Nonstandard Analysis, Axiomatically

Vladimir Kanovei • Michael Reeken

Nonstandard Analysis, Axiomatically

Vladimir Kanovei

IITP, Institute for Information Transmission
Bol. Karetny 19, 127994 Moscow

Russian Federation

Michael Reeken

Bergische Universität Wuppertal
FB C Mathematik
Gaußstr. 20, 42119 Wuppertal

Germany

Mathematics Subject Classification (2000): Primary: 03E, 03C, 03H05, 03E70
Secondary: 00A30, 00A35, 26E35, 28E05, 54J05

ISBN 978-3-642-06077-9

Springer-Verlag Berlin Heidelberg New York
a member of Springer Science+Business Media GmbH

Printed in Germany

Cover Design: *design & production GmbH*, Heidelberg

Printed on acid-free paper 41/3142XT 5 4 3 2 1 0

Preface

In the aftermath of the discoveries in foundations of mathematics there was surprisingly little effect on mathematics as a whole. If one looks at standard textbooks in different mathematical disciplines, especially those closer to what is referred to as applied mathematics, there is little trace of those developments outside of mathematical logic and model theory. But it seems fair to say that there is a widespread conviction that the principles embodied in the Zermelo – Fraenkel theory with Choice (**ZFC**) are a correct description of the set theoretic underpinnings of mathematics.

In most textbooks of the kind referred to above, there is, of course, no discussion of these matters, and set theory is assumed informally, although more advanced principles like Choice or sometimes Replacement are often mentioned explicitly. This implicitly fixes a point of view of the mathematical universe which is at odds with the results in foundations. For example most mathematicians still take it for granted that the real number system is uniquely determined up to isomorphism, which is a correct point of view as long as one does not accept to look at "unnatural" interpretations of the membership relation.

One of the crucial discoveries in foundations was that the structures studied in mathematics do have nonstandard models. Starting with A. Robinson, this gave rise to a new mathematical discipline, *nonstandard analysis*, which has made important contributions to different branches of pure mathematics. The earliest practice of nonstandard analysis, concentrated on a variety of different nonstandard extensions of "standard" mathematical structures, was codified by Robinson and Zakon in the late 60s. They suggested (in [RobinZ 69] or in an improved version [Zak 74]) a system, of type-theoretic character, which reduced methods of nonstandard analysis to a few principles, which, once established, allow to develop nonstandard analysis without paying much attention to details related to the construction of nonstandard extensions. The key concept of this direction of foundations of nonstandard analysis [1], which we shall call "model-theoretic foundations", is that of a *nonstandard superstructure*, or *nonstandard universe*, actually, a type-theoretic

[1] It has several modifications which differ, in particular, in terms of the level of formalization.

superstructure over a nonstandard extension of a chosen mathematical structure.

Despite its great success in applications since the early 70s [2] , the model-theoretic approach has, from a philosophical point of view, the disadvantage that set theory becomes in some sense dissociated from mathematical practice. It becomes the substrate for the construction of a variety of non-isomorphic "nonstandard" models related to various fields of mathematics. This leads naturally to the idea to modify the axiomatic system for set theory in such a way that it restores the essential uniqueness of fundamental mathematical structures. [3] Such a system can be intended for two distinct purposes. It may be conceived either as a subject of foundational studies, which is a legitimate purpose in itself, or as a tool for the "working mathematician" which would be used in much the same way as **ZFC** is used.

Given the remarkable unity of mathematics, such a system should be equivalent to **ZFC** in a metamathematical sense. Usually this is understood as the requirement that the theory should possess, within the "universe of discourse" of all sets, a class — let us denote it by $\mathbb{S}$ and call *the standard universe* — informally seen as a copy of the "real universe of sets" described by **ZFC**, and that the theory should prove those and only those theorems about $\mathbb{S}$ which **ZFC** proves about all sets. In this sense, the theory may be viewed as equivalent to **ZFC**.

Notice that the latter requirement, called *conservativity*, only ensures that "provable" truths are reproduced. But we would like to relate all true sentences of the nonstandard theory to the truth in **ZFC**. This is a stronger requirement; indeed, a conservative extension of **ZFC**, although it proves nothing about $\mathbb{S}$ beyond the theorems of **ZFC**, may well shut some "windows" which are implicitly open in **ZFC**. This could be dealt with by stipulating that the extended theory be interpretable in the **ZFC** universe in such a way that the class $\mathbb{S}$ in the sense of interpretation is isomorphic to the ground **ZFC** universe. Such an interpretation may be thought of as an *extension* of the ground **ZFC** universe to the "universe of discourse" of the theory. (This extension is rather an embedding of the **ZFC** universe **V** into a definable class–size structure in **V**, which is a model of the extended theory.) We call "realistic" [4] theories which admit such an interpretation in **ZFC**. This is a reasonable demarcation line between nonstandard theories which respect mathematical reality as described by **ZFC** from purely syntactical deduction schemes, be the latter even conservative.

[2] See footnote 12 on page 2 for a (rather incomplete) list of related references.

[3] Keisler wrote in [Keis 94] that the unique existence of the pair consisting of the real and hyperreal number systems should be ensured by the underlying set theory.

[4] The meaning which we associate with this concept is not the same as in the recent paper of Hrbaček [Hr 01] entitled "Realism, nonstandard set theory, and large cardinals".

Although speculations in this direction of modifying the axiomatic set-up itself can be traced to the late 60s [5], it was only in mid-70s when Hrbaček and Nelson [6] independently proposed satisfactory nonstandard set theories of this kind [7], of which Nelson's [Nel 77] *internal set theory* **IST** gained a lot of support among practitioners of nonstandard methods.

The universe of **IST** is arranged in a remarkably simple way. It includes a class $\mathbb{S}$, called *the universe of all standard sets*, distinguished by the standardness predicate st, an atomic predicate. $\mathbb{S}$ is postulated to fulfill the

[5] Kreisel was, perhaps, one of the first who discussed axiomatization of nonstandard methods, as the following citation from [Kr 69, p. 93] shows: *"Is there a simple formal system ... in which existing practice of nonstandard analysis can be codified? And if the answer is positive: is this formal system a conservative extension of the current systems of analysis ... ?"* Yet Kreisel's interest was focused on type theoretic rather than set theoretic systems. An outstanding contribution to this research line was made by Henson and Keisler [HenK 86], by demonstrating that nonstandard systems of type n can be compared with standard systems of type $n+1$ rather than type n.

[6] Names listed alphabetically.

[7] ***Historical note.*** Nelson's paper [Nel 77] appeared in the Nov. 1977 issue of *Bull. Amer. Math. Soc.*, and was received by the journal Nov. 19, 1976, being based on a lecture to AMS in the Summer of 1976. Hrbaček's paper [Hr 78] in *Fund. Math.*, 1978, was accepted by the journal in 1975 (editorial note: "accepté par la Rédaction le 16. 5. 1975"), and based on an abstract in *J. Symbolic Logic* 1976, 41, no 1, p. 285, submitted for the January 1975 meeting of ASL.

Another nonstandard axiomatic system, *alternative set theory* **AST** of Vopenka [Vop 79], appeared, in published form, even somewhat earlier, Sochor [Soc 76] is, perhaps, the first really relevant publication, although the origins go back to [VopH 72] and earlier. **AST** is closer, in its spirit, to Hrbaček's approach, in particular, in that it directly involves "external sets" (called *semisets* in the **AST** vocabulary), but differs from both Nelson's and Hrbaček's theories in that its "standard universe" consists of hereditarily finite sets rather than all sets of **ZFC**, and in this sense **AST** is a nonstandard extension of Peano arithmetic (or **ZFC** where the Infinity axiom is replaced by its negation) rather than full **ZFC**, which implies affinities rather with the model-theoretic version of foundations.

These three initial attempts to fully axiomatize nonstandard mathematics can hardly be linearly ordered in any reasonable sense, with any sort of preference assigned in some sound manner. It is fair to assert, on the base of available records, that all three were undertaken independently of each other and led to results of comparable quality (although not of comparable impact on the practice of nonstandard mathematics, where **IST** has preference), in addition, all three were based upon earlier development in foundations of nonstandard analysis.

It must be pointed out that at least some axioms of Nelson's, Hrbaček's, and other systems of the same kind have direct or implicit predecessors elaborated within model-theoretic foundations. "Bounded internal set theory" of Diener and Stroyan [DienS 88] demonstrates transmission of ideas from principles valid in nonstandard superstructures to axioms of **IST**.

axioms of **ZFC** while the "universe of discourse" is postulated, by a special axiom schema called Transfer, to be an elementary extension of $\mathbb{S}$ in the $\in$-language. Additional axioms of Standardization and Idealization provide some regularity for the interactions between standard and nonstandard sets, as well as the existence of truly "nonstandard" objects like infinitesimals.

The price one has to pay for this nice image is very high indeed. The predicate `st`, being independent from the membership $\in$, does not always form sets in **IST**. For instance the collection of all standard natural numbers is not a set in **IST**. Thus certain things which were done in model theory and its nonstandard analysis branch cannot be reproduced in **IST**, at least not directly. But as Keisler remarked in [Keis 94] "... mathematics in **IST** looks more like traditional mathematics ...". This was probably the reason that quite a number of mathematicians outside of logic and foundations took up this approach.

Hrbaček [Hr 78, Hr 79] proposed several theories which incorporate more of the set theoretic instrumentarium one is accustomed to from the model-theoretic development of nonstandard analysis. In these systems the predicate `st` can participate in definitions of sets, subsequently, in addition to the standard universe $\mathbb{S}$ and the internal universe $\mathbb{I}$ (quite similar to the set universe of **IST**) there is now a larger underlying universe $\mathbb{H}$ of "external sets", including in particular standard and internal sets, sets like the collection of all standard natural numbers, and many more. Unlike the situation in **IST**, $\mathbb{I}$ consists of all sets which are elements of standard sets. This means that the standard universe exerts a tighter control over internal sets than is possible in **IST**. As axioms there are again the **ZFC** axioms in $\mathbb{S}$, Transfer between the standard and internal universes, an axiom which says that $\mathbb{I}$ is a transitive class in $\mathbb{H}$. In addition, $\mathbb{H}$ is postulated to be well-founded over $\mathbb{I}$ [8] and to satisfy quite a big fragment of **ZFC** including Replacement for all formulas (even those in which the predicate `st` occurs). Saturation for families (of internal sets) of standard size [9] is included: as usual, it is a general source of getting various internal nonstandard objects like infinitesimal or infinitely large numbers.

Later Kawaï [Kaw 81, Kaw 83] extended **IST** to a theory similar to those introduced by Hrbaček, and very well equipped technically.

Hrbaček, Nelson, and Kawaï showed that their theories are equiconsistent and conservative extensions of **ZFC**. Thus as far as proving theorems is concerned, they would be admissible as a tool replacing **ZFC**. But, as one of the authors (Kanovei [Kan 94b]) has shown, **IST** transcends **ZFC** in another sense. For instance, **IST** does not admit an interpretation in **ZFC** of the kind

[8] But not well-founded in proper sense: it contains $\in$-decreasing infinite chains of sets, which is a characteristic property of all nonstandard set theories.

[9] Being of standard size means to be in $1-1$ correspondence with the set of all standard elements of a standard set. In Hrbaček's system this is equivalent to being well-orderable.

discussed above, moreover, not every model of **ZFC** can be embedded, as the class of all standard sets, in a model of **IST**. Furthermore there exists a sentence in the st-$\in$-language which is not equivalent in **IST** to an $\in$-sentence. Thus, **IST** "knows" something about the standard universe which **ZFC** itself does not "know" about sets. (This "something" is not a first–order property, of course.) We can conclude that **IST** and Kawaï's theories considered as a working tool represent a definite step away from **ZFC**.

As the authors have shown [KanR 95, KanR 97], a minor modification of one of the theories of Hrbaček is a perfect candidate for the stronger requirements mentioned above. In some sense this theory also improves upon **IST** as it is an external extension of *bounded set theory* **BST**, essentially, the theory of those sets of **IST** which are elements of standard sets. If one scrutinizes known applications of **IST** none of them makes any essential use of those internal sets which are not of this type. Moreover, **BST** admits without any restriction a very useful algorithm of conversion of formulas (known as *Nelson's reduction algorithm*), valid in **IST** only for special types of formulas, and has some other advantages, especially in a fair treatment of "external sets" (*i.e.*, definable subclasses of sets, which are not necessarily sets themselves in internal theories like **IST** or **BST**) causing so much trouble for **IST** practitioners.

We call this nonstandard set theory *Hrbaček set theory*, **HST**. Its axioms, stated in a very convenient way, allow to deduce, in quite a straightforward manner and even using the same technical notation ("asterisks") known from the model-theoretic development of nonstandard analysis, many substantial results in nonstandard mathematics. On the other hand, since the internal universe $\mathbb{I}$ of **HST** models bounded set theory, an improved version of **IST**, "internal" forms of reasoning well known to **IST** practitioners can also be carried out in **HST**. Metamathematically, unlike **IST** and **IST**-based theories, **HST** admits an interpretation in the **ZFC** universe of the kind mentioned above. In addition, it poses deep set theoretic problems and admits such advanced foundational tools as constructibility and forcing to find solutions.

This book was written with the goal to systematize recent development in the domain of axiomatic foundations of nonstandard analysis, including metamathematical, applied, and, to less extent, philosophical issues, in particular, with the following three general aims in mind:

1) to present Hrbaček set theory **HST**, including its principal metamathematical properties and connections with other nonstandard set theories — in this aspect the exposition is based mainly on the research papers in this field written since the late 70s, including our own recent papers of the 90s;
2) to show that "internal" methods in axiomatic nonstandard analysis go, in foundational issues, far beyond what is usually presented even in manuals

based on **IST** [10]: essentially, the diversity of "external" sets maintained by **HST** begins in the internal universe;

3) to demonstrate how "nonstandard" arguments (in domains which vary from traditional topics like calculus to those which attract attention nowadays like "hyperfinite" descriptive set theory) can be maintained on the base of **HST** — where we build upon research papers and, occasionally, books (with respect to more traditional material) in nonstandard analysis.

Accordingly, the reader envisaged has at least an idea of (if not some experience with) nonstandard analysis, either in the model-theoretic or the **IST**-based version, and wants to support it with a universal axiomatic background, or specializes rather in set theoretic foundations in general and looks for new fields of foundational research. Some knowledge in foundations, that is, set theory (including basics of descriptive set theory – for Chapter 9), and (to lesser extent) model theory is expected.

Acknowledgements The authors are thankful to David Ballard and Karel Hrbaček for their patience in reading through several consecutive drafts and many remarks and corrections. One of the authors (V. Kanovei) acknowledges the support of RFFI in 1998–2000 and 2003–2004. Both authors are grateful to the DFG for the continued support of their cooperation without which this book would not have been written.

Wuppertal,
March 2004

Vladimir Kanovei
Michael Reeken

[10] Such as Lutz and Gose [LutG 81], van den Berg [vdBerg 87, vdBerg 92], Robert [Rob 88], F. and M. Diener [DienD 95].

Table of Contents

Introduction ... 1
Basic notation ... 10

1 Getting started ... 11
1.1 The axiomatical system of Hrbaček set theory ... 12
1.1a The universe of **HST** ... 12
1.1b Axioms for the external universe ... 14
1.1c Axioms for standard and internal sets ... 14
1.1d Well-founded sets ... 16
1.1e The $\in$-structure of internal and well-founded sets ... 17
1.1f Axioms for sets of standard size ... 19
1.1g Putting it all together ... 20
1.1h Zermelo – Fraenkel theory **ZFC** ... 20
1.2 Basic elements of the nonstandard universe ... 22
1.2a How to define fundamental set theoretic notions in **HST** ... 22
1.2b Closure properties and absoluteness ... 22
1.2c Ordinals and cardinals ... 24
1.2d Natural numbers, finite and $*$-finite sets ... 25
1.2e Hereditarily finite sets ... 28
1.3 Sets of standard size ... 29
1.3a Cardinalities of sets of standard size ... 29
1.3b Saturation and the Hrbaček paradox ... 30
1.3c The principle of Extension ... 32
1.4 The class $\boldsymbol{\Delta}_2^{ss}$ ... 34
1.4a Basic properties of $\boldsymbol{\Delta}_2^{ss}$ ... 34
1.4b Cuts (initial segments) of $*$-ordinals ... 35
1.4c Monads and transversals ... 37
1.4d On non-well-founded cardinalities ... 38
1.4e Small and large sets ... 40
1.5 Some finer points ... 42
1.5a Von Neumann hierarchy and Reflection in **ZFC** ... 42
1.5b Von Neumann hierarchy over internal sets in **HST** ... 44
1.5c Classes and structures ... 45
1.5d Interpretations ... 47
1.5e Models ... 48
1.5f Simulation of models of **ZFC** ... 49
1.5g Asterisk is an elementary embedding ... 50
Historical and other notes to Chapter 1 ... 52

2 Elementary real analysis in the nonstandard universe 53
2.1 Hyperreal line 54
2.1a Hyperreals 54
2.1b Fundamentals of nonstandard real analysis 56
2.1c Directed Saturation 57
2.1d Nonstandard characterization of closed and compact sets 58
2.2 Sequences and functions 59
2.2a Limits 60
2.2b Continuous functions 61
2.2c Intermediate value theorem 62
2.2d Robinson's lemma and uniform limits 62
2.3 Topics in nonstandard real analysis 64
2.3a Shadows and equivalences 64
2.3b Near-standard elements 66
2.3c Topology 69
2.4 Two special applications 73
2.4a Euler factorization of the sine function 73
2.4b Jordan curve theorem 76
Historical and other notes to Chapter 2 81

3 Theories of internal sets 83
3.1 Introduction to internal set theories 84
3.1a Internal set theory 84
3.1b Bounded set theory 86
3.1c Internal sets interpret **BST** in the external universe 87
3.1d Basic internal set theory 88
3.1e Standard natural numbers and standard finite sets 90
3.1f Remarks on Basic Idealization and Saturation 92
3.2 Development of bounded set theory 93
3.2a Half-bounded forms of Idealization 93
3.2b Reduction to two "external" quantifiers 94
3.2c Finite axiomatizability of **BST** and other corollaries 95
3.2d Collection in **BST** 97
3.2e Other basic theorems of **BST** 99
3.2f Introduction to the problem of external sets 101
3.2g More on "external sets" in **BST** 104
3.3 Internal theories with partial Saturation 105
3.3a Two schemes of partially saturated internal theories 105
3.3b κ-deep Basic Idealization scheme 106
3.3c κ-size Basic Idealization scheme 109
3.4 Development of Nelson's internal set theory 111
3.4a Bounded sets in **IST** 111
3.4b Bounded formulas: reduction to two "external" quantifiers 113
3.4c Collection in **IST** 114
3.4d Uniqueness in **IST** 117
3.5 Truth definition in internal set theory 118
3.5a Truth definition for the standard universe 118
3.5b Connection with the ordinary truth 120
3.5c Extension of the definition of formal truth 122

3.6 Second edition of **IST** 124
3.6a Standard and nonstandard theories of Nelson's system 124
3.6b The background nonstandard universe 125
3.6c Three "myths" of **IST** 127
Historical and other notes to Chapter 3 129

4 Metamathematics of internal theories 131
4.1 Outline of metamathematical properties 132
4.1a Nonstandard extensions of structures 132
4.1b Nonstandard extensions of theories 133
4.1c Comments 134
4.1d Metamathematics of internal theories: the main results 136
4.2 Ultrapowers and saturated extensions 138
4.2a Saturated structures and nonstandard set theories 138
4.2b Quotient power extensions 140
4.2c Adequate and good ultrafilters and ultrapowers 142
4.2d Elementary chains of structures 144
4.3 Metamathematics of **BST** 146
4.3a Warmup: several examples 146
4.3b Infinite Fubini products of adequate ultrafilters 148
4.3c Standard core interpretation of **BST** in **ZFC** 150
4.3d Saturated standard core interpretation 152
4.4 The conservativity and equiconsistency of **IST** 154
4.4a Good extensions of von Neumann sets in **ZFC** universe 154
4.4b Iterated adequate extensions of von Neumann sets 155
4.4c Iterated adequate extensions in the ϑ-version of **ZFC** 156
4.4d Long iterated quotient power chains 156
4.4e Conservativity of **IST** by inner models 157
4.5 Non-reducibility of **IST** 159
4.5a The minimality axiom 159
4.5b The source of counterexamples 160
4.5c The ultrafilter 161
4.5d "Definable" adequate quotient power 163
4.5e Corollaries and remarks 164
4.6 Interpretability of **IST** in a standard theory 166
4.6a Standard theory with a global choice and a truth predicate 166
4.6b Formally definable classes 168
4.6c A nonstandard theory extending **IST** 169
4.6d The ultrafilter 170
4.6e The interpretation 173
4.6f Extendibility of standard models 175
Historical and other notes to Chapter 4 176

5 Definable external sets and metamathematics of HST 179
5.1 Introduction to metamathematics of **HST** 180
5.1a Internal core embeddings and interpretability 180
5.1b Metamathematics of **HST** : an overview 181
5.2 From internal to elementary external sets 184
5.2a Interpretation of **EEST** in **BST** 184

5.2b Elementary external sets in external theories 186
5.2c Some basic theorems of **EEST** 188
5.2d Standard size, natural numbers, finiteness in **EEST** 189
5.3 Assembling of external sets in **HST** 191
5.3a Well-founded trees 191
5.3b Coding of the assembling construction 192
5.3c Examples of codes... 193
5.3d Regular codes... 195
5.4 From elementary external to all external sets 196
5.4a The domain of the interpretation......................... 196
5.4b Basic relations between codes 198
5.4c The structure of basic relations 200
5.4d The interpretation and the embedding 202
5.4e Verification of the **HST** axioms 204
5.4f Superposition of interpretations 207
5.4g The problem of external sets revisited 209
5.5 The class $\mathbb{L}[\mathbb{I}]$: sets constructible from internal sets 211
5.5a Sets constructible from internal sets 211
5.5b Proof of the theorem on I-constructible sets 212
5.5c The axiom of $\mathbb{I}$-constructibility 214
5.5d Transfinite constructions in $\mathbb{L}[\mathbb{I}]$ 215
Historical and other notes to Chapter 5 217

6 Partially saturated universes and the Power Set problem 219
6.1 Internal subuniverses 220
6.1a Some basic definitions and results 220
6.1b Relative standardness.................................... 221
6.1c Simple relative standardness 222
6.1d Gordon classes 224
6.1e Associated structures 225
6.1f More on internal subuniverses 228
6.1g Appendix: Kunen's theorem 229
6.2 Partially saturated internal universes 230
6.2a Partially saturated classes $\mathbb{I}_\kappa$ 230
6.2b Good internal subuniverses 232
6.2c Internal universes over complete sets.................... 233
6.3 External universes 237
6.3a External universes and internal core extensions 237
6.3b Von Neumann construction over non-transitive classes 239
6.3c Absoluteness for external subuniverses 240
6.4 Partially saturated external universes.......................... 241
6.4a Partially saturated external theories 241
6.4b Extensions of thin classes 243
6.4c Constructible extensions 244
6.4d Constructible extensions of self-definable classes.......... 246
6.4e The classes $\mathbb{L}[\mathbb{I}_\kappa]$ 248
6.4f External universes over complete sets 249
6.4g Collapse onto a transitive class......................... 251
6.4h Outline of applications: subuniverses satisfying Power Set.. 252

Historical and other notes to Chapter 6 254

7 Forcing extensions of the nonstandard universe 257
7.1 Generic extensions of models of **HST** 258
7.1a Ground model 258
7.1b Regular extensions 259
7.1c Forcing notions and names 260
7.1d Adding a set 261
7.1e Forcing relation 263
7.1f Generic extensions and the truth lemma 266
7.1g The extension models **HST** 267
7.2 Applications: collapse maps and isomorphisms 270
7.2a Making two internal sets equinumerous 270
7.2b Internal preserving bijections 272
7.2c Making elementarily equivalent structures isomorphic 273
7.2d The forcing notion 274
7.2e Key lemma 276
7.2f Generic isomorphisms 278
7.3 Consistency of the isomorphism property 279
7.3a The product forcing notion 280
7.3b Externalization 281
7.3c Restricted forcing relations 282
7.3d Automorphisms and the restriction property 283
7.3e The product generic extension 284
Historical and other notes to Chapter 7 287

8 Other nonstandard theories 289
8.1 Nonstandard set theory of Kawaï 290
8.1a The axioms of Kawaï's theory 290
8.1b Metamathematical properties 292
8.1c Special model axiom 293
8.2 "Nonstandard set theory" of Hrbaček 295
8.2a Axioms 295
8.2b Additional axioms of Collection 297
8.2c Conservativity and consistency 298
8.2d Remarks and exercises 301
8.3 Non-well-founded set theories 303
8.3a Boffa's non-well-founded set theory 303
8.3b Extensions of proper classes 305
8.3c Applications to nonstandard analysis 306
8.3d Alpha theory 307
8.3e Interpretation of Alpha theory in **ZFBC** 311
8.4 Miscellanea: some other theories 312
8.4a A theory with "definable" Saturation 312
8.4b Stratified nonstandard set theories 313
8.4c Nonstandard class theories 314
Historical and other notes to Chapter 8 315

9 "Hyperfinite" descriptive set theory 317
9.1 Introduction to "hyperfinite" DST 319
9.1a General set-up 319
9.1b Comments on notation 320
9.1c Borel and projective sets in a nonstandard domain 321
9.1d Some applications of countable Saturation 323
9.1e Operation A and Souslin sets 324
9.2 Operations, countably determined sets, shadows 325
9.2a Operations and quantifiers 325
9.2b Countably determined sets 327
9.2c Shadows or standard part maps 329
9.3 Structure of the hierarchies 331
9.3a Operations associated with Borel and projective classes 331
9.3b The "shadow" theorem 332
9.3c Closure properties of the classes 335
9.4 Some classical questions 338
9.4a Separation and reduction 338
9.4b Countably determined sets with countable cross-sections 340
9.4c Countably determined sets with internal and $\mathbf{\Sigma}^0_1$ cross-sections 343
9.4d Uniformization 344
9.4e Variations on Louveau's theme 347
9.4f On sets with $\mathbf{\Pi}^0_1$ cross-sections 350
9.5 Loeb measures 351
9.5a Definitions and examples 351
9.5b Loeb measurability of projective sets 353
9.5c Approximations almost everywhere 354
9.5d Randomness in a hyperfinite domain 356
9.5e Law of Large Numbers 358
9.5f Random sequences and hyperfinite gambling 359
9.6 Borel and countably determined cardinalities 362
9.6a Preliminaries 362
9.6b Borel cardinals and cuts 364
9.6c Proof of the theorem on Borel cardinalities 366
9.6d Complete classification of Borel cardinalities 368
9.6e Countably determined cardinalities 368
9.7 Equivalence relations and quotients 370
9.7a Silver's theorem for countably determined relations 371
9.7b Application: nonstandard partition calculus 373
9.7c Generalization 375
9.7d Transversals of "countable" equivalence relations 376
9.7e Equivalence relations of monad partitions 378
9.7f Borel and countably determined reducibility 380
9.7g Reducibility structure of monad partitions 382
Historical and other notes to Chapter 9 386

References 389

Index 397

Introduction

For the convenience of the reader, we give a short and informal resumé of the content, without going into technical details.

Chapter 1 presents *Hrbaček set theory* **HST** [11] and demonstrates some examples of reasoning in **HST**. We begin with the description of the system of axioms which govern the relations between three main categories of sets in the nonstandard set universe $\mathbb{H}$ of **HST** : standard, internal, well-founded.

The class $\mathbb{I}$ of all *internal* sets forms the basement of the $\in$-structure of the **HST** set universe $\mathbb{H}$: the axioms of Transitivity of $\mathbb{I}$ and Regularity over $\mathbb{I}$ postulate $\mathbb{I}$ to be transitive (*i.e.*, elements of internal sets are themselves internal), while the whole universe is well-founded over $\mathbb{I}$, so that, in particular, any infinite $\in$-decreasing chain $x_1 \ni x_2 \ni x_3 \ni \dots$ necessarily contains an internal set (but $\mathbb{I}$ itself is not well-founded!).

Internal sets are formally defined as elements of *standard* sets, which form a subclass $\mathbb{S} \subsetneqq \mathbb{I}$, distinguished by an atomic predicate of standardness $\mathtt{st}$, thus, $\mathbb{S} = \{x : \mathtt{st}\, x\}$. Special axioms guarantee that $\langle \mathbb{S}; \in \rangle$ is an elementary substructure of $\langle \mathbb{I}; \in \rangle$ and both structures obey **ZFC**. A very important axiom of Standardization postulates that for any $X \subseteq \mathbb{S}$ there is a standard set Y such that $X = Y \cap \mathbb{S}$. Finally, *well-founded* sets are those sets x which do not admit an infinite $\in$-decreasing chain $x \ni x_1 \ni x_2 \ni \dots$, they form a transitive class $\mathbb{WF}$. The classes $\mathbb{WF}$ and $\mathbb{I}$, as well as $\mathbb{WF}$ and $\mathbb{S}$, have the class of all hereditarily finite sets as intersection.

A special group of axioms of **HST** is related to *sets of standard size* (those equinumerous to subsets of $\mathbb{S}$ — this turns out to be the same in **HST** as well-orderable sets), in particular, Saturation for all standard size families of internal sets, and Standard Size Choice, an axiom of choice for standard size families. As for the whole universe $\mathbb{H}$ of external (that is, all) sets, the theory **HST** contains all axioms of **ZFC** (the schemata of Separation and Collection admitted in the $\mathtt{st}$-$\in$-language), with the exception of the axioms of Power Set, Choice, and Regularity which are incompatible with **HST** (but Choice and Regularity are added in useful partial versions).

It turns out that **HST** allows to view things in a manner highlighting its intrinsic analogy with the model theoretic treatment of nonstandard mathe-

[11] A slightly improved version of the theory $\mathbf{NS}_1(\mathbf{ZFC})$ in [Hr 78].

matics. [12] Indeed, given any standard set we may remove all its nonstandard elements. We are left with a set containing exclusively standard elements. With these elements, we can repeat the argument. Doing things correctly by $\in$-induction we thus define, after the removal of all subsequent nonstandard elements, a new set called *the condensed set*. The class of all condensed sets is equal to the class $\mathbb{WF}$ of all well-founded sets in $\mathbb{H}$. Furthermore, it follows from the axiom of Standardization that for any $x \in \mathbb{WF}$ there exists a unique standard set ${}^*x$, the $*$-*extension* of x, the condensed form of which is equal to x, so that $\mathbb{WF}$ is $\in$-isomorphic to the class $\mathbb{S}$ of all standard sets via the map $x \mapsto {}^*x$ having pretty much the same properties as "star" embeddings in the model-theoretic version of nonstandard analysis.

The universe $\mathbb{WF}$ satisfies the axioms of **ZFC** (because so does $\mathbb{S}$) but is a more convenient version of the conventional set universe of "standard" mathematics than $\mathbb{S}$ because besides transitivity it is closed under subset formation in $\mathbb{H}$: any set $x \subseteq \mathbb{WF}$ belongs to $\mathbb{WF}$. It follows that $\mathbb{WF}$ and $\mathbb{H}$ contain the same natural numbers, reals, ordinals, cardinals *etc.*, while for instance $\mathbb{S}$-ordinals are "stars" of $\mathbb{WF}$-ordinals. This specific point of view (so that $\mathbb{WF}$ rather than $\mathbb{S}$ is the principal "copy" of the standard set universe) allows to develop nonstandard mathematics in close analogy (even using the same notation) to the model-theoretic version of nonstandard analysis: $\mathbb{H}$, the **HST** "universe of discourse", corresponds to the ground **ZFC** universe, $\mathbb{WF}$ to a standard structure, $\mathbb{I}$ to its nonstandard extension in which $\mathbb{WF}$ is embedded by the "star" map, and $\mathbb{S} = \{{}^*x : x \in \mathbb{WF}\} \subseteq \mathbb{I}$ is a copy of $\mathbb{WF}$ (formally, different from $\mathbb{WF}$) and an elementary substructure of $\mathbb{I}$.

Chapter 2 considers some applications. We show how the classics of nonstandard mathematics, nonstandard real analysis, can be developed in **HST**. In fact this does not much differ from the model-theoretic approach, the "star" embedding and Saturation work quite the same way. But we don't have to construct nonstandard models in **HST**: they are already given; essentially, $\mathbb{I}$, the class of all internal sets, is a nonstandard extension of the whole mathematical universe (viewed as $\mathbb{WF}$), whose level of Saturation exceeds any given standard cardinal. It turns out that the axiomatic approach of **HST** makes it easier to apply syntactical methods to shorten and formalize proofs. One such method allows to obtain proofs of many basic facts, related to sequences and limits, in the form of simple and formal transformations of formulas.

Together with rather traditional topics in the foundations of real analysis, we consider in **Chapter 2** several more elaborate applications of nonstandard methods. One of those is the famous factorization of the sine function, orig-

[12] We refer to Chang and Keisler [CK 92], Henson and Keisler [HenK 86], Stroyan and Bayod [StrB 86], Lindstrøm [Lin 88], Keisler [Keis 94], Henson [Hen 97], Cutland [Cut 97], Goldblatt [Gol 98], Loeb and Wolff [LoebW 00] in matters of nonstandard analysis in "superstructures" or "nonstandard universes", and to Chang and Keisler [CK 92] in matters of model theory in general.

inally obtained by Leonhard Euler through an argument which deals with infinitesimals and infinitely large numbers (understood naïvely, of course) in such a liberal manner that it takes quite a bit of work to properly recover all obscure aspects of the proof on the rigorous basis of (modern) nonstandard analysis. Another application concerns the Jordan curve theorem, where we present a quite elementary nonstandard proof of this (rather difficult) result.

Then we come back to foundational topics. The foundational study of **HST** is concentrated on the following issues:

1) metamathematical properties of **HST**, in particular, in relation to **ZFC** and some nonstandard set theories;
2) structure of the **HST** universe, which includes both set theoretic aspects and development of more applied parts of nonstandard mathematics.

The ultimate metamathematical goal will be to show that the "standard" set universe of conventional mathematics (assumed to obey the axioms of **ZFC**) admits a nonstandard extension, in a certain mathematically rigorous sense, which interprets the **HST** axioms and has the original universe as the class of all standard sets (or, in a different but equivalent version, as the class of all well-founded sets). The possibility of such an extension is of great importance for the philosophy of nonstandard mathematics. Indeed, it shows that **HST** is not only a useful syntactical tool to prove "standard" theorems, but it also admits a mechanism which allows us to consistently extend the "standard" universe of classical mathematics to a universe of **HST**, with a semblance of, *e.g.*, the extension of the reals to the complex numbers such that all properties and features of the extension are adequately traceable in the ground universe.

This extension will have two distinct steps, with the class $\mathbb{I}$ of all internal sets, or, more exactly, the structure $\langle \mathbb{I}; \in, \mathtt{st} \rangle$, to be created first.

Clearly $\mathbb{I}$ has to satisfy everything that **HST** says about internal sets. This turns out to be *bounded set theory* **BST**, explicitly introduced by one of the authors (Kanovei [Kan 91]), a simpler version of Nelson's *internal set theory* **IST** : actually, **BST** is a theory of those sets in **IST** which are members of standard sets. **Chapter 3** presents a development of **IST** and **BST**, in comparison with each other. As a matter of fact even the common part **BIST** of **IST** and **BST** allows to consistently treat basic notions like natural numbers and finite sets. However the theories **BST** and **IST** offer much more for foundational studies than just that. For instance, we prove that any st-$\in$-formula is equivalent in **BST** to a $\Sigma_2^{\mathtt{st}}$ formula, *i.e.*, a formula of the form $\exists^{\mathtt{st}} x \, \forall^{\mathtt{st}} y \, \Phi(x, y, ...)$, Φ being an $\in$-formula, a result known for **IST** only for st-$\in$-formulas with quantifiers bounded by standard sets. A special feature of **IST** is that it provides a uniform truth definition (by a st-$\in$-formula) for all $\in$-formulas, or, what is the same by Transfer, a uniform truth definition for the class $\mathbb{S}$ of all standard sets. This appears somewhat paradoxical, because, usually, to provide a truth definition one has to employ a stronger theory than the theory associated with the object of the truth definition, yet **IST** is a conservative and equiconsistent extension of **ZFC**, a theory naturally associated with $\mathbb{S}$.

Chapter 4 proves the main metamathematical properties of **BST** and **IST**. We prove that both theories are equiconsistent and conservative extensions of **ZFC**. In addition, **BST** is a *reducible* extension, that is for any st-$\in$-sentence φ there is an $\in$-sentence φ' with the equivalence $\varphi \Longleftrightarrow \varphi'$ provable in **BST**. On the other hand, **IST** fails to satisfy this property. This is seen by extending standard structures, for instance those of the form $\mathbf{V}_\vartheta$, ϑ being a cardinal, to nonstandard structures. The question, which models of full **ZFC** can be extended to models of **BST** or **IST**, turns out to be interesting and difficult, especially for **IST**. The **BST** case is simpler: it turns out that any model of **ZFC** can be extended to a model of **BST**.

The extension method can be applied to obtain nonstandard extensions even of the whole set universe. For instance, the "standard" universe of **ZFC** can be extended to a **BST**-like world. Saying it differently, **BST** admits an interpretation in **ZFC** such that the class of all standard sets is definably isomorphic to the ground **ZFC** universe! Nonstandard theories which admit this sort of interpretation are called "realistic" (Definition 4.1.8). This principal property is somewhat stronger than the conservativity or equiconsistency with **ZFC**. For instance, **IST**, known to be a conservative nonstandard extension of **ZFC**, is not "realistic". Actually **IST** needs more than **ZFC** in the ground "standard" set universe to define an interpretation.

Chapter 5 (especially Sections 5.2, 5.4) demonstrates how a **BST** universe $\mathbb{I}$ can be extended to a **HST** universe, where the former remains the class of all internal sets. To explain the idea behind this construction, recall that **BST** (as well as **IST** and any other theory which postulates the set universe to be an elementary extension of the standard universe in the $\in$-language) does not include the axiom schema of Separation in the st-$\in$-language, for instance, collections like $\mathbb{N} \cap \mathbb{S}$ (all *standard* natural numbers) are usually not sets in **BST**. On the contrary, **HST** does include Separation in the st-$\in$-language, therefore, to obtain an **HST**-like extension of a **BST** universe, we have, at least, to adjoin to $\mathbb{I}$ all st-$\in$-definable (with parameters) subobjects (not yet subsets) of sets in $\mathbb{I}$, which we call *elementary* [13] *external sets*, with the understanding that they are **HST**-sets but, generally, not **BST**-sets. The following startling property of **BST** supports this extension mechanism: despite the multitude of st-$\in$-formulas, elementary external sets admit a uniform description as $\mathbf{\Delta}_2^{\mathrm{ss}}$ sets, *i.e.*, those of the form $\bigcup_{a\in A\cap\mathbb{S}}\bigcap_{b\in B\cap\mathbb{S}} X_{ab}$, where A, B are standard sets while $\langle a,b\rangle \mapsto X_{ab}$ is a map (a set) in $\mathbb{I}$, defined on $A\times B$, as well as of the dual form, so that we have a *parametrization* of all elementary external sets by internal sets. From the point of view of the internal universe of **BST**, this is a parametrization of all definable external

[13] "Elementary" here means that they form the simplest class of non-internal sets, because, first, all of them contain only internal elements, and second, they are st-$\in$-definable in $\mathbb{I}$.

"non-sets" by true sets, which immediately attaches a definite meaning to formulas containing quantifiers over (definable) external sets, thus, solving a major foundational problem ("the problem of external sets") for **IST** practitioners.

To develop an **HST**-like structure of external sets, we have to incorporate definable collections of elementary external sets, definable collections of those collections, *etc.*, where "*etc.*" means that the procedure may be iterated along all ordinals. Technically, this is realized by a coding of sets by well-founded trees, so that a set $F(t) \in \mathbb{I}$ is assigned, by an elementary external function F to every endpoint t of a well-founded elementary external tree T, then each point of the tree assembles the sets obtained at immediate successors, and the set assembled at the root is the result. We call sets obtained this way *sets constructible from internal sets*, because a certain version of the Gödel constructibility relative to $\mathbb{I}$ leads exactly to the class $\mathbb{L}[\mathbb{I}]$ of all those sets.

There are two important issues related to this construction. First, it results in a universe $\mathbb{L}[\mathbb{I}]$ which interprets the **HST** axioms: we prove this in Section 5.5. Second, using the fact that the construction of $\mathbb{L}[\mathbb{I}]$ can be fully coded in $\mathbb{I}$, we observe in Sections 5.2 and 5.4 that the basic relations ($\in$, $=$, standardness) between coded sets can be determined in $\mathbb{I}$ as relations between the codes. This allows to define, still on the **BST** platform, a coded version of the class $\mathbb{L}[\mathbb{I}]$ within $\mathbb{I}$, and prove that it satisfies **HST**. Saying the same differently, we obtain an **HST** extension of the **BST** universe. This demonstrates a large scale of flexibility of the **BST**/**HST**–based foundations of nonstandard mathematics: indeed, a minimal kit of tools provided by **BST** (where no external sets exist) is essentially sufficient to extend the **BST** universe to a much wider universe of **HST** containing plenty of external sets (but the same internal sets).

Yet another aspect of $\mathbb{L}[\mathbb{I}]$ deserves attention. We prove that the class $\mathbf{\Delta}_2^{\mathtt{ss}}$ is equal to the collection of all sets $X \subseteq \mathbb{I}$ which belong to $\mathbb{L}[\mathbb{I}]$. It follows that $\mathbf{\Delta}_2^{\mathtt{ss}}$ is connected with a larger set universe which models **HST** — namely, the class $\mathbb{L}[\mathbb{I}]$. This explains why "external sets", informally introduced essentially as $\mathbf{\Delta}_2^{\mathtt{ss}}$ sets in some papers written by **IST**-followers (see, *e.g.*, van den Berg [vdBerg 87, vdBerg 92]) do not lead to concrete problems: the naïve reasoning actually stands on the firm platform given by $\mathbb{L}[\mathbb{I}]$.

Then we concentrate on the problem of absence of the Power Set axiom in **HST**. It was discovered by Hrbaček [Hr 78] that full Saturation is incompatible with Power Set in the presence of Replacement in the st-$\in$-language, in particular, in **HST**, infinite internal sets do not have power sets (of all external subsets; the "internal" power set does exist in $\mathbb{I}$, of course). This leads to a problem: how to perform in **HST** mathematical constructions which involve power sets ? The absence of Choice in **HST** (*e.g.*, the set ${}^*\mathbb{N}$ of all natural numbers in the sense of $\mathbb{I}$, is not well-orderable) also causes problems, but to less extent: we still have Choice for well-orderable families.

Fortunately, the problem can be reasonably resolved. Given an infinite cardinal κ, we define in **Chapter 6** the class $\mathbb{I}_\kappa$ of all internal sets that belong to standard sets of cardinality $\leq {}^*\kappa$ in $\mathbb{S}$, and show that the class $\mathbb{L}[\mathbb{I}_\kappa] \subseteq \mathbb{L}[\mathbb{I}]$ of all sets constructible in $\mathbb{H}$ from sets in $\mathbb{I}_\kappa$ models an appropriate κ-version of **HST** (with Saturation restricted by κ) together with the Power Set axiom. Thus, practically, nonstandard arguments which involve power sets can be adequately modeled in a subuniverse of the form $\mathbb{L}[\mathbb{I}_\kappa]$, where κ can be chosen big enough to provide the amount of Saturation required in any given nonstandard construction. Another system of κ-saturated subuniverses of the form $\mathbb{WF}[\mathscr{I}]$ satisfying both Power Set and Choice will be proposed: each class $\mathbb{WF}[\mathscr{I}]$ is a kind of von Neumann hull over a class $\mathscr{I} \subseteq \mathbb{I}$.

Classes $\mathbb{I}_\kappa$ and $\mathscr{I}$ involved in this construction belong to the family of subclasses of the internal universe $\mathbb{I}$ having the common property of being *standard–closed*, that is closed under application of any standard function. All of them are elementary substructures of $\mathbb{I}$ in the $\in$-language. For instance if $w \in \mathbb{I}$ then $\mathbb{S}[w] = \{{}^*\!f(w) : f \in \mathbb{WF}\}$ is such a class, and actually an ultrapower of $\mathbb{WF}$. It is an attractive idea to replace the standardness predicate by its relative version, *i.e.* the predicate $w\text{-}\mathtt{st}\,x$ iff $x \in \mathbb{S}[w]$. Unfortunately the structure $\langle \mathbb{I}; \in, w\text{-}\mathtt{st}\rangle$ obtained this way does not satisfy **BST** (unless w is standard, of course: then $\mathbb{S}[w] = \mathbb{S}$). To be more exact, we demonstrate in **Chapter 6** that Standardization fails.

Chapter 7 shows how *forcing* can be used to explore metamathematical properties of **HST**. Normally forcing needs the ground universe to be $\in$-well-founded, but it can be easily adjusted to the case when the ground universe is well-founded over an ill-founded transitive set or class of "urelements", however in this case the "urelements" cannot gain new members in the extension. As the universe $\mathbb{H}$ of all sets is well-founded over the transitive class $\mathbb{I}$ of all internal sets, forcing can be employed to build up extensions of models of **HST** which preserve internal and standard sets. (Special arguments will be used to verify that those axioms of **HST** like Standardization, which have no direct analogies in **ZFC**, remain true in the extension).

Our first application of forcing deals with the following question: is it possible that internal sets X, Y of different cardinality in $\mathbb{I}$ become equinumerous in the external universe $\mathbb{H}$? We prove that the universe $\mathbb{L}[\mathbb{I}]$ is regular enough to prohibit such a collapse, with some motivated exceptions among hyperfinite sets. In the other direction, we define a generic extension of $\mathbb{L}[\mathbb{I}]$ (still a model of **HST**) in which two given infinite internal sets (perhaps of different $\mathbb{I}$-cardinalities, *e.g.*, ${}^*\mathbb{N}$ and ${}^*\mathbb{R}$) are equinumerous. This example leads to a more complicated construction of a generic extension of $\mathbb{L}[\mathbb{I}]$ where **IP**, Henson's *isomorphism property* (saying, in this context, that all elementarily equivalent internally presented structures are isomorphic) holds: then, in particular, all internal elementary extensions of $\mathbb{R}$ are isomorphic, which represents a very strong solution of the uniqueness problem. This construction causes a lot of work, its elementary part is a generic isomorphism between two given elementarily equivalent structures $\mathfrak{A} = \langle A; ... \rangle$ and $\mathfrak{B} = \langle B; ... \rangle$, obtained with a forcing which consists of all internal partial $1 - 1$ maps $p : A \to B$ such that every $b = p(a)$ satisfies in $\mathfrak{B}$ the same formulas as a satisfies in $\mathfrak{A}$, and a back-and-forth argument.

We end the foundational part of the book in **Chapter 8** with a review of metamathematical properties of some other nonstandard set and class theories, including those by Kawaï, Hrbaček, Ballard, Gordon, Di Nasso. Some of them give alternative solutions of the Hrbaček paradox, and, generally, propose different views, sometimes very different from **HST**, of "the" nonstandard set universe, based on different principles. But it seems as if **HST** is the only theory which has the following reasonable and philosophically motivated combination of properties:

1°. Unique existence of a $*$-extension ${}^*\mathfrak{A}$ for any "standard" mathematical structure $\mathfrak{A}$ (for instance, ${}^*\mathbb{R}$ and ${}^*\mathbb{N}$) which is standard size saturated rather then simply κ-saturated for a chosen standard cardinal κ. (See Footnote 3 on p. VI on the uniqueness in this context.)

2°. The property of being "realistic", *i.e.*, interpretability in **ZFC** in such a way that the class of all standard sets in the sense of the interpretation coincides with the **ZFC** set universe — this implies the properties of conservativity and equiconsistency with respect to **ZFC**.

3°. $*$-methods of model-theoretic nonstandard analysis and those elaborated by **IST** practitioners are both fully available.

The final **Chapter 9** is devoted to *"hyperfinite"* descriptive set theory, a branch of nonstandard mathematics concerned with structures, quite similar to those known in classical descriptive set theory, but built upon internal subsets of a certain internal set rather than on open sets of a separable metric space. This leads to Borel and projective hierarchies, quite similar to hierarchies of classical, or Polish, descriptive set theory according to their definitions, but sometimes having rather different properties, and quite often involving very different arguments even if the properties look similar. In addition, a larger class of *countably determined* sets (*i.e.*, those obtained by δs-operations over internal sets) appears; it has no direct "Polish" analogy. This part of the book is not intended to be a really comprehensive exposition, our aim will rather be to show by selected examples that the topic admits a unified development on the base of **HST**. This will include well established branches like the structure of the hierarchies and Loeb measures, together with several distinguished results of recent time, including reducibility and cardinality properties of Borel and countably determined sets and quotient structures, with generalizations of Jin's results on transversals for monad equivalence relations.

Table of logical dependencies.

In the following table, which displays major logical connections between different parts of the book, thick lines mean a strong dependence while thin lines mean a circumstantial but still recognizable dependence. A figure like **1**:1,2,3 in the framebox means: Sections 1.1, 1.2, 1.3.

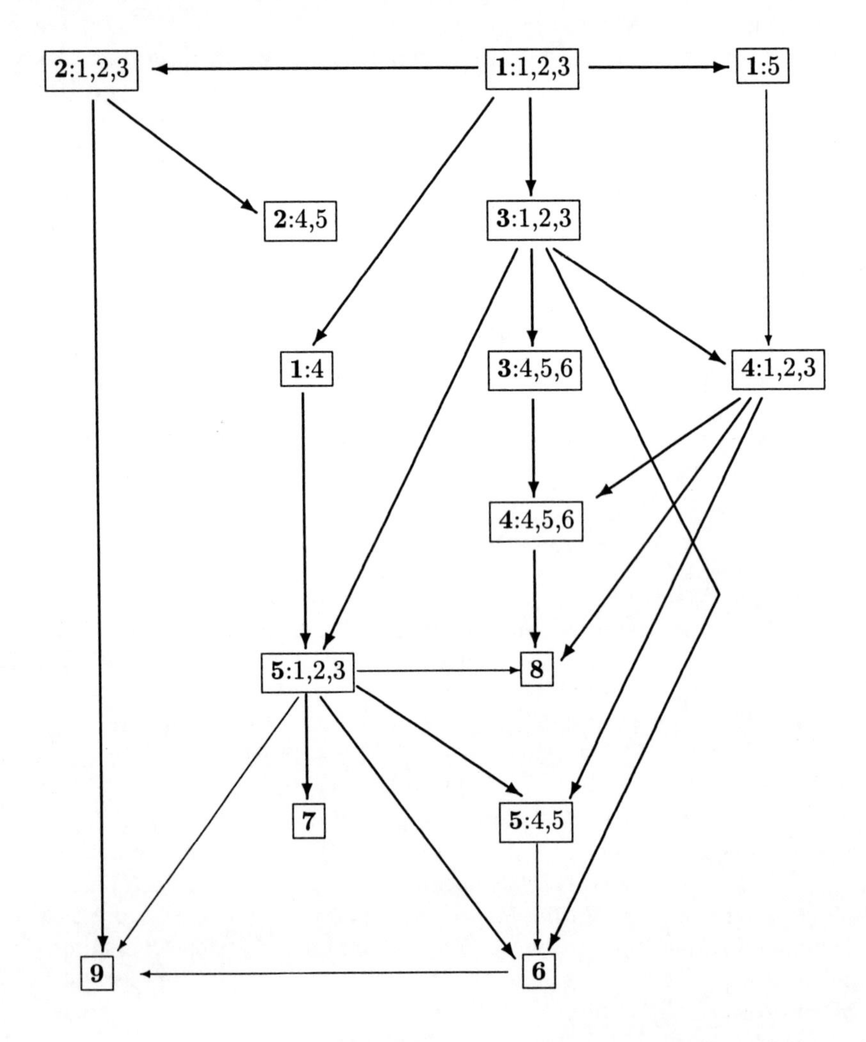

How to read this book

The **main course** consists of Sections 1.1 – 1.4, 3.1, 3.2, 4.1, 4.2, 4.3 (with an excursion to 1.5, if necessary, where notions like *model* or *interpretation* are considered), and **Chapter 5** as a whole. This is a course on the axioms of **HST**, the structure of the **HST** set universe, including the class $\mathbb{L}[\mathbb{I}]$ of all sets constructible from internal sets (an inner model of **HST**), the main mathematical properties of **HST** (which includes the metamathematics of *bounded set theory* **BST**, essentially, the theory of the internal part of the **HST** universe), and finally a complete solution of the question how to deal with external sets (of any kind) within the frameworks of an internal set universe.

Additional choices include

(1) the study of metamathematical properties of **IST** in Sections 3.4 – 3.6 and 4.4, 4.5, 4.6, including the truth definition for $\in$-formulas in **IST** and the nature of **IST**-extendible standard models of **ZFC**;
(2) the study of "internal subuniverses" — subclasses of $\mathbb{I}$ closed under the application of standard functions, and their external extensions which (unlike the whole set universe of **HST**) can satisfy the Power Set axiom, in Chapter 6;
(3) forcing in **HST** in Chapter 7, in particular, the consistency of the following statement (false in $\mathbb{L}[\mathbb{I}]$): *all infinite internal sets are externally equinumerous.*

These issues are largely independent of each other over the main course, and so is the following one, except for its dependence on (1):

(4) a review of metamathematical properties of some other nonstandard set theories, in comparison with **HST**, in Chapter 8.

Finally, there is a **special course** concentrated on an application: "hyperfinite", or "nonstandard" descriptive set theory, in Chapter 9, available independently of the foundational issues, after a superficial reading of Sections 1.1 – 1.4 and Chapter 2. In principle, the content of the two first sections of Chapter 2 suffices to make sure that the $*$-tools developed in the model theoretic version of nonstandard analysis work as well (and in the same notational system) on the axiomatic base of **HST**.

Basic notation

An index of notation is given in the end, but here we would like to precede the exposition with a review of more general set theoretic notation. Some of the following notions will be specified and explained in more detail below.

class: a collection of sets defined by a formula [14]

$\exists$, $\forall$: *quantifiers* (there exists... , for all...)

$\mathtt{card}\, X$ is the *cardinality* of a set X

sets X, Y are *equinumerous* (or have the same cardinality) if there is a bijection $f : X$ onto Y

$\#X = \mathtt{card}\, X$ for finite sets X (the number of elements)

$\mathtt{dom}\, R = \{x : \exists y\, (\langle x, y\rangle \in R)\}$ for R being a set of pairs

$\mathtt{ran}\, R = \{y : \exists x\, (\langle x, y\rangle \in R)\}$ for R being a set of pairs

$\mathtt{dom}\, R$, $\mathtt{ran}\, R$ also have the meaning of the *projection* of R onto the first, resp., second co-ordinate

$f"X = \{f(x) : x \in \mathtt{dom}\, f \cap X\}$ the f-image of a set X (f being a function)

$X \Delta Y = (X \smallsetminus Y) \cup (Y \smallsetminus X)$ the *symmetric difference* of sets X, Y

$X^{\mathtt{C}} = C \smallsetminus X$: the *complement* of X to a given domain C

$\mathscr{P}(X)$, the *power set* or *power class* of X (also a set or class): the collection of all subsets or all subclasses of X depending on the context

X^Y (sometimes ${}^Y X$): the family of all functions $f : Y \to X$

$\bigcup X = \{y : \exists x \in X\, (y \in x)\}$ the *union* of all elements of X

$\bigcap X = \{y : \forall x \in X\, (y \in x)\}$ the *intersection* of all elements of X

$X \cup Y$, $X \cap Y$ the union, resp., intersection of two sets

natural number: formally the set of all smaller natural numbers, so that $n = \{0, 1, 2, \ldots, n-1\}$

finite set: a set equinumerous to a natural number

$\mathbb{N}$ is the set of all natural numbers

ordinal: formally the set of all smaller ordinals, $\alpha = \{\gamma : \gamma < \alpha\}$

$\mathtt{Ord}$ is the class of all ordinals

ω is the least (infinite) limit ordinal, formally $\omega = \mathbb{N}$

cardinal: an initial ordinal, *i.e.* an ordinal not equinumerous to any smaller ordinal

$\aleph_0$, $\aleph_1$, $\mathfrak{c} = 2^{\aleph_0}$: in the ordinary sense

$s^\wedge a$: a is added as a rightmost term to a sequence s

$s^\wedge t$: *concatenation* of sequences s, t

$M \models \Phi$ means that a closed formula Φ is true in a structure M [15]

Φ^M denotes the *relativization* of a formula Φ to a structure M

---

[14] Theories which study classes as formal objects will not be considered.

[15] In the sence of the existence of a validation function defined on all subformulas and satisfying certain known rules.

1 Getting started

This chapter introduces **HST**, *Hrbaček set theory*.

We present the system of axioms (Section 1.1), outline important patterns of reasoning in **HST**, and explore the structure of the set universe under the axioms of **HST**. This will be based on an asterisk map $x \mapsto {}^*x$ (§ 1.1d), an $\in$-isomorphism of the class $\mathbb{WF}$ of well-founded sets onto the standard universe $\mathbb{S} \subseteq \mathbb{I}$, so that the followers of the model-theoretic version of foundations of nonstandard analysis may feel comfortable. We introduce in Section 1.2 ordinals, cardinals, natural numbers, finite sets *etc.*, and their $*$-versions. This continues in Section 1.3 where we discuss sets of standard size, Saturation, and the principle of Extension. The next Section 1.4 studies the class $\mathbf{\Delta}_2^{\mathtt{ss}}$ of all sets which can be obtained from internal sets by standard size unions and intersections; we prove, for instance, that there is no $\mathbf{\Delta}_2^{\mathtt{ss}}$ surjection of an internal set X onto a $*$-infinite internal set Y of bigger $*$-cardinality. We finish in Section 1.5 with some topics related rather to model theory and metamathematics, like the notions of *model* and *interpretation*.

Before the technical part starts, it is worth to note that the **HST** universe resembles a **ZFC**-like universe with urelements (atoms), the urelements being internal sets (*i. e.*, elements of standard sets). It is clear that internal sets do not really behave like urelements, in particular, they are true sets containing elements (themselves internal sets) and form a proper class $\mathbb{I}$ which is itself a model of **ZFC**. However, similarly to urelement-based set theories, the rest of the set universe will be postulated to be a von Neumann–like well-founded superstructure over $\mathbb{I}$, the height of which is equal to the ordinals in the well-founded universe $\mathbb{WF}$.

Unfortunately, some intrinsic properties of this construction discovered by Hrbaček (see Theorem 1.3.9 below) do not allow us to keep all of **ZFC** in the **HST** universe: the axioms of Power Set and Choice (and Regularity, of course) are to be sacrificed, if we want that the amount of Saturation available is not restricted by a cardinal. Chapter 6 of this book will show how to cope with this problem in **HST**. More radical methods to save Power Set and Choice in the nonstandard set-up at the cost of some other axioms (for instance, Replacement) will be considered in Chapter 8.

Now let us come to the details.

1.1 The axiomatical system of Hrbaček set theory

This section presents the axioms of Hrbaček set theory, or **HST**, and describes the basic structure of the **HST** set universe. Syntactically, **HST** is a theory in the st-∈-*language*, which contains a binary predicate of membership ∈ and a unary predicate of *standardness* st (and equality of course) as the primary notions. Formula $x \in y$ reads: x *belongs to* y, or x *is an element of* y, with the usual set theoretic understanding of membership. The formula $\mathtt{st}\, x$ reads: x *is standard*, its meaning will be explained below.

A st-∈-*formula* is a formula of the st-∈-language.

An ∈-*formula* is a formula of the ∈-*language* having ∈ as the only atomic predicate. Thus an ∈-formula is a st-∈-formula in which the standardness predicate does not occur. ∈-formulas are also called *internal formulas*, in opposition to *external formulas*, *i.e.*, those st-∈-formulas containing st.

1.1a The universe of HST

Hrbaček set theory **HST** deals with four major types of sets: external, internal, standard, and well-founded.

First of all, *standard* sets are those sets x which satisfy $\mathtt{st}\, x$. *Internal* sets are those sets y which satisfy $\mathtt{int}\, y$, where $\mathtt{int}\, y$ is the formula $\exists^{\mathtt{st}} x\,(y \in x)$ (saying: y belongs to a standard set). Thus,

$$\mathbb{S} = \{x : \mathtt{st}\, x\} \quad \text{is the class of all standard sets;}$$

$$\mathbb{I} = \{y : \mathtt{int}\, y\} = \{y : \exists^{\mathtt{st}} x\,(y \in x)\} \quad \text{is the class of all internal sets.}$$

The class $\mathbb{I}$ is the source of some typical objects of "nonstandard" mathematics like hyperintegers and hyperreals.

Blanket agreement 1.1.1. Thus, internal sets are precisely all sets which are elements of standard sets. This understanding of the notion of internality and the associated notions like $\mathbb{I}$, $\exists^{\mathtt{int}}$, $\forall^{\mathtt{int}}$ is default throughout the book. All exceptions (*e.g.*, when **IST** is considered) will be explicitly indicated. □

External sets are simply all sets in the nonstandard universe of **HST**. We shall use $\mathbb{H}$ (a tribute to Hrbaček) to denote the class of all external sets. Thus, $\mathbb{H}$ is the "universe of discourse", the universe of all sets considered by the theory, including the class $\mathbb{WF}$ of all well-founded sets which can be informally identified with the "conventional" mathematical universe (in which all sets are well-founded); $\mathbb{WF}$ will satisfy all axioms of **ZFC**. The class $\mathbb{S}$ of all standard sets (determined by the predicate st, as above) will be shown to be ∈-isomorphic to $\mathbb{WF}$. In a sense, $\mathbb{S}$ is an "isomorphic expansion" of $\mathbb{WF}$ into $\mathbb{H}$. Given that $\mathbb{S}$ is not transitive, $\mathbb{I}$ arises naturally as the class of all elements of sets in $\mathbb{S}$. It is viewed as an elementary extension of $\mathbb{S}$ (in ∈-language), and thereby also of $\mathbb{WF}$. Finally, $\mathbb{H}$ is a comprehensive universe in which all these classes coexist in a reasonable common set theoretic structure, with $\in$ having the natural meaning in all mentioned universes.

An additional advantage of the identification of $\mathbb{WF}$ with the "conventional" mathematical universe is that it allows to carry out "nonstandard" arguments in close analogy and using the same $*$-notation as it is customary in the model theoretic version of nonstandard analysis.

Correspondingly, there are alternative interpretations in which either $\mathbb{S}$ or $\mathbb{I}$ is informally considered as incarnation of the "conventional" mathematical universe. Identifying the latter with $\mathbb{S}$ [1] is well in line with model theoretic ideas but strange from a "realistic" point of view of set theory which fixes the meaning of $\in$ to be the natural one. Assuming this point of view forces one to accept that traditional mathematics talks about sets with "invisible" (*i.e.*, nonstandard) elements. Identifying the "conventional" universe with $\mathbb{I}$ [2] is more compatible with a "realistic" point of view but has the irritating effect that the standardness predicate destroys the intrinsic well-foundedness of the sets in $\mathbb{I}$ in the $\in$-language, because standardness defines a "smaller" infinity than given in the original theory. Thus, from a "realistic" point of view, identifying $\mathbb{WF}$ with the "conventional" universe is the most appealing (see also § 1.2a below), but technically the class $\mathbb{S}$ will play a crucial role in streamlining certain arguments.

The axioms of **HST** are assembled in three groups, presented in §§ 1.1b, 1.1c, 1.1f together with a few introductory results. For convincing technical

[1] "... intuitively, standard sets should be identified with the members of the 'universe of discourse' of $\mathfrak{T}$", this is from [Hr 78], where $\mathfrak{T}$ is a standard theory, in our context, **ZFC**.

[2] This scheme, based on Nelson's ideas [Nel 77, Nel 88], views $\mathbb{I}$, the internal universe, as the universe of conventional mathematics. For the reader's convenience, we cite here a few phrases from Nelson's unpublished book [Nel **].

"I want to begin by introducing a new predicate 'standard' to ordinary mathematics without defining it. The reason for not defining 'standard' is that it plays a syntactical, rather than semantic, role in the theory. It is similar to the use of 'fixed' in informal mathematical discourse. One does not define this notion [...]. But the predicate 'standard' – unlike 'fixed' – will be part of the formal language of our theory [...]. We shall introduce axioms for handling this new predicate 'standard' in a relatively consistent way. In doing so, we do not enlarge the world of mathematical objects in any way, we merely construct a richer language to discuss the same objects as before."

In other words, Nelson views $\mathbb{I}$ as the same old "standard" set universe, satisfying **ZFC**, to which mathematicians are accustomed, where a new predicate of standardness $\mathtt{st}$ is defined by a list of axioms (of *internal set theory* **IST**, see Chapter 3) which govern its interaction with $\in$, the basic predicate of membership in **ZFC**. As long as we deal only with standard and internal sets (as in **IST**), this approach can be appropriate and defensible, but in the **HST** environment the major inconvenience is that $\mathbb{I}$-notions are different from the $\mathbb{H}$-notions, for instance, there exist much more $\mathbb{I}$-natural numbers than natural numbers. Anyway, this is very different from our attitude.

reasons it is more convenient to arrange the axioms in such a way that the pair $\mathbb{S} \subseteq \mathbb{I}$ instead of $\mathbb{WF}$ occupies the central place.

1.1b Axioms for the external universe

This group includes the **ZFC** Extensionality, Pair, Union, Infinity axioms [3] and the schemata of Separation and Collection (therefore also Replacement, which is a consequence of Collection, as usual) for all st-$\in$-formulas.

The Power Set, Choice, and Regularity axioms known from **ZFC** are not included: they happen to contradict the axioms of **HST** see Exercise 1.2.15(3) and Remark 1.3.10. Still we'll be able to incorporate Regularity and Choice in weaker but useful forms. However the included axioms do support most of basic set theoretic notions in $\mathbb{H}$, like ordered and unordered pairs, function, domain, range, relation, Cartesian products, unions and intersections *etc.*, as long as power sets are not involved. In addition, the axioms of this group legalize the use of the standardness predicate in definitions of sets; for instance, ${}^{\sigma}X = \{x \in X : \mathtt{st}\, x\} = X \cap \mathbb{S}$, the collection of all standard elements of X, is a legitimate set for any set X in **HST**.

Exercise 1.1.2. Prove, using the axioms of this group, that the cartesian product $X \times Y$ of sets X, Y is a set. (*Hint.* By Pair, $f_x(y) = \langle x, y\rangle$ is a set. By Replacement, $P_x = \{\langle x, y\rangle : y \in Y\}$ is a set. Again by Replacement and Union, $X \times Y = \bigcup_{x \in X} P_x$ is a set.) □

1.1c Axioms for standard and internal sets

Let us introduce convenient notation. Let quantifiers $\exists^{\mathtt{st}}$ and $\forall^{\mathtt{st}}$ be shortcuts meaning: *there exists a standard* ... , *for all standard* ... , formally,

$$\exists^{\mathtt{st}} x\, \varphi(x) \text{ means } \exists\, x\, (\mathtt{st}\, x \wedge \varphi(x)), \quad \forall^{\mathtt{st}} x\, \varphi(x) \text{ means } \forall\, x\, (\mathtt{st}\, x \Longrightarrow \varphi(x)).$$

Quantifiers $\exists^{\mathtt{int}}$ and $\forall^{\mathtt{int}}$ (meaning *there exists an internal* ... , *for all internal* ...) are introduced similarly.

If Φ is an $\in$-formula then $\Phi^{\mathtt{st}}$, the *relativization* of Φ to $\mathbb{S}$, is the formula obtained by restriction of all quantifiers in Φ to the class $\mathbb{S}$, so that all occurrences of $\exists\, x \dots$ are changed to $\exists^{\mathtt{st}} x \dots$ while all occurrences of $\forall\, x \dots$ are changed to $\forall^{\mathtt{st}} x \dots$. In other words, $\Phi^{\mathtt{st}}$ says that Φ is true in $\mathbb{S}$. Relativization $\Phi^{\mathtt{int}}$, which displays the truth of an $\in$-formula Φ in the universe $\mathbb{I}$, is defined similarly: the quantifiers $\exists$, $\forall$ change to $\exists^{\mathtt{int}}$, $\forall^{\mathtt{int}}$.

The following axioms specify the behaviour of standard and internal sets.

ZFC$^{\mathtt{st}}$: The collection of all formulas of the form $\Phi^{\mathtt{st}}$, where Φ is an $\in$-statement which is an axiom of **ZFC**.

[3] We present, for the reader's convenience, the axioms of **ZFC**, the "standard" Zermelo–Fraenkel axiomatic set theory, in § 1.1h below. Yet some knowledge of set theory is assumed.

In other words, it is postulated that the universe $\mathbb{S}$ is a **ZFC** universe. (Note that the **ZFC** axioms are assumed to be formulated as certain closed $\in$-formulas in this definition.) This is enough to prove the following result:

Lemma 1.1.3. $\mathbb{S} \subseteq \mathbb{I}$.

Proof. Let $x \in \mathbb{S}$. The formula $\exists y\,(x \in y)$ is a theorem of **ZFC**, therefore $(\exists y\,(x \in y))^{\mathtt{st}}$, that is the formula $\exists^{\mathtt{st}} y\,(x \in y)$, is true. In other words, x is an element of a standard set, which means $x \in \mathbb{I}$. □

Transfer: $\Phi^{\mathtt{int}} \Longleftrightarrow \Phi^{\mathtt{st}}$, where Φ is an arbitrary closed $\in$-formula containing only standard sets as parameters [4].

Transitivity of $\mathbb{I}$: $\forall^{\mathtt{int}} x\, \forall y \in x\, (\mathtt{int}\, y)$.

Regularity over $\mathbb{I}$: For any nonempty set X there exists $x \in X$ such that $x \cap X \subseteq \mathbb{I}$. (The full Regularity of **ZFC** requires $X \cap x = \varnothing$.)

Standardization: $\forall X\, \exists^{\mathtt{st}} Y\, (X \cap \mathbb{S} = Y \cap \mathbb{S})$. (Such a standard set Y, unique by Transfer and Extensionality, is sometimes denoted by ${}^{\mathrm{S}}X$.)

Transfer can be considered as saying that $\mathbb{I}$, the universe of all internal sets, is an elementary extension of $\mathbb{S}$ in the $\in$-language. It follows, by **ZFC**$^{\mathtt{st}}$, that the class $\mathbb{I}$ of all internal sets satifies **ZFC** (in the $\in$-language), in fact, we can replace **ZFC**$^{\mathtt{st}}$ by **ZFC**$^{\mathtt{int}}$, with relativization to $\mathbb{I}$, in the list of **HST** axioms. See also Theorem 1.1.9 below.

Transitivity of $\mathbb{I}$ postulates internal sets to form the basement of the $\in$-structure of the universe $\mathbb{H}$. This axiom is very important since it implies that some set operations in $\mathbb{I}$ retain their sense in the whole universe $\mathbb{H}$.

Regularity over $\mathbb{I}$ organizes the **HST** set universe $\mathbb{H}$ in a sort of hierarchy over the internal universe $\mathbb{I}$, in the same way as the Regularity axiom organizes the universe in the von Neumann hierarchy over the empty set $\varnothing$ in **ZFC**. (The **ZFC** Regularity fails in $\mathbb{H}$: the set of all nonstandard $\mathbb{I}$-natural numbers does not contain an $\in$-minimal element, see Exercise 1.2.15(3).) Yet there is an essential difference with the **ZFC** setting: we shall see that $\mathbb{I}$, the ground level, implicitly contains a sufficient amount of information on all ordinals involved in the cumulative construction of $\mathbb{H}$ from $\mathbb{I}$.

Standardization postulates that $\mathbb{H}$ does not contain collections of standard sets other than those of the form $S \cap \mathbb{S}$ for a standard set S.

---

[4] To be more exact, Transfer is the collection of all statements of the form

$$\forall^{\mathtt{st}} x_1 \ldots \forall^{\mathtt{st}} x_n \left(\Phi^{\mathtt{st}}(x_1, \ldots, x_n) \Longleftrightarrow \Phi^{\mathtt{int}}(x_1, \ldots, x_n) \right),$$

where $\Phi(x_1, \ldots, x_n)$ is an arbitrary $\in$-formula. The quantifiers $\forall^{\mathtt{st}} x_1 \ldots \forall^{\mathtt{st}} x_n$ express the requirement that "all parameters are standard". Yet this is too cumbersome, especially in some other cases, so we prefer to use semi-formal phrases like: "for any formula with such-and-such parameters" as long as an exact meaning is clear and automatically recoverable.

1.1d Well-founded sets

Now we can consistently introduce the last principal class: well-founded sets. Recall the following notions from general set theory.

Definition 1.1.4. A binary relation $\prec$ on a set or class X is *well-founded* if any nonempty set $Y \subseteq X$ contains a $\prec$-minimal element, that is there exists $x \in Y$ such that no $y \in Y$ satisfies $y \prec x$.

A set or class X is *transitive* if any $x \in X$ satisfies $x \subseteq X$, *i.e.*, elements of elements of X are elements of X, and is $\subseteq$*-complete* if we have $y \in X$ whenever $y \subseteq x \in X$, that is subsets of elements of X are elements of X.

A set x is *well-founded* if there is a transitive set X such that $x \subseteq X$ and the restriction $\in \restriction X$ is a well-founded relation. □

It is known that all sets are well-founded in **ZFC** by the Regularity axiom. This is not the case in **HST** : the set ${}^*\mathbb{N}$ of all $\mathbb{I}$-natural numbers is ill-founded. But well-founded sets play a special role in **HST**!

Definition 1.1.5 (HST). Let $\mathtt{wf}\, x$ mean that x is well-founded [5].

We put $\mathbb{WF} = \{x : \mathtt{wf}\, x\}$, the class of all well-founded sets. □

We introduce quantifiers $\exists^{\mathtt{wf}}$ and $\forall^{\mathtt{wf}}$ (meaning: there is a well-founded ... , for any well-founded ...) and the relativization $\Phi^{\mathtt{wf}}$ to $\mathbb{WF}$ similarly to $\exists^{\mathtt{st}}$, $\forall^{\mathtt{st}}$, $\Phi^{\mathtt{st}}$ in § 1.1c. In other words, $\Phi^{\mathtt{wf}}$ says that Φ is true in $\mathbb{WF}$.

The main property of the class $\mathbb{WF}$ in **HST** is that it admits a definable $\in$-isomorphism $w \longmapsto {}^*w$ onto the class $\mathbb{S}$ of all standard sets.

Definition 1.1.6. Define ${}^*w \in \mathbb{S}$ (the $*$*-extension* of w) for any set $w \in \mathbb{WF}$ by $\in$-induction as follows:

($*$) If all sets ${}^*u \in \mathbb{S}$, $u \in w$, are defined, let, by Standardization, ${}^*w = {}^{\mathbb{S}}\{{}^*u : u \in w\}$ be the only standard set x such that $x \cap \mathbb{S} = \{{}^*u : u \in w\}$.

Exercise: prove that ${}^*\varnothing = \varnothing$. □

Remark 1.1.7. Definition 1.1.6 is a definition by $\in$-*induction*, with $\mathbb{WF}$ as the domain of induction: we define a set ${}^*w$ for any $w \in \mathbb{WF}$, assuming that ${}^*u$ has been defined for every $u \in w$. It is known from basic studies in set theory (see, *e.g.*, [Jech 78, 1.9]) that to make such a definition legitimate the relation $\mathsf{E} = \in \restriction \mathbb{WF}$ has to be well-founded (then we say: well-founded induction), which will be proved below in Theorem 1.1.9, and, for any x in the domain, the collection $\{y : y \mathsf{E}\, x\}$ must be a set, which is in our case vacuous. □

It follows from Theorem 1.1.9 below that for any standard x there is a unique set $w \in \mathbb{WF}$, denoted by $\hat{x}$, such that $x = {}^*w$. The construction of $\hat{x}$

[5] Note that the formula $\mathtt{wf}\, x$ is $\in$-definable in **HST**, that is it does not involve $\mathtt{st}$, the other atomic element of the language of **HST**.

can be explained through the following process, a kind of *Mostowski collapse*. Any given nonempty standard set x can contain standard and nonstandard elements. Let us remove all nonstandard elements. Each of the remaining standard elements of x can also contain nonstandard elements, so we remove them as well. *Et cetera*. The result is a set $\hat{x} \in \mathbb{WF}$, the *condensed set* of x.

Exercise 1.1.8. Prove that $\widehat{X} = \{\hat{x} : x \in X \cap \mathbb{S}\}$ for any standard X. □

1.1e The $\in$-structure of internal and well-founded sets

The following theorem shows that the axioms introduced in §§ 1.1b, 1.1c turn the classes $\mathbb{S}$, $\mathbb{I}$, $\mathbb{WF}$ into rather convenient set universes which satisfy the axioms of **ZFC** in $\in$-language. For instance, we prove that $\mathbb{WF}$ and $\mathbb{I}$ are transitive classes while $\mathbb{S}$ is a non-transitive $\in$-icomorphic "copy" of $\mathbb{WF}$ with $\mathbb{I}$ being its "transitive closure" and an elementary extension of $\mathbb{S}$ in $\in$-language, in the sense of the following principle:

$*$-Transfer: $\Phi(x, y, ...)^{\mathrm{wf}} \iff \Phi({}^*x, {}^*y, ...)^{\mathrm{int}}$, where Φ is an arbitrary parameter-free $\in$-formula and x, y, ... any elements of $\mathbb{WF}$.

Theorem 1.1.9. *The classes* $\mathbb{WF}$ *and* $\mathbb{S} \subseteq \mathbb{I}$ *have the following properties*:

(i) *The relation* $\in \restriction \mathbb{S}$ *is well-founded. The class* $\mathbb{S}$ *interprets* **ZFC** [6].

(ii) *The relation* $\in \restriction \mathbb{WF}$ *is well-founded. The class* $\mathbb{WF}$ *is transitive,* $\subseteq$*-complete (moreover,* $X \subseteq \mathbb{WF} \Longrightarrow X \in \mathbb{WF}$*), and interprets* **ZFC** [7]. *The map* $x \mapsto {}^*x$ *is an* $\in$*-isomorphism of* $\mathbb{WF}$ *onto* $\mathbb{S} = \{{}^*w : w \in \mathbb{WF}\}$.

(iii) *The class* $\mathbb{I}$ *is transitive* [8] *and interprets* **ZFC**. *The map* $x \mapsto {}^*x$ *is an* $\in$*-elementary embedding of* $\mathbb{WF}$ *in* $\mathbb{I}$*, so that* $*$*-Transfer holds.*

Proof. (i) Suppose that $\varnothing \neq Y \subseteq \mathbb{S}$. By Standardization, there is a standard set X with $Y = X \cap \mathbb{S}$. By $\mathbf{ZFC}^{\mathrm{st}}$, there is an element $x \in X \cap \mathbb{S}$ which is $\in$-minimal in the $\mathbb{S}$-sense, *i.e.*, no set $y \in \mathbb{S}$ satisfies $y \in x \cap X$. Then y is $\in$-minimal in Y. That $\mathbb{S}$ interprets **ZFC** is asserted by $\mathbf{ZFC}^{\mathrm{st}}$.

(ii) To prove that $\mathbb{WF}$ is transitive show that any $X \in \mathbb{WF}$ satisfies $X \subseteq \mathbb{WF}$. By definition $X \subseteq u$ for a transitive set u with $\in \restriction u$ well-founded. Then any $x \in X$ satisfies $x \subseteq u$ since u is transitive, thus $x \in \mathbb{WF}$.

Now suppose that $X \subseteq \mathbb{WF}$. By the axiom of Collection there is a set U such that for any $x \in X$ there is a transitive set $u \in U$ with $x \subseteq u$ and

[6] By this word "interprets" we mean here that, for any axiom Φ of **ZFC**, Φ^{st} is a theorem of **HST**, or, less formally, **HST** proves that Φ is true in $\mathbb{S}$, *i.e.*, the structure $\langle \mathbb{S}; \in \restriction \mathbb{S}\rangle$. (In this case Φ^{st} is even an axiom of **HST**.) See §§ 1.5d, 1.5e on the difference between the notions of "interpretation" and "model".

[7] Generally, it follows from the axioms of **ZF** minus Regularity that the class of all well-founded sets interprets **ZF**. This easy result belongs to the folklore of set theory and has no direct connection with nonstandard mathematics.

[8] The relation $\in \restriction \mathbb{I}$ is <u>not</u> well-founded and $\mathbb{I}$ is <u>not</u> $\subseteq$-complete, see below.

$\in\restriction u$ well-founded. Let Z be the union of all transitive sets $u \in U$ with $\in\restriction u$ well-founded. We leave it as a simple **exercise** for the reader to prove that Z is transitive, $\in\restriction Z$ is well-founded, and $X \subseteq Z$, and hence $X \in \mathbb{WF}$.

To see that $\in\restriction\mathbb{WF}$ is well-founded, let $\varnothing \neq X \subseteq \mathbb{WF}$. Take any $x \in X$. There is a transitive set Y with $x \subseteq Y$ and $\in\restriction Y$ well-founded, and hence there is an $\in$-minimal element in the set $Y \cap X$. This element is $\in$-minimal in X as well.

Prove that $w \longmapsto {}^*w$ is an $\in$-isomorphism $\mathbb{WF}$ onto $\mathbb{S}$. (By the way, it follows that $\mathbb{WF}$ interprets **ZFC** since so does $\mathbb{S}$.) According to the well-foundedness of $\in\restriction\mathbb{WF}$, we can prove that ${}^*w = {}^*v$ implies $w = v$, by $\in$-induction, that is, assuming that ${}^*a = {}^*b$ implies $a = b$ for all $a \in v$ and $b \in w$ (the inductive hypothesis). Now, if ${}^*v = {}^*w$ then, by definition, $\{{}^*a : a \in v\} = \{{}^*b : b \in w\}$, and hence $v = w$ by the inductive hypothesis.

Prove that any standard x has the form ${}^*w$ for some $w \in \mathbb{WF}$. Otherwise, by the well-foundedness of $\in\restriction\mathbb{S}$ (see above), there exists an $\in$-minimal standard set y not of this form, but such that every standard $z \in y$ is of this form, thus there is a set $w_z \in \mathbb{WF}$, unique by the above, such that $z = {}^*w_z$. We have $w = \{w_z : z \in y\} \in \mathbb{WF}$ by the above. Then clearly $y = {}^*w$, contradiction.

That $u \in v \Longleftrightarrow {}^*u \in {}^*v$ is left as an easy **exercise** for the reader.

(iii) Apply Transitivity of $\mathbb{I}$, Transfer, and the results of (ii), (i). □

Corollary 1.1.10. (i) *Suppose that an $\in$-formula $\varphi(x_1, ..., x_n, y)$ defines a* **ZFC** *operator, i.e.* **ZFC** *proves $\forall x_1 ... \forall x_n \exists! y\, \varphi(x_1, ..., x_n, y)$. Then for any $x_1, ..., x_n \in \mathbb{I}$ there is a unique set $y \in \mathbb{I}$ satisfying $\varphi^{\text{int}}(x_1, ..., x_n, y)$.*

(ii) *In particular, if $\psi(x_1, ..., x_n, y)$ is an $\in$-formula and $x_1, ..., x_n, X \in \mathbb{I}$ then the set $Y = \{y \in X : \psi^{\text{int}}(x_1, ..., x_n, y)\}$ belongs to $\mathbb{I}$.*

(iii) *The same holds for the classes $\mathbb{WF}$ and $\mathbb{S}$ and relativizations φ^{wf}, φ^{st}.*

Thus each of the three classes $\mathbb{S}$, $\mathbb{WF}$, $\mathbb{I}$ is closed under $\in$-definitions.

(iv) *Moreover if $x_1, ..., x_n \in \mathbb{S}$ and $y \in \mathbb{I}$ is a unique set in $\mathbb{I}$ satisfying $\varphi^{\text{int}}(x_1, ..., x_n, y)$ then y is standard.*

Proof. The classes interpret **ZFC** by Theorem 1.1.9. Claim (ii) follows because $\langle x_1, ..., x_n, X\rangle \mapsto \{y \in X : \psi(x_1, ..., x_n, y)\}$ is a **ZFC** operator. Claim (iv) is an easy consequence of Transfer. □

Exercise 1.1.11. Prove the following: (1) There is no set x with $x \in x$.

(2) None of the classes $\mathbb{WF}$, $\mathbb{S}$, $\mathbb{I}$ is a set: they are proper classes in **HST**.

(3) (*Boundedness*) If $X \subseteq \mathbb{I}$ then $X \subseteq S$ for a standard S.

(4) If $X, Y \in \mathbb{S}$ and $\mathtt{card}^{\mathbb{S}} X < \mathtt{card}^{\mathbb{S}} Y$ then there is no surjection $f : X \cap \mathbb{S} \xrightarrow{\text{onto}} Y \cap \mathbb{S}$. [9]

[9] Here, $\mathtt{card}^{\mathbb{S}} X$ is the cardinal of X defined in $\mathbb{S}$, a **ZFC** universe. Note that there still can exist a surjection of X onto Y in this case, see § 1.4d.

Hints. (1) By Regularity over $\mathbb{I}$, $x \cap \{x\} \subseteq \mathbb{I}$, hence, if $x \in x$ then $x \in \mathbb{I}$, a contradiction as $\mathbb{I}$ satisfies **ZFC**.

(2) If $\mathbb{WF}$ is a set then $\mathbb{WF} \in \mathbb{WF}$ by Theorem 1.1.9(ii), contradiction.

(3) Each $x \in X$ belongs to a set ${}^*y$, $y \in \mathbb{WF}$. By Collection, we have a set $Y \subseteq \mathbb{WF}$ such that any $x \in X$ belongs to some ${}^*y$, $y \in Y$. However $Y \in \mathbb{WF}$ by Theorem 1.1.9, hence, $U = \bigcup Y \in \mathbb{WF}$ as well. Put $S = {}^*U$.

(4) By Standardization, there is a standard set $F \subseteq X \times Y$ with $F \cap \mathbb{S} = f$ (maps are identified with their graphs). Show that, by Transfer, F maps X onto Y, which contradicts the assumption $\mathtt{card}^{\mathbb{S}}\, X < \mathtt{card}^{\mathbb{S}}\, Y$. □

1.1f Axioms for sets of standard size

We complete the list of **HST** axioms with three axioms leading to typically "nonstandard" ways of reasoning. Their common feature is a special role of sets of standard size.

Definition 1.1.12. *Sets of standard size* are sets of the form $\{f(x) : x \in X \cap \mathbb{S}\}$, where X is any set and f is any function with $X \cap \mathbb{S} \subseteq \mathtt{dom}\, f$. □

We shall see in § 1.3a that, in **HST**, to be of standard size, to be equinumerous to a well-founded set, and to be well-orderable are equivalent notions. As far as **HST** is concerned, we would, perhaps, use the well-orderability, or the property of being equinumerous to a well-founded set, as the primary property to distinguish this class of sets. However, we would like to have a uniform definition, applicable to some other relevant nonstandard theories, where the equivalence does not hold any more because the class $\mathbb{WF}$ is too small (as in **EEST**, a theory introduced in § 5.2a).

Note that the axioms introduced in §§ 1.1b, 1.1c are satisfied, for instance, in **ZFC** if we define $\mathtt{st}\, x$ to be true for all x — then $\mathbb{WF} = \mathbb{S} = \mathbb{I} = \mathbb{H}$ and ${}^*x = x$. The next axiom yields internal sets which are not standard, and, implicitly, non-internal sets, leading to the correct picture $\mathbb{S} \subsetneqq \mathbb{I} \subsetneqq \mathbb{H}$.

Say that a set $\mathscr{X}$ is $\cap$-*closed* if $X \cap Y \in \mathscr{X}$ holds for any $X, Y \in \mathscr{X}$.

Saturation: The class $\mathbb{I}$ is *standard size saturated*, i. e., if $\mathscr{X} \subseteq \mathbb{I}$ is a $\cap$-closed set of standard size and every $X \in \mathscr{X}$ is nonempty then $\bigcap \mathscr{X} \neq \varnothing$. [10]

Standard Size Choice: Choice in the case when the domain of the choice function is a set of standard size: formally, if X is a set of standard size, F is a function defined on X, and $F(x) \neq \varnothing$ for any $x \in X$, then there is a function f defined on X so that $f(x) \in F(x)$ for all x.

Dependent Choice: An ω-sequence of choices exists in the case when the domain of the n-th choice depends on the result of the $(n-1)$-th choice. Equivalently: any nonempty partially ordered set without maximal elements includes a nonempty linearly ordered subset where any element has its immediate successor.

[10] A more traditional form of saturation will be proved below, see Theorem 1.3.5.

1.1g Putting it all together

To conclude, we define **HST** as follows:

Definition 1.1.13. **HST** is the theory in the st-$\in$-language which consists of the axioms listed in §1.1b (Extensionality, Pair, Union, Infinity, and the schemata of Separation and Collection for all st-$\in$-formulas), §1.1c (**ZFC**$^{\mathtt{st}}$, Transfer, Regularity over $\mathbb{I}$, Transitivity of $\mathbb{I}$, Standardization), and §1.1f (Saturation, Standard Size Choice, Dependent Choice). □

The following theorem will be proved in Section 5 (see Theorem 5.1.4 and Corollary 5.1.5). It summarizes most important metamathematical properties of **HST**, including its relationships with the "standard" set theory **ZFC**.

Theorem 1.1.14. **HST** *is equiconsistent with* **ZFC**. *In addition*:

Conservativity: *Any* $\in$*-sentence* Φ *is a theorem of* **ZFC** *iff* $\Phi^{\mathtt{wf}}$ *is a theorem of* **HST** *iff* $\Phi^{\mathtt{st}}$ *is a theorem of* **HST**.

Standard core Interpretability: *There is an interpretation of* **HST** *in the theory* **ZFC** *such that the ground set universe of* **ZFC** *is isomorphic to the class of all standard sets of the interpretation.*

This theorem is of principal importance, especially the final statement which shows that a **ZFC**-oriented mathematician can assume that the ("standard") universe of ordinary mathematics is just a standard or well-founded part of a bigger universe governed by the **HST** axioms. There are interesting and important additional issues related to the interpretability part of the theorem, which will be commented upon in §4.1.

Blanket agreement 1.1.15. Below, all theorems, lemmas, *etc.* (as well as results of §1.1e above) are results in **HST** unless clearly stated otherwise. This does not apply for "metatheorems", *i.e.*, statements which explicitly assert provability, consistency *etc.*, like, for example, Theorem 1.1.14. □

1.1h Zermelo – Fraenkel theory ZFC

The following list of the axioms of **ZFC** is given for reference only: in principle we assume some knowledge of set theory. We give [Jech 78] and [Kun 80] as broad references in general set theory.

Extensionality: $\forall X\, \forall Y\, \big(X = Y \iff \forall x\, (x \in X \iff x \in Y)\big)$, or: less formally, sets are equal if and only if they have the same elements.

Pair: $\forall x\, \forall y\, \exists Z\, \forall z \big(z \in Z \iff (z = x \vee z = y)\big)$, or: for any sets x, y there is a set Z whose elements are precisely x, y. This set Z is denoted by $Z = \{x, y\}$ and called the (unordered) pair of x, y.

Separation (or *Comprehension*): $\forall X \exists Y \forall x \big(x \in Y \iff (x \in X \wedge \Phi(x))\big)$: any collection of elements of a set, definable by a formula, is a set.

If we really consider **ZFC** then $\Phi(x)$ can be any $\in$-formula with arbitrary sets as parameters [11]. When Separation is considered as an axiom schema in **HST**, Φ can be any $\mathtt{st}$-$\in$-formula (with any parameters). This also applies for the formula $\Phi(x,y)$ in the two following schemata. Note that the Replacement schema follows from Collection and Separation.

Collection: For any set X there is a set Y such that
$$\forall X \,\exists Y \,\forall x \in X \,(\exists y \,\Phi(x,y) \Longrightarrow \exists y \in Y \,\Phi(x,y)).$$

Replacement: For any set X, if $\forall x \in X \,\exists !\, y \,\Phi(x,y)$ then there is a function f defined on X such that $\forall x \in X \,\Phi(x, f(x))$.

Union: For every set X there exists a set $U = \bigcup X = \{y : \exists x \in X \,(y \in x)\}$ (the union of all elements of X).

Infinity: There is a set X such that $\varnothing \in X$ and $x \in X \Longrightarrow x \cup \{x\} \in X$.

(The existence of the *empty set* $\varnothing$ follows from Separation.)

In the presence of the Regularity axiom, any set X as in Infinity is infinite in any natural sense: indeed, the sets $0 = \varnothing$, $1 = \{\varnothing\}$, $2 = \{\varnothing, \{\varnothing\}\}$, $3 = \{\varnothing, \{\varnothing\}, \{\varnothing, \{\varnothing\}\}\}$, ... , which belong to X, are paiwise different.

Recall that the following **ZFC** axioms do not belong to **HST**.

Regularity[12] : For any set $X \neq \varnothing$ there is $x \in X$ such that $x \cap X = \varnothing$.

Power Set: For any set x, the *power class* $\mathscr{P}(x) = \{y : y \subseteq x\}$ is a set.

Choice: For any set X, there is a function f (called a *choice function*) defined on X, such that $f(x) \in x$ for all $x \in X \smallsetminus \{\varnothing\}$.

The axiom of Power Set postulates the existence of a set $\mathscr{P}(x)$, called *the power set of* x, which consists of all subsets of a given set x.

Remark 1.1.16. (1) Some of the set theories considered in this book do contain the Power Set axiom, some don't, but independently of this $\mathscr{P}(X)$ will be used to denote the collection $\{x : x \subseteq X \text{ is a set}\}$, the "power class" of any given set (or, occasionally, proper class) X.

(2) Many notable applications of the Choice axiom, for instance, the principle of Well-Ordering (*i.e.*, the statement that every set can be well-ordered) work properly only in the presence of Power Set. □

To conclude, the theory **ZFC** consists of the axioms and axiom schemata of Extensionality, Pair, Separation (or Comprehension), Collection, Replacement, Union, Infinity, Regularity, Power Set, Choice.

A weaker system of Zermelo **ZC** contains the same axioms except for Replacement and Collection.

---

[11] Which means, formally, that Φ can contain free variables other than x, but Y is not allowed to occur in $\Phi(x)$ as a variable.

[12] Also called the axiom of Foundation.

1.2 Basic elements of the nonstandard universe

The first question for any set theoretic foundational system is how it models the most fundamental notions like natural numbers, ordinals, finiteness, *i.e.*, the basis of everything to which we are accustomed in the "standard" **ZFC** universe. The universe of **HST** is comparatively more complicated, which forces us first of all to find some clues to its structure.

1.2a How to define fundamental set theoretic notions in HST

Since we decided to identify (informally) $\mathbb{WF}$ with the set universe of "conventional" mathematics, it is also natural to take $\mathbb{WF}$ as the basic universe for main set theoretic notions, with a reasonable hope of their good, **ZFC**-like behaviour. We call this approach the **scheme** "$\mathbb{WF} \xrightarrow{*} \mathbb{I}$ [in $\mathbb{H}$]".

We may observe that, although $\mathbb{S}$ and $\mathbb{WF}$ are $\in$-isomorphic, $\mathbb{WF}$ is a better version of the "conventional" set universe than $\mathbb{S}$, in particular because $\mathbb{WF}$, unlike $\mathbb{S}$, is transitive and $\subseteq$-complete (*i.e.*, $x \subseteq \mathbb{WF} \Longrightarrow x \in \mathbb{WF}$) — it follows that many basic set theoretic notions have one and the same meaning in $\mathbb{WF}$ and $\mathbb{H}$. (See below on absoluteness.)

Taking $\mathbb{WF}$ as the universe where, intuitively, the objects of "standard" mathematics live, we can then proceed in close analogy with the model theoretic version of nonstandard analysis: $\mathbb{WF}$ corresponds to a "standard" mathematical structure, $\mathbb{I}$ to its nonstandard extension via the map $*$ and the store of nonstandard $*$-extensions and their elements (which include such typically nonstandard objects as infinitesimals), and $\mathbb{H}$ to the ground set universe (of **ZFC**). Two differences with the model theoretic set-up can be mentioned: first, the membership relations in both $\mathbb{WF}$ and $\mathbb{I}$ are of one and the same nature, namely restrictions of the basic membership relation $\in$ given in the set universe $\mathbb{H}$ of **HST**; second, $\mathbb{I}$ is saturated with respect to all cardinals in $\mathbb{WF}$ rather than with respect to a particular cardinal.

Following this scheme, we consider the notions of ordinals, cardinals, natural numbers, finite sets, and their $*$-versions in this section.

The alternative **scheme** "$\mathbb{S} \subseteq \mathbb{I} \subseteq \mathbb{H}$" [13], which intuitively identifies the set universe of "conventional" mathematics with the class $\mathbb{S}$ of all standard sets, will be useful in some metamathematical studies, in particular, those related to nonstandard set theories of internal type like **IST**.

1.2b Closure properties and absoluteness

According to Corollary 1.1.10, each of the classes $\mathbb{I}$, $\mathbb{WF}$, $\mathbb{S}$ is closed under set theoretic operations definable in **ZFC**. The most meaningful applications of the corollary happen in the case when the formula φ is absolute, in other words, the result of the operation does not depend on the choice of the uni-

[13] The schemes "$\mathbb{WF} \xrightarrow{*} \mathbb{I}$ [in $\mathbb{H}$]" and "$\mathbb{S} \subseteq \mathbb{I} \subseteq \mathbb{H}$" are called, resp., *the external picture* and *the internal picture* in Hrbaček [Hr 01].

verse where the operation is carried out, as, for instance, the operation of pair $\{x, y\}$ and many more. This leads to the following

Corollary 1.2.1. (i) *The classes* $\mathbb{WF}$, $\mathbb{S}$, $\mathbb{I}$ *are closed under pairs, unions, intersections, Cartesian products.*

(ii) *If* $X, Y \in \mathbb{WF}$ *then the collection* X^Y *of all functions* $f : Y \to X$ *and the "power class"* $\mathscr{P}(X) = \{y : y \subseteq X\}$ *are sets and belong to* $\mathbb{WF}$.

(iii) *For any* $X \in \mathbb{I}$ *there is a set* $\mathscr{P}_{\mathtt{int}}(X) = \mathscr{P}(X) \cap \mathbb{I} \in \mathbb{I}$ (called: *the internal power set* of X). *If, moreover,* X *is a standard set then so is* $\mathscr{P}_{\mathtt{int}}(X)$.

Unlike **ZFC**, $\mathscr{P}(X)$ (as well as X^Y) is not necessarily a set in **HST**, in fact, $\mathscr{P}(X)$ is not a set and $\mathscr{P}_{\mathtt{int}}(X) \subsetneqq \mathscr{P}(X)$ for all infinite internal sets X in **HST** (Theorem 1.3.9 below). However, it follows from (ii) of the corollary that $\mathscr{P}(X)$ is a set, hence the power set for any well-founded set X.

Proof. (i) Prove, for instance, that $\mathbb{I}$ is closed under $\bigcup$. Let $X \in \mathbb{I}$. By Corollary 1.1.10 there is a set $U \in \mathbb{I}$ which is the union of all elements of X from the $\mathbb{I}$-point of view, that is, $y \in U$ iff $\exists^{\mathtt{int}} x \in X\ (y \in x)$ holds for any $y \in \mathbb{I}$. However $\mathbb{I}$ is transitive, hence, U coincides with $\bigcup X$ taken in the universe $\mathbb{H}$ of all sets. The result for $\mathbb{WF}$ can be proved similarly. The result for $\mathbb{S}$ follows from the result for $\mathbb{I}$ and Corollary 1.1.10(iv).

(ii) Let $X \in \mathbb{WF}$. By Corollary 1.1.10 there is a power set $\mathscr{P}_{\mathtt{wf}}(X) \in \mathbb{WF}$ defined in $\mathbb{WF}$, *i.e.*, $\mathscr{P}_{\mathtt{wf}}(X) = \mathscr{P}(X) \cap \mathbb{WF}$. However this includes all subsets of X by Theorem 1.1.9(ii), *i.e.*, $\mathscr{P}_{\mathtt{wf}}(X) = \mathscr{P}(X) \in \mathbb{WF}$.

(iii) As above, there is a set $P \in \mathbb{I}$ which is the power set of X in $\mathbb{I}$, that is, we have $y \in P$ iff $y \subseteq X$ for any internal y. Then $P = \mathscr{P}_{\mathtt{int}}(X)$ because any $y \in P$ is internal by Transitivity of $\mathbb{I}$. If $X \in \mathbb{S}$ then $P = \mathscr{P}_{\mathtt{int}}(X) \in \mathbb{S}$ by Corollary 1.1.10(iv). (**Exercise**. Prove that if $X = {}^*u \in \mathbb{S}$, where $u \in \mathbb{WF}$, then $p = \mathscr{P}(u) \in \mathbb{WF}$ and ${}^*p = \mathscr{P}_{\mathtt{int}}(X)$.)

Other parts of (ii) and (iii) are left as an easy **exercise** for the reader. □

Results of this type admit reformulation in terms of a general notion of absoluteness, useful every time one has to argue in different universes.

Definition 1.2.2. Let $\mathbb{K} \subseteq \mathbb{H}$ be a class of sets.

- A $\mathtt{st}$-$\in$-formula $\Phi(x_1, ..., x_n)$ is ***absolute for*** $\mathbb{K}$ if, for any sets $x_1, ..., x_n \in \mathbb{K}$, $\Phi(x_1, ..., x_n)$ holds in $\mathbb{K}$ iff it is true in $\mathbb{H}$, the universe of all sets, *i.e.*,

$$\forall x_1, ..., x_n \in \mathbb{K}\ \left(\Phi^{\mathbb{K}}(x_1, ..., x_n) \Longleftrightarrow \Phi(x_1, ..., x_n)\right).$$

- A set X definable by a $\mathtt{st}$-$\in$-formula $\varphi(x)$ (*i.e.*, X is the only set satisfying $\varphi(X)$) is ***absolute for*** $\mathbb{K}$ if the formulas $\varphi(x)$ and $\exists!x\, \varphi(x)$ are absolute.
- An operation $\langle x_1, ..., x_n \rangle \mapsto y = F(x_1, ..., x_n)$ [14] is ***absolute for*** $\mathbb{K}$ if so are the formulas $y = F(x_1, ..., x_n)$ and $\exists!y\, (y = F(x_1, ..., x_n))$. □

Exercise 1.2.3. Prove that operations $\{x_1, ..., x_k\}$ and $\langle x_1, ..., x_k \rangle$ (unordered and ordered tuple), $\bigcup X$, $\bigcap X$, $\mathscr{P}(X)$, X^Y, $\mathtt{dom}\, f$, $\mathtt{ran}\, f$ (domain and range), $X \times Y$ (Cartesian product) are absolute for $\mathbb{WF}$.

Hint. The class $\mathbb{WF}$ is transitive, $\subseteq$-complete, and satisfies **ZFC**. □

Corollary 1.2.4. *Any* $\in$*-formula* $\Phi(x_1, ..., x_n)$ *with all quantifiers bounded by* $\tau(x_1, ..., x_n)$*, where* τ *is a superposition of operations mentioned in Exercise 1.2.3, e.g.,* $\tau(x, y) = \mathscr{P}(x \cup \{\mathscr{P}(y)\})$*, is absolute for* $\mathbb{WF}$. □

A formula non-absolute for $\mathbb{WF}$ is: "there exists an ill-founded set". Furthermore, the class $\mathbb{I}$ is not $\subseteq$-complete, so that the operation $\mathscr{P}$ is not absolute for $\mathbb{I}$. Generally speaking, $\mathscr{P}_{\mathtt{int}}(X) \subsetneqq \mathscr{P}(X)$ for internal sets X. However the transitivity of $\mathbb{I}$ and the fact that $\mathbb{I}$ interprets **ZFC** allow us to save some amount of absoluteness, which is as follows:

Exercise 1.2.5. Let $\Phi(x_1, ..., x_n)$ be an $\in$-formula all quantifiers of which are bounded by $\tau(x_1, ..., x_n)$, where τ is a superposition of operations mentioned in 1.2.3, except for $\mathscr{P}$ and X^Y. Prove that Φ is absolute for $\mathbb{I}$. □

This still includes operations $\mathtt{dom}$, $\mathtt{ran}$, $\times$, $\bigcup$, $\bigcap$ and many other.

As for the classes $\mathbb{S} \subseteq \mathbb{I}$ and the $*$-embedding $\mathbb{WF} \to \mathbb{I}$, these combinations of universes admit rather full absoluteness in the $\in$-language, in the frameworks described by the principles of Transfer and $*$-Transfer.

1.2c Ordinals and cardinals

It is known that **ZFC** admits several equivalent definitions of ordinals. Since not all of them remain equivalent in **HST**, let us specify the definition. First of all, sets X, Y are *equinumerous* iff there is a bijection $f : X \xrightarrow{\text{onto}} Y$. In **ZFC** this is the same as sets of equal cardinality, but in **HST** we prefer to have a definition independent of the notion of cardinality.

Definition 1.2.6. An *ordinal* is a transitive set well-ordered by $\in$. A *cardinal* is an initial ordinal, *i.e.*, an ordinal not equinumerous to a smaller ordinal. $\mathtt{Ord}$, resp. $\mathtt{Card}$ is the class of all ordinals, resp. all cardinals. □

This definition is explicitly related to the universe $\mathbb{H}$, which is not in all respects a **ZFC** universe, so we should take some time to see if the normal **ZFC**-like behaviour of the ordinals and cardinals persists in **HST**.

Lemma 1.2.7. $\mathtt{Ord} = (\mathtt{Ord})^{\mathbb{WF}}$, *thus, the ordinals and the* $\mathbb{WF}$*-ordinals* (ordinals defined in $\mathbb{WF}$) *are one and the same. Similarly,* $\mathtt{Card} = (\mathtt{Card})^{\mathbb{WF}}$.

---

[14] We view an *operation* as just an $\in$-formula, say, $\varphi(x_1, ..., x_n, y)$, so that the equality $y = F(x_1, ..., x_n)$ is considered simply as a shortcut for the formula "y is the unique set satisfying $\varphi(x_1, ..., x_n, y)$".

Proof. If $\xi \in (\mathtt{Ord})^{\mathbb{WF}}$ then ξ remains an ordinal in $\mathbb{H}$ because all subsets of ξ belong to $\mathbb{WF}$ by Theorem 1.1.9(ii). Conversely if $\xi \in \mathtt{Ord}$ then by definition $\xi \in \mathbb{WF}$. Similarly, if $\kappa \in (\mathtt{Card})^{\mathbb{WF}}$ then at least $\kappa \in \mathtt{Ord}$ by the above. A possible bijection onto a smaller ordinal in $\mathbb{H}$ is effectively coded by a subset of $\kappa \times \kappa$, therefore it would belong to $\mathbb{WF}$ by Theorem 1.1.9(ii). □

Thus an ordinal (resp., cardinal) is a set $\alpha \in \mathbb{WF}$ such that it is true in $\mathbb{WF}$ that α is an ordinal (resp., cardinal). In other words, the formula "α is an ordinal" is absolute for $\mathbb{WF}$. Since $\mathbb{WF}$ models **ZFC**, the ordinals satisfy all usual theorems, for instance $\mathtt{Ord}$ is well-ordered by the relation: $\alpha < \beta$ iff $\alpha \in \beta$, an ordinal is the set of all smaller ordinals, $0 = \varnothing$ is the least ordinal, there exist limit ordinals, *etc.* In general we may rely on the **ZFC**-like behaviour of ordinals and cardinals in **HST**. However, $\mathtt{card}\, X \in \mathtt{Card}$ cannot be defined for every set X; see § 1.3a below on cardinalities in **HST**.

The common aleph-notation $\aleph_0 = \omega$, $\aleph_1 = \omega_1, \dots, \aleph_\xi = \omega_\xi$ $(\xi \in \mathtt{Ord}), \dots$ will be used to denote infinite cardinals (in $\mathbb{WF}$).

Definition 1.2.8. Ordinals and cardinals in the sense of $\mathbb{I}$ will be called resp. $*$-*ordinals* and $*$-*cardinals*. Let $^*\mathtt{Ord}$ = all $*$-ordinals, and correspondingly $^*\mathtt{Card}$ = all $*$-cardinals. For any internal set X, let $^*\mathtt{card}\, X \in {}^*\mathtt{Card}$ be the cardinality of X in $\mathbb{I}$, or the $*$-*cardinal* (or $*$-*cardinality*) of X. □

Thus $*$-ordinals is the same as $\mathbb{I}$-ordinals. By Transfer, $\mathbb{S}$-*ordinals* (that is standard sets that are ordinals in the sense of $\mathbb{S}$) is then the same as *standard* $*$-ordinals. Similarly for cardinals.

Exercise 1.2.9. Prove, using Transfer and $*$-Transfer, that $^*\alpha \in {}^*\mathtt{Ord}$ whenever $\alpha \in \mathtt{Ord}$, and in general $*$ maps ordinals onto $\mathbb{S}$-ordinals, the same for cardinals. As a consequence show that standard $*$-ordinals are well-ordered by $\in$. Why does this fail for the class $^*\mathtt{Ord}$ of all $*$-ordinals ? □

Exercise 1.2.10. Let a *pseudo-ordinal* be a transitive set linearly ordered by $\in$. Thus, ordinals (in the sense above) are well-founded pseudo-ordinals. (In **ZFC** all pseudo-ordinals are well-founded !) Prove the following:

(1) $*$-ordinals are also pseudo-ordinals;

(2) if α is a pseudo-ordinal and ξ a (well-founded) ordinal then $\alpha + \xi$, defined, as usual, by transfinite induction on ξ, is a pseudo-ordinal;

(3) (difficult !) any pseudo-ordinal has the form $\alpha + \xi$, where α is a $*$-ordinal and ξ a (well-founded) ordinal. □

1.2d Natural numbers, finite and $*$-finite sets

The next definition is an ordinary definition of natural numbers and the concept of finiteness by means of the $\in$-language.

Definition 1.2.11. $\mathbb{N} = \omega =$ the least limit ordinal. Elements of $\mathbb{N}$ are called *natural numbers*.

A *finite set* is a set equinumerous to a natural number. In other words, a set X is finite if and only if there is a number $n = \#X \in \mathbb{N}$ (easily seen to be unique if it exists) and a bijection $b : n = \{0, 1, ..., n-1\} \xrightarrow{\text{onto}} X$. Note that if $X \in \mathbb{WF}$ then any such bijection b belongs to $\mathbb{WF}$ by Theorem 1.1.9(ii). It follows that a well-founded set X is finite in $\mathbb{WF}$ iff it is finite in $\mathbb{H}$. □

It is known from **ZFC** that $n = \{0, 1, ..., n-1\}$ for any $n \in \mathbb{N}$.

Thus, "natural number" means: a $\mathbb{WF}$-natural number, or an $\mathbb{H}$-natural number, both being one and the same by the above. By definition $\mathbb{N} \in \mathbb{WF}$; moreover, any $n \in \mathbb{N}$ and any $X \subseteq \mathbb{N}$ belong to $\mathbb{WF}$ by Theorem 1.1.9(ii).

Remark 1.2.12. Note that $\mathbb{N} \subseteq \mathtt{Ord}$, therefore, $\mathbb{N}$ is well-ordered. Thus we can freely use definitions and proofs by induction, as well as definitions and proofs by *transfinite* induction (on ordinals). □

To see how $\mathbb{N}$ relates to the universe $\mathbb{I}$, consider the $*$-extension ${}^*\mathbb{N} \in \mathbb{I}$ of the set $\mathbb{N} \in \mathbb{WF}$. It follows from $*$-Transfer that it is true in $\mathbb{I}$ that "${}^*\mathbb{N}$ is the set of all natural numbers". This does not imply that the elements of ${}^*\mathbb{N}$ are precisely the natural numbers. Passing from $\mathbb{WF}$ to $\mathbb{I}$, the "set of natural numbers" acquires new elements beyond the true natural numbers.

Definition 1.2.13. Elements of ${}^*\mathbb{N}$ will be called *$*$-natural numbers* (all of them belong to $\mathbb{I}$ by the axiom of Transitivity of $\mathbb{I}$). Internal sets X which are formally finite in the sense of $\mathbb{I}$ will be called *$*$-finite*, or *hyperfinite*. In other words, a set $X \in \mathbb{I}$ is $*$-finite iff there is a number $n = \#X \in {}^*\mathbb{N}$ and an internal [15] bijection $b : n = \{0, 1, ..., n-1\} \xrightarrow{\text{onto}} X$. □

Exercise 1.2.14. Show that the values $\#X$ given by definitions 1.2.11 and 1.2.13 coincide when both apply, *i.e.*, X is both finite and internal $*$-finite.

Hint. Prove by induction on n that if $n \in \mathbb{N}$ and there is a bijection $b : n$ onto $h \in {}^*\mathbb{N}$ then $n = h$. □

Note that all finite internal sets are $*$-finite, but not conversely: any number $n = \{0, 1, ..., n-1\} \in {}^*\mathbb{N} \smallsetminus \mathbb{N}$ is $*$-finite but infinite (prove this !).

Exercise 1.2.15. Prove the following:

(1) ${}^*n = n$ for all $n \in \mathbb{N}$, thus, natural numbers = $\mathbb{S}$-natural numbers;

(2) $\mathbb{N} = {}^*\mathbb{N} \cap \mathbb{S}$, thus, natural numbers = standard $*$-natural numbers;

(3) $\mathbb{N} \subsetneqq {}^*\mathbb{N}$, moreover, $\mathbb{N}$ is a proper initial segment of ${}^*\mathbb{N}$, and ${}^*\mathbb{N} \smallsetminus \mathbb{N}$ does not contain a least element, therefore, the **ZFC** axiom of Regularity fails.

[15] *External* bijections between internal sets need some care, *e.g.*, for any $n \in {}^*\mathbb{N} \smallsetminus \mathbb{N}$ there is an external bijection n onto $n+1$, just put $b(i) = i$ for all $i \in \mathbb{N}$ and $b(i) = i + 1$ for all $i < n$, $i \notin \mathbb{N}$. See more about this phenomenon in § 1.4d.

Elements of ${}^*\mathbb{N} \smallsetminus \mathbb{N}$, *i.e.*, nonstandard $*$-natural numbers, are called *infinitely large*, each of them is non-well-founded and bigger than any $n \in \mathbb{N}$.

(4) $\mathbb{S} \subsetneqq \mathbb{I} \subsetneqq \mathbb{H}$;

(5) for any set X the collection $X^{<\omega}$ of all finite sequences of elements of X and the collection $\mathscr{P}_{\mathtt{fin}}(X)$ of all finite subsets of X are sets;

(6) if $X \in \mathbb{WF}$ and $n \in \mathbb{N}$ then $({}^*X)^n = {}^*(X^n)$.

Hints. (1) We have **ZFC** in $\mathbb{WF}$, hence, formally $0 = \varnothing$ and $n+1 = n \cup \{n\}$ for any n. To see that ${}^*0 = \varnothing = 0$ note that ${}^*\varnothing \in \mathbb{S}$ and, by $*$-Transfer, ${}^*\varnothing \cap \mathbb{I} = \varnothing$, hence, ${}^*\varnothing = \varnothing$ by Transitivity of $\mathbb{I}$. Now, arguing by induction on n we have ${}^*(n+1) = {}^*(n \cup \{n\}) = {}^*n \cup \{{}^*n\} = ({}^*n) + 1$ by $*$-Transfer.

(2) Direction $\subseteq$ follows from (1). Conversely if $a \in {}^*\mathbb{N} \cap \mathbb{S}$ then $a = {}^*n$ for some $n \in \mathbb{N}$, but ${}^*n = n$.

(3) That $\mathbb{N}$ is an initial segment of ${}^*\mathbb{N}$ easily follows from (1). To find a number in ${}^*\mathbb{N} \setminus \mathbb{N}$, note that the set $X_n = \{H \in {}^*\mathbb{N} : H > n\}$ is internal for each $n \in \mathbb{N}$ because it is defined in $\mathbb{I}$ via an $\in$-formula with $n, {}^*\mathbb{N} \in \mathbb{I}$ as parameters. Since $\mathbb{N}$ is a set of standard size (as any well-founded set), the intersection $\bigcap_{n \in \mathbb{N}} X_n = {}^*\mathbb{N} \smallsetminus \mathbb{N}$ is nonempty by Saturation.

(4) Any $n \in {}^*\mathbb{N} \smallsetminus \mathbb{N}$ belongs to $\mathbb{I} \smallsetminus \mathbb{S}$ while $X = {}^*\mathbb{N} \smallsetminus \mathbb{N} \notin \mathbb{I}$.

(5) The cartesian power X^n is a set, by induction on $n \in \mathbb{N}$ (Exercise 1.1.2 is applied). Now $X^{<\omega} = \bigcup_n X^n$ is a set by Replacement and Union, and $\mathscr{P}_{\mathtt{fin}}(X)$ is a set by Replacement via the map $\langle x_1, ..., x_n \rangle \mapsto \{x_1, ..., x_n\}$. □

We are already equipped enough to obtain less trivial results.

Lemma 1.2.16. (i) *Every finite set $X \subseteq \mathbb{S}$ is standard, hence, internal;*

(ii) *every internal set $I \subseteq \mathbb{S}$ is finite;*

(iii) *every $*$-finite set $X \in \mathbb{S}$ is finite and satisfies $X \subseteq \mathbb{S}$;*

(iv) *every finite set $X \subseteq \mathbb{I}$ is internal, hence, $\mathbb{I}$ is closed under finite unions and intersections, and Cartesian products.*

Proof. (i) Use induction on the number $\#X$ of elements in X.

(ii) $F = \mathscr{P}_{\mathtt{fin}}(I)$ is a set by (5) of Exercise 1.2.15, and $F \subseteq \mathbb{S}$ by (i), hence, F is a set of standard size. The collection $\mathscr{X}$ of all sets $X_w = I \smallsetminus w$, $w \in F$, is $\cap$-closed, and, if I is not finite, consists of nonempty internal sets; then, by Saturation, $\bigcap_{w \in F} X_w \neq \varnothing$, easily leading to contradiction.

(iii) Let $\#X = n+1 > 0$. By Transfer, X contains an element $x \in \mathbb{S}$, and $X' = X \smallsetminus \{x\}$ is also standard. Then $X' \subseteq \mathbb{S}$ by the induction hypothesis.

(iv) Induction on the number $\#X \in \mathbb{N}$ of elements of X proves the first part. To prove the "hence" part, use Transitivity of $\mathbb{I}$ to make sure that $\mathbb{I}$-unions *etc.* are equal to unions in the whole universe. □

1.2e Hereditarily finite sets

Which sets are both internal and well-founded ? To answer this question, recall that, in **ZFC**, a set x is called *hereditarily finite* if its transitive closure $\mathtt{TC}(x)$ [16] is finite.

Alternatively, the collection $\mathbb{HF}$ of all hereditarily finite sets can be defined, both in **ZFC** and in **HST**, as $\mathbb{HF} = \bigcup_{n \in \mathbb{N}} \mathbb{HF}_n$, where $\mathbb{HF}_0 = \varnothing$ and $\mathbb{HF}_{n+1} = \mathscr{P}(\mathbb{HF}_n)$ by induction on n. All $\mathbb{HF}_n$ are finite sets (induction on n: $\mathbb{HF}_{n+1}$ has $2^{\# \mathbb{HF}_n}$ elements). Moreover, $\mathbb{HF}_n \subseteq \mathbb{WF}$, and hence $\mathbb{HF}_n \in \mathbb{WF}$ by Theorem 1.1.9(ii). Finally, $\mathbb{HF}$ is a set by Collection and $\mathbb{HF} \in \mathbb{WF}$ still by Theorem 1.1.9(ii). Clearly every finite set $x \subseteq \mathbb{HF}$ belongs to $\mathbb{HF}$.

Exercise 1.2.17. Prove that there is a bijection $\pi : \mathbb{N} \xrightarrow{\text{onto}} \mathbb{HF}$. In addition prove the following:

(1) $\mathbb{HF} = \{x \in \mathbb{WF} : {}^*x = x\}$;

(2) $\mathbb{HF} = \mathbb{S} \cap \mathbb{WF} = \mathbb{I} \cap \mathbb{WF} = {}^*\mathbb{HF} \cap \mathbb{S} = {}^*\mathbb{HF} \cap \mathbb{WF}$;

(3) $\mathbb{N} = {}^*\mathbb{N} \cap \mathbb{WF} = {}^*\mathbb{N} \cap \mathbb{HF} = \mathbb{I} \cap \mathtt{Ord} = \mathbb{S} \cap \mathtt{Ord}$.

Hints. Put $\pi'(0) = \pi'(1) = \varnothing$ and $\pi'(n) = \{\pi'(k) : k \in u_n\}$ for $n \geq 2$, where u_n is the set of all numbers $i < n$ such that p_i (i-th prime; $p_0 = 2$) divides n. the function $\pi' : \mathbb{N} \xrightarrow{\text{onto}} \mathbb{HF}$ is not a bijection, of course, but it does not take much effort to convert it to a bijection.

(1) To show that ${}^*x = x$ for any $x \in \mathbb{HF}$ argue by induction on n, where $x \in \mathbb{HF}_n$. If x belongs to $\mathbb{HF}_{n+1}$ and ${}^*y = y$ for all $y \in x$ then $x = \{{}^*y : y \in x\} \in \mathbb{S}$ by Lemma 1.2.16(ii), hence, $x = {}^*x$ by Definition 1.1.6. The converse statement follows from (2).

(2) If $x \in \mathbb{HF}$ then $x = {}^*x \in \mathbb{S}$. Prove that any $x \in \mathbb{WF} \cap \mathbb{I}$ belongs to $\mathbb{HF}$ by $\in$-*induction*, *i.e.*, we assume that the result holds for all $y \in x$ and derive the result for x itself. (The method is applicable because the relation $\in \restriction \mathbb{WF}$ is well-founded, see Remark 1.1.7.) If $x \in \mathbb{I} \cap \mathbb{WF}$ then any $y \in x$ belongs to $\mathbb{I} \cap \mathbb{WF}$ because the classes are transitive, hence, $y = {}^*y \in \mathbb{HF}$ by the inductive hypothesis and (1), hence, $x \subseteq \mathbb{HF}$. In addition, x is finite by Lemma 1.2.16(ii), thus, $x \in \mathbb{HF}$.

Other equalities in (2) easily follow from $\mathbb{HF} = \mathbb{S} \cap \mathbb{WF} = \mathbb{I} \cap \mathbb{WF}$.

Finally, (3) is a consequence of (2). □

We observe that, by (3), for a $*$-natural number n to be well-founded, to be standard, and to belong to $\mathbb{N}$ is one and the same.

---

[16] The transitive closure $\mathtt{TC}(x)$ of a set x is the $\subseteq$-least transitive set X satisfying $x \subseteq X$. It can be defined by $X = x \cup \bigcup x \cup \bigcup\bigcup x \cup \ldots$. See § 1.5b below on the existence in **HST** and other properties.

1.3 Sets of standard size

We show that sets of standard size, well-orderable sets, and sets equinumerous to a cardinal is one and the same in **HST**. This enables us to consistently define cardinalities of sets of standard size. Then we prove several corollaries of the axiom of Saturation, including a Saturation theorem for families with the finite intersection property, Hrbaček's theorem of refutation of the axioms of Power Set and Choice (to be more exact, the Well-Ordering principle), and the principle of Extension.

1.3a Cardinalities of sets of standard size

The next theorem proves a result cited above.

Theorem 1.3.1. *Sets in* $\mathbb{WF}$ *and sets* $X \subseteq \mathbb{S}$ *are of standard size. In addition, the following properties are equivalent for any set* x :

(1) *x is a set of standard size*; (2) *x is well-orderable*;

(3) *x is equinumerous with a cardinal κ, i.e., there is a bijection* $\kappa \overset{\text{onto}}{\longrightarrow} x$.

Proof. If $U \in \mathbb{WF}$ then U is the image of $X = {}^*U \cap \mathbb{S}$ via the map $f({}^*u) = u$, hence, both U and any functional image of U are sets of standard size. (Note that f is a set since **HST** contains Replacement in the st-$\in$-language.)

(1) $\Longrightarrow$ (3). Suppose that $x = \{f(a) : a \in A\}$, where $A \subseteq \mathbb{S}$. Let, by Standardization, $U \in \mathbb{WF}$ be a set satisfying $A = {}^*U \cap \mathbb{S}$. Then we have $x = \{g(u) : u \in U\}$, where $g(u) = f({}^*u)$. As $\mathbb{WF}$, a transitive and $\subseteq$-complete class, obeys **ZFC**, there is a well-ordering $\prec$ of U. Let V be the set of all $v \in U$ such that $g(u) \neq g(v)$ for any $u \prec v$. Then V is well-founded, $g \restriction V$ is a bijection, and still $x = \{g(u) : u \in V\}$. Finally, u is in $1-1$ correspondence with a (unique) cardinal $\kappa = \{\xi : \xi < \kappa\} \in \mathtt{Card}$ in $\mathbb{WF}$.

(3) $\Longrightarrow$ (2). It suffices to check that any $X \in \mathbb{WF}$ can be well-ordered (in $\mathbb{H}$). Let $\prec$ be a well-ordering of X in $\mathbb{WF}$: recall that the axiom of choice holds in $\mathbb{WF}$. Then $\prec$ still well-orders X in the whole universe $\mathbb{H}$ because any subset of X belongs to $\mathbb{WF}$ by Theorem 1.1.9(ii).

(2) $\Longrightarrow$ (1). Let x be well-ordered by a relation $\prec$. As the class $\mathtt{Ord}$ of all ordinals is well-ordered by $\in$, either there is an order preserving map of $\mathtt{Ord}$ onto an initial segment of x or there is an order preserving map f of x onto a *proper* initial segment of $\mathtt{Ord}$. The "either" case is impossible: $\mathtt{Ord}$ would be a set, contradiction. In the "or" case, let λ be the least ordinal which does not belong to the initial segment of $\mathtt{Ord}$ on which f maps x. We have a $1-1$ map of the set $\lambda = \{\xi : \xi < \lambda\} \in \mathbb{WF}$ onto x. □

Now we can introduce cardinalities of sets of standard size.

Definition 1.3.2. For a set X of standard size let $\mathtt{card}\, X$, the *cardinality* of X, be the only cardinal $\kappa \in \mathtt{Card}$ equinumerous with X. □

This is well-defined by the theorem. On the other hand, we cannot define $\mathtt{card}\, X$ this way for any set X not of standard size, as long as cardinals are understood as in § 1.2c. Yet we defined $^*\mathtt{card}\, X$ for any internal set X (Definition 1.2.8) to be a $*$-cardinal equal to the cardinality of X in $\mathbb{I}$.

Exercise 1.3.3. Prove that subsets, functional images, power sets, cartesian products, standard size unions of sets of standard size are still sets of standard size. Also, if X, Y are sets of standard size then so is X^Y.

Hint: apply Theorem 1.3.1. Prove that, *e.g.*, the power set $\mathscr{P}(X)$ of a standard size set X is a set of standard size. (If X is not of standard size then $\mathscr{P}(X)$ may not even be a set.) By Theorem 1.3.1 we can assume that $X \in \mathbb{WF}$. Then any $Y \subseteq X$ also is well-founded, hence, $\mathscr{P}(X)$ coincides with the $\mathbb{WF}$-power set of X, existing in $\mathbb{WF}$ as this universe interprets **ZFC**.

The results for unions and images also require Standard Size Choice. □

1.3b Saturation and the Hrbaček paradox

The axiom of Saturation was formulated for $\cap$-closed families, in order to avoid premature reference to the notion of finiteness. The latter was defined in § 1.2d, thus we can now derive the ordinary, and often more convenient, form of Saturation.

Definition 1.3.4. A family of sets $\mathscr{X}$ satisfies the *finite intersection property*, or is a *f. i. p. family*, if $\bigcap \mathscr{X}'$ is nonempty for any finite $\mathscr{X}' \subseteq \mathscr{X}$. □

For instance, any family $\mathscr{X}$ of nonempty sets, $\cap$-*directed* in the sense that for any $X, Y \in \mathscr{X}$ there exists $Z \in \mathscr{X}$, $Z \subseteq X \cap Y$, is a f. i. p. family.

Theorem 1.3.5 (Saturation). *Suppose that a set $\mathscr{X} \subseteq \mathbb{I}$ of standard size is a f. i. p. family. Then $\bigcap \mathscr{X} \neq \varnothing$.*

Proof. Let $\mathscr{X}'$ be the set of all finite intersections of sets in $\mathscr{X}$. (To see that $\mathscr{X}'$ is a set use the result of Exercise 1.3.3.) By Lemma 1.2.16, all sets in $\mathscr{X}'$ are internal. Finally, $\mathscr{X}'$ is $\cap$-closed, and all sets in $\mathscr{X}'$ are nonempty by the choice of $\mathscr{X}$. It remains to apply the Saturation axiom for $\mathscr{X}'$. □

The following corollary was implicitly used in the proof of the existence of infinitely large numbers above.

Corollary 1.3.6 (Compactness). *If $\mathscr{X}, \mathscr{Y} \subseteq \mathbb{I}$ are two nonempty families of standard size, and $\bigcap \mathscr{X} \subseteq \bigcup \mathscr{Y}$, then there exist finite nonempty subfamilies $\mathscr{X}' \subseteq \mathscr{X}$ and $\mathscr{Y}' \subseteq \mathscr{Y}$ such that still $\bigcap \mathscr{X}' \subseteq \bigcup \mathscr{Y}'$.*

In particular, if $X \in \mathbb{I}$ and $X \subseteq \bigcup \mathscr{Y}$, then there is a finite nonempty subfamily $\mathscr{Y}' \subseteq \mathscr{Y}$ such that still $X \subseteq \bigcup \mathscr{Y}'$.

Proof. Otherwise the standard size collection $\mathscr{Z}$ of all sets of the form $Z = X \setminus Y$, where $X \in \mathscr{X}$, $Y \in \mathscr{Y}$, is a f. i. p. collection. Theorem 1.3.5 implies $\bigcap \mathscr{Z} \neq \varnothing$, that is, $\bigcap \mathscr{X} \setminus \bigcup \mathscr{Y} \neq \varnothing$, contradiction. □

Exercise 1.3.7 (Compare with Idealization of **IST** or **BST**). Prove that if $\Phi(a,x)$ is an $\in$-formula with parameters in $\mathbb{I}$ and $A_0 \subseteq \mathbb{S}$ is any set then

$$\forall^{\mathtt{fin}} A \subseteq A_0 \; \exists^{\mathtt{int}} x \; \forall a \in A \; \Phi^{\mathtt{int}}(a,x) \iff \exists^{\mathtt{int}} x \; \forall a \in A_0 \; \Phi^{\mathtt{int}}(a,x) .$$

Hint. To prove the nontrivial direction $\Longrightarrow$, show, using Collection and Exercise 1.1.11(3), that there is an internal set X such that

$$\forall^{\mathtt{fin}} A \subseteq A_0 \; \exists x \in X \; \forall a \in A \; \Phi^{\mathtt{int}}(a,x) .$$

Apply 1.3.5 to the family of all sets $X_a = \{x \in X : \Phi^{\mathtt{int}}(a,x)\}$, $a \in A_0$. □

Exercise 1.3.8. Prove the following:

(1) if $X \subseteq \mathbb{I}$ is a set of standard size and $n \in {}^*\mathbb{N} \smallsetminus \mathbb{N}$ then there is a $*$-finite internal set I with $X \subseteq I$ and $\#I \le n$;

(2) any internal set X of standard size is finite, and hence by Lemma 1.2.16 a standard size set $X \subseteq I$ is internal iff it is finite;

(3) If $X \subseteq {}^*\mathbb{N}$ is a set of standard size then X is is neither cofinal in ${}^*\mathbb{N}$ nor coinitial in ${}^*\mathbb{N} \smallsetminus \mathbb{N}$.

Hints. (1) Cover X by an internal set Y. Apply Theorem 1.3.5 to the f. i. p. family $\mathscr{X}$ of all sets $C_x = \{z \in \mathbb{I} : x \in z \subseteq Y \wedge \#z \le n\}$, $x \in X$. (It is a separate **exercise** to show that each C_x is an internal set.)

(2) If X is infinite then the family of all sets $X \smallsetminus \{x\}$, $x \in X$, is a f. i. p. family, leading to contradiction with Theorem 1.3.5.

(3) If, on the contrary, X is cofinal in ${}^*\mathbb{N}$ then, applying 1.3.6 to the family $\mathscr{Y}$ of all intervals $[0,n)$, $n \in X$, we have a contradiction. To prove the non-coinitiality claim, apply 1.3.6 to the family $\mathscr{Y}$ of all intervals $[0,n)$, $n \in \mathbb{N}$, and the family $\mathscr{X}$ of all intervals $[0,n)$, $n \in x \smallsetminus \mathbb{N}$. □

Theorem 1.3.9 (*Hrbaček paradox*). *No infinite internal set X can be well-ordered nor does it have a power set. Moreover, there is no set P containing all sets $Y \subseteq X$ of standard size.*

Proof. First of all X is not a set of standard size (Exercise 1.3.8), hence, it is not well-orderable (Theorem 1.3.1). Prove the "moreover" assertion. Suppose on the contrary that there is a set P such that any $Y \subseteq X$ of standard size belongs to X. Note that X, an internal set, is either $*$-infinite or $*$-finite and then $\#X = n \in {}^*\mathbb{N} \smallsetminus \mathbb{N}$ as X is not finite. In both cases there is a $*$-number $n \in {}^*\mathbb{N} \smallsetminus \mathbb{N}$ such that X contains an internal subset X' of $\#X' = n$. Thus we can assume that simply $X = n \in {}^*\mathbb{N} \smallsetminus \mathbb{N}$, in other words, $X = \{0,1,...,n-1\}$.

To get a contradiction show that any ordinal $\xi = \{\alpha : \alpha < \xi\} \in \mathtt{Ord}$ admits an order preserving $1-1$ embedding into X. As any set in $\mathbb{WF}$ is of standard size, this yields a map $F : P \xrightarrow{\text{onto}} \mathtt{Ord}$, so that $\mathtt{Ord}$ is a set, contradiction. Thus let $\xi \in \mathtt{Ord}$. For any finite $x \subseteq \xi$, the (internal) set F_x of all internal maps $f : {}^*\xi \to X$, order–preserving on ${}^*x$, is nonempty. (**Exercise**:

use induction on $\#x$ to show that there is a map $f \in F_x$ with $f({}^*\alpha) \in \mathbb{N}$ for any $\alpha \in x$.) On the other hand the collection of all sets F_x, $x \in \mathscr{P}_{\text{fin}}(\xi)$, is of standard size because so is the set $\mathscr{P}_{\mathtt{fin}}(\xi) \in \mathbb{WF}$ by Lemma 1.3.1. Finally, if $X \subseteq \mathscr{P}_{\text{fin}}(\xi)$ is finite then $y = \bigcup X$ is finite and $F_y = \bigcap_{x \in X} F_x$ is nonempty. It follows by Theorem 1.3.5 that there is $f \in \bigcap_{x \subseteq \xi \text{ finite}} F_x$. Now $g(\alpha) = f({}^*\alpha)$ is an order preserving map $\xi \to X$. □

Remark 1.3.10. (1) It follows from the theorem that the Power Set axiom contradicts **HST**. Yet the internal power set $\mathscr{P}_{\mathtt{int}}(X) = \mathscr{P}(X) \cap \mathbb{I}$ exists and is an internal set for any internal X. Note also that the full power $\mathscr{P}(X)$ is a set (of standard size) for any set X of standard size (Exercise 1.3.3).

(2) It follows that the Well-Ordering principle (*i.e.*, the statement that every set can be well-ordered) also contradicts **HST**. As for the Choice axiom as given in § 1.1h, its status in **HST** is not completely clear. At least its negation is consistent with **HST**, see Theorem 5.5.8(ii). Yet the axiom of Choice in this form is rather useless in the absence of Power Set, *e.g.*, it does not support typical transfinite Dependent Choice-like constructions. □

1.3c The principle of Extension

Theorems and axioms gathered under this title allow to extend a function defined on a "small" (standard size) set to an internal function (hence, defined on a "big", internal set). We define it as follows.

Definition 1.3.11. Let $\mathscr{I}$ be a class of sets satisfying $\mathbb{S} \subseteq \mathscr{I} \subseteq \mathbb{I}$.

A function $f : S \to \mathscr{I}$ defined on a set $S \subseteq \mathscr{I}$ is *$\mathscr{I}$-extendible* if there is a function $f^\circ \in \mathscr{I}$ (hence internal) satisfying $S \subseteq \mathtt{dom}\, f^\circ$ and $f = f^\circ \restriction S$.

The class $\mathscr{I}$ is said to satisfy Extension if any function $f : S \to \mathscr{I}$ defined on a set $S \subseteq \mathscr{I}$ of standard size is $\mathscr{I}$-extendible. □

Theorem 1.3.12. *The class $\mathbb{I}$ satisfies Extension.*

Proof. By (3) of Exercise 1.1.11, there is a set $R \in \mathbb{S}$ with $\mathtt{ran}\, f \subseteq R$. Let

$$G_w = \{h \in \mathbb{I} : h \text{ is a map } S \to R \text{ and } h \restriction w = f \restriction w\}$$

for any finite set $w \subseteq S$. Note that all sets G_w are internal: indeed, both the set w and the map $f \restriction w$ defined on w are internal by Lemma 1.2.16, so that we can define G_w in $\mathbb{I}$ (which is possible, because $\mathbb{I}$ models **ZFC** in the $\in$-language), getting just the same set. Prove that $G_w \neq \varnothing$, by induction on the number $\#w$ of elements in w. If w is empty there is nothing to prove. To carry out the induction step, prove the statement for a set $w' = w \cup \{z\}$, assuming that $G_w \neq \varnothing$. Fix any $g \in G_w$, so that $g \in \mathbb{I}$ is a function $S \to R$ and $g \restriction w = f \restriction w$. Define, in $\mathbb{I}$, a function h on S so that $h(z) = g(z)$ and $h(y) = f(y)$ for all $y \neq z$ in S : clearly $h \in G_{w'}$.

To prove the theorem, apply the **HST** axiom of Saturation to the standard size family $\mathscr{X} = \{G_w : w \in \mathscr{P}_{\mathtt{fin}}(S)\}$: any $f^\circ \in \bigcap \mathscr{X}$ is as required. □

It occurs that, conversely, Extension implies Saturation modulo a weaker, "internal" form of Saturation: see e.g. the proof of Theorem 8.2.10 below.

Corollary 1.3.13. (i) *If $S \subseteq \mathbb{S}$ and $f : S \to \mathbb{I}$ is any function then there is an internal function f° such that $S \subseteq \mathtt{dom}\, f^\circ$ and $f = f^\circ \restriction S$.*

(ii) *If $W \in \mathbb{WF}$ and $f : W \to \mathbb{I}$ is any function then there is an internal function $f^\circ : {}^*W \to \mathbb{I}$ such that $f(w) = f^\circ({}^*w)$ for all $w \in W$.*

Proof. To prove (i) note that by definition any set $S \subseteq \mathbb{S}$ is a set of standard size, and apply Theorem 1.3.12.

To prove (ii) apply the theorem to the set $S = \{{}^*w : w \in W\}$ and the map $g({}^*w) = f(w)$. We obtain an internal function g° such that $S \subseteq D = \mathtt{dom}\, g^\circ$ and $g = g^\circ \restriction S$. Define, in $\mathbb{I}$, a function $f^\circ : {}^*W \to \mathbb{I}$ so that $f^\circ(x) = g^\circ(x)$ for all $x \in {}^*W \cap D$ and $f^\circ(x) = \varnothing$ otherwise. □

To show how Extension works let us prove a useful generalization of (5) of Exercise 1.2.15. Let, for any set X and ordinal λ:

$$X^\lambda = \text{all functions } f : \lambda \to X, \quad \text{whenever } \lambda \text{ is an ordinal};$$

$$[X]^\lambda = \text{all sets } Y \subseteq X \text{ of } \mathtt{card}\, Y = \lambda, \quad \text{whenever } \lambda \text{ is a cardinal};$$

in addition, $X^{<\lambda} = \bigcup_{\xi<\lambda} X^\xi$ and $[X]^{<\lambda} = \bigcup_{\xi<\lambda} [X]^\xi$. In **ZFC**, these collections are sets by simple arguments based on the Power Set axiom. This argument fails in **HST** due to the absense of Power Set, however we have

Theorem 1.3.14. *If $X \subseteq \mathbb{I}$ then X^λ, $X^{<\lambda}$ are sets for any $\lambda \in \mathtt{Ord}$ and $[X]^\lambda$, $[X]^{<\lambda}$ are sets for any $\lambda \in \mathtt{Card}$.*

Proof. It suffices to show that X^λ is a set, the rest of the lemma follows by Collection and Separation: for instance, $[X]^\lambda \subseteq \mathtt{ran}\, \Phi$, where Φ is defined on X^λ by $\Phi(f) = \mathtt{ran}\, f$. We can assume that X is standard by (3) of Exercise 1.1.11. Then $X = {}^*W$ for some $W \in \mathbb{WF}$. Let, in $\mathbb{WF}$, $F = W^\lambda$, the set of all functions $\varphi : \lambda \to W$. It follows from Corollary 1.3.13(ii) that for any $f \in X^\lambda$ there is a function $f^\circ \in {}^*F$ with $f^\circ({}^*w) = f(w)$ for any $w \in W$. Thus, $f^\circ \longmapsto f$ maps ${}^*F$ onto X^λ, and hence X^λ is a set by Collection. □

The extended function can sometimes even be standard. Then the principal tool involved is Standardization rather than Saturation, as in the proof of the next lemma.

Lemma 1.3.15. *If $X \subseteq \mathbb{S}$ and $\phi : X \to \mathbb{S}$ is any function then there is a standard function f such that $X \subseteq \mathtt{dom}\, f$ and $f(x) = \phi(x)$ for all $x \in X$.*

Proof. By Standardization, there exist standard sets U, f such that $X = U \cap \mathbb{S}$ and the only standard elements of f are pairs of the form $\langle x, \phi(x) \rangle$, where $x \in X$. It follows from Transfer that f is a function with $\mathtt{dom}\, f = U$. **Exercise**: prove that f is as required. □

1.4 The class $\mathbf{\Delta}_2^{\mathtt{ss}}$

The idea to obtain new sets applying some operations, for instance, of union and intersection to sets already defined, is quite common in set theory, and often leads to meaningful collections of sets, like Borel, projective, or Gödel-constructible sets. In **HST**, standard size unions and intersections, applied to internal sets, lead to an important class $\mathbf{\Delta}_2^{\mathtt{ss}}$ of subsets of $\mathbb{I}$.

Definition 1.4.1. Define the following classes of sets [17]:

$\mathbf{\Sigma}_1^{\mathtt{ss}}$ is the collection of all sets of the form $x = \bigcup_{\xi<\kappa} x_\xi$;

$\mathbf{\Pi}_1^{\mathtt{ss}}$ is the collection of all sets of the form $x = \bigcap_{\xi<\kappa} x_\xi$;

$\mathbf{\Sigma}_2^{\mathtt{ss}}$ is the collection of all sets of the form $x = \bigcup_{\xi<\kappa} \bigcap_{\eta<\lambda} x_{\xi\eta}$;

$\mathbf{\Pi}_2^{\mathtt{ss}}$ is the collection of all sets of the form $x = \bigcap_{\xi<\kappa} \bigcup_{\eta<\lambda} x_{\xi\eta}$;

where $\kappa, \lambda \in \mathtt{Ord}$ and all sets x_ξ, $x_{\xi\eta}$ are internal.

Finally put $\mathbf{\Delta}_1^{\mathtt{ss}} = \mathbf{\Sigma}_1^{\mathtt{ss}} \cap \mathbf{\Pi}_1^{\mathtt{ss}}$ and $\mathbf{\Delta}_2^{\mathtt{ss}} = \mathbf{\Sigma}_2^{\mathtt{ss}} \cap \mathbf{\Pi}_2^{\mathtt{ss}}$. □

The goal of this section is to present several rather nontrivial properties of sets in $\mathbf{\Delta}_2^{\mathtt{ss}}$, in particular, related to cuts, external maps between internal sets, and cardinalities of sets not of standard size. Another related example will be given below (a Dependent Choice-like construction in Theorem 5.5.12).

1.4a Basic properties of $\mathbf{\Delta}_2^{\mathtt{ss}}$

It turns out that as far as the external universe **HST** is concerned the class $\mathbf{\Delta}_2^{\mathtt{ss}}$ somehow exhausts the domain of applicability of classical methods of nonstandard analysis based on Standardization and Saturation: this follows, for instance, from Theorem 1.4.3. The next theorem presents some rather technical properties of this class.

Theorem 1.4.2. (i) *Theorem 1.3.5 (Saturation) holds also for f. i. p. families $\mathscr{X}$ of standard size, which consist of $\mathbf{\Pi}_1^{\mathtt{ss}}$ sets. In addition,*

(ii) $\mathbf{\Delta}_1^{\mathtt{ss}} = \mathbb{I}$, *moreover if X is $\mathbf{\Sigma}_1^{\mathtt{ss}}$, Y is $\mathbf{\Pi}_1^{\mathtt{ss}}$, and $Y \subseteq X$, then there is an internal set Z with $Y \subseteq Z \subseteq X$;*

(iii) $\mathbb{N}$ *is a $\mathbf{\Sigma}_1^{\mathtt{ss}}$ non-internal set, while ${}^*\mathbb{N} \setminus \mathbb{N}$ is a $\mathbf{\Pi}_1^{\mathtt{ss}}$ non-internal set, hence there are sets in $\mathbf{\Sigma}_1^{\mathtt{ss}} \setminus \mathbf{\Pi}_1^{\mathtt{ss}}$ and vice versa, in addition $\mathbf{\Sigma}_1^{\mathtt{ss}} \cup \mathbf{\Pi}_1^{\mathtt{ss}} \subsetneqq \mathbf{\Delta}_2^{\mathtt{ss}}$;*

(iv) $\mathbf{\Sigma}_2^{\mathtt{ss}} = \mathbf{\Pi}_2^{\mathtt{ss}}$, *thus the class $\mathbf{\Delta}_2^{\mathtt{ss}} = \mathbf{\Sigma}_2^{\mathtt{ss}} = \mathbf{\Pi}_2^{\mathtt{ss}}$ is closed under operations of standard size union and standard size intersection;*

(v) *the class $\mathbf{\Delta}_2^{\mathtt{ss}}$ is also closed under projections: if P is $\mathbf{\Delta}_2^{\mathtt{ss}}$ then $\mathrm{dom}\, P = \{x : \exists y\, P(x,y)\}$ and $\mathrm{ran}\, P = \{y : \exists x\, P(x,y)\}$ are $\mathbf{\Delta}_2^{\mathtt{ss}}$.*

---

[17] The upper index ss accounts for the fact that the operations used to define the families are standard size unions and intersections, by Theorem 1.3.1.

Proof. (i) Let $\mathscr{X} = \{X_\xi : \xi < \kappa\}$, where $\kappa \in \mathtt{Ord}$ and all sets X_ξ are $\mathbf{\Pi}_1^{\mathtt{ss}}$. According to Standard Size Choice, we can simultaneously present sets X_ξ in the form $X_\xi = \bigcap_{\eta<\kappa_\xi} X_{\xi\eta}$, where all κ_ξ belong to $\mathtt{Card}$ while all sets $X_{\xi\eta}$ are internal. Apply Theorem 1.3.5 to the family $\mathscr{X}' = \{X_{\xi\eta} : \xi < \kappa \wedge \eta < \kappa_\xi\}$.

(ii) Let $Y = \bigcap_{\xi<\kappa} Y_\xi \subseteq X = \bigcup_{\xi<\kappa} X_\xi$, all sets X_ξ, Y_ξ being internal. Apply Corollary 1.3.6. $\mathbb{I}$ is closed under finite unions by Lemma 1.2.16(iv).

(iii) That ${}^*\mathbb{N}$ is $\mathbf{\Sigma}_1^{\mathtt{ss}}$ is obvious, that it is not in $\mathbf{\Pi}_1^{\mathtt{ss}}$ follows from (2) of Exercise 1.3.8. To prove the additional statement take $X = \mathbb{N} \times ({}^*\mathbb{N} \smallsetminus \mathbb{N})$ to show that $\mathbf{\Delta}_2^{\mathtt{ss}}$ is strictly bigger.

(iv) We have $\bigcap_{\xi<\kappa} \bigcup_{\eta<\lambda} x_{\xi\eta} = \bigcup_{f\in F} \bigcap_{\xi<\kappa} x_{\xi,f(\xi)}$ by the axiom of Standard Size Choice, where $F = \lambda^\kappa$, the collection of all functions $f : \kappa \to \lambda$, is a set in $\mathbb{WF}$ by Theorem 1.1.9(ii), which easily leads to $\mathbf{\Sigma}_2^{\mathtt{ss}}$ presentation. The "hence" statement follows because the classes $\mathbf{\Sigma}_2^{\mathtt{ss}}$ and $\mathbf{\Pi}_2^{\mathtt{ss}}$ are obviously closed under standard size resp. unions and intersections.

(v) Starting with a $\mathbf{\Sigma}_2^{\mathtt{ss}}$-presentation of P, apply the syntactical transformation given by Exercise 1.3.7. □

It follows from the closure property (iv) that the index $_2$ in $\mathbf{\Delta}_2^{\mathtt{ss}}$ is deceptive: we could write something like $\mathbf{\Delta}_\infty^{\mathtt{ss}}$ as well. Even more, we have

Theorem 1.4.3 (= Theorem 5.2.10 below). *Any set $X \subseteq \mathbb{I}$, $\mathtt{st}$-$\in$-definable in $\mathbb{I}$ (with parameters in $\mathbb{I}$ allowed) is $\mathbf{\Delta}_2^{\mathtt{ss}}$.*

This will be the reason to call sets in $\mathbf{\Delta}_2^{\mathtt{ss}}$ *elementary external*. Despite such a general character, the class $\mathbf{\Delta}_2^{\mathtt{ss}}$ still admits a meaningful study, as demonstrated by the examples below in this section.

It will be shown below (Section 5.5) that $\mathbf{\Delta}_2^{\mathtt{ss}}$ can be inflated to an even bigger class $\mathbb{L}[\mathbb{I}]$, which satisfies **HST** and generally admits a fruitful study, (but all sets $X \subseteq \mathbb{I}$ in $\mathbb{L}[\mathbb{I}]$ already belong to $\mathbf{\Delta}_2^{\mathtt{ss}}$), and because of this it is consistent with **HST** that <u>all</u> sets $X \subseteq \mathbb{I}$ belong to $\mathbf{\Delta}_2^{\mathtt{ss}}$.

1.4b Cuts (initial segments) of $*$-ordinals

It follows from Theorem 1.4.2(iii) that there exist sets "properly" $\mathbf{\Delta}_2^{\mathtt{ss}}$. Theorem 1.4.6 will show that such sets do not exist in a certain category of sets of $*$-ordinals, namely initial segments of ${}^*\mathtt{Ord}$ commonly called *cuts*.

Definition 1.4.4. (1) Let $X \subseteq {}^*\mathtt{Ord}$. A set $Y \subseteq X$ is *cofinal*, resp., *coinitial* in X if $\forall \xi \in X\, \exists \eta \in Y\, (\eta \geq \xi)$, resp., $\forall \xi \in X\, \exists \eta \in Y\, (\eta \leq \xi)$.

(2) A cut (initial segment) $C \subseteq {}^*\mathtt{Ord}$ is *standard size cofinal (*resp., *coinitial)* iff C contains a cofinal (resp., ${}^*\mathtt{Ord} \smallsetminus C$ contains coinitial) subset of standard size.

(3) In particular a cut C is *countably cofinal (*resp., *coinitial)* iff there exists a strictly increasing sequence $\{c_n\}_{n\in\mathbb{N}}$ cofinal in C (resp., there exists a strictly decreasing sequence $\{c_n\}_{n\in\mathbb{N}}$ coinitial in ${}^*\mathtt{Ord} \smallsetminus C$). □

Note that if a cut C has a maximal element c then trivially C is standard size cofinal while ${}^*\mathtt{Ord} \smallsetminus C$ is standard size coinitial.

Exercise 1.4.5. (i) Let $X \subseteq {}^*\mathtt{Ord}$ be a set of standard size. Prove that there is an ordinal $\vartheta \in \mathtt{Ord}$ and an **increasing** (resp., **decreasing**) map $f : \vartheta \to X$ such that $\mathtt{ran}\, f$ is cofinal (resp., coinitial) in X.

(ii) Let C be a cut in ${}^*\mathtt{Ord}$. Prove the following claims (1) and (2):

(1) *If C is internal then $C \in {}^*\mathtt{Ord}$, and if C is then a limit $*$-ordinal then it is not standard size cofinal while ${}^*\mathtt{Ord} \smallsetminus C$ contains a least element.* *Hint*: if $Y \subseteq C$ is a set of standard size then apply Saturation to the family of sets $\{\xi \in C : \eta < \xi\}$, $\eta \in Y$.

(2) *If C is non-internal then C is a gap* (that is C has no maximal element and ${}^*\mathtt{Ord} \smallsetminus C$ has no minimal element), *and it is impossible that C is both standard size cofinal and standard size coinitial.* *Hint*: if $Y \subseteq C$ and $Y' \subseteq {}^*\mathtt{Ord} \smallsetminus C$ are sets of standard size then apply Saturation to the family of sets $\{\xi : \eta < \xi < \eta'\}$, $\eta \in Y$, $\eta' \in Y'$.) □

Theorem 1.4.6. (i) *Suppose that $C \subseteq {}^*\mathtt{Ord}$ is a $\mathbf{\Delta}_2^{\mathrm{ss}}$ cut. Then we have exactly one of the following:*

(1) *C is internal, hence a $*$-ordinal, and either C is a successor ordinal or it is not standard size cofinal;*

(2) *C is a gap and standard size cofinal, hence $\mathbf{\Sigma}_1^{\mathrm{ss}}$;*

(3) *C is a gap and standard size coinitial, hence C is $\mathbf{\Pi}_1^{\mathrm{ss}}$.*

(ii) *If $\gamma \in {}^*\mathtt{Ord}$, $X \subseteq \gamma$ is a $\mathbf{\Delta}_2^{\mathrm{ss}}$ set cofinal in $\gamma = [0, \gamma)$, and $X \cap \xi$ is internal for any $\xi \in X$ (equivalently, for any $\xi < \gamma$) then X is internal.*

Proof. (i) Let $C = \bigcup_{a \in A} \bigcap_{b \in B} C_{ab}$, where $A, B \in \mathbb{WF}$ while all sets C_{ab} are internal. If none of the sets $C_a = \bigcap_{b \in B} C_{ab}$ is cofinal in C, *i.e.*, for any $a \in A$ there exists $\xi_a \in C \smallsetminus C_a$, then $Y = \{\xi_a : a \in A\}$ is cofinal in C, therefore, C is internal provided Y has a maximal element, and standard size cofinal with no maximal element otherwise.

Assume that some C_a is cofinal in C. If for any finite $F \subseteq B$ there is a $*$-ordinal $\xi_F \in C_F = \bigcap_{b \in F} C_{ab}$, $\xi_F \notin C$, then the set $Y = \{\xi_F : F \in \mathscr{P}_{\mathtt{fin}}(B)\}$ is coinitial in ${}^*\mathtt{Ord} \smallsetminus C$. (Indeed, if $\xi \in {}^*\mathtt{Ord} \smallsetminus C$ satisfies $\xi < \xi_F$ for any finite $F \subseteq B$ then every set C_F, $F \in \mathscr{P}_{\mathtt{fin}}(B)$ contains a $*$-ordinal $\geq \xi$. It follows by Theorem 1.3.5 or Corollary 1.3.6 that C_a itself contains an element $\geq \xi$, contradiction.) If Y contains a minimal element ξ then $C = \xi$ is internal, otherwise C is standard size coinitial with no minimal element in ${}^*\mathtt{Ord} \smallsetminus C$.

To accomplish the proof of (i) apply (ii) of Exercise 1.4.5.

(ii) The set $U = \{X \cap \xi : \xi < \gamma\}$ consists of internal subsets of γ by the choice of X. Moreover, U is $\mathbf{\Delta}_2^{\mathrm{ss}}$. (It needs some care to derive this using (iv), (v) of Theorem 1.4.2. We leave this as an **exercise**.) Let $U = \bigcup_{a \in A} \bigcap_{b \in B} U_{ab}$, where $A, B \in \mathbb{WF}$ while all sets U_{ab} are internal subsets of $\mathscr{P}_{\mathtt{int}}(\gamma)$.

Note that U is linearly ordered by the relation: $u \preccurlyeq v$ iff u is an initial segment of $v \subseteq {}^*\mathtt{Ord}$. If none of the sets $U_a = \bigcap_{b \in B} U_{ab}$ is $\preccurlyeq$-cofinal in U then for any $a \in A$ there exists $\xi_a \in X$ such that $u_a = X \cap \xi_a$ belongs to $U \smallsetminus U_a$ and $u \preccurlyeq u_a$ for all $u \in U_a$. Then the set $Y = \{\xi_a : a \in A\}$ is cofinal in X, hence in γ, contrary to (i) (the incompatibility of (i)(1) and (i)(2)).

Thus at least one set U_a is $\preccurlyeq$-cofinal in U. Note that $U_a \times U_a = \bigcap_{b,b' \in B}(U_{ab} \times U_{ab'})$ is a subset of $U \times U$, hence, a subset of the internal set of all pairs $\langle u, v\rangle$ of internal sets $u, v \subseteq \gamma$ such that $u \preccurlyeq vu \vee v \preccurlyeq u$. It follows by Compactness (Corollary 1.3.6) that there exists a finite set $B' \subseteq B$ such that the set $U' = \bigcap_{b \in B'} U_{ab}$ (internal by Lemma 1.2.16) still is pairwise $\preccurlyeq$-comparable. Note that $U_a \subseteq U' \subseteq \mathscr{P}_{\mathtt{int}}(\gamma)$, and hence easily $U' \subseteq U$ is $\preccurlyeq$-cofinal in U. It follows that $X = \bigcup U'$ is internal. □

1.4c Monads and transversals

Here we concentrate on cuts (initial segments) in ${}^*\mathbb{N}$. Cuts $U \subseteq {}^*\mathbb{N}$ closed under $+$ are called *additive*. For instance, if $h \in {}^*\mathbb{N} \smallsetminus \mathbb{N}$ then $h\mathbb{N} = \bigcup_{n \in \mathbb{N}}[0, hn)$ and $h/\mathbb{N} = \bigcap_{n \in \mathbb{N}}[0, h/n)$ are additive cuts.

Any additive cut $U \subseteq {}^*\mathbb{N}$ defines an equivalence relation on ${}^*\mathbb{N}$, $x \mathsf{M}_U y$ iff $|x - y| \in U$, and subsequently a partition of ${}^*\mathbb{N}$ into *U-monads*, *i.e.*, sets of the form $[x]_U = \{y : x \mathsf{M}_U y\} = \{y : |x - y| \in U\}$. (**Exercise**: prove that any two different monads are disjoint.) A *transversal* for such an equivalence relation M_U is any set having exactly one element in each U-monad.

A part of the following result will be employed to demonstrate that Choice is unprovable in **HST** (Theorem 5.5.8 below). See §§ 9.7e, 9.7f, 9.7g on further studies of monad partitions.

Theorem 1.4.7. *Suppose that $\{\varnothing\} \neq U \subsetneqq {}^*\mathbb{N}$ is an additive cut. Then*

(i) *if U is countably cofinal then M_U admits a $\mathbf{\Delta}_2^{\mathrm{ss}}$ transversal if and only if U has the form $h\mathbb{N}$, $h \in {}^*\mathbb{N}$;*

(ii) *if U is countably coinitial then M_U admits a $\mathbf{\Delta}_2^{\mathrm{ss}}$ transversal if and only if U has the form $h/\mathbb{N}$, $h \in {}^*\mathbb{N} \smallsetminus \mathbb{N}$.*

Proof. (i) *The "if" part.* It suffices to consider the case $U = \mathbb{N}$. In $\mathbb{WF}$, let, for any natural number n, $s(n) \in 2^{\mathbb{N}}$ be the full binary expansion extended by zeros, for instance $s(13) = \langle 1, 1, 0, 1, 0, 0, 0, \ldots\rangle$. Then, in $\mathbb{I}$, ${}^*s(x)$ is an internal sequence in ${}^*(2^{\mathbb{N}})$ for any $x \in {}^*\mathbb{N}$. Still in $\mathbb{WF}$, define for $f, g \in 2^{\mathbb{N}}$: $f \mathsf{E}_0 g$ iff $\{k : f(k) \neq g(k)\}$ is finite. Let $F \subseteq 2^{\mathbb{N}}$ be any transversal for the partition of $2^{\mathbb{N}}$ into E_0-classes in $\mathbb{WF}$. The following set is $\mathbf{\Delta}_2^{\mathrm{ss}}$:

$$T = \{x \in {}^*\mathbb{N} : {}^*s(x) \restriction \mathbb{N} \in F\} = \bigcup_{f \in F} \bigcap_{n \in \mathbb{N}} \{x \in {}^*\mathbb{N} : {}^*s(x)(n) = f(n)\}$$

Exercise: show that T is a transversal for $\mathsf{M}_{\mathbb{N}}$.

The "only if" part. If U is not of the form $h\mathbb{N}$ then it admits a cofinal sequence $\{d_n\}_{n \in \mathbb{N}}$ with $d_{n+1} \notin d_n\mathbb{N}$, $\forall n$. Suppose that $T = \bigcup_{a \in A} \bigcap_{b \in B} T_{ab}$

is a transversal for M_U, where $A, B \in \mathbb{WF}$ and all sets $T_{ab} \subseteq {}^*\mathbb{N}$ are internal. By Saturation, for any $a \in A$ and n there is a finite set $B_a(n) \subseteq B$ such that any two elements $x \neq y$ in $T_a(n) = \bigcap_{b \in B_a(n)} T_{ab}$ satisfy $|x - y| > d_{n+1}$. We have $T_a \subseteq \bigcap_n T_a(n)$, hence, as T intersects any U-monad,

$$ {}^*\mathbb{N} = \bigcup_n T + [-d_n, d_n] = \bigcup_{a \in A} \bigcup_n T_a(n) + [-d_n, d_n] , $$

where $X + [-d, d] = \bigcup_{x \in X} [x - d, x + d]$ for any $X \subseteq {}^*\mathbb{N}$. Saturation yields a finite set $P \subseteq \mathbb{N} \times A$ such that ${}^*\mathbb{N} = \bigcup_{\langle n,a \rangle \in P} T_a(n) + [-d_n, d_n]$. Now let $h \in {}^*\mathbb{N} \smallsetminus U$. Each set $(T_a(n) + [-d_n, d_n]) \cap [0, h)$ contains at most $2hd_n/d_{n+1}$ elements, where d_n/d_{n+1} is infinitesimal, hence, no finite union of sets $T_a(n) + [-d_n, d_n]$ can cover $[0, h)$, contradiction.

(ii) *The "if" part*. Suppose that $U = h/\mathbb{N}$. There is a unique $c \in {}^*\mathbb{N}$ with $2^c \le h < 2^{c+1}$, then $U = 2^c/\mathbb{N}$, and hence we can assume that $h = 2^c$ in $\mathbb{I}$. Let ${}^{\text{int}}2^c$ denote the (internal) set of all internal maps $\xi : c \to 2$ (that is, all $*$-finite dyadic sequences of length c).

Put $\varphi(\xi) = \xi \restriction \mathbb{N}$ for each $\xi \in {}^{\text{int}}2^c$, so that φ is a map from ${}^{\text{int}}2^c$ onto $2^{\mathbb{N}}$. (Maps of this sort will be called *shadows* below.) Define, in $\mathbb{I}$, $b(\xi) = \sum_{k=0}^{c-1} 2^k \xi(k)$ for any $\xi \in {}^{\text{int}}2^c$. Then b is an internal bijection of ${}^{\text{int}}2^c$ onto $[0, h)$, and we have $\varphi(\xi) = \varphi(\eta) \iff b(\xi) \; \mathsf{M}_U \; b(\eta)$. However φ has all its values in the set $2^{\mathbb{N}} \in \mathbb{WF}$. Thus the quotient $[0, h)/\mathsf{M}_U$ is a set of standard size together with $2^{\mathbb{N}}$. Standard Size Choice yields a transversal $T \subseteq [0, h)$ for M_U, which is also a set of standard size, hence, a $\mathbf{\Delta}_2^{\mathbf{ss}}$ set. It remains to uniformly reproduce T in every interval in ${}^*\mathbb{N}$ of the form $[\nu h, \nu h + h)$.

The "only if" part. There is a decreasing sequence $\{d_n\}_{n \in \mathbb{N}}$ such that $U = \bigcap_{n \in \mathbb{N}} [0, d_n)$ and $d_{n+1} \in d_n/\mathbb{N}$ for any n. Suppose towards the contrary that $T = \bigcup_{a \in A} \bigcap_{b \in B} T_{ab}$ is a transversal for M_U, where A, $B \in \mathbb{WF}$ and all sets T_{ab} are internal. For any $a \in A$, if $x \neq y$ belong to $T_a = \bigcap_{b \in B} T_{ab} \subseteq T$ then $|x - y| >$ some d_n. By Saturation there exist a finite set $B_a \subseteq B$ and some $n = n(a) \in \mathbb{N}$ such that $|x - y| > d_n$ whenever $x \neq y$ belong to $T'_a = \bigcap_{b \in B_a} T_{ab}$. For any $a \in A$, the set $N_a = \bigcup_{c \in T'_a} [c - d_{n(a)+1}, c + d_{n(a)+1}]$ is internal, and the set $N = \bigcup_{a \in A} N_a$ coincides with ${}^*\mathbb{N}$ since $T \subseteq \bigcup_{a \in A} T'_a$. Now take any $h \in {}^*\mathbb{N}$ with $h > d_n^2, \forall n$. Any intersection $T'_a \cap [0, h)$ has $\le h/d_{n(a)}$ elements, Thus $N_a \cap [0, h)$ has $\le 2hd_{n(a)+1}/d_{n(a)}$ elements, infinitesimally small w. r. t. h. Thus no finite union of sets N_a can cover $[0, h)$, and hence, by Saturation, the whole union N does not cover $[0, a)$, a contradiction to the above. □

1.4d On non-well-founded cardinalities

We proved in § 1.3a that cardinalities of sets of standard size are identic to cardinals in $\mathbb{WF}$, a transitive **ZFC**-universe, hence, we may trust their **ZFC**-like behaviour in **HST**. Much less is known about cardinalities of sets not of standard size, for instance, infinite internal sets. The latter suggest what may look as an easy entry into the non-standard size cardinalities, because $\mathbb{I}$ has its own *scala* of $*$-cardinals (*i. e.*, cardinals in the sense of $\mathbb{I}$).

Exercise 1.4.8. Show that no $*$-cardinal is a (well-founded) cardinal in the sense of Definition 1.2.6 unless it is finite. □

Despite of this, can we trust a **ZFC**-like behaviour of $*$-cardinals in the external universe of **HST**? For instance, are internal sets X, Y of different $*$-cardinalities necessarily not equinumerous in the external universe? An entirely positive answer fails, *e.g.*, for $X = n$ and $Y = n + 1$, where $n \in {}^*\mathbb{N} \smallsetminus \mathbb{N}$, see footnote 15 on p. 26. There is a much better example, essentially due to Keisler *et. al.* [KKML 89], which can be converted to the following result in **HST** : if $a \in {}^*\mathbb{N} \smallsetminus \mathbb{N}$ then there is a $\mathbf{\Delta}_2^{\mathtt{ss}}$ surjection of $a = [0, a)$ onto $a\mathbb{N} = \bigcup_{k \in \mathbb{N}} [0, ak)$, hence, for any $k \in \mathbb{N}$ there is a $\mathbf{\Delta}_2^{\mathtt{ss}}$ surjection of a onto ak, see Lemma 9.6.12 in Section 9.6. The next theorem shows that in any other case different $*$-cardinalities cannot be "glued" by $\mathbf{\Delta}_2^{\mathtt{ss}}$ maps. Recall that ${}^*\mathtt{card}\, X$ is the $*$-cardinality of an internal set X in $\mathbb{I}$.

Theorem 1.4.9. *If* $X, Y \in \mathbb{I}$, *and* $f : X \to Y$ *is a* $\mathbf{\Delta}_2^{\mathtt{ss}}$ *function then*:

(i) *for any* $h \in {}^*\mathbb{N} \smallsetminus \mathbb{N}$, *the set* $\mathtt{ran}\, f$ *can be covered by a set* $R \in \mathbb{I}$ *with* ${}^*\mathtt{card}\, R \le h\, {}^*\mathtt{card}\, X$ *in* $\mathbb{I}$, *in particular, we have* ${}^*\mathtt{card}\, R \le {}^*\mathtt{card}\, X$ *whenever* X *is* $*$-*infinite*;

(ii) *if* $\mathtt{ran}\, f = Y$ *then* ${}^*\mathtt{card}\, Y \le n\, {}^*\mathtt{card}\, X$ *for some* $n \in \mathbb{N}$, *in particular, we have* ${}^*\mathtt{card}\, Y \le {}^*\mathtt{card}\, X$ *whenever* X *is* $*$-*infinite.*

By the above-mentioned result of Lemma 9.6.12, it cannot be required in (i) that ${}^*\mathtt{card}\, R \le n\, {}^*\mathtt{card}\, X$ for some $n \in \mathbb{N}$.

Proof. (i) There exist sets $A, B \in \mathbb{WF}$ and a family $\{W_{ab}\}_{a \in A,\, b \in B}$ of internal sets, such that the graph of f is equal to $\bigcup_{a \in A} \bigcap_{b \in B} W_{ab}$. Corollary 1.3.13 (Extension) yields an internal function φ defined on ${}^*A \times {}^*B$ such that $\varphi({}^*a, {}^*b) = W_{ab}$ for all $a \in A$ and $b \in B$. Then we have

$$f(x) = y \iff \exists^{\mathtt{st}} a \in {}^*A\, \forall^{\mathtt{st}} b \in {}^*B\; W(x, y, a, b)$$

where $W = \{\langle x, y, a, b\rangle \in X \times Y \times {}^*A \times {}^*B : \langle x, y\rangle \in \varphi(a, b)\}$ is internal.

By (1) of Exercise 1.3.8 there is a $*$-finite internal set Z containing all standard elements of ${}^*A \cup {}^*B$ and satisfying $\#Z < \sqrt{\log_2 h}$ in $\mathbb{I}$. Put

$$F(x, y) = \{\langle a, b\rangle \in Z \times Z : W(x, y, a, b)\}$$

for all $x \in X$ and $y \in Y$. Then obviously $f(x) = y \iff f(x') = y'$, whenever $x, x' \in X$ and $y, y' \in Y$ satisfy the equality $F(x, y) = F(x', y')$. On the other hand, F is an internal function into $\mathscr{P}_{\mathtt{int}}(Z \times Z)$, a $*$-finite set with $< 2^{(\#Z)^2} \le h$ elements in $\mathbb{I}$ by the choice of Z. Arguing in $\mathbb{I}$, we obtain a set Y of cardinality $h\, {}^*\mathtt{card}\, X$ in $\mathbb{I}$ such that for all $x \in X$ and $y \in Y$ there is $y' \in Y$ with $F(x, y) = F(x, y')$, and hence $\mathtt{ran}\, f \subseteq Y$.

(ii) Let, by (i), n be the least number in ${}^*\mathbb{N}$ such that Y is covered by an internal set Y' with ${}^*\mathtt{card}\, Y' \le n\, {}^*\mathtt{card}\, X$. (Thus n is legitimally defined by an $\in$-formula in $\mathbb{I}$.) We observe that $n \in \mathbb{N}$ because otherwise $h = n - 1$ gives a contradiction to the choice of n. □

Thus, $*$-infinite $*$-cardinals are preserved at least under mappings of class $\mathbf{\Delta}_2^{\mathrm{ss}}$. That this may be not the case for external mappings of more complicated nature, will be shown in Chapter 7.

1.4e Small and large sets

Let us call *small* any set of standard size, and *large* any set containing a subset equinumerous to an infinite internal set, in particular, all infinite internal sets are large. Thus subsets of small sets are small and supersets of large sets are large. We have even more:

Exercise 1.4.10. Prove that small sets are not large, but if X is small and Y large then Y contains a subset equinumerous to X.

Hint. Internal sets of standard size are finite (Exercise 1.3.8). If $Y \in \mathbb{I}$ is infinite then any $X \subseteq \mathbb{S}$ admits an injection to Y by the same argument, based on the Saturation Theorem 1.3.5, as in the proof of Theorem 1.3.9. □

Are there sets neither small nor large? The following theorem shows that at least not in the class $\mathbf{\Delta}_2^{\mathrm{ss}}$.

Theorem 1.4.11. *Any $\mathbf{\Delta}_2^{\mathrm{ss}}$ set $X \subseteq \mathbb{I}$ is either large or of standard size. Moreover, if E is a $\mathbf{\Delta}_2^{\mathrm{ss}}$ equivalence relation on a $\mathbf{\Delta}_2^{\mathrm{ss}}$ set X then the quotient X/E is either large or of standard size.*

Proof. An elementary proof of the first claim is as follows. By definition, $X = \bigcup_{a\in A} \bigcap_{b\in B} X_{ab}$, where all X_{ab} are internal and $A, B \in \mathbb{WF}$. If all $\mathbf{\Pi}_1^{\mathrm{ss}}$ sets $X_a = \bigcap_{b\in B} X_{ab}$ are finite, then, using Standard Size Choice, we obtain a surjection $f : A \times \mathbb{N} \xrightarrow{\text{onto}} X$, thus X is of standard size. Suppose that some X_a is infinite. Then, for each finite $u \subseteq B$, the partial intersection $Y_u = \bigcap_{b\in u} X_{ab}$ is infinite as well, and internal, hence, for any $m \in \mathbb{N}$, the internal set P_{um} of all internal sets $Y \subseteq Y_u$ with ${}^*\mathtt{card}\, Y \geq m$ is nonempty. The family of all sets P_{um}, where $u \in \mathscr{P}_{\mathtt{fin}}(B)$ and $m \in \mathbb{N}$, is of standard size and f. i. p., therefore, there is $Y \in \bigcap_{u,m} P_{um}$, an internal infinite set.

Now prove the "moreover" assertion.

First of all, it can be assumed that X is internal. (Indeed, any set $X \subseteq \mathbb{I}$ can be covered by an internal set Y, Exercise 1.1.11. We can extend E to Y so that all elements of $Y \smallsetminus X$ are E-equivalent. The extended relation is still $\mathbf{\Delta}_2^{\mathrm{ss}}$ and has just one more equivalence class.) Thus, E has the form $\mathsf{E} = \bigcup_{a\in A} \bigcap_{b\in B} C_b^a$, where A, $B \in \mathbb{WF}$ while $C_b^a \subseteq X \times X$ are internal.

Below, a, b are always assumed to belong to resp. A, B and x, y, z to X, unless explicitly indicated otherwise. **Important:** we can w. l. o. g. assume that the sets C_b^a are symmetric, that is $C_b^a = (C_b^a)^{-1}$: indeed, in any case

$$\mathsf{E} = \mathsf{E} \cup \mathsf{E}^{-1} = \bigcup_a \bigcap_{b,b'} C_b^a \cup (C_{b'}^a)^{-1} = \bigcup_a \bigcap_{b,b'} S_{b,b'}^a\,,$$

where the sets $S_{b,b'}^a = (C_b^a \cup (C_{b'}^a)^{-1}) \cap ((C_b^a)^{-1} \cup C_{b'}^a)$ satisfy $S_{b,b'}^a = {S_{b,b'}^a}^{-1}$.

It follows from the transitivity of E that, for any x and y,

$$\exists\, a\, \exists\, z\, \forall\, b\, (x\, C_b^a\, z \wedge y\, C_b^a\, z) \implies x\, \mathsf{E}\, y\,.$$

Using **Saturation**, in the form of the equivalence in Exercise 1.3.7, we obtain

$$\exists\, a\, \forall^{\mathtt{fin}} B' \subseteq B\, \exists\, z\, (x\, C_{B'}^a\, z \wedge y\, C_{B'}^a\, z) \implies x\, \mathsf{E}\, y\,,$$

where $C_{B'}^a = \bigcap_{b \in B'} C_b^a$. Applying **Choice** in $\mathbb{WF}$, we transform this to

$$\forall\, \varphi \in \varPhi\, \exists\, a\, \exists\, z\, (x\, C_{\varphi(a)}^a\, z \wedge y\, C_{\varphi(a)}^a\, z) \implies x\, \mathsf{E}\, y\,, \tag{1}$$

where $\varPhi \in \mathbb{WF}$ is the set of all functions $\varphi : A \to \mathscr{P}_{\mathtt{fin}}(B)$.

Now let us approach the problem from another angle. Assume that X/E is not large, in particular, there is no infinite internal pairwise E-inequivalent set $Y \subseteq X$. Our goal is to show that the quotient X/E is then a set of standard size. The assumption can be formally expressed as

$$\forall\, Y\, \big(\forall\, n\, (\mathtt{card}\, Y > n) \Longrightarrow \exists\, x \neq y \in Y\, \exists\, a\, \forall\, b\, (x\, C_b^a\, y)\big),$$

where (as well as below) Y is assumed to be an internal subset of X. We have, by **Saturation** (as in Exercise 1.3.7),

$$\forall\, Y\, \big(\forall\, n\, (\mathtt{card}\, Y > n) \Longrightarrow \exists\, a\, \forall\, B' \in \mathscr{P}_{\mathtt{fin}}(B)\, \exists\, x \neq y \in Y\, (x\, C_{B'}^a\, y)\big),$$

which can be transformed, by **Choice** in $\mathbb{WF}$, to

$$\forall\, \varphi \in \varPhi\, \forall\, Y\, \big(\forall\, n\, (\mathtt{card}\, Y > n) \Longrightarrow \exists\, a\, \exists\, x \neq y \in Y\, (x\, C_{\varphi(a)}^a\, y)\big),$$

so that, still by **Saturation**, for any $\varphi \in \varPhi$,

$$\exists\, n\, \exists^{\mathtt{fin}} A' \subseteq A\, \forall\, Y\, \big(\, \mathtt{card}\, Y > n \Longrightarrow \exists\, x \neq y \in Y\, \exists\, a \in A'\, (x\, C_{\varphi(a)}^a\, y)\big),$$

therefore, for any $\varphi \in \varPhi$ there exist a finite set $A_\varphi \subseteq A$ and a finite set $Z_\varphi \subseteq X$ with at most n elements such that

$$\forall\, x\, \exists\, z \in Z_\varphi\, \exists\, a \in A_\varphi\, (x\, C_{\varphi(a)}^a\, z)\,. \tag{2}$$

(*Hint*: take as Z_φ a maximal subset of X such that $\neg\, x\, C_{\varphi(a)}^a\, y$ whenever $x \neq y \in Z_\varphi$ and $a \in A_\varphi$; Z_φ contains $\leq n$ elements. Recall that the sets C_b^a are symmetric.) Let, for any $x \in X$, ζ_x be a function defined on $\varPhi$ by

$$\zeta_x(\varphi) = \{\langle z, a\rangle : z \in Z_\varphi \wedge a \in A_\varphi \wedge x\, C_{\varphi(a)}^a\, z\}\,.$$

Since $\varPhi$ is a set of standard size, while Z_φ and A_φ are finite, there are only standard size many possible maps of the form ζ_x. Thus, it remains to show that, for all x, $y \in X$, $\zeta_x = \zeta_y$ implies $x\, \mathsf{E}\, y$.

Suppose that x, y satisfy $\zeta_x = \zeta_y$, and prove $x\, \mathsf{E}\, y$. According to (1), it suffices to find, for any given $\varphi \in \varPhi$, elements $a \in A$ and $z \in X$ satisfying both $x\, C_{\varphi(a)}^a\, z$ and $y\, C_{\varphi(a)}^a\, z$. It follows from (2) that the set $\zeta_x(\varphi) = \zeta_y(\varphi)$ is nonempty. Let $\langle z, a\rangle$ be any of its elements, thus, $z \in Z_\varphi$, $a \in A_\varphi$, and both $\langle x, z\rangle$ and $\langle y, z\rangle$ belong to $C_{\varphi(a)}^a$, as required. □

1.5 Some finer points

The issues considered here belong to the "folklore" of set theory, proof theory, and model theory (except for the content of §1.5b). Our intention is not to give a systematic introduction to the topics considered, which we assume to be in principle known to the reader, at least to some extent, but rather to present those particular versions of basic definitions which are used in this book, as well as to pay some attention to certain points where a deeper knowledge is necessary to consciously understand metamathematical results and arguments in this book, for instance, the difference between the notions of *interpretation* and *model*. These mainly metamathematical issues are more substantially considered, for instance, in Shoenfield [Shoen 67] and, from the point of view of **ZFC**, in Kunen's book [Kun 80], Appendices to Chapter I and Chapter IV.

The section ends with a theorem saying that, in **HST**, the $*$-extension ${}^*\mathfrak{M} \in \mathbb{I}$ of any structure $\mathfrak{M} \in \mathbb{WF}$ is also an elementary extension, in the model theoretic sense.

We begin with the von Neumann hierarchy in **ZFC** and **HST**.

1.5a Von Neumann hierarchy and Reflection in ZFC

The Regularity axiom organizes the **ZFC** set universe $\mathbf{V}$ in the form $\mathbf{V} = \bigcup_{\xi\in\mathtt{Ord}} \mathbf{V}_\xi$ ($\mathtt{Ord}$ is the class of all ordinals), where sets $\mathbf{V}_\xi$ (levels of *the von Neumann hierarchy*) are defined by transfinite induction on ξ as follows:

$$\mathbf{V}_0 = \varnothing\,, \quad \mathbf{V}_{\xi+1} = \mathscr{P}(\mathbf{V}_\xi) \text{ for each } \xi\,, \quad \mathbf{V}_\lambda = \textstyle\bigcup_{\xi<\lambda}\mathbf{V}_\xi \text{ for limit ordinals } \lambda\,.$$

An associated notion, *the von Neumann rank* $\mathtt{rank}\, x \in \mathtt{Ord}$, is defined, for any set x, as the least ordinal ξ with $x \in \mathbf{V}_{\xi+1}$, so that $\mathbf{V}_\xi = \{x : \mathtt{rank}\, x < \xi\}$.

Exercise 1.5.1 (ZFC). Prove that all sets $\mathbf{V}_\xi$ are transitive and $\subseteq$-complete, and $\mathtt{rank}\, x = \mathtt{sup}_{y\in x}\mathtt{rank}\, y$ for any set x, where $\mathtt{sup}\, O$ is the least ordinal strictly bigger than each ordinal in a set $O \subseteq \mathtt{Ord}$. [18] □

One of the most important applications of the sets $\mathbf{V}_\xi$ is that we can find among them models of arbitrarily big finite fragments of **ZFC**, and, in a sense, arbitrarily close approximations of the whole universe of **ZFC**.

Recall that Σ_n *formulas* and Π_n *formulas* are $\in$-formulas of the form

$$(\Sigma_n)\ \exists x_1\, \forall x_2\, \exists x_3 \ldots \forall(\exists)\, x_n\, \psi\,, \qquad (\Pi_n)\ \forall x_1\, \exists x_2\, \forall x_3 \ldots \exists(\forall)\, x_n\, \psi\,,$$

where ψ is a *bounded* $\in$-formula, *i.e.*, an $\in$-formula having no quantifiers except for those of the form $\exists x \in y$ or $\forall x \in y$ (*bounded quantifiers*).

[18] This is why the $\in$-induction (see Remark 1.1.7) is often called, in **ZFC**, *induction on the von Neumann rank*: a property $P(x)$ is proved, or an object $F(x)$ is defined, for all sets x by induction on $\mathtt{rank}\, x$, *i.e.*, on the assumption that $P(y)$ has been established, or, resp., $F(y)$ defined, for all sets y with $\mathtt{rank}\, y < \mathtt{rank}\, x$.

Exercise 1.5.2 (ZFC). Show that for any limit $\lambda \in \mathtt{Ord}$ the set $\mathbf{V}_\lambda$ is an $\in$-model of **ZC**. □

Now for any n we define the following subtheory of **ZFC**:

Σ_n-**ZFC**: the theory **ZC** + Σ_n-Collection, where Σ_n-Collection is the Collection schema of § 1.1h for Σ_n formulas Φ.

Definition 1.5.3. A set V is an *elementary submodel* of the universe with respect to an $\in$-formula $\varphi(v_1, ..., v_n)$, or V *reflects* φ, iff $\varphi(x_1, ..., x_n) \Longleftrightarrow \varphi^V(x_1, ..., x_n)$ for all $x_1, ..., x_n \in V$, where φ^V is the relativization to V, that is, all quantifiers $\exists\, x$, $\forall\, x$ in φ are substituted by $\exists\, x \in V$, $\forall\, x \in V$. □

Theorem 1.5.4. (i) (*Reflection*) *If Φ is a finite list of $\in$-formulas then* **ZFC** *proves that for every $\xi \in \mathtt{Ord}$ there is an ordinal $\vartheta > \xi$ such that $\mathbf{V}_\vartheta$ is an elementary submodel of the universe w. r. t. all formulas in Φ.*

(ii) *For each n,* **ZFC** *proves that for every $\xi \in \mathtt{Ord}$ there is an ordinal $\vartheta > \xi$ such that $\mathbf{V}_\vartheta$ is an elementary submodel of the universe w. r. t. all Σ_n formulas, hence, $\mathbf{V}_\vartheta$ models Σ_n-***ZFC**.

Proof (*Sketch*). (i) This is the basic Reflection theorem, we can refer, for instance, to [Kun 80, pp. 136–137].

(ii) It is known [Jech 78, pp. 125–126] that the multitude of Σ_n formulas $\Phi(\cdot,\cdot)$ for any given $n \geq 1$ can be reduced to a single Σ_n formula $\Phi_n(k,\cdot,\cdot)$, *universal* in the sense that for any Σ_n formula $\Phi(\cdot,\cdot)$ there is $k = k(\Phi)$ such that **ZFC** minus Collection proves $\forall\, x, y\, (\Phi(x,y) \Longleftrightarrow \Phi_n(k,x,y))$. Subsequently, Σ_n-Collection is equivalent to $\forall\, k\, (\Sigma_n$-Collection for the formula $\Phi_n(k,\cdot,\cdot))$. Now the main part of (ii) becomes a corollary of (i).

Prove the "hence" assertion. Suppose that $X \in \mathbf{V}_\vartheta$ and $\Phi(x,y)$ is a Σ_n formula with parameters in $\mathbf{V}_\vartheta$. We can assume that $\forall\, x \in X\, \exists\, y\, \Phi(x,y)$ (otherwise consider $X' = \{x \in X : \exists\, y\, \Phi(x,y)\} \in \mathbf{V}_\vartheta$ instead). Then we have

$$\exists\, Y\, \forall\, x \in X\, \exists\, y \in Y\, \Phi(x,y) \tag{1}$$

in the universe. Now it suffices to show that (1) is "essentially" a Σ_n formula: then (1) holds in $\mathbf{V}_\vartheta$, hence, there is $Y \in \mathbf{V}_\vartheta$ with $\forall\, x \in X\, \exists\, y \in Y\, \Phi(x,y)$, as required. The claim clearly follows from the two observations:

(A) Any formula of the form $\exists\, x\, \Phi$ or $\exists\, x \in z\, \Phi$, where Φ is Σ_n and $n \geq 1$, is equivalent, in **ZC**, to a Σ_n formula;

(B) Any formula of the form $\forall\, x \in z\, \Phi$, where Φ is Σ_n and $n \geq 1$, is equivalent, in Σ_n-**ZFC** (perhaps not in **ZC**!), to a Σ_n formula.

To prove (A), let Φ be $\exists\, y\, \mathbf{Q}\, \mathbf{v}\, \varphi(x,y,\mathbf{v},\mathbf{u})$, where $\mathbf{Q}$ is a Π_{n-1} prefix, φ is a bounded formula, $\mathbf{v}$ is the list of $n-1$ free variables bounded by $\mathbf{Q}$, and $\mathbf{u}$ the list of all other free variables of φ. Then $\exists\, x\, \Phi(x,y)$ is equivalent to

$$\exists\, p\, \mathbf{Q}\, \mathbf{v}\, \left[\exists\, x \in\in p\, \exists\, y \in\in p\, \left(p = \langle x,y\rangle \wedge \varphi(x,y,\mathbf{v},\mathbf{u})\right)\right],$$

where $x \in\in p$ means $\exists a \in p\,(x \in a)$, $p = \langle x, y\rangle$ is the bounded formula $\exists a, b \in p\,(p = \{a, b\} \wedge a = \{x\} \wedge b = \{x, y\})$, and $p = \{a, b\}$ is the bounded formula $a \in p \wedge b \in p \wedge \forall c \in p\,(c = a \vee c = b)$, therefore, $[...]$ in the displayed line above is a bounded formula.

To prove (B) let Φ, φ be as above. Then $\forall x \in z\, \Phi(x, y)$ is equivalent to

$$\exists f\, \mathbf{Q}\, \mathbf{v}\, \big[\mathtt{Fun}(f, z) \wedge \forall x \in z\, \exists a, b \in\in f(x)\, \big(f(x) = \langle a, b\rangle \wedge \varphi(x, y, \mathbf{v}, \mathbf{u})\big)\big], \quad (2)$$

where $\mathtt{Fun}(f, z)$ means that f is a function defined on z. We leave it as an **exercise** for the reader to check that $[...]$ in (2) (including $\mathtt{Fun}(f, z)$) can be written as a bounded formula, and that the equivalence of (2) and $\forall x \in z\, \Phi(x, y)$ is provable in Σ_n-**ZFC** (why do we need Σ_n-Collection ?). □

1.5b Von Neumann hierarchy over internal sets in HST

Exercise 1.5.5 (HST). Prove that all levels $\mathbf{V}_\xi$, $\xi \in \mathtt{Ord}$ of the von Neumann hierarchy are still sets in **HST**, and $\mathbb{WF} = \bigcup_{\xi \in \mathtt{Ord}} \mathbf{V}_\xi$. □

Thus, in **HST**, only well-founded sets are in the scope of the von Neumann hierarchy as defined in § 1.5a: However the definition can be modified to accomodate all sets. The idea is straightforward. The standard von Neumann construction, based on the Regularity axiom of **ZFC**, begins with $\varnothing$, the empty set. Therefore, if all, not only well-founded, sets are to be incorporated, we have to begin with internal sets and apply Regularity over $\mathbb{I}$.

For any set or class $U \subseteq \mathbb{I}$ define $\mathbf{V}_\xi[U]$ (which can be a proper class even if U is a set) by induction on $\xi \in \mathtt{Ord}$ as follows:

$$\mathbf{V}_0[U] = U\,, \quad \mathbf{V}_{\xi+1}[U] = U \cup \mathscr{P}(\mathbf{V}_\xi[U]) \quad \text{for each ordinal } \xi\,,$$
$$\mathbf{V}_\lambda[U] = \textstyle\bigcup_{\xi<\lambda} \mathbf{V}_\xi[U] \quad \text{for limit ordinals } \lambda\,.$$

This is not immediately a definition by transfinite induction, because, due to the absence of the axiom of Power Set in **HST**, the classes $\mathbf{V}_\xi[U]$ are not necessarily sets. Yet the decision whether a set x belongs to $\mathbf{V}_\xi[U]$ can be made on the base of those sets which belong to the *transitive closure*

$$\mathtt{TC}(x) = x \cup (\textstyle\bigcup x) \cup (\bigcup\bigcup x) \cup \ldots = \bigcup_{n \in \mathbb{N}} (\bigcup^n x)\,.$$

of x. Thus $\mathtt{TC}(x)$ is a set by the axioms of Replacement and Union, and obviously the least transitive set containing x as a subset.

Exercise 1.5.6. Suppose that $U \subseteq \mathbb{I}$, x is any set, X a transitive set, and $U \subseteq X$, $x \in X$. Put $v_0 = U$ and then, by induction, $v_{\xi+1} = U \cup (X \cap \mathscr{P}(v_\xi))$ and $v_\lambda = \bigcup_{\xi<\lambda} v_\xi$ for limit ordinals λ. Prove that all v_ξ are sets and $x \in v_\xi$ iff $x \in \mathbf{V}_\xi[U]$. This validates the inductive definition of $\mathbf{V}_\xi[U]$. □

Put $\mathbb{WF}[U] = \bigcup_{\xi \in \mathtt{Ord}} \mathbf{V}_\xi[U]$ (the class of all sets *well-founded over* U) and, for $x \in \mathbb{WF}[U]$, define $\mathtt{irk}_U\, x$, *the rank over* U, to be the least ordinal ξ with $x \in \mathbf{V}_\xi[U]$. In particular, define $\mathtt{irk}\, x = \mathtt{irk}_{\mathbb{I}}\, x$, the *rank over* $\mathbb{I}$.

Exercise 1.5.7 (HST). Prove the following:

(1) $\mathbb{H} = \mathbb{WF}[\mathbb{I}] = \bigcup_{U \in \mathbb{I}} \mathbb{WF}[U] = \bigcup_{\xi \in \mathtt{Ord}} \mathbf{V}_\xi[(\mathbf{V}_{{}^*\xi})^{\mathbb{I}}]$, where $(\mathbf{V}_{{}^*\xi})^{\mathbb{I}}$ is the ${}^*\xi$-th von Neumann level defined in the internal universe $\mathbb{I}$.
(*Hint*. By definition, $\mathbb{I} \subseteq \mathbb{WF}[\mathbb{I}]$. To show that each non-internal z belongs to $\mathbb{WF}[\mathbb{I}]$ argue by $\in$-induction: $\in \restriction (\mathbb{H} \smallsetminus \mathbb{I})$ is a well-founded relation.)

(2) $\mathtt{irk}_U\, x = 0$ for $x \in U$, while $\mathtt{irk}_U\, x = \mathtt{sup}_{y \in x} \mathtt{irk}_U\, y$ for $x \in \mathbb{WF}[U] \smallsetminus U$, in particular, $\mathtt{irk}\, x = 0$ for all internal x, while $\mathtt{irk}\, x = \mathtt{sup}_{y \in x} \mathtt{irk}\, y$ for $x \notin \mathbb{I}$.

(3) If U is internal and $x \in \mathbb{WF}[U]$ then $\mathtt{irk}\, x \le \mathtt{irk}_U\, x$ and (using Lemma 1.5.8 below) $\mathtt{irk}_U\, x < \mathtt{irk}\, x + \omega$. □

Lemma 1.5.8. *Let x, U be internal sets and α, ξ be ordinals. If $x \in \mathbf{V}_\xi[U]$ then $x \in \mathscr{P}_{\mathtt{int}}{}^n(U)$* [19] *for some $n \in \mathbb{N}$. If $\alpha \in \mathbf{V}_\xi[U]$ then $\alpha < \omega + \xi$.*

Proof. Argue by induction on ξ. If $\xi = 0$ then $\mathbf{V}_0[U] = U$, hence, if $\alpha \in U$ then $\alpha \in \mathbb{N}$ by Exercise 1.2.17(3). The limit step is trivial. If $x \in \mathbf{V}_{\xi+1}[U]$ then any $y \in x$ belongs to $\mathbf{V}_\xi[U]$, hence, as y is still internal by Transitivity of $\mathbb{I}$, $y \in \mathscr{P}_{\mathtt{int}}{}^n(U)$ for some $n = n_y \in \mathbb{N}$. Arguing in $\mathbb{I}$, let N be the set of all $n \in {}^*\mathbb{N}$ such that there is $y \in x$ with $y \in \mathscr{P}_{\mathtt{int}}{}^n(U) \smallsetminus \mathscr{P}_{\mathtt{int}}{}^{n-1}(U)$. This is an internal set, and $N \subseteq \mathbb{N}$ by the above, thus, $n = \mathtt{sup}\, N \in \mathbb{N}$, and then we easily obtain $x \in \mathscr{P}_{\mathtt{int}}{}^{n+1}(U)$. If $\alpha \in \mathbf{V}_{\xi+1}[U]$ then any $\beta < \alpha$ belongs to $\mathbf{V}_\xi[U]$, and hence $\beta < \omega + \xi$ and $\alpha < \omega + \xi + 1$. □

1.5c Classes and structures

It is a common approach in set theory to understand **classes** as collections of sets defined by formulas. For instance, in **HST**, a class is any collection of sets $X = \{x : \varphi(x)\}$, where φ is a formula which may contain arbitrary sets as parameters. For instance, $X_y = \{x : y \in x\}$ is a class for any set y. A *proper* class is a class which is not a set. Classes like Ord, Card, and, for instance, X_y just defined, are proper in **HST** (and in **ZFC**). Normally a proper class is just a collection too big to be a set (however see § 3.2f below).

Classes are routinely used in a way which simulates class theories, that is, not necessarily with formulas explicitly tagged to classes. This is a legitimate use assuming that 1) all classes appear only through expressions of the form $x \in X$ where x means a set, 2) there are no quantifiers over classes.

Structures. Let $\mathcal{L}$ be a language [20] containing $\nu \in \mathbb{N}$ atomic predicates, $P_j(x_1, ..., x_{a(j)})$, $j = 1, ..., \nu$, where each P_j has arity $a(j) \in \mathbb{N}$. An *invariant $\mathcal{L}$-structure* [21] is any structure of the form $\mathfrak{q} = \langle D; R_1, ..., R_\nu; \equiv \rangle$, where D, the *domain* or *universe* of $\mathfrak{q}$ can be a set or proper class, $R_j \subseteq D^{a(j)}$ are *basic relations*, $\equiv$ is an equivalence relation on D, and the relations R_j are *$\equiv$-invariant*, that is for any $x_1, y_1, ..., x_{a(j)}, y_{a(j)} \in D$:

[19] $\mathscr{P}_{\mathtt{int}}{}^n$ is n-th iteration of the power set operation $\mathscr{P}_{\mathtt{int}}(X) = \mathscr{P}(X) \cap \mathbb{I}$ in $\mathbb{I}$.
[20] We consider only 1st order languages with finitely many atomic predicates.
[21] Close to what Shoenfield defines as interpretation in [Shoen 67, 9.5].

$$x_1 \equiv y_1 \wedge ... \wedge x_{a(j)} \equiv y_{a(j)} \text{ implies } R_j(x_1, ..., x_{a(j)}) \Longleftrightarrow R_j(y_1, ..., y_{a(j)}).$$

In this case, for any $\mathcal{L}$-formula Φ we define its **q-relativization** Φ^{q} as follows:

1) all quantifiers are relativized to D, in other words, $\exists y$, $\forall y$ are changed to resp. $\exists y \in D$, $\forall y \in D$;
2) any occurrence of $P_j(x_1, ..., x_{a(j)})$ in Φ is changed to $R_j(x_1, ..., x_{a(j)})$ (which is a shorthand for $\langle x_1, ..., x_{a(j)} \rangle \in R_j$).
3) any occurrence of $x = y$ in Φ is changed to $x \equiv y$.

If we replace all occurrences of the classes D and R_j in Φ^{q} by formulas which define these classes, the result will be a formula of the language of the underlying set theory (for instance, a st-$\in$-formula if we argue in **HST**).

If $\equiv$ is the equality on D then an $\mathcal{L}$-structure $\mathsf{q} = \langle D; R_1, ..., R_\nu; = \rangle$ is called *an $\mathcal{L}$-structure with true equality*, and $=$ is dropped, so that a typical description of such a structure looks like $\mathbf{s} = \langle D; R_1, ..., R_\nu \rangle$. The relativization $\Phi^{\mathbf{s}}$ is defined in this case by means of 1), 2) only.

Structures of the $\in$-language and the st-$\in$-language are called *$\in$-structures* and *st-$\in$-structures* respectively. For instance $\mathbf{s} = \mathbb{S} = \langle \mathbb{S}; \in \restriction \mathbb{S} \rangle$, routinely truncated to $\langle \mathbb{S}; \in \rangle$, is an $\in$-structure (with true equality) and $\Phi^{\mathbf{s}}$ is the same as $\Phi^{\mathtt{st}}$. Similarly, $\langle \mathbb{I}; \in, \mathtt{st} \rangle$ is a st-$\in$-structure (in **HST**).

Any relativized formula Φ^{q} is informally considered as saying "Φ holds in q", or "Φ is true in q". Example: $y = \mathscr{P}(x)$ says that y consists of all subsets of x (in the universe of all sets), while, for any transitive class K (considered as the $\in$-structure $\langle K; \in \rangle$ with true equality), the relativized formula $(y = \mathscr{P}(x))^K$ (where $x, y \in K$) says that $y = \mathscr{P}(x)$ holds in K, which is in fact equivalent to $y = \mathscr{P}(x) \cap K$. Yet this is not exactly the same as the model theoretic notion of truth, see below.

Reduction to true equality. Most structures used below as interpretations of nonstandard set theories will arise as invariant structures with the equivalence relation very far from being the equality. This leads to the problem whether a given invariant structure can be replaced by one with true equality.

Definition 1.5.9. A structure $\mathsf{q}' = \langle D'; R'_1, ..., R'_\nu \rangle$ with true equality is said to *reduce* an invariant structure $\mathsf{q} = \langle D; R_1, ..., R_\nu; \equiv \rangle$ if there exists a map $r : D \xrightarrow{\text{onto}} D'$ (called a *reduction* of q to q') such that
$x \equiv y \Longleftrightarrow r(x) = r(y)$ for all $x, y \in D$, and
$R_j(x_1, ..., x_{a(j)}) \Longleftrightarrow R'_j(r(x_1), ..., r(x_{a(j)}))$ for all j and $x_1, ..., x_{a(j)} \in D$. □

Proposition 1.5.10. *If $r : D \xrightarrow{\text{onto}} D'$ reduces q to q' then for any $\mathcal{L}$-formula $\Phi(x_1, ..., x_n)$ and any $x_1, ..., x_n$ in the domain D of q we have:*

$$\Phi(x_1, ..., x_n)^{\mathsf{q}} \quad \Longleftrightarrow \quad \Phi(r(x_1), ..., r(x_n))^{\mathsf{q}'} .$$

Proof. A routine proof by induction on the complexity of formulas is left as an **exercise** for the reader. □

An obvious way to get such a reduction is to define a quotient structure. If $\mathfrak{q} = \langle D; R_1, ..., R_\nu; \equiv\rangle$ is an invariant structure then the *quotient structure* $\mathfrak{q}/{\equiv} = \langle D/{\equiv}; R_1, ..., R_\nu\rangle$ has the universe $D/{\equiv}$ consisting of $\equiv$*-classes* $[x]_\equiv = \{y \in D : x \equiv y\}$ of elements x of the *underlying domain* D while the relations R_j on $D/{\equiv}$ are naturally defined so that $R_j([x_1]_\equiv, ..., [x_{a(j)}]_\equiv)$ iff just $R_j(x_1, ..., x_{a(j)})$. The $\equiv$-invariance of the relations R_j validates the consistency of this definition. The original invariant structure $\mathfrak{q} = \langle D; R_1, ..., R_\nu, \equiv\rangle$ will be called the *underlying* structure.

It is clear that the quotient structure $\mathfrak{q}/{\equiv}$ is a structure with true equality, and the map $x \mapsto [x]_\equiv$ reduces $\mathfrak{q}$ to $\mathfrak{q}/{\equiv}$. Yet this may not be a plausible solution of the reduction problem: if the domain D of $\mathfrak{q}$ is a proper class then the domain of $\mathfrak{q}/{\equiv}$ appears to be a collection of proper classes, which leaves, strictly speaking, such a structure out of any direct consideration in a set theory like **ZFC**. Yet the problem is resolved positively in **ZFC**:

Theorem 1.5.11 (ZFC). *Any invariant structure* $\mathfrak{q} = \langle D; R_1, ..., R_\nu; \equiv\rangle$ *can be reduced to a structure* $\mathfrak{q}' = \langle D'; R'_1, ..., R'_\nu\rangle$ *with true equality (and with the universe that consists of sets).*

Proof. If there is a definable well-ordering $<$ of the universe then, choosing the $<$-least element c_x in each equivalence class $[x]_\equiv = \{y : y \equiv x\}$ we obtain a reduced domain $D' = \{c_x : x \in D\}$, on which $\equiv$ is the equality.

In the general case, put $C_x = [x]_\equiv \cap V_{\alpha(x)} \subseteq [x]_\equiv$ for any $x \in D$, where $\alpha(x)$ is the least ordinal such that $[x]_\equiv \cap \mathbf{V}_{\alpha(x)} \neq \varnothing$, and $\mathbf{V}_\alpha$ are sets of the von Neumann hierarchy. Unlike $[x]_\equiv$, C_x is obviously a set. Consider the class $D' = \{C_x : x \in D\}$ as the new domain and define $R'_j(C_{x_1}, ..., C_{x_{a(j)}})$ iff just $R_j(x_1, ..., x_{a(j)})$, for any $j = 1, ..., \nu$. The map $x \mapsto C_x$ reduces $\mathfrak{q}$ to $\mathfrak{q}' = \langle D'; R'_1, ..., R'_\nu\rangle$. □

Exercise 1.5.12. Prove, using (1) of Exercise 1.5.7, that the theorem remains true for **HST**. □

1.5d Interpretations

Suppose that T_1 and T_2 are set theories, like **ZFC** or **HST**, whose languages (see footnote 20 on page 45) are $\mathcal{L}_1$ and $\mathcal{L}_2$. An interpretation of T_2 in T_1 is, roughly, an $\mathcal{L}_2$-structure, defined in $\mathcal{L}_1$, and satisfying, provably in T_1, all axioms of T_2. This (together with the notion of a *model* considered below) is one of the most useful metamathematical notions.

Suppose that $\mathcal{L}_2$ contains ν atomic predicates, $P_j(x_1, ..., x_{a(j)})$, $j = 1, ..., \nu$, where each P_j has arity $a(j)$. An *interpretation* of $\mathcal{L}_2$ in $\mathcal{L}_1$ is an $\mathcal{L}_2$-structure $\mathfrak{q} = \langle D; R_1, ..., R_\nu; \equiv\rangle$, defined by means of $\mathcal{L}_1$, such that $\Phi^\mathfrak{q}$ is a theorem of T_1 for any axiom Φ of T_2. To be more exact, it is assumed here that D, $\equiv$, and each of $R_j \subseteq D^{a(j)}$ are classes definable by certain parameter-free formulas of $\mathcal{L}_1$. Then, for any $\mathcal{L}_2$-formula Φ, the relativization Φ^{q} is a formula of $\mathcal{L}_1$, moreover, the transformation $\Phi \mapsto \Phi^{q}$ is recursive. In these rigorous terms, the notion of interpretation has a clear metamathematical meaning and understanding.

According to § 1.5c, there exist invariant interpretations (the general case), interpretations with true equality, and quotient interpretations.

Example 1.5.13. $\mathsf{w} = \langle \mathbb{WF}\,; \in \restriction \mathbb{WF}\rangle$ is an $\in$-structure with true equality defined by means of the $\mathtt{st}$-$\in$-language. It follows from Theorem 1.1.9 that $\langle \mathbb{WF}\,; \in \restriction \mathbb{WF}\rangle$ is an interpretation of **ZFC** in **HST**. Obviously Φ^{w} is the same as $\Phi^{\mathtt{wf}}$ for any $\in$-formula Φ. □

1.5e Models

Generally, a *model* is a *set-size* structure $\mathfrak{M} = \langle M\,; R_1, ..., R_\nu\rangle$, whose universe M and relations R_j are sets rather than proper classes (then $\mathfrak{M}$ is a set itself). Model theory provides us with a certain $\in$-formula, say, $\mathtt{Form}(\mathcal{L}, \mathfrak{M}, \Phi)$ which says that $\mathcal{L}$ is a language, $\mathfrak{M} = \langle M\,; R_1, ..., R_n\rangle$ is a set-size $\mathcal{L}$-structure, and Φ is an $\mathcal{L}$-formula [22] containing only sets in M as parameters, and with another $\in$-formula, say, $\mathrm{TRUE}(\mathfrak{M}, \Phi)$, mostly written as $\mathfrak{M} \models \Phi$, which says that $\mathfrak{M}$ is a structure of the language of Φ and Φ is true in $\mathfrak{M}$, in the sense that there is a function τ, defined on the set of all subformulas of Φ (including Φ) in which all free variables are replaced by sets in M as parameters, and taking values in $\{\mathtt{true}, \mathtt{false}\}$ [23], with $\tau(\Phi) = \mathtt{true}$, which is a *validation function*, so that

1) τ respects every relation R_j, in the sense that $\tau(P_j(x_1, ..., x_{a(j)})) = \mathtt{true}$ iff $\langle x_1, ..., x_{a(j)}\rangle \in R_j$,
2) τ obeys rules of first-order logic [24], for instance $\tau(\varphi \wedge \psi) = \mathtt{true}$ iff both $\tau(\varphi) = \mathtt{true}$ and $\tau(\psi) = \mathtt{true}$, $\tau(\neg\,\varphi) = \mathtt{true}$ iff $\tau(\varphi) = \mathtt{false}$, and $\tau(\exists\, x\, \varphi(x)) = \mathtt{true}$ iff $\tau(\varphi(x)) = \mathtt{true}$ for some $x \in M$.

Remark 1.5.14. The truth of a formula Φ in a structure $\mathfrak{M}$ can be expressed in two ways, $\Phi^{\mathfrak{M}}$ (interpretation) and $\mathfrak{M} \models \Phi$, applicable in different cases: $\Phi^{\mathfrak{M}}$ needs that Φ is a metamathematically given formula (perhaps, with sets as parameters) and $\mathfrak{M}$ a fixed structure which can be a proper class, while $\mathfrak{M} \models \Phi$ assumes that $\mathfrak{M}$ and Φ are just set variables (Φ denotes a finite sequence of symbols and sets as parameters).

Fortunately $\Phi^{\mathfrak{M}}$ and $\mathfrak{M} \models \Phi$ are equivalent on the domain of common applicability. Namely, if $\Phi(v_1, ..., v_n)$ is a formula of a certain fixed (finite, as above) language $\mathcal{L}$, then it is a theorem of **HST** (and of **ZFC**, of course) that for any set-size $\mathcal{L}$-structure $\mathfrak{M} = \langle M\,; R_1, ..., R_\nu\rangle$ we have, for all $x_1, ..., x_n \in M$, $\Phi(x_1, ..., x_n)^{\mathfrak{M}} \iff (\mathfrak{M} \models \Phi(x_1, ..., x_n))$. □

[22] Here, a finite sequence of symbols which codes, in certain way, a formula of $\mathcal{L}$. We assume some acquaintance with model theoretic notions.

[23] $\mathtt{true}$ and $\mathtt{false}$ are usually identified with numbers, resp., 1 and 0.

[24] Typically called *Tarski rules* in this context.

Exercise 1.5.15. It is a good exercise, useful for a general understanding of related issues, to prove the result just mentioned. See [Kun 80, Chapter IV], especially Lemma 10.1 of Appendix 3, for additional explanations. □

Accordingly, *a model of a theory* T in a language $\mathcal{L}$ is any set-size $\mathcal{L}$-structure $\mathfrak{M}$ such that we have $\mathfrak{M} \models \Phi$ for every axiom Φ of T.

In this definition, the property "to be a model of T" is expressed by a single formula $\forall \Phi \in T\,(\mathfrak{M} \models \Phi)$. This is, in general, impossible for interpretations whose domains are proper classes, because then a function τ as above is also a proper class, the existence of which cannot be expressed by a set theoretic formula. Thus, an interpretation of T_2 in T_1, in the sense of § 1.5d, is, generally speaking, a metamathematical notion as it needs infinitely many formulas (one for each axiom of T_2) to be adequately presented.

1.5f Simulation of models of ZFC

It is known from the Gödel incompleteness theorem that the existence of a model of **ZFC** is not provable in **ZFC**. This obstacle can be circumvented to great practical effect by the following trick which yields *interpretations* rather than *models* of **ZFC** and finds typical applications in conservativity proofs, where it allows to avoid inconveniences sometimes connected with finite subtheories. (See examples in Sections 4.4 and 4.5.)

Let **ZFC**$\boldsymbol{\vartheta}$ be a theory in the $\in$-language enriched by a constant symbol $\boldsymbol{\vartheta}$, containing all of **ZFC** ($\boldsymbol{\vartheta}$ can occur in the schemata), the axiom "$\boldsymbol{\vartheta}$ is an ordinal", and the following schema (see § 1.5a on $\mathbf{V}_\xi$)

(∗) all sentences of the form

$$\forall x_1, \dots, x_n \in \mathbf{V}_{\boldsymbol{\vartheta}} \left(\Phi(x_1, \dots, x_n) \Longleftrightarrow \Phi(x_1, \dots, x_n)^{\mathbf{V}_{\boldsymbol{\vartheta}}}\right),$$

where $\Phi(x_1, \dots, x_n)$ is an $\in$-formula (perhaps with free variables, but $\boldsymbol{\vartheta}$ not allowed to occur) while ${}^{\mathbf{V}_{\boldsymbol{\vartheta}}}$ means the relativization to $\mathbf{V}_{\boldsymbol{\vartheta}}$.

Definition 1.5.16. A set theory $\mathfrak{T}$ in a language which includes the $\in$-language is said to be a *conservative* extension of **ZFC** if any $\in$-formula Φ is a theorem of $\mathfrak{T}$ if and only if Φ is a theorem of **ZFC**. □

Exercise 1.5.17. (i) Prove that **ZFC**$\boldsymbol{\vartheta}$ is a conservative extension of **ZFC**, in the sense of Definition 1.5.16, that is, if φ is an $\in$-formula ($\boldsymbol{\vartheta}$ not allowed) then **ZFC** proves φ iff **ZFC**$\boldsymbol{\vartheta}$ proves φ iff **ZFC**$\boldsymbol{\vartheta}$ proves $\varphi^{\mathbf{V}_{\boldsymbol{\vartheta}}}$.

Hint. If **ZFC**$\boldsymbol{\vartheta}$ proves φ then φ is provable in a subtheory T of the form **ZFC**+ "$\boldsymbol{\vartheta}$ is an ordinal" + a finite list L of equivalences of the form (∗). Theorem 1.5.4 yields an ordinal ϑ such that $\langle \mathbf{V}\,; \vartheta, \in\rangle$ is an elementary submodel of the universe w. r. t. φ and all $\in$-formulas Φ which occur in L, hence, an interpretation of T in **ZFC**. Thus, φ "is true", and the whole argument is a proof of φ in **ZFC**.

(ii) Infer that **ZFC**ϑ and **ZFC** are equiconsistent.

(iii) By definition $\langle \mathbf{V}_\vartheta\,; \in\rangle$ is an interpretation of **ZFC** in **ZFC**ϑ. Assuming `Consis` **ZFC** show that **ZFC**ϑ does not prove that $\langle \mathbf{V}_\vartheta\,; \in\rangle$, or just $\mathbf{V}_\vartheta$, for brevity, is a model of **ZFC**.

Hint. Otherwise **ZFC**ϑ would prove `Consis` **ZFC**, hence **ZFC** would prove the same by the above, which is a contradiction to the Gödel incompleteness theorem. □

By definition **ZFC**ϑ proves that $\mathbf{V}_\vartheta$ is a model of **ZC** and, in addition, for any particular case of Collection **ZFC**ϑ proves that this case holds in $\mathbf{V}_\vartheta$ in spite of 1.5.17(iii). This phenomenon is often used to "simulate" transitive models of **ZFC** in **ZFC**ϑ.

Exercise 1.5.18 (ZFCϑ). Prove that ϑ is a cardinal and $\mathtt{card}\,\mathbf{V}_\vartheta = \vartheta$. Prove that ϑ is a strong limit cardinal, that is, $\kappa < \vartheta \Longrightarrow 2^\kappa < \vartheta$.

Hint. To see that ϑ is a cardinal note that in the universe for any ordinal there is a bigger cardinal, hence this is true in $\mathbf{V}_\vartheta$. Similarly we have $\mathtt{card}\,\mathbf{V}_\xi < \vartheta$ for any $\xi < \vartheta$, with $\mathtt{card}\,\mathbf{V}_\vartheta = \vartheta$ being an easy consequence. □

1.5g Asterisk is an elementary embedding

A key fact in the model-theoretic version of nonstandard mathematics is that the "asterisk" map elementarily embeds any well-founded structure $\mathfrak{M}$ in ${}^*\mathfrak{M}$. Therefore the status of this phenomenon in **HST** merits a brief review.

Arguing in **HST**, consider a finite [25] language $\mathcal{L} \in \mathbb{WF}$, containing ν atomic predicates, $P_j(x_1, ..., x_{a(j)})$, $j = 1, ..., \nu$, of arities $a(j) \in \mathbb{N}$, and an $\mathcal{L}$-structure $\mathfrak{M} = \langle M\,; R_1, ..., R_\nu\rangle \in \mathbb{WF}$, where, accordingly, $R_j \subseteq M^{a(j)}$ are sets in $\mathbb{WF}$. Then, ${}^*\mathfrak{M} = \langle {}^*M\,; {}^*R_1, ..., {}^*R_\nu\rangle \in \mathbb{I}$ is an $\mathcal{L}$-structure, too.

According to Remark 1.5.14, the "key fact" above can be interpreted, in **HST**, in two ways: in terms of relativization and in terms of $\models$. The first approach leads us to the following formulation of the result:

Proposition 1.5.19 (HST). *If $\mathcal{L}$ is a finite language and $\Phi(v_1, ..., v_n)$ any $\mathcal{L}$-formula then* (it is a theorem of **HST** that) *for any $\mathcal{L}$-structure $\mathfrak{M} \in \mathbb{WF}$ and any $x_1, ..., x_n \in M$, we have $\Phi(x_1, ..., x_n)^{\mathfrak{M}} \Longleftrightarrow \Phi({}^*x_1, ..., {}^*x_n)^{{}^*\mathfrak{M}}$.*

Proof. Suppose, for the sake of simplicity, that $\mathcal{L}$ contains only one atomic predicate, $P(\cdot,\cdot,\cdot)$, and $\Phi(x)$ is $\exists\, y\, \forall\, z\, P(x, y, z)$. Then any $\mathcal{L}$-structure $\mathfrak{M}$ has the form $\mathfrak{M} = \langle M\,; R\rangle$, where $R \subseteq M^3$, and $\Phi(x)^{\mathfrak{M}}$ is the $\in$-formula

[25] If $\mathcal{L}$ is an infinite language, containing, say, predicates $\{P_j\}_{j\in\mathbb{N}}$, then any well-founded $\mathcal{L}$-structure has the form $\mathfrak{M} = \langle M\,; \{R_j\}_{j\in\mathbb{N}}\rangle \in \mathbb{WF}$, accordingly, ${}^*\mathfrak{M} = \langle {}^*M\,; \{{}^*R_j\}_{j\in{}^*\mathbb{N}}\rangle \in \mathbb{I}$, which is rather an ${}^*\mathcal{L}$-structure. To view it as an $\mathcal{L}$-structure, drop all relations ${}^*R_j$ with nonstandard indices $j \in {}^*\mathbb{N} \smallsetminus \mathbb{N}$. With this understanding, Theorem 1.5.20 remains true for infinite languages.

$\forall y \in M \, \exists z \in M \; R(x,y,z)$ with x, M, R as free variables, which we can denote by $\varphi(x, M, R)$. We observe that, by $*$-Transfer,

$$\varphi(x, M, R)^{\mathtt{wf}} \iff \varphi({}^*x, {}^*M, {}^*R)^{\mathtt{int}}.$$

It follows from the results of 1.2.4 and 1.2.5 that the relativization superscipts can be removed, so that we have $\varphi(x, M, R) \iff \varphi({}^*x, {}^*M, {}^*R)$. But this is just the equivalence to be proved. □

This result is usually sufficient for all practical reasons, yet we may note that both $\mathcal{L}$ and Φ are "metamathematically given" objects in the proposition, *i. e.*, any concrete language and any concrete formula which are not under **HST** quantifiers. The following theorem gives a stronger result, in terms of $\models$, in which $\mathcal{L}$ and Φ are **HST** variables.

Theorem 1.5.20 (HST). *If $\mathcal{L} \in \mathbb{WF}$ is a finite language and $\mathfrak{M} \in \mathbb{WF}$ is an $\mathcal{L}$-structure then the map $x \longmapsto {}^*x$ restricted to M is an elementary embedding of $\mathfrak{M}$ in ${}^*\mathfrak{M}$ in the sense of $\mathcal{L}$. Furthermore, ${}^*\mathfrak{M}$ is κ-saturated, for any cardinal $\kappa \in \mathtt{Card}$, that is, any f. i. p. collection $\{X_\xi\}_{\xi<\kappa}$ of sets $\varnothing \neq X_\xi \subseteq M$, $\mathcal{L}$-definable in ${}^*\mathfrak{M}$, has nonempty intersection.*

Proof. Besides the saturation property, we have to prove that:

$$\forall \text{ closed } \mathcal{L}\text{-formula } \Phi\big[\, \mathtt{Form}(\mathcal{L}, \mathfrak{M}, \Phi) \implies (\mathfrak{M} \models \Phi \text{ iff } {}^*\mathfrak{M} \models {}^*\Phi)\,\big],$$

where ${}^*\Phi$ is the result of the replacement of any parameter $x \in M$ in Φ by ${}^*x$, while $\mathtt{Form}$ and $\models$ are defined as in § 1.5e.

Thus suppose $\mathtt{Form}(\mathcal{L}, \mathfrak{M}, \Phi)$ and prove (in **HST**) that $\mathfrak{M} \models \Phi$ iff ${}^*\mathfrak{M} \models {}^*\Phi$. Suppose that $\mathfrak{M} \models \Phi$. Since TRUE is by the construction absolute for $\mathbb{WF}$ (we can refer to Corollary 1.2.4) $\mathfrak{M} \models \Phi$ is true in $\mathbb{WF}$, subsequently, ${}^*\mathfrak{M} \models {}^*\Phi$ holds in $\mathbb{I}$ by $*$-Transfer, that is, by definition, there is $\tau \in \mathbb{I}$ with $\tau({}^*\Phi) = \mathtt{true}$ which is a validation function in the sense of $\mathbb{I}$. It remains to note that τ still is a validation function in the whole universe $\mathbb{H}$. (The key observation is that any assignment of parameters from ${}^*M$ to free variables of subformulas of Φ which we can maintain in $\mathbb{H}$ is already in $\mathbb{I}$ because this class contains all its finite subsets by Lemma 1.2.16.) Conversely, suppose that ${}^*\mathfrak{M} \models {}^*\Phi$ is witnessed by a validation function τ with $\tau({}^*\Phi) = \mathtt{true}$. Then ${}^*\mathfrak{M} \models {}^*\Phi$ is also true in $\mathbb{I}$. (If not then we have ${}^*\mathfrak{M} \models \neg\, {}^*\Phi$ in $\mathbb{I}$, hence, ${}^*\mathfrak{M} \models \neg\, {}^*\Phi$ in the universe $\mathbb{H}$ of all sets, as above, which is a contradiction.) Then $\mathfrak{M} \models \Phi$ holds in $\mathbb{WF}$ by $*$-Transfer, hence, $\mathfrak{M} \models \Phi$ by the absoluteness.

To derive the "furthermore" part of the theorem from Theorem 1.3.5, it suffices to show that any set $X \subseteq {}^*M$, definable in ${}^*\mathfrak{M}$ by a formula of $\mathcal{L}$ with parameters in ${}^*M$, belongs to $\mathbb{I}$. Suppose that $X = \{x \in {}^*\mathfrak{M} : {}^*\mathfrak{M} \models \varphi(x)\}$, where $\varphi(x)$ is such a formula. Then $\varphi(x)$ belongs to $\mathbb{I}$ since $\mathbb{I}$ contains all finite subsets of $\mathbb{I}$, moreover, as above, we can rewrite the definition of X as

$$X = \{x \in {}^*\mathfrak{M} : \big({}^*\mathfrak{M} \models \varphi(x)\big)^{\mathbb{I}}\,\}.$$

It remains to use Corollary 1.1.10. □

Historical and other notes to Chapter 1

Section 1.1. **HST** is Hrbaček's theory $\mathbf{NS}_1(\mathbf{ZFC})$ of [Hr 78] plus the axioms of Regularity over $\mathbb{I}$, Standard Size Choice, and Dependent Choice.

Definition of the $*$-extension map in the context of axiomatical nonstandard set theories (Definition 1.1.6) was given in [Hr 79]. Hrbaček called sets x for which ${}^*x$ can be defined (*i.e.* well-founded sets here) "external standard".

Note that Nelson [Nel 77] defined $*$ differently: if $\langle V\,;{}^*\!\!\in, \mathtt{st}\rangle$ is a model of **IST** then, for any set $x \in V$, ${}^*x = \{y \in V : y\ {}^*\!\!\in x\}$, the true set of all ${}^*\!\!\in$-elements of x.

Di Nasso [DiN 99] suggested an equivalent method of axiomatization of **HST**-like theories: the language contains $\in$ and $*$, with appropriate axioms for $*$ (in particular, that it is a function defined on all well-founded sets), in this modification, $\mathbb{S}$ and $\mathtt{st}$ are definable notions.

Section 1.2. The origins of the scheme "$\mathbb{WF} \xrightarrow{*} \mathbb{I}$ [in $\mathbb{H}$]" go back to the model theoretic version of nonstandard analysis. The scheme was introduced in the context of nonstandard set theories in [Hr 79, Kaw 83]. Systematic treatment was presented in our [KanR 97], see also Di Nasso [DiN 97, DiN 99]. "$\mathbb{S} \subseteq \mathbb{I} \subseteq \mathbb{H}$", the other scheme, is due to [Hr 78, Hr 79] and, in its truncated form "$\mathbb{S} \subseteq \mathbb{I}$", this is the famous internal set theory of Nelson [Nel 77]. The content of this section is to prove, in **HST**, a variety of little things mostly quite trivial in the context of model theoretic nonstandard analysis.

Section 1.3. Saturation was first considered, in the context of nonstandard set theories, in [Hr 78] and, in the form of Idealization, in [Nel 77]. Hrbaček paradox was observed in [Hr 78]. The principle of Extension, in the context of nonstandard set theories, appeared in [Hr 79]. Similar principles have been known in model theoretic nonstandard analysis under different names, *e.g.* *Comprehension* in [Hen 97] or *Comprehensiveness* in [Gol 98, 15.4].

Section 1.4. "External", mainly countable operations over internal sets naturally appeared in the theory of Loeb measures in 70s. External sets in $\mathbf{\Sigma}_1^{\mathtt{ss}}$, $\mathbf{\Pi}_1^{\mathtt{ss}}$ were studied by v. d. Berg [vdBerg 87] in the frameworks of **IST** under the names: generalized (pre)galaxies, generalized (pre)halos, the prefix "pre" means that internal objects are not excluded, "generalized" means that the operations are not necessarily countable.

Theorem 1.4.6(i) was proved in [vdBerg 87]; part (ii) appears in [AnH 04]. Theorem 1.4.7 is essentially from [Jin 01], where the result is obtained for countably determined cuts in the context of model theoretic nonstandard analysis. Our results in §§ 1.4d, 1.4e also have some parallel forms in "hyperfinite" descriptive set theory discussed in Chapter 9.

Section 1.5. Theorem 1.5.20 was proved in [KanR 97]. The rest of the content of this section is rather well known in foundations since while ago and mostly external with respect to axiomatic nonstandard analysis.

2 Elementary real analysis in the nonstandard universe

Our main subject in this Chapter will be the development of nonstandard real analysis in the frameworks of the foundational scheme "$\mathbb{WF} \xrightarrow{*} \mathbb{I}$ [in $\mathbb{H}$]" of **HST** (as explained in § 1.2a). Of course, by no means can we hope to prove any new mathematical fact this way: indeed, if Φ is an $\in$-sentence then Φ^{wf}, the relativization of Φ to $\mathbb{WF}$, is provable in **HST** if and only if Φ is a theorem of **ZFC** (Theorem 1.1.14). Yet a broader "external" view brings us new insights into the nature of very common mathematical objects, or rather restores, at the level of full mathematical rigor, mathematical ideas and constructions once successfully employed by the masters of early calculus but then abandoned as too vague to admit rigorous treatment.

The main issues can be described as follows:

1) we consider real analysis in its "standard" **ZFC** form as relativized to the well-founded universe $\mathbb{WF}$: this is fully validated by Theorem 1.1.14;
2) we study $*$-extended versions of some fundamental notions of real analysis in $\mathbb{WF}$, that is, definitions which refer rather to the internal universe $\mathbb{I}$ than to $\mathbb{WF}$;
3) we give alternative proofs of "standard" results, based on connections between the concepts mentioned in 1) and 2).

However the intrinsic logic of the study will immediately lead us to results, like Robinson's lemma, which are not of the form 1), 2), 3), but rather are related to nonstandard structures themselves in such a manner that it is not immediately clear what they do mean in the universe $\mathbb{WF}$ chosen as the implementation of "standard" mathematics. This can be qualified as "true nonstandard analysis".

The first two sections, containing rather traditional material, are intended to show that both the spirit and, to great extent, even the letter of model theoretic nonstandard real analysis are respected in **HST**. Section 2.3 contains more general considerations, while the final Section 2.4 presents a couple of less known applications of nonstandard methods, Euler's decomposition of the sine function, and the Jordan curve theorem.

Blanket agreement 2.0.1. We argue in **HST** in this Chapter. Some of the theorems are indicated as theorems of **ZFC**, but the proofs, of course, proceed in **HST**. That this gives a correct proof follows from Theorem 1.1.14. □

2.1 Hyperreal line

Recall that $\mathbb{N}$ was defined in § 1.2 as the set of all natural numbers, *i.e.*, formally, all ordinals smaller than the first limit ordinal — and this has one and the same meaning in both the **HST** universe $\mathbb{H}$ and the class $\mathbb{WF}$ of all well-founded sets, in particular, $\mathbb{N} \in \mathbb{WF}$.

Definition 2.1.1. $\mathbb{Q}$, $\mathbb{R} \in \mathbb{WF}$ are resp. the set of all $\mathbb{WF}$-rationals and the set of all $\mathbb{WF}$-reals. Elements of $\mathbb{Q}$ and $\mathbb{R}$ are called *rationals* and *reals*. □

To be more exact, note that $\mathbb{Q}$ can be formally defined as a collection of triples $\langle p, m, n\rangle$, where $p = 0, 1$ and m, $n \in \mathbb{N}$, viewed as fractions $(1-2p)\frac{m}{n}$, where $n \neq 0$, with obvious provisions made to guarantee the uniqueness and $\mathbb{N} \subseteq \mathbb{Q}$ (thus, any triple $\langle 0, m, 1\rangle$ changes to m). Then we can explicitly define the usual order on the rationals and define, following Dedekind, $\mathbb{R}$ to be the set of all proper cuts in $\mathbb{Q}$ in the sense of this order. Usual definition of the operations $+$, $\times$, x^y ends the construction of the structure of $\mathbb{R}$. Identifying rationals with certain cuts in $\mathbb{R}$, we have $\mathbb{Q} \subseteq \mathbb{R}$,

Exercise 2.1.2. With these definitions, prove that the structures of $\mathbb{Q}$ and $\mathbb{R}$ are absolute for $\mathbb{WF}$, using the results of § 1.2b. □

This allows us to legitimately call elements of $\mathbb{R}$ "reals" rather than "$\mathbb{WF}$-reals" in Definition 2.1.1, and similarly for $\mathbb{Q}$.

2.1a Hyperreals

It follows from $*$-Transfer that the $*$-extensions ${}^*\mathbb{R}$, ${}^*\mathbb{Q} \in \mathbb{I}$ of the sets $\mathbb{R}$, ${}^*\mathbb{Q} \in \mathbb{WF}$ are (internal) sets of resp. $\mathbb{I}$-reals and $\mathbb{I}$-rational numbers. As a certain relational/operational structure, containing, *e.g.*, $<$, $+$, $\times$, x^y, is associated with $\mathbb{R}$, we should consider their $*$-extensions, say, ${}^*\!<$, yet it is common practice to drop $*$ here and use the same symbols $<$, $+$, $\times$, x^y to denote the order and the operations, as this never leads to confusion. Similarly, say, the sine function on ${}^*\mathbb{R}$ is denoted still $\sin x$ rather than ${}^*\!\sin x$, and the same with other typical functions.

Elements of ${}^*\mathbb{R}$ will be called *hyperreals* or $*$-*reals*, elements of ${}^*\mathbb{Q}$ *hyperrationals* or $*$-*rationals*. It follows from Theorem 1.5.20 that the map $x \mapsto {}^*x$ is an elementary embedding of $\mathbb{R}$ in ${}^*\mathbb{R}$ and $\mathbb{Q} \subseteq \mathbb{R}$ in ${}^*\mathbb{Q} \subseteq {}^*\mathbb{R}$.

Thus we have defined the $*$-versions ${}^*\mathbb{N} \subsetneqq {}^*\mathbb{Q} \subsetneqq {}^*\mathbb{R}$ of well-founded structures $\mathbb{N} \subsetneqq \mathbb{Q} \subsetneqq \mathbb{R}$, so that the former have the same mathematical properties (*i.e.*, those expressible in the $\in$-language) in $\mathbb{I}$ as the latter in $\mathbb{WF}$. The goal, however, is to study ${}^*\mathbb{R}$ from the $\mathbb{H}$-point of view.

Exercise 2.1.3. Prove that a $*$-real $x \in {}^*\mathbb{R}$ is standard iff $x = {}^*a$ for some $a \in \mathbb{R}$. (*Hint*: otherwise $x = {}^*u$ for $u \in \mathbb{WF} \smallsetminus \mathbb{R}$, leading to contradiction.) □

The map $x \mapsto {}^*x$ is a $1-1$ elementary embedding $\mathbb{R} \to {}^*\mathbb{R}$, hence, reals $a \in \mathbb{R}$ could be identified with ${}^*a \in {}^*\mathbb{R}$, getting $\mathbb{R} \subseteq {}^*\mathbb{R}$. Yet we shall not go this way, except for natural (and rational, if defined in the form of triples as above) numbers: recall that ${}^*n = n$ for any $n \in \mathbb{N}$.

Exercise 2.1.4. It follows from $*$-Transfer and Transitivity of $\mathbb{I}$ that ${}^*\mathbb{R}$ consists of all <u>internal</u> proper cuts in $\mathbb{Q}$. Show that ${}^*\mathbb{Q}$ has external (non-internal) cuts which, therefore, are not presented in ${}^*\mathbb{R}$: in particular, the cut with the left set $\{x \in {}^*\mathbb{R} : \exists^{\mathtt{wf}} a\,(x < {}^*a)\}$ is not internal. □

Theorem 2.1.5. *${}^*\mathbb{R}$ is standard size saturated, so that if $\mathscr{F}$ is a standard size family of internal subsets of ${}^*\mathbb{R}$, and $\bigcap \mathscr{F}' \neq \varnothing$ for any finite $\mathscr{F}' \subseteq \mathscr{F}$, then $\bigcap \mathscr{F} \neq \varnothing$.* (Proof: use Theorem 1.3.5.) □

Thus ${}^*\mathbb{R}$ is saturated but Dedekind-incomplete in $\mathbb{H}$ (*e.g.*, by the result of Exercise 2.1.4). On the contrary, $\mathbb{R}$ is Dedekind-complete, but not saturated, of course: we leave this as an **exercise**.

Definition 2.1.6. A $*$-real $x \in {}^*\mathbb{R}$ is:

- *standard* if $x = {}^*a$ for some $a \in \mathbb{R}$ (in fact this is a claim rather than a definition, see Exercise 2.1.3);
- *infinitesimal*, or *infinitely small*, if $|x| < \varepsilon$ for <u>all</u> standard $\varepsilon > 0$;
- *bounded*, or *limited*, if $|x| < a$ for <u>some</u> standard $a > 0$;
- *unbounded*, *unlimited*, or *infinitely large*, if $|x| > a$ for <u>all</u> standard a;
- *appreciable*, if it is limited but not infinitesimal;
- *near-standard*, if there is a standard $a \in {}^*\mathbb{R}$ such that $x \simeq a$, where $x \simeq y$ means that $|x - y|$ is infinitesimal. □

By definition $0 = {}^*0$ is infinitesimal. To obtain a non-0 infinitesimal $x \in {}^*\mathbb{R}$, note that the collection of sets $X_n = \{x \in {}^*\mathbb{R} : 0 < x < n^{-1}\}$, $n \in \mathbb{N}$, has standard size (because any well-founded set, in particular, $\mathbb{N}$ has standard size) and $\varnothing \neq X_{n+1} \subseteq X_n$ for each n, hence, $X = \bigcap_{n \in \mathbb{N}} X_n$ is nonempty by Saturation (Theorem 2.1.5). But any $x \in X$ is a positive infinitesimal.

A similar Saturation-based argument proves the existence of infinitely large $*$-reals, basically, if $\varepsilon > 0$ is infinitesimal then ε^{-1} is infinitely large.

Exercise 2.1.7. Prove that a finite sum of infinitesimals is infinitesimal, *i.e.*, if $n \in \mathbb{N}$ and $x_k \simeq 0$ for any $1 \le k \le n$ then $\sum_{k=1}^n x_k \simeq 0$. To begin with, note that the sequence $\{x_k\}_{k=1}^n$ belongs to $\mathbb{I}$ by Lemma 1.2.16, hence, the sum $\sum_{k=1}^n x_k$ is a legitimate finite sum from the $\mathbb{I}$-point of view. □

Exercise 2.1.8. Prove that $\simeq$ is an equivalence relation. Prove that if $x, y \in {}^*\mathbb{R}$ are standard and $x \simeq y$ then $x = y$. Prove that if $a, b \in \mathbb{R}$ and ${}^*a \simeq x \le y \simeq {}^*b$ (where $x, y \in {}^*\mathbb{R}$) then $a \le b$. □

Lemma 2.1.9. *Near-standard ∗-reals is the same as bounded ones.*

Proof. If $x \in {}^*\mathbb{R}$ is near-standard, say, $x \simeq {}^*a$, where $0 < a \in \mathbb{R}$, then $|x - {}^*a| < 1$, hence, $|x| < {}^*a + 1 = {}^*(a+1)$. Suppose that $x \in {}^*\mathbb{R}$ is bounded, say, $|x| < {}^*b$, where $b \in \mathbb{R}$, and prove the near-standardness. Note that the set $L = \{q \in \mathbb{Q} : {}^*q < x\}$ belongs to $\mathbb{WF}$ as a subset of a well-founded set. Clearly L is non-empty (for instance $-b \in L$) and bounded from above (any $q \in L$ satisfies $q < b$ – why ?), actually, L is a proper initial segment of $\mathbb{R}$. Define (in $\mathbb{H}$) $a = \sup L$, thus, $a \in \mathbb{R}$. **Exercise**: show that $x \simeq {}^*a$. □

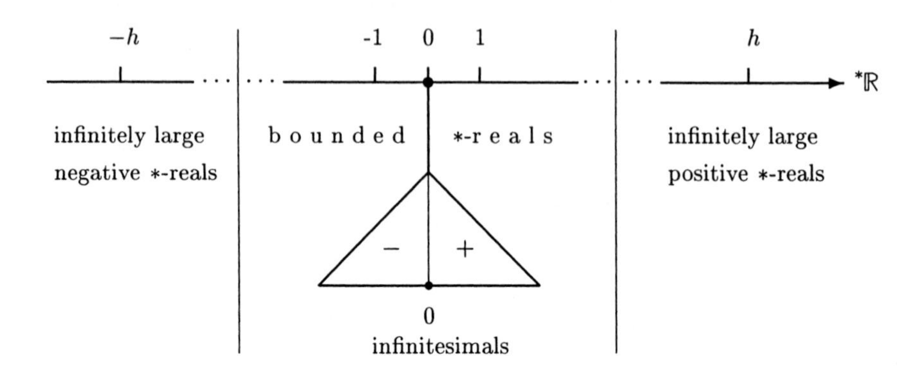

Definition 2.1.10. For any bounded $x \in {}^*\mathbb{R}$, the *shadow* [1] ${}^\circ x$ is the unique real $a \in \mathbb{R}$ such that $x \simeq {}^*a$. (The existence of such an a follows from Lemma 2.1.9, its uniqueness from the second claim in Exercise 2.1.8.)

For any $x \in {}^*\mathbb{R}$, $\mathtt{mon}\, x = \{y \in {}^*\mathbb{R} : x \simeq y\}$, the *monad*, or *halo* of x. □

2.1b Fundamentals of nonstandard real analysis

Here we display several principal results about relationships between the real and ∗-real structures, which belong to the folklore of model-theoretic nonstandard analysis. Each of them needs quite a bit of effort to be established in a superstructure, here we shall see that they comfortably follow from the axioms of **HST**, some of them being just tautologies in the **HST** set-up.

Internal Definitions: Any set defined in $\mathbb{I}$ by an $\in$-formula (internal parameters allowed) is internal.

To prove this, suppose that $Y \subseteq \mathbb{I}$, $\varphi(x)$ is an internal formula, and $Y = \{x \in \mathbb{I} : \varphi^{\mathtt{int}}(x)\}$, we have to show that $Y \in \mathbb{I}$. According to (3) of Exercise 1.1.11, there is a standard, hence, internal set X such that $Y \subseteq X$. Now apply Corollary 1.1.10.

[1] In papers following the model theoretic version of nonstandard analysis, it is more customary to call shadows *standard parts* and denote them by $\mathtt{st}\, x$, however this is impossible for the axiomatic version because here $\mathtt{st}$ is reserved for the standardness predicate. Shadows are sometimes also denoted by ${}^\circ x$.

Internal Induction: If $X \subseteq {}^*\mathbb{N}$ is an internal set containing 0, and we have $n \in X \Longrightarrow n+1 \in X$ for any $*$-integer n, then $X = {}^*\mathbb{N}$.

This is because $\mathbb{I}$ is an $\in$-interpretation of **ZFC**.

Overflow: If $X \subseteq {}^*\mathbb{N}$ is internal, $k_0 \in \mathbb{N}$, and each $k \in \mathbb{N}$, $k \geq k_0$, belongs to X, then there is $K \in {}^*\mathbb{N} \setminus \mathbb{N}$ such that all $k \in [k_0, K]$ belong to X.

Underflow: If $X \subseteq {}^*\mathbb{N}$ is internal, $K \in {}^*\mathbb{N} \setminus \mathbb{N}$ and each $k \in {}^*\mathbb{N} \setminus \mathbb{N}$, $k \leq K$, belongs to X, then there is $k_0 \in \mathbb{N}$ such that all $k \in [k_0, K]$ belong to X.

Permanence: If $X \subseteq {}^*\mathbb{R}$ is internal, $x \in {}^*\mathbb{R}$, and $\mathtt{mon}\, x \subseteq X$ then there is a real $r > 0$ such that $[x - {}^*r\,,\, x + {}^*r] \subseteq X$.

Exercise 2.1.11. Prove Overflow, Underflow, and Permanence. For instance, to prove Permanence, define, in $\mathbb{I}$, ε to be the largest real (*i.e.*, $*$-real from outside) satisfying $[x - \varepsilon,\, x + \varepsilon] \subseteq X$. Obviously ε cannot be infinitesimal because $\mathtt{mon}\, x \subseteq X$. Thus there is a real $r > 0$ such that ${}^*r < \varepsilon$. □

2.1c Directed Saturation

We start with a technical tool, a couple of theorems which allow to reduce some basic nonstandard arguments to simple transformations of formulas.

Definition 2.1.12. A relation $R(a, x_1, ..., x_n)$ is *directed in* a if and only if one of the following two conditions holds:

$$\forall x_1 \dots \forall x_n\, \forall a\, \forall a' > a\, \big(R(a, x_1, ..., x_n) \Longrightarrow R(a', x_1, ..., x_n)\big)$$

$$\forall x_1 \dots \forall x_n\, \forall a\, \forall a' < a\, \big(R(a, x_1, ..., x_n) \Longrightarrow R(a', x_1, ..., x_n)\big)$$

(resp. *increasingly* and *decreasingly* directed). □

In this definition, it is assumed that the variables a, a', x_i range over ${}^*\mathbb{R}$, accordingly, $<$ is the usual real number order on ${}^*\mathbb{R}$. (We could also write ${}^*\!<$.) Yet as a rule the definition will be applied when the domain of a, a' is restricted to ***positive*** $*$-reals.

Theorem 2.1.13 (*D-Saturation*, i.e., *Directed Saturation*). *If $R(x, a_1, ..., a_n)$ is an internal relation, the domain of each variable a_i is ${}^*\mathbb{R}^+$, and R is directed in each a_i, then we have*

$$\forall^{\mathtt{wf}} a_1 \dots \forall^{\mathtt{wf}} a_n\, \exists x\, R(x, {}^*a_1, ..., {}^*a_n) \iff \exists x\, \forall^{\mathtt{wf}} a_1 \dots \forall^{\mathtt{wf}} a_n\, R(x, {}^*a_1, ..., {}^*a_n)\,,$$

$$\exists^{\mathtt{wf}} a_1 \dots \exists^{\mathtt{wf}} a_n\, \forall x\, R(x, {}^*a_1, ..., {}^*a_n) \iff \forall x\, \exists^{\mathtt{wf}} a_1 \dots \exists^{\mathtt{wf}} a_n\, R(x, {}^*a_1, ..., {}^*a_n)\,.$$

Proof. We prove $\Longrightarrow$, the nontrivial direction, in the first equivalence. Suppose that R is just $R(x, a, b)$, decreasingly directed in a and increasingly on b; both a and b having ${}^*\mathbb{R}^+$ as the domain. The left-hand side takes the form $\forall^{\mathtt{wf}} a\, \forall^{\mathtt{wf}} b\, \exists x\, R(x, {}^*a, {}^*b)$, hence, implies $\forall^{\mathtt{wf}} n\, \exists x\, R(x, n, n^{-1})$. By Overflow, there exists an infinitely large $*$-integer H such that $\exists x\, R(x, H, H^{-1})$. Let x satisfy $R(x, H, H^{-1})$. To derive $\forall^{\mathtt{wf}} a\, \forall^{\mathtt{wf}} b\, R(x, {}^*a, {}^*b)$, the right-hand side, let $a, b \in \mathbb{R}^+$. Then ${}^*a, {}^*b^{-1} < H$, hence ${}^*b > H^{-1}$. This implies $R(x, {}^*a, {}^*b)$ by the assumption that R is directed. □

Let $\exists^{\infty\mathtt{lg}} H$ and $\forall^{\infty\mathtt{lg}} H$ mean, resp., "there exists an infinitely large $H \in {}^*\mathbb{N}$ (such that ...)" and "for all infinitely large $H \in {}^*\mathbb{N}$". These are shorthands, of course, meaning, resp.:

$$\exists H \left(\forall^{\mathtt{wf}} h\, (H \geq {}^*h) \wedge \Phi(H)\right) \quad \text{and} \quad \forall H \left(\forall^{\mathtt{wf}} h\, (H \geq {}^*h) \Longrightarrow \Phi(H)\right).$$

The following theorem is perhaps not so useful as the first one, but it will still allow us to carry out some arguments.

Theorem 2.1.14 (*Inf. Large Exchange*). *For any set $A \subseteq \mathbb{R}$, if $R \subseteq {}^*A \times {}^*\mathbb{N}$ satisfies $R(x, n) \Longrightarrow R(x, k)$ whenever $k, n \in {}^*\mathbb{N} \smallsetminus \mathbb{N}$ and $k \leq n$, then*

$$\forall^{\mathtt{wf}} a \in A\, \exists^{\infty\mathtt{lg}} H\, R({}^*a, H) \Longleftrightarrow \exists^{\infty\mathtt{lg}} H\, \forall^{\mathtt{wf}} a \in A\, R({}^*a, H).$$

Proof. Prove $\Longrightarrow$, the nontrivial direction. As A is a set of standard size (Theorem 1.3.1), there is, by Standard Size Choice, a map $f : A \to {}^*\mathbb{N} \smallsetminus \mathbb{N}$ such that $R({}^*a, f(a))$ holds for any $a \in A$. Applying Saturation to the family of all intervals $[n, f(a)]$, where $n \in \mathbb{N}$ and $a \in A$ (this is a family of standard size since so are A and $\mathbb{N}$), we obtain a number $H \in {}^*\mathbb{N} \smallsetminus \mathbb{N}$ with $H < f(a)$ for any $a \in A$. Then $R({}^*a, H)$ holds for any $a \in A$ by the choice of R. □

2.1d Nonstandard characterization of closed and compact sets

As one of most notable applications of D-Saturation, we prove the following theorem, which presents a characterization of closed and compact sets of reals $X \subseteq \mathbb{R}$ in terms of their $*$-extensions ${}^*X$.

Recall that any set $X \subseteq \mathbb{R}$ (generally, any $X \subseteq \mathbb{WF}$) is itself well-founded by Theorem 1.1.9.

Lemma 2.1.15. *Suppose that $X \subseteq \mathbb{R}$. Then*

(i) *X is closed* $\Longleftrightarrow$ $\forall^{\mathtt{wf}} a\, \forall x \in {}^*X\, (x \simeq {}^*a \Longrightarrow a \in X)$;

(ii) *X is compact* $\Longleftrightarrow$ $\forall x \in {}^*X\, \exists a \in X\, (x \simeq {}^*a)$.

Exercise 2.1.16. Why is the statement "X is closed" in the left-hand side of (i) absolute for $\mathbb{WF}$ (§ 1.2b), *i.e.*, expresses one and the same property in $\mathbb{WF}$ and in $\mathbb{H}$? The same for the compactness in (ii). □

Proof. The left-hand side of (i) means, by definition, that

$$\forall a \left(\forall \varepsilon > 0\, \exists x \in X\, (|x - a| < \varepsilon) \Longrightarrow a \in X\right)$$

holds in $\mathbb{WF}$, so, applying $*$-Transfer, we obtain

$$\forall^{\mathtt{wf}} a\, \big(\forall^{\mathtt{wf}} \varepsilon > 0\, \exists x \in {}^*X\, (|x - {}^*a| < {}^*\varepsilon) \implies {}^*a \in {}^*X\big)\,, \text{ that is,}$$
$$\forall^{\mathtt{wf}} a\, \exists^{\mathtt{wf}} \varepsilon\, \forall x \in {}^*X\, \big((|x - {}^*a| < {}^*\varepsilon) \implies a \in X\big)\,.$$

As the relation in brackets is obviously directed in ε, D-Saturation shows that the last displayed formula is equivalent to

$$\forall^{\mathtt{wf}} a\, \forall x \in {}^*X\, \exists^{\mathtt{wf}} \varepsilon\, \big((|x - {}^*a| < {}^*\varepsilon) \implies a \in X\big)\,, \text{ hence, to}$$
$$\forall^{\mathtt{wf}} a\, \forall x \in {}^*X\, \big(\forall^{\mathtt{wf}} \varepsilon\, (|x - {}^*a| < {}^*\varepsilon) \implies a \in X\big)\,,$$

which is equivalent to the right-hand side of (i).

To prove (ii), let X be compact. Then X is bounded, so there exists $b \in \mathbb{R}$ such that $|x| < b$ for any $x \in X$. To prove the right-hand side of (ii), let $x \in {}^*X$. Then $|x| < {}^*b$ by $*$-Transfer, hence, x is bounded, and, by Lemma 2.1.9, there is a real $a = {}^\circ x \in \mathbb{R}$ such that $x \simeq {}^*a$. However $a \in X$ by (i).

To prove the direction $\Longleftarrow$ in (ii), we note that the right-hand side of (ii) implies $\forall x \in {}^*X\, \exists^{\mathtt{wf}} c\, (|x| < {}^*c)$, and hence $\exists^{\mathtt{wf}} c\, \forall x \in {}^*X\, (|x| < {}^*c)$ by D-Saturation, and $\exists c\, \forall x \in X\, (|x| < c)$ by $*$-Transfer, so X is bounded. To prove that X is closed it is sufficient to verify the right-hand side of (i). Let $a \in \mathbb{R}$, $x \in {}^*X$, and $x \simeq {}^*a$. By the right-hand side of (ii), there is an element $b \in X$ with $x \simeq {}^*b$. Then ${}^*a \simeq {}^*b$, hence, $a = b \in X$ (Exercise 2.1.8). □

Exercise 2.1.17. Show that the right-hand sides of (i) and (ii) in Theorem 2.1.15 can be expressed using shadows resp. as:

$$\forall x \in {}^*X\, (x \text{ is bounded} \implies {}^\circ x \in X) \quad \text{and} \quad \forall x \in {}^*X\, ({}^\circ x \in X)\,.$$

Thus, in particular, a set of reals X is compact if and only if the shadow map sends ${}^*X$ onto X. □

2.2 Sequences and functions

By a *sequence* we shall here understand any $\mathbb{N}$-sequence $\{x_n\}_{n\in\mathbb{N}}$, where x_n are objects of any kind, in particular, reals or $*$-reals. A $*$*-sequence* will mean any ${}^*\mathbb{N}$-sequence $\{x_n\}_{n\in{}^*\mathbb{N}}$, with x_n as above. We shall consider mainly *internal* $*$-sequences, *i.e.*, those which belong to $\mathbb{I}$.

It follows from Theorem 1.1.9(ii) that any sequence $\{x_n\}_{n\in\mathbb{N}}$ of reals (generally, of any well-founded sets) belongs to $\mathbb{WF}$, hence, ${}^*\{x_n\} \in \mathbb{S}$ is a standard $*$-sequence of the form $\{{}^*x_n\}_{n\in{}^*\mathbb{N}}$, where ${}^*x_n = {}^*(x_n)$ for any $n \in \mathbb{N}$ by $*$-Transfer and because $n = {}^*n$ for any $n \in \mathbb{N}$. Thus, any sequence $\{x_n\}_n$ of reals has a unique standard $*$*-extension* $\{{}^*x_n\}_{n\in{}^*\mathbb{N}}$, so that ${}^*x_n \in {}^*\mathbb{R}$ is defined for any $*$-integer n. For instance if $x_n = \frac{1}{n}$ for all $n \in \mathbb{N}$ then the $*$-extension still satisfies ${}^*x_n = \frac{1}{n}$ for all $n \in {}^*\mathbb{N}$ since $n = {}^*n$ for $n \in \mathbb{N}$.

Exercise 2.2.1. Put $x_n = 0$ for $n \in \mathbb{N}$ and $x_n = 1$ for $n \in {}^*\mathbb{N} \setminus \mathbb{N}$. Fix $h \in \mathbb{N}$ and $H \in {}^*\mathbb{N} \setminus \mathbb{N}$ and put $y_n = 0$ for $n < H$ and $z_n = 0$ for $n < h$, and $y_n = 1$ for $n \geq H$ and $z_n = 1$ for $n \geq h$. Prove that $\{x_n\}_{n \in {}^*\mathbb{N}}$ is not internal, $\{y_n\}_{n \in {}^*\mathbb{N}}$ is internal nonstandard, and $\{z_n\}_{n \in {}^*\mathbb{N}}$ is standard. □

2.2a Limits

Recall that $\forall^{\infty\mathrm{lg}} H$ means: for all infinitely large $H \in {}^*\mathbb{N}$.

Theorem 2.2.2. *Let $x \in \mathbb{R}$ and $\{x_n\}_{n \in \mathbb{N}}$ be a sequence of reals. Then*

$$\lim_{n \to \infty} x_n = x \iff \forall^{\infty\mathrm{lg}} H\, ({}^*x_H \simeq {}^*x).$$

Remark 2.2.3. The equality in the left-hand side is understood in the $\mathbb{WF}$-sense, *i. e.*, as "it is true in $\mathbb{WF}$ that $\lim_{n\to\infty} x_n = x$": this will be our general agreement in such cases. (Recall that any sequence of reals $\{x_n\}_{n \in \mathbb{N}}$ is well-founded, see above.) Yet the equality $\lim_{n\to\infty} x_n = x$ expresses exactly the same meaning in $\mathbb{H}$, the **HST** universe of all sets, as it has in $\mathbb{WF}$, because $\mathbb{N}$ and $\mathbb{R}$ defined in $\mathbb{WF}$ remain unchanged in $\mathbb{H}$. □

Proof. To shorten formulas, let ε denote positive reals or hyperreals, while n and H denote elements of $\mathbb{N}$ or ${}^*\mathbb{N}$. Then $\lim_{n\to\infty} x_n = x \iff$

$$\iff \text{in } \mathbb{WF}\text{: } \forall \varepsilon\, \exists n\, \forall H \geq n\, (|x_H - x| < \varepsilon) \iff (\text{by } *\text{-Transfer})$$
$$\iff \forall^{\mathtt{wf}} \varepsilon\, \exists^{\mathtt{wf}} n\, \forall H \geq n\, (|{}^*x_H - {}^*x| < {}^*\varepsilon) \quad ,$$

which we can re-write as $\forall^{\mathtt{wf}} \varepsilon\, \exists^{\mathtt{wf}} n\, \forall H\, (H \geq n \Longrightarrow |{}^*x_H - {}^*x| < {}^*\varepsilon)$. As the formula $H \geq n \Longrightarrow |{}^*x_H - {}^*x| < {}^*\varepsilon$ is obviously (incresingly) directed in n, we can apply D-Saturation to the formula in the last displayed line, obtaining

$$\iff \forall H\, \forall^{\mathtt{wf}} \varepsilon\, \exists^{\mathtt{wf}} n\, \big(H \geq n \Longrightarrow |{}^*x_H - {}^*x| < {}^*\varepsilon\big) \iff$$
$$\iff \forall H\, \big(\forall^{\mathtt{wf}} n\, (H \geq n) \Longrightarrow \forall^{\mathtt{wf}} \varepsilon\, (|{}^*x_H - {}^*x| < {}^*\varepsilon)\big) \quad ;$$

which is equivalent to $\forall^{\infty\mathrm{lg}} H\, ({}^*x_H \simeq {}^*x)$, as required. □

The following exercises can be carried out using D-Saturation similarly.

Exercise 2.2.4. Prove that if $a, b \in \mathbb{R}$ and f is a function $\mathbb{R} \to \mathbb{R}$ then

$$\lim_{x \to a} f(x) = b \iff \forall x \in {}^*\mathbb{R}\, \big(x \simeq {}^*a \Longrightarrow {}^*f(x) \simeq {}^*b\big). \quad \square$$

Exercise 2.2.5 (*Cauchy criterion*). Prove that a sequence of reals $\{x_n\}_{n \in \mathbb{N}}$ converges (in $\mathbb{WF}$) if and only if ${}^*x_m \simeq {}^*x_n$ for any $m, n \in {}^*\mathbb{N} \setminus \mathbb{N}$. □

Exercise 2.2.6. Prove that for a sequence of of reals $\{x_n\}_{n \in \mathbb{N}}$ to converge to x (in $\mathbb{WF}$) it is still sufficient that $\exists^{\infty\mathrm{lg}} H\, \forall^{\infty\mathrm{lg}} K < H\, ({}^*x_K \simeq {}^*x)$. □

Exercise 2.2.7. Prove that a sequence of reals $\{x_n\}_{n \in \mathbb{N}}$ is bounded (*i. e.*, $|x_n| < c$ for all n, for some real c) iff ${}^*x_n$ is a bounded $*$-real for each $n \in {}^*\mathbb{N} \setminus \mathbb{N}$. Prove that a real function $f : \mathbb{R} \to \mathbb{R}$ is bounded iff ${}^*f(x)$ is bounded for each infinitely large hyperreal x. □

2.2b Continuous functions

Note that any set $X \subseteq \mathbb{R}$ and any function $f : \mathbb{R} \to \mathbb{R}$ are well-founded by Theorem 1.1.9(ii). Recall that a real function f is *continuous*, resp. *uniformly continuous* on a set $X \subseteq \mathbb{R}$ iff resp.

(1) $\forall \varepsilon\, \forall x \in X\, \exists \delta\, \forall x' \in X\, \big(|x - x'| < \delta \Longrightarrow |f(x) - f(x')| < \varepsilon\big)$;

(2) $\forall \varepsilon\, \exists \delta\, \forall x \in X\, \forall x' \in X\, \big(|x - x'| < \delta \Longrightarrow |f(x) - f(x')| < \varepsilon\big)$.

(Variables ε and δ are assumed to range over positive reals.)

Theorem 2.2.8. *If* $X \subseteq \mathbb{R}$ *and* $f : \mathbb{R} \to \mathbb{R}$ *then* f *is continuous, resp., uniformly continuous on* X *in* $\mathbb{WF}$ *iff resp.*

(1′) $\forall x \in X\, \forall x' \in {}^*X\, \big({}^*x \simeq x' \Longrightarrow {}^*f({}^*x) \simeq {}^*f(x')\big)$;

(2′) $\forall x, x' \in {}^*X\, \big(x \simeq x' \Longrightarrow {}^*f(x) \simeq {}^*f(x')\big)$.

Proof. Variables x and x' are supposed to range over resp. X and ${}^*X$. Applying $*$-Transfer and D-Saturation, we have: (1) holds in $\mathbb{WF}$ $\Longleftrightarrow$

$$\Longleftrightarrow \forall x\, \forall^{\text{wf}}\varepsilon\, \exists^{\text{wf}}\delta\, \forall x'\, \big(|{}^*x - x'| < {}^*\delta \Longrightarrow |{}^*f({}^*x) - {}^*f(x')| < {}^*\varepsilon\big) \Longleftrightarrow$$

$$\Longleftrightarrow \forall x\, \forall^{\text{wf}}\varepsilon\, \forall x'\, \exists^{\text{wf}}\delta\, \big(|{}^*x - x'| < {}^*\delta \Longrightarrow |{}^*f({}^*x) - {}^*f(x')| < {}^*\varepsilon\big) \Longleftrightarrow$$

$$\Longleftrightarrow \forall x\, \forall x'\, \big(\forall^{\text{wf}}\delta\, (|{}^*x - x'| < {}^*\delta) \Longrightarrow \forall^{\text{wf}}\varepsilon\, (|{}^*f({}^*x) - {}^*f(x')| < {}^*\varepsilon)\big) \Longleftrightarrow$$

$\Longleftrightarrow$ (1′), as required. The uniform case differs in some details: (2) $\Longleftrightarrow$

$$\Longleftrightarrow \forall^{\text{wf}}\varepsilon\, \exists^{\text{wf}}\delta\, \forall x\, \forall x'\, \big(|x - x'| < {}^*\delta \Longrightarrow |{}^*f(x) - {}^*f(x')| < {}^*\varepsilon\big) \Longleftrightarrow$$

$$\Longleftrightarrow \forall x\, \forall x'\, \forall^{\text{wf}}\varepsilon\, \exists^{\text{wf}}\delta\, \big(|x - x'| < {}^*\delta \Longrightarrow |{}^*f(x) - {}^*f(x')| < {}^*\varepsilon\big) \Longleftrightarrow$$

$\Longleftrightarrow$ (2′), as required. (Here both x, x' range over ${}^*X$.) □

The following classical theorem is now a simple consequence.

Corollary 2.2.9 (ZFC). *A continuous function* f *on a compact set of reals* X *is uniformly continuous.*

Proof. According to Theorem 1.1.14, it suffices to show, in **HST**, that the statement is true in $\mathbb{WF}$. But this easily follows from 2.2.8 and 2.1.15. □

There is a curious question related to the proof of Lemma 2.2.8 and Corollary 2.1.15. The usual proof of the Corollary is based on another definition of compactness (open covers), and involves the theorem which claims that a closed bounded set is compact. The latter is implied by the basic principle of Dedekind completeness of the real line. So it seems that we are able to avoid the Dedekind completeness by nonstandard reasoning.

In fact, of course, we are not. The Dedekind completeness is hidden in Lemma 2.1.9. Indeed, consider a non-closed set of reals, e.g. the set $\mathbb{Q}$ of all rationals. Lemma 2.1.9 fails for ${}^*\mathbb{Q}$ because there are nonstandard $*$-rationals with irrational standard parts.

2.2c Intermediate value theorem

Let us apply "nonstandard" technique to give a remarkably elementary proof of the following important theorem of analysis.

Theorem 2.2.10 (ZFC). *If a continuous function f is defined on a real interval $[a, b]$ with $a < b$, and $f(a) < 0$ but $f(b) > 0$, then there is a real $a < x < b$ such that $f(x) = 0$.*

Proof. Again it suffices to prove, in **HST**, that the statement is true in $\mathbb{WF}$. Thus we assume that $a < b$ belong to $\mathbb{R}$ and $f : [a, b] \to \mathbb{R}$ is a continuous function in $\mathbb{WF}$. Take any infinitely large $H \in {}^*\mathbb{N}$. Then any $x_n = {}^*a + \frac{n}{H}({}^*b - {}^*a)$ is a $*$-real, more exactly, a member of the $*$-real interval $[{}^*a, {}^*b]$, and $x_0 = {}^*a$ but $x_H = {}^*b$, hence, every $y_n = {}^*f(x_n)$ is also a $*$-real, and $y_0 < 0$ while $y_H > 0$ (say, by $*$-Transfer). Moreover the mapping $n \mapsto y_n$ is internal. It follows from Internal Induction that there is $n < H$ such that $y_n < 0$ but $y_{n+1} \geq 0$. (Let n be simply the least number with $y_{n+1} \geq 0$. Why is the argument not applicable if we define $y_n = -1$ for all $n \in \mathbb{N}$ but $y_n = 1$ for infinitely large n?) As the $*$-reals x_n, x_{n+1} are bounded (indeed, ${}^*a \leq x_n < x_{n+1} \leq {}^*b$), and $x_n \simeq x_{n+1}$ (because $x_{n+1} - x_n = \frac{1}{H} \simeq 0$) there exists, by Lemma 2.1.9, a real $x \in \mathbb{R}$ such that $x_n \simeq x_{n+1} \simeq {}^*x$. We claim that $f(x) = 0$. Suppose, towards the contrary, that $y = f(x) \neq 0$. However the $*$-real ${}^*y = {}^*f({}^*x)$ satisfies ${}^*y \simeq y_n \simeq y_{n+1}$ by Theorem 2.2.8, hence, obviously ${}^*y \simeq 0$, which contradicts the assumption $y \neq 0$ (Exercise 2.1.8). □

Exercise 2.2.11. Prove, using the same arguments, the maximal value theorem (*i.e.*, a real function, defined and continuous on a real interval $[a, b]$, has a maximum and minimum on $[a, b]$). □

2.2d Robinson's lemma and uniform limits

A number of theorems of the foundations of Analysis are of the following kind: two limits can be interchanged provided the inner limit is uniform on the variable of the outer limit. To prove theorems of this kind, we shall apply the following tool of "nonstandard" classics.

Theorem 2.2.12 (*Robinson's Lemma*). *Let $\{a_n\}_{n \in {}^*\mathbb{N}}$ be an internal $*$-sequence of $*$-reals. Assume that $a_n \simeq 0$ for all $n \in \mathbb{N}$. Then there exists an infinitely large number $H \in {}^*\mathbb{N}$ such that $a_m \simeq 0$ for all $m \leq H$.*

Proof. We proceed the same way: $\forall^{\mathrm{wf}} n\, (a_n \simeq 0) \Longrightarrow$

$$\Longrightarrow \quad \forall^{\mathrm{wf}} n\, \exists H > n\, \forall m \leq H\, \forall^{\mathrm{wf}} \varepsilon\, (|a_m| < {}^*\varepsilon) \quad \Longleftrightarrow$$

$$\Longleftrightarrow \quad \forall^{\mathrm{wf}} n\, \forall^{\mathrm{wf}} \varepsilon\, \exists H\, \forall m \leq H\, (H > n \wedge |a_m| < {}^*\varepsilon) \quad \Longleftrightarrow$$

$$\Longleftrightarrow \quad \exists H\, \forall^{\mathrm{wf}} n\, \forall^{\mathrm{wf}} \varepsilon\, \forall m \leq H\, (H > n \wedge |a_m| < {}^*\varepsilon) \quad \Longleftrightarrow$$

$$\Longleftrightarrow \exists H\, \big(\forall^{\mathrm{wf}} n\, (H > n) \wedge \forall m \leq H\, \forall^{\mathrm{wf}} \varepsilon\, (|a_m| < {}^*\varepsilon)\big) \quad ,$$

which is equivalent to $\exists^{\infty \lg} H \, \forall m \le H \, (a_m \simeq 0)$. □

(Note that D-Saturation was applied twice in the proof of the theorem !) Now we are equipped well enough to consider uniform limits.

Theorem 2.2.13. *Assume that $X \subseteq \mathbb{R}$, $f : X \to \mathbb{R}$, and $\{f_n\}$ is a sequence of functions $f_n : X \to \mathbb{R}$. Then*

$$f_n \to f \text{ in } \mathbb{WF}, \text{ uniformly on } X \quad \text{iff} \quad \forall^{\infty \lg} H \, \forall x \in {}^*X \, ({}^*f_H(x) \simeq {}^*f(x)).$$

Proof. $\{f_n\}$ converges to f uniformly on $X \iff$

$$\iff \forall \varepsilon \, \exists n \, \forall H \ge n \, \forall x \in X \, (|f_H(x) - f(x)| < \varepsilon) \text{ holds in } \mathbb{WF} \iff$$
$$\iff \quad \forall^{\mathtt{wf}} \varepsilon \, \exists^{\mathtt{wf}} n \, \forall H \ge n \, \forall x \in {}^*X \, (|{}^*f_H(x) - {}^*f(x)| < {}^*\varepsilon) \quad \iff$$
$$\iff \forall H \, \forall x \in {}^*X \, \forall^{\mathtt{wf}} \varepsilon \, \exists^{\mathtt{wf}} n \, \big(H \ge n \Longrightarrow |{}^*f_H(x) - {}^*f(x)| < {}^*\varepsilon\big) \iff$$
$$\iff \forall H \, \forall x \in {}^*X \, \forall^{\mathtt{wf}} \varepsilon \, \big(\forall^{\mathtt{wf}} n \, (H \ge n) \Longrightarrow |{}^*f_H(x) - {}^*f(x)| < {}^*\varepsilon\big) \quad ,$$

which is equivalent to $\forall^{\infty \lg} H \, \forall x \in {}^*X \, ({}^*f_H(x) \simeq {}^*f(x))$. □

Exercise 2.2.14. Find the places where $*$-Transfer and D-Saturation were applied in the proof of Theorem 2.2.13. □

Theorem 2.2.15 (ZFC). *Assume that $X \subseteq \mathbb{R}$, $f : X \to \mathbb{R}$, and $\{f_n\}$ is a sequence of continuous (on X) functions $f_n : X \to \mathbb{R}$. Suppose that $f_n \to f$ uniformly on X. Then f is continuous on X.*

Proof. As in the proof of Corollary 2.2.9, it suffices to show, in **HST**, that the statement holds in $\mathbb{WF}$. It follows from Theorem 2.2.8 that

$$\forall^{\mathtt{wf}} n \, \forall x \in X \, \forall x' \in {}^*X \, \big(x' \simeq {}^*x \Longrightarrow {}^*f_n({}^*x) \simeq {}^*f_n(x')\big)$$

because all f_n are continuous. We apply Robinson's lemma:

$$\forall x \in X \, \forall x' \in {}^*X \, \exists^{\infty \lg} H \, \big(x' \simeq {}^*x \Longrightarrow {}^*f_H({}^*x) \simeq {}^*f_H(x')\big). \tag{1}$$

To complete the proof, assume that $x \in X$ and $x' \in {}^*X$, $x' \simeq {}^*x$; we have to show that ${}^*f({}^*x) \simeq {}^*f(x')$. Choose H as in (1). Then

$${}^*f({}^*x) \simeq {}^*f_H({}^*x) \simeq {}^*f_H(x') \simeq {}^*f(x').$$

(The left and the right relations $\simeq$ are guaranteed by Theorem 2.2.13 while the middle $\simeq$ holds by the choice of H. □

The following is another consequence of Robinson's lemma. Suppose that $\nu \in {}^*\mathbb{N}$, possibly, $\nu \notin \mathbb{N}$. We consider <u>internal</u> sequences of the form $\{x_k\}_{k=1}^{\nu}$, where $\nu \in {}^*\mathbb{N}$ and, typically, $x_k \in {}^*\mathbb{R}$. Any such a sequence belongs to $\mathbb{I}$ and is $\mathbb{I}$-finite, hence, it admits computation of $\sum_{k=1}^{\nu} x_k$ and $\prod_{k=1}^{\nu} x_k$ in $\mathbb{I}$, the results of which are $*$-reals.

Corollary 2.2.16. *Suppose that* $1 \le \nu \in {}^*\mathbb{N}$, *and* $\{x_k\}_{k=1}^{\nu}$ *is an internal sequence of* $*$-*reals*, $x_k \simeq 0$ *for each* $k \in \mathbb{N}$, *and there is a real sequence* $\{r_n\}_{n\in\mathbb{N}}$ *with* $|x_k| \le {}^*r_k$ *for all* $k \le \nu$, $k \notin \mathbb{N}$ *and* $\sum_{k=1}^{\infty} r_k < \infty$. *Then* $\sum_{k=1}^{\nu} |x_k| \simeq 0$.

Proof. Note that $s_n = \sum_{k=1}^{n} |x_k| \simeq 0$ for any $n \in \mathbb{N}$ (Exercise 2.1.7), hence, by Robinson's lemma, $s_\mu \simeq 0$ for some $\mu \le \nu$, $\mu \notin \mathbb{N}$. It remains to show that $\sum_{k=\mu}^{\nu} |x_k| \simeq 0$. Since $|x_k| \le {}^*r_k$, we have to prove that $\sum_{k=\mu}^{\nu} {}^*r_k \simeq 0$, or, what is the same, that $\sum_{k=\mu}^{\nu} {}^*r_k < {}^*\varepsilon$ for any real $\varepsilon > 0$. As the series $\sum r_k$ converges (in $\mathbb{WF}$), for any real $\varepsilon > 0$ there is a number $l \in \mathbb{N}$ such that $\sum_{k=m}^{n} r_k < \varepsilon$ whenever $l < m < n$ in $\mathbb{WF}$. It follows, by $*$-Transfer, that $\sum_{k=m}^{n} {}^*r_k < {}^*\varepsilon$ whenever ${}^*l < m < n$, but ${}^*l < \mu$ since $\nu \notin \mathbb{N}$. □

2.3 Topics in nonstandard real analysis

There is more to real analysis than the real number system. Sets of reals, functions of reals, and spaces of functions are the core of real analysis. We now want to show that these matters can be studied with equal ease and success in **HST** as it is the case for the model theoretic approach. In passing by we shall also point out that it needs not more than a twist of language to give the arguments the shape the followers of **IST** are accustomed to.

The historic source of nonstandard analysis are ideas formulated by Leibniz and later abandoned as apparently inconsistent. This "Leibnizian intuition" which he himself qualified as a "façon de parler" can be revived naturally. We present these results in terms of the scheme "$\mathbb{WF} \xrightarrow{*} \mathbb{I}$ [in $\mathbb{H}$]". The most concise exposition is probably the following: to do the classical intuitive arguments in the "style of Leibniz" in $\mathbb{H}$ on a suitable set of internal objects and then to pass to a quotient identifying objects which are "infinitesimally close" to each other. This defines (in $\mathbb{H}$) a quotient. The quotient turns out to be small (in $\mathbb{H}$) and is bijectively mapped onto the classical structure (in $\mathbb{WF}$) thus explaining classical analysis as "Leibnizian intuition" factored by the notion of being "infinitesimally close". The "factoring" is the technical part being taken care of by **HST** while the intuitive content of the analytic theorem is properly expressed by the "Leibnizian intuition". This separation of intuitive content and set theoretical technicalities is remarkable, the latter being reduced to straightforward deductive reasoning in the axiomatic set-up. As these arguments always follow one and the same general pattern it makes sense to describe this pattern in some detail.

2.3a Shadows and equivalences

Definition 2.3.1. In the most general sense, a *shadow map* is any map φ such that $\mathtt{dom}\,\varphi \subseteq \mathbb{I}$ and $\mathtt{ran}\,\varphi \subseteq \mathbb{WF}$. □

For a "standard mathematician" this is bound to appear odd. Why should a rather unspecified map merit a special name? Well, in **HST** maps with large domains of definition and small range are special. Recall that the axiom of Choice takes the form of Standard Size Choice in **HST**. The customary source of maps in the form of "for all $x \in X$ there is a y in ..." here only works for domains X of standard size. Shadows are, in this respect, special. The name tries to convey that the rich structure of a set with exclusively internal elements is being projected onto the much simpler structure of a well-founded set. In Chapter 9 specific shadows will play an important role in the study of "hyperfinite" descriptive set theory.

A shadow always, trivially, defines an equivalence relation on its domain: $x \approx_\varphi y$ iff $\varphi(x) = \varphi(y)$. We can turn this around and ask whether an equivalence relation $\approx$ on a set $X \subset \mathbb{I}$ defines a shadow. This, clearly, depends upon the size of the quotient $X/{\approx}$. If this quotient is a set of standard size, *i.e.* a small set (see § 1.4e) then the quotient map can be easily converted in a shadow. Thus we retain that "shadow" and "equivalence relation with small quotient" are equivalent notions.

We fix some notation. Let us consider an arbitrary equivalence relation (a set of pairs) $\mathsf{E} \subset \mathbb{I}^2$; equivalence relation is here understood as a set of pairs which is symmetric and transitive. Its *domain* $\operatorname{dom} \mathsf{E} = \{x : \exists y\, (\langle x, y\rangle \in \mathsf{E})\}$ is obviously equal to the *domain of reflexivity* $\{x : \langle x, x\rangle \in \mathsf{E}\}$. The *quotient* $\operatorname{dom} \mathsf{E}/\mathsf{E}$ of E consists of the equivalence classes $[x]_{\mathsf{E}} = \{y : x\,\mathsf{E}\,y\}$, $x \in \operatorname{dom} \mathsf{E}$, that can also be called *monads*.

Theorem 1.4.11 says that in case that E is of class $\mathbf{\Delta}_2^{\mathtt{ss}}$ the quotient $\operatorname{dom} \mathsf{E}/\mathsf{E}$ is either large or small (of standard size). We shall see in § 5.5c that in **HST** it is consistent to assume that all subsets of $\mathbb{I}$ are of this class. This shows why the notion of shadow implies a lot of structure, at least in the case the associated equivalence is of class $\mathbf{\Delta}_2^{\mathtt{ss}}$. As was pointed out in Section 1.4, this class somehow exhausts the classical methods of nonstandard analysis based on Saturation and Standardization.

We illustrate this notion of shadow with a simple example. We recall that a $*$-real is bounded iff its absolute value is bounded by a standard real. We also recall that for any $*$-reals x and y we say that x *nearly equals* y, written $x \sim y$, iff $\forall^{\mathtt{st}} \varepsilon > 0\, (|x - y| < \varepsilon)$. For each bounded $*$-real x there is a unique real r such that $x \sim {}^*r$. This allows to define: the *shadow* of a bounded $*$-real x, denoted by ${}^\circ x$, is that unique real number r the $*$-image of which nearly equals x: $x \sim {}^*r$. The equivalence relation related to this shadow is evidently the relation of being bounded and nearly equal:

$$x \simeq y \iff \exists^{\mathtt{st}} z\, \forall^{\mathtt{st}} \varepsilon > 0\, (|z - x| + |z - y| < \varepsilon). \qquad (\dagger)$$

From now on only $\simeq$ will be of interest. If we now consider the structure $\langle {}^*\mathbb{R}_{\mathsf{bd}}; 0, 1, +, \cdot, <, = \rangle$, where ${}^*\mathbb{R}_{\mathsf{bd}}$ is the bounded part of ${}^*\mathbb{R}$, then it is a theorem of **HST** that it is $\simeq$-invariant and that the quotient is isomorphic to $\langle \mathbb{R}; 0, 1, +, \cdot, <, = \rangle$. This set theoretic description is literally correct in $\mathbb{H}$ but not in $\mathbb{I}$ because it depends upon the availability of non-internal sets.

Exercise 2.3.2. For the proof of $\simeq$-invariance of $\langle {}^*\mathbb{R}_{bd}\,; 0, 1, +, \cdot, <, =\rangle$, it has, among other things, to be proved that for bounded x, y the implication $x < y \wedge x \not\simeq y \Longleftrightarrow {}^\circ x < {}^\circ y$ holds. Give a proof. □

We now explore equivalences a little further. The *canonical quotient map* $\mathtt{sh}_\mathsf{E}$ from the domain $\mathtt{dom}\,\mathsf{E}$ to the equivalence classes is a map in $\mathbb{H}$ because for each $x \in \mathtt{dom}\,\mathsf{E}$ there is precisely one uniquely determined set of elements equivalent to x. (Define the set of pairs by Separation and prove that the so defined set is a graph.) But what about transversals?

A *transversal* for the quotient map $\mathtt{sh}_\mathsf{E}$ is a map $\tau : \mathtt{dom}\,\mathsf{E}/\mathsf{E} \longrightarrow \mathtt{dom}\,\mathsf{E}$ which is a right inverse of $\mathtt{sh}_\mathsf{E}$. A transversal τ always satisfies, by definition, $x\,\mathsf{E}\,\tau(\mathtt{sh}_\mathsf{E}(x))$ for all $x \in \mathtt{dom}\,\mathsf{E}$, which just says that τ is a left inverse of $\mathtt{sh}_\mathsf{E}$ modulo the equivalence.

Lemma 2.3.3. *A small quotient has transversal maps.*

Proof. For each class from the quotient there is an x from this class and the quotient being small we get by Standard Size Choice a transversal. By definition this map is a right inverse of the quotient map. □

2.3b Near-standard elements

We are naturally led to the following notion: an element $x \in D = \mathtt{dom}\,\mathsf{E}$ is E-*near-standard* iff its class $[x]_\mathsf{E}$ contains a standard element. [2]

We put $\mathsf{E}_{\mathbf{ns}} = \{\langle x, y\rangle \in \mathsf{E} : \exists^{\mathrm{st}} z\, (x\,\mathsf{E}\,z)\}$: the *near-standard part* of E.

Exercise 2.3.4. Prove that the quotient $\mathtt{dom}\,\mathsf{E}_{\mathbf{ns}}/\mathsf{E}_{\mathbf{ns}}$ is a set of standard size (that is, small). □

If we assume, for simplicity, that each $\mathsf{E}_{\mathbf{ns}}$-class contains precisely one standard element then we get a bijective map from the quotient onto the set $R = \{r \in \mathbb{WF} : {}^*r \in \mathtt{dom}\,\mathsf{E}_{\mathbf{ns}}\} \in \mathbb{WF}$. This motivates a narrower notion of shadow.

Definition 2.3.5. Given a set $R \in \mathbb{WF}$, a $*$-*compatible shadow* is a map $\varphi : \mathtt{dom}\,\varphi \subseteq {}^*R \xrightarrow{\text{onto}} R$ such that ${}^*r \in \mathtt{dom}\,\varphi$ and $\varphi({}^*r) = r$ for all $r \in R$, *i.e.* the shadow is a left inverse of the restriction of the $*$-map to R. □

---

[2] Note that it is only due to particular properties of $\mathbb{R}$, the reals, that $\simeq$-near standard elements of ${}^*\mathbb{R}$ are the same as bounded elements. For instance if we consider $\mathbb{N}$ as a discrete space with the distance $\rho(k, \ell) = 1$ for any $k \neq \ell$ then any $x \in {}^*\mathbb{N}$ will be bounded, that is, being at a finite distance from a standard element, but every $x \in {}^*\mathbb{N} \setminus \mathbb{N}$ is not near-standard. There are much more meaningful counterexamples, of course.

With a $*$-compatible shadow φ there is always associated a projection map $\widetilde{\varphi}(x) = {}^*(\varphi(x))$ from the domain of φ to the standard core of the domain. Example: the standard part $x \longmapsto {}^{*\circ}x$ related to ${}^\circ x$. This is the map sending each bounded $*$-real to the standard $*$-real almost equal to it.

Let us now consider $\langle {}^*\mathbb{Q}; 0, 1, +, \cdot, <, = \rangle$ and the relation $\simeq$ of near equality, defined by ($\dagger$) on the previous page. We have $\mathtt{dom}\simeq\, = {}^*\mathbb{Q}_{\mathtt{bd}}$ (the bounded part of ${}^*\mathbb{Q}$), but not all classes contain a standard $*$-rational element. Indeed take the recursively defined sequence $x_0 = 1$, $x_{n+1} = (\frac{x_n}{2} + \frac{1}{x_n})$. All of its elements with nonstandard indices are in one $\simeq$-class but there is no standard $*$-rational in that class (the class representing $\sqrt{2}$). **HST** offers a very simple way to deal with this problem. Knowing classical analysis we can argue that there is a shadow mapping of ${}^*\mathbb{Q}_{\mathtt{bd}}$ onto $\mathbb{R}$. But even when we give up all previous knowledge of $\mathbb{R}$ we can deduce from general abstract considerations that the structure $\langle \mathbb{R}; 0, 1, +, \cdot, <, = \rangle$ exists as a shadow (or a quotient, see § 1.5c) of $\langle {}^*\mathbb{Q}_{\mathtt{bd}}; 0, 1, +, \cdot, <, = \rangle$.

Definition 2.3.6. Let us call an equivalence E *complete* iff $\mathsf{E}_{\mathtt{ns}} = \mathsf{E}$ and *simple* iff no class contains more than one standard element. □

It follows from the preceding considerations that a complete equivalence E has a small quotient. Its quotient map is a $*$-compatible shadow if it is, in addition, simple. Conversely, equivalences defined from $*$-compatible shadows are complete and simple. The next lemma can be rendered as the existence of a "completion" of any small quotient. The name is intended to convey that this process of completion "fills" lost classes with new elements taking the place of the missing standard elements. But the name also recalls the classical notion of completion which, of course, is a special case of this nonstandard notion.

Lemma 2.3.7. *For any equivalence* $\mathsf{E} \subset \mathbb{I}$ *with a small quotient* $\mathtt{dom}\,\mathsf{E}/\mathsf{E}$ *there is a complete simple equivalence such that the quotients* $\mathtt{dom}\,\mathsf{E}/\mathsf{E}$ *and* $\mathtt{dom}\,\mathsf{E}'/\mathsf{E}'$ *are in a natural bijection.*

Proof. By Theorem 1.3.1, the smallness of the quotient implies the existence of a bijection β from $\mathtt{dom}\,\mathsf{E}/\mathsf{E}$ onto a set $R \in \mathbb{WF}$. Take any transversal τ for E. Then consider the map $f({}^*r) = \tau(\beta^{-1}(r))$ from ${}^*R \cap \mathbb{S}$ into $\mathtt{dom}\,\mathsf{E}$. By Extension there is an internal map π, $\mathtt{dom}\,\pi = {}^*R$, extending f. Consider an equivalence relation $x \mathrel{\mathsf{E}_\pi} y$ iff $\pi(x)$, $\pi(y)$ belong to $\mathtt{dom}\,\mathsf{E}$ and $\pi(x) \mathrel{\mathsf{E}} \pi(y)$. Note that $\mathtt{dom}\,\mathsf{E}_\pi$ is a subset of ${}^*R$ with ${}^*R \cap \mathbb{S} \subseteq \mathtt{dom}\,\mathsf{E}_\pi$. By definition both quotients are in bijection. (The classes of E_π are the inverse images of the classes of E under the map π.) Each class of E_π is a subset of ${}^*R$ containing a standard element, in addition, E_π is simple because β is a bijection from the quotient onto R. □

This allows us to get the full structure $\langle \mathbb{R}; 0, 1, +, \cdot, <, = \rangle$ from the quotient $\langle {}^*\mathbb{Q}_{\mathtt{bd}}; 0, 1, +, \cdot, <, = \rangle/\simeq$. The intuition behind this is reals as rational approximations with "infinitesimal error". The whole story of Dedekind cuts or Cantor's equivalence classes of convergent rational sequences as realizations of the reals disappears into the general set theoretic background and the "Leibnizian intuition" becomes literally correct. While the preceding arguments concerning $\langle \mathbb{R}; 0, 1, +, \cdot, <, = \rangle$ as the $*$-compatible shadow of $\langle {}^*\mathbb{R}_{\mathsf{bd}}; 0, 1, +, \cdot, <, = \rangle / \simeq$ rested on comparing two versions of the continuum ($\mathbb{R}$ and ${}^*\mathbb{R}$), this approach via completion does not use any previous knowledge of $\mathbb{R}$, only of $\mathbb{Q}$ (and ${}^*\mathbb{Q}$). There is evidently a problem with this second approach: when we drop any previous knowledge about $\mathbb{R}$ then we also do not dispose of the shadow mapping $\langle {}^*\mathbb{Q}_{\mathsf{bd}}; 0, 1, +, \cdot, <, = \rangle$ onto $\langle \mathbb{R}; 0, 1, +, \cdot, <, = \rangle$. We then have to verify independently that the quotient ${}^*\mathbb{Q}_{\mathsf{bd}}/\simeq$ is small. Once we know this we apply the preceding theorem and just choose any small set in bijection with the quotient and christen it $\mathbb{R}$. The structure on this set, which is the main point we are interested in, is being imported from $\langle {}^*\mathbb{Q}_{\mathsf{bd}}; 0, 1, +, \cdot, <, = \rangle / \simeq$ via the bijection. But proving smallness of the quotient is easy because it just needs some bound however crude it may be. The obvious guess is something of the type of $\mathbb{Q}^{\mathbb{N}}$.

Theorem 2.3.8. *The set of strictly increasing and bounded sequences from* $\mathbb{Q}^{\mathbb{N}}$ *is mapped onto the quotient* ${}^*\mathbb{Q}_{\mathsf{bd}}/\simeq$.

Proof. Take any strictly increasing but bounded sequence $\{x_i\}_{i \in \mathbb{N}}$ in $\mathbb{Q}^{\mathbb{N}}$. (It belongs to $\mathbb{WF}$.) Choose a nonstandard index h. The element ${}^*x_h$ falls into a class in ${}^*\mathbb{Q}_{\mathsf{bd}}/\simeq$. It remains to show that this map is onto. Take any class $[x]_\simeq$. But because of Exercise 2.3.2 $\forall^{\mathsf{st}}\varepsilon > 0\, \exists^{\mathsf{st}} y\, (|x - y| < \varepsilon)$, and therefore, by Standard Size Choice, there is a standard strictly increasing sequence $\{{}^*x_i\}_{i \in {}^*\mathbb{N}}$ below x such that $\forall^{\mathsf{st}} n (0 < x - {}^*x_n < \frac{1}{n+1})$. This implies, by Overflow, that up to some hyperfinite index M the element ${}^*x_h \in [x]_\simeq$ for all hyperfinite $h \leq M$. If $M < h$ the assertion follows because the sequence is strictly increasing and below any ${}^*y > x$. If $M \geq h$ the assertion follows directly. Thus the set of all classes is covered by the range of the map taking standard strictly increasing bounded sequences $\{x_i\}_{i \in \mathbb{N}}$ to $[{}^*x_h]_\simeq$. □

This shows how technicalities can be separated from the intuitive content very neatly. In the classical approach the construction of $\langle \mathbb{R}; 0, 1, +, \cdot, <, = \rangle$ must work through a mess of tedious technicalities to exhibit explicitly that a suitable quotient of the rational Cauchy sequences or the set of Dedekind cuts in the rationals does have the structure of $\langle \mathbb{R}; 0, 1, +, \cdot, <, = \rangle$. In the nonstandard approach the structure comes at almost no cost in form of $\langle {}^*\mathbb{Q}_{\mathsf{bd}}; 0, 1, +, \cdot, <, = \rangle / \simeq$, and it only remains to show that the quotient is of standard size. To highlight the importance of smallness of the quotient recall that $\simeq$ extends to the whole of ${}^*\mathbb{R}$ by $x \sim y$ iff $\forall^{\mathsf{st}}\varepsilon > 0\, (|x - y| < \varepsilon)$. The quotient ${}^*\mathbb{Q}/\sim$ is not of standard size, so it is not anything we can find in $\mathbb{WF}$ – a truly nonstandard object! (**Exercise**: show that ${}^*\mathbb{Q}/\sim$ is not of standard size!) We can condense these results into the following theorem.

Theorem 2.3.9. *Let* $\mathfrak{D} = \langle D; f_1, \ldots, f_k, R_1, \ldots, R_m \rangle$ *be an* E*-invariant internal structure, where* E *is a complete equivalence on* D*. Then* $\mathfrak{D}/\mathsf{E}$ *is isomorphic to a structure* $\widetilde{\mathfrak{D}} \in \mathbb{WF}$. □

This shows that there is a rather automatic formal machinery in place depending on some hypotheses of a more algebraic kind (invariance of $\mathfrak{D}$) delivering the structures of classical analysis. We conclude this with the remark that in **HST** we can either consistently assume that all subsets of $\mathbb{I}$ are of class $\mathbf{\Delta}_2^{\mathtt{ss}}$ (see Theorem 5.5.8 below) or restrict our attention to equivalence relations which are of this kind. Therefore all arguments in **IST** involving standard parts (defined by bounded st-∈-formulas) are fully covered in **HST** either by axiomatically incorporating this assumption (all sets $X \subseteq \mathbb{I}$ are in $\mathbf{\Delta}_2^{\mathtt{ss}}$) into the set-up or by considering only equivalence relations in $\mathbf{\Delta}_2^{\mathtt{ss}}$ (which, by the way, is the case for any of the concrete structures considered in classical analysis).

The notion of standard part is used extensively in the model theoretic approach where it corresponds to what we call shadow. The notion has also been used by **IST** followers where it denotes essentially the same thing as here, although it is not a set in the sense of that theory.

Discretization

It remains to note that $*$-compatible shadows describe locally the interaction between well-founded sets and their $*$-image. This allows to find new characterizations of notions of classical analysis. Another interesting fact is that Saturation allows to reduce the $*$-images of sets in $\mathbb{WF}$ of arbitrary cardinality to $*$-finite (discrete) sets.

Lemma 2.3.10. *For every internal set* $D \in \mathbb{I}$ *there is a* $*$*-finite (internal) set* $D' \subseteq D$ *such that both sets have the same standard core:* $D' \cap \mathbb{S} = D \cap \mathbb{S}$.

Proof. Applying 1.3.8(1) (which is a consequence of Saturation) with $X = D \cap \mathbb{S}$ we find a $*$-finite set F with $D \cap \mathbb{S} \subseteq F$. Further, by Standardization there is a standard set $G = {}^{\mathsf{S}}D$ satisfying $D \cap \mathbb{S} = G \cap \mathbb{S}$. It remains to define $D' = D \cap F \cap G$. □

It follows that given a $*$-compatible shadow $\varphi : \mathtt{dom}\,\varphi \to R$ (where $R \in \mathbb{WF}$ and ${}^*R \cap \mathbb{S} \subseteq \mathtt{dom}\,\varphi \subseteq {}^*R$) we can find a $*$-finite (internal) set $F \subseteq {}^*R$ such that the restriction $\psi = \varphi \restriction (F \cap \mathtt{dom}\,\varphi)$ is a $*$-compatible shadow still with $\mathtt{ran}\,\psi = R$, and hence the associated quotients are in a natural bijection. Such a $*$-discretization can be employed in many contexts. It is an essential tool in reducing classical calculus to "Leibnizian intuition". It was employed in this sense in § 2.2c.

2.3c Topology

Given a $*$-compatible shadow φ, it should be clear by now that the classes $[x]_\varphi$ are just the classes of "infinitesimally close" objects we were referring to in the introductory remarks. The intuitive reference of this notion suggests a connection with classical topology which is considered to be the mathematical formalization of the intuitive notion of "nearness". By the shadow the elements of $\mathtt{dom}\,\varphi$ (where ${}^*R \cap \mathbb{S} \subseteq \mathtt{dom}\,\varphi \subseteq {}^*R$ and $R \in \mathbb{WF}$) are collected into subsets of elements "near" to a standard element. Thus, in **HST**, we expect the standard part to contain all the information about some topology on R which somehow encodes the notion of "nearness" defined by the shadow.

The relation is actually more subtle than it might be expected to be. Classical topology is a large and complex field and we have no intention, here, to study these questions in any detail, nor do we present any new results. The aim is rather to show that in **HST** we do have a sufficient instrumentarium to successfully and efficiently study such questions without attention to the technical details of the model theoretic approach but, contrary to the **IST** environment, in the more familiar environment of a sufficiently well-equipped set theory in which the meaning of $\in$ is fixed to be the ordinary meaning of membership. In the model theoretic set-up the meaning of $\in$ has to be varied in various artificial model constructions. This invites the "Leibnizian" comment that set theory itself becomes a "façon de parler". Axiomatic set theories avoid this by keeping $\in$ fixed.

Our exposition in this subsection follows the model theoretic way rather closely but the **IST** followers just have to remove references to sets in $\mathbb{WF}$ and replace them by their standard counterparts in the internal universe to see the proof they would naturally think of.

Getting a shadow out of a topology

We now ask whether a topology on a set $R \in \mathbb{WF}$, given by its family $\mathcal{O} \subseteq \mathscr{P}(R)$ of open sets, defines a shadow in a natural way. Define the following:

- for any $r \in R$, $\mathcal{O}_r \subseteq \mathcal{O}$ denotes the subfamily of open sets containing x;
- for $x \in {}^*R$ and $r \in R$, define $\mathtt{sh}_{\mathcal{O}}(x) = r$ iff $x \in \bigcap_{U \in \mathcal{O}_r} {}^*U$.
- for $x, y \in {}^*R$, $x \simeq_{\mathcal{O}} y$ means that $x \in {}^*U \iff y \in {}^*U$ for any set $U \in \mathcal{O}$, $U \subseteq R$.
- $x \in {}^*R$ is <u>near-standard</u> iff $x \simeq_{\mathcal{O}} {}^*r$ for some $r \in R$.

Theorem 2.3.11. *Under these conditions, the map* $\mathtt{sh}_{\mathcal{O}}$ *is well-defined* (that is, we have $r = r'$ whenever $\mathtt{sh}_{\mathcal{O}}(x) = r$ and $\mathtt{sh}_{\mathcal{O}}(x) = r'$ for one and the same $x \in {}^*R$) *if and only if* $\mathcal{O}$ *is a* T_2 *topology.*[3]

If $\mathcal{O}$ *is indeed a* T_2 *topology then* $\mathtt{sh}_{\mathcal{O}}$ *is a* $*$*-compatible shadow* $D \xrightarrow{\text{onto}} R$, *where the domain* $D = \mathtt{dom}\,\mathtt{sh}_{\mathcal{O}}$ *is equal to the set of all near-standard elements of* ${}^*R$.

Proof. If $\mathcal{O}$ is T_2 and $r \neq r' \in R$ then there exists a pair of disjoint open sets $U, V \subseteq R$ containing resp. r, r'. Then ${}^*U \cap {}^*V = \varnothing$ by $*$-Transfer,

[3] Recall that T_2 means that any two distinct points have disjoint open nbhds.

and hence no element $x \in {}^*R$ can belong to both ${}^*U$ and ${}^*V$. To prove the converse suppose that points $r \neq r' \in R$ are not separable as required by T_2. This means that $\mathcal{O}_r \cup \mathcal{O}_{r'}$ is a f. i. p. family. It follows by $*$-Transfer that $\mathscr{F} = \{{}^*U : U \in \mathcal{O}_r \cup \mathcal{O}_{r'}\}$ also is a f. i. p. family (of internal sets), and hence there exists $x \in \bigcap \mathscr{F}$. Thus we have both $\mathtt{sh}_{\mathcal{O}}(x) = r$ and $\mathtt{sh}_{\mathcal{O}}(x) = r'$.

That $\mathtt{sh}_{\mathcal{O}}({}^*r) = r$ for any $r \in R$ follows by $*$-Transfer.

If $r \in \mathbb{R}$ and $x \simeq_{\mathcal{O}} {}^*r$ then clearly $\mathtt{sh}_{\mathcal{O}}(x)$ is defined and equal to r. □

Getting a topology out of a shadow

We now ask whether a topology can be extracted from a shadow given as an independent map. Recall that a topology on a set $R \in \mathbb{WF}$ can be defined by five different (but equivalent) methods: a neighbourhood system, the families of resp. open or closed sets, and, finally, by a closure operator or an interior operator. Recall that a map $\boldsymbol{\alpha} : \mathscr{P}(R) \to \mathscr{P}(R)$ is a *closure operator* iff the following four conditions are satisfied:

1) $\boldsymbol{\alpha}(\varnothing) = \varnothing$, 2) $\boldsymbol{\alpha}(X_1 \cup X_2) = \boldsymbol{\alpha}(X_1) \cup \boldsymbol{\alpha}(X_2)$,
3) $X \subseteq \boldsymbol{\alpha}(X)$, 4) $\boldsymbol{\alpha}(\boldsymbol{\alpha}(X)) = \boldsymbol{\alpha}(X)$.

Thus a closure operator is a special projection map on the power set of R. The family of closed sets defining the topology is the family of sets invariant under $\boldsymbol{\alpha}$ or, equivalently, the range of $\boldsymbol{\alpha}$. If one drops requirement 4) this leads to what is sometimes called a generalized topology.

Proposition 2.3.12. (i) *Given an arbitrary $*$-compatible shadow $\varphi : D = \mathtt{dom}\,\varphi \subseteq {}^*R \to \mathtt{ran}\,\varphi = R \in \mathbb{WF}$, the map $\boldsymbol{\alpha}(Y) = \varphi"{}^*Y$ $(Y \subseteq R)$ defines a generalized* T_1 *topology: properties 1), 2), 3) are satisfied and* (a special case of 4) !) *all singletons are invariant under $\boldsymbol{\alpha}$.*

(ii) *If moreover φ is $\mathtt{sh}_{\mathcal{O}}$, where $\mathcal{O}$ is a* T_2 *topology on R, then the operator $\boldsymbol{\alpha}_{\mathcal{O}}(Y) = \mathtt{sh}_{\mathcal{O}}"{}^*Y = \{\mathtt{sh}_{\mathcal{O}}(x) : x \in {}^*Y \cap \mathtt{dom}\,\mathtt{sh}_{\mathcal{O}}\}$ $(Y \subseteq R)$ is equal to the $\mathcal{O}$-closure operator on R.*

Proof. (i) We look at 1), 2). By definition, $\boldsymbol{\alpha}(\varnothing) = \varnothing$ and

$$\boldsymbol{\alpha}(Y_1 \cup Y_2) = \varphi"{}^*(Y_1 \cup Y_2) = \varphi"({}^*Y_1 \cup {}^*Y_2) = \boldsymbol{\alpha}(Y_1) \cup \boldsymbol{\alpha}(Y_2).$$

We have $Y \subseteq \boldsymbol{\alpha}(Y)$, and, generally speaking, $\subsetneqq$ because $\varphi"{}^*Y$ may pick up new elements as shadows of nonstandard elements of ${}^*Y$. This takes care of 3). Singletons are necessarily invariant: $\boldsymbol{\alpha}(\{x\}) = \varphi"\{{}^*x\} = \{\varphi({}^*x)\} = \{x\}$.

(ii) Consider any $Y \subseteq R$ and prove that $F = \boldsymbol{\alpha}_{\mathcal{O}}(Y)$ coincides with the $\mathcal{O}$-closure of Y. First of all $Y \subseteq F$ because ${}^*r \in {}^*Y$ and $\mathtt{sh}_{\mathcal{O}}({}^*r) = r$ by Theorem 2.3.11 for any $r \in Y$. To prove that F is closed suppose that $r \in R$ and any nbhd $U \in \mathcal{O}_r$ has a non-empty intersection with F, and hence the (internal) set $X_U = {}^*Y \cap {}^*U$ is non-empty. As the sets X_U, $U \in \mathcal{O}_r$, form a f. i. p. family of standard size, their intersection is non-empty. Thus there is $x \in {}^*Y$ such that $x \in {}^*U$ whenever $U \in \mathcal{O}_r$, thus $r = \mathtt{sh}_{\mathcal{O}}(x) \in \boldsymbol{\alpha}_{\mathcal{O}}(Y)$.

Finally, suppose towards the contrary that some $r \in F$ has a nbhd U satisfying $U \cap Y = \varnothing$. By definition $r = \mathbf{sh}_{\mathcal{O}}(x)$ for some $x \in {}^*Y$, and $x \in {}^*U$. Thus ${}^*Y \cap {}^*U \neq \varnothing$, and hence $Y \cap U \neq \varnothing$ by $*$-Transfer, contradiction. □

The proofs of the two preceding results show that the notion of $*$-compatible shadow is more general than that of a topology. We always get a generalized T_1-topology from a $*$-compatible shadow but if we want 4) as well (to get a true topology) then the shadow must possess additional properties. Such a shadow may be called *topological*. Yet we don't know any reasonable property of a $*$-compatible shadow φ which implies that the map $\boldsymbol{\alpha} = \boldsymbol{\alpha}_\varphi$ defined as in 2.3.12 is a closure operator satisfying 4).

Interior and accumulation points

When we consider an arbitrary topology $\mathcal{O}$ (open sets) on a set $R \in \mathbb{WF}$ then an equivalence can still be defined on the near-standard domain as follows:

$$x \simeq_{\mathcal{O}} y \iff \forall U \in \mathcal{O}\,(x \in {}^*U \iff y \in {}^*U)\,.$$

It is evidently complete (all standard elements of ${}^*R$ are in the domain of $\simeq_{\mathcal{O}}$), but not necessarily simple (some $\simeq_{\mathcal{O}}$-classes may have more than one standard element) unless the topology provides enough separation, and hence it does not necessarily have a naturally associated $*$-compatible shadow. In this context the question then becomes whether a set $U \subseteq R$ for which $\forall^{\mathtt{st}} x \in {}^*U\ ([x]_{\simeq_{\mathcal{O}}} \subseteq {}^*U)$ holds is an open set in the given topology. The answer is positive.

Lemma 2.3.13. *Given a topological space $\langle R; \mathcal{O}\rangle$, a set $U \subseteq R$ is open iff all of the standard points of ${}^*U$ are s-interior points, that is, formally $\forall^{\mathtt{st}} x \in {}^*R\,(x \in {}^*U \Longrightarrow [x]_{\simeq_{\mathcal{O}}} \subseteq {}^*U)$.*

A set $F \subseteq R$ is closed iff all of the standard points of ${}^*F$ are s-accumulation points, that is, formally $\forall x \in {}^*R\,([x]_{\simeq_{\mathcal{O}}} \cap {}^*F \neq \varnothing \Longrightarrow x \in {}^*F)$.

Proof. If $U \subseteq R$ is open and ${}^*r \in {}^*U$ then $U \in \mathcal{O}_r$ and $[{}^*r]_{\simeq_{\mathcal{O}}} \subseteq {}^*U$. If $[{}^*r]_{\simeq_{\mathcal{O}}} \subseteq {}^*U$ for all $r \in U$ then, by Saturation, there is an open set $V \in \mathcal{O}_r$ with ${}^*r \in {}^*V \subseteq {}^*U$, which proves (by $*$-Transfer) that U is open.

The second assertion follows by duality. □

The more interesting question is whether any given equivalence $\approx$ can serve to define a topology which then produces the original equivalence relation. But as in the case of $*$-compatible shadows this is, in general, not the case. An equivalence which has this property may again be called *topological*.

Compactness

We may take up compactness again and consider it from a different angle. Instead of the syntactic proof in Lemma 2.1.15 we can prove the same result in a purely set theoretic style.

Theorem 2.3.14 (Robinson). *A topology* $\mathcal{O}$ *a set* $R \in \mathbb{WF}$ *is a compact topology iff all elements of* $^*R$ *are near-standard.*

Proof. First, let us assume that all elements of $^*R$ are near-standard. Let $\mathcal{U}$ be an arbitrary open cover of R. If $x \in {}^*R$ then there is $r \in R$ with $x \simeq_{\mathcal{O}} {}^*r$. We have $r \in U$ for some $U \in \mathcal{U}$, and hence $^*r \in {}^*U$ by $*$-Transfer, and finally $x \in {}^*U$ because $x \simeq_{\mathcal{O}} {}^*r$. Thus $^*R$ is covered by $\{{}^*U : U \in \mathcal{U}\}$. By Saturation there is a finite subcover, say, $\{{}^*U : U \in \mathcal{U}'\}$, $\mathcal{U}' \subseteq \mathcal{U}$ being finite. Then $\mathcal{U}'$ obviously covers R by $*$-Transfer.

Now assume that each open cover of R can be reduced to a finite subcover. By $*$-Transfer every open standard cover of $^*R$ can be reduced to a standard finite subcover. Assume there is a non-near-standard element $x \in {}^*R$. Then for any $r \in R$ there is a nbhd $U_r \in \mathcal{O}$ of r such that $x \notin {}^*U_r$. Under our assumptions there is a finite subcover $\mathcal{U}' \subseteq \mathcal{U}$ of R. It follows by $*$-Transfer that $\{{}^*U : U \in \mathcal{U}'\}$ covers $^*R$, and hence $x \in {}^*U$ for some $U \in \mathcal{U}'$, which is a contradiction. □

2.4 Two special applications

This section presents two applications of nonstandard methods. We call them special because they clearly demonstrate how the nonstandard framework liberates analytic or geometric ideas from the technical web of ε-δ-methods, limits, and other machinery of "standard" methods.

2.4a Euler factorization of the sine function

Euler's decomposition of the sine function into an infinite product,

$$\mathtt{sinh}\, x = x\left(1+\frac{x^2}{\pi^2}\right)\left(1+\frac{x^2}{4\pi^2}\right)\left(1+\frac{x^2}{9\pi^2}\right)\cdots, \tag{1}$$

$$\mathtt{sin}\, x = x\left(1-\frac{x^2}{\pi^2}\right)\left(1-\frac{x^2}{4\pi^2}\right)\left(1-\frac{x^2}{9\pi^2}\right)\cdots, \tag{2}$$

provides us with an excellent advanced example of "nonstandard" reasoning in analysis. We begin with Euler's own arguments, short and brilliant, albeit not looking rigorous from any point of view, then proceed with a recovery of Euler's proof in **HST**.

Part I: Euler's argument

Euler begins his proof of (2) with the observation that

$$2\,\mathtt{sinh}\, x = e^x - e^{-x} = \left(1+\frac{x}{\omega}\right)^\omega - \left(1-\frac{x}{\omega}\right)^\omega, \tag{3}$$

where ω is an infinitely large natural number. Now, Euler uses the fact that $a^\omega - b^\omega$ is the product of factors

$$T_k = a^2 + b^2 - 2ab\cos\frac{2k\pi}{\omega}, \quad \text{where} \quad k \geq 1, \tag{4}$$

together with the factor $a - b$ and, if ω is an even number, the factor $a + b$, too. Taking $a = 1 + \frac{x}{\omega}$ and $b = 1 - \frac{x}{\omega}$, Euler converts (4) to

$$T_k = 2 + 2\frac{x^2}{\omega^2} - 2\left(1 - \frac{x^2}{\omega^2}\right)\cos\frac{2k\pi}{\omega}, \tag{5}$$

and further to

$$T_k' = \frac{4k^2\pi^2}{\omega^2}\left(1 + \frac{x^2}{k^2\pi^2} - \frac{x^2}{\omega^2}\right), \tag{6}$$

applying the replacement

$$\cos\frac{2k\pi}{\omega} \longmapsto 1 - \frac{2k^2\pi^2}{\omega^2}, \tag{7}$$

which he motivates by the relative smallness of the remainder of the Maclaurin series for cos. Therefore, argues Euler, $e^x - e^{-x}$ can be divided by $1 + \frac{x^2}{k^2\pi^2} - \frac{x^2}{\omega^2}$, where the term $\frac{x^2}{\omega^2}$ can be omitted by the substitution

$$1 + \frac{x^2}{k^2\pi^2} - \frac{x^2}{\omega^2} \longmapsto 1 + \frac{x^2}{k^2\pi^2}, \tag{8}$$

since even after multiplying by ω it remains infinitesimal. As still there is a factor $a - b = 2x/\omega$, Euler produces the final expression (1), arguing that the numerical factor must be equal to 1, in order to be consistent with the Maclaurin series for `sinh`. After that, (2) is obtained by substitution $x \mapsto ix$.

The whole argument is quite short of looking rigorous from any point of view, so it is rather surprising that it gives the right result. We shall see that in fact Euler's proof can be explained in **HST** on the basis of some analytic phenomena of a general nature.

Part II: Recovered proof

To prove (1), let us fix $\nu \in {}^*\mathbb{N} \setminus \mathbb{N}$ and put $\omega = 2\nu + 1$. As $\mathtt{lim}_{n\to\infty}(1 + \frac{\alpha}{n})^n = e^\alpha$, and on the other hand the product in the right-hand side of (1) converges, it suffices, by Theorem 2.2.2, to prove that

$$\left(1 + \frac{x}{\omega}\right)^\omega - \left(1 - \frac{x}{\omega}\right)^\omega \simeq x\prod_{k=1}^{\nu}\left(1 + \frac{x^2}{k^2\pi^2}\right) \tag{9}$$

whenever $x = {}^*a \in {}^*\mathbb{R}$, $\nu \in {}^*\mathbb{N} \setminus \mathbb{N}$, and $\omega = 2\nu + 1$. Let us fix x, ν, ω as indicated, and assume, in addition, that $x > 0$. To prove (9), we begin with the following algebraic identity of polynomials known to Euler:

$$\left(1+\frac{x}{\omega}\right)^{\omega}-\left(1-\frac{x}{\omega}\right)^{\omega}=\frac{2x}{\omega}\prod_{k=1}^{\nu}\underbrace{\left[2+2\frac{x^2}{\omega^2}-2\left(1-\frac{x^2}{\omega^2}\right)\cos\frac{2k\pi}{\omega}\right]}_{T_k}. \quad (10)$$

Now Euler changes $\cos\frac{2k\pi}{\omega}$ to $1-\frac{2k^2\pi^2}{\omega^2}$ in T_k, arguing that the remainder of the expansion of cos is relatively small. Let us see how this argument works out. Let $2\alpha_k=\cos\frac{2k\pi}{\omega}-1+\frac{2k^2\pi^2}{\omega^2}$, then, by easy transformations,

$$T_k=\left(\frac{4k^2\pi^2}{\omega^2}-4\alpha_k\right)\left(1+\frac{x^2}{k^2\pi^2}-\frac{x^2}{\omega^2}+h_k x^2\right), \quad (11)$$

where $h_k=\frac{\omega^2\alpha_k}{k^2\pi^2(k^2\pi^2-\omega^2\alpha_k)}$, thus,

$$\left(1+\frac{x}{\omega}\right)^{\omega}-\left(1-\frac{x}{\omega}\right)^{\omega}=Cx\prod_{k=1}^{\nu}\left(1+\frac{x^2}{k^2\pi^2}-\frac{x^2}{\omega^2}+h_k x^2\right), \quad (12)$$

where C and h_k do not depend on x. Following Euler, let us infer that

$$\left(1+\frac{x}{\omega}\right)^{\omega}-\left(1-\frac{x}{\omega}\right)^{\omega}\simeq Cx\prod_{k=1}^{\nu}\left(1+\frac{x^2}{k^2\pi^2}-\frac{x^2}{\omega^2}\right). \quad (13)$$

As obviously $u_k=1+\frac{x^2}{k^2\pi^2}-\frac{x^2}{\omega^2}>\frac{1}{2}$ (indeed, $x>0$ is standard while ω infinitely large), for (12) $\Longrightarrow$ (13) it suffices to show that $\sum_{k=1}^{\nu}|h_k|$ is infinitesimal, which would follow if we find a standard constant γ such that

$$0\le h_k<\frac{\gamma}{\omega^2}, \quad \text{in all assumptions made}. \quad (14)$$

To find such a γ, let us analyse Euler's vague reference to the expansion

$$\cos\frac{2k\pi}{\omega}=1-\frac{2k^2\pi^2}{\omega^2}+\underbrace{\frac{2k^4\pi^4}{3\omega^4}-\frac{2^6k^6\pi^6}{6!\,\pi^6}+\ldots}_{2\alpha_k}. \quad (15)$$

in (7). The terms of the expansion clearly decrease in absolute values, beginning with $\frac{2k^2\pi^2}{\omega^2}$, in the domain $0<k\le\nu=\frac{\omega-1}{2}$, moreover, they decrease faster than the geometric series with denominator $\frac{(2k\pi)^2}{\omega^2\cdot 3\cdot 4}=\frac{\pi^2}{12}<0.9$, hence,

$$0<\alpha_k<\frac{k^4\pi^4}{3\omega^4}<0.9\cdot\frac{k^2\pi^2}{\omega^2}, \quad (16)$$

which immediately implies (14) with, say, $\gamma=4$. This is how the derivation of (13) from (12), or Euler's passage from (5) to (6) can be justified: at the bottom of things we have the fact that terms of the Maclaurin expansion of cos uniformly decrease to 0 as fast as a convergent geometric series.

In continuation of the argument, we obtain

$$\left(1+\frac{x}{\omega}\right)^{\omega}-\left(1-\frac{x}{\omega}\right)^{\omega}\simeq Cx\prod_{k=1}^{\nu}\left(1+\frac{x^2}{k^2\pi^2}\right) \tag{17}$$

from (13) using the same argument but with much less effort, because now $\sum_{k=1}^{\nu} x^2/\omega^2 = \nu x^2/\omega^2 \le x^2/2\omega \simeq 0$. Finally, to show that $C=1$, note that (12) is an identity of polynomials in x, hence, they have equal coefficients, in particular, 1, the coefficient of x in the left-hand side is equal to C, the coefficient of x in the right-hand side of (12), so that $C=1$ in (12), (13), (17), as required. □

Exercise 2.4.1. Prove that

$$e^x - 1 = e^{x/2}\cdot x\cdot\prod_{k=1}^{\infty}\left(1+\frac{x^2}{4k^2\pi^2}\right),$$

using the a similar "nonstandard" argument. The result follows from (1) by an elementary substitution, so this is only a technical exercise. Factorizing $(1+\frac{1}{\omega})^{\omega}-1$, the first transformation of factors yields those of the form $1+\frac{x^2}{4k^2\pi^2}+\frac{x}{\omega}$, to be replaced by $(1+\frac{x}{\omega})\left(1+\frac{x^2}{4k^2\pi^2}\right)$. □

2.4b Jordan curve theorem

The following theorem is known under this name.

Theorem 2.4.2. *A simple*[4] *closed continuous curve $\mathscr{K}$ in the plane separates its complement into two open sets of which it is the common boundary.*

One of these open sets, the **outer** *or* **exterior** *region $\mathscr{K}_{\text{ext}}$ is an open, unbounded, path-connected set, while another, the* **inner** *or* **interior** *region $\mathscr{K}_{\text{int}}$ is an open, simply path-connected, bounded set.*

This theorem (referred to as JCT below) was one of the starting points in the development of topology (originally called *Analysis Situs*). Although the result appears natural to an analyst, it is far not so easy to prove. Jordan's original proof in his *Cours d'analyse* [Jor 1893] is rather elementary as to the tools employed: Jordan considered the assertion to be evident for polygons and reduced the case of a continuous curve to that of a polygon by approximating the curve by a sequence of suitable simple polygons. Yet the argument extends over nine pages and, as a matter of fact, was not viewed as complete by mathematicians of his time.

We are interested here in this proof.

It is certainly not true that all classical arguments can be replaced in some useful or reasonable way by simpler "nonstandard" arguments. But as

[4] *Simple* (polygon, curve) will always mean: having no self-intersections.

we shall show it is possible to simplify the approximation argument specific to Jordan's proof. We shall follow the proof quite closely but take a somewhat different approach when proving path-connectedness.

Arguing in **HST**, we are going to prove that JCT is true in the well-founded universe $\mathbb{WF}$. According to Theorem 1.1.14, this suffices to claim that JCT is then a theorem of **ZFC**.

Beginning the proof, we fix a simple closed curve $\mathscr{K} = \{K(t) : 0 \le t < 1\}$ in $\mathbb{WF}$, where $K : \mathbb{R} \to \mathbb{R}^2$ is a continuous 1-periodic function, injective modulo 1 (*i.e.*, $K(t) = K(t')$ iff $t - t' \equiv 0 \bmod 1$). Then ${}^*\mathscr{K} \in \mathbb{I}$.

Exercise 2.4.3. Prove, using $*$-Transfer, that ${}^*\mathscr{K}$ is a simple $*$-curve, *i.e.*, a simple curve from the point of view of the internal universe $\mathbb{I}$, and ${}^*K(t) \simeq {}^*K(t')$ iff $t \simeq t'$ mod 1. □

We outline the proof we are going to give:

Step 1. We shall infinitesimally approximate ${}^*\mathscr{K}$ by a simple $*$-polygon Π, using a construction, essentially due to Jordan, of consecutively cutting off loops from an originally self-intersecting approximation.

Step 2. Taking JCT for polygons for granted, we define the interior region $\mathscr{K}_{\text{int}}$ as the open set of all points A such that ${}^*A$ belongs to Π_{int} but does not belong to the monad of Π. $\mathscr{K}_{\text{ext}}$ is defined accordingly.

Step 3. To prove that $\mathscr{K}_{\text{int}}$ is path-connected we define a simple $*$-polygon Π' which lies entirely within Π_{int}, does not intersect $\mathscr{K}$, and contains all "asteriscs" of points of $\mathscr{K}_{\text{int}}$. This will imply the path-connectedness.

First step

We want to approximate the curve ${}^*\mathscr{K}$ by a simple $*$-polygon (that is, an internal object which is a polygon from the point of view of $\mathbb{I}$).

Definition 2.4.4. A $*$-polygon $\Pi = P_1P_2\dots P_nP_1$ (where $n \in {}^*\mathbb{N}$) *approximates* ${}^*\mathscr{K}$ if there is an internal sequence of $*$-reals $0 \le t_1 < \dots < t_n < 1$ such that

1°. $P_i = {}^*K(t_i)$ for $1 \le i < n$, and

2°. $t_n - t_1 \ge \frac{1}{2}$, and $t_{i+1} - t_i \le \frac{1}{2}$ for all $1 \le i < n$.

In addition, Π *approximates* ${}^*\mathscr{K}$ *infinitesimally* if

3°. $\Delta(\Pi) \simeq 0$, where $\Delta(\Pi) = \max_{1 \le k \le n} |P_kP_{k+1}|$.

It is understood in 3° that $P_{n+1} = P_1$. □

Lemma 2.4.5. *Suppose that a $*$-polygon $\Pi = P_1 \dots P_nP_1$ approximates the curve ${}^*\mathscr{K}$ infinitesimally. Then*

(a) *n is infinitely large, $t_{i+1} \simeq t_i$ for all $1 \le i < n$, $t_1 \simeq 0$, and $t_n \simeq 1$;*

(b) *there is an infinitesimal* $\varepsilon > 0$ *such that* ${}^*\mathscr{K}$ *is in the* ε*-neighbourhood of* Π *and* Π *is in the* ε*-neighbourhood of* ${}^*\mathscr{K}$;

(c) *if points* $P \simeq Q$ *are on* Π *then precisely one of the two arcs* Π *is decomposed into by these points must be included in the monad of* P.

Proof. (a) Suppose towards the contrary that $t_{i+1} \not\simeq t_i$. Then the reals $r_i = {}^\circ t_i$ and $r_{i+1} = {}^\circ t_{i+1}$ in $\mathbb{R}$ (the shadows, see Definition 2.1.10) satisfy $r_i < r_{i+1}$ strictly. By definition, we have $t_i \simeq {}^*r_i$, and hence ${}^*K({}^*r_i) \simeq {}^*K(t_i)$ in the plane by Theorem 2.2.8. Similarly, ${}^*K({}^*r_{i+1}) \simeq {}^*K(t_{i+1})$. Thus, by 3°, ${}^*K({}^*r_i) \simeq {}^*K({}^*r_{i+1})$, or, that is the same, ${}^*(K(r_i)) \simeq {}^*(K(r_{i+1}))$. It follows, by $*$-Transfer, that $K(r_i)$ coincides with $K(r_{i+1})$. However it follows from 2° that $|r_{i+1} - r_i| \le \frac{1}{2}$, in particular, it is not true that $r_i = 0$ and $r_{i+1} = 1$, and on the other hand $r_i < r_{i+1}$ strictly. [5] Thus we have a self-intersection in $\mathscr{K}$, contradiction.

The rest of (a) is proved similarly; note that the injectivity modulo 1 of K is used in the proof that $t_1 \simeq 0$ and $t_n \simeq 1$.

(b) It follows from (a) and Theorem 2.2.8 that for each i, $1 \le i \le n$ the $*$-real $\delta_i = \max_{t_i \le t \le t_{i+1}} |{}^*K(t) - {}^*K(t_i)|$ is infinitesimal, and therefore $\varepsilon = 2\max_{1 \le i \le n} \delta_i$ is infinitesimal and proves the assertion.

(c) Since all edges of Π are infinitesimal by (a), we may assume that P and Q are vertices, say $P = P_i$ and $Q = P_j$. Then either $t_i \simeq t_j$ or $t_i \simeq 0$ while $t_j \simeq 1$ (otherwise $\mathscr{K}$ would have a self-intersection as in the proof of (a) above). Consider the first case. The arc determined by $t_i \le t \le t_j$ is clearly within the monad of P. To see that the other arc is not included in the monad consider any t_k which is $\not\simeq$ any of t_i, 0, 1. Then $P_k \not\simeq P$ as otherwise $\mathscr{K}$ would have a self-intersection. □

Lemma 2.4.6. ${}^*\mathscr{K}$ *is infinitesimally approximable by a* **simple** $*$*-polygon.*

Proof. To get a $*$-polygon Π which infinitesimally approximates ${}^*\mathscr{K}$ just take $t_i = \frac{i}{n}$ for some infinitely large n. But Π may have self-intersections.

Assume two non-adjacent sides intersect, that is, P_iP_{i+1} intersects P_jP_{j+1} for some $1 \le i < j - 1 < n$. By the triangle inequality the shorter of the segments P_iP_j and $P_{i+1}P_{j+1}$ is not longer than the longer of the segments P_iP_{i+1} and P_jP_{j+1} which is bounded in length by $\Delta(\Pi)$.

Let us assume that $|P_iP_j| \le |P_{i+1}P_{j+1}|$. We now replace in $P_1 \dots P_nP_1$ the arc $P_i \dots P_j$ by a new side P_iP_j if $t_j - t_i \le \frac{1}{2}$. This ensures that 2° is satisfied for the new $*$-polygon. If $t_j - t_i > \frac{1}{2}$ then we replace the complementary arc $P_j \dots P_nP_1 \dots P_i$ by a new side P_jP_i such that again 2° is satisfied. The case $|P_iP_j| > |P_{i+1}P_{j+1}|$ is treated in the same way, so that the resulting $*$-polygon Π_{new} still infinitesimally approximates ${}^*\mathscr{K}$ because 1°

[5] Note the role of 2° in this argument: it does not allow the hyperreals t_k to collapse into a sort of infinitesimal "cluster" or into a pair of them around 0 and 1, which would be compatible with $\Delta(\Pi) \simeq 0$ alone.

and 2° are satisfied (for the accordingly reduced system of parameter values t_i) and $\Delta(\Pi_{\text{new}}) \leq \Delta(\Pi)$, so that 3° is also satisfied.

This (internal) procedure does not necessarily reduce the number of self-intersections because for the one which is removed there may be others appearing on the newly introduced side of the reduced polygon Π_{new}. But the number of vertices of Π_{new} is obviously strictly less than that of Π. Therefore the internal sequence of $*$-polygons arising from Π by iterated applications of this procedure eventually ends with a simple $*$-polygon Π' which approximates ${}^*\mathscr{K}$ infinitesimally. □

Second step

We now define the interior and exterior regions.

Let us fix for the remainder a simple $*$-polygon $\Pi = P_1P_2 \dots P_nP_1$ which approximates ${}^*\mathscr{K}$ infinitesimally. In $\mathbb{WF}$, let $\mathscr{K}_{\text{int}}$, resp., $\mathscr{K}_{\text{ext}}$ be the set of all points $A \in \mathbb{R}^2$ such that ${}^*A \in \Pi_{\text{int}}$, resp., ${}^*A \in \Pi_{\text{ext}}$, and ${}^*A$ has a non-infinitesimal distance from Π.

Exercise 2.4.7. Prove that, in $\mathbb{WF}$, $\mathscr{K}_{\text{int}}$ and $\mathscr{K}_{\text{ext}}$ are open sets, $\mathscr{K}_{\text{int}}$ is bounded, $\mathscr{K}_{\text{ext}}$ is unbounded, and $\mathscr{K}_{\text{int}} \cup \mathscr{K}_{\text{ext}} = \mathbb{R}^2 \setminus \mathscr{K}$. □

Show that for $A \in \mathscr{K}_{\text{int}}$ and $B \in \mathscr{K}_{\text{ext}}$ any continuous arc α from A to B in $\mathbb{WF}$ intersects $\mathscr{K}$. Indeed ${}^*\alpha$ intersects Π in some point P since it starts in Π_{int} and ends in Π_{ext}. (The $*$-form of JCT for polygons is applied.) By Lemma 2.4.5 and the compactness of $\mathscr{K}$ there is, by Lemma 2.1.15, a point $B \in \mathscr{K}$ such that ${}^*B \simeq P$. As $\mathscr{K}$ and the arc are closed, B is in $\mathscr{K} \cap \alpha$.

Now prove that, in $\mathbb{WF}$, each point of $\mathscr{K}$ is a limit point for both the interior region $\mathscr{K}_{\text{int}}$ and the exterior region $\mathscr{K}_{\text{ext}}$. By the choice of Π and the definition of $\mathscr{K}_{\text{int}}$ and $\mathscr{K}_{\text{ext}}$, it suffices to prove the following in $\mathbb{I}$, the internal universe: *given a vertex A on Π, then for any square S with center in A and positive size the domain ${}^*S_{\text{int}}$ contains points in both Π_{int} and Π_{ext} which have non-infinitesimal distance from Π.* We prove this assertion for Π_{int} only; the proof for the exterior region is similar.

Let B be another vertex of Π chosen such that the distance $|AB|$ is non-infinitesimal. We can assume that B lies in ${}^*S_{\text{ext}}$ and has non-infinitesimal distance from ${}^*S$, and in addition ${}^*S$ itself does not contain any vertex of Π. The interior region Π_{int} is decomposed by ${}^*S$ into a number of polygonal domains. Let Π' be the polygon which bounds that domain among them the boundary of which contains A. Let α and β be the simple broken lines – connecting A with B – into which Π is partitioned by the vertices A and B. Then Π' consists of parts of α and β and connected parts of ${}^*S$. Since A is the only common point of α and β except for B (which is far away from ${}^*S$), going around Π' we find a connected broken "interval" C_1C_2 of ${}^*S$ (which may occasionally contain one or more of the four vertices of ${}^*S$) such that the points C_1 and C_2 belong to different curves among α, β. Since C_1C_2 is also a part of Π', any inner point E of C_1C_2 belongs to Π_{int}.

Consider a point E in C_1C_2 which has equal distance $d = \mathrm{d}(E, \alpha) - \mathrm{d}(E, \beta)$ from both α and β. Note that d is not infinitesimal: otherwise there are points $A' \in \alpha$ and $B' \in \beta$ such that $A' \simeq E \simeq B'$, which is impossible by Lemma 2.4.5(c) as ${}^*S$ has non-infinitesimal distance from both A and B.

Thus $E \in \Pi_{\text{int}}$ has a non-infinitesimal distance from Π, as required.

Third step

We prove the path-connectedness of $\mathscr{K}_{\text{int}}$; the result for $\mathscr{K}_{\text{ext}}$ is a minor modification. Prove that, in $\mathbb{WF}$, any two points A, B in $\mathscr{K}_{\text{int}}$ are connected by a broken line which does not intersect $\mathscr{K}$. This is based on the following

Lemma 2.4.8. *There exists a simple polygon Π' which lies entirely within Π_{int}, contains no point of ${}^*\mathscr{K}$ in Π'_{int}, and ${}^*A \in \Pi'_{\text{int}}$ holds for any $A \in \mathscr{K}_{\text{int}}$.*

Proof. Let an infinitesimal $\varepsilon > 0$ be defined as in (b) of Lemma 2.4.5, so that $\mathscr{K}$ and ${}^*\mathscr{K}$ are included in the ε-neighbourhood of Π in ${}^*\mathbb{R}^2$. For any side PQ of Π (which is infinitesimal by definition) we draw a rectangle of the size $(|PQ| + 4\varepsilon) \times (4\varepsilon)$ so that the side PQ lies within the rectangle at equal distance 2ε from each of the four sides of the rectangle.

Let us say that a point E is the *inner intersection* of two straight segments σ and σ' iff E is an inner point of both σ and σ', and $\sigma \cap \sigma' = \{E\}$. For any point $C \in \Pi_{\text{int}}$ which is either a vertex of some of the rectangles above, or an inner intersection of sides of two different rectangles in this family let CC' be a shortest straight segment which connects C with a point C' on Π; obviously each CC' is infinitesimal.

Let us fix a point A in $\mathscr{K}_{\text{int}}$. The parts of the rectangles lying within Π and the segments CC' decompose the interior region Π_{int} into a (possibly hyperfinite) number of polygonal domains. Let the polygon Π' be the boundary of the domain containing ${}^*A$. (Note that all the lines involved lie in the monad of Π, hence none of them contains ${}^*A$.) It remains to prove that Π'_{int} also contains ${}^*B$, where B is any other point of $\mathscr{K}_{\text{int}}$.

Let $\Pi' = C_1C_2 \dots C_n$. We observe that by construction, for any $k = 1, ..., n$, there is a shortest segment $\sigma_k = C_kC'_k$, connecting C_k with a point C'_k in Π which does not intersect Π'_{int}. Moreover, by the triangle equality, the segments σ_k have no inner intersections. Therefore, any two of them intersect each other only in such a way that either the only intersection point is the common endpoint $C'_k = C'_l$ or one of them is an end-part of the other one. Then the segments σ_k decompose the ring-like polygonal region $\mathscr{R}$ between Π and Π' into n open domains $\mathscr{D}_k$ $(k = 1, ..., n)$ defined as follows.

If σ_k and σ_{k+1} are disjoint (σ_{n+1} equals σ_1) then the border of $\mathscr{D}_k$ consists of σ_k, σ_{k+1}, the side C_kC_{k+1} of Π', and that arc $\widehat{C'_kC'_{k+1}}$ of Π which does not contain any of the points C'_l as an inner point. If σ_k and σ_{k+1} have the common endpoint $C_k = C_{k+1}$ and no more common points then the border shrinks to σ_k, σ_{k+1}, and C_kC_{k+1}. If, finally, one of the segments is included in the other then $\mathscr{D}_k$ is empty.

If now $^*B \in \Pi'_{\text{ext}}$ then $^*B$ belongs to one of the domains $\mathscr{D}_k$. If this is a domain of the first type then the infinitesimal simple arc $C'_k C_k C_{k+1} C'_{k+1}$ separates $^*A$ from $^*B$ within Π, which easily implies, by Lemma 2.4.5(c), that either $^*A$ or $^*B$ belongs to the monad of Π, which contradicts the choice of the points. If $\mathscr{D}_k$ is a domain of second type then the barrier accordingly shrinks, leading to the same contradiction. □

The lemma implies the the path-connectedness of $\mathscr{K}_{\text{int}}$: indeed, by the JCT for polygons in $\mathbb{I}$, $^*A$ connects to $^*B$ by a broken line which lies within Π'_{int} therefore does not intersect $^*\mathscr{K}$. By $*$-Transfer there is, in $\mathbb{WF}$, a broken line which connects A and B and does not intersect $\mathscr{K}$ as required.

The lemma also implies the *simple* path-connectedness of $\mathscr{K}_{\text{int}}$. Indeed to prove that, in $\mathbb{WF}$, every simple closed kurve $\mathscr{K}_1 \subseteq \mathscr{K}_{\text{int}}$ can be appropriately contracted into a point, note that $^*\mathscr{K}_1$ is evidently situated within Π'_{int}, the interior of a simple polygon, so that $^*\mathscr{K}_1$ has the required property in the nonstandard domain by the JCT for polygons. It remains to apply $*$-Transfer.

□ (*Theorem 2.4.2*)

Historical and other notes to Chapter 2

Sections 2.1, 2.2.

We present classics of nonstandard analysis in the frameworks of scheme "$\mathbb{WF} \xrightarrow{*} \mathbb{I}$ [in $\mathbb{H}$]" (see § 1.2a).

The principle of D-Saturation (Theorem 2.1.13), used here as a general method to codify proofs in what can be characterized as *elementary* nonstandard analysis, was explicitly formulated in Kanovei [Kan 94b] (a remote predecessor of this book), but definitely it goes back to various studies since 60s, in particular, in the **IST** setting, in Nelson [Nel 77] and Reeken [R 92].

Section 2.3.

See [Cut 97, Loeb 97] on topics in nonstandard topology and [Lux 77] on early studies in this area.

Section 2.4.

Euler's proof of the decomposition formulas is given in [E:Inf, § 156]. Luxemburg presented his famous recovered proof, by means of modern nonstandard analysis, in [Lux 73]. [6] However the recovered proof in [Lux 73] in fact de-

[6] Luxemburg motivated his interest in recovery of Euler's mysterious ideas as follows (p. 63 in [Lux 73]): "Nowadays this representation of the sine function belongs to that part of function theory that studies the behavior of entire functions whenever its zeros are given. [...] There are many proofs known for this result. Some of the proofs are even elementary. But all of these proofs are somewhat artificial in the sense that they rely on some analytical trick. It is therefore not without interest to examine how Euler proved his formula."

viates from Euler's original argument: the transformation from (12) to (17) is carried out circumpassing (13) as an intermediate step and with different estimates of the remainder. Kanovei [Kan 88] gave a more faithful recovery of Euler's decomposition argument; it is presented in § 2.4a (part II).

Criticisms of Jordan's original proof of the Jordan curve theorem can be found, for instance, in [Veb 05, Os 12]. As a curiosity we note in passing that Jordan speaks of *infinitesimals* in his proof but it is only a figure of speech for a number which may be chosen as small as one wishes or for a function which tends to zero. Veblen [Veb 05] is considered the first to have given a (by modern standards) rigorous proof of JCT.

Narens [Nar 71] gave a nonstandard proof of JCT which rather closely follows Jordan's argument, yet some part of this proof has been criticized in mid-90s as inconclusive, in the course of an informal discussion within GALAXY, an email exchange list devoted to nonstandard analysis. This criticism initiated our own research, which resulted in article [KanR 99a].

A totally different nonstandard proof of JCT, which avoids polygons and approximations entirely by looking at a nonstandard discretization of the plane and reducing the problem to a combinatorial version of the JCT, due to Stout, was given by Bertoglio and Chuaqui [BertC 94].

3 Theories of internal sets

The class $\mathbb{I}$ of all internal sets, or, more exactly, the structure $\langle \mathbb{I}; \in, \mathtt{st}\rangle$, is a very important substructure of the nonstandard set universe of **HST** because it contains many typically nonstandard objects like infinitely large or infinitesimal numbers (see Chapter 2). It will be demonstrated (Theorem 3.1.8) that $\langle \mathbb{I}; \in, \mathtt{st}\rangle$ satisfies the axioms of *bounded set theory* **BST**, a variant of Nelson's internal set theory **IST**.

This Chapter is devoted to these two theories. We present what can be called "internal methods" of nonstandard analysis, or *the scheme* "$\mathbb{S} \subseteq \mathbb{I}$".

Some attention will be paid to rather traditional issues well known to **IST** practitioners. In particular, we show in Section 3.1 that a rather weak theory **BIST**, essentially the common part of **BST** and **IST**, allows to prove typical results about standard and nonstandard natural numbers, finite sets, *etc.*, in an "internal" manner, that is without any use of non-internal sets.

Then we present an interesting and sometimes difficult foundational study related to the internal theories **BST** and **IST**. The axioms of Inner Boundedness of **BST** and (unbounded) Idealization of **IST**, absent in **BIST**, add quite a bit of regularity to the structure of the internal nonstandard universe.

In particular, we prove in § 3.2b that, in **BST**, any $\mathtt{st}$-$\in$-formula is equivalent to a $\Sigma_2^{\mathtt{st}}$ formula, a result of great importance, which will lead to the proof of Collection and Uniqueness in **BST** for all $\mathtt{st}$-$\in$-formulas, and also to the construction of "external" extensions of a **BST** universe in Chapter 5. The list of basic theorems of **BST** also includes some forms of Extension, Standard Size Choice, and Uniqueness.

Partially saturated versions of **BST** will be considered in Section 3.3.

The axiom of Idealization of **IST** (with unbounded quantifiers) also leads to Collection and Uniqueness in **IST** for all $\mathtt{st}$-$\in$-formulas, see Section 3.4. In addition, it is demonstrated in Section 3.5 that **IST** provides a uniform truth definition for closed $\in$-formulas in $\mathbb{S}$. This rather surprising fact can lead to new insights into the axiomatics of nonstandard theories. For instance the truth definition enables us to obtain counterexamples showing that some theorems of **BST**, for instance Reduction to $\Sigma_2^{\mathtt{st}}$ or some forms of Extension, are not, generally speaking, provable in **IST** despite of rather common beliefs to the contrary among **IST** practitioners based on a misinterpretation of some results in Nelson [Nel 88]. This issue is considered in Section 3.6.

3.1 Introduction to internal set theories

This type of nonstandard set theories is based on an idea of the nonstandard set universe quite different from **HST**. We have a much simpler picture: similarly to the class $\mathbb{I}$ in **HST**, the whole universe of an internal theory is an elementary extension, in the $\in$-language, of the class $\mathbb{S} = \{x : \mathtt{st}\, x\}$ of all standard sets. This property characterizes *internal* nonstandard set theories.

Blanket agreement 3.1.1. In this Chapter, in the course of our study of internal theories like **IST** and **BST**, some **HST** notions will be temporarily modified, in particular, $\mathbb{I}$ will denote the universe of all sets. However, sets in $\mathbb{I}$ (that is, all sets) will sometimes be called *internal*, especially in order to make distinction with "external sets" (see § 3.2f below).

Following the practice of **IST** studies, we take $\mathbb{I}$ (rather than $\mathbb{S}$) as the basic universe for general set theoretic notions like ordinals, natural numbers, finite sets. This does not lead to any problems because $\mathbb{I}$ will always satisfy the axioms of **ZFC** in the $\in$-language (see Exercise 3.1.3).

In particular, *a finite set* will mean: a finite set in the sense of the "universe of discourse", that is an $\mathbb{I}$-finite set, and the same for other basic notions. (See also Definition 3.1.17 below.) □

3.1a Internal set theory

Nelson's famous *internal set theory* **IST** [1] is defined here as follows:

Definition 3.1.2. **IST** is a theory in $\mathtt{st}$-$\in$-language, containing the axiom schemata of $\mathbf{ZFC}^{\mathtt{st}}$, Inner Transfer, Idealization, Inner Standardization described below in this Subsection. □

$\mathbf{ZFC}^{\mathtt{st}}$: exactly as in § 1.1c.

Inner Transfer: [2] $\Phi^{\mathtt{st}} \Longleftrightarrow \Phi$,

where Φ is any closed $\in$-formula with standard sets as parameters. (Compare with Transfer in § 1.1c.)

[1] Introduced in [Nel 77]. Nelson's subsequent paper [Nel 88] introduced a system different from and incompatible with the "canonical" **IST**. See Section 3.6.

[2] The word "inner" here, and in some other principles below, means only that the statements express *inner properties* of the class $\mathbb{I}$ and are thought of either to be considered in an internal theory like **IST** where all sets are internal, or, if considered in **HST**, to be formally relativized to $\mathbb{I}$, to distinguish from synonymous axioms and theorems of **HST**. We would like to use the word "internal" instead of "inner" here, but that would rather mean the restriction to internal formulas, *i.e.*, $\in$-formulas, which is not really applicable here.

In papers on **IST**, Inner Transfer and Inner Standardization are called just Transfer and Standardization (which we reserved for the **HST** axioms of the same name). Idealization is referred to as *Internalization* in [Keis 94].

Exercise 3.1.3. Show that, in the presence of Inner Transfer, $\mathbf{ZFC}^{\mathtt{st}}$ is equivalent to **ZFC** (in the $\in$-language). In particular, $\mathbf{ZFC}^{\mathtt{st}}$ and Inner Transfer imply all **ZFC** axioms (in the $\in$-language) in the universe $\mathbb{I}$ of all sets. □

In fact **IST** is usually formulated with **ZFC** instead of $\mathbf{ZFC}^{\mathtt{st}}$ [3].

Idealization: $\forall^{\mathtt{stfin}} A\, \exists x\, \forall a \in A\, \Phi(a,x) \Longleftrightarrow \exists x\, \forall^{\mathtt{st}} a\, \Phi(a,x)$,
where $\Phi(a,x)$ is any $\in$-formula with arbitrary parameters.
(Compare with the equivalence in Exercise 1.3.7. However Idealization is somewhat stronger: here the domain A of a is not bounded by any set.)

The quantifier $\forall^{\mathtt{stfin}}$ in Idealization means: "for all standard finite", while the adjective "standard finite" refers to any set x with $\mathtt{st}\, x$, finite in the sense of the usual **ZFC** definition of finiteness in the universe of all sets (now denoted by $\mathbb{I}$), in accordance with Blanket agreement 3.1.1. See more on finiteness in internal theories in § 3.1e, especially Definition 3.1.17.

Inner Standardization: $\forall^{\mathtt{st}} X\, \exists^{\mathtt{st}} Y\, \forall^{\mathtt{st}} x\, (x \in Y \Longleftrightarrow x \in X \wedge \Phi(x))$,
where $\Phi(x)$ is any $\mathtt{st}$-$\in$-formula with arbitrary parameters.
(Compare with Standardization in § 1.1c.)

Definition 3.1.4. Let ${}^{\sigma}X = X \cap \mathbb{S}$ for any set X.
For any standard X and any $\mathtt{st}$-$\in$-formula $\Phi(x)$, ${}^{\mathbb{S}}\{x \in X : \Phi(x)\}$ is the unique set $Y \in \mathbb{S}$ satisfying $\forall^{\mathtt{st}} x\, (x \in Y \Longleftrightarrow x \in X \wedge \Phi(x))$. □

Exercise 3.1.5. Prove using Inner Transfer that if standard sets X, Y satisfy ${}^{\sigma}X = {}^{\sigma}Y$ then $X = Y$. Infer the uniqueness of Y in Definition 3.1.4. □

Comments. Note that in purely syntactical terms the schema of Inner Transfer consists of all formulas of the form

$$\forall^{\mathtt{st}} x_1 \ldots \forall^{\mathtt{st}} x_n\, \big(\Phi(x_1, ..., x_n)^{\mathtt{st}} \Longleftrightarrow \Phi(x_1, ..., x_n)\big),$$

where $\Phi(x_1, ..., x_n)$ is any $\in$-formula (with all free variables indicated) while the prefix $\forall^{\mathtt{st}} x_1 \ldots \forall^{\mathtt{st}} x_n$ accounts for the condition that the parameters must be standard. Accordingly, Inner Standardization consists of all formulas

$$\forall x_1 \ldots \forall x_n\, \big(\forall^{\mathtt{st}} X\, \exists^{\mathtt{st}} Y\, \forall^{\mathtt{st}} x\, (x \in Y \Longleftrightarrow x \in X \wedge \Phi(x, x_1, ..., x_n))\big),$$

where $\Phi(x, x_1, ..., x_n)$ is any $\mathtt{st}$-$\in$-formula (all free variables indicated). Finally Idealization consists of all formulas of the form

$$\forall x_1 \ldots \forall x_n\, \big(\forall^{\mathtt{stfin}} A\, \exists x\, \forall a \in A\, \Phi(a, x, x_1, ..., x_n) \Longleftrightarrow \exists x\, \forall^{\mathtt{st}} a\, \Phi(a, x, x_1, ..., x_n)\big),$$

where $\Phi(a, x, x_1, ..., x_n)$ is any $\in$-formula (all free variables indicated).
Changing the quantifier prefix to $\forall^{\mathtt{st}} x_1 \ldots \forall^{\mathtt{st}} x_n$ in the last displayed formula, we obtain a somewhat weaker schema:

[3] This approach, due to Nelson, allows to view $\mathbb{I}$ rather than $\mathbb{S}$ (or $\mathbb{WF}$, as in **HST**) as the universe of traditional mathematics, see Footnote 2 on page 13.

Enlargement: $\forall^{\text{stfin}} A \, \exists x \, \forall a \in A \, \Phi(a,x) \iff \exists x \, \forall^{\text{st}} a \, \Phi(a,x)$,
where $\Phi(a,x)$ is any $\in$-formula with **standard** parameters.

The main feature and the main problem of **IST** (and the "internal" approach as a whole) is that st-$\in$-definable "parts" of sets are not necessarily sets themselves. For example, ${}^{\sigma}X = X \cap \mathbb{S}$ is not necessarily a set in **IST**, see Exercise 3.1.21. Yet this set-back can be overcome for most practical purposes: the "non-sets" like ${}^{\sigma}X$ can be successfully treated in the same manner as definable proper classes in **ZFC**. The real problem appears when collections of "non-sets" and quantifiers over those are to be considered, as this is the case in some branches of nonstandard analysis. We shall address this issue below.

On the other hand, **IST** has gained a lot of adherents due to its remarkable simplicity and capacity to successfully provide an adequate basis for many branches of nonstandard mathematics. **IST** is arguably the only nonstandard set theory which has been extensively used so far as a working foundation for nonstandard mathematics.

3.1b Bounded set theory

Inner Transfer and Inner Standardization of **IST** are direct counterparts of Transfer and Standardization in **HST**. On the other hand, Idealization of **IST** is somewhat stronger than Saturation: for instance it implies the existence of a set x such that $\mathbb{S} \subseteq x$ (take $a \in x$ as Φ), which is impossible in **HST** — hence, it is not true that the class $\mathbb{I}$ of all internal sets in **HST** interprets **IST**. The apparently unnecessary strength of Idealization led one of the authors in [Kan 91] to the idea to weaken Idealization to a form both reasonable for practical purposes and compatible with the structure of internal sets in **HST**. This plan was realised in *bounded set theory* **BST**.

Definition 3.1.6. **BST** is a theory in st-$\in$-language, containing $\mathbf{ZFC^{st}}$, Inner Transfer, and Inner Standardization (as defined in § 3.1a) together with the two following axioms of Inner Boundedness and Basic Idealization. □

Inner Boundedness: *every set is an element of a standard set.*
(Compare with the definition of internal sets in **HST**.)

Basic Idealization:
$\forall^{\text{stfin}} A \subseteq A_0 \, \exists x \in X \, \forall a \in A \, (x \in \psi(a)) \iff \exists x \in X \, \forall^{\text{st}} a \in A_0 \, (x \in \psi(a))$
— for any standard sets A_0, X and any (internal) map $\psi : A_0 \to \mathscr{P}(X)$.

Basic Idealization looks too special, but in fact it implies much more general idealization schemata, see § 3.2a. Still it is weaker than the full **IST** Idealization, but Inner Boundedness is a good compensation. It is not immediately clear what we gain by Inner Boundedness at the cost of Idealization, yet it is demonstrated below that the step from **IST** to **BST** provides us with the possibility to define external extensions of the internal universe, while further restrictions lead to the Power Set axiom in the relevant external universes.

It follows from Exercise 3.1.3 that $\mathbf{ZFC}^{\mathtt{st}}$ can be replaced by **ZFC** (in the $\in$-language). Thus **BST** can be presented as **ZFC** plus the axioms of Basic Idealization and Inner Boundedness and the schemata of Inner Transfer and Inner Standardization. Surprisingly, the schemata can be replaced by suitable finite sub-schemata in this case, hence **BST** turns out to be a finitely axiomatizable extension of **ZFC**, see § 3.2c.

Similarly to the case of **IST**, one can consider the following weaker version of Basic Idealization:

Basic Enlargement:
$\forall^{\mathtt{stfin}} A \subseteq A_0 \, \exists x \in X \, \forall a \in A \, (x \in \psi(a)) \iff \exists x \in X \, \forall^{\mathtt{st}} a \in A_0 \, (x \in \psi(a))$
— for any standard sets A_0, X and any **standard** map $\psi : A_0 \to \mathscr{P}(X)$.

Exercise 3.1.7. (i) Prove that Inner Boundedness contradicts **IST**.

(ii) Prove that Basic Idealization implies the following:

$$\forall^{\mathtt{stfin}} A \subseteq A_0 \, \exists x \in X \, \forall a \in A \, \Phi(a, x) \iff \exists x \in X \, \forall^{\mathtt{st}} a \in A_0 \, \Phi(a, x).$$

(X, A_0 standard, Φ an $\in$-formula.) *Hint*: let $\psi(a) = \{x \in X : \Phi(a, x)\}$.

(iii) Prove that, in **BST**, for any set X there is a standard U with $X \subseteq U$. (*Hint*. Take $Y \in \mathbb{S}$ with $X \in Y$. Then $X \subseteq U = \bigcup Y \in \mathbb{S}$.) □

As far as the metamathematical side is concerned, we shall prove in Chapter 4 (Theorem 4.1.10) that **BST** fully satisfies the conclusion of Theorem 1.1.14, while **IST** satisfies Theorem 1.1.14 except for the interpretability. In general **BST** and **HST** are somewhat closer to **ZFC** than **IST**.

The ideas behind **IST** and **BST** can be applied not only to **ZFC** as the basic "standard" theory. For any theory T in the $\in$-language, let **IST**[T] be a version of **IST** in which $\mathbf{ZFC}^{\mathtt{st}}$ is replaced by $T^{\mathtt{st}}$ (all sentences $\Phi^{\mathtt{st}}$, $\Phi \in T$), for instance, the theory **IST**[**ZC**] will be of some use. **BST**[**ZC**] can be defined similarly, but this version is somewhat deceptive, see § 3.6b.

3.1c Internal sets interpret BST in the external universe

Temporarily coming back to **HST**, we prove that, in **HST**, the class $\mathbb{I}$ of all internal sets satisfies the axioms of **BST**. Recall that $\Phi^{\mathtt{int}}$ denotes the relativization of an $\in$-formula Φ to the class $\mathbb{I} = \{x : \mathtt{int}\, x\}$ of all internal sets in **HST**, see § 1.1c.

Theorem 3.1.8. $\mathbf{I} = \langle \mathbb{I}; \in, \mathtt{st} \rangle$ *is an interpretation of* **BST** *in* **HST** : *if* Φ *is an axiom of* **BST** *then* $\Phi^{\mathtt{int}}$ *is a theorem of* **HST**.

Proof. *We argue in* **HST**. Verification of all **BST** axioms in $\mathbb{I}$, except for Basic Idealization, is left as an easy **exercise**. To approach Basic Idealization, we first note the following.

Remark 3.1.9. *Finite* in internal theories corresponds to $*$-*finite* in **HST** (see § 1.2d) while *standard finite* in **BST** corresponds to *finite* in **HST**. (See also § 3.1e.) More formally, suppose in **HST** that $X \in \mathbb{I}$. Then

(i) X is finite in the sense of $\mathbb{I}$ iff X is $*$-finite (or, which is the same, hyperfinite) in the **HST** universe $\mathbb{H}$ – see Definition 1.2.13;

(ii) if moreover X is standard then X is finite in $\mathbb{I}$ iff X is finite in $\mathbb{H}$, and each side implies $X \subseteq \mathbb{S}$ – by Lemma 1.2.16. □

Coming back to the theorem, we observe that the formula of Basic Idealization begins with the quantifier $\forall^{\mathtt{stfin}} A$, *i. e.*, "for every standard and finite". By definition (§ 1.1c), $(\mathtt{st}\, x)^{\mathtt{int}}$ is $\mathtt{st}\, x$ while $(x \text{ finite})^{\mathtt{int}}$ is "x is internal and $*$-finite" by (i) of Remark 3.1.9. Moreover, by (ii) of Remark 3.1.9, given a standard set A_0, a set $A \subseteq A_0$ is standard and $*$-finite iff it is finite in the **HST** sense and $A \subseteq \mathbb{S}$. This allows us to present (Basic Idealization)$^{\mathtt{int}}$ as

$$\forall^{\mathtt{fin}} A \subseteq A_0 \cap \mathbb{S} \left(\bigcap_{a \in A} \psi(a) \neq \varnothing \right) \iff \bigcap_{a \in A \cap \mathbb{S}} \psi(a) \neq \varnothing \right),$$

where A_0, X are standard sets and $\psi : A_0 \to X$ is an internal map. It remains to apply the result of Exercise 1.3.7. □ (*Theorem 3.1.8*)

Theorem 3.1.8 is the easier direction, related to the necessity in the next fundamental result, showing that **BST** is essentially the theory of the internal universe in **HST**, the "internal part" of **HST** in brief. The sufficiency in Theorem 3.1.10 will be established in Section 5: see 7° of Corollary 5.1.5.

Theorem 3.1.10. *The theory* **HST** *is a conservative extension of* **BST**, *in the sense that for a* $\mathtt{st}$-$\in$-*sentence* Φ *to be a theorem of* **BST** *it is necessary and sufficient that* $\Phi^{\mathtt{int}}$ *is a theorem of* **HST**.

According to Theorem 3.1.8 any theorem of **BST** is true for internal sets in **HST**. In practice this is not the most profitable method to study internal sets in **HST** because the convenient "envelope" of external sets in **HST** offers more opportunities to work with internal sets. On the other hand, the results obtained in Section 3.2 (for **BST**) are of great foundational importance as they will allow us to construct an **HST** "envelope" over any **BST** universe in Chapter 5.

3.1d Basic internal set theory

Coming back to internal theories, we introduce *basic internal set theory* **BIST**, a subtheory of both **BST** and **IST**, containing the axioms of $\mathbf{ZFC}^{\mathtt{st}}$, Inner Transfer, Inner Standardization, Basic Idealization (but not Inner Boundedness of **BST**). Such a reduced theory can be reasonably characterized as representing *elementary methods* of "internal" nonstandard set theories.

Let us present several rather simple theorems of **BIST**, leaving the special issue of finiteness and natural numbers to § 3.1e.

Lemma 3.1.11 (BIST: a very useful form of Transfer!). *If $\Phi(x)$ is an $\in$-formula with standard parameters then $\exists\, x\, \Phi(x) \Longrightarrow \exists^{\mathtt{st}} x\, \Phi(x)$.*

Proof. We have $(\exists\, x\, \Phi(x))^{\mathtt{st}}$ by Inner Transfer, so that there is a standard x satisfying $\Phi(x)^{\mathtt{st}}$. Then $\Phi(x)$ holds by Inner Transfer. □

Exercise 3.1.12 (BIST). Prove that if f is a standard function and $w \in \mathtt{dom}\, f$ is standard then the value $f(w)$ is also standard. (*Hint.* Apply Lemma 3.1.11 for the formula $x = f(w)$.) □

Definition 3.1.13 (BIST: relative standardness). For any set w, $\mathbb{S}[w]$ is the class of all sets of the form $f(w)$, where f is a standard function and $w \in \mathtt{dom}\, f$. Sets in $\mathbb{S}[w]$ will be called *standard relative to w*, or *w-standard.*

Generally, $\mathbb{S}[c_1, ..., c_n] = \{f(c_1, ..., c_n) : f \in \mathbb{S} \wedge \langle c_1, ..., c_n\rangle \in \mathtt{dom}\, f\}$.

Exercise. Prove that if a set w does not belong to a standard set then $\mathbb{S}[w] = \varnothing$. (*Hint.* If $f \in \mathbb{S}$ then $\mathtt{dom}\, f \in \mathbb{S}$ by Lemma 3.1.14.) □

Classes $\mathbb{S}[w]$ will be extensively studied below in the context of **HST**.

Lemma 3.1.14 (BIST). *Let $\Phi(x)$ be an $\in$-formula with standard parameters and a parameter $w \in W$, where W is standard. If there is a unique set x with $\Phi(x)$ then this x belongs to $\mathbb{S}[w]$. If w is standard then so is x.*

Proof. Let Φ be $\Phi(x, w)$. Applying Choice of **ZFC** (we refer to Exercise 3.1.3) and Inner Transfer, we find a standard function f defined on W so that $\exists\, y\, \Phi(y, w') \Longrightarrow \Phi(f(w'), w')$ holds for all $w' \in W$, in particular, $x = f(w)$. To prove the last claim employ the result of 3.1.12. □

The stronger theories **BST** and **IST** will allow us to prove the lemma for all $\mathtt{st}$-$\in$-formulas (see Uniqueness in Theorems 3.2.11 and 3.4.16); it is not known whether **BIST** is strong enough to prove such a generalization.

Note that collections of the form ${}^{\sigma}X = X \cap \mathbb{S}$ are not necessarily sets in "internal" nonstandard theories: Separation can be used only in $\in$-language ! Thus, occurrences of objects like ${}^{\sigma}X$ in "internal" reasoning are nothing but shortcuts for the corresponding longer legitimate expressions.

The next lemma shows the strength of different versions of Idealization. Recall Blanket agreement 3.1.1 regarding the notion of finiteness in **BST**.

Lemma 3.1.15 (BIST). *For any set A there is a finite set $C \subseteq A$ such that ${}^{\sigma}A \subseteq C$. If, in addition, the (unbounded) schema of Idealization* (as in **IST**) *holds then there is a finite set C with $\mathbb{S} \subseteq C$.*

Proof. The following equivalence is a case of Basic Idealization:

$$\forall^{\mathtt{stfin}} A' \subseteq A\, \exists^{\mathtt{fin}} C \subseteq A\, (A' \subseteq C) \Longleftrightarrow \exists^{\mathtt{fin}} C \subseteq A\, ({}^{\sigma}A \subseteq C)\,.$$

(Take $\psi(a) = \{C \subseteq A : a \in C\}$.) The left-hand side is true (take $C = A'$), therefore, the right-hand side holds as well. The proof of the second claim is a minor modification. □

The following lemma is a useful application of Inner Standardization.

Lemma 3.1.16 (BIST). *Let $\Phi(x,y)$ be a* st-$\in$-*formula with any parameters. Then, for any standard sets A, B, we have:*

$$\forall^{\mathtt{st}} a \in A\, \exists^{\mathtt{st}} b \in B\, \Phi(a,b) \iff \exists^{\mathtt{st}} f \in B^A\, \forall^{\mathtt{st}} a \in A\, \Phi(a, f(a))\,. \qquad (*)$$

Proof. The set $F = B^A$ of all functions $f : A \to B$ is standard by 3.1.14. The direction $\Longleftarrow$ is easy (Exercise 3.1.12). To prove $\Longrightarrow$ we apply Inner Standardization to obtain a standard set $P = {}^{\mathsf{S}}\{\langle a,b\rangle \in A \times B : \Phi(a,b)\} \subseteq A \times B$ such that $\langle a,b\rangle \in P \iff \Phi(a,b)$ for all standard $a \in A$, $b \in B$. The left-hand side of $(*)$ takes the form: $\forall^{\mathtt{st}} a \in A\, \exists^{\mathtt{st}} b \in B\, (\langle a,b\rangle \in P)$. This implies $\forall\, a \in A\, \exists\, b \in B\, (\langle a,b\rangle \in P)$, by Inner Transfer. Now Choice and Inner Transfer prove the existence of a standard function $f : A \to B$ such that $\langle a, f(a)\rangle \in P$ for any $a \in A$, leading to the right–hand side of $(*)$. □

3.1e Standard natural numbers and standard finite sets

It has been observed (Remark 3.1.9) that finite sets in **HST** correspond to standard finite sets in internal theories. Let us consider the latter notion in more detail.

Definition 3.1.17 (For "internal" theories like **BIST**, **IST**, **BST**).
According to Blanket agreement 3.1.1, $\mathbb{N}$ denotes the set of all $\mathbb{I}$-natural numbers while ${}^\sigma\mathbb{N} = \mathbb{N} \cap \mathbb{S}$ is the collection of all *standard* natural numbers. A *finite* set is any set equinumerous to some $n = \{0, 1, ..., n-1\} \in \mathbb{N}$. A *standard finite set* is a set which is both standard and finite. □

Thus we can expect that the pair ${}^\sigma\mathbb{N} \subseteq \mathbb{N}$ has the same basic properties in internal theories as the pair $\mathbb{N} \subseteq {}^*\mathbb{N}$ in **HST**. This turns out to be the case, moreover some important results do not use Basic Idealization. The next lemma presents several startling consequences of Inner Standardization, in particular it justifies induction in ${}^\sigma\mathbb{N}$ for any st-$\in$-formula.

Lemma 3.1.18 (BIST without Basic Idealization**).** *Suppose that $\Psi(x)$ is a* st-$\in$-*formula with any parameters. Then*

(i) *if there are numbers $n \in {}^\sigma\mathbb{N}$ satisfying $\Psi(n)$ then there is a least one among such numbers n* (that is, ${}^\sigma\mathbb{N}$ is "externally" well-ordered);

(ii) (Induction) *if $\Psi(0)$ holds and $\Psi(n) \Longrightarrow \Psi(n+1)$ for all $n \in {}^\sigma\mathbb{N}$ then $\Psi(n)$ holds for all $n \in {}^\sigma\mathbb{N}$;*

(iii) *if X is a finite set of a standard finite number of elements then the collection $\{x \in X : \Psi(x)\}$ is a set;*

(iv) *${}^\sigma\mathbb{N}$ is an initial segment of $\mathbb{N}$;*

(v) *if $X \subseteq \mathbb{N}$ is a finite set of a standard finite number of elements then $X \cap {}^\sigma\mathbb{N}$ is not cofinal in ${}^\sigma\mathbb{N}$ and $X \smallsetminus {}^\sigma\mathbb{N}$ is not coinitial in $\mathbb{N} \smallsetminus {}^\sigma\mathbb{N}$.*

Proof. (i) By Inner Standardization, put $X = {}^{\mathsf{S}}\{n \in \mathbb{N} : \Phi(n)\} \subseteq \mathbb{N}$ (a standard set such that $n \in X \iff \Phi(n)$ for any standard n). Then X is nonempty, hence, by **ZFC** in the universe, it contains a least element, which is standard by Lemma 3.1.14 (note that the latter depends on Inner Transfer).

To prove (ii) apply (i) for $\neg \Psi$.

(iv) Apply Induction to the formula "all numbers $x < n$ are standard".

(iii), (v) Induction on the number of elements of X. □

Exercise 3.1.19. Why will (i) of the lemma be wrong if we replace ${}^{\sigma}\mathbb{N}$ by $\mathbb{N}$? Why are (i), (ii), (iii) trivial for $\in$-formulas Ψ? □

It follows from Inner Transfer that a set is standard finite iff it is equinumerous to some $n = \{0, 1, ..., n-1\} \in {}^{\sigma}\mathbb{N}$. There is another characterization (compare to Lemma 1.2.16!):

Lemma 3.1.20. (i) (**BIST** without Basic Idealization) *Any standard finite set contains only standard elements;*

(ii) (**IST** or **BST**) *conversely, any set $X \subseteq \mathbb{S}$ is standard and finite.*

Proof. (i) Let X be standard finite. By Inner Transfer, there is a standard map $f : n \overset{\text{onto}}{\longrightarrow} X$ for some $n \in {}^{\sigma}\mathbb{N}$. Any $k < n$ is standard by (iv) of Lemma 3.1.18, hence, $f(k) \in \mathbb{S}$ (Exercise 3.1.12), as required.

(ii) Assume that $X \subseteq \mathbb{S}$. Consider first the case of **IST**. The equivalence

$$\forall^{\mathtt{stfin}} A\, \exists x \in X\, \forall a \in A\, (x \neq a) \iff \exists x \in X\, \forall^{\mathtt{st}} a\, (x \neq a) \qquad (\dagger)$$

is an example of Idealization, where the right-hand side is false as $X \subseteq \mathbb{S}$. Let a standard finite A witness that the left-hand side is false, so that $X \subseteq A$. By Inner Standardization there is a standard set $Y = {}^{\mathsf{S}}X \subseteq A$ such that X and Y contain the same standard elements, which means that $X = Y$ because $A \subseteq \mathbb{S}$ by the first part of the proof. Thus X is standard and finite.

Now consider the case of **BST**. By Inner Boundedness, there is a standard set S such that $X \subseteq S$. It suffices to replace (†) by

$$\forall^{\mathtt{stfin}} A \subseteq S\, \exists x \in X\, \forall a \in A\, (x \neq a) \iff \exists x \in X\, \forall^{\mathtt{st}} a \in S\, (x \neq a)$$

and apply Basic Idealization with $\psi(a) = \{x \in X : x \neq a\}$. □

Exercise 3.1.21. (1) Prove 3.1.20(ii) in **BIST** plus the following "axiom": if f is a function, ${}^{\sigma}\mathbb{N} \subseteq \mathtt{dom}\, f$, and $f(n) \in \mathbb{S}$ for any standard $n \in \mathbb{N}$ then there is a standard set X such that $f(n) \in X$ for any standard n.

(2) (**BIST**) Prove that ${}^{\sigma}\mathbb{N}$ is a proper initial segment of $\mathbb{N}$ and is <u>not</u> a set. (*Hint.* To prove the properness apply Lemma 3.1.15. Now if ${}^{\sigma}\mathbb{N}$ were a set then it would contain a maximal element n; consider $n+1$.)

(3) (**BST** or **IST**) Prove, using Lemma 3.1.20, that for any (internal) X, if ${}^{\sigma}X$ is a set then it is standard and finite. Prove that if X is standard then ${}^{\sigma}X$ is a set iff X is finite (and in this case $X = {}^{\sigma}X$). □

3.1f Remarks on Basic Idealization and Saturation

Remark 3.1.22. Note that the proof of (i) of Lemma 3.1.20 does not use Basic Idealization. It follows that the implications $\Longleftarrow$ in Idealization–like axioms are consequences of Inner Standardization. □

The following result is a Saturation-like corollary of Basic Idealization.

Theorem 3.1.23 (Saturation in BIST). *Let ψ be a function defined on a set $A_0 \in \mathbb{S}$, such that $\psi(a) \neq \varnothing$ for all $a \in A_0 \cap \mathbb{S}$, and the family $\mathscr{X} = \{\psi(a) : a \in A_0 \cap \mathbb{S}\}$ is closed under intersection of two sets. Then the intersection $\bigcap \mathscr{X} = \bigcap_{a \in A_0 \cap \mathbb{S}} \psi(a)$ is nonempty.*

Here $A_0 \cap \mathbb{S}$ and $\mathscr{X}$ are not necessarily sets, but ψ is a set (internal) !

Proof. It suffices to check that the left-hand side of Basic Idealization holds in this case. For this, it suffices to prove that if $A \subseteq A_0 \cap \mathbb{S}$ is standard finite then there is $a \in A_0 \cap \mathbb{S}$ with $\psi(a) = \bigcap_{b \in A} \psi(b)$. This is proved by induction on the (standard finite) number $n(A) \in {}^\sigma\mathbb{N}$ of elements of A, based on Lemma 3.1.18(ii). If $A = \{b\}$ is a one-element set put $a = b$. The induction step utilizes the fact that $\mathscr{X}$ is closed under pairwise intersections. □

Exercise 3.1.24 (BIST). Prove arguing as in 3.1.18(i) the following generalization: for any st-$\in$-formula $\Psi(x)$ with any parameters, if there are standard ordinals $\alpha \in \mathtt{Ord}$ satisfying $\Psi(\alpha)$ then there is a least one among such ordinals α. (In other words $\mathtt{Ord} \cap \mathbb{S}$ is "externally" well-ordered.) □

The assumption ${}^\sigma\mathbb{N} = \mathbb{N}$ in the following result has little to do with nonstandard analysis in its traditional forms (and is incompatible with weakest forms of Idealization), yet the result is interesting because it shows unexpected potentialities of Standardization.

Theorem 3.1.25 (BST without Basic Idealization). *Assume that $\mathbb{S} \subsetneqq \mathbb{I}$ but ${}^\sigma\mathbb{N} = \mathbb{N}$. Then there exists a measurable cardinal $\kappa > \omega$ such that $\kappa \not\subseteq \mathbb{S}$.*

Proof. Let $x \notin \mathbb{S}$. Let, by Inner Boundedness, X be a standard set of least possible cardinality containing x. We can assume that $X = \kappa$ is a (standard) cardinal – then $\kappa > \omega$ because ${}^\sigma\mathbb{N} = \mathbb{N}$. (As usual, ω and $\mathbb{N}$ is one and the same.) We apply Inner Standardization: the standard set $U = {}^{\mathrm{S}}\{X \subseteq \kappa : x \in X\}$ is an ultrafilter on X. (Indeed, otherwise, by Inner Transfer, the opposite is witnessed by standard counterexamples, e.g., there exist standard sets $Y, Z \subseteq X$ with $Y, Z \in U$ but $Y \cap Z \notin U$ – clearly nonsense.) Prove that U is κ-complete. By Inner Transfer, it suffices to prove this for standard sequences. Thus, let $\gamma < \kappa$ be a standard cardinal, and $f : \gamma \to U$ a standard map; prove that $X = \bigcap_{\xi < \gamma} f(\xi) \in U$. By definition we have to show that $x \in X$, i.e., $x \in f(\xi)$ for all $\xi < \gamma$. Suppose that $x \notin f(\xi)$. But $\xi < \gamma$, hence, $\xi \in \mathbb{S}$ by the choice of κ. It follows that $f(\xi) \in U$ is also standard by Lemma 3.1.14, contradiction. □

3.2 Development of bounded set theory

This section presents several fundamental theorems of bounded set theory **BST** (as defined in § 3.1b) which fully exploit Inner Boundedness and do not seem to be available in **BIST** alone. The list of theorems include a very important theorem of Reduction to $\Sigma_2^{\mathtt{st}}$ (Theorem 3.2.3), saying that, in **BST**, any st-∈-formula can be transformed to a certain special form ($\Sigma_2^{\mathtt{st}}$ form), finite axiomatizability of **BST** over **ZFC**, and the theorems of Collection, Extension, Uniqueness, Dependent Choice. Some of them also have foundational applications to **HST**.

Blanket agreement 3.2.1. We argue in **BST** in Section 3.2. □

3.2a Half-bounded forms of Idealization

We begin with a technical result which shows that Basic Idealization gains additional strength from the interaction with the rest of the **BST** axioms.

Recall that the (unbounded) axiom of Idealization of **IST** is incompatible with **BST**. However, bounding any of the two active variables by a standard set, we obtain a compatible form, moreover, a consequence of Basic Idealization (which bounds both of them).

Local Idealization: for any standard set X,

$$\forall^{\mathtt{stfin}} A\, \exists x \in X\, \forall a \in A\, \Phi(a,x) \iff \exists x \in X\, \forall^{\mathtt{st}} a\, \Phi(a,x).$$

Inner Saturation: for any standard set A_0,

$$\forall^{\mathtt{stfin}} A \subseteq A_0\, \exists x\, \forall a \in A\, \Phi(a,x) \iff \exists x\, \forall^{\mathtt{st}} a \in A_0\, \Phi(a,x).$$

(Compare this with the equivalence in Exercise 1.3.7.)

In both cases, $\Phi(a,x)$ is any ∈-formula with arbitrary parameters.

Lemma 3.2.2 (BST). *Local Idealization and Inner Saturation hold.*

Proof. Assume the left–hand side of Local Idealization and prove the right–hand side. (See Remark 3.1.22 in matters of the other direction.) The case of many parameters in Φ can be reduced to one parameter, thus, let Φ contain a single parameter p_0, so that $\Phi(a,x)$ is $\Phi(a,x,p_0)$. Let, by Inner Boundedness, P be a standard set containing p_0. Put $Z_a = \{\langle p,x\rangle \in P \times X : \Phi(a,x,p)\}$ for any a. By Collection (see Exercise 3.1.3) and Inner Transfer, there is a standard set A_0 such that $\forall a'\, \exists a \in A_0\, (Z_a = Z_{a'})$. Basic Idealization yields $x \in X$ satisfying $\Phi(a,x,p_0)$ for all standard $a \in A_0$. It remains to prove $\Phi(a',x,p_0)$ for any standard a', not necessarily in A_0. By Inner Transfer and the choice of A_0 we have $Z_a = Z_{a'}$ for some standard $a \in A_0$. Then

$$\Phi(a,x,p_0) \implies \langle p_0,x\rangle \in Z_a = Z_{a'} \implies \Phi(a',x,p_0)\,, \quad \text{as required}.$$

Now assume the left–hand side of Inner Saturation and prove the right–hand side. Let p_0 and P be as above, and $F = \mathscr{P}_{\mathtt{fin}}(A_0)$ (the set of all finite subsets of A_0); F is standard (Exercise 3.1.12). By the axiom of **ZFC** Collection, there is a set X such that, for all $p \in P$ and $A \in F$,

$$\exists x\, \forall a \in A\, \Phi(a,x,p) \implies \exists x \in X\, \forall a \in A\, \Phi(a,x,p)\,.$$

One can choose a standard set X of this kind by Transfer. It remains to apply Basic Idealization to X and the function $\psi(a) = \{x \in X : \Phi(a,x,p_0)\}$. □

Lemma 3.2.2, especially in the part related to Local Idealization, will be of great importance below, because several basic theorems of **BST** need really Local Idealization rather than the weaker Basic Idealization.

3.2b Reduction to two "external" quantifiers

Let $\Sigma_2^{\mathtt{st}}$ indicate the class of all $\mathtt{st}$-$\in$-formulas of the form $\exists^{\mathtt{st}} a\, \forall^{\mathtt{st}} b\, \varphi$, where φ is an $\in$-formula. The following key result is a syntactical, or "metamathematical" theorem; it is assumed that $x_1, ..., x_m$ is the full list of free variables of Φ and that Φ does not contain sets as parameters.

Theorem 3.2.3 (*Reduction to* $\Sigma_2^{\mathtt{st}}$). *Suppose that* $\Phi(x_1, ..., x_m)$ *is a* $\mathtt{st}$-$\in$-*formula. Then there exist* $\in$-*formulas* $\varphi(x_1, ..., x_m, a, b)$ *and* $\psi(x_1, ..., x_m)$ *such that the following are theorems of* **BST** :

(i) $\forall x_1 ... \forall x_m \,(\Phi(x_1, ..., x_m) \iff \exists^{\mathtt{st}} a\, \forall^{\mathtt{st}} b\, \varphi(x_1, ..., x_m, a, b))\,;$

(ii) $\forall^{\mathtt{st}} x_1 ... \forall^{\mathtt{st}} x_m \,(\Phi(x_1, ..., x_m) \iff \psi(x_1, ..., x_m) \iff \psi^{\mathtt{st}}(x_1, ..., x_m))\,.$

Thus every $\mathtt{st}$-$\in$-formula Φ is provably equivalent in **BST** to a $\Sigma_2^{\mathtt{st}}$ formula with the same list of free variables, and even to an $\in$-formula ψ assuming that only standard values of the arguments are of interest.

The proof employs a generalization of *Nelson's algorithm* of transformation of $\mathtt{st}$-$\in$-formulas to $\in$-form. The algorithm (see § 3.6b) works under the assumption that all quantifiers in a given formula Φ are restricted by standard sets and there is no occurrence of $\mathtt{st}$ except for those in quantifiers $\exists^{\mathtt{st}}$, $\forall^{\mathtt{st}}$ — then the step for $\neg$ is rather easy. The method we use preassumes only that any (internal) set in the nonstandard set universe belongs to a standard set, which is implied by Inner Boundedness.

Proof. First of all if φ satisfieses (i) then by Inner Transfer the formula $\exists a\, \forall b\, \varphi(x_1, ..., x_m, a, b)$ can be taken as ψ to witness (ii).

We prove (i) by induction on the length of Φ. Let us write $\mathbf{x}$ instead of $x_1, ..., x_m$. As any occurrence of $\mathtt{st}\, z$ can be replaced by $\exists^{\mathtt{st}} w\, (z = w)$, it suffices to carry out induction steps for $\wedge$, $\exists^{\mathtt{st}}$, $\neg$, $\exists$. The first two of them are quite routine (for instance, we have to transform the conjunction $\Phi \wedge \Psi$ of two $\Sigma_2^{\mathtt{st}}$ formulas to $\Sigma_2^{\mathtt{st}}$ form), and are left as an **exercise** for the reader.

Induction step for $\neg$ [4]. We search for a $\Sigma_2^{\mathtt{st}}$ formula equivalent to the formula $\forall^{\mathtt{st}} a\, \exists^{\mathtt{st}} b\, \varphi(\mathbf{x}, a, b)$ (where φ is an $\in$-formula), taken as $\Phi(\mathbf{x})$. For any set U, let us define $\mathtt{mon}\, U = \bigcap\{u \in U : \mathtt{st}\, u\}$ (the *monad* of U). Let $\mathtt{Ult}\, U$ be a $\in$-formula saying U *is an ultrafilter*. The following equivalence

$$\Phi(\mathbf{x}) \iff \exists^{\mathtt{st}} U \left(\mathtt{Ult}\, U \wedge \mathbf{x} \in \mathtt{mon}\, U \wedge \forall a\, \exists b\, \exists u \in U\, \forall \mathbf{y} \in u\, \varphi(\mathbf{y}, a, b) \right)$$

suffices to complete the step since the rigth–hand side is $\Sigma_2^{\mathtt{st}}$ (**exercise**: show this !). To prove the equivalence, note that, by Inner Boundedness and Inner Standardization, for every $\mathbf{x}$ there is a standard ultrafilter U with $\mathbf{x} \in \mathtt{mon}\, U$. (Let, by Inner Boundedness, X be a standard set containing $\mathbf{x}$. Standard elements of U are all standard $u \subseteq X$ with $\mathbf{x} \in u$.) It remains to verify

$$\Phi(\mathbf{x}) \iff \forall a\, \exists b\, \exists u \in U\, \forall \mathbf{y} \in u\, \varphi(\mathbf{y}, a, b)$$

for every U of this type. By Inner Transfer, we have

$$\forall a\, \exists b\, \exists u \in U\, \forall \mathbf{y} \in u\, \varphi(\mathbf{y}, a, b) \iff \forall^{\mathtt{st}} a\, \exists^{\mathtt{st}} b\, \exists^{\mathtt{st}} u \in U\, \forall \mathbf{y} \in u\, \varphi(\mathbf{y}, a, b)\,,$$

thus it suffices to check that, for all standard a and b,

$$\varphi(\mathbf{x}, a, b) \iff \exists^{\mathtt{st}} u \in U\, \forall \mathbf{y} \in u\, \varphi(\mathbf{y}, a, b)\,.$$

Let a, b be standard and $u = u_{ab} = \{\mathbf{y} \in X : \varphi(\mathbf{y}, a, b)\}$, where $X = \bigcup U$. Both X and u are standard by Inner Transfer. If $u \in U$ then both sides of the last displayed equivalence are true, otherwise both of them are false.

Induction step for $\exists$. Let $\varphi(\mathbf{x}, w, a, b)$ be an $\in$-formula. We need a $\Sigma_2^{\mathtt{st}}$ formula equivalent to the formula $\exists w\, \exists^{\mathtt{st}} a\, \forall^{\mathtt{st}} b\, \varphi(\mathbf{x}, w, a, b)$ taken as $\Phi(\mathbf{x})$. Applying Inner Boundedness and Local Idealization (Lemma 3.2.2) we get

$$\begin{aligned} \Phi(\mathbf{x}) &\iff \exists^{\mathtt{st}} a\, \exists^{\mathtt{st}} W\, \exists w \in W\, \forall^{\mathtt{st}} b\, \varphi(\mathbf{x}, w, a, b) \iff \\ &\iff \exists^{\mathtt{st}} a\, \exists^{\mathtt{st}} W\, \forall^{\mathtt{stfin}} B\, \exists w \in W\, \forall b \in B\, \varphi(\mathbf{x}, w, a, b)\,. \end{aligned}$$

□

3.2c Finite axiomatizability of BST and other corollaries

The results of this subsection, including Theorem 3.2.6, are mainly due to Andreev and Hrbaček [AnH 04]. Most of them, grouped in the following Exercise, are variations on the themes of §§ 3.2a, 3.2b.

Exercise 3.2.4. (1) (**BST**) Suppose that $p \in X$, where X is standard. Prove that there is a unique standard ultrafilter U_p on X such that $p \in \mathtt{mon}\, U_p = \bigcap_{Y \in U_p \cap \mathbb{S}} Y$. Prove that $\Phi(p) \iff \Phi(q)$ whenever $p, q \in X$ satisfy $U_p = U_q$ and $\Phi(y)$ is a $\mathtt{st}$-$\in$-formula with standard parameters. [5]

[4] This version of the argument for the step related to $\neg$ is due to P. Andreev.

[5] A remarkable corollary: in **BST** any property of a set $p \in X$ expressible by a $\mathtt{st}$-$\in$-formula with standard parameters depends only on U_p, a standard set!

(2) Prove that any st-$\in$-formula $\Phi(x)$ is equivalent in **BST** to a formula of the form $\exists^{st} U\,(\mathtt{Ult}\,U \wedge x \in \mathtt{mon}\,U \wedge \psi(U))$, where ψ is an $\in$-formula.

(3) Prove the following "class form Standardization" in **BST** : for any st-$\in$-formula $\Phi(p, x)$ with an arbitrary parameter p, there exists an $\in$-formula $\varphi(x)$ with standard parameters such that $\forall^{st} x\,(\Phi(p, x) \Longleftrightarrow \varphi(x))$.

(4) Prove that in **BST** a class X is *standard*, *i.e.* definable by an $\in$-formula with standard parameters, iff $X \cap S$ is standard for any standard set S.

Hints. (1) Put $U_p = {}^{\mathrm{S}}\{Y \subseteq X : p \in Y\}$, where, for any collection $\mathscr{X} \subseteq \mathscr{P}(X)$, ${}^{\mathrm{S}}\mathscr{X}$ is the unique standard set $U \subseteq \mathscr{P}(X)$ such that $U \cap \mathbb{S} = \mathscr{X} \cap \mathbb{S}$. To prove the second part assume, by Theorem 3.2.3, that Φ is $\exists^{st} a\, \forall^{st} b\, \varphi(a, b, x_1, ..., x_n, y)$, where φ is an $\in$-formula and $x_1, ..., x_n$ are standard. Then $\Phi(p)$ is obviously equivalent to $\exists^{st} a\, \forall^{st} b\,(X_{ab} \in U_p)$, where $X_{ab} = \{p \in P : \varphi(a, b, x_1, ..., x_n, p)\}$.)

(2) It follows from (1) that the formula $\psi'(U)$ saying "U is an ultrafilter over a set of the form $X = X_1 \times \ldots \times X_n$, and $\Phi(x)$ holds for all $x = \langle x_1, ..., x_n\rangle \in \mathtt{mon}\,U$" satisfies (2) except that it is not an $\in$-formula. Apply Theorem 3.2.3(ii) to obtain an $\in$-formula ψ equivalent to ψ' for standard U.

(3) Let $p \in X$, X standard. Then $\Phi(p, x) \Longleftrightarrow \forall q \in U_p\, \Phi(q, x)$ holds for any standard x by (1). Use 3.2.3(ii) to convert the right-hand side to $\in$-form.

(4) By (3) there is an $\in$-formula $\sigma(S)$ with standard parameters such that $\sigma(S) \Longleftrightarrow S \subseteq X$ for any standard S. Prove that if $X \cap S$ is standard for any standard X then the formula $\exists\, S\,(x \in S)$ defines X. □

Problem 3.2.5. Let, in **BST**, X be a class (defined by any st-$\in$-formula with arbitrary parameters) such that $X \cap x$ is internal for any (internal) set x. Is then X necessarily definable by an $\in$-formula?

The positive answer is known for classes X defined by a st-$\in$-formula with standard parameters: this case is a good **exercise** for the reader. □

Recall that $\mathbf{ZFC}^{st}$ can be replaced by simply **ZFC** (in the $\in$-language) in the definition of **BST** in the presence of Inner Transfer. Somewhat surprisingly, the schemata of Inner Transfer and Inner Standardization are then reducible to appropriate finite fragments. Consider the following axioms:

(I) $\forall^{st}\xi \in \mathtt{Ord}\,\forall^{st}\varphi$ (if φ is a closed $\in$-formula [6] with standard sets in $\mathbf{V}_\xi$ as parameters then $(\mathbf{V}_\xi \models \varphi) \Longleftrightarrow (\mathbf{V}_\xi \models \varphi)^{st}$).

(II) $\forall X\, \exists^{st} Y\, \forall^{st} x (x \in X \Longleftrightarrow x \in Y)$ – Inner Standardization for internal sets.

The axiom (I) is obviously Inner Transfer for the $\in$-formula $\mathbf{V}_\xi \models \varphi$ (with ξ, φ as free variables or parameters). It follows that (I) and (II) are theorems of **BST**. Conversely, let **BST**$'$ be the theory **ZFC** (in the $\in$-language) plus the axioms (I), (II), Inner Boundedness, and Basic Idealization.

[6] Here a finite sequence of symbols satisfying certain conditions, see §1.5e for further explanations, also regarding the model theoretic notion of truth $\models$.

Theorem 3.2.6. BST′ *implies* **BST**.

Proof. Inner Transfer is the key issue. Consider a closed $\in$-formula φ of the form $\mathsf{Q}_1 y_1\, \mathsf{Q}_2 y_2\, \dots\, \mathsf{Q}_n y_n\, \psi(y_1, \dots, y_n)$, where Q_k are quantifiers $\exists$ and $\forall$ while ψ is a bounded $\in$-formula with standard parameters. Prove that

$(*)\ \varphi \Longleftrightarrow \varphi^{\mathtt{st}}$.

We argue by induction on n. For $n = 0$ the result follows from (I): indeed take as ξ any standard ordinal ξ such that $\mathbf{V}_\xi$ contains all parameters in φ, and use the obvious absoluteness of bounded formulas under the relativization to any transitive set. To carry out the step, note that, by **ZFC**, there exists a sequence $Y_1 \subseteq \dots \subseteq Y_n$ of sets $Y_k = \mathbf{V}_{\xi_k}$, $\xi \in \mathtt{Ord}$, such that for all $k \geq 0$ and $y_1 \in Y_1, \dots, y_{k-1} \in Y_{k-1}$ the following holds:

$(\dagger)\ \mathsf{Q}_k y_k\, \dots\, \mathsf{Q}_n y_n\, \varphi(y_1, \dots, y_{k-1}, y_k, \dots, y_n) \Longleftrightarrow$

$\Longleftrightarrow \mathsf{Q}_k y_k \in Y_k\, \mathsf{Q}_{k+1} y_{k+1}\, \dots\, \mathsf{Q}_n y_n\, \varphi(y_1, \dots, y_{k-1}, y_k, \dots, y_n) \Longleftrightarrow$

$\Longleftrightarrow \mathsf{Q}_k y_k \in Y_k\, \mathsf{Q}_{k+1} y_{k+1} \in Y_{k+1}\, \dots\, \mathsf{Q}_n y_n \in Y_n\, \varphi(y_1, \dots, y_{k-1}, y_k, \dots, y_n)$.

We can w. l. o. g. assume, by Inner Boundedness, that the sets Y_j are standard.

Let, for the sake of definiteness, Q_1 be $\exists$. Suppose that φ is true. Then

$$\exists y_1 \in Y_1\, \mathsf{Q}_2 y_2 \in Y_2\, \dots\, \mathsf{Q}_n y_n \in Y_n\, \varphi(y_1, y_2, \dots, y_n) \qquad (\ddagger)$$

also holds by $(\dagger)$, hence, $(\ddagger)^{\mathtt{st}}$ holds by $(*)$ for bounded formulas. Thus, there is $y_1 \in Y_1 \cap \mathbb{S}$ such that

$$(\mathsf{Q}_2 y_2 \in Y_2\, \dots\, \mathsf{Q}_n y_n \in Y_n\, \varphi(y_1, y_2, \dots, y_n))^{\mathtt{st}}.$$

We can drop $^{\mathtt{st}}$ by $(*)$ for bounded formulas. This yields

$$\mathsf{Q}_2 y_2\, \dots\, \mathsf{Q}_n y_n\, \varphi(y_1, y_2, \dots, y_n)$$

by $(\dagger)$. Here $^{\mathtt{st}}$ can be added by the inductive hypothesis, therefore, we obtain $\varphi^{\mathtt{st}}$ since y_1 is standard. [7]

This ends the proof of $(*)$ and Inner Transfer. To see that in **BST** without Inner Standardization the latter follows from (II) [8], first check that all applications of Inner Standardization in the proof of Theorem 3.2.3 and Exercise 3.2.4(3) can be substituted by (II). Then apply 3.2.4(3). □

3.2d Collection in BST

Recall that Collection is an axiom of **ZFC**, and hence it is automatically true in **BST** for all $\in$-formulas Φ. However, Theorem 3.2.3 (Reduction to $\Sigma_2^{\mathtt{st}}$) will help us to derive Collection even for all $\mathtt{st}$-$\in$-formulas!

[7] Inner Transfer is derived in [AnH 04] from its restriction to formulas connected with Gödel operations, the idea goes back to Robinson and Zakon [RobinZ 69].

[8] The result was first observed by Nelson [Nel 88] in a somewhat different setting.

Inner Collection: [9] $\forall X\, \exists Y\, \forall x \in X\, \big(\exists y\, \Phi(x,y) \Longrightarrow \exists y \in Y\, \Phi(x,y)\big)$,
where $\Phi(x)$ is any $\mathtt{st}$-$\in$-formula with arbitrary parameters.

To prove Inner Collection, we need a special form to be proved beforehand.

Lemma 3.2.7 (BST). *Assume that $\varphi(a,b,x)$ is a parameter–free $\in$-formula. For any standard set X there exist standard sets A and B of cardinality resp. $\le 2^\kappa$ and $\le 2^{2^\kappa}$, where $\kappa = \operatorname{card} X$, such that for all $x \in X$:*

$$\exists^{\mathtt{st}} a\, \forall^{\mathtt{st}} b\, \varphi(a,b,x) \Longleftrightarrow \exists^{\mathtt{st}} a \in A\, \forall^{\mathtt{st}} b\, \varphi(a,b,x) \Longleftrightarrow$$
$$\Longleftrightarrow \exists^{\mathtt{st}} a \in A\, \forall^{\mathtt{st}} b \in B\, \varphi(a,b,x)\,.$$

Proof. We define, for all a and b, $W_{ab} = \{x \in X : \varphi(a,b,x)\}$,

$$W_a = \{W_{a,b} : b \text{ is an arbitrary set}\} \subseteq \mathscr{P}(X) \quad , \quad \text{and}$$
$$W = \{W_a : a \text{ is an arbitrary set}\} \subseteq \mathscr{P}^2(X)\,.$$

The set W has cardinality at most $\lambda = 2^{2^\kappa}$ while every set W_a has cardinality at most 2^κ. Using the **ZFC** axioms of Collection and Choice, and then Inner Transfer, we obtain standard sets A and B of cardinality resp. $\le 2^\kappa$ and $\le 2^{2^\kappa}$ with $\forall a'\, \exists a \in A\, (W_a = W_{a'})$ and $\forall b'\, \exists b \in B\, (W_{ab} = W_{ab'})$ for any $a \in A$. We claim that A and B are as required.

Let (1), (2), (3) denote the parts of the asserted equivalences from left to right. It is clear that (2) implies both (1) and (3).

To prove (1) $\Longrightarrow$ (2), let a standard set a satisfy $\forall^{\mathtt{st}} b\, \varphi(a,b,x)$. By the choice of A and Inner Transfer, $W_a = W_{a'}$ holds for a standard $a' \in A$. We claim now that $\varphi(a',b',x)$ is true for every standard b'. Indeed, $W_{a'b'}$ is a standard member of the set $W_a = W_{a'}$, therefore, by Inner Transfer, we have $W_{a'b'} = W_{ab}$ for a suitable standard b. Then $\varphi(a,b,x)$ by the choice of a, thus $x \in W_{ab} = W_{a'b'}$, and finally $\varphi(a',b',x)$, as required.

To prove (3) $\Longrightarrow$ (2), let a standard $a \in A$ satisfy $\forall^{\mathtt{st}} b \in B\, \varphi(a,b,x)$. We claim that $\varphi(a,b',x)$ is true for every standard b'. Notice that $W_{ab'}$ is a standard member of $X[a]$, therefore, $W_{ab'} = W_{ab}$ for a standard $b \in B$ by Inner Transfer and the choice of B. Then we have $\varphi(a,b,x)$ by the choice of a, so $x \in W_{ab} = W_{ab'}$, and finally $\varphi(a,b',x)$. □

Theorem 3.2.8 (BST). *Inner Collection holds.* [10]

Proof. Let X be any set and $\Phi(x,y)$ any $\mathtt{st}$-$\in$-formula. We can assume that X is standard (Exercise 3.1.7(iii)), and that Φ is $\Phi(x,y,p)$, with a set p as the only parameter. By Inner Boundedness, $\exists y\, \Phi(x,y,p)$ implies $\exists^{\mathtt{st}} z\, \exists y \in z\, \Phi(x,y,p)$. Let $\Psi(x,z,p)$ denote the formula $\exists y \in z\, \Phi(x,y,p)$.

[9] See footnote 2 on page 84 regarding the meaning of the word “inner”.

[10] It is rather surprising that Collection, unlike the principles of Reduction to $\Sigma_2^{\mathtt{st}}$ and Extension, also holds in **IST** for all $\mathtt{st}$-$\in$-formulas, see below.

By Theorem 3.2.3 (Reduction to Σ_2^{st}), the formula $\Psi(x,z,p)$ is equivalent to a Σ_2^{st} formula, say, $\exists^{\text{st}} a\, \forall^{\text{st}} b\, \varphi(x,z,p,a,b)$. Covering the parameter p by a standard set P and applying Lemma 3.2.7, we get a standard set Z such that

$$\exists^{\text{st}} z\, \Psi(x,z,p) \iff \exists^{\text{st}} z \in Z\, \Psi(x,z,p) \quad - \quad \text{for all } x \in X\,.$$

The set $Y = \bigcup Z = \{y : \exists\, z \in Z\, (y \in z)\}$ is as required. □

Exercise 3.2.9. Prove, following Theorem 1.4.9, that if $H \in \mathbb{N}$ is nonstandard then $\operatorname{card} Y \le H \cdot \operatorname{card} X$ can be required in Inner Collection. □

Exercise 3.2.10. Recall that in **ZFC** Collection implies an important principle of Reflection (Theorem 1.5.4). To see that Theorem 3.2.8 does not draw such a consequence in **BST** for all st-∈-formulas, let Φ say that

$$\forall\, x\, \exists\, y = \{x\} \;\wedge\; \forall\, x\, \forall\, y\, \exists\, z = x \cup y \;\wedge\; \forall\, x\, \exists^{\text{st}} y\, (x \cap \mathbb{S} = y \cap \mathbb{S})\,.$$

Show, in **BST**, that Φ holds but there is no transitive set X such that Φ is true in X. (By Lemma 3.1.15, there is a finite set $C \subseteq X$ such that $X \cap \mathbb{S} \subseteq C$. If Φ is true in X then X is closed under finite set formation, hence, $C \in X$, and, applying the last part of Φ, we find a standard set $Y \in X$ with $Y \cap \mathbb{S} = C \cap \mathbb{S} = X \cap \mathbb{S}$, contradiction.) □

3.2e Other basic theorems of BST

To demonstrate more of the power and significance of Theorem 3.2.3 (Reduction to Σ_2^{st}), we show that **BST** is strong enough to prove the following important principles. In the definitions of the first four of them, $\Phi(x,y)$ is any st-∈-formula with arbitrary parameters.

Map-Standardization: (Compare with Lemma 1.3.15 in **HST**!)
For any standard set X there is a standard function f defined on X satisfying $\forall^{\text{st}} x \in X\, (\exists^{\text{st}} y\, \Phi(x,y) \Longrightarrow \Phi(x, f(x)))$.

Inner Extension: (Compare with Theorem 1.3.12 in **HST**!)
For any standard set X there is a function f defined on X satisfying $\forall^{\text{st}} x \in X\, (\exists\,!\, y\, \Phi(x,y) \Longrightarrow \Phi(x, f(x)))$.

Inner S. S. Choice: (Compare with Standard Size Choice in **HST**, § 1.1f !)
For any standard set X there is a function f defined on X satisfying $\forall^{\text{st}} x \in X\, (\exists\, y\, \Phi(x,y) \Longrightarrow \Phi(x, f(x)))$.

Inner Dependent Choice: (Compare with Dependent Choice in **HST**, § 1.1f !)
For any standard X, if $\forall\, x \in X\, \exists\, y \in X\, \Phi(x,y)$ then there is a function h with $\operatorname{dom} h = \mathbb{N}$ and $\Phi(h(k), h(k+1))$ for all standard $k \in \mathbb{N}$.

Uniqueness: If $\Phi(p,x)$ is a parameter-free st-∈-formula, p is any set, and $\exists\,!\, x\, \Phi(p,x)$, then this unique x belongs to $\mathbb{S}[p]$ (see Definition 3.1.13). In particular if p is standard then x also is standard.

Theorem 3.2.11 (BST). *Map-Standardization, Uniqueness, Inner S. S. Choice (hence Inner Extension), and Inner Dependent Choice hold.*

Proof. Map-Standardization. We can assume that $\forall^{\mathrm{st}} x \in X\, \exists^{\mathrm{st}} y\, \Phi(x,y)$. (Otherwise consider the formula $\Phi(x,y) \vee \neg\, \exists z\, \Phi(x,z)$.) Inner Collection yields a standard set Y with $\forall^{\mathrm{st}} x \in X\, \exists^{\mathrm{st}} y \in Y\, \Phi(x,y)$. Apply Lemma 3.1.16.

Uniqueness. We can assume, by Theorem 3.2.3, that Φ is a Σ_2^{st} formula, for instance, $\exists^{\mathrm{st}} a\, \forall^{\mathrm{st}} b\ \varphi(a,b,p,x)$. There is a standard a such that x is still the unique set satisfying $\forall^{\mathrm{st}} b\, \varphi(a,b,p,x)$. Let X, P be standard sets containing resp. x, p. The uniqueness of x can be expressed by

$$\neg\, \exists y \in X\, \forall^{\mathrm{st}} b\, (y \neq x \wedge \varphi(a,b,p,y))\,.$$

Applying Local Idealization, we get a standard finite set B such that

$$\neg\, \exists y \in X\, \forall b \in B\, (y \neq x \wedge \varphi(a,b,p,y))\,,$$

thus, as $B \subseteq \mathbb{S}$ by Lemma 3.1.20, x is still the unique element of X satisfying $\forall b \in B\, \varphi(a,b,p,x)$, an $\in$-formula with parameters $a \in \mathbb{S}$ and p. Now $x = f(p)$, where $f : P \to X$ is the (standard) function defined as follows: if $p' \in P$ and there is a unique $x' \in X$ with $\forall b \in B\, \varphi(a,b,p',x')$, then $f(p') = x'$, otherwise $f(p') = x_0$, where x_0 is a fixed standard element of X.

Inner S. S. Choice. As above, we may assume that Φ is a Σ_2^{st} formula $\exists^{\mathrm{st}} a\, \forall^{\mathrm{st}} b\, \varphi(x,y,a,b)$, where φ is an $\in$-formula, and $\forall^{\mathrm{st}} x \in X\, \exists y\, \Phi(x,y)$. By Theorem 3.2.8, there is a set Y with $\forall^{\mathrm{st}} x \in X\, \exists y \in Y\, \Phi(x,y)$, moreover, a standard set of this kind (Exercise 3.1.7(iii)). Applying Theorem 3.2.8, we obtain standard sets A, B such that

$$\Phi(x,y) \iff \exists^{\mathrm{st}} a \in A\, \forall^{\mathrm{st}} b \in B\, \varphi(x,y,a,b) \quad - \quad \text{for all } x \in X,\ y \in Y.$$

Thus we have $\forall^{\mathrm{st}} x \in X\, \exists y \in Y\, \exists^{\mathrm{st}} a \in A\, \forall^{\mathrm{st}} b \in B\, \varphi(x,y,a,b)$.

Changing the quantifiers over y and a and applying Lemma 3.1.16, we derive $\forall^{\mathrm{st}} x \in X\, \exists y \in Y\, \forall^{\mathrm{st}} b \in B\, \varphi(x,y,\widetilde{a}(x),b)$ for a standard function $\widetilde{a} : X \to A$. Apply Basic Idealization to the pair of quantifiers $\exists y\, \forall^{\mathrm{st}} b$, use the fact that the quantifier $\forall^{\mathrm{st}} x \in X$ is equivalent to the combination of quantifiers $\forall^{\mathrm{stfin}} X' \subseteq X \forall x \in X'$ by Lemma 3.1.20, and obtain consecutively

$$\forall^{\mathrm{stfin}} X' \subseteq X\ \forall^{\mathrm{stfin}} B' \subseteq B\, \forall x \in X'\, \exists y \in Y\, \forall b \in B'\ \varphi(x,y,\widetilde{a}(x),b)\,,$$

$$\forall^{\mathrm{stfin}} X' \subseteq X\ \forall^{\mathrm{stfin}} B' \subseteq B\, \exists \hat{y} \in \widehat{Y}\, \forall x \in X'\, \forall b \in B'\ \varphi(x,\hat{y}(x),\widetilde{a}(x),b)\,,$$

where $\widehat{Y}$ is the (standard) set of all functions $\hat{y}$ such that $\mathtt{dom}\,\hat{y}$ is a finite subset of X while $\mathtt{ran}\,\hat{y} \subseteq Y$. Converting the pair of variables x, b into one variable and applying Basic Idealization backwards, we obtain

$$\exists \hat{y}\, \forall^{\mathrm{st}} x \in X\ \forall^{\mathrm{st}} b \in B\ \varphi(x,\hat{y}(x),\widetilde{a}(x),b)\,,$$

that is, $\exists \hat{y}\, \forall^{\mathrm{st}} x \in X\, \Phi(x,\hat{y}(x))$, as required.

Inner Dependent Choice. First of all, by theorems 3.2.3, 3.2.8 (Reduction to $\Sigma_2^{\mathtt{st}}$, Inner Collection) Φ may be assumed to be a formula of the form $\exists^{\mathtt{st}} a \in A\, \forall^{\mathtt{st}} b \in B\, \varphi(a, b, x, y)$, where A and B are standard while φ is an $\in$-formula. Let $\mathbf{a} = \langle a_0, ..., a_{n-1} \rangle$ be a finite sequence of elements of A. Put

$$X(\mathbf{a}) = \{\langle x_0, ..., x_n \rangle \in X^{n+1} : \forall k < n\, \forall^{\mathtt{st}} b \in B\, \varphi(a_k, b, x_k, x_{k+1})\}\,.$$

We say that $\mathbf{a}$ is *good* if $X(\mathbf{a}) \neq \varnothing$. The empty sequence Λ is evidently good: $n = 0$ and $X(\Lambda) = X$. If $\mathbf{a} = \langle a_0, ..., a_{n-1} \rangle$ is good then by the assumption there exists a *standard* $a_n \in A$ such that $\mathbf{a}^\wedge a_n = \langle a_0, ..., a_{n-1}, a_n \rangle$ is also good. By Inner Standardization, there is a standard set S whose elements are finite sequences of elements of A, and whose *standard* elements are all standard good sequences and nothing more. By what is said above and by Inner Transfer, every sequence in S can be extended to a sequence in S by adding one more element. Therefore there exists an infinite sequence $\alpha = \{a_n\}_{n \in \mathbb{N}}$ such that $\langle a_0, ..., a_{n-1} \rangle \in S$ for all n. By Transfer again, there exists a *standard* sequence α of this type. Then, for any standard n, a_n is standard and $\langle a_0, ..., a_{n-1} \rangle$ is (standard and) good. Thus,

$$\forall^{\mathtt{st}} n\, \exists \langle x_0, ..., x_n \rangle \in X^{<\omega}\, \forall k < n\, \forall^{\mathtt{st}} b \in B\, \varphi(a_k, b, x_k, x_{k+1})\,,$$

where $X^{<\omega}$ is the set of all finite sequences of elements of X. We obtain

$$\forall^{\mathtt{st}} n\, \forall^{\mathtt{stfin}} B' \subseteq B\, \exists \langle x_0, ..., x_n \rangle \in X^{<\omega}\, \forall k < n\, \forall b \in B'\, \varphi(a_k, b, x_k, x_{k+1})\,.$$

by Local Idealization. Applying Local Idealization backwards, we get

$$\exists \langle x_0, ..., x_n \rangle \in X^{<\omega}\, \forall^{\mathtt{st}} k\, \forall^{\mathtt{st}} b \in B\, \varphi(a_k, b, x_k, x_{k+1})\,;$$

in particular, n, the length of this sequence $\langle x_0, ..., x_n \rangle$ minus 1, is nonstandard. Put $h(k) = x_k$ for $k \leq n$ and $h(k) = x_0$ (not essential) for $k > n$; then h is as required since a_k is standard provided k is standard. □

This Theorem and other results of Section 3.2 characterize **BST** as a rather complete nonstandard theory; for instance, **IST** lacks such a degree of completeness, see below. It is quite difficult to find a meaningful "nonstandard" sentence of clear foundational value which **BST** does not prove or disprove. Problem 3.2.5 is a possible candidate.

3.2f Introduction to the problem of external sets

Users of internal theories have to pay some price for the simplicity of the set-up: not all objects of interest turn out to be sets!

Definition 3.2.12 (informal, for internal theories like **BST** or **IST**). "External sets" (will be used, in this sense, only with quotation marks) are $\mathtt{st}$-$\in$-definable subclasses of (internal) sets.[11] □

Recall that, in nonstandard set theories of "internal" type, like **IST** or **BST**, *internal set* means just any set. While any (internal) set is an "external set" (by means of the formula $x = x$), there exists plenty of "external sets" which fail to be really sets! For instance, the collection ${}^{\sigma}\mathbb{N} = \mathbb{N} \cap \mathbb{S}$ is not a set (Exercise 3.1.21). This contradicts the **ZFC** idea of sets according to which anything "smaller" than a set is also a set, while classes (*i.e.*, definable collections of sets) are not sets just because they are "too big" to be sets. In internal theories, some objects are non-sets not because they are big but because they are too complicated!

Yet this does not mean that one cannot deal with "external sets" in **BST**. On the contrary, as long as we consider individual "external sets", any reasonable claim related to them can be properly transformed to a legitimate st-$\in$-statement which can be studied in **BST** the same way as proper classes are studied in **ZFC**. The real problem comes from another angle:

Problem 3.2.13 (The problem of external sets). How to quantify over "external sets", for instance, if $P(\cdot,\cdot)$ is a st-$\in$-formula, what could be the meaning, in **BST** (or **IST**) of a statement like:

for any "external set" A there is an "external set" B such that $P(A, B)$. □

To solve the problem, one has to find a **parametrization**, that is, a st-$\in$-definable surjection from the universe of internal sets onto the variety of all "external sets". This seems to be quite a difficult task: indeed, the multitude of all st-$\in$-formulas which define "external sets" is to be eliminated by a single st-$\in$-formula that defines all of them, varying only an internal parameter. To see that this is a nontrivial problem note that there definitely does not exist an $\in$-definable parametrization of all (unbounded) $\in$-definable classes in **ZFC**. Yet "external sets" do admit a parametrization in **BST**. This is perhaps the most surprising and important feature of this theory.

Definition 3.2.14. An *E-code* [12] is any (internal) [13] function p whose domain has the form $A \times B$ where A, B are arbitrary standard sets. Put

$$\mathsf{E}_p = \bigcup_{a \in A \cap \mathbb{S}} \bigcap_{b \in B \cap \mathbb{S}} p(a,b) = \{x : \exists^{\text{st}} a \in A \, \forall^{\text{st}} b \in B \, (x \in p(a,b))\}.$$

for any such p. If p is not an E-code then put $\mathsf{E}_p = \varnothing$.

We put $\mathbb{E} = \{\mathsf{E}_p : p \text{ is a E-code}\}$. □

[11] From a broader perspective, introduced in Chapter 5 below, these "external sets" may be considered as only the first level of external objects in **BST**. The next level would contain definable collections of "external sets", and so on. Accordingly, Problem 3.2.13 presents only a part of the full problem related to "external sets" of all levels.

[12] In this definition, E refers both to "external" and "elementary", see § 5.2b on the latter aspect.

[13] The word "internal" in brackets is irrelevant as long as we argue in **BST** where all sets are internal. However, this definition will also be used in **HST**, where it will really mean that only functions $p \in \mathbb{I}$ are to be considered.

Thus E_p is a part of the (internal) set $\bigcup \operatorname{ran} p$, but, generally speaking, E_p is not a set in **BST**, see (ii) of the next exercise. It follows that the family $\mathbb{E}$ can be considered in **BST** only informally.

Exercise 3.2.15. (i) Let for any (internal) x, ${}^{\mathsf{e}}x$ be a function defined on the set $\{\varnothing\} \times \{\varnothing\}$ by the equality ${}^{\mathsf{e}}x(\varnothing, \varnothing) = x$; technically ${}^{\mathsf{e}}x = \{\langle\langle\varnothing, \varnothing\rangle, x\rangle\}$. Prove that ${}^{\mathsf{e}}x$ is a E-code and $\mathsf{E}_{{}^{\mathsf{e}}x} = x$.

(ii) Define $p(m, n) = \{m\}$ for all $m, n \in {}^*\mathbb{N}$. Prove that p is a E-code but $\mathsf{E}_p = \mathbb{S} \cap \mathbb{N}$ and hence E_p is not a set (internal) by 3.1.21. □

Collections of the form E_p ($\mathbf{\Sigma}_2^{\mathrm{ss}}$ sets in **HST**, § 1.4) seem to be a rather particular type of all possible "external sets", yet according to the next theorem they exhaust all of them! In other words, the map $p \mapsto \mathsf{E}_p$ is a *parametrization* of all "external sets" in **BST**. By the way, it is unknown whether any similar parametrization exists in **IST**.

Theorem 3.2.16. (i) *If $\Phi(x, y)$ is a* st-$\in$*-formula then it is a theorem of* **BST** *that for any sets C, y_0 there is p such that $\mathsf{E}_p = \{x \in C : \Phi(x, y_0)\}$. In other words all "external sets" have the form E_p in* **BST**.

(ii) (**BST**) *Any "external set" X definable* (in $\mathbb{I}$) *by a* st-$\in$*-formula with parameters $c_1, ..., c_n \in \mathbb{I}$ has the form $X = \mathsf{E}_p$ for a E-code $p \in \mathbb{S}[c_1, ..., c_n]$.*

(iii) (**BST**) *Moreover if X is internal in* (ii) *then $X = \mathsf{E}_p$, where $p = {}^{\mathsf{e}}X$, and both X and p belong to $\mathbb{S}[c_1, ..., c_n]$.*

Proof. (i) We can assume that Φ is $\exists^{\mathrm{st}} a\, \forall^{\mathrm{st}} b\, \varphi(a, b, x, y)$, where φ is an $\in$-formula, by Theorem 3.2.3. By Inner Boundedness, there is a standard set Y containing y_0. Then, by Theorem 3.2.8 (Collection), there exist standard sets A, B such that

$$\text{for all } x \in C,\ y \in Y: \quad \Phi(x, y) \iff \exists^{\mathrm{st}} a \in A\, \forall^{\mathrm{st}} b \in B\, \varphi(a, b, x, y).$$

Put $p(a, b) = \{x \in C : \varphi(a, b, x, y)\}$ for all $a \in A$, $b \in B$. Then p is a E-code and $\mathsf{E}_p = \{x \in C : \Phi(x, y)\}$.

(ii) For brevity, consider the case of a single parameter. Thus let a set $X = \{x \in C : \Phi(x, y_0)\}$ be definable (in $\mathbb{I}$) by a st-$\in$-formula $\Phi(x, y_0)$ with the only parameter y_0, where C is a standard set. Note that in the proof for (i) p is defined by a formula containing, as parameters, only standard sets A, B, C and an internal set y_0. It follows that $p \in \mathbb{S}[y_0]$ by Lemma 3.1.14, as required.

(iii) If X is internal then both X and $p = {}^{\mathsf{e}}X$ belong to $\mathbb{S}[c_1, ..., c_n]$ by Uniqueness in **BST** (Theorem 3.2.11) and still $X = \mathsf{E}_p$. □

Theorem 3.2.16 will play the key role in our study of the **HST** set universe and metamathematics of **HST** (Chapter 5). It will be shown (Theorem 5.2.6) that "external sets" satisfy axioms of a certain nonstandard set theory. Generally, this will lead to a complete solution of the problem of external sets in the broadest sense outlined in Footnote 11.

3.2g More on "cxternal sets" in BST

Thus **BST** makes it possible to effectively study "external sets" on the base of an appropriate coding considered in §3.2f. As the following exercise shows, this coding can be easily extended to the next level of "external" objects: collections of "external sets". But we shall see in Chapter 9.5.20 that in fact the whole von Neumann hierarchy of "external" objects of higher types can be suitably coded by internal sets!

Exercise 3.2.17. Informally, a *set-size collection of "external sets"* is any collection of the form $\{\Omega_p : p \in P\}$, where $\Omega_p = \{x : \langle x, p\rangle \in \Omega\}$, P is internal, and Ω is an "external set". Let X be a set containing more than a standard finite number of elements. Prove in **BST**, arguing as in the proof of Theorem 1.3.9, that the family $\mathscr{P}_{\mathbf{ext}}(X)$ of all "external subsets" of X (the "external power set") is not a set-size collection. □

Recall that well-founded sets were very instrumental in the development of **HST**. Is there a reasonable parallel in **BST**? Definition 1.1.4 gives nothing in **BST**: indeed, as **BST** proves all of **ZFC** in the $\in$-language, all sets appear to be well-founded. The following version of the notion of well-foundedness can be more meaningful.

Definition 3.2.18. A binary relation $\prec$ on X is *"externally"* well-founded if any non-empty "external set" $Y \subseteq X$, *i.e.*, any collection of the form $\mathsf{E}_p \subseteq X$, contains a $\prec$-minimal element. A set or "external set" x is *"externally"* well-founded if there is a transitive set X such that $x \subseteq X$ and the restriction $\in \restriction X$ is "externally" well-founded. □

Exercise 3.2.19 (BST). Prove that a set x is "externally" well-founded iff there is no function f defined on $\mathbb{N}$ so that $f(0) = x$ and $f(n+1) \in f(n)$ for any $n \in {}^{\sigma}\mathbb{N}$. □

It follows that, for instance, $\mathbb{N}$ is not "externally" well-founded, for take any $n_0 \in \mathbb{N} \setminus {}^{\sigma}\mathbb{N}$ and define $f(n) = n_0 - n$ for $n \le n_0$ and $f(n) = 0$ for $n > n_0$. But which sets are "externally" well-founded in **BST**? Let $\mathbb{HF}$ be the set of all hereditarily finite sets (in the $\mathbb{I}$-sense, see Blanket agreement 3.1.1). Then $\mathbb{N} \subseteq \mathbb{HF} \in \mathbb{S}$. Put ${}^{\sigma}\mathbb{HF} = \mathbb{HF} \cap \mathbb{S}$ (an "external set").

Exercise 3.2.20 (BST). Prove that a set x is "externally" well-founded iff $x \in {}^{\sigma}\mathbb{HF}$, while an "external set" $E = \mathsf{E}_p$ is "externally" well-founded iff $E \subseteq {}^{\sigma}\mathbb{HF}$.

Hint. The result is analogous to the equality $\mathbb{I} \cap \mathbb{S} = \mathbb{HF}$ in **HST**, see Exercise 1.2.17(2), and admits a similar proof. □

3.3 Internal theories with partial Saturation

One of the priorities of nonstandard set theory is to keep the amount of Saturation, or Idealization in the "internal" notational system, not restricted by any fixed cardinal. However, to incorporate the Power Set and Choice axioms into the **HST**-type of reasoning, we will have to consider partially saturated theories and universes. We begin with partially saturated versions of **BST** which, unlike **BST**, imply that every set has a set-size "external power set"; this will lead us to versions of **HST** compatible with the Power Set axiom.

3.3a Two schemes of partially saturated internal theories

Recall that the formulation of Basic Idealization in § 3.1b contains two set parameters, standard sets A_0 and X. The restriction of their cardinalities leads to two different partially saturated versions of **BST**.

Definition 3.3.1. $\mathbf{BST}_{\kappa}$, resp., $\mathbf{BST}'_{\kappa}$ be modifications of **BST** obtained by 1) adding a symbol κ to the language, together with the axiom: "κ is an infinite standard cardinal", 2) strengthening Inner Boundedness to, resp.,

Inner κ-Boundedness: every set belongs to a standard set of cardinality $\leq \kappa$,

Inner Strong κ-Boundedness: every set belongs to a standard set of cardinality $\leq \kappa$ and κ is a set of standard size in the sense of Definition 1.1.12,

and 3) weakening Basic Idealization to, resp.,

κ-deep BI: Basic Idealization of § 3.1b in the case $\operatorname{card} X \leq \kappa$,

κ-size BI: Basic Idealization of § 3.1b in the case $\operatorname{card} A_0 \leq \kappa$. [14] □

Intuitively, if κ increases then the strength of κ-deep BI and κ-size BI also increases while the strength of the κ-Boundedness axioms decreases.

The notion of relative standardness (Definition 3.1.13) allows us to obtain reasonable equivalent formulations of the κ-Boundedness axioms:

Lemma 3.3.2. *Inner κ-Boundedness and Inner Strong κ-Boundedness are equivalent, in* **BST** *minus Basic Idealization, to, resp., the following:*

(i) *for any x there exists $\xi < \kappa$ such that $x \in \mathbb{S}[\xi]$;*

(ii) *there exists $\xi < \kappa$ such that every set x belongs to $\mathbb{S}[\xi]$.*

In addition, (ii) implies that all sets are sets of standard size.

Proof. If $\xi < \kappa$ and $x \in \mathbb{S}[\xi]$ then $x = f(\xi)$ for a suitable standard map f defined on κ, hence, x belongs to the f-image X of κ, which is a standard

[14] In the ordinary model theoretic nomenclature, κ-size BI should be denoted by κ^+-size BI: indeed, for instance, Saturation for families of cardinality κ is κ^+-Saturation. We prefer to keep κ here in order to preserve uniformity of the whole system of definitions.

set of cardinality $\leq \kappa$. Conversely, if X is a standard set of cardinality κ then, by Inner Transfer, there is a standard bijection $f : \kappa \xrightarrow{\text{onto}} X$, hence, if $x \in X$ then $x \in \mathbb{S}[\xi]$ where $\xi = f^{-1}(x)$.

If κ is of standard size then there is a standard set Z and a map (internal) h defined on Z such that κ is the h-image of $Z \cap \mathbb{S}$. Then $\kappa \subseteq \mathbb{S}[h]$, hence, all sets belong to $\mathbb{S}[h]$ by the above. But h belongs to $\mathbb{S}[\xi]$ for some $\xi < \kappa$.

It remains to prove the additional statement. Let X be any (internal) set. Take a standard S with $X \subseteq S$. The set F of all (internal) maps $f : \kappa \to S$ is standard. Put $H(f) = f(\xi)$ for all $f \in F$. The H-image of $F \cap \mathbb{S}$ is equal to S. Using Standardization, we can find a standard set $F' \subseteq F$ such that the H-image of $F' \cap \mathbb{S}$ is equal to X. □

Exercise 3.3.3. (1) Prove in $\mathbf{BST}_\kappa$ that κ-size BI holds, but Inner Strong κ-Boundedness fails — and hence $\mathbf{BST}_\kappa$ contradicts $\mathbf{BST}'_\kappa$ [15].

(2) Prove that if $\lambda < \kappa$ are standard cardinals then κ-size BI contradicts Inner λ-Boundedness.

Hints. (1) To infer κ-size BI argue as in the proof of Inner Saturation in Lemma 3.2.2, taking P, A_0 (then F, too) to be standard sets of cardinality $\leq \kappa$. By **ZFC** Collection, X can be chosen to be a standard set of cardinality $\leq \kappa$. [16] That κ is not a set of standard size under κ-deep BI is elementary.

(2) Even if κ is singular, there is a regular [17] cardinal κ', $\lambda < \kappa' \leq \kappa$. By Inner λ-Boundedness, any $\xi < \kappa'$ belongs to a set $X = [0, \gamma)$, where $\gamma < \kappa$ is a standard ordinal, easily leading to a contradiction with κ'-size BI. □

Thus, theories $\mathbf{BST}_\kappa$ for different (standard) cardinals κ are pairwise incompatible, and the same for $\mathbf{BST}'_\kappa$.

Interpretations of the partially saturated theories $\mathbf{BST}_\kappa$ and $\mathbf{BST}'_\kappa$ in **BST** will be defined in Theorem 3.4.5(ii) ($\mathbf{BST}_\kappa$) and Theorem 6.2.6 ($\mathbf{BST}'_\kappa$). It will follow that, similarly to **BST** or **IST**, the theories $\mathbf{BST}_\kappa$ and $\mathbf{BST}'_\kappa$ are equiconsistent and conservative extensions of **ZFC** (see § 4.1d).

3.3b κ-deep Basic Idealization scheme

For any E-code p (by definition, an internal function defined on a set of the form $A \times B$, A, B being standard), put

$$\|p\| = \{\{p(a,b) : b \in B \cap \mathbb{S}\} : a \in A \cap \mathbb{S}\}$$

and define $p \doteq q$ iff $\|p\| = \|q\|$. Obviously $\|p\|$ may not be a legitimate set in **BST** (generally, it is a collection of "external sets"), but the equality

---

[15] We shall see below (§ 6.2b) that κ-deep BI is really stronger than κ-size BI.

[16] The proof of Local Idealization in Lemma 3.2.2 has a similar κ-version: Inner κ-Boundedness plus 2^κ-size BI imply κ-deep BI. Yet 2^κ-size BI contradicts Inner κ-Boundedness by (2) of this Exercise, so that this version is vacuous.

[17] A cardinal λ is regular if no set $X \subseteq \lambda$ of smaller cardinality is cofinal in λ, and singular otherwise.

$||p|| = ||q||$ can be adequately expressed by a suitable st-$\in$-formula (we leave this as an **exercise** for the reader).

Exercise 3.3.4. Show that $p \doteq q$ implies $\mathsf{E}_p = \mathsf{E}_q$. Actually, $p \doteq q$ can be considered as an effective form of the equality $\mathsf{E}_p = \mathsf{E}_q$. □

Theorem 3.3.5 (BST$_\kappa$). *We have Inner Collection, Uniqueness, Map-Standardization, Inner Extension, Inner Dependent Choice, the parametrization theorem (Theorem 3.2.16), Inner S. S. Choice in the case when* $\mathtt{card}\, X \le 2^\kappa$, *and Reduction to* $\Sigma_2^{\mathtt{st}}$ *with the correction that the formulas* φ, ψ *in Theorem 3.2.3 can contain* κ *as a parameter. In addition,*

(∗) *for every (internal) set* X *there is a standard set* P *such that for any E-code* q *with* $\mathsf{E}_q \subseteq X$ *there exists an E-code* $p \in P$ *with* $p \doteq q$.

It follows that many of **BST** theorems (3.2.3, 3.2.8, 3.2.16, 3.2.11) remain true in **BST**$_\kappa$ (Inner S. S. Choice in a κ-form). The additional claim (∗) is extremely important: it implies that the "external power set" of any set X is a set-size collection in **BST**$_\kappa$: take $\Omega = \{\langle p, x\rangle : p \in P \wedge x \in X \cap \mathsf{E}_p\}$ in the definition in Exercise 3.2.17. Recall that in **BST** "external power sets" are not necessarily set-size collections.

Proof. The step for $\exists$ in the proof of Theorem 3.2.3: simply change $\exists^{\mathtt{st}} W$ to $\exists^{\mathtt{st}} W$, $\mathtt{card}\, W \le \kappa$. Proof of Uniqueness in Theorem 3.2.11: take X to be a standard set of cardinality $\le \kappa$. It remains to consider Inner Extension, Inner S. S. Choice ($\mathtt{card}\, X \le 2^\kappa$), and Inner Dependent Choice, because the rest of the theorems do not depend at all on Idealization in any form.

Inner Extension. We can assume that Φ contains only one parameter, say p, *i.e.* Φ is $\Phi(p, x, y)$. By Uniqueness, for any standard $x \in X$, if $\exists !\, y\, \Phi(p, x, y)$ then there is a standard function ξ defined on X such that $\Phi(p, x, \xi(p))$. Applying Map-Standardization, we find a standard function g defined on X and satisfying $\Phi(p, x, g(x)(p))$ for any standard $x \in X$ with $\exists !\, y\, \Phi(p, x, y)$. It remains to define $f(x) = g(x)(p)$.

Inner S. S. Choice in the case $\mathtt{card}\, X \le 2^\kappa$. [18] Let, for instance, $X = {}^\kappa 2$, the (standard) set of all (internal) functions $x : \kappa \to 2 = \{0, 1\}$).

Assuming $\forall^{\mathtt{st}} x \in X\, \exists y\, \Phi(x, y)$, as above, we have, by Inner κ-Boundedness, $\forall^{\mathtt{st}} x \in X\, \exists^{\mathtt{st}} f\, \exists \alpha < \kappa\, \Phi(x, f(\alpha))$, hence, by Map-Standardization, there is a standard map g, defined on X, such that $\forall^{\mathtt{st}} x \in X\, \exists \alpha < \kappa\, \Phi(x, g(x, \alpha))$. Now, it obviously suffices to prove the theorem for the formula $\Phi(x, g(x, \alpha))$, with x, α as variables. In other words, we can suppose, from the beginning, that $\forall^{\mathtt{st}} x \in X\, \exists y \in Y\, \Phi(x, y)$ holds for a standard set Y of cardinality $\le \kappa$.

In this case, a fragment of the proof of Theorem 3.2.11 can be carried out. Assuming w. l. o. g. that $\Phi(x, y)$ is $\exists^{\mathtt{st}} a \in A\, \forall^{\mathtt{st}} b \in B\, \varphi(x, y, a, b)$, where A, B are standard sets and φ an $\in$-formula, we advance to

[18] The case $\mathtt{card}\, X \le \kappa$ needs only a slight and obvious correction of the proof of Inner S. S. Choice in Theorem 3.2.11. The case we consider is more complicated.

$$\forall^{\text{stfin}} X' \subseteq X \, \forall^{\text{stfin}} B' \subseteq B \, \exists \hat{y} \subset \widehat{Y} \, \forall x \in X' \, \forall b \in B' \, \varphi(x, \hat{y}(x), \tilde{a}(x), b) \quad (\dagger)$$

for a standard function $\tilde{a} : X \to A$, where we stumble because the set $\widehat{Y}$ of all partial maps $\hat{y} : X \to Y$ with finite domain $\texttt{dom}\,\hat{y} \subseteq X$ is of cardinality $> \kappa$. To overcome this difficulty, let, by κ-deep BI, $\Omega \subseteq \kappa$ be a finite set such that $\kappa \cap \mathbb{S} \subseteq \Omega$. For any $x \in {}^{\kappa}2$ the restriction $x \restriction \Omega$ belongs to the (standard) set F of all (internal) maps f with $\texttt{dom}\, f \subseteq \kappa$ finite and $\texttt{ran}\, f \subseteq \{0,1\}$; obviously $\texttt{card}\, F = \kappa$. We observe that $x \restriction \Omega \neq y \restriction \Omega$ whenever $x \neq y \in {}^{\kappa}2$ are standard: indeed, by Inner Transfer, there is a standard $\alpha < \kappa$ with $x(\alpha) \neq y(\alpha)$. Then $\alpha \in \Omega$, and hence $x \restriction \Omega \neq y \restriction \Omega$.

Let H be the set of all internal partial maps $h : F \to Y$, with finite domains. It follows from $(\dagger)$ that

$$\forall^{\text{stfin}} X' \subseteq X \, \forall^{\text{stfin}} B' \subseteq B \, \exists h \in H \, \forall x \in X' \, \forall b \in B' \, \varphi(x, h(x \restriction \Omega), \tilde{a}(x), b).$$

Since H is a standard set with $\texttt{card}\, H = \kappa$, κ-deep BI can be applied, and we obtain a function $h \in H$ such that

$$\forall^{\text{st}} x \in X \, \forall^{\text{st}} b \in B \, \varphi(x, h(x \restriction \Omega), \tilde{a}(x), b).$$

Put $f(x) = h(x \restriction \Omega)$, then $\forall^{\text{st}} x \in X \, \Phi(x, f(x))$, as required.

Inner Dependent Choice. We have to slightly amend the argument in the proof of Theorem 3.2.11. For any $\mathbf{a} = \langle a_0, ..., a_{n-1} \rangle \in A^n$ and any sequence $\mathbf{f} = \langle f_0, ..., f_n \rangle$ of standard functions $f_k : \kappa \to X$, put

$$\Xi_{\mathbf{a},\mathbf{f}} = \{ \langle \xi_0, ..., \xi_n \rangle \in \kappa^{n+1} : \forall k < n \, \forall^{\text{st}} b \in B \, \varphi(a_k, b, f_k(\xi_k), f_{k+1}(\xi_{k+1})) \} .$$

Then we obtain two infinite standard sequences $\{a_n\}_{n \in \mathbb{N}}$ and $\{f_n\}_{n \in \mathbb{N}}$ such that $\Xi_{\langle a_0, ..., a_{n-1} \rangle, \langle f_0, ..., f_n \rangle} \neq \varnothing$ for any standard $n \geq 1$. The argument ends as in the proof of Theorem 3.2.11 (with κ instead of X).

Finally prove the additional claim $(*)$.[19] We can assume that X is standard. Prove that the standard set $P = \{p : p \text{ maps } \lambda \times \lambda \text{ into } \mathscr{P}(X)\}$, where $\lambda = 2^{2^{\vartheta}}$ and $\vartheta = \texttt{Max}\{\texttt{card}\, X, \kappa\}$, is as required. Consider any E-code q for a subset of X, thus, $q : U \times V \to \mathscr{P}(X)$, where U, V are standard. Then

$$x \in \mathsf{E}_q \iff \exists^{\text{st}} a \in U \, \forall^{\text{st}} b \in V \, \big(x \in q(a,b)\big) .$$

By Inner κ-Boundedness there is a standard set $Q \subseteq \mathscr{P}(X)^{U \times V}$ of cardinality $\leq \kappa$ containing q. As $\texttt{card}\, X \leq \vartheta$, too, the proof of Lemma 3.2.7 yields standard sets $A \subseteq U$ and $B \subseteq V$ of cardinality at most $\lambda = 2^{2^{\vartheta}}$ such that

$$\{\{W_{ab} : b \in B\} : a \in A\} = \{\{W_{ab} : b \in V\} : a \in U\} ,$$

where $W_{ab} = \{\langle x, r \rangle \in X \times Q : x \in r(a,b)\}$. It follows, by Inner Transfer, that

$$\{\{W_{ab} : b \in B \cap \mathbb{S}\} : a \in A \cap \mathbb{S}\} = \{\{W_{ab} : b \in V \cap \mathbb{S}\} : a \in U \cap \mathbb{S}\}$$

[19] Note that the proof does not involve κ-deep BI.

(see arguments in the proof of Lemma 3.2.7), so that $q \doteq r$, where $r = q \upharpoonright (A \times B)$. It remains to note that A and B are sets of cardinality at most λ, which allows to suitably convert r into an E-code $p \in P$, $p \doteq r$. □

Remark 3.3.6. It is consistent with the practice of nonstandard analysis that Saturation for κ-size families (which takes the form of κ-size BI in $\mathbf{BST}'_\kappa$) is matched with 2^κ-size Choice. For instance, typical model theoretic instrumentarium for nonstandard analysis includes "countable" Saturation and, mostly, only continuum-size Choice. Theorem 3.3.5 reveals another factor: a somewhat stronger κ-form of Basic Idealization implies Inner 2^κ-size Choice.

It turns out that the axiom of Inner S. S. Choice fails in $\mathbf{BST}_\kappa$ in the case when $\mathtt{card}\, X > 2^\kappa$, so that this part of Theorem 3.2.11 is optimal.

Indeed, arguing in $\mathbf{BST}_\kappa$, let $X \subseteq \mathscr{P}({}^\kappa 2)$ be any standard set of cardinality $\lambda > 2^\kappa$. Let $\mathscr{U}$ be the least class of (internal) subsets of ${}^\kappa 2$ which contains all sets of the form $\{x \in {}^\kappa 2 : x(\alpha) = 0\}$, $\alpha < \kappa$ and is closed under complements and finite unions and intersections; clearly $\mathscr{U}$ is a standard set of cardinality κ. Let $\Phi(S, U)$ say that $S \in X$, $U \in \mathscr{U}$, all standard elements of S belong to U but each standard $x \in {}^\kappa 2 \setminus S$ does not belong to U. If Y, $Z \subseteq {}^\kappa 2$ are disjoint finite sets then there is a set $U \in \mathscr{U}$ with $Y \subseteq U$ and $Z \cap U = \varnothing$. It follows, by κ-deep BI, that $\forall S \in X\, \exists U \in \mathscr{U}\, \Phi(S, U)$.

On the other hand, there is no (internal) map $\xi : X \to \mathscr{U}$ such that $\Phi(S, \xi(S))$ holds for any standard $S \in X$. Indeed, suppose that there is such a map ξ. Then, by Inner κ-Boundedness, ξ belongs to a standard set Ξ of cardinality κ. Let $\Xi = \{\xi_\alpha : \alpha < \kappa\}$ be a standard enumeration. We can assume that each ξ_α is a map $X \to \mathscr{U}$. Let $r(S) = \{\xi_\alpha(S)\}_{\alpha<\kappa}$: Then r is a standard map from X to $\mathscr{U}^\kappa$, a standard set of cardinality $\kappa^\kappa < \lambda = \mathtt{card}\, X$, hence, by Inner Transfer, there are two different $S \neq S' \in X$ such that $\xi_\alpha(S) = \xi_\alpha(S')$ for all $\alpha < \kappa$. In other words, we have $\xi(S) = \xi(S')$ for any $\xi \in \Xi$. Thus $\Phi(S, U)$ and $\Phi(S', U)$ hold for one and the same set $U = \xi(S) = \xi(S') \in \mathscr{U}$ and two different standard $S, S' \in X$. Yet by Inner Transfer there is a standard $x \in S \triangle S'$. Let, for instance, $x \in S \setminus S'$. Then $\Phi(S, U)$ implies $x \in U$, but $\Phi(S', U)$ implies $x \notin U$, contradiction. □

3.3c κ-size Basic Idealization scheme

Theorem 3.3.5 shows that $\mathbf{BST}_\kappa$ is a theory of the same type as $\mathbf{BST}$: the difference is related only to the amount of Saturation and Inner Boundedness available while the principal theorems remain roughly the same. This is not the case for $\mathbf{BST}'_\kappa$, whose universe looks very different. For instance by Lemma 3.3.2 there is an ordinal $\xi < \kappa$ such that $f \mapsto f(\xi)$ is a definable map from $\mathbb{S}$ onto the class of all sets, which is impossible in $\mathbf{BST}$ or $\mathbf{BST}_\kappa$. Nevertheless, $\mathbf{BST}'_\kappa$ proves everything which $\mathbf{BST}$ proves without Basic Idealization, *e.g.* some results in §§ 3.1d, 3.1e and, most notably, Lemma 3.2.7.

Theorem 3.3.7 ($\mathbf{BST}'_\kappa$). *We have Inner Collection, Inner S. S. Choice, Inner Dependent Choice, Inner Extension, Map-Standardization.*

Proof. Inner Collection. Fix any $\xi < \kappa$ such that all sets belong to $\mathbb{S}[\xi]$. Then $\exists\, x\, \varphi(x)$ is equivalent to $\exists^{\text{st}} f\ (f$ is a function $\wedge\ \xi \in \operatorname{dom} f\ \wedge\ x = f(\xi))$. It follows that any st-$\in$-formula $\Phi(x,y)$ can be converted to the form

$$\mathsf{Q}_1^{\text{st}} a_1 \dots \mathsf{Q}_n^{\text{st}} a_n\, \varphi(a_1, \dots, a_n, x, y, \xi)\,, \tag{1}$$

where Q_i are quantifiers while φ is an $\in$-formula. In this case, let X be a standard set and $\psi(a_1, ..., a_n, x, f, v)$ be the formula (with v a free variable)

$$f \text{ is a function} \wedge \xi \in \operatorname{dom} f \wedge \varphi(a_1, \dots, a_n, x, f(v), v)\,.$$

By the choice of ξ, $\exists\, y\, \Phi(x,y) \iff \exists^{\text{st}} f\, \Psi(x, f, \xi)$, where $\Psi(x, f, v)$ is

$$\mathsf{Q}_1^{\text{st}} a_1 \dots \mathsf{Q}_n^{\text{st}} a_n\, \psi(a_1, \dots, a_n, x, f, v)\,.$$

Arguing as in the proof of Lemma 3.2.7 (but with n quantifiers instead of two), we can find standard sets $F, A_1, ..., A_n$ (which depend on X and φ) such that, for any $x \in X$ and $\eta < \kappa$,

$$\begin{aligned}\exists^{\text{st}} f\, \Psi(x, f, v) &\iff \exists^{\text{st}} f \in F\, \mathsf{Q}_1^{\text{st}} a_1 \dots \mathsf{Q}_n^{\text{st}} a_n\, \psi(a_1, \dots, a_n, x, f, \eta) \\ &\iff \exists^{\text{st}} f \in F\, \mathsf{Q}_1^{\text{st}} a_1 \in A_1 \dots \mathsf{Q}_n^{\text{st}} a_n \in A_n\, \psi(a_1, ..., a_n, x, f, \eta)\,,\end{aligned}$$

that is, $\exists^{\text{st}} f\, \Psi(x, f, \xi) \iff \exists^{\text{st}} f \in F\, \Psi(x, f, \xi)$. Let $Y = \{f(\xi) : f \in F\}$; then, by definition, we have $\exists\, y\, \Phi(x,y) \iff \exists\, y \in Y\, \Phi(x,y)$ for any $x \in X$.

Map-Standardization: follows from Inner Collection.

Inner S. S. Choice. Let X be any set and $\Phi(x,y)$ a st-$\in$-formula. Assume, for the sake of simplicity, that $\forall^{\text{st}} x \in X\, \exists\, y\, \Phi(x,y)$. By Collection and Lemma 3.3.2 there exist standard sets Y and A and an (internal) map $g : A \to Y$ such that $\forall^{\text{st}} x \in X\, \exists\, y \in Y\, \Phi(x,y)$ and $Y = \{g(a) : a \in A \cap \mathbb{S}\}$, thus, we have $\forall^{\text{st}} x \in X\, \exists^{\text{st}} a \in A\, \Phi(x, g(a))$. By Lemma 3.1.16 (true in $\mathbf{BST}'_\kappa$), there is a standard map $h : X \to A$ such that $\forall^{\text{st}} x \in X\, \Phi(x, g(h(x)))$.

Inner Extension: follows from Inner S. S. Choice.

Inner Dependent Choice. Lemma 3.3.2 reduces the task to the case when the domain of Dependent Choice is a set of the form $X \cap \mathbb{S}$, where X is standard, and the binary relation considered is $R \cap \mathbb{S}$, where $R \subseteq X \times X$ is standard. Now apply Dependent Choice in **ZFC** and use Inner Transfer. □

The additional claim $(*)$ of Theorem 3.3.5 is also provable in $\mathbf{BST}'_\kappa$, but here the fact that the family of all "external subsets" of any set is a set-size collection can be obtained by a different and much simpler argument:

Exercise 3.3.8 ($\mathbf{BST}'_\kappa$). Let X be any set. According to Lemma 3.3.2, there exists a standard set Y and an internal map f defined on Y sich that $X = \{f(y) : y \in Y \cap \mathbb{S}\}$. Prove, using Standardization, that any "external set" (*i.e.*, a st-$\in$-definable subclass, not necessarily an internal set) $X' \subseteq X$ is equal to $\{f(y) : y \in Y' \cap \mathbb{S}\}$ for a suitable standard $Y' \subseteq Y$. Infer that the "external power set" $\mathscr{P}_{\text{ext}}(X)$ is a set-size collection. □

Problem 3.3.9. Does $\mathbf{BST}'_\kappa$ prove Reduction to Σ_2^{st} and Uniqueness? □

3.4 Development of Nelson's internal set theory

Despite the similarity of the axiomatic systems and many theorems common to them (mainly those provable in **BIST**), the absence of Inner Boundedness in **IST** leaves nonstandard sets in a looser connection with standard sets than **BST** does. Subsequently, despite the presence of Idealization, not all of the results of §§ 3.2b – 3.2g remain true in **IST**, and if something indeed remains true, as Inner Collection and Uniqueness, then this is due to arguments rather different from those in **BST** (see §§ 3.4c and 3.4d).

Reduction to $\Sigma_2^{\mathtt{st}}$ as in Theorem 3.2.3 fails in **IST** (this will be demonstrated in § 3.6c), subsequently, we don't know whether the problem of external sets can be solved in **IST** (see the end of § 3.4c).

As for Inner Extension, Inner S. S. Choice, Map-Standardization (as in § 3.2e), these principles turn out to be *undecidable* in **IST**, *i. e.*, as we shall see in Chapter 4, **IST** does not prove them, nor does it prove their negations. In addition, there are other interesting principles independent of **IST**, which we discuss in § 3.5c.

The other issue is that **BST** theorems turn out to be partial results in **IST**, because **BST** admits a model in the **IST** universe, namely, the class $\mathbb{B}$ of all *bounded* sets. In particular, Reduction to $\Sigma_2^{\mathtt{st}}$ still has a provable form sufficient for all practical purposes in Theorem 3.4.7.

Blanket agreement 3.4.1. We argue in **IST** in § 3.4 unless specified otherwise. $\mathbb{S} = \{x : \mathtt{st}\, x\}$ is the class of all standard sets, but $\mathbb{I}$ will denote the universe of *all sets in* **IST**. Elements of $\mathbb{I}$ are called *internal sets.* □

3.4a Bounded sets in IST

Our goal in this subsection is to reveal natural semantical connections between the theories **IST** and **BST**, and also the partially saturated version **BST**$_\kappa$ of **BST**.

Definition 3.4.2 (IST). A set x is *bounded*, briefly $\mathtt{bd}\, x$, iff there exists a standard set X containing x. The class of all bounded sets is denoted by

$$\mathbb{B} = \{x : x \text{ is bounded}\} = \{x : \exists^{\mathtt{st}} X\,(x \in X)\}\,.$$

Similarly to some definitions above, $\Phi^{\mathtt{bd}}$ indicates the relativization to $\mathbb{B}$, obtained by changing in Φ all quantifiers $\exists$, $\forall$ to resp. $\exists^{\mathtt{bd}}$, $\forall^{\mathtt{bd}}$, where $\exists^{\mathtt{bd}}$ and $\forall^{\mathtt{bd}}$ mean: *there is bounded...* , *for all bounded...* . (Occurrences of $\mathtt{st}$, in particular, in $\exists^{\mathtt{st}}$, $\forall^{\mathtt{st}}$, do not change.)

For any standard cardinal κ, let $\mathbb{I}_\kappa = \{x : \exists^{\mathtt{st}} X\,(x \in X \wedge \mathtt{card}\, X \le \kappa)\}$. The definition of $\mathbb{I}_\kappa$ is equally meaningful in **BST**. [20] □

[20] The classes $\mathbb{I}_\kappa$ will be redefined in **HST** and extensively studied in Chapter 6, in connection with the Power Set problem in **HST**.

Exercise 3.4.3 (IST). (1) Show that each nonstandard integer is bounded while any set C with $\mathbb{S} \subseteq C$ (Lemma 3.1.15) is unbounded.

(2) Prove that $\mathbb{S} \subseteq \mathbb{I}_\kappa \subseteq \mathbb{B}$. (*Hint*: if $x \in \mathbb{S}$ then $X = \{x\} \in \mathbb{S}$.)

(3) Show that $\mathbb{I}_\kappa \subsetneqq \mathbb{I}_{\kappa'}$ whenever $\kappa < \kappa'$ are standard cardinals.
Hint: use Idealization to get an ordinal $\xi < \kappa'$ which does not belong to any standard subset of κ' of cardinality κ. □

Thus, $\mathbb{S} \subsetneqq \mathbb{B} \subsetneqq \mathbb{I}$: both inclusions are strict in **IST**. Note that in **BST** the axiom of Inner Boundedness implies that every set is bounded.

Remark 3.4.4. The formulas $\mathtt{bd}\,x$ and $\mathtt{int}\,x$ (see § 1.1c) is just one and the same st-$\in$-formula that defines the class $\mathbb{I}$ of all internal sets in **HST** and the class $\mathbb{B}$ of all bounded sets in **IST**. This analogy goes quite far: in both cases the class separated by this formula satisfies the axioms of **BST**. This is established by Theorem 3.1.8 in **HST** and by (i) of the next theorem in **IST**.

Moreover it follows from Theorem 3.1.10 (for **HST**) and Theorem 3.4.5(iii) (for **IST**) that whatever **HST** proves to be true in $\mathbb{I}$ or **IST** proves to be true in $\mathbb{B}$ is a theorem of **BST**. Thus **BST** is, in an exact and rather strong sense, the theory of the class $\mathbb{I}$ in **HST** and the theory of the class $\mathbb{B}$ in **IST**.

Likewise $\mathbf{BST}_\kappa$ turns out to be the theory of the classes $\mathbb{I}_\kappa$, κ an infinite standard cardinal, in **BST** and hence in **HST** (with "cardinal" replaced by "$*$-cardinal") and in **IST** by Theorems 3.4.5(ii), (iii). [21] □

Theorem 3.4.5. (i) $\mathbb{B}$ *is an interpretation of* **BST** *in* **IST**;

(ii) *for any standard infinite cardinal* κ, $\mathbb{I}_\kappa$ *is an interpretation of* $\mathbf{BST}_\kappa$ *in* **IST** *and in* **BST**. [22]

(iii) *for any* st-$\in$-*sentence* Φ, **BST** *proves* Φ *iff* **IST** *proves* $\Phi^{\mathtt{bd}}$;

(iv) *for any* st-$\in$-*sentence* Φ, $\mathbf{BST}_\kappa$ *proves* Φ *iff* **BST** *proves that* $\Phi^{\mathbb{I}_\kappa}$ (the relativization to $\langle \mathbb{I}_\kappa ; \in, \mathtt{st}\rangle$) *holds for any standard infinite cardinal* κ.

Proof. (i) We have to check that all axioms of **BST** hold in $\langle \mathbb{B}; \in, \mathtt{st}\rangle$. Let us begin with Inner Transfer. To prove $\Phi^{\mathtt{bd}} \Longleftrightarrow \Phi^{\mathtt{st}}$ for any $\in$-formula Φ with standard parameters, it suffices to show that

(1) $\exists^{\mathtt{bd}} x\, \varphi^{\mathtt{bd}}(x) \Longrightarrow \exists^{\mathtt{st}} x\, \varphi^{\mathtt{bd}}(x)$, for any $\in$-formula φ with parameters in $\mathbb{S}$.

This would immediately follow from the **IST** Inner Transfer could we prove that $\varphi^{\mathtt{bd}} \Longleftrightarrow \varphi$ holds for any $\in$-formula φ with parameters in $\mathbb{B}$, that is, a sort of Inner Transfer between $\mathbb{B}$ and $\mathbb{I}$, which is a consequence of

(2) $\exists\, x\, \varphi(x) \Longrightarrow \exists^{\mathtt{bd}} x\, \varphi(x)$, for any $\in$-formula φ with parameters in $\mathbb{B}$.

Let us concentrate on this assertion.

Suppose w. l. o. g. that φ contains a single parameter p_0. We fix a standard set P such that $p_0 \in P$. The implication in (2) takes the form

[21] See § 6.2b on the other partially saturated theory $\mathbf{BST}'_\kappa$ in this respect.

[22] Precisely it is asserted that the structures $\langle \mathbb{B}; \in, \mathtt{st}\rangle$ and $\langle \mathbb{I}_\kappa ; \in, \mathtt{st}, \kappa\rangle$ are interpretations of, resp., **BST** and $\mathbf{BST}_\kappa$ (the cardinal κ interprets the constant $\boldsymbol{\kappa}$).

$$\exists z\, \varphi(z, p_0) \implies \exists z \in \mathbb{B}\, \varphi(z, p_0),$$

where $\varphi(z,p)$ is a parameter-free $\in$-formula. The idea of the proof is to forget about the special parameter p_0 and consider the problem for *all* relevant p. By the **ZFC** Collection axiom, there is a set Z such that

$$\forall p \in P\, \big(\exists z\, \varphi(z,p) \implies \exists z \in Z\, \varphi(z,p)\big)\,.$$

We can choose a *standard* Z with this property by the **IST** Transfer. Setting $p = p_0$, we obtain the required result.

This ends the proof of Inner Transfer in $\mathbb{B}$. Now we immediately have $\mathbf{ZFC}^{\mathtt{st}}$ in $\mathbb{B}$. Inner Standardization, Inner Boundedness, and Basic Idealization (since standard sets retain their elements in $\mathbb{B}$) are left as an **exercise**.

(ii) If $p_0 \in \mathbb{I}_\kappa$ in the argument for (i) then we can choose the sets P and Z so that $\mathtt{card}\, P \le \kappa$ and $\mathtt{card}\, Z \le \kappa$. Then all elements of Z still belong to $\mathbb{I}_\kappa$. This modification of the argument for $\mathbb{B}$ yields Inner Transfer in $\mathbb{I}_\kappa$. That the other axioms of $\mathbf{BST}_\kappa$ hold in $\mathbb{I}_\kappa$ is now a matter of a short routine verification, which we leave as an **exercise** for the reader. (This includes $\boldsymbol{\kappa}$-deep BI, where we have to keep in mind that any standard set X of cardinality $\le \kappa$ retains all its elements in $\mathbb{I}_\kappa$.)

(iii), (iv): the proof follows in § 4.1d. □

Corollary 3.4.6. *Generally, all results of §§ 3.2b–3.2g remain true in* **IST**, *provided the formulas involved are relativized to the class* $\mathbb{B}$ *of all bounded sets, and all sets involved belong to* $\mathbb{B}$. □

We leave it as an **exercise** for the reader to "concretize" this Corollary for all particular results of §§ 3.2b–3.2g. As for Reduction to $\Sigma_2^{\mathtt{st}}$ and its corollaries in § 3.2b, this issue deserves special attention.

3.4b Bounded formulas: reduction to two "external" quantifiers

The results of § 3.2b, generally speaking, fail for **IST**, in particular, there is a $\mathtt{st}$-$\in$-sentence not provably equivalent in **IST** to any $\in$-sentence (Exercise 3.5.8), and there is a $\mathtt{st}$-$\in$-formula provably not equivalent in **IST** to any $\Sigma_2^{\mathtt{st}}$ formula (see § 3.6c). Yet there is a large family of *bounded* $\mathtt{st}$-$\in$-formulas which admit reduction to $\Sigma_2^{\mathtt{st}}$. Informally, a $\mathtt{st}$-$\in$-formula is bounded if all its variables and quantifiers are restricted by standard sets, and there are no occurrences of $\mathtt{st}$ except for those in the quantifiers $\exists^{\mathtt{st}}$, $\forall^{\mathtt{st}}$. Surprisingly it is not so easy to convert this into a formal definition, yet fortunately it is clear that any such formula is automatically relativized to the class $\mathbb{B}$ of all bounded sets. This leads to the following reduction theorem.

Theorem 3.4.7. *Any* $\mathtt{st}$-$\in$*-formula* $\Phi(x_1, ..., x_m)$ *relativized to* $\mathbb{B}$ *is equivalent in* **IST** *to a* $\Sigma_2^{\mathtt{st}}$ *formula. Thus it is asserted that for any* $\mathtt{st}$-$\in$*-formula* $\Phi(x_1, ..., x_m)$ *there is an* $\in$*-formula* $\varphi(x_1, ..., x_m, a, b)$ *such that* **IST** *proves:*

$$\forall^{\mathtt{bd}} x_1 \dots \forall^{\mathtt{bd}} x_m\, \big(\Phi^{\mathtt{bd}}(x_1, ..., x_m) \iff \exists^{\mathtt{st}} a\, \forall^{\mathtt{st}} b\, \varphi(x_1, ..., x_m, a, b)\big)\,.$$

Proof. It follows from Theorem 3.2.3 that there is an $\in$-formula φ such that

$$\forall x_1 \dots \forall x_m \left(\Phi(x_1, \dots, x_m) \iff \exists^{\mathtt{st}} a\, \forall^{\mathtt{st}} b\, \varphi(x_1, \dots, x_m, a, b)\right)$$

is a theorem of **BST**. In particular the displayed sentence holds in $\mathbb{B}$ because this class satisfies **BST** by Theorem 3.4.5(i). □

Corollary 3.4.8. *Let $\Phi(x_1, \dots, x_m)$ be any* st-$\in$*-formula. Then there exists an $\in$-formula $\psi(x_1, \dots, x_m)$ such that the following is a theorem of* **IST** :

$$\forall^{\mathtt{st}} x_1 \dots \forall^{\mathtt{st}} x_m \left(\Phi^{\mathtt{bd}}(x_1, \dots, x_m) \iff \psi^{\mathtt{st}}(x_1, \dots, x_m) \iff \psi(x_1, \dots, x_m)\right).$$

Thus, in **IST**, any st-$\in$-formula with standard parameters and relativized to $\mathbb{B}$ is provably equivalent to an $\in$-formula with the same list of parameters.

Proof. Let φ be the formula given by Theorem 3.4.7. By Inner Transfer, the formula $\exists\, a\, \forall\, b\, \varphi(x_1, \dots, x_m, a, b)$ can be taken as ψ. □

3.4c Collection in IST

The following theorem is analogous to Theorem 3.2.8 + Exercise 3.2.9, and to Theorem 1.4.9 in **HST**, in particular in matters of the inequality $\mathtt{card}\, Y \le H \cdot \mathtt{card}\, X$, and the proof is based on the same idea (similar, in a sense, to Scott's analysis of infinitary sentences), yet here the technical realization is somewhat more complicated.

Theorem 3.4.9 (IST). *Inner Collection* (see § 3.2d) *holds.*

If, in addition, $h \in \mathbb{N}$ is nonstandard (i.e., infinitely large) *then Y in Inner Collection can be chosen so that* $\mathtt{card}\, Y \le h \cdot \mathtt{card}\, X$.

Proof. Let X be any set and $\Phi(x, y)$ any st-$\in$-formula. We argue in **IST**. Using the fact that the class $\mathbb{S}$ of all standard sets can be covered by a finite set of an arbitrary $*$-finite number of elements, we'll define an internal map F taking values in a finite set of cardinality $\le h$, such that $\Phi(x, y) \iff \Phi(x, y')$ whenever x, y, y' satisfy $F(x, y) = F(x, y')$. This will allow to use the **ZFC** Choice to get an internal function $\widetilde{Y}$ defined on X and assigning a finite set $\widetilde{Y}(x)$ of cardinality $\le h$ to each $x \in X$ so that for any y there is $y' \in \widetilde{Y}(x)$ with $F(x, y) = F(x, y')$. The proof will be completed by $Y = \bigcup_{x \in X} \widetilde{Y}(x)$.

Now let us consider details. We can assume that $\Phi(x, y)$ has the form

$$\mathsf{Q}_2\, x_2 \dots \mathsf{Q}_m\, x_m\, \varphi(x, y, x_2, \dots, x_m),$$

where φ is a quantifier-free $\in$-formula and quantifiers Q_i, $i = 2, \dots, m$, are among $\exists$, $\forall$, $\exists^{\mathtt{st}}$, $\forall^{\mathtt{st}}$, independently of each other. Define Q_1 to be $\exists$, to correspond to the formula $\exists\, y\, \Phi(x, y)$.

It will be convenient to replace x by x_0 and y by x_1.

The first part of the proof contains a construction carried out in the **ZFC** part of **IST**. Let Z be an arbitrary set. Induction on $k = m, m-1, ..., 2, 1$ is used to define a set Z_k and a map

$$x_0, ..., x_k \longmapsto F_Z(x_0, ..., x_k) \in Z_k .$$

Thus $F_Z(x_0, ..., x_k)$ is being defined for *all* sets $x_0, ..., x_k$.

Base of induction: $k = m$. Put $Z_m = \{0,1\}$ and, for all $x_0, ..., x_m$, define

$$F_Z(x_0, ..., x_m) = \begin{cases} 1, \text{ if } \varphi(x_0, ..., x_m) \text{ is true} \\ 0, \text{ if } \varphi(x_0, ..., x_m) \text{ is false} \end{cases} .$$

Induction step. If Q_{k+1} is either $\exists^{\text{st}}$ or $\forall^{\text{st}}$ then we put $Z_k = {Z_{k+1}}^Z = \{f : Z \to Z_{k+1}\}$ and, given $x_{k+1} \in Z$ and arbitrary sets $x_0, ..., x_k$, define

$$F_Z(x_0, ..., x_k)(x_{k+1}) = F_Z(x_0, ..., x_k, x_{k+1}).$$

Thus, for all $x_0, ..., x_k$, $F_Z(x_0, ..., x_k)$ is a function defined on Z (by the last displayed formula) and taking values among elements of Z_{k+1}. If Q_{k+1} is either $\exists$ or $\forall$ we put $Z_k = \mathscr{P}(Z_{k+1})$ and

$$\begin{aligned} F_Z(x_0, ..., x_k) &= \{F_Z(x_0, ..., x_k, x_{k+1}) : x_{k+1} \text{ is an arbitrary set}\} = \\ &= \{c \in Z_{k+1} : \exists x_{k+1}\ (c = F_Z(x_0, ..., x_k, x_{k+1}))\} \quad . \end{aligned}$$

Thus, for all $x_0, ..., x_k$, $F(x_0, ..., x_k)$ is a subset of Z_{k+1}.

Lemma 3.4.10. *Assume that Z contains all standard sets. Suppose that the sets $x_0, ..., x_k$, $x'_0, ..., x'_k$ satisfy $F_Z(x_0, ..., x_k) = F_Z(x'_0, ..., x'_k)$. Then*

$$\begin{aligned} &\mathsf{Q}_{k+1}\, x_{k+1} \ldots \mathsf{Q}_m\, x_m\ \varphi(x_0, ..., x_k, x_{k+1}, ..., x_m) \iff \\ &\qquad \iff \mathsf{Q}_{k+1}\, x'_{k+1} \ldots \mathsf{Q}_m\, x'_m\ \varphi(x'_0, ..., x'_k, x'_{k+1}, ..., x'_m) . \end{aligned}$$

Proof. We proceed by descending induction on k. The case $k = m$ is evident. As for the step, we consider only the case when Q_{k+1} is either $\exists$ or $\exists^{\text{st}}$ (universal quantifiers do not differ significantly) and the direction $\Longrightarrow$.

Let Q_{k+1} be $\exists^{\text{st}}$. Thus assume that there is a standard x_{k+1} such that

$$\mathsf{Q}_{k+2}\, x_{k+2} \ldots \mathsf{Q}_m\, x_m\ \varphi(x_0, ..., x_k, x_{k+1}, x_{k+2}, ..., x_m) . \qquad (\dagger)$$

We note that $x_{k+1} \in Z$ by the choice of Z, thus

$$\begin{aligned} F_Z(x_0, ..., x_k, x_{k+1}) &= F_Z(x_0, ..., x_k)(x_{k+1}) = \\ &= F_Z(x'_0, ..., x'_k)(x_{k+1}) = F_Z(x'_0, ..., x'_k, x_{k+1}) \end{aligned}$$

by the definition of F_Z. Hence, by the induction hypothesis,

$$\mathsf{Q}_{k+2}\, x'_{k+2} \ldots \mathsf{Q}_m\, x'_m\ \varphi(x'_0, ..., x'_k, x_{k+1}, x'_{k+2}, ..., x'_m) .$$

This completes the proof of the right-hand side.

We consider the case when Q_{k+1} is $\exists$. Assume that there exists x_{k+1} satisfying (†). By the definition of F_Z there is x'_{k+1} such that

$$F_Z(x_0, ..., x_k, x_{k+1}) = F_Z(x'_0, ..., x'_k, x'_{k+1}).$$

To complete the proof apply the induction hypothesis. □ *(Lemma)*

The case $k = 1$ is especially interesting. Recall that x_1 is y and x_0 is x.

Corollary 3.4.11. *If Z contains all standard sets, $F_Z(x, y) = F_Z(x, y')$, and $\Phi(x, y)$ holds, then $\Phi(x, y')$ holds, too.* □

Let, for any natural ℓ and $k \le m$, $g(k, \ell)$ be defined by induction on $k = m, m-1, ..., 2, 1$ by $h(m, \ell) = 2$ and

$$g(k,\ell) = \begin{cases} g(k+1,\ell)^\ell & \text{, if } Q \text{ is either } \exists^{st} \text{ or } \forall^{st} \\ 2^{g(k+1,\ell)} & \text{, if } Q \text{ is either } \exists \text{ or } \forall \end{cases} \quad : \text{ for all } k < m.$$

Let, further, $g(\ell) = g(1, \ell)$. Then $g(k, \ell)$ is exactly the number $\#(Z_k)$ of elements in Z_k provided Z has ℓ elements. In particular $g(\ell) = \#(Z_1)$.

We introduce the following internal formula:

$\boldsymbol{\theta}(Z, x, Y)$: Z is finite, Y is finite and contains $\le g(\#Z)$ elements, and for any y there is $y' \in Y$ such that $F_Z(x, y) = F_Z(x, y')$.

Lemma 3.4.12. *For any x and any finite Z there is Y with $\boldsymbol{\theta}(Z, x, Y)$.*

Proof. Let $\ell = \#Z$. By definition, $F_Z(x, y) \in Z_1$, and Z_1 contains at most $g(\ell)$ elements. Therefore there is a set Y, $\#Y = g(\ell)$, such that each value $F_Z(x, y)$ belongs to $\{F_Z(x, y) : y \in Y\}$. □

Lemma 3.4.13. *If Z contains all standard sets, then $\boldsymbol{\theta}(Z, x, Y)$ implies*

$$\exists y\, \Phi(x, y) \Longrightarrow \exists y \in Y\, \Phi(x, y).$$

Proof. Apply Corollary 3.4.11 and the definition of the formula $\boldsymbol{\theta}$. □

We come back to the proof of Theorem 3.4.9. Since g is an increasing function, there exists a natural number L such that $g(L) \le h < g(L+1)$. Then L is infinitely large because h is infinitely large while g is a standard function. By Idealization there exists a finite set Z of cardinality L containing all standard sets. As $\boldsymbol{\theta}$ is an internal formula, by Lemma 3.4.12 and the **ZFC** Choice there is a function $\widetilde{Y}$ such that $\widetilde{Y}(x)$ is defined and satisfies $\boldsymbol{\theta}(Z, x, \widetilde{Y}(x))$ for all $x \in X$. Lemma 3.4.13 implies that the set $Y = \bigcup_{x \in X} \widetilde{Y}(x)$ is as required.

□ *(Theorem 3.4.9)*

Unfortunately Theorem 3.4.9 does not seem to lead to a full solution of the problem of "external sets" in **IST** (see § 3.2f): the restricted character of Theorem 3.4.7 implies the restricted character of the solution.

Problem 3.4.14. Is there a parametrization of all "external sets" in **IST**, *i. e.*, a st-∈-formula $\Psi(x,p)$ such that for any st-∈-formula $\Phi(x,y)$ we can prove in **IST** that $\forall X\,\forall y\,\exists p\,\forall x \in X\;(\Phi(x,y) \Longleftrightarrow \Psi(x,p))$? □

Theorems 3.2.16 and 3.4.5(i) enable us to solve the problem in the positive, provided we consider only those formulas Φ which are relativized to the class $\mathbb{B}$ of all bounded sets. As the latter is the domain of all reasonable applications of **IST**, we can claim a pragmatically sufficient positive solution of the problem of "external sets" in **IST**, yet the problem in principle remains open. We would expect the negative answer.

3.4d Uniqueness in IST

The principle of Uniqueness as defined in § 3.2e fails in **IST** for trivial reasons: there exist sets X which do not belong to any standard set, in particular, do not belong to the domain of any standard function. For instance any set X with $\mathbb{S} \subseteq X$ is of this sort. However a partial result survives. Recall that $\mathbb{B} = \{x : \exists^{\text{st}} y\,(x \in y)\}$, the class of all bounded sets in **IST**, and, for $x \in \mathbb{B}$, $\mathbb{S}[x]$ is the class of all sets of the form $f(x)$, where f is a standard function with $x \in \operatorname{dom} f$ (see Definition 3.1.13).

Exercise 3.4.15 (IST). Why is $\mathbb{S}[x] = \varnothing$ for any $x \notin \mathbb{B}$? □

Theorem 3.4.16 (IST, *Uniqueness*). *Suppose that $\Phi(x,y)$ is a parameter-free* st-∈-*formula. If $x \in \mathbb{B}$ and there is a unique set y with $\Phi(x,y)$ then this y belongs to $\mathbb{S}[x]$. In particular if x is standard then y is standard.*

Proof. We use the notation from the proof of Theorem 3.4.9. Assume that a finite set Z contains all standard sets, and Y satisfies $\boldsymbol{\theta}(Z,x,Y)$ (see above). Then $y \in Y$ by Lemma 3.4.13. Thus, formally,

$$\forall Z\,\forall Y\,\big(\forall^{\text{st}} z\,(z \in Z) \wedge \boldsymbol{\theta}(Z,x,Y) \Longrightarrow y \in Y\big).$$

Applying Idealization, we obtain a standard finite set Z such that

$$\forall Y\,\big(\boldsymbol{\theta}(Z,x,Y) \Longrightarrow y \in Y\big).$$

Now, let X be a standard set containing x. According to Lemma 3.4.12, for any $x' \in X$ there is a set Y with $\boldsymbol{\theta}(Z,x',Y)$, hence, $\#Y = g(\#Z)$. As X and Z are standard, we can apply Choice and Inner Transfer, getting a standard function $\widetilde{Y}$ defined on X so that $\boldsymbol{\theta}(Z,x',\widetilde{Y}(x'))$ holds for all $x' \in X$, in particular, we have $\boldsymbol{\theta}(Z,x,\widetilde{Y}(x))$. Moreover, as any $\widetilde{Y}(x')$, $x' \in X$, contains $\le n$ elements, where $n = g(\#Z)$ is standard, there is another standard function η defined on $X \times \{0,1,...,n-1\}$ so that $\widetilde{Y}(x) = \{\eta(x,k) : k < n\}$. It remains to note that $\eta(x,k) \in \mathbb{S}(x)$ whenever k is standard. □

3.5 Truth definition in internal set theory

This subsection is devoted to a remarkable property of the class $\mathbb{S}$ of all standard sets in **IST**. We prove that there exists a uniform truth definition for all $\in$-formulas with standard parameters, or, what amounts to the same by Transfer, a uniform truth definition for $\in$-formulas in $\mathbb{S}$. This is rather surprising because the existence of a truth definition usually witnesses that the theory that gives the definition is essentially stronger than the underlying theory; but **ZFC** and **IST** are equiconsistent.

These considerations lead (see § 3.5c) to new principles, in st-$\in$-language, which have no reasonable analogies for **BST** yet naturally strengthen **IST** in the direction of more regularity of the class of all standard sets.

We shall argue mainly in **IST** below in Section 3.5.

3.5a Truth definition for the standard universe

We have to provide here a careful distinction between set theoretic ("mathematical") and metamathematical notions related to the $\in$-language, truth of $\in$-formulas, and so on. To simplify the arguments, we restrict the $\in$-language at the formal level to the symbols $\neg$, $\wedge$, $\exists$, $\in$, $=$, brackets (and), variables v_i, $i \in \mathbb{N}$, and arbitrary sets as parameters. We define, at the formal level, Form to be the class of all finite sequences, whose elements are among

- integers 0, 1, 2, 3, 4, 5, 6, identified with the logical signs $\neg$, $\wedge$, $\exists$, $\in$, $=$, and the brackets (and) respectively;
- pairs $\langle 0, i\rangle$, $i \in \mathbb{N}$, identified with variables v_i; and
- pairs $\dot{x} = \langle 1, x\rangle$, where x is an arbitrary set; $\dot{x}$ is identified with the set x when the latter is used as a parameter;

and which satisfy the usual conditions of being an $\in$-formula in the sense of the mentioned identification. Lower case φ, ψ are used in this section to denote elements of Form, called *coded formulas*.

The Gödel numbering symbol $\ulcorner ... \urcorner$ will be used to denote the function restoring the correct coding as sequences, so that, for instance, $\ulcorner x = y \urcorner$ is the sequence $\langle\langle 1, x\rangle, 4, \langle 1, y\rangle\rangle$, $\ulcorner \varphi \wedge \psi \urcorner$ is $\langle 5, \varphi, 6, 1, 5, \psi, 6\rangle$ while $\ulcorner \exists\, v_i\, \varphi \urcorner$ is the sequence $\langle 2, \langle 0, i\rangle, 5, \varphi, 6\rangle$. A writing like $\varphi(v_i) \in \mathtt{Form}$ means that the pair $\langle 0, i\rangle$ occurs in φ in a certain proper manner, and if, in this case, x is any set then $\ulcorner \varphi(x) \urcorner$ means that the pair $\langle 0, i\rangle$ is replaced by $\dot{x} = \langle 1, x\rangle$.

A coded formula $\psi \in \mathtt{Form}$ is *subordinate* to φ if ψ is a subformula of φ in which some (maybe none or all) free variables are replaced by arbitrary sets as parameters. For instance, φ itself is subordinate to φ; $\ulcorner \varphi(x) \urcorner$ for all x, and $\varphi(v_i)$ (v_i is free) are subordinate to $\exists\, v_i\, \varphi(v_i)$. We put

$$\mathtt{Form}[\varphi] = \{\psi \in \mathtt{Form} : \ \psi \text{ is subordinate to } \varphi\}.$$

For example $\ulcorner \varphi(x) \urcorner \in \mathtt{Form}[\exists\, v_i\, \varphi(v_i)]$ for all x. Distinguish closed formulas:

$$\mathtt{ClForm} = \{\varphi \in \mathtt{Form} : \varphi \text{ is a closed formula}\},$$

and $\mathtt{ClForm}[\varphi]$ the same way.

One can easily determine in **IST** whether a given sequence belongs to `Form` and, if so, compute its length (the number of symbols) and distinguish the set of involved parameters.

Exercise 3.5.1. Prove that a coded formula $\varphi \in \mathtt{Form}$ is standard iff it has standard length and standard parameters. □

The following definition establishes the formal notion of a set of coded formulas satisfying what should be considered as a truth definition with respect to truth within the universe $\mathbb{S}$ of all standard sets. The idea is clear: the set has to meet several evident rules (*Tarski rules*) which connect the truth of a formula with the truth of subformulas.

Definition 3.5.2. $\mathtt{Truth}\,T$ is the conjunction of the following $\mathtt{st}$-$\in$-formulas:

(1) $T \subseteq \mathtt{ClForm}$;

(2) $\forall^{\mathtt{st}} p\, \forall^{\mathtt{st}} q\, \big((\ulcorner p = q \urcorner \in T \iff p = q) \wedge (\ulcorner p \in q \urcorner \in T \iff p \in q)\big)$;

(3) for any standard φ, ψ: $\ulcorner \varphi \wedge \psi \urcorner \in T \implies \varphi \in T \wedge \psi \in T$, and
$\ulcorner \neg\, (\varphi \wedge \psi) \urcorner \in T \implies \ulcorner \neg\, \varphi \urcorner \in T \vee \ulcorner \neg\, \psi \urcorner \in T$;

(4) for any standard φ : $\ulcorner \neg\, \varphi \urcorner \in T \implies \varphi \notin T$ and
$\ulcorner \neg \neg\, \varphi \urcorner \in T \implies \varphi \in T$;

(5) for any standard $\varphi(v_i)$: $\ulcorner \exists\, v_i\, \varphi(v_i) \urcorner \in T \Longrightarrow \exists^{\mathtt{st}} x\, (\ulcorner \varphi(x) \urcorner \in T)$ and
$\ulcorner \neg\, \exists\, v_i\, \varphi(v_i) \urcorner \in T \Longrightarrow \forall^{\mathtt{st}} x\, (\ulcorner \neg\, \varphi(x) \urcorner \in T)$.

Sets T satisfying $\mathtt{Truth}\,T$ are called *truth sets for* $\mathbb{S}$. [23]

A coded formula $\varphi \in \mathtt{Form}$ is *formally true* (f. true in brief) in $\mathbb{S}$ if there is a truth set T for $\mathbb{S}$ containing φ, and is *formally false* (f. false) in $\mathbb{S}$ iff $\ulcorner \neg\, \varphi \urcorner$ is f. true. Thus "φ is f. true (false) in $\mathbb{S}$" are $\mathtt{st}$-$\in$-formulas with φ as the unique free variable. □

Proposition 3.5.3. *The union of two truth sets is a truth set.*

Proof. We first note that if T_1, T_2 satisfy $\mathtt{Truth}\,T_1$ and $\mathtt{Truth}\,T_2$, but $\mathtt{Truth}(T_1 \cup T_2)$ fails, then there is a standard $\varphi \in \mathtt{ClForm}$ such that $\varphi \in T_1$ and $\ulcorner \neg\, \varphi \urcorner \in T_2$ or *vice versa*. Prove that this cannot happen, by induction on the (standard) number of symbols in φ.

If φ is an elementary coded formula then (2) and (4) of Definition 3.5.2 give the result immediately. If φ is $\ulcorner \neg\, \psi \urcorner$ then, $\psi \in T_2$ by (4) of Definition 3.5.2, but still $\ulcorner \neg\, \psi \urcorner \in T_1$, so that we have the controversy for a simpler

[23] Note that sets T satisfying $\mathtt{Truth}\,T$ are, generally speaking, **partial** truth sets, in the sense that it is not required that for any relevant formula φ T contains φ or its negation. This is by necessity: most likely the existence of such **complete** truth sets is too strong to be provable in **IST**.

coded formula ψ, which contradicts the induction hypothesis. If φ is $\ulcorner \psi \wedge \chi \urcorner$ then, by (3) of Definition 3.5.2, both ψ and χ belong to T_1 but at least one of $\ulcorner \neg\, \psi \urcorner$, $\ulcorner \neg\, \chi \urcorner$ belongs to T_2, so that still the controversy is reproduced at a lower level. Finally, if φ is $\exists\, v_i\, \psi(v_i)$ then, by (5) of Definition 3.5.2, $\ulcorner \psi(x) \urcorner \in T_1$ for a standard x, but $\ulcorner \neg\, \psi(x) \urcorner \in T_2$ for all standard x, thus, the controversy is again reproduced. □

The next theorem shows that the notion of formal truth, based on truth sets, satisfies all reasonable requirements which any definition of truth in $\mathbb{S}$ has to satisfy.

Theorem 3.5.4. *The following statements are theorems of* **IST**.

(i) *Let* $x,\, y \in \mathbb{S}$. *Then* $\ulcorner x \in y \urcorner$ *is f. true in* $\mathbb{S}$ *iff* $x \in y$. *The same for* $x = y$.

(ii) *A standard* $\varphi \in \mathtt{ClForm}$ *cannot be both f. true and f. false in* $\mathbb{S}$.

(iii) $\ulcorner \varphi \wedge \psi \urcorner \in \mathtt{ClForm}$ *is f. true in* $\mathbb{S}$ *iff both* φ *and* ψ *are f. true in* $\mathbb{S}$.

(iv) $\ulcorner \exists\, v_i\, \varphi(v_i) \urcorner \in \mathtt{ClForm}$ *is f. true in* $\mathbb{S}$ *iff* $\exists^{\mathrm{st}} x\, (\ulcorner \varphi(x) \urcorner$ *is f. true in* $\mathbb{S})$.

Proof. (i) To prove the nontrivial direction take a set X satisfying $\mathbb{S} \subseteq X$ (Lemma 3.1.15). The set T of all coded formulas $\ulcorner p = p \urcorner$, where $p \in X$, and $\ulcorner p \in q \urcorner$, where $p,\, q \in X$ and $p \in q$, is a truth set.

(iv) The nontrivial direction is to show that if $\ulcorner \varphi(x) \urcorner$ is f. true in $\mathbb{S}$ for some standard x then $\ulcorner \exists\, v_i \varphi(v_i) \urcorner$ also is f. true. To see this take a set T with $\mathtt{Truth}\, T$ containing $\ulcorner \varphi(x) \urcorner$; then $T' = T \cup \{\ulcorner \exists\, v_i\, \varphi(v_i) \urcorner\}$ satisfies $\mathtt{Truth}\, T'$.

(ii) immediately follows from Proposition 3.5.3.

(iii) The nontrivial direction is to prove that if $\varphi,\, \psi \in \mathtt{ClForm}$ are f. true in $\mathbb{S}$ then $\ulcorner \varphi \wedge \psi \urcorner$ also is f. true. We can assume, by Proposition 3.5.3, that one and the same set T with $\mathtt{Truth}\, T$ contains both φ and ψ. Then the set $T' = T \cup \{\ulcorner \varphi \wedge \psi \urcorner\}$ satisfies $\mathtt{Truth}\, T'$ and witnesses that $\varphi \wedge \psi$ is f. true. □

3.5b Connection with the ordinary truth

The definition of being f. true or f. false might cause some trouble since we have already had the notion of truth in $\mathbb{S}$ given by the relativization Φ^{st}. Fortunately this is not the case. To formulate the relevant theorem, note that the function $\ulcorner ... \urcorner$ can also be applied for metamathematically given formulas. For instance if $\Phi(x,y)$ is $\exists\, v_3\, (v_3 \in x \wedge v_3 \notin y)$ then

$$\ulcorner \Phi(x,y) \urcorner = \langle 2, \langle 0,3 \rangle, 5, \langle 0,3 \rangle, 3, \langle 1,x \rangle, 1, 0, 5, \langle 0,3 \rangle, 3, \langle 1,y \rangle, 6, 6 \rangle \in \mathtt{Form}.$$

Theorem 3.5.5. *Let* $\Phi(v_1, ..., v_n)$ *be an* $\in$*-formula. Then* **IST** *proves that:*

$$\forall^{\mathrm{st}} x_1 \ldots \forall^{\mathrm{st}} x_n \big(\Phi^{\mathrm{st}}(x_1, ..., x_n) \iff (\ulcorner \Phi(x_1, ..., x_n) \urcorner \text{ is f. true in } \mathbb{S}) \iff$$
$$\iff (\ulcorner \Phi(x_1, ..., x_n) \urcorner \text{ is not f. false in } \mathbb{S}) \big).$$

Proof. We leave it as a routine **exercise** to prove the theorem by straightforward induction on the length of formulas using Theorem 3.5.4 and the following lemma:

Lemma 3.5.6. *Let $\Phi(v_1, ..., v_n)$ be an $\in$-formula. Then* **IST** *proves that: there is a truth set T for $\mathbb{S}$ such that, for any tuple of standard $x_1, ..., x_n$, one of the formulas $\ulcorner \Phi(x_1, ..., x_n) \urcorner$, $\ulcorner \neg \Phi(x_1, ..., x_n) \urcorner$ belongs to T.*

To prove the lemma, let $\Phi_i(v_1, ..., v_{n(i)})$, $1 \le i \le m$, be the list of all subformulas (including Φ itself) of the formula $\Phi(v_1, ..., v_n)$. Let C be a set that contains all standard sets. We define

$$T_i = \{\ulcorner \Phi_i(x_1, ..., x_{n(i)}) \urcorner : x_1, ..., x_{n(i)} \in C \wedge \Phi_i(x_1, ..., x_{n(i)})\} \cup \\ \cup \{\ulcorner \neg \Phi_i(x_1, ..., x_{n(i)}) \urcorner : x_1, ..., x_{n(i)} \in C \wedge \neg \Phi_i(x_1, ..., x_{n(i)})\}.$$

The set $T = \bigcup_{1 \le i \le m} T_i$ is as required: that it satisfies $\texttt{Truth}\,T$ easily follows from Inner Transfer. □ (*Lemma and Theorem 3.5.5*)

Remark 3.5.7. It will be of crucial importance in an argument in § 4.5d that **IST** can be replaced by **IST**[**ZC**], a weaker theory [24], in Theorem 3.5.5 — in other words the schemata Collection$^{\mathtt{st}}$ and Replacement$^{\mathtt{st}}$ of **IST** are irrelevant to the proof.

Detailed analysis of the proof of Theorem 3.5.5 shows that Idealization is involved insofar that it implies the existence of a set C with $\mathbb{S} \subseteq C$, while Inner Standardization can be fully eliminated. Inner Transfer is *sine qua non* in the proof of Lemma 3.5.6, of course. □

The following result shows that the notion of being f. true can be used to obtain a formula not equivalent in **IST** to any $\Sigma_2^{\mathtt{st}}$ formula.

Exercise 3.5.8. Let $\Theta(\varphi)$ be the $\mathtt{st}$-$\in$-formula saying: $\varphi \in \mathtt{ClForm}$ is standard and f. true in $\mathbb{S}$. Prove that, in **IST**, Θ is not equivalent to any $\Sigma_2^{\mathtt{st}}$ formula.

Hint. Otherwise there is an $\in$-formula $\sigma(\cdot)$ such that $\Theta(\varphi) \iff \sigma(\varphi)$ for any standard $\varphi \in \mathtt{ClForm}$. For any $\in$-formula $\Phi(x)$ we have:

$$\Phi(x) \iff \Phi(x)^{\mathtt{st}} \iff \Theta(\ulcorner \Phi(x) \urcorner) \iff \sigma(\ulcorner \Phi(x) \urcorner)^{\mathtt{st}} \iff \sigma(\ulcorner \Phi(x) \urcorner).$$

for any $x \in \mathbb{S}$ by Theorem 3.5.5, hence, $\Phi(x) \iff \sigma(\ulcorner \Phi(x) \urcorner)$ for any x. Thus any $\in$-formula Φ is reducible to σ by a simple substitution $x \mapsto \ulcorner \Phi(x) \urcorner$. But this can be shown to be impossible. For instance if σ is a Σ_n formula then take any properly Σ_{n+1} formula Φ. □

---

[24] Recall that **IST**[**ZC**] consists of Inner Transfer, Idealization, Inner Standardization, and **ZC**$^{\mathtt{st}}$ instead of **ZFC**$^{\mathtt{st}}$, see § 3.1b.

3.5c Extension of the definition of formal truth

Let us study the notion of "formal truth", introduced by Definition 3.5.2, with respect to standard formulas (*i.e.*, those with standard length and standard parameters) considered as internal objects of **IST**.

It is not known whether the following hypothesis is provable in **IST**:

Formal Truth Completeness: If $\varphi \in \mathtt{ClForm}$ is standard then either φ or $\ulcorner\neg\,\varphi\urcorner$ is f. true in $\mathbb{S}$.

Yet we shall see that Formal Truth Completeness is a consequence of

$\mathbb{S}$-*Separation*: $\exists\, X\, \forall^{\mathtt{st}} x\, \big(x \in X \Longleftrightarrow \Phi(x)\big)$,
where $\Phi(x)$ is any $\mathtt{st}$-$\in$-formula with arbitrary parameters.

(One cannot, in general, replace $\exists\, X$ by $\exists^{\mathtt{st}} X$, for take $\Phi(x)$ to be the formula $x = x$). The following is a similar version of Choice:

$\mathbb{S}$-*Size Choice*: $\forall^{\mathtt{st}} x\, \exists\, y\, \Phi(x,y) \Longrightarrow \exists\, F\, \forall^{\mathtt{st}} x\, \Phi(x, F(x))$, [25]
where $\Phi(x,y)$ is any $\mathtt{st}$-$\in$-formula with arbitrary parameters and F postulated to exist is assumed to be a function with $\mathbb{S} \subseteq \operatorname{\mathtt{dom}} F$.

Exercise 3.5.9. Prove that $\mathbb{S}$-Size Choice implies $\mathbb{S}$-Separation. □

Both $\mathbb{S}$-Size Choice and $\mathbb{S}$-Separation are independent of **IST** (Corollary 4.5.14 below) and are trivially false in the **BST** setting.

Theorem 3.5.10 (IST + $\mathbb{S}$-Separation). *Formal Truth Completeness holds.*

Proof. Say that a truth set T is *good* for $\varphi(v_1, ..., v_n) \in \mathtt{Form}$ if, whenever $x_1, ..., x_n$ are standard, at least one of the coded formulas $\ulcorner\varphi(x_1, ..., x_n)\urcorner$, $\ulcorner\neg\,\varphi(x_1, ..., x_n)\urcorner$ belongs to T. To prove the theorem it suffices to show that for any standard $\varphi(v_1, ..., v_n) \in \mathtt{Form}$ there exists a good truth set T.

The proof goes by formal external induction in **IST** on the length of φ. Only the step for $\exists$ needs some care. Prove the existence of a good truth set T for any standard coded formula of the form $\exists\, v_0\, \varphi(v_0, v_1, ..., v_n)$, provided a good truth set T' exists for the coded formula $\varphi(v_0, v_1, ..., v_n)$. Under the assumption of $\mathbb{S}$-Separation, there exists a set Z satisfying:

$$\forall^{\mathtt{st}} x_1 \ldots \forall^{\mathtt{st}} x_n\, \big(\langle x_1, ..., x_n\rangle \in Z \Longleftrightarrow \exists^{\mathtt{st}} x\, (\ulcorner\varphi(x, x_1, ..., x_n)\urcorner \in T')\big).$$

We obtain the required set T by adjoining to T':

1) all coded formulas $\ulcorner\exists\, v_0\, \varphi(v_0, x_1, ..., x_n)\urcorner$ such that $\langle x_1, ..., x_n\rangle \in Z$, and
2) all coded formulas $\ulcorner\neg\exists\, v_0\, \varphi(v_0, x_1, ..., x_n)\urcorner$ such that $\langle x_1, ..., x_n\rangle \notin Z$ but $x_1, ..., x_n \in C$. (Here, as above, C contains all standard sets.) □

[25] This does not imply Map-Standardization of § 3.2e !

Thus $\mathbb{S}$-Separation provides a formal definition in **IST** for the truth of all (coded) $\in$-formulas of standard finite length and with parameters in $\mathbb{S}$. One may ask whether, say, all standard **ZFC** axioms are f. true in $\mathbb{S}$ in this sense. The positive answer can be obtained by adjoining one more additional hypothesis, Map-Standardization introduced in § 3.2e above. Map-Standardization is also consistent with **IST**, but to prove the consistency we need to assume something beyond **ZFC**, *e. g.*, the existence of an inaccessible cardinal. (These questions will be briefly considered in § 4.4c.)

Theorem 3.5.11 (IST + $\mathbb{S}$-Separation + Map-Standardization). *If $\Phi \in$ `ClForm` is a standard axiom of* **ZFC** *then Φ is f. true in $\mathbb{S}$.*

It is quite important to understand that this theorem gives much more than the truth of all *metamathematically given* **ZFC** axioms.

Proof (Exercise). There are exactly two relevant axioms, Separation and Replacement. (All the other are single metamathematically given statements true in $\mathbb{S}$ by Inner Transfer.) Prove, applying Theorem 3.5.10 and Inner Standardization, that all standard cases of Separation are f. true in $\mathbb{S}$. Prove the same for Replacement, using Map-Standardization. □

Corollary 3.5.12 (IST + $\mathbb{S}$-Separation + Map-Standardization). **ZFC** *is consistent.*

Proof (Exercise). Arguing in the theory indicated, assume on the contrary that **ZFC** is inconsistent. This means that there exists a formal deduction of, say, $0 = 1$ in **ZFC**. By Inner Transfer, there is a deduction of $0 = 1$ which proceeds entirely inside the collection of all coded formulas of standard length with standard parameters, that is, those covered by the truth definition.

Note that, by Theorem 3.5.4, the usual rules of deduction transform f. true coded formulas in `ClForm` into f. true ones. Thus we obtain a contradiction since axioms are f. true by Theorem 3.5.11 while $0 = 1$ is false. □

There is an important case when Formal Truth Completeness can be obtained for free: models whose "standard parts" are really standard !

Theorem 3.5.13 (ZFC). *Suppose that $\langle {}^*V\,; {}^*\!\in, \mathtt{st}\rangle$ is a model of* **IST** *whose standard part $V = \{x \in {}^*V : \mathtt{st}\,x\}$ is a transitive set and ${}^*\!\in \restriction V = \,\in \restriction V$. Then Formal Truth Completeness holds in ${}^*V$.*

Proof. As V is a transitive model, the metamathematics within V is based on natural numbers of the **ZFC** universe. In other words, a standard formula in `Form` from the point of view of ${}^*V$ is just an ordinary $\in$-formula with sets in V as parameters. It remains to apply Lemma 3.5.6 or Theorem 3.5.5. □

Problem 3.5.14. Study the truth definition in **IST** with respect to the universe $\mathbb{B}$ of all bounded sets. □

Problem 3.5.15. Does **IST** prove Formal Truth Completeness ? □

3.6 Second edition of IST

Somewhat surprisingly, it has gone unnoticed that the system of nonstandard deduction introduced in Nelson's second paper [Nel 88] on foundations of nonstandard analysis, is different from **IST**, to be more exact, it is equivalent in its $\in$-part but is essentially stronger in its $\mathtt{st}$-$\in$-part. The scheme proposed in [Nel 88] can be summarized as follows:

Part 1: a model (or an interpretation, as in the case $T = \mathbf{ZFC}$ [Nel 88, § 6]) V of a given (standard) theory T is studied by means of a (standard) type-theoretic superstructure, say $S(V)$ over V;

Part 2: $S(V)$ is studied by means of its nonstandard extension, say, ${}^*S(V)$;

Part 3: those results of this study which can be relativized to V are considered as relevant to T.

Let us consider the deductive scheme of [Nel 88] in more detail.

3.6a Standard and nonstandard theories of Nelson's system

In the case of $T = \mathbf{ZFC}$ as the basic theory, Part 1 of Nelson's scheme can be adequately formalized as the theory of the type theoretic hull of a transitive interpretation of **ZFC** in the **ZFC** universe. Arguing in the theory $\mathbf{ZFC}\boldsymbol{\vartheta}$ (see § 1.5f), we take $\mathbf{V}_{\boldsymbol{\vartheta}}$ as the given transitive interpretation of **ZFC**. (Nelson uses V to denote $\mathbf{V}_{\boldsymbol{\vartheta}}$.) The type theoretic superstructure over $\mathbf{V}_{\boldsymbol{\vartheta}}$ is essentially the same as the set $\mathbf{V}_{\boldsymbol{\vartheta}+\omega} = \mathbf{V}_{\boldsymbol{\vartheta}} \cup \mathscr{P}(\mathbf{V}_{\boldsymbol{\vartheta}}) \cup \mathscr{P}^2(\mathbf{V}_{\boldsymbol{\vartheta}}) \cup \dots$.

Let **ZCN** (N from "Nelson") be a theory, in the $\in$-language enriched by a constant symbol $\boldsymbol{\vartheta}$, containing all sentences φ in this language such that the relativization $\varphi^{\mathbf{V}_{\boldsymbol{\vartheta}+\omega}}$ is a theorem of $\mathbf{ZFC}\boldsymbol{\vartheta}$.

Problem 3.6.1. Is **ZCN** recursively axiomatizable ? □

Exercise 3.6.2. Prove that **ZCN** implies $\mathbf{ZC}\boldsymbol{\vartheta}$, the theory which includes

(1) all axioms of **ZC** (with $\boldsymbol{\vartheta}$ allowed to occur in the schemata),

(2) the sentence "$\boldsymbol{\vartheta}$ is an ordinal", and

(3) all sentences of the form $\Phi^{\mathbf{V}_{\boldsymbol{\vartheta}}}$, Φ being a **ZFC** axiom not containing $\boldsymbol{\vartheta}$,

(compare with $\mathbf{ZFC}\boldsymbol{\vartheta}$ of § 1.5f!), in addition, **ZCN** implies

(4) the universe of all sets is equal to $\mathbf{V}_{\boldsymbol{\vartheta}+\omega}$, *i.e.*, $\forall x\, \exists n\, (x \in \mathbf{V}_{\boldsymbol{\vartheta}+n})$. □

Even $\mathbf{ZC}\boldsymbol{\vartheta} + (4)$ is weaker than **ZCN**, for instance, it does not prove $\mathtt{Consis}\ \mathbf{ZC}\boldsymbol{\vartheta}$ which **ZCN** does, but on the other hand $\mathbf{ZC}\boldsymbol{\vartheta} + (4)$ tends to prove the same meaningful sentences as **ZCN** does, while $\mathbf{ZC}\boldsymbol{\vartheta}$ tends to prove the same meaningful sentences expressed by *bounded* $\in$-formulas as **ZCN** does. Moreover, the theories $\mathbf{ZC}\boldsymbol{\vartheta} + (4)$ and $\mathbf{ZC}\boldsymbol{\vartheta}$ can substitute **ZCN** in all results below in Section 3.6.

Proposition 3.6.3. **ZCN** *is a conservative extension of* **ZFC** *in the sense that* **ZFC** *proves an* $\in$*-formula* φ *iff* $\varphi^{\mathbf{V}_\vartheta}$ *is a theorem of* **ZCN**.

Proof. If **ZCN** proves $\varphi^{\mathbf{V}_\vartheta}$ then **ZFC**ϑ proves $(\varphi^{\mathbf{V}_\vartheta})^{\mathbf{V}_{\vartheta+\omega}}$, hence proves $\varphi^{\mathbf{V}_\vartheta}$. It follows (Exercise 1.5.17) that φ is a theorem of **ZFC**. □

The nonstandard side of Nelson's system is not presented in [Nel 88] as a first-order theory. Yet the transformation rules (T), (I), (S′) in [Nel 88, § 2], along with Nelson's commitment to consider only those (internal) sets which belong to standard sets, fit perfectly within the theory ***ZCN** containing

1°: Inner Boundedness, Inner Standardization, Basic Idealization,

2°: Bounded Inner Transfer, that is, the schema of Inner Transfer restricted to bounded $\in$-formulas (as defined in § 1.5a),

3°: $\mathbf{ZCN}^{\mathtt{st}}$ along with the axiom: "ϑ is standard";

4°: an axiom saying that "for any standard x, the standard power set of x (*i.e.*, a standard set P satisfying $(P = \mathscr{P}(x))^{\mathtt{st}}$; it exists by $\mathbf{ZCN}^{\mathtt{st}}$) is the true power set of x".

Axiom 4° implies that $\mathscr{P}(x)$ exists and is standard for any standard x.

The next theorem belongs to the same class of conservativity results as the theorems 1.1.14 and 4.1.10.

Theorem 3.6.4 (see Exercise 4.4.12(3) below). ***ZCN** *is a conservative extension of* **ZCN** *in the sense that an* $\in$-ϑ*-formula* φ *is a theorem of* ***ZCN** *iff* $\varphi^{\mathtt{st}}$ *is a theorem of* **ZCN**. □

3.6b The background nonstandard universe

We rather consider a universe described by the ϑ-unrelated part ***ZC** of ***ZCN**, *i.e.*, the theory obtained by the reduction of $\mathbf{ZCN}^{\mathtt{st}}$ to $\mathbf{ZC}^{\mathtt{st}}$ in 3° above. The theory ***ZC** differs in two principal aspects from either of **BST** and **IST** : first, we have only **ZC** rather than **ZFC** in the standard universe, second, we have only Bounded Inner Transfer of 2° in § 3.6a instead of the full schema of Inner Transfer. The latter issue is motivated by the following

Proposition 3.6.5. *The theory* ***ZC** *plus Inner Transfer, that is,* **BST**[**ZC**] *implies* **BST**, *therefore, it is not a conservative extension of* **ZC**.

Proof. To derive (Collection)$^{\mathtt{st}}$ suppose that X is standard and $\varphi(x,y)$ is an $\in$-formula with standard parameters. To find a standard Y satisfying

$$\forall^{\mathtt{st}} x \in X \left(\exists^{\mathtt{st}} y \, \varphi(x,y) \Longrightarrow \exists^{\mathtt{st}} y \in Y \, \varphi(x,y)\right), \qquad (*)$$

let X' be a finite set such that $\mathbb{S} \cap X \subseteq X'$ (Lemma 3.1.15). As φ is an $\in$-formula and X' is finite, there is a set Y such that $\exists y \, \varphi(x,y) \Longrightarrow \exists y \in Y \, \varphi(x,y)$ for every $x \in X'$. By Inner Boundedness, there is a standard set Y of this sort. That this Y satisfies $(*)$ follows from Inner Transfer. □

Remark 3.6.6. We conclude that **ZC** does not have a reasonable conservative nonstandard **BST**-like extension with the full schema of Inner Transfer.

The key ingredients of the proof are Inner Transfer and Inner Boundedness; the schema of Inner Standardization is not involved, and as far as Basic Idealization is concerned, only the fact that $\mathbb{S} \cap X$ can be covered by a finite set (Lemma 3.1.15) is used. It follows that, unlike **ZFC**, the theory **ZC** does not have a conservative nonstandard **BST**-like (*i.e.*, with Inner Boundedness) extension which satisfies the full schema of Inner Transfer and just a little bit of Basic Idealization. □

Due to the restriction imposed on Inner Transfer, the full internal set universe of ***ZC** is less organized than its standard universe. Let $\boldsymbol{\Delta}_0$-**ZC** be **ZC** with the schema of Separation restricted to bounded $\in$-formulas.

Lemma 3.6.7. ***ZC** *implies* $\boldsymbol{\Delta}_0$-**ZC** *in the universe of all sets.*

Proof. (1) The equality $x = \bigcup y$ is a bounded formula, therefore, for any standard y, the "standard" $\bigcup y$ (*i.e.*, a standard set x satisfying $(x = \bigcup y)^{\mathtt{st}}$; it exists by $\mathbf{ZCN}^{\mathtt{st}}$) is equal to the "true" $\bigcup y$. Saying it differently, $\bigcup y$ is a standard set for any standard y. Similarly for $\{x, y\}$ and $x \cup y$. [26]

(2) For any x there is a standard u with $x \subseteq u$: take $u = \bigcup y$, where y is a standard set containing x.

(3) Let X be any set and $\varphi(x, a)$ any bounded $\in$-formula with a set a as the only parameter. Prove that $Y = \{x \in X : \varphi(x, a)\}$ is a set. By (2), it can be assumed that X is standard. Then $P = \mathscr{P}(X)$ is standard by 4°. Let A be a standard set containing a. Apply Bounded Inner Transfer to the formula

$$\forall a \in A \, \exists Y \in P \, \forall x \in X \, \big(x \in Y \iff \varphi(x, a)\big) .$$

(4) It easily follows from the above that $\bigcup x$ is a set for any (not necessarily standard) set x, and similarly for $\mathscr{P}(x)$, $\{x, y\}$. The rest of the axioms of $\boldsymbol{\Delta}_0$-**ZC** also are rather obvious on the base of (3). □

Exercise 3.6.8 (*ZC). Prove that B^A and $\mathscr{P}_{\mathtt{fin}}(B)$ are sets for any sets A, B, in addition, B^A and $\mathscr{P}_{\mathtt{fin}}(B)$ are standard provided so are A, B. □

The theory ***ZC** contains Inner Boundedness, hence, is of **BST** type rather than of **IST** type, but is much weaker, so that we hardly can expect much of the results estalished for **BST** in § 3.2 to survive in ***ZC**. Yet in fact certain forms of Reduction to $\Sigma_2^{\mathtt{st}}$ and Inner S. S. Choice remain true in ***ZC**.

The notion of a bounded $\in$-formula was introduced in § 1.5a. Similarly to that, let a *bounded* $\mathtt{st}$-$\in$-*formula* be any $\mathtt{st}$-$\in$-formula which has only quantifiers of the form $\exists x \in y$, $\forall x \in y$. (The quantifiers $\exists^{\mathtt{st}}$, $\forall^{\mathtt{st}}$ are considered as shortcuts, hence, they still have to have the form $\exists^{\mathtt{st}} x \in y$, $\forall^{\mathtt{st}} x \in y$.)

[26] This argument does not work for the power set because $P = \mathscr{P}(X)$ is not a bounded formula. This is why 4° is included into the axioms of **ZCN**.

Lemma 3.6.9. *For any bounded* st-$\in$-*formula* $\Phi(y_1,...,y_n)$ *there is an* $\in$-*formula* $\varphi(a,b,y_1,...,y_n)$ *such that* ***ZC** *proves the following: for any standard* $Y_1,...,Y_n$ *there exist standard sets* A, B *satisfying*

$$\forall y_1 \in Y_1 \ldots \forall y_n \in Y_n \; \Phi(y_1,...,y_n) \iff \exists^{\mathtt{st}} a \in A \, \forall^{\mathtt{st}} b \in B \; \varphi(a,b,y_1,...,y_n) \, .$$

Proof. If $\Phi(y)$ is $\mathtt{st}\, y$ then let $\varphi(a,b,y)$ be $y = a$. Now, in ***ZC**, for any standard Y take $A = Y$, then the equivalence $\mathtt{st}\, y \iff \exists^{\mathtt{st}} a \in A \, \varphi(a,b,y)$ holds for any $y \in Y$. The cases of other atomic formulas are trivial.

The induction steps for $\neg$ and $\exists$ are based on the procedure called *Nelson's algorithm*; it is rather similar to the transformations used in the proof of Theorem 3.2.3, but the step for $\neg$ is much easier. The algorithm is based on the following equivalences provable in ***ZC**:

$$\forall^{\mathtt{st}} a \in A \, \exists^{\mathtt{st}} b \in B \; \psi(a,b) \iff \exists^{\mathtt{st}} f \in F \, \forall^{\mathtt{st}} a \in A \; \psi(a, f(a)) \, , \quad \text{and}$$

$$\exists w \, \exists^{\mathtt{st}} a \, \forall^{\mathtt{st}} b \in B \; \psi(w,a,b) \iff \exists^{\mathtt{st}} a \, \forall^{\mathtt{st}} B' \in P \, \exists w \, \forall b \in B' \; \psi(w,a,b) \, ,$$

where $P = \mathscr{P}_{\mathtt{fin}}(B)$, $F = B^A$, $\exists w$ and $\exists^{\mathtt{st}} a$ mean $\exists w \in W$ and $\exists^{\mathtt{st}} a \in A$ in the second line, the sets A, B, W are standard, and φ is an $\in$-formula. Note that P and F are standard sets by 3.6.8. □

Exercise 3.6.10. Prove the *restricted* Inner S. S. Choice schema in ***ZC** :

$$\forall^{\mathtt{st}} x \in X \, \exists y \in Y \; \Phi(x,y) \implies \exists f \in Y^X \, \forall^{\mathtt{st}} x \in X \; \Phi(x, f(x)) \, ,$$

whenever X, Y are standard and $\Phi(x,y)$ is a bounded st-$\in$-formula. (*Hint.* According to 3.6.9 we assume that $\Phi(x,y)$ is $\exists^{\mathtt{st}} a \in A \, \forall^{\mathtt{st}} b \in B \; \varphi(x,y,a,b)$, where A, B are standard and φ an $\in$-formula. Apply the argument used in the proof of Theorem 3.2.11.) □

3.6c Three "myths" of IST

Part 3 is the most interesting part of Nelson's scheme. Let **IST**$'$ be the set of all closed st-$\in$-formulas φ such that the theory ***ZCN** proves $\varphi^{\mathbf{V}_\vartheta}$. Note that **IST**$'$ is deductively closed.

Problem 3.6.11. Is **IST**$'$ recursively axiomatizable ? □

Exercise 3.6.12. Prove that $\langle \mathbf{V}_\vartheta \, ; \in, \mathtt{st} \rangle$ is an interpretation of **IST** in ***ZCN**, therefore, **IST** $\subseteq$ **IST**$'$. (See § 1.5f for the reasons why we do not claim that $\langle \mathbf{V}_\vartheta \, ; \in, \mathtt{st} \rangle$ is a *model* of **IST**.) □

Thus, if Φ is a theorem of **IST** then Φ belongs to **IST**$'$, in other words, ***ZCN** proves that Φ is true in $\mathbf{V}_\vartheta$. Conversely, does **IST** imply all of **IST**$'$? The answer is "yes" as long as $\in$-formulas are considered:

Exercise 3.6.13. Prove, using 3.6.3, 3.6.4, that any $\in$-formula φ which belongs to **IST**$'$ is a theorem of **ZFC**, hence, **IST** proves both φ and $\varphi^{\mathtt{st}}$. □

But in general the answer is in the negative: the theory **IST**$'$ is definitely stronger than **IST**. In fact ***ZCN** makes the structure $\langle \mathbf{V}_\vartheta \,; \in \rangle$ more saturated than **IST** does for its nonstandard set universe. For instance, Claim 3 below, a theorem of **IST**$'$, is not provable in **IST** (see the fourth remark below on this page).

The misunderstanding of the difference between **IST**$'$ and **IST**, and also of the difference (much more transparent, of course) between **IST** and (what we define here as) ***ZC** and ***ZCN** led **IST** practitioners to incorrect attribution of certain results to **IST** proper. In particular, the following assertions have become parts of the spoken, and sometimes written folklore of **IST**. (We refer to [vdBerg 87, pp. 118,182], [Nel 88, Thm 5], [vdBerg 92, pp. 74,80], to mention a few.)

Claim 1. *In* **IST**, *every* st-$\in$-*formula can be transformed to an equivalent* $\Sigma_2^{\mathtt{st}}$ *formula, hence, any external set belongs to either* $\mathbf{\Delta}_2^{\mathtt{ss}}$, *or* $\mathbf{\Sigma}_1^{\mathtt{ss}}$, *or* $\mathbf{\Pi}_1^{\mathtt{ss}}$.

Claim 2. *In* **IST**, *every* st-$\in$-*sentence is a subject of a certain algorithm of* [Nel 77, Nel 88] *which transforms it to an equivalent* $\in$-*sentence.*

Claim 3. *In* **IST**, *restricted Inner S. S. Choice* (as in Exercise 3.6.10) *is true.*

The following should be said regarding these claims.

First, each of the three claims is true in **BST**, hence, true in **IST** as long as we consider only those st-$\in$-formulas which are relativized to the class $\mathbb{B}$ of all bounded sets, in particular, formulas with quantifiers restricted by standard sets. Needless to say that this covers all applications meaningful for practitioners, so that the claims are "practically true".

Second, each of the three claims is, generally speaking, false (in **IST**). Leaving the falsity of Claims 2 and 3 to Section 4.5 (Corollary 4.5.14(ii) and Exercise 4.5.15), we refer to Exercise 3.5.8 regarding Claim 1.

Third, appropriate reservations (like the restriction of all quantifiers by standard sets) are not always tagged to the claims, even in **IST** papers of foundational character. For instance, there is no applicable comments in [Nel 88] regarding Theorem 5 there (= Claim 3), which is the source of subsequent unscrupulous references of **IST** practitioners.

Fourth, Claim 3 is true in **IST**$'$ (Exercise 3.6.10), Claim 1 is false in **IST**$'$ because it is refutable even in **IST** by the result of Exercise 3.5.8. As for Claim 2, we don't know whether it is true in **IST**$'$ that every closed st-$\in$-formula is provably equivalent to an $\in$-formula.

Fifth, Claims 1 and 3 are true in ***ZC**, hence, in ***ZCN**, for bounded st-$\in$-formulas (Lemma 3.6.9 and Exercise 3.6.10), Claim 2 should also be true in ***ZC** (under the same restriction).

Historical and other notes to Chapter 3

Sections 3.1, 3.6. Nelson's paper [Nel 77] introduced **IST**. The nonstandard deduction scheme given in his subsequent note [Nel 88] is not equivalent to **IST** (as demonstrated in Section 3.6). **BST** was explicitly formulated in [Kan 91], but implicitly it is equivalent to the theory of internal sets in Hrbaček's theories $\mathfrak{NS}_1$(**ZFC**) and $\mathfrak{NS}_2$(**ZFC**) [Hr 78].

Basic Idealization looks somewhat weaker than Local Idealization (Bounded Idealization of [Kan 91, KanR 97]). Yet it implies Local Idealization, see § 3.2a. The advantage of Basic Idealization is that it is a single axiom.

The content of §§ 3.1d, 3.1e: mainly Nelson [Nel 77]. See comments on Chapter 6 on relative standardness. Theorem 3.1.25: Prohorova [Pr 98]. Lemma 3.6.9 (Nelson's algorithm) and Exercise 3.6.10: Nelson [Nel 88].

Section 3.2. Fundamentals of **BST** are due to Kanovei [Kan 91]. Some of the results (for instance, Reduction to $\Sigma_2^{\mathtt{st}}$ in Theorem 3.2.3, Map-Standardization, Uniqueness, Inner S. S. Choice in Theorem 3.2.11) were suggested by earlier similar theorems in **IST**, due to Nelson [Nel 77, Nel 88] (in particular, Theorem 2.2 in [Nel 77], the Uniqueness principle), proved by him in the case when all quantifiers and variables are bounded by fixed standard sets.

The content of § 3.2c is due to Andreev and Hrbaček [AnH 04].

That "external sets" cause a problem in internal theories was clear to Nelson, who proposed in [Nel 77] a solution: any statement about "individual" external sets can be unambiguously encoded into a proper $\mathtt{st}$-$\in$-formula. Nelson also observed that the solution does not cover the case of quantification over external sets, but this rarely occurs in (elementary) nonstandard analysis. This approach was detailed by **IST** followers, most notably v. d. Berg [vdBerg 87, vdBerg 92], but these attempts fell short of a proper treament of quantifiers over external sets, let alone sets of external sets and higher levels. In [R 92] a very rudimentary metamathematical treatment of "external sets" is described, which is largely sufficient for a discussion of external quotients and "external subsets" thereof, but does not lead to any sort of set theoretic structure. Theorem 3.2.16, first established in [Kan 91] led to a complete solution for **BST** in [KanR 95]. We present the solution in Chapter 5.

Section 3.3. Partially saturated theories were introduced in [KanR 95, Part 3]. The counterexample in 3.3.6 is due to Hrbaček; generally his remarks were very useful to elaborate the final layout.

Section 3.4. Theorems of Collection and Uniqueness 3.4.9, 3.4.16 are due to Kanovei [Kan 94b, Kan 95]. That bounded sets interpret **BST** was proved in [Kan 91]. Weaker versions of Theorems 3.4.7 (Reduction to $\Sigma_2^{\mathtt{st}}$) and 3.4.16 (Uniqueness in **IST**), for formulas with all quantifiers and variables bounded by standard sets, were obtained by Nelson [Nel 77, Nel 88].

Section 3.5. This material is due to Kanovei [Kan 91, Kan 94b].

4 Metamathematics of internal theories

One of the most important metamathematical issues related to any formal theory is the question of *consistency*: that is, a theory should not imply a contradiction. As long as minimally reasonable set theories are considered, Gödel's famous incompleteness theorems make it impossible to prove the consistency in any absolute sense, so that usually the results are given in terms of *equiconsistency* with some other theory, for instance, **ZFC**. In this Chapter, we prove that the internal theories **IST** and **BST** considered above are equiconsistent with **ZFC**, that is, consistency of **ZFC** logically implies consistency of both **BST** and **IST**. (For the opposite direction, if **IST** or **BST** is consistent then obviously so is **ZFC** as a subtheory of each of **IST**, **BST**, see Exercise 3.1.3.)

Three other metamathematical issues related to nonstandard theories will be considered: conservativity, reducibility, interpretability. A special kind of interpretation of a nonstandard set theory $\mathfrak{T}$ in a standard theory $\mathfrak{U}$, called *standard core interpretation*, will be defined: roughly, this property means that $\mathfrak{U}$ is strong enough to define an extension of its whole set universe to a universe of $\mathfrak{T}$. The importance of standard core interpretability in **ZFC** for the evaluation of the metamathematical "quality" of a nonstandard theory, will be stressed. Our main goal in this Chapter will be to investigate the theories **BST** and **IST** with respect to these properties. Theorem 4.1.10 presents the main results, in particular, both **BST** and **IST** are conservative extensions of **ZFC**, but of these two theories only **BST** admits a standard core interpretation in **ZFC**. The results related to **BST** will be the base of the subsequent study of metamathematical properties of **HST** in Chapter 5.

The key technical tool used in this Chapter is the quotient power construction which naturally includes ultrapowers, ultralimits, iterated ultrapowers. It is applied, in the form of adequate ultralimit construction, in Section 4.3 to define a standard core interpretation of **BST** in **ZFC**, and in Section 4.6 to define a standard core interpretation of **IST** in **ZFGT**, a "standard" theory extending the **ZFC** universe by a global choice function and a truth predicate for the $\in$-language. A "definable" version of this construction is used in Section 4.5 to show that certain meaningful sentences, like Inner Extension, are undecidable in **IST**.

4.1 Outline of metamathematical properties

Our study will be concentrated around several principal relations between standard and nonstandard set theories (mainly **ZFC**, **IST**, **BST**, and their variants) and also between their models. We consider nonstandard extensions of standard theories and standard models, in particular, extensions of the whole set universe (in the form of interpretations, see § 1.5d).

We shall use the word *extension* both for extensions of structures and extensions of theories; it will always be clear from the context what is the intended meaning. After a few precise definitions related to these notions, and some remarks, we formulate a theorem (Theorem 4.1.10) which presents the main metamathematical properties of the theories **IST** and **BST**.

4.1a Nonstandard extensions of structures

According to our general definitions in § 1.5c, an $\in$-*structure* (with true equality [1]) is a structure of the form $\mathbf{v} = \langle V\,; \varepsilon\rangle$ where V is a set or a class and ε is a binary relation on V. The truth of an $\in$-formula Φ in such a structure $\mathbf{v}$ is naturally understood in the sense that the atomic predicate $\in$ is interpreted as ε . Formally, the *relativization* $\Phi^{\mathbf{v}}$ is defined as follows: all occurrences of $\in$ in Φ are changed to ε and all quantifiers relativized to V. Then $\Phi^{\mathbf{v}}$ expresses the truth of Φ in $\mathbf{v}$.

Similarly, a $\mathtt{st}$-$\in$-*structure* (with true equality) is a structure of the form $\mathbf{v} = \langle V\,; \varepsilon, {}^*\mathtt{st}\rangle$ where V and ε are as above and ${}^*\mathtt{st}$, a standardness relation, is a unary relation on V. ($*$ is added to distinguish this from the standardness predicate $\mathtt{st}$ of the $\mathtt{st}$-$\in$-language.) The *relativization* $\Phi^{\mathbf{v}}$ of a $\mathtt{st}$-$\in$-formula Φ to such a structure $\mathbf{v}$ is defined as follows: all occurrences of $\in$ and $\mathtt{st}$ in Φ are changed to resp. the relations ε and ${}^*\mathtt{st}$ and all quantifiers relativized to V. Then $\Phi^{\mathbf{v}}$ expresses the truth of Φ in $\mathbf{v}$.

If $\mathbf{v} = \langle V\,; \varepsilon, {}^*\mathtt{st}\rangle$ is a $\mathtt{st}$-$\in$-structure then $\mathbb{S}^{(\mathbf{v})} = \{x \in V : {}^*\mathtt{st}\, x\}$ is *the standard core* (or standard universe) of $\mathbf{v}$.

Definition 4.1.1. Suppose that $\mathbf{v} = \langle V\,; \varepsilon\rangle$ is an $\in$-structure and ${}^*\mathbf{v} = \langle {}^*V\,; {}^*\varepsilon, ...\rangle$ is an $\in$-structure or a $\mathtt{st}$-$\in$-structure (... indicates the presence or absence of a standardness relation). A map $* : V \to {}^*V$ is

(1) an $\in$-*embedding* if $*$ is $1-1$ and we have $x\,\varepsilon\, y \iff {}^*x\; {}^*\varepsilon\; {}^*y$ for all x, $y \in V$. [2] In this case we say that ${}^*\mathbf{v}$ is an *extension of* $\mathbf{v}$ *via* $*$;

(2) a *standard core embedding*, or an $\in$-*isomorphism* of $\mathbf{v}$ onto the standard core of ${}^*\mathbf{v}$, if in addition ${}^*\mathbf{v} = \langle {}^*V\,; {}^*\varepsilon, {}^*\mathtt{st}\rangle$ is a $\mathtt{st}$-$\in$-structure and the standard core $\mathbb{S}^{({}^*\mathbf{v})} = \{z \in {}^*V : {}^*\mathtt{st}\, z\}$ coincides with $\{{}^*x : x \in V\}$. In this case we say that ${}^*\mathbf{v}$ is a *standard core extension of* $\mathbf{v}$ *via* $*$;

---

[1] By the reasons related to Theorem 1.5.11 there will be no need for invariant structures in this Chapter.

[2] Notation: ${}^*x$ is typically used instead of $*(x)$.

(3) an *elementary* $\in$-embedding if we have $\varphi^{\mathbf{v}} \iff ({}^*\varphi)^{{}^*\mathbf{v}}$ whenever φ is a closed $\in$-formula with sets in V as parameters and ${}^*\varphi$ is obtained by the substitution of ${}^*x$ for any $x \in V$ occurring in φ as a parameter. In this case we say that ${}^*\mathbf{v}$ is an *elementary* extension of $\mathbf{v}$ via $*$. □

Exercise 4.1.2. Let ${}^*\mathbf{v} = \langle {}^*V\,;\, {}^*\varepsilon, {}^*\mathtt{st}\rangle$ be a standard core extension of $\mathbf{v} = \langle V\,;\,\varepsilon\rangle$. Prove that if φ and ${}^*\varphi$ are as in 4.1.1(3) then φ is true in $\mathbf{v}$ iff $({}^*\varphi)^{\mathtt{st}}$ is true in ${}^*\mathbf{v}$. □

4.1b Nonstandard extensions of theories

By a *standard set theory* we understand a theory in a language which includes the $\in$-language but does not contain $\mathtt{st}$. A *nonstandard set theory* will be a theory in a language which includes the $\mathtt{st}$-$\in$-language. Thus **ZFC** is a standard theory while **HST**, **BST**, **IST** are nonstandard theories.

Any nonstandard theory $\mathfrak{T}$ distinguishes the *standard core* $\mathbb{S} = \{x : \mathtt{st}\, x\}$ of the set universe, and hence for any $\in$-formula φ, $\varphi^{\mathtt{st}}$ is a formula of the language of $\mathfrak{T}$. The next definition introduces several important notions which characterize the relationships between a nonstandard theory $\mathfrak{T}$ and a standard theory $\mathfrak{U}$, in terms of the standard core of the universe of $\mathfrak{T}$.

Definition 4.1.3. (1) $\mathfrak{T}$ is a *standard core extension* of a theory $\mathfrak{U}$ in the $\in$-language if for any axiom Φ of $\mathfrak{U}$, $\Phi^{\mathtt{st}}$ is a theorem of $\mathfrak{T}$.[3]

(2) $\mathfrak{T}$ is a *conservative standard core extension* of $\mathfrak{U}$ if for any $\in$-formula Φ, $\mathfrak{U}$ proves Φ if and only if $\mathfrak{T}$ proves $\Phi^{\mathtt{st}}$.[3]

(3) $\mathfrak{T}$ is a *reducible* nonstandard theory if for any sentence Φ of the language of $\mathfrak{T}$ there is an $\in$-sentence ψ such that $\mathfrak{T}$ proves $\Phi \iff \psi^{\mathtt{st}}$.

(4) $\mathfrak{T}$ is *standard core interpretable* in $\mathfrak{U}$ if there exist:
   1) an interpretation ${}^*\mathbf{v} = \langle {}^*\mathbf{V}\,;\, {}^*\!\in, {}^*\mathtt{st}, \ldots\rangle$ of $\mathfrak{T}$ in $\mathfrak{U}$, where ${}^*\mathtt{st}$ interprets the atomic predicate $\mathtt{st}$ while ... denotes classes which interpret other possible atomic symbols of the language of $\mathfrak{T}$, and
   2) a standard core embedding $* : \mathbf{V} \to {}^*\mathbf{V}$ of the $\mathfrak{U}$-universe $\mathbf{v} = \langle \mathbf{V}\,;\,\in\rangle$ of all sets into ${}^*\mathbf{v}$.[4]

   Such an interpretation is called: a *standard core* interpretation. □

Remark 4.1.4. The notion of conservativity in 4.1.3(2) is different from Definition 1.5.16. Yet both notions obviously coincide for nonstandard theories containing Inner Transfer like **BST** or **IST**. □

---

[3] In (1), (2) $\mathfrak{U}$ is supposed to be a standard theory in the $\in$-language like **ZFC**. If $\mathfrak{U}$ is a theory in a language properly extending the $\in$-language, like **ZFC**$\boldsymbol{\vartheta}$ of § 1.5f or theories with global choice like **ZFGT** below, then the definitions become more complicated, see Remark 4.6.13 on a suitable example.

[4] As usual, both the interpretation and the embedding must be defined by formulas of the language of $\mathfrak{U}$, and their indicated properties provable in $\mathfrak{U}$.

Remark 4.1.5. Nonstandard theories with a meaningful class $\mathbb{WF}$ of well-founded sets, like **HST**, admit modifications of these definitions oriented towards $\mathbb{WF}$ rather than $\mathbb{S}$. A nonstandard theory $\mathfrak{T}$ is a *conservative wf-core extension* of $\mathfrak{U}$ if, for any $\in$-sentence Φ, $\mathfrak{U}$ proves Φ iff $\mathfrak{T}$ proves $\Phi^{\mathtt{wf}}$. The notion of a *wf-core interpretable* theory, along with an associated notion of the *well-founded core* $\mathbb{WF}^{({}^*\mathbf{V})}$ of a $\mathtt{st}$-$\in$-structure ${}^*\mathbf{V}$ is defined similarly.

For $\mathfrak{T} = \mathbf{HST}$, these concepts are equivalent to the standard core notions since the classes $\mathbb{S}$ and $\mathbb{WF}$ are $\in$-isomorphic in **HST**. □

Standard theories may transcend **ZFC** by both new axioms (like the continuum-hypothesis) and new elements of the language (so does **ZFC**ϑ of § 1.5f or theories with global choice in Definition 4.3.4), in both cases the set universe satisfies **ZFC**. The language of a standard core extension of **ZFC** is at least the $\mathtt{st}$-$\in$-language, hence, a standard universe (*core*) $\mathbb{S} = \{x : \mathtt{st}\, x\}$ is defined, and the requirement 4.1.3(1) means that $\langle \mathbb{S} ; \in \rangle$ is postulated, by $\mathfrak{T}$, to interpret $\mathfrak{U}$.

Standard core interpretability of a nonstandard theory $\mathfrak{T}$ in a standard theory $\mathfrak{U}$ means that $\mathfrak{U}$ is strong enough to define a structure that interprets $\mathfrak{T}$ in $\mathfrak{U}$ (§ 1.5d), along with an isomorphism of the universe of all sets onto the standard core of the structure — and thus the set universes of both theories must be connected in a certain way, in addition to the general requirements contained in the definition of interpretation in § 1.5d. Many examples of interpretation of nonstandard theories in **ZFC** (and some other standard theories) will be given below.

The properties of conservativity, reducibility, interpretability, together with equiconsistency with **ZFC**, will be the main issues of the metamathematical study of nonstandard theories below. The following definition introduces a property of structures rather than theories, but still it contains a certain indirect characterization of nonstandard set theories.

Definition 4.1.6 (in **ZFC**). Let $\mathfrak{T}$ be a theory in the $\mathtt{st}$-$\in$-language. A set M is *$\mathfrak{T}$-extendible* if $\langle M ; \in \restriction M \rangle$ admits a standard core embedding into a $\mathtt{st}$-$\in$-structure which models $\mathfrak{T}$. □

Exercise 4.1.7. (1) Prove that every conservative standard core extension of **ZFC** is equiconsistent with **ZFC**. (Reducibility does not imply equiconsistency, moreover, every inconsistent extension of **ZFC** is reducible.)

(2) Prove that if $\mathfrak{T}$ is a standard core extension of **ZFC** then any $\mathfrak{T}$-extendible set is a model of **ZFC**. □

4.1c Comments

Why are these properties important and deserve attention besides just an interest related to a purely foundational study ? Suppose that one is going to "work" in a nonstandard set theory $\mathfrak{T}$, that is, to prove theorems in $\mathfrak{T}$ and interpret the results as mathematically true. Naturally, $\mathfrak{T}$ is a theory whose language contains both $\in$ and st. The standard core $\mathbb{S} = \{x : \mathtt{st}\, x\}$ of the set universe of $\mathfrak{T}$ can be identified with the convenional mathematical set universe [5]. Since the legitimate kit of mathematical tools is almost universally identified with **ZFC**, it is reasonable to require that the theory $\mathfrak{T}$ proves those and only those $\in$-statements about standard sets which **ZFC** proves about all sets, which is exactly the standard core conservativity requirement.

Objects outside of $\mathbb{S}$ can be viewed in two ways. We can see them as auxiliary objects which do not possess the same mathematical reality as those in $\mathbb{S}$, or, saying it differently, as objects invented by $\mathfrak{T}$, which appear in the beginning of a proof and die with QED, when we forget about them.

However, if mathematics is not merely a formal game for us we should consider it as a principle that the "nonstandard" objects ought to have some kind of reality too, perhaps a "relativized" reality with respect to $\mathbb{S}$ which is taken as "real". Then $\mathfrak{T}$ would only provide a kind of $\mathfrak{T}$-*envelope* of $\mathbb{S}$. In this situation we may want the "envelope" to fit tightly to $\mathbb{S}$ such that all st-$\in$-properties sets in $\mathbb{S}$ do have in the envelope are traceable down to $\mathbb{S}$. This is where the property of reducibility appears.

But at the end of the day a direct definition of the "envelope" within $\mathbb{S}$ is the best thing! (Compare with the definition of complex numbers as pairs of reals.) Here we face an obstacle: it is literally impossible to extend the universe of all sets since everything is already here. This is where the notion of *interpretable extension* appears: standard core interpretability of $\mathfrak{T}$ in $\mathfrak{U}$ means, informally, that a theory $\mathfrak{U}$ is strong enough to extend the universe $\mathbf{V}$ of all sets to a structure satisfying $\mathfrak{T}$ where $\mathbf{V}$ becomes the class of all standard sets.

The distinguished role of **ZFC** in the foundations of "standard" mathematics leads us to the following definition:

Definition 4.1.8. A nonstandard set theory is *"realistic"* [6] iff it admits a standard core interpretation in **ZFC**. □

We consider the property of being "realistic" as a principal property which separates nonstandard theories that reflect mathematical reality (as long as the latter is based on the Zermelo – Fraenkel system **ZFC**) from schemes of a purely syntactical nature. It will be our goal to prove that amongst the nonstandard theories considered in this book, **BST** and **HST** are "realistic" while **IST** and some theories considered in Chapter 8 are not.

Proposition 4.1.9. *Any "realistic" nonstandard theory* $\mathfrak{T}$ *is a conservative (hence, equiconsistent) standard core extension of* **ZFC**.

[5] A special feature of **HST** is that it allows to consider the class $\mathbb{WF}$ of well-founded sets, an $\in$-isomorphic, and transitive copy of $\mathbb{S}$ as a more convenient domain of objects of "standard" mathematics than $\mathbb{S}$. Note that Nelson considers things differently, see Footnote 2 on page 13.

[6] The meaning of this word here is not the same as in Hrbaček [Hr 01].

Proof. That conservativity implies equiconsistency is easy: if $\mathfrak{T}$ proves $0 = 1$ then $(0 = 1)^{\mathrm{st}}$ is also provable, thus **ZFC** proves $0 = 1$. To establish conservativity, suppose that $\mathfrak{T}$ proves Φ^{st}, where Φ is an $\in$-sentence. We have to prove Φ in **ZFC**. Consider a standard core interpretation q of $\mathfrak{T}$ in **ZFC**. Then Φ^{st} is true in q, and hence Φ is true in the standard core of q. Therefore, as the latter is $\in$-isomorphic to the **ZFC** universe $\mathbf{V}$, Φ is true in $\mathbf{V}$ as well, which is a proof of Φ in **ZFC**. □

4.1d Metamathematics of internal theories: the main results

Claim (iv) of the next theorem involves **ZFGT**, a theory (defined in § 4.6a) in the language $\mathcal{L}_{\in,\mathbf{G},\mathbf{T}}$ with symbols $\mathbf{G}$ for a global choice function and $\mathbf{T}$ for the truth predicate for formulas of $\mathcal{L}_{\in,\mathbf{G}}$. Note that **ZFGT** contains Separation in $\mathcal{L}_{\in,\mathbf{G},\mathbf{T}}$, but the schemata of Replacement and Collection are included in the $\in$-language only. It will be shown (Theorem 4.6.3) that **ZFGT** is a conservative (in the sense of Definition 1.5.16) standard extension of **ZFC**.

Theorem 4.1.10. (i) **BST** *is a "realistic" theory — hence, it is an equiconsistent and conservative standard core extension of* **ZFC**.

(ii) **BST** *is a reducible theory* — this follows from Theorem 3.2.3(ii).

(iii) **IST** *is an equiconsistent and conservative standard core extension of* **ZFC**. *However* **IST** *is not a reducible theory and* **IST** *is not standard core interpretable in* **ZFC** *— hence it is not "realistic".*

(iv) *On the other hand* **IST** *is standard core interpretable in* **ZFGT**.

Claim (i) of Theorem 4.1.10 will be established in § 4.3c, claims (iii), (iv) related to **IST** — in Sections 4.4, 4.5, 4.6. Note that the conservativity and equiconsistency of **BST** in (i) easily follow from these properties of **IST** via the inner model of bounded sets, but the standard core interpretability of **BST** does not seem to follow from any property of **IST** whatsoever.

Theorem 4.1.10, in its **BST** part, will be an essential precondition in our study of metamathematical properties of **HST** in Chapter 5.

Applying the conservativity in Theorem 4.1.10 and Inner Transfer, we have

Corollary 4.1.11. *Any of the four following conditions is necessary and sufficient for an $\in$-sentence Φ to be a theorem of* **ZFC** :

(1) Φ *is a theorem of* **BST** ; (3) Φ^{st} *is a theorem of* **BST** ;
(2) Φ *is a theorem of* **IST** ; (4) Φ^{st} *is a theorem of* **IST** . □

Let us draw several further consequences. Claims (ii) and (iii) in the next corollary is our backlog from § 3.4a: (iii) and (iv) of Theorem 3.4.5. They show that **BST** is *the theory of the class* $\mathbb{B}$ *of all bounded sets in* **IST**, while $\mathbf{BST}_\kappa$ [7] is *the theory of classes* $\mathbb{I}_\kappa$ *in* **IST**. The claims can also be viewed as a sort of conservativity of **IST** over **BST** and of **BST** over $\mathbf{BST}_\kappa$.

[7] See § 3.3a on partially saturated theories $\mathbf{BST}'_\kappa$ and $\mathbf{BST}_\kappa$.

Corollary 4.1.12. (i) $\mathbf{BST}_{\boldsymbol{\kappa}}$ *is a "realistic" theory — and hence it is an equiconsistent and conservative standard core extension of* **ZFC**. [8]

(ii) *If Φ is a* $\mathtt{st}$-$\in$*-sentence then* **BST** *proves Φ iff* **IST** *proves* $\Phi^{\mathbf{bd}}$.

(iii) *If Φ is a* $\mathtt{st}$-$\in$*-sentence then* $\mathbf{BST}_{\boldsymbol{\kappa}}$ *proves Φ iff* **BST** *proves that* $\Phi^{\mathbb{I}_\kappa}$ (the relativization to $\mathbb{I}_\kappa$) *holds for any standard infinite cardinal κ.*

Proof. (i) Recall that by Theorem 3.4.5(i) any structure of the form $\mathbb{I}_\kappa = \langle \mathbb{I}_\kappa ; \in, \mathtt{st}, \kappa \rangle$, where κ is an infinite standard cardinal, is an interpretation of $\mathbf{BST}_{\boldsymbol{\kappa}}$ in **BST**. (Note: the cardinal κ is an interpretation of the constant $\boldsymbol{\kappa}$ of the language of $\mathbf{BST}_{\boldsymbol{\kappa}}$.) Obviously $\mathbb{I}_\kappa$ contains all standard sets. Taking, for instance, $\kappa = \aleph_0$ (or $\aleph_1$, *etc.*), we find the following: any standard core interpretation of **BST** in **ZFC** can be reduced to a standard core interpretation of $\mathbf{BST}_{\boldsymbol{\kappa}}$ in **ZFC**. [9] It remains to apply Theorem 4.1.10(i).

(ii) The direction "only if" follows from (i) of Theorem 3.4.5: indeed if Φ is a theorem of **BST** then it must be true in $\mathbb{B}$ because this class interprets **BST**. To establish the claim "if", suppose that **IST** proves $\Phi^{\mathbf{bd}}$ where, we recall, $^{\mathbf{bd}}$ indicates relativization to the class $\mathbb{B} = \{x : \exists^{\mathtt{st}} y\, (x \in y)\}$ of all bounded sets in **IST**. By Theorem 3.2.3(ii) (Reduction to $\Sigma_2^{\mathtt{st}}$) there is an $\in$-sentence φ such that **BST** proves $\varphi \iff \Phi$. As $\mathbb{B}$ is an interpretation of **BST** in **IST** by Theorem 3.4.5(i), and **IST** proves $\Phi^{\mathbf{bd}}$, **IST** also proves $\varphi^{\mathbf{bd}}$, and hence proves $\varphi^{\mathtt{st}}$ by Inner Transfer of **BST**. Then **ZFC** proves φ by Corollary 4.1.11, thus **BST** proves $\varphi^{\mathtt{st}}$ and φ itself by $\mathbf{ZFC}^{\mathtt{st}}$ and Inner Transfer of **BST**. It follows that **BST** proves Φ by the choice of φ.

(iii) By Theorem 3.4.5(ii), we can concentrate on the claim "if". We shall assume that $\mathtt{Card}$ indicates only infinite cardinals in the course of the proof.

Suppose that **BST** proves $\forall^{\mathtt{st}} \kappa \in \mathtt{Card}\ \Phi^{\mathbb{I}_\kappa}$. It follows from Theorem 3.3.5 (Reduction to $\Sigma_2^{\mathtt{st}}$ in $\mathbf{BST}_{\boldsymbol{\kappa}}$) that there is an $\in$-formula $\varphi(\boldsymbol{\kappa})$ containing the constant $\boldsymbol{\kappa}$ such that $\mathbf{BST}_{\boldsymbol{\kappa}}$ proves $\Phi \iff \varphi(\boldsymbol{\kappa})$. Then, as $\mathbb{I}_\kappa$ is an interpretation of $\mathbf{BST}_{\boldsymbol{\kappa}}$ in **BST** by Theorem 3.4.5(ii), **BST** also proves $\forall \kappa \in \mathtt{Card}\ \varphi(\boldsymbol{\kappa})^{\mathbb{I}_\kappa}$ and hence proves $\forall \kappa \in \mathtt{Card}\, \varphi(\kappa)$ by Inner Transfer of **BST** and of $\mathbf{BST}_{\boldsymbol{\kappa}}$. But then **ZFC** proves $\forall \kappa \in \mathtt{Card}\, \varphi(\kappa)$ by the conservativity of **BST** and $\mathbf{BST}_{\boldsymbol{\kappa}}$ proves $\forall \kappa \in \mathtt{Card}\, \varphi(\kappa)$ by the conservativity of $\mathbf{BST}_{\boldsymbol{\kappa}}$. (Inner Transfer also works in this argument.)

However $\mathbf{BST}_{\boldsymbol{\kappa}}$ postulates $\boldsymbol{\kappa}$ to be an infinite cardinal. It follows that $\mathbf{BST}_{\boldsymbol{\kappa}}$ proves $\varphi(\boldsymbol{\kappa})$, and hence proves Φ by the choice of φ. □

Exercise 4.1.13. Study the reducibility of $\mathbf{BST}_{\boldsymbol{\kappa}}$. In this case **ZFC** cannot serve as a ground standard theory because Reduction to $\Sigma_2^{\mathtt{st}}$ in Theorem 3.3.5 leads to $\in$-formulas containing $\boldsymbol{\kappa}$. However we can enrich **ZFC** by $\boldsymbol{\kappa}$ as a constant, with an associated axiom saying that $\boldsymbol{\kappa}$ is an infinite cardinal. □

---

[8] Proposition 4.3.2 below proves that $\mathbf{BST}'_{\boldsymbol{\kappa}}$, the other partially saturated theory, is also "realistic", in a somewhat modified sense. Yet no result like 4.1.12(iii) is known for $\mathbf{BST}'_{\boldsymbol{\kappa}}$. See also § 6.2b.

[9] A more direct construction of a standard core interpretation of $\mathbf{BST}_{\boldsymbol{\kappa}}$ in **ZFC** is outlined in Exercise 4.3.15.

4.2 Ultrapowers and saturated extensions

Technically, the proof of Theorem 4.1.10 will consist of a series of nonstandard extensions of different standard structures. In this section we review the basic tools involved. The focal point will be the saturation properties of extensions.

We begin in § 4.2a with a brief introduction into saturated extensions and enlargements as nonstandard structures; Theorem 4.2.4 will show which parts of nonstandard theories are satisfied in these nonstandard structures.

All nonstandard extensions used below belong to a certain general category called quotient powers in § 4.2b. This class includes ordinary ultrapowers, ultralimits, iterated and "definable" ultrapowers in a uniform and natural fashion. Then we consider two particular classes of ultrafilters in § 4.2c, adequate and good ultrafilters: they naturally lead to saturated quotient powers. Limits of transfinite elementary chains of extensions are considered in § 4.2d.

Blanket agreement 4.2.1. We argue in **ZFC** in this section. □

4.2a Saturated structures and nonstandard set theories

The property of saturation is considered here in less generality than in model theory but more in line with its applications in this book. See 4.2.5 for a more general concept.

Definition 4.2.2. Let κ be an infinite cardinal.

An $\in$-structure $\langle {}^*V ; {}^*\varepsilon \rangle$ is *κ-saturated* iff any family $\mathscr{X} \subseteq {}^*V$ with $\mathtt{card}\, \mathscr{X} < \kappa$ and satisfying ${}^*\varepsilon$-f. i. p. (the finite intersection property w. r. t. ${}^*\varepsilon$, meaning that any finite subfamily $\mathscr{X}' \subseteq \mathscr{X}$ has a common ${}^*\varepsilon$-element) in $\langle {}^*V ; {}^*\varepsilon \rangle$ has an ${}^*\varepsilon$-element in ${}^*V$ common for the whole family $\mathscr{X}$.

Suppose that $\langle V ; \varepsilon \rangle$ is another $\in$-structure and $* : V \to {}^*V$ is an $\in$-embedding. Then $\langle {}^*V ; {}^*\varepsilon \rangle$ is a *κ-enlargement* of $\langle V ; \varepsilon \rangle$ via $*$ iff any family $\mathscr{X} \subseteq \{{}^*X : X \in V\}$ with $\mathtt{card}\, \mathscr{X} < \kappa$, ${}^*\varepsilon$-f. i. p. in $\langle {}^*V ; {}^*\varepsilon \rangle$, has an ${}^*\varepsilon$-element in ${}^*V$ common for the whole family $\mathscr{X}$. □

Exercise 4.2.3. Suppose that $* : V \to {}^*V$ is an $\in$-embedding of an $\in$-structure $\mathbf{v} = \langle V ; \varepsilon \rangle$ in ${}^*\mathbf{v} = \langle {}^*V ; {}^*\varepsilon \rangle$. Prove the following:

(1) if ${}^*\mathbf{v}$ is κ-saturated then it is a κ-enlargement of $\mathbf{v}$ via $*$;

(2) for ${}^*\mathbf{v}$ to be a κ-enlargement of $\mathbf{v}$ via $*$ the following is necessary and, *in the case when $*$ is an elementary embedding*, also sufficient: for any family $\mathscr{X} \subseteq V$ with $\mathtt{card}\, \mathscr{X} < \kappa$, satisfying ε-f. i. p. in $\mathbf{v}$, the family ${}^*\mathscr{X} = \{{}^*X : X \in \mathscr{X}\}$ has a common ${}^*\varepsilon$-element in ${}^*V$. □

Thus the property of κ-enlargement essentially requires that any f. i. p. family of size $< \kappa$ in the original structure gains an element in the extension.

The next theorem contains sufficient conditions for a standard core extension of a standard structure to satisfy certain axioms. Recall that κ-size BI is Basic Idealization of § 3.1b in the case $\mathtt{card}\, A_0 \le \kappa$ (see § 3.3a). Similarly let κ-size BE be Basic Enlargement of § 3.1b in the case $\mathtt{card}\, A_0 \le \kappa$.

Theorem 4.2.4. *Suppose that V is either a set of the form $\mathbf{V}_\vartheta$, ϑ being a limit ordinal, or else the universe $\mathbf{V}$ of all sets, and $*$ is an elementary standard core embedding of $\mathbf{v} = \langle V ; \in \restriction V\rangle$ in a* st-$\in$*-structure* $^*\mathbf{v} = \langle {}^*V ; {}^*\!\in, {}^*\mathtt{st}\rangle$. *Let finally $\kappa \in V$ be a cardinal in V.* [10] *Then:*

(i) $^*\mathbf{v}$ *satisfies* [11] *Inner Transfer, Inner Standardization,* **ZC**, *and* $\mathbf{ZC}^{\mathtt{st}}$;

(ii) *if* $\mathbf{v}$ *satisfies* **ZFC** *then* $^*\mathbf{v}$ *satisfies* $\mathbf{ZFC}^{\mathtt{st}}$;

(iii) *if for any $x \in {}^*V$ there is an element $a \in V$ such that $x \,{}^*\varepsilon\, {}^*a$ then $^*\mathbf{v}$ satisfies Inner Boundedness;*

(iv) *if there exists $\xi \in {}^*V$ such that $\xi \,{}^*\varepsilon\, {}^*\kappa$ and for any $x \in {}^*V$ there is a function $f \in V$ defined on κ such that $x = {}^*f(\xi)$ then $^*\mathbf{v}$ satisfies Inner Strong κ-Boundedness* (see § 3.3a);

(v) *if $^*\mathbf{v}$ is κ^+-saturated then it satisfies κ-size BI;*

(vi) *if $^*\mathbf{v}$ is a κ^+-enlargement of $\mathbf{v}$ then it satisfies κ-size BE.*

Proof. (i) and (ii). Inner Transfer follows from the result of Exercise 4.1.2 and the elementarity of the embedding $*$. $\mathbf{ZC}^{\mathtt{st}}$ holds by the same reasons, and also because V itself obviously satisfies **ZC** (any set $\mathbf{V}_\vartheta$, ϑ limit, does).

To check Inner Standardization, let $\varphi(x)$ be an $\in$-formula with sets in $^*V$ as parameters. Suppose that $X \in V$; then $^*X$ is a standard set in $^*\mathbf{v}$ while $Y = \{x \in X : \varphi^{^*\mathbf{v}}({}^*x)\}$ is a set in V by the choice of V. Thus $^*Y \in {}^*V$ is a standard set in $^*\mathbf{v}$. We claim that $\forall^{\mathtt{st}} x \in {}^*X\, (x \in {}^*Y \iff \varphi(x))$ holds in $^*\mathbf{v}$. Since standard sets in $^*\mathbf{v}$ are those of the form $^*y$, it suffices to show $^*x \,{}^*\!\in\, {}^*Y \iff \varphi^{^*\mathbf{v}}({}^*x)$ for every $x \in X$. Yet either side is equivalent to $x \in Y$.

(iii) is obvious. (iv) follows from Lemma 3.3.2(ii)

(v) The notion of finiteness is obviously absolute for V. Thus any set $B \in {}^*V$ such that "B is standard and finite" holds in $^*\mathbf{v}$ has the form $B = {}^*A$ for a unique finite $A \in V$. This observation reduces κ-size BI in $^*\mathbf{v}$ to

$$\forall^{\mathtt{fin}} A \subseteq A_0\, \exists x \,{}^*\varepsilon\, {}^*X\, \forall a \in A\, (x \,{}^*\varepsilon\, X_a) \iff \exists x \,{}^*\varepsilon\, {}^*X\, \forall a \in A_0\, (x \,{}^*\varepsilon\, X_a) \quad (\ddagger)$$

where A and X_0 are sets in V, $\mathtt{card}\, A_0 \le \kappa$, and $X_a \in {}^*V$ for any $a \in A$. (The sets X_a arise as follows. Take, as in Basic Idealization, $\psi \in {}^*V$ such that "ψ is a map ${}^*A_0 \to \mathscr{P}({}^*X)$" holds in $^*\mathbf{v}$. Let X_a be the unique element of $^*V$ such that $\psi({}^*a) = X_a$ holds in $^*\mathbf{v}$. The argument is validated by the fact that $^*\mathbf{v}$ satisfies **ZC** by (i).)

[10] That is, $\kappa \in \mathtt{Ord}$ and V does not contain a bijection from κ onto any $\xi < \kappa$.

[11] The word "satisfies" in the theorem means either *models*, in the form $\mathbf{v} \models \varphi$ or $^*\mathbf{v} \models \varphi$, — provided $V = \mathbf{V}_\vartheta$ and $^*V$ are sets, or *interprets*, in the form $\varphi^{\mathbf{v}}$ or $\varphi^{^*\mathbf{v}}$, — provided $V = \mathbf{V}$. Accordingly, this theorem splits into two theorems of rather different logical structure but essentially identical proofs which we combined in a common proof. This is a bit too complicated, but we'll have to consider extensions of both sets $\mathbf{V}_\vartheta$ and the universe $\mathbf{V}$ below!

The left-hand side of (‡) implies that the family $\mathscr{X} = \{X_a : a \in A_0\}$ satisfies $^*\!\in$-f. i. p. in $^*\mathbf{v}$ and obviously $\mathtt{card}\,\mathscr{X} \le \mathtt{card}\,A_0 \le \kappa$. Then all X_a have a common $^*\!\in$-element $x \in {}^*V$ by the κ^+-saturation, getting the right-hand side of (‡).

(vi) If $\psi \in {}^*V$ and "ψ is a standard map $^*A_0 \to \mathscr{P}(^*X)$" holds in $^*\mathbf{v}$ then easily $\psi = {}^*h$ for a (unique) map $h \in V$, $h : A_0 \to \mathscr{P}(X)$, and further the sets $X_a \in {}^*V$ defined as above coincide with $^*Y_a$, where $Y_a = h(a)$. Now employ the same argument as in (v). □

Exercise 4.2.5. Say that an $\in$-structure $^*\mathbf{v} = \langle {}^*V\,;\, {}^*\varepsilon \rangle$ is *strongly κ-saturated* iff for any $\in$-formula $\varphi(a, x, y_1, ..., y_n)$ and any elements $y_1, ..., y_n \in {}^*V$ the following holds:

(†) for any set $A_0 \subseteq {}^*V$ with $\mathtt{card}\,A_0 < \kappa$, if for every finite $A \subseteq A_0$ there is $x \in {}^*V$ satisfying $\forall\, a \in A\, \varphi^{^*\mathbf{v}}({}^*a, x, y_1, ..., y_n)$ then there is $x \in {}^*V$ such that $\forall\, a \in A_0\, \varphi^{^*\mathbf{v}}({}^*a, x, y_1, ..., y_n)$.

Note that the κ-saturation as in Definition 4.2.2 is the case $\varphi(a, x) := x \in a$ of this definition. Prove under the conditions of Theorem 4.2.4 that if V is a set and $^*\mathbf{v}$ is strongly $(\mathtt{card}\,V)^+$-saturated then $^*\mathbf{v}$ satisfies Idealization. □

4.2b Quotient power extensions

The scheme of quotient power extensions considered here is somewhat more general than the ordinary ultrapower construction in the following two aspects. First, it employs a filter which is not necessarily an ultrafilter. Second, the family of functions the equivalence classes of which form the ultrapower not necessarily contains all functions mapping the index set into the ground structure. Such a generality will be crucial in several applications below.

A quotient power extension begins with the following objects:

1°. An $\in$-structure $\mathbf{v} = \langle V\,;\, \varepsilon \rangle$ where ε is a binary relation on a set or class [12] V. For instance V can be a transitive set or class while $\varepsilon = \in \restriction V$.

2°. An infinite set or class I, the *index set (class)*. A *filter* U over I. A set or class of functions $\mathscr{F} \subseteq V^I$, the *underlying set (class)* of the quotient power. Thus $\mathscr{F}$ consists of functions $f : I \to V$. It is required that for any $x \in V$ the *constant function* $\mathbf{f}_x(i) = x$, $\forall i$, belongs to $\mathscr{F}$.

By definition a *filter over I* is any $U \subseteq \mathscr{P}(I)$ closed under finite intersections and supersets (that is, $X \in U \Longrightarrow Y \in U$ whenever $X \subseteq Y \subseteq I$) and not containing $\varnothing$. An *ultrafilter* is a filter U containing exactly one element of any pair of complementary sets (classes) X, $I \setminus X$ in $\mathscr{P}(I)$. In those cases below when I is a proper class, accordingly, U is a collection of proper classes, suitable provisions will be taken to fix a *parametrization by sets* of the classes involved, to keep the arguments within legitimate frameworks.

[12] *Classes* in **ZFC** are collections defined by formulas, like *e.g.* $\mathtt{Ord}$. We'll have $V = \mathbf{V}$, the **ZFC** universe of all sets, in the most important applications.

The following requirements 3°, 4° will be instrumental in the proof of the Łoś theorem below. Note that both requirements are satisfied for obvious reasons provided U is an ultrafilter and $\mathscr{F} = V^I$.

3°. If $f_1, ..., f_n \in \mathscr{F}$ and $\varphi(x_1, ..., x_n)$ is an $\in$-formula then the set $X = \{i \in I : \varphi^{\mathbf{v}}(f_1(i), ... f_n(i))\}$ is *U-measurable*, *i.e.* it belongs to U or to the complementary ideal $U^{\complement} = \{I \setminus X : X \in U\}$.

4°. If $f_1, ..., f_n \in \mathscr{F}$, $\varphi(x, x_1, ..., x_n)$ is an $\in$-formula, and $\psi(x_1, ..., x_n)$ is $\exists\, x\, \varphi(x, x_1, ..., x_n)$ then

$$\forall\, i \in I\; \psi^{\mathbf{v}}(f_1(i), ..., f_n(i)) \implies \exists\, f \in \mathscr{F}\; \forall\, i \in I\; \varphi^{\mathbf{v}}(f(i), f_1(i), ..., f_n(i)).$$

Recall that $\varphi^{\mathbf{v}}$ means that the $\in$-formula φ is relativized to $\mathbf{v}$, that is all quantifiers $\exists\, z$, $\forall\, z$ are replaced by $\exists\, z \in V$, $\forall\, z \in V$ and $\in$ changed to ε . In different words $\varphi^{\mathbf{v}}$ means that φ is true in $\mathbf{v} = \langle V\, ;\, \varepsilon\rangle$, see § 4.1a. If V is a set rather than a proper class then $\varphi^{\mathbf{v}}$ can be replaced by $\mathbf{v} \models \varphi$.

Definition 4.2.6. If U, I are as indicated then $U i\, \Phi(i)$, $U i \in I\, \Phi(i)$ mean that the set $\{i \in I : \Phi(i)\}$ belongs to U. (The quantifier: "U-many".) □

Under the assumptions 1° – 4° put for all $f, g \in \mathscr{F}$

$$f \mathrel{{}^*{=}} g \quad \text{iff} \quad U i\, (f(i) = g(i))\, ; \qquad {}^*\mathtt{st}\, f \quad \text{iff} \quad f \mathrel{{}^*{=}} \mathbf{f}_x \;\text{ for some } x \in V\, ;$$
$$f\; {}^*\varepsilon\; g \quad \text{iff} \quad U i\, (f(i)\; \varepsilon\; g(i))\, .$$

Exercise 4.2.7. Prove that ${}^*{=}$ is an equivalence relation on $\mathscr{F}$ and the relations ${}^*\varepsilon$ and ${}^*\mathtt{st}$ are ${}^*{=}$-invariant. □

Thus $\langle \mathscr{F}\, ;\, {}^*\varepsilon, {}^*\mathtt{st}\, ;\, {}^*{=}\rangle$ is an invariant structure in the sense of § 1.5c. Yet we are more interested in the associated quotient structure $\langle \mathscr{F}/{}^*{=}\, ;\, {}^*\varepsilon, {}^*\mathtt{st}\rangle$.

Definition 4.2.8. Put $[f] = \{g \in \mathscr{F} : f \mathrel{{}^*{=}} g\}$ for any $f \in \mathscr{F}$, and further

$$[f]\; {}^*\varepsilon\; [g] \;\text{ iff }\; f\; {}^*\varepsilon\; g\, , \qquad {}^*x \;=\; [\mathbf{f}_x] \qquad \text{for each } x \in V,$$
$${}^*\mathtt{st}\, [f] \;\text{ iff }\; {}^*\mathtt{st}\, f\, , \qquad {}^*V \;=\; \mathscr{F}/({}^*{=}) \;=\; \{[f] : f \in \mathscr{F}\}\, .$$

If V is a proper class – a rather typical case below – then the definition of $[f]$ is to be amended so that the classes $[f]$ become sets. We define $[f] = \{g \in \mathscr{F} \cap \mathbf{V}_{\alpha(f)} : f \mathrel{{}^*{=}} g\}$ for any $f \in \mathscr{F}$, where $\alpha(f)$ is the least ordinal α such that $\mathbf{V}_\alpha$, the von Neumann set, contains some $g \in \mathscr{F}$ with $f \mathrel{{}^*{=}} g$.

The structure ${}^*\mathbf{v} = \langle \mathscr{F}/{}^*{=}\, ;\, {}^*\varepsilon, {}^*\mathtt{st}\rangle = \langle {}^*V\, ;\, {}^*\varepsilon, {}^*\mathtt{st}\rangle$, also denoted by $\mathscr{F}/U$ and often truncated to $\langle {}^*V\, ;\, {}^*\varepsilon\rangle$, is the *$U, \mathscr{F}$-quotient power* of $\mathbf{v} = \langle V\, ;\, \varepsilon\rangle$. The map $x \mapsto {}^*x : V \to {}^*V$ is *the natural embedding*.

If I is a set then the quotient power is called *set-indexed*. □

Exercise 4.2.9. Prove that $\{{}^*x : x \in V\}$ is the standard core of ${}^*\mathbf{v}$ in the sense of § 4.1a, that is the collection of all ${}^*\mathtt{st}$-standard elements of ${}^*V$. □

Clearly $\langle {}^*V ; {}^*\varepsilon \rangle$ is a usual ultrapower of $\langle V ; \varepsilon \rangle$ provided U is an ultrafilter and $\mathscr{F} = V^I$. Conditions 3° and 4° are obvious in this case.

We need another definition to formulate the Łoś theorem. If Φ is any $\in$-formula Φ with parameters in $\mathscr{F}$ then for any $i \in I$ we let $\Phi[i]$ indicate the result of the substitution of $f(i)$ for every $f \in \mathscr{F}$ in Φ, and let $[\Phi]$ indicate the result of the substitution of $[f]$ for every $f \in \mathscr{F}$ in Φ. Thus $\Phi[i]$ and $[\Phi]$ are $\in$-formulas having sets in resp. V and ${}^*V$ as parameters.

Lemma 4.2.10 (Łoś Theorem). *Under the assumptions 1° – 4°, suppose that Φ is a closed $\in$-formula with parameters in $\mathscr{F}$. Then $[\Phi]^{{}^*\mathbf{v}} \iff Ui\,(\Phi^{\mathbf{v}}[i])$.*

Proof. We argue by induction on the length of Φ. The case of elementary formulas $f = g$, $f \in g$ easily follows from the definition. It suffices to consider only the induction steps for $\wedge$, $\neg$, $\exists$. The step for $\wedge$ is trivial. The step for $\neg$ follows from the equivalence $Ui\,\neg\,\Phi[i] \iff \neg\,Ui\,\Phi[i]$ by standard arguments. The equivalence itself is a consequence of 3°.

The step for $\exists$. Prove the lemma for a formula $\Psi := \exists\, x\, \Phi(x)$ assuming that the result holds for $\Phi(f)$ whenever $f \in \mathscr{F}$. The direction $\Longrightarrow$ is trivial: $[\Psi]^{{}^*\mathbf{v}}$ implies $[\Phi(f)]^{{}^*\mathbf{v}}$ for some $f \in \mathscr{F}$, hence $Ui\,\Phi^{\mathbf{v}}(f)[i]$ by the induction hypothesis. This obviously implies $Ui\,\Psi^{\mathbf{v}}[i]$.

The direction $\Longleftarrow$. Suppose that $Ui\,\Psi^{\mathbf{v}}[i]$. Let $\varphi(x)$ be the formula $\exists\, y\, \Phi(y) \Longrightarrow \Phi(x)$. Then obviously $\forall\, i\, (\exists\, x\, \varphi(x))^{\mathbf{v}}[i]$, hence by 4° there is a function $f \in \mathscr{F}$ such that $\forall\, i\, \varphi^{\mathbf{v}}(f)[i]$. Then $\Psi^{\mathbf{v}}[i]$ implies $\Phi^{\mathbf{v}}(f)[i]$ for any i, hence in our assumptions we have $Ui\,\Phi^{\mathbf{v}}(f)[i]$. This implies $[\Phi(f)]^{{}^*\mathbf{v}}$ by the inductive hypothesis, hence $[\Psi]^{{}^*\mathbf{v}}$, as required. □

Corollary 4.2.11. *Under the assumptions 1° – 4°, the natural embedding $*$ is an elementary standard core $\in$-embedding of $\mathbf{v} = \langle V ; \varepsilon \rangle$ into ${}^*\mathbf{v} = \langle {}^*V ; {}^*\varepsilon, {}^*\mathtt{st} \rangle$, thus ${}^*\mathbf{v}$ is an elementary standard core extension of $\mathbf{v}$.*

Proof. Since $\{{}^*x : x \in V\}$ coincides with the standard core of ${}^*\mathbf{v}$ by 4.2.9, it suffices to show that $\varphi^{\mathbf{v}} \iff {}^*\varphi^{{}^*\mathbf{v}}$ whenever φ and ${}^*\varphi$ are as in 4.1.1(3). To prove this claim let Φ be obtained by changing each $z \in V$ in φ to $\mathbf{f}_z$. Thus ${}^*\varphi$ is $[\Phi]$ while $\Phi[i]$ coincides with φ for all i. Apply Lemma 4.2.10. □

Remark 4.2.12. It is assumed above that I, U are sets. Yet the crucial applications will be those in which they are proper classes, together with V. Such a class-size modification will be explained in due course. □

4.2c Adequate and good ultrafilters and ultrapowers

The general quotient power construction outlined in § 4.2b yields elementary standard core extensions of standard models, thereby structures satisfying Inner Transfer. An appropriate choice of the ultrafilter is needed to obtain quotient powers being saturated extensions and enlargements. Two types of ultrafilters lead to this goal: adequate and good ultrafilters.

Definition 4.2.13. For any set or class C put $C^{\mathtt{fin}} = \mathscr{P}_{\mathtt{fin}}(C)$ (all finite subsets of C). A filter or ultrafilter U on $C^{\mathtt{fin}}$ is *C-adequate* iff it contains all sets of the form $I(C,j) = \{i \in C^{\mathtt{fin}} : j \subseteq i\}$, where $j \in C^{\mathtt{fin}}$.

In terms of the "U-many" quantifier, this means $Ui\,(j \subseteq i)$ for any finite $j \subseteq C$ or equivalently $Ui\,(c \in i)$ for any $c \in C$. □

It is somewhat confusing that the index set $C^{\mathtt{fin}}$ itself consists of sets, *i.e.* finite subsets of C. If X belongs to a C-adequate filter U then X consists of finite subsets of C, subsequently $U \subseteq \mathscr{P}(C^{\mathtt{fin}})$.

Exercise 4.2.14. Let C be an infinite set. Using Choice prove that:

(1) C-adequate ultrafilters exist, moreover, any C-adequate filter, for instance, $\{I(C,j) : j \in C^{\mathtt{fin}}\}$, can be extended to a C-adequate ultrafilter;

(2) every C-adequate filter is nonprincipal (contains only infinite sets) and *κ-regular* where $\kappa = \mathtt{card}\, C$ (the intersection of any infinite family of sets of the form $I(C,j) = \{i \in C^{\mathtt{fin}} : j \subseteq i\}$, $j \in C$, is empty). □

Good ultrafilters form a somewhat more complicated species.

Definition 4.2.15. Let κ be an infinite cardinal. Suppose that f, g are maps defined on the set $\kappa^{\mathtt{fin}}$ of all finite sets $u \subseteq \kappa$. We say that:

$$\begin{aligned} f \text{ is } \textit{monotone} \quad &\text{if} \quad u \subseteq v \text{ implies } f(u) \supseteq f(v) \text{ for all } u, v \in \kappa^{\mathtt{fin}};\\ g \text{ is } \textit{additive} \quad &\text{if} \quad g(u \cup v) = g(v) \cap g(u) \text{ for all } u, v \in \kappa^{\mathtt{fin}};\\ g \le f \quad &\text{if} \quad g(u) \subseteq f(u) \text{ for all } u \in \kappa^{\mathtt{fin}}. \end{aligned}$$

(Note the inversion in lines 1 and 2!) Clearly any additive map is monotone. An ultrafilter U on κ is *κ^+-good* if for any monotone map $f : \kappa^{\mathtt{fin}} \to U$ there exists an additive map $g : \kappa^{\mathtt{fin}} \to U$, $g \le f$.

An ultrafilter U is *countably incomplete*, or $\aleph_0$-regular, if there is a family $\{X_n : n \in \mathbb{N}\} \subseteq U$ with $\bigcap_n X_n = \varnothing$. □

Proposition 4.2.16 (ZFC; see [CK 92, 6.1.4]**).** *For any infinite cardinal κ there exists a κ^+-good countably incomplete ultrafilter U on κ.* □

This known existence result is cited here without a proof.

To observe the effect of adequate and good ultrafilters we prove

Theorem 4.2.17. *Suppose that $\mathbf{v} = \langle V ; \varepsilon \rangle$ is an ε-structure and κ is an infinite cardinal. Then*

(i) *if $I = \kappa^{\mathtt{fin}}$ and U is a κ-adequate ultrafilter on I then the ultrapower ${}^*\mathbf{v} = V^I/U$ is a κ^+-enlargement of $\mathbf{v}$ via the natural embedding;*

(ii) *if U is a κ^+-good countably incomplete ultrafilter on $I = \kappa$ then the ultrapower ${}^*\mathbf{v} = V^I/U$ is κ^+-saturated, and even strongly κ^+-saturated in the sense of 4.2.5.*

Proof. (i) [13] Let ${}^*\mathbf{v} = V^I/U = \langle {}^*V ; {}^*\varepsilon\rangle$. Consider a set $\mathscr{X} = \{X_\xi : \xi < \kappa\} \subseteq V$. Suppose that $\mathscr{X}$ is ε-f. i. p., so that for any finite $i \subseteq \kappa$ there is $f(i) \in V$ such that $f(i) \;\varepsilon\; X_\xi$ for all $\xi \in i$. Thus $f \in V^I$. We claim that $[f] \;{}^*\varepsilon\; {}^*X_\xi$ in ${}^*\mathbf{v}$ for any $\xi < \kappa$, so that $z = [f]$ is a common ${}^*\varepsilon$-element of all sets ${}^*X_\xi$, as required. By Lemma 4.2.10 it suffices to show that $Ui\,(f(i) \in X_\xi)$. But this is clear because by definition $\xi \in i \Longrightarrow f(i) \in X_\xi$ and U is adequate.

(ii) Let ${}^*\mathbf{v} = V^I/U = \langle {}^*V ; {}^*\varepsilon\rangle$. Consider a set $\{X_\alpha : \alpha < \kappa\} \subseteq {}^*V$ satisfying ${}^*\varepsilon$-f. i. p.. Then $X_\alpha = [h_\alpha]$, $h_\alpha \in V^I$. We put, for $\alpha, \xi < \kappa$, $H_\alpha(\xi) = \{x \in V : x \;\varepsilon\; h_\alpha(\xi)\}$. It follows from the choice of U that there is a decreasing countable chain $\kappa = I_0 \supseteq I_1 \supseteq I_2 \supseteq ...$ of sets $I_n \in U$ with $\bigcap_{n\in\mathbb{N}} I_n = \varnothing$. For any finite $s \subseteq \kappa$ define $H_s(\xi) = \bigcap_{\alpha\in s} H_\alpha(\xi)$. It follows from the f. i. p. assumption and the Łoś theorem (Lemma 4.2.10) that $D_s = \{\xi < \kappa : h_s(\xi) \neq \varnothing\} \in U$, and hence the set $f(s) = I_n \cap D_s$, where n is the number of elements in s, belongs to U.

The map $f : \kappa^{\mathtt{fin}} \to U$ is obviously monotone, thus by the choice of U there exists an additive map $g : \kappa^{\mathtt{fin}} \to U$, $g \le f$. Put $s_\xi = \{\alpha < \kappa : \xi \in g(\{\alpha\})\}$ for each $\xi < \kappa$. Note that any s_ξ is finite. Indeed if s_ξ contains at least n elements, say, ordinals $\alpha_1 < \cdots < \alpha_n$, then $\xi \in g(\{\alpha_i\})$, $\forall i$, hence $\xi \in g(\{\alpha_1, ..., \alpha_n\}) \subseteq f(\{\alpha_1, ..., \alpha_n\}) \subseteq I_n$ by the additivity, but $\bigcap_n I_n = \varnothing$.

Further, given $\xi < \kappa$, we have $\xi \in \bigcap\{g(\{\alpha\}) : \alpha \in s_\xi\} = g(s_\xi) \subseteq f(s_\xi)$, thus $\xi \in f(s_\xi)$, so that $H_{s_\xi}(\xi) \neq \varnothing$. Choose any $x(\xi) \in H_{s_\xi}(\xi)$; thus x is a map in V^I. We claim that $[x] \in X_\alpha$ for any $\alpha < \kappa$.

We have by definition $x(\xi) \in H_{s_\xi}(\xi) = \bigcap_{\alpha\in s_\xi} H_\alpha(\xi) = \bigcap_{\xi\in g(\{\alpha\})} H_\alpha(\xi)$, thus $U\xi\,(x(\xi) \;\varepsilon\; h_\alpha(\xi))$ for any $\alpha < \kappa$ because all values of g belong to U. It follows by the Łoś theorem that $[x] \in [h_\alpha]$ as required.

A similar proof of the strong saturation is left as an **exercise**. □

4.2d Elementary chains of structures

In some cases below a saturated structure appears as the result of a transfinite sequence of enlargements. Here we introduce an appropriate notation. As sometimes `Ord`-long chains will be considered, let us reserve ∞ for an object larger than any ordinal, and put $\mathtt{Ord}^+ = \mathtt{Ord} \cup \{\infty\}$.

Definition 4.2.18. Let $\Omega \subseteq \mathtt{Ord}^+$ be an initial segment. An *elementary continuous chain of* $\in$*-structures of length* Ω is any sequence of $\in$-structures $\langle V_\xi ; \varepsilon_\xi\rangle$ and embeddings $e_{\eta\xi} : V_\eta \to V_\xi$ $(\eta \le \xi \in \Omega)$ satisfying the following:

1*. Any $e_{\eta\xi}$ $(\eta \le \xi \in \Omega)$ is an elementary embedding of $\langle V_\eta ; \varepsilon_\eta\rangle$ into $\langle V_\xi ; \varepsilon_\xi\rangle$, $e_{\zeta\xi} = e_{\eta\xi} \circ e_{\zeta\eta}$ (the superposition) whenever $\zeta \le \eta \le \xi$, and finally $e_{\xi\xi}$ is the identity on V_ξ for any ξ.

[13] The result follows from the κ-regularity of U as in 4.2.14 by a general argument, see [CK 92, 4.3.14]. Nevertheless we present a more transparent argument here.

2*. For any $\lambda \in \Omega$ either a limit ordinal or ∞, $\langle V_\lambda ; \varepsilon_\lambda \rangle$ is the *direct limit* of $\langle V_\xi ; \varepsilon_\xi \rangle$, $\xi < \lambda$, in the sense that $V_\lambda = \{e_{\xi\lambda}(x) : \xi < \lambda \wedge x \in V_\xi\}$. □

Note that any initial segment $\Omega \subseteq \mathtt{Ord}^+$ is either an ordinal or the class $\mathtt{Ord}$ of all ordinals or $\mathtt{Ord}^+$ itself.

Proposition 4.2.19. *Suppose that $\langle V_\xi ; \varepsilon_\xi \rangle$ and $e_{\eta\xi} : V_\eta \to V_\xi$ $(\eta \le \xi \le \gamma)$ is an elementary continuous chain of $\in$-structures, γ being either a limit ordinal or ∞. If κ is a cardinal, $\mathtt{cof}\,\gamma > \kappa$, and any $\langle V_{\xi+1} ; \varepsilon_{\xi+1} \rangle$ is a κ-enlargement of $\langle V_\xi ; \varepsilon_\xi \rangle$ then $\langle V_\gamma ; \varepsilon_\gamma \rangle$ is κ-saturated.*

Proof. Consider any ε_γ-f. i. p. set $\mathscr{X} \subseteq V_\gamma$ of cardinality $< \kappa$. (Recall that ε_γ-f. i. p. indicates the finite intersection property w. r. t. ε_ξ as the membership, see Definition 4.2.2.) Since $\mathtt{cof}\,\gamma > \kappa$, there is an ordinal $\xi < \kappa$ such that every $X \in \mathscr{X}$ has the form $X = e_{\eta\gamma}(Y)$ for some $\eta < \xi$ and $Y \in V_\eta$. In this case the set $Z = e_{\eta\xi}(Y)$ belongs to V_ξ and still $X = e_{\xi\gamma}(Z)$. It follows that there is a set $\mathscr{Z} \subseteq V_\xi$ still with $\mathtt{card}\,\mathscr{Z} = \mathtt{card}\,\mathscr{X} < \kappa$ such that $\mathscr{X} = \{e_{\xi\gamma}(Z) : z \in \mathscr{Z}\}$.

As $e_{\xi\gamma}$ is an elementary embedding, $\mathscr{Z}$ is ε_ξ-f. i. p.. However $\langle V_{\xi+1} ; \varepsilon_{\xi+1} \rangle$ is a κ-enlargement of $\langle V_\xi ; \varepsilon_\xi \rangle$, and hence there is an element $z \in V_{\xi+1}$ with $z \;\varepsilon_{\xi+1}\; e_{\xi,\xi+1}(Z)$ for any $Z \in \mathscr{Z}$. Then $x = e_{\xi+1,\gamma}(z)$ satisfies $x \in e_{\xi\gamma}(Z)$ for any $Z \in \mathscr{Z}$ simply because $e_{\xi\gamma}(Z) = e_{\xi+1,\gamma}(e_{\xi,\xi+1}(Z))$. □

There are different methods to maintain the step $\xi \to \xi + 1$ in the construction of an elementary continuous chain so that the next structure is an enlargement of the previous one or even a saturated structure for a suitable cardinal — for instance adequate or good ultrapowers. As for the limit step, there is a simple universal construction.

Lemma 4.2.20. *If γ is a limit ordinal or ∞ then any elementary continuous chain of $\in$-structures of length γ can be extended to an elementary continuous chain of length $\gamma \cup \{\gamma\}$ $(= \gamma + 1$ in the case when $\gamma \in \mathtt{Ord})$.*

Proof. Consider a elementary continuous chain which consists of structures $\langle V_\xi ; \varepsilon_\xi \rangle$ $(\xi < \gamma)$ and elementary embeddings $e_{\eta\xi} : V_\eta \to V_\xi$ $(\eta \le \xi < \gamma)$. Define V_γ to be the collection of all pairs of the form $\langle \xi + 1, x \rangle$, where $\xi < \gamma$ and $x \in V_{\xi+1} \smallsetminus \mathtt{ran}\, e_{\xi,\xi+1}$, along with all pairs of the form $\langle 0, x \rangle$, $x \in V_0$. Define $\langle \xi', x \rangle \;\varepsilon_\gamma\; \langle \eta', y \rangle$ (ξ', η' being 0 or successor ordinals) iff $e_{\xi'\zeta}(x) \;\varepsilon_\zeta\; e_{\eta'\zeta}(y)$ where $\zeta = \mathtt{max}\{\xi', \eta'\}$. To define embeddings $e_{\xi\gamma}$, $\xi < \gamma$, suppose that $\xi < \gamma$ and $x \in V_\xi$. There is a least ordinal $\zeta \le \xi$ such that $x = e_{\zeta\xi}(y)$ for some (unique) $y \in V_\zeta$. (For instance if $\xi = \eta + 1$ and $x \notin \mathtt{ran}\, e_{\eta\xi}$ then $\zeta = \xi$ and $y = x$.) Put $e_{\xi\gamma}(x) = \langle \zeta, y \rangle$. □

In the most elementary case when $V_\eta \subseteq V_\xi$, $\varepsilon_\eta = \varepsilon_\xi \restriction V_\eta$, and $e_{\eta\xi} =$ the identity on V_η (for all $\eta \le \xi < \gamma$) the direct limit is isomorphic to the union $\langle \bigcup_{\xi<\gamma} V_\xi ; \bigcup_{\xi<\gamma} \varepsilon_\xi \rangle$ of the chain.

4.3 Metamathematics of BST

The goal of this section is to prove claim (i) of Theorem 4.1.10. Our plan is as follows. Working in **ZFC**, we define a standard core interpretation of **BST** in **ZFC** in the form of a quotient power of the universe which belongs to the type of *iterated ultrapowers* (with finite support), introduced in § 4.3b. The ultrafilters involved in the iteration belong to the category of *adequate* ultrafilters (§ 4.2c). (The same effect can be achieved by *good* ultrafilters, a more complicated type.) The rest of Theorem 4.1.10(i) is a simple corollary.

Finally, we show in § 4.3d how to modify the construction in order to obtain an interpretation that is κ^+-saturated for any cardinal κ.

Blanket agreement 4.3.1. We argue in **ZFC** in this section unless explicitly specified otherwise. □

4.3a Warmup: several examples

The following examples are immediate applications of adequate and good quotient powers to the metamathematics of internal theories. Albeit inconclusive in the sense of our final goal, they give a good introduction into more complicated constructions of nonstandard extensions below.

Example 1: a κ^+-saturated extension of the universe

Consider the **ZFC** universe of all sets $\mathbf{v} = \langle \mathbf{V}\,;\in\rangle$ as the initial structure. Let κ be an infinite cardinal and U_κ be a κ^+-good countably incomplete ultrafilter on $I = \kappa$. (Proposition 4.2.16 is applied.)

Proposition 4.3.2. *The ultrapower* $^*\mathbf{v} = \mathbf{V}^I/U_\kappa$ *is a standard core interpretation of* $\mathbf{BST}'_{^*\kappa}$ [14] *in* **ZFC**. *Thus* $\mathbf{BST}'_\kappa$ *is a "realistic"* [15] *theory, and hence it is an equiconsistent and conservative standard core extension of* **ZFC**.

Proof. Let $^*\mathbf{v} = \langle {}^*\mathbf{V}\,; {}^*\!\in, {}^*\mathtt{st}\rangle$ Corollary 4.2.11 enables us to apply Theorem 4.2.4. Thus we immediately have $\mathbf{ZFC}^{\mathbf{st}}$, Inner Transfer, Inner Standardization in $^*\mathbf{v}$, as well as κ-size BI by Theorem 4.2.17(ii). Finally to prove Inner Strong κ-Boundedness we have to verify the requirement of 4.2.4(iv). Consider any element $F = [f] \in {}^*\mathbf{V}$; $f \in \mathbf{V}^I$. Put $g(\xi) = \xi$ for any $\xi < \kappa$. Then easily $G = [g] \in {}^*\mathbf{V}$, $G \;{}^*\!\in {}^*\kappa$, and $F = {}^*\!f(G)$ in $^*\mathbf{v}$ by the Łoś theorem. □

It will be demonstrated in § 6.2b that essentially the same κ^+-good ultrapower of a **ZFC** universe can be represented as a class of the form $\mathbb{S}[x]$ in **BST**. This gives an interpretation of $\mathbf{BST}'_\kappa$ in the form of an inner structure in **BST** (and thereby in **IST** and **HST**). See Theorem 3.4.5(i) on interpretations of $\mathbf{BST}_\kappa$, another partially saturated theory, by means of classes $\mathbb{I}_\kappa$.

[14] Meaning that $^*\mathbf{v}$ is an interpretation of $\mathbf{BST}_\kappa$ such that the constant κ is interpreted as $^*\kappa$. See § 3.3a on partially saturated theories $\mathbf{BST}'_\kappa$ and $\mathbf{BST}_\kappa$.

[15] Albeit in a slightly modified sense which allows papameters, like U_κ here, to participate in the definition of an interpretation.

Example 2: a κ^+-saturated extension of the universe, 2nd version

We can build up a κ^+-saturated extension of the universe $\mathbf{V}$ also by means of an elementary continuous chain of adequate ultrapowers.

Fix a κ-adequate ultrafilter U over $I = \kappa^{\mathtt{fin}}$. Define an elementary continuous chain of $\in$-structures $\mathbf{v}_\xi = \langle V_\xi ; \varepsilon_\xi \rangle$ $(\xi \le \kappa^+)$ along with associated embeddings $e_{\eta\xi} : V_\eta \to V_\xi$ $(\eta \le \xi \le \kappa^+)$ so that every $\mathbf{v}_{\xi+1}$ is the U-ultrapower of $\mathbf{v}_\xi$, and every $e_{\xi,\xi+1}$ is the natural embedding of $\mathbf{v}_\xi$ in $\mathbf{v}_{\xi+1}$. The definition goes on by transfinite induction on ξ. Put $\mathbf{v}_0 = \langle \mathbf{V} ; \in \rangle$. At limit steps (including κ^+) apply Lemma 4.2.20. At successor steps $\xi \to \xi + 1$ define $e_{\xi,\xi+1}$ as indicated and then put $e_{\eta,\xi+1} = e_{\xi,\xi+1} \circ e_{\eta\xi}$ for any $\eta < \xi$.

Define a unary relation ${}^*\mathtt{st}$ on V_{κ^+} so that ${}^*\mathtt{st}\, x$ iff $x \in \mathtt{ran}\, e_{0\kappa^+}$. This accomplishes the definition of the *final structure* ${}^*\mathbf{v} = \langle V_{\kappa^+} ; \varepsilon_{\kappa^+}, {}^*\mathtt{st} \rangle$. With $\mathbf{v}_{\xi+1}$ being a κ^+-enlargement of $\mathbf{v}_\xi$ for every ξ by Theorem 4.2.17(i), the final structure ${}^*\mathbf{v}$ is a κ^+-saturated standard core extension of $\mathbf{V}$ via the embedding $e_{0\kappa^+}$ by Proposition 4.2.19. Yet unlike Example 1 Inner Strong κ-Boundedness fails in $\mathbf{v}_{\kappa^+}$. This shows that sometimes good ultrapowers cannot be substituted by chains of adequate ultrapowers.

Example 3: a putative interpretation of BST

We can try to employ elementary continuous chains of successive ultrapowers to obtain a standard core interpretation of **BST**. Indeed suppose that U_ξ is a ξ-adequate ultrafilter on $I_\xi = \xi^{\mathtt{fin}}$ for any ordinal ξ. (Good ultrafiltes do not yield anything better in this construction.) Define, as in Example 2, an $\mathtt{Ord} \cup \{\infty\}$-long elementary continuous chain of $\in$-structures $\mathbf{v}_\xi = \langle V_\xi ; \varepsilon_\xi \rangle$ $(\xi \le \infty)$ along with associated embeddings $e_{\eta\xi} : V_\eta \to V_\xi$ $(\eta \le \xi \le \infty)$ so that $\mathbf{v}_0$ is $\langle \mathbf{V} ; \in \rangle$, every $\mathbf{v}_{\xi+1}$ is the U_ξ-ultrapower of $\mathbf{v}_\xi$, and every $e_{\xi,\xi+1}$ is equal to the natural embedding of $\mathbf{v}_\xi$ in $\mathbf{v}_{\xi+1}$.

Define a unary relation ${}^*\mathtt{st}$ on V_∞ so that ${}^*\mathtt{st}\, x$ iff $x \in \mathtt{ran}\, e_{0\infty}$. The *final structure* ${}^*\mathbf{v} = \langle V_\infty ; \varepsilon_\infty, {}^*\mathtt{st} \rangle$ is then a κ^+-saturated standard core extension of $\mathbf{v} = \mathbf{v}_0 = \langle \mathbf{V} ; \in \rangle$ via the embedding $e_{0\infty}$ for **any** cardinal κ by the same reasons as in Example 2. It immediately follows from Theorem 4.2.4 that $\mathbf{ZFC}^{\mathtt{st}}$, Inner Transfer, Inner Standardization, Basic Idealization hold in ${}^*\mathbf{v}$.

To prove Inner Boundedness we have to verify the requirement of 4.2.4(iii). The case $\xi = \infty$ in the next lemma gives the result required.

Lemma 4.3.3. *For any $x \in V_\xi$ there is $s \in \mathbf{V}$ such that $x \;\varepsilon_\xi\; e_{0\xi}(s)$.*

Proof. We argue by induction on $\xi \in \mathtt{Ord} \cup \{\infty\}$. The case $\xi = 0$ is trivial.

The step $\xi \to \xi + 1$. Suppose that $x \in V_{\xi+1}$. Then $x = [f]$ for some $f \in V_\xi^I$. Thus $\mathtt{ran}\, f \subseteq V_\xi$. By the inductive hypothesis for any $i \in I_\xi = \mathtt{dom}\, f$ there is a set $x_i \in \mathbf{V}$ such that $f(i) \;\varepsilon_\xi\; e_{0\xi}(x_i)$. Consider the set $X = \bigcup_{i \in I_\xi} x_i$. Then $x_i \subseteq X$. However $e_{0\xi}$ is an elementary embedding. It follows that $e_{0\xi}(x_i) \subseteq_\xi X'$, where $X' = e_{0\xi}(X)$ and $\subseteq_\xi$ means the inclusion induced by ε_ξ as the membership. Thus we have $f(i) \;\varepsilon_\xi\; X'$ for any $i \in I_\xi$. Now Lemma 4.2.10 implies $x \;\varepsilon_{\xi+1}\; e_{\xi,\xi+1}(X') = e_{0,\xi+1}(X)$, as required.

The limit step. If λ is a limit ordinal or ∞ and $x \in V_\lambda$ then by definition $x = e_{\xi\lambda}(z)$ for some $\xi < \lambda$ and $z \in V_\xi$, and hence by the inductive hypothesis $z \,\varepsilon_\xi\, e_{0\xi}(s)$ for some $s \in \mathbf{V}$, thus $x \,\varepsilon_\lambda\, e_{\xi\lambda}(e_{0\xi}(s)) = e_{0\lambda}(s)$. □

Thus ${}^*\mathbf{v}$ *is a standard core interpretation of* **BST**, with $e_{0\infty}$ being a standard core embedding $\mathbf{V} \to V_\infty$.

But unfortunately this argument contains a gap!

Example 4: an interpretation of BST in ZF with global choice

The problem can be easily observed in the very beginning of the argument: there seem to be no way to assign a ξ-adequate or a ξ-good ultrafilter to every $\xi \in \mathtt{Ord}$ by a single formula in **ZFC**.

Definition 4.3.4. **ZFGC** is the "global choice" extension of **ZFC**, *i.e.*, a theory in the language $\mathcal{L}_{\in,\mathbf{G}}$, with the membership $\in$ and an additional unary functional symbol **G**. The axioms of **ZFGC** include all of **ZFC** in $\mathcal{L}_{\in,\mathbf{G}}$ (so that **G** may occur in the axiom schemata of Separation, Collection, and Replacement) and the Global Choice axiom which says that **G** is a global choice function, that is, $\mathbf{G}(x) \in x$ for any $x \neq \varnothing$.

$\mathbf{ZFGC}^{\text{weak}}$ is a weaker version of **ZFGC** : the symbol **G** can occur in the schema of Separation but not in Collection and not in Replacement. □

The construction of ${}^*\mathbf{v}$ in Example 3 works perfectly in **ZFGC**, in fact even in $\mathbf{ZFGC}^{\text{weak}}$: it suffices to define U_ξ as the **G**-choice in the (non-empty) set of all ξ-adequate ultrafilters on $\xi^{\mathtt{fin}}$. This proves:

Theorem 4.3.5. **BST** *is standard core interpretable in* $\mathbf{ZFGC}^{\text{weak}}$. □

It is known (Felgner [Fel 71]) that **ZFGC** and $\mathbf{ZFGC}^{\text{weak}}$ are conservative (and hence equiconsistent) extensions of **ZFC**: any $\in$-sentence provable in **ZFGC** is a theorem of **ZFC** as well. (See Theorem 4.6.3 for a stronger result.) Thus Theorem 4.3.5 implies that **BST** is a conservative (and hence equiconsistent) standard core extension of **ZFC**, as in Theorem 4.1.10(i).

As for interpretation in **ZFC** (without global choice), we will have to modify the construction. The successful modification involves *Fubini products* of ultrafilters with finite support, such that the family of ultrafilters involved is ordered by a linear order which is not a well-order. We present the "adequate" version of the construction; the "good" version is pretty much analogous.

4.3b Infinite Fubini products of adequate ultrafilters

Suppose that C_1, C_2 are disjoint sets and U_1, U_2 are resp. a C_1-adequate and a C_2-adequate ultrafilter. The *Fubini product* of them is defined by

$$U_1 \times U_2 = \{X \subseteq (C_1 \cup C_2)^{\mathtt{fin}} : U_2\, i_2\ U_1\, i_1\ (i_1 \cup i_2 \in X)\}\,.$$

Thus in terms of generalized quantifiers "U-many" (see Definition 4.2.6) the product $U = U_1 \times U_2$ can be characterized by the equivalence $U i\, \Phi(i) \iff U_2\, i_2\ U_1\, i_1\ \Phi(i_1 \cup i_2)$. (Note the order of quantifiers!)

Exercise 4.3.6. (1) Prove that $U_1 \times U_2$ is a $(C_1 \cup C_2)$-adequate ultrafilter.

(2) Why is $C_1 \cap C_2 = \varnothing$ required in the definition?

(3) Prove that the Fubini product is associative but non-commutative.

Hint for (1). That $U = U_1 \times U_2$ is closed under supersets in $\mathscr{P}(I)$, where $I = (C_1 \cup C_2)^{\mathtt{fin}}$, is clear. To check that U is closed under intersection use the equivalences $U_\nu\, i_\nu\, \Phi(i_\nu) \wedge U_\nu\, i_\nu\, \Psi(i_\nu) \iff U_\nu\, i_\nu\, (\Phi(i_\nu) \wedge \Psi(i_\nu))$, $\nu = 1, 2$. The equivalences $\neg\, U_\nu\, i_\nu\, \Phi(i_\nu) \iff U_\nu\, i_\nu\, \neg\, \Phi(i_\nu)$, $\nu = 1, 2$, prove that U is "ultra". The equivalences follow from the fact that U_ν are ultrafilters. □

See more on the Fubini products of not necessarily adequate ultrafilters, including an example of non-commutativity, in [CK 92, 6.5].)

The Fubini product admits an immediate generalization to any finite number of factors. What about infinite products?

Suppose that $P = \langle P; < \rangle$ is a linearly ordered set or class, and $U_p \subseteq \mathscr{P}(I_p)$ is a C_p-adequate ultrafilter over $I_p = C_p{}^{\mathtt{fin}}$ for any $p \in P$. We assume that C_p are pairwise disjoint sets; this does not restrict generality because we can define $C'_p = \{p\} \times C_p$ and replace each U_p by its shift copy $U'_p \subseteq (C'_p)^{\mathtt{fin}}$ obtained by means of the map $c \mapsto \langle p, c \rangle$.

Put $C = \bigcup_{p \in P} C_p$ and $I = C^{\mathtt{fin}}$ (can be proper classes). For any finite $s \subseteq P$ we set $C^s = \bigcup_{p \in s} C_p$, $I^s = (C^s)^{\mathtt{fin}}$, $i \restriction s = i \cap C^s$ for any $i \in I$,

$$X{\Uparrow} = \{ i \in I : i \restriction s \in X \} \quad \text{for any } X \subseteq I^s\,;$$

$$U{\Uparrow} = \{ X{\Uparrow} : X \in U \} \qquad \text{for any } U \subseteq \mathscr{P}(I^s)\,;$$

and finally $U^s = U_{p_1} \times \ldots \times U_{p_n}$ whenever $s = \{p_1 < \ldots < p_n\}$, the Fubini product of the ultrafilters $U_{p_1}, \ldots, U_{p_n}$.

Say that $X \subseteq I$ is a set or class *of finite support* if there is a finite set $s \subseteq P$ such that $i \in X \iff i \restriction s \in X$ holds for all $i \in I$. In this case obviously there is a least finite set s of this kind, which will be denoted by $\|X\|$ and called the *support* of X, and we have $X = Y{\Uparrow}$, where $Y = X \cap I^s$.

Lemma 4.3.7. (i) *If* $t = s \cup \{p\}$, $p > \sup s$, *and* $X \subseteq I^t$, *then*

$$U_p\, j \in I_p\;\; U^s\, i \in I^s\, (i \cup j \in X) \iff U^t\, i \in I^t\, (i \in X)\,.$$

(ii) *Any* U^s *is a* C^s*-adequate ultrafilter.*

(iii) *If* $s \subseteq t$ *and* $X \subseteq I^t$, $\|X\| \subseteq s$, *then* $U^s\, i\, (i \in X) \iff U^t\, i\, (i \in X)$, *and hence* $U^s{\Uparrow} \subseteq U^t{\Uparrow}$.

Proof. (iii) If, say, $t = \{p_1 < \ldots < p_n < p\}$, $s = t \setminus \{p\}$, then easily

$$U^s\, i \in I^s\, (i \in X) \iff U_p\, j \in I_p\;\; U^s\, i \in I^s\, (i \cup j \in X)$$

for any $X \subseteq I^t$ with $\|X\| \subseteq s$, simply because in this case whether $i \in X$ holds (for $i \in I^t$) does not depend on $i \restriction p$. It remains to apply (i). (See [CK 92, 6.5.3] for a complete accurate proof.) □

Obviously $\mathbf{U}' = \bigcup_{s \subseteq P \text{ finite}} (U^s \Uparrow)$ is a collection of sets or classes of finite support, and hence it is not even a filter as it is not closed under the superset formation. Nevertheless we define $\mathbf{U} = \bigtimes_{p \in P} U_p$, the collection of all $Y \subseteq I$ containing a subset or subclass that belongs to $\mathbf{U}'$.

Exercise 4.3.8. Prove that $\mathbf{U}'$ is an ultrafilter in the algebra $\mathscr{A}$ of all $X \subseteq I$ of finite support, $\mathbf{U} = \bigtimes_{p \in P} U_p$ is a C-adequate filter, and

(*) if $X \subseteq I$ and $\|X\| \subseteq s$, then $\mathbf{U}\, i \in I\ (i \in X) \iff U^s i \in I^s\ (i \in X)$. □

Definition 4.3.9. A function f on I is a function *of finite support* if there is a finite set $s \subseteq P$ such that $f(i) = f(i \restriction s)$ for all $i \in I$. In this case there is a least such finite set s, it denoted by $\|f\|$, the *support* of f. For a finite $s \subseteq P$ let $\mathscr{F}^s$ be the class of all functions defined on I with finite support $\|f\| \subseteq s$. (Values of $f \in \mathscr{F}^s$ can be arbitrary.) Put $\mathscr{F}^P = \bigcup_{s \subseteq P \text{ finite}} \mathscr{F}^s$.

Exercise: prove that $\operatorname{ran} f$ is always a set for $f \in \mathscr{F}^P$, even in the case when $I = \operatorname{dom} f$ is a proper class. □

Exercise 4.3.10. Prove that 3°, 4° of § 4.2b are satisfied with $V = \mathbf{V}$ (the universe of all sets), $U = \bigtimes_{p \in P} U_p$, and $\mathscr{F} = \mathscr{F}^P$. (*Hint.* $\mathscr{F}^P$ and $\bigtimes_{p \in P} U_p$ match each other: the former consists of functions with finite support while the latter is an ultrafilter in the algebra of all $X \subseteq I$ with finite support.) □

Remark 4.3.11. C and I will be proper classes in principal applications below. Accordingly, U^s, $\mathbf{U}'$, $\mathscr{F}^P$ and $\mathbf{U} = \bigtimes_{p \in P} U_p$ itself, will be collections of proper classes. Yet this can be considered as merely a figure of speech since only classes $X \subseteq I$ and maps f defined on I *of finite support* will be involved, a category which admits a fully effective coding by sets. □

4.3c Standard core interpretation of BST in ZFC

It follows from 4.3.10 that we can consider quotient powers of different structures by means of filters of type $\bigtimes_{a \in A} U_a$. Extensions of this type are often called *iterated ultrapowers*. The first application of this construction is a standard core interpretation of **BST** in **ZFC** based on the two following results.

Theorem 4.3.12 (ZFC). *One can define a linearly ordered class* $P = \langle P; < \rangle$*, and for any* $p \in P$ — *a set* $C_p \neq \varnothing$ *and a* C_p*-adequate ultrafilter* U_p *such that* P *has no largest element, the sets* C_p *are pairwise disjoint, and* $\{p \in P : \operatorname{card} C_p > \kappa\}$ *is cofinal in* P *for any cardinal* κ.

Proof. The idea is to define $\{U_p : p \in P\}$ equal to the class of all adequate ultrafilters. For any ordinal ξ, define a wellordering $<''_\xi$ on the set $\xi^{\mathtt{fin}}$ of all finite subsets of ξ so that $i <''_\xi j$ whenever $\operatorname{card} i < \operatorname{card} j$, and $<''_\xi$ acts lexicographically on finite sets $i, j \subseteq \xi$ of equal size. Let $<'_\xi$ will denote the induced lexicographical *linear* ordering of $\mathscr{P}(\xi^{\mathtt{fin}})$ obtained by means of the identification of any $X \subseteq \xi^{\mathtt{fin}}$ with its characterictic function. Finally, let $<_\xi$

indicate the induced lexicographical *linear* ordering of the set $P_\xi = \mathscr{P}(\xi^{\mathtt{fin}})^{2^\xi}$ of all functions $p : 2^\xi \to \mathscr{P}(\xi^{\mathtt{fin}})$ such that $\mathtt{ran}\, p = \{p(\nu) : \nu < 2^\xi\}$ is a ξ-adequate ultrafilter over $\xi^{\mathtt{fin}}$. (Here 2^ξ denotes the cardinal $\mathtt{card}\, \mathscr{P}(\xi)$.)

Let $C_p = \xi$ and $U_p = \mathtt{ran}\, p$ for $p \in P_\xi$. Then $\{U_p : p \in P_\xi\}$ is the set of all ξ-adequate ultrafilters; note that $U_p = U_q$ is possible for $p \neq q$.

We finally put $P = \bigcup_{\xi \in \mathtt{Ord}} P_\xi$ and define $p < q$ if either $p \in P_\xi$, $q \in P_\eta$, $\xi < \eta$ or $p, q \in P_\xi$, $p <_\xi q$. The transformation mentioned in the beginning of § 4.3b converts the sets C_p to pairwise disjoint form $C'_p = \{p\} \times \xi$. □

Theorem 4.3.13 (ZFC). *If P, C_p, U_p are as in Theorem 4.3.12, $\mathbf{U} = \mathsf{X}_{p \in P} U_p$, then the $\mathbf{U}, \mathscr{F}^P$-quotient power $^*\mathbf{v} = \mathscr{F}^P/\mathbf{U} = \langle {}^*\mathbf{V}\,; {}^*\!\in, {}^*\mathtt{st}\rangle$ of the universe $\langle \mathbf{V}\,; \in\rangle$ is a standard core interpretation of* **BST** *in* **ZFC**, *with the natural embedding $*$ being a standard core embedding $\mathbf{V} \to {}^*\mathbf{V}$.*

Proof. The proof of Theorem 4.3.12 results in a concrete unambiguous definition of P, C_p, U_p, with the sets C_p pairwise disjoint, and hence the derived objects like $\mathbf{U} = \mathsf{X}_{p \in P} U_p$ and $\mathscr{F}^P$ defined in § 4.3b as well as the $\mathbf{U}, \mathscr{F}^P$-quotient power $^*\mathbf{v} = \langle {}^*\mathbf{V}\,; {}^*\!\in, {}^*\mathtt{st}\rangle$ of $\langle \mathbf{V}\,; \in\rangle$ and the natural embedding $*$ also admit concrete unambiguous $\in$-definitions in **ZFC**.

We have to prove that any **BST** axiom holds in $^*\mathbf{v}$. (To be more exact it is asserted that $\Phi^{^*\mathbf{v}}$ *is a theorem of* **ZFC** *whenever Φ is an axiom of* **BST**.) We observe that Inner Transfer, $\mathbf{ZFC}^{\mathtt{st}}$, Inner Standardization follow from Theorem 4.2.4.

Inner Boundedness. Suppose that $F = [f] \in {}^*\mathbf{V}$, $f \in \mathscr{F}^P$. By definition there is a finite set $s \subseteq P$ such that $f(i) = f(i \restriction s)$ for all $i \in I$. (In the notation of § 4.3b.) It follows that the range $r = \mathtt{ran}\, f$ is a set. Thus $f(i) \in r$ for all i. We conclude that $F = [f] \;{}^*\!\!\in {}^*r$ by Lemma 4.2.10 (the Łoś theorem).

Basic Idealization. Unfortunately we cannot hope to obtain Basic Idealization as a consequence of Saturation because $^*\mathbf{v}$ is even not countably saturated, see Exercise 4.3.16 below. Yet there is a more direct argument. We have to prove that the following is true in $^*\mathbf{v}$:

$$\forall^{\mathtt{stfin}} A \subseteq {}^*\!A_0\, \exists x\, \forall a \in A\, \Phi(a,x) \implies \exists x\, \forall^{\mathtt{st}} a \in {}^*\!A_0\, \Phi(a,x), \qquad (1)$$

where $\Phi(a,x)$ is an $\in$-formula with parameters in $^*\mathbf{V}$ while $A_0 \in \mathbf{V}$. Suppose that Φ has just one parameter $[h] \in {}^*\mathbf{V}$; the general case is similar. Thus we let Φ be $\Phi(a,x,[h])$, where $h \in \mathscr{F}^P$, that is, $h \in \mathscr{F}^s$ for a finite set $s = \{p_1 < \dots < p_n\} \subseteq A_0$. Then $h(i) = h(i \restriction s)$ for any $i \in I$.

Assuming the left-hand side of (1), we have $\exists x\, \forall a \in {}^*\!A\, \Phi(a,x,[h])$ in $^*\mathbf{v}$ for any finite $A \subseteq A_0$. This is equivalent to $\mathbf{U}\, i\, (\exists x\, \forall a \in A\, \Phi(a,x,h(i)))$ by the Łoś theorem (Lemma 4.2.10), and hence to

$$U^s\, i\, (\exists x\, \forall a \in A\, \Phi(a,x,h(i))) \qquad (2)$$

by 4.3.8($*$). (Note that the ground standard structure is $\mathbf{v} = \langle \mathbf{V}\,; \in\rangle$, where $\mathbf{V}$ is the universe of all sets, thus Φ and $\Phi^{\mathbf{v}}$ is one and the same.)

Choose any $p > p_n$ in P with $\mathtt{card}\, C_p \geq \mathtt{card}\, A_0$. Let $\beta : A_0 \to C_p$ be any bijection. Put $g(i) = \beta^{-1}(i \cap C_p)$ for any $i \in I$; this is a finite subset of A_0. Define a function $f \in \mathscr{F}^{s\cup\{p\}}$ as follows. Consider any $i \in I$. If there is x such that $\forall a \in g(i)\, \Phi(a, x, h(i \restriction s))$ then let $f(i)$ denote any such an x; otherwise put $f(i) = \varnothing$. Easily $\|f\| \subseteq t = s \cup \{p\}$ and, for any $i' \in I_p$,

$$U^s i \in I^s\, \forall a \in g(i')\, \Phi(a, f(i \cup i'), h(i)) \quad \text{by (2),} \quad \text{and hence}$$

$$U_p\, i'\;\; U^s i \in I^s\, \forall a \in g(i \cup i')\, \Phi(a, f(i \cup i'), h(i)),$$

and finally $U^s i \in I^t\, \forall a \in g(i)\, \Phi(a, f(i), h(i))$ by Lemma 4.3.7(i). Here the quantifier $U^s i \in I^t$ can be replaced to $\mathbf{U}\, i \in I$ by 4.3.8(*). This yields $\forall a \in [g]\, \Phi(a, [f], [h])$ in ${}^*\mathbf{v}$ still by the Łoś theorem.

To infer the right-hand side of (1) it now suffices to prove that ${}^*a\; {}^*\varepsilon\; [g]$ for any $a \in \mathbf{V}$. Still by the Łoś theorem and 4.3.8(*) this is equivalent to $U^t i\; (a \in g(i))$, then to $U_p\, j\; (a \in g(j))$ by Lemma 4.3.7(iii) because obviously $\|g\| = \{p\}$, and hence to $U_p\, j\; (c \in j)$, where $c = \beta(a) \in C_p$. Yet the last formula is a true sentence since U_p is an adequate ultrafilter. □

Thus, by theorems 4.3.12 and 4.3.13, **BST** admits a standard core interpretation in **ZFC**, and hence **BST** is a "realistic" theory.

□ (*Theorem 4.1.10*(i))

Corollary 4.3.14. *Every model of* **ZFC** *is* **BST***-extendible in the sense of Definition 4.1.6.*

Proof. Apply Theorem 4.1.10(i) (or 4.3.12 + 4.3.13) within the model. □

It follows that basically the whole set universe of **ZFC** is **BST**-extendible! We shall see in § 4.6f that the result badly fails for **IST**.

Exercise 4.3.15. Let κ be a fixed infinite cardinal. Modify the construction of quotient power in Theorem 4.3.13 so that $\mathscr{F}^P$ consists only of those functions f with finite support which satisfy $\mathtt{card}\,(\mathtt{ran}\, f) \leq \kappa$. Prove that in this case the quotient power will be a standard core interpretation of $\mathbf{BST}_{\boldsymbol{\kappa}}$ in **ZFC**, with $\boldsymbol{\kappa}$ naturally interpreted as ${}^*\kappa$. (Some problems here begin with the verification of 3°, 4° of § 4.2b.) □

4.3d Saturated standard core interpretation

The standard core extension involved in Theorem 4.3.13 satisfies a suitable form of Idealization but easily fails to fulfill even countable Saturation!

Exercise 4.3.16. Prove that the structure ${}^*\mathbf{v} = \langle {}^*\mathbf{V}; {}^*\!\in \rangle$ considered by Theorem 4.3.13 is not even ω_1-saturated. (*Hint.* For $i \in I$ let $f_n(i)$ be the number of elements of the finite set $i_n = \{x : \langle n, x\rangle \in i\}$. Prove using the finite support requirement that there is no $H \in {}^*\mathbf{V}$ such that "H is a function, ${}^*n \in \mathtt{dom}\, H$, and $H({}^*n) = [f_n]$" holds in ${}^*\mathbf{v}$ for any n.) □

To improve the interpretation construction in Theorem 4.3.13 so that the result is a κ^+-saturated structure for any cardinal κ, we apply a modified version of the elementary continuous chain construction described in § 4.3a, Example 3. We define an $\mathtt{Ord} \cup \{\infty\}$-long elementary continuous chain of $\in$-structures $\mathbf{v}_\xi = \langle V_\xi ; \varepsilon_\xi \rangle$ $(\xi \le \infty)$ along with associated embeddings $e_{\eta\xi}$: $V_\eta \to V_\xi$ $(\eta \le \xi \le \infty)$ so that $\mathbf{v}_0$ is $\langle \mathbf{V} ; \in \rangle$ and for any $\xi \in \mathtt{Ord}$ the structure $\mathbf{v}_{\xi+1} = \langle V_{\xi+1} ; \varepsilon_{\xi+1} \rangle$ is the (truncated) $U_\xi, \mathscr{F}_\xi$-quotient power $\mathscr{F}_\xi / U_\xi$ in the sense of Definition 4.2.8, where U_ξ, $\mathscr{F}_\xi$ are defined as follows.

Put $U_\xi = \bigtimes_{p \in P_\xi} U_p$, where P_ξ and U_p are defined as in the proof of Theorem 4.3.12, in particular, each $U_p \subseteq C_p{}^{\mathtt{fin}}$ is a C_p-adequate ultrafilter where $C_p = \{p\} \times \xi$, and hence U_ξ is a C_ξ-adequate filter where $C_\xi = \bigcup_{p \in P_\xi} C_p = P_\xi \times \xi$, by 4.3.8. By definition U_ξ contains only those subsets of $I_\xi = C_\xi{}^{\mathtt{fin}}$ that include a subset of finite support. Accordingly, $\mathscr{F}_\xi$ is the set of all maps $f : I_\xi \to V_\xi$ of finite support. (See § 4.3b on notation.) That 1° – 4° of § 4.2b hold with such a choice follows from 4.3.10.

We finally let $e_{\xi,\xi+1}$ be the natural embedding of $\mathbf{v}_\xi$ in $\mathbf{v}_{\xi+1}$. This completes the step $\xi \to \xi + 1$.

Define an unary relation $^*\mathtt{st}$ on V_∞ so that $^*\mathtt{st}\, x$ iff $x \in \mathtt{ran}\, e_{0\infty}$. This accomplishes the definition of the *final structure* $^*\mathbf{v} = \langle V_\infty ; \varepsilon_\infty, {}^*\mathtt{st} \rangle$.

Theorem 4.3.17. *The structure* $^*\mathbf{v} = \langle V_\infty ; \varepsilon_\infty, {}^*\mathtt{st} \rangle$ *is a standard core interpretation of* **BST** *in* **ZFC**, κ^+*-saturated for any cardinal* κ. *The map* $e_{0\infty}$ *is an elementary standard core* $\in$*-embedding of* $\langle \mathbf{V} ; \in \restriction \mathbf{V} \rangle$ *into* $^*\mathbf{v}$.

Proof. Coming back to the construction of Example 3 in § 4.3a, we observe that to prove the theorem it suffices to check the following:

(∗) For any $\xi \in \mathtt{Ord}$ the structure $\mathbf{v}_{\xi+1}$ is a ξ^+-enlargement of $\mathbf{v}_\xi$.

To prove (∗) consider a set $\mathscr{X} = \{X_\alpha : \alpha < \xi\} \subseteq V_\xi$. Suppose that $\mathscr{X}$ is ε_ξ-f. i. p., so that for any finite $A \subseteq \xi$ there is $z_A \in V_\xi$ such that $z_A \;\varepsilon_\xi\; X_\alpha$ for all $\alpha \in A$. Fix an arbitrary $p \in P_\xi$. For any $i \in I_\xi$ put $A(i) = \{\alpha < \xi : \langle p, \alpha \rangle \in i\}$ – this is a finite subset of ξ. Put $f(i) = z_{A(i)}$. Then $f : I_\xi \to V_\xi$ is a function of finite support, actually $\|f\| = \{p\}$. Thus $f \in \mathscr{F}_\xi$ and by definition we have $\forall\, a \in A(i)\; (f(i) \;\varepsilon_\xi\; X_\alpha)$ for every $i \in I_\xi$.

However $U_\xi\, i\, (\alpha \in A(i))$ holds for each $\alpha < \xi$ since U_ξ is C_ξ-adequate. Thus $U_\xi\, i\, (f(i) \;\varepsilon_\xi\; X_\alpha)$ for any $\alpha < \xi$. It follows by the Łoś theorem that $[f] \in V_{\xi+1}$ satisfies $[f] \;\varepsilon_{\xi+1}\; e_{\xi,\xi+1}(X_\alpha)$ for all $\alpha < \xi$, as required. □

Problem 4.3.18. The construction of the structure $^*\mathbf{v}$ above can be altered in minor details, leading to formally different saturated standard core interpretations of **BST** in **ZFC**. All of them are elementarily equivalent to each other w. r. t. $\mathtt{st}$-$\in$-formulas with standard parameters by Theorem 3.2.3(ii). Are all of them definably $\mathtt{st}$-$\in$-isomorphic in **ZFC**?

One might seek to answer this question in the positive by a back-and-forth argument, but this method does not seem to work here because an $\mathtt{Ord}$-long sequence of choices is to be carried out to get an isomorphism this way. □

4.4 The conservativity and equiconsistency of IST

This part of Theorem 4.1.10(iii) can be proved by several different methods, among them 1) saturated extensions of transitive models of fragments of **ZFC** in the **ZFC** universe, 2) saturated extensions of the set $\mathbf{V}_\vartheta$ in **ZFC**ϑ, 3) inner models in a ϑ-version of **BST**, 4) saturated extensions of the universe of **ZFGT**, a standard theory somewhat stronger than **ZFC**. The methods have many common features, in particular, essentially the same technique of saturated extensions based on adequate quotient powers (good quotient powers can also be used), so that we can rather speak of metamathematically different versions of one and the same argument.

We present constructions 1), 2), 3) in this section, and 4) in Section 4.6.

4.4a Good extensions of von Neumann sets in ZFC universe

Suppose that ϑ is an infinite ordinal. Let us consider two examples of saturated extensions of the set $\mathbf{V}_\vartheta$. The first of them is essentially Example 1 in § 4.3a, but with $\mathbf{V}_\vartheta$ instead of the whole universe as the ground standard structure, the other one resembles the extension considered in § 4.3c.

Let U be any ϑ^+-good countably incomplete ultrafilter on the set $I = \vartheta = \{\xi : \xi < \vartheta\}$ (Proposition 4.2.16). Consider the ultrapower ${}^*\mathbf{v} = {\mathbf{V}_\vartheta}^I/U = \langle {}^*\mathbf{V}_\vartheta\,;\, {}^*{\in}, {}^*\mathtt{st}\rangle$, along with the natural elementary standard core embedding $* : \mathbf{V}_\vartheta \to {}^*\mathbf{V}_\vartheta$.

Proposition 4.4.1. *If* $\vartheta = \mathtt{card}\,\mathbf{V}_\vartheta$ *then* ${}^*\mathbf{v}$ *is an elementary standard core extension of* $\mathbf{v} = \langle \mathbf{V}_\vartheta\,;\, \in\rangle$ *via* $*$, *and a model of* **IST**[**ZC**]. [16]

Proof. The structure ${}^*\mathbf{v}$ is ϑ^+-saturated, and even strongly ϑ^+-saturated by Theorem 4.2.17(ii), and hence ${}^*\mathbf{v}$ satisfies Idealization by 4.2.5. That ${}^*\mathbf{v}$ satisfies Inner Transfer, Inner Standardization, and $\mathbf{ZC}^{\mathtt{st}}$ follows from Theorem 4.2.4. □

Corollary 4.4.2 (a part of Theorem 4.1.10(iii)). *The theory* **IST** *is a conservative and hence equiconsistent standard core extension of* **ZFC**.

Proof. Assume that **IST** proves $\varphi^{\mathtt{st}}$, where φ is an $\in$-sentence. Then $\mathbf{IST}[\mathbf{ZC}] + \Psi^{\mathtt{st}}$ proves $\varphi^{\mathtt{st}}$, where Ψ is an appropriate finite list of axioms of **ZFC** (in the $\in$-language) and $\Psi^{\mathtt{st}} = \{\psi^{\mathtt{st}} : \psi \in \Psi\}$. Suppose towards the contrary that $\neg\,\varphi$ is compatible with **ZFC**. Applying Theorem 1.5.4 in the consistent theory $T = \mathbf{ZFC} + \Psi$, we find an ordinal γ such that $\mathbf{V}_\gamma$ reflects φ and all formulas $\psi \in \Psi$ together with and all their subformulas. Moreover, there is an infinite sequence of ordinals γ_n, $n \in \mathbb{N}$, of this kind such that $\gamma_{n+1} \geq \mathtt{card}\,{\mathbf{V}_{\gamma_n}}^+$ for all n. Then $\vartheta = \sup_n \gamma_n$ is a cardinal, $\vartheta = \mathtt{card}\,\mathbf{V}_\vartheta$,

[16] Recall that **IST**[**ZC**] consists of $\mathbf{ZC}^{\mathtt{st}}$, Idealization, Inner Standardization, and Inner Transfer. **IST** itself contains a stronger schema $\mathbf{ZFC}^{\mathtt{st}}$ instead of $\mathbf{ZC}^{\mathtt{st}}$.

and still $\mathbf{V}_\vartheta$ reflects φ and all formulas $\psi \in \Psi$ by a Löwenheim – Skolem argument. Therefore, as all formulas in Ψ are **ZFC** axioms, all of them are true in $\mathbf{V}_\vartheta$, and by the same reason φ fails in $\mathbf{V}_\vartheta$.

It follows from Proposition 4.4.1 that $\mathbf{v} = \langle \mathbf{V}_\vartheta ; \in \rangle$ admits a standard core embedding $*$ in a model $^*\mathbf{v} = \langle {}^*\mathbf{V}_\vartheta ; {}^*\!\in, {}^*\mathtt{st} \rangle$ of **IST**[**ZC**]. Note that if ψ is a formula in Ψ then $\mathbf{V}_\vartheta \models \psi$ by the above, and hence $\mathbf{V}_\vartheta \models \psi^{\mathtt{st}}$ by 4.1.2. Thus $^*\mathbf{v}$ models $\Psi^{\mathtt{st}}$. It follows that $^*\mathbf{v} \models \varphi^{\mathtt{st}}$ by the choice of Ψ, therefore φ holds in $\mathbf{V}_\vartheta$ still by 4.1.2, in contradiction to the above. □

4.4b Iterated adequate extensions of von Neumann sets

The following construction does not give anything beyond Corollary 4.4.2, but will be used as a pattern for more complicated quotient power extensions.

We still assume that ϑ is a limit ordinal. Let U be any $\mathbf{V}_\vartheta$-adequate ultrafilter over the set $J = {\mathbf{V}_\vartheta}^{\mathtt{fin}}$.

Within the general scheme of §4.3b, put $P = \mathbb{N}$ and $C_p = \{p\} \times \mathbf{V}_\vartheta$ for any $p \in \mathbb{N}$ and let U_p be the C_p-adequate ultrafilter over $I_p = {C_p}^{\mathtt{fin}}$ obtained by the action of the shift $x \mapsto \langle p, x \rangle$ on U. Let

$$C = \textstyle\bigcup_{p \in \mathbb{N}} C_p = \mathbb{N} \times \mathbf{V}_\vartheta\,, \quad I = \bigcup_{p \in \mathbb{N}} I_p = C^{\mathtt{fin}}, \quad \mathbf{U} = \bigtimes_{p \in \mathbb{N}} U_p\,.$$

The definition of $\mathscr{F}$ suitably changes because the ground structure now is $\mathbf{V}_\vartheta$ rather than the universe $\mathbf{V}$ of all sets as in the proof of Theorem 4.3.13. For any finite $s \subseteq \mathbb{N}$ let $\mathscr{F}^s$ be the set of all functions $f : I \to \mathbf{V}_\vartheta$ such that $\|f\| \subseteq s$, that is, $f(i) = f(i \restriction s)$ for all $i \in I$. Put $\mathscr{F} = \bigcup_{s \subseteq \mathbb{N} \text{ finite}} \mathscr{F}^s$.

It follows from 4.3.8, 4.3.10 that $\mathbf{U}$ is a C-adequate filter containing $U_p\Uparrow$ for all $p \in \mathbb{N}$, where $C = \bigcup_{p \in \mathbb{N}} C_p = \mathbb{N} \times \mathbf{V}_\vartheta$, and 1° – 4° of §4.2b are satisfied with $V = \mathbf{V}_\vartheta$, $U = \mathbf{U}$, and $\mathscr{F}$. Thus we can define the quotient power $^*\mathbf{v} = \langle {}^*\mathbf{V}_\vartheta ; {}^*\!\in, {}^*\mathtt{st} \rangle = \mathscr{F}/\mathbf{U}$ and the natural elementary standard core embedding $* : \mathbf{V}_\vartheta \to {}^*\mathbf{V}_\vartheta$.

Proposition 4.4.3. *If* $\vartheta = \mathsf{card}\,\mathbf{V}_\vartheta$ *then* $^*\mathbf{v}$ *is an elementary standard core extension of* $\mathbf{v} = \langle \mathbf{V}_\vartheta ; \in \rangle$ *via* $*$, *and a model of* **IST**[**ZC**].

Proof. Inner Transfer, Inner Standardization, and $\mathbf{ZC}^{\mathtt{st}}$: see Proposition 4.4.3. As for Idealization, we have to prove

$$\forall^{\mathtt{stfin}} A\, \exists\, x\, \forall\, a \in A\, \Phi(a, x, [h]) \implies \exists\, x\, \forall^{\mathtt{st}} a\, \Phi(a, x, [h]), \qquad (\ddagger)$$

where $\Phi(a, x, [h])$ is an $\in$-formula with a parameter $[h] \in {}^*\mathbf{V}_\vartheta$, that is $h \in \mathscr{F}^s$ for a finite $s = \{p_1 < \dots < p_n\} \subseteq \mathbb{N}$. The reasoning proceeds as in the Basic Idealization part of the proof of Theorem 4.3.13, with the following minor differences. We replace Φ by $\Phi^{\mathbf{V}_\vartheta}$ wherever necessary because $\mathbf{V}_\vartheta$ is now the standard model. Obviously $\mathbf{V}_\vartheta$ now also plays the role of A_0, and hence A is an arbitrary finite subset (or, equivalently, element) of $\mathbf{V}_\vartheta$. We let $p = p_n + 1$. Finally, β is the map $\beta(x) = \langle p, x \rangle$ for all $x \in \mathbf{V}_\vartheta$. □

Corollary 4.4.2 follows from 4.4.3 as well as from 4.4.1, of course.

4.4c Iterated adequate extensions in the ϑ-version of ZFC

Here we specify the construction of § 4.4b, applying it to the set $\mathbf{V}_\vartheta$, in the frameworks of **ZFC**ϑ, a theory defined in § 1.5f. We are going to make use of the fact that **ZFC**ϑ postulates $\mathbf{V}_\vartheta$ to be an interpretation of **ZFC**.

Arguing in **ZFC**ϑ, let us fix a $\mathbf{V}_\vartheta$-adequate ultrafilter U over the set $J = \mathbf{V}_\vartheta{}^{\mathtt{fin}}$. Define P, C_p, U_p, I_p, C, I, $\mathbf{U}$, $\mathscr{F}^s$, $\mathscr{F}$, and finally the quotient power ${}^*\mathbf{v} = \langle {}^*\mathbf{V}_\vartheta\,;\, {}^*\!\in, {}^*\mathtt{st}\rangle = \mathscr{F}/\mathbf{U}$ and the natural elementary standard core embedding $* : \mathbf{V}_\vartheta \to {}^*\mathbf{V}_\vartheta$ exactly as in § 4.4b, for $\vartheta = \boldsymbol{\vartheta}$.

Proposition 4.4.4. (i) (**ZFC**ϑ) *The structure* ${}^*\mathbf{v}$ *is an elementary standard core extension of* $\mathbf{v} = \langle \mathbf{V}_\vartheta\,;\, \in \restriction \mathbf{V}_\vartheta\rangle$ *via* $*$ *and a model of* **IST**[**ZC**].
(ii) *The structure* ${}^*\mathbf{v}$ *is an interpretation of* **IST** *in* **ZFC**ϑ.

Proof. (i) follows from Proposition 4.4.3. Indeed **ZFC**ϑ proves that $\boldsymbol{\vartheta} = \mathtt{card}\,\mathbf{V}_\vartheta$ by 1.5.18, and hence **ZFC**ϑ proves for $\boldsymbol{\vartheta}$ everything which **ZFC** proves for any ordinal ϑ satisfying $\vartheta = \mathtt{card}\,\mathbf{V}_\vartheta$, in particular, 4.4.3.

(ii) The exact content of the interpretability claim here is as follows. As the construction of ${}^*\mathbf{v}$ obviously depends on $\boldsymbol{\vartheta}$ and the choice of an ultrafilter U, let us write ${}^*\mathbf{v}(\boldsymbol{\vartheta}, U)$. Then, it is asserted that *for any axiom* Φ *of* **IST**, *it is a theorem of* **ZFC**ϑ *that* $\Phi^{{}^*\mathbf{v}(\boldsymbol{\vartheta},U)}$ *for any* $\mathbf{V}_\vartheta$*-adequate ultrafilter* U *over* $\mathbf{V}_\vartheta{}^{\mathtt{fin}}$. The result follows from (i) — for all axioms except those in $\mathbf{ZFC}^{\mathtt{st}}$, and from Theorem 4.2.4(ii) for the latter category. □

Exercise 4.4.5. Why is ${}^*\mathbf{v}$ **not** a standard core interpretation in Proposition 4.4.4? Why can we not claim that ${}^*\mathbf{v}$ is a **model** of **IST** in Proposition 4.4.4? (See § 1.5f regarding the same question w. r. t. $\mathbf{V}_\vartheta$ itself.) □

4.4d Long iterated quotient power chains

We continue to argue in **ZFC**ϑ *in this subsection.*

Our goal is to define a more complicated standard core extension of $\mathbf{V}_\vartheta$ that satisfies addititional principles considered in § 3.5c and is $\boldsymbol{\vartheta}^+$-saturated. (The model ${}^*\mathbf{v}$ of § 4.4b is not even countably saturated: see 4.3.16 for an entirely analogous fact.) To obtain such a model we employ a modification of the elementary continuous chain construction introduced in § 4.3d.

We put $\gamma = \boldsymbol{\vartheta}^+$ (the length of the elementary chain in our construction), choose any $\mathbf{V}_\vartheta$-adequate ultrafilter U over $I = \mathbf{V}_\vartheta{}^{\mathtt{fin}}$, let $\langle V_0\,;\, \varepsilon_0\rangle = \langle \mathbf{V}_\vartheta\,;\, \in \restriction \mathbf{V}_\vartheta\rangle$, and specify the successor step in the elementary chain construction so that for any $\xi < \gamma$ the structure $\langle V_{\xi+1}\,;\, \varepsilon_{\xi+1}\rangle$ is the (truncated) $U, \mathscr{F}_\xi$-quotient power $\mathscr{F}_\xi/U$ (in this case a true ultrapower of $\langle V_\xi\,;\, \varepsilon_\xi\rangle$) in the sense of Definition 4.2.8, where $\mathscr{F}_\xi = V_\xi{}^I$.

Define an unary relation ${}^*\mathtt{st}$ on V_γ so that ${}^*\mathtt{st}\,x$ iff $x \in \mathtt{ran}\,e_{0\gamma}$. Thus the structure ${}^*\mathbf{v} = \langle V_\gamma\,;\, \varepsilon_\gamma, {}^*\mathtt{st}\rangle$ has been defined.

See §§ 3.5c, 3.2e on the additional schemata in the next theorem.

Theorem 4.4.6. (i) (**ZFC**ϑ) ${}^*\mathbf{v}$ *is a* ϑ^+*-saturated elementary standard core extension of* $\mathbf{v} = \langle \mathbf{V}_\vartheta \,; \in \restriction \mathbf{V}_\vartheta \rangle$ *via the elementary standard core* $\in$*-embedding* $e_{0\gamma}$*, and a model of* **IST**[**ZC**] *plus the schemata of* $\mathbb{S}$-Size Choice, $\mathbb{S}$-Separation, Inner S. S. Choice, Inner Extension.

(ii) *The structure* ${}^*\mathbf{v}$ *is an interpretation of* **IST** *in* **ZFC**ϑ.

Proof (Sketch). (i) The first part: follow the proof of Theorem 4.3.17. Note that $\mathtt{card}\,\mathbf{V}_\vartheta = \vartheta$ by 1.5.18 in the proof of the saturation property. To see that $\mathbb{S}$-Size Choice holds in ${}^*\mathbf{v}$ show that for any set $\{h_\xi : \xi < \vartheta\} \subseteq V_\gamma$ there is an $h \in V_\gamma$ such that "h is a function $\wedge$ ${}^*\xi \in \mathtt{dom}\, h$ $\wedge$ $h({}^*\xi) = h_\xi$" is true in ${}^*\mathbf{v}$ for any $\xi < \vartheta$. The rest of the schemata follows from $\mathbb{S}$-Size Choice.

(ii) Follow the proof of Proposition 4.4.4(ii). □

Similarly to Corollary 4.4.2, we obtain

Corollary 4.4.7. *The theory* **IST** *with the schemata of* $\mathbb{S}$-Size Choice, $\mathbb{S}$-Separation, Inner S. S. Choice, Inner Extension *is a conservative and equiconsistent standard core extension of* **ZFC**. *The schemata mentioned are not refutable in* **IST** *provided* **IST** *(or, equivalently,* **ZFC***) is consistent.* □

Exercise 4.4.8. (i) Prove that ${}^*\mathbf{v}$ interprets Map-Standardization iff ϑ is a strongly inaccessible cardinal. Infer that **IST** + Map-Standardization is consistent provided so is **ZFC** plus "there is an inaccessible cardinal".

(ii) Prove that ${}^*\mathbf{v}$ satisfies $\mathbb{S}$-Separation whenever γ is any limit ordinal, not necessarily ϑ^+.

(iii) Take as $\mathscr{F}_\xi$ the set of all functions $f : I \to V_\xi$ bounded in the sense that there is an element $r \in V_\xi$ such that $f(i) \mathrel{\varepsilon_\xi} r$ for all $i \in I$. Prove that then ${}^*\mathbf{v}$ is an interpretation of **BST** rather than **IST**.

Hints. (i) If ϑ is strongly inaccessible then any $g : X \to \mathbf{V}_\vartheta$ defined on a set $X \in \mathbf{V}_\vartheta$ belongs to $\mathbf{V}_\vartheta$, easily leading to Map-Standardization in ${}^*\mathbf{v}$. If ϑ is not strongly inaccessible then there is a function $g \notin \mathbf{V}_\vartheta$, $g : X \to \mathbf{V}_\vartheta$, $\mathtt{dom}\, g = X \in \mathbf{V}_\vartheta$. Define $F \in V_\gamma$ such that $F({}^*x) = {}^*(g(x))$ for any $x \in X$.

(ii) It suffices to show that for any set $X \subseteq \mathbf{V}_\vartheta$ there is an element $F = [f] \in V_1$ such that $x \in X \iff {}^*x \mathrel{\varepsilon} F$ for all $x \in \mathbf{V}_\vartheta$. Define $f : \mathbf{V}_\vartheta{}^{\mathtt{fin}} \to \mathbf{V}_\vartheta$ so that $f(i) = i \cap X$ for all i. □

Problem 4.4.9. Is 4.4.8(i) provable on the base of the consistency of **ZFC** alone? A reason to expect the negative answer is that **IST**+ $\mathbb{S}$-Separation + Map-Standardization is not an equiconsistent extension of **ZFC** by 3.5.12. □

4.4e Conservativity of IST by inner models

Here we present another and quite different method to prove the conservativity of **IST**. Consider a theory **BST**ϑ in the $\mathtt{st}$-$\in$-language enriched by a constant ϑ, whose axioms include all of **BST** (ϑ can freely occur in the schemata), the axiom "ϑ is a standard ordinal", and the elementary submodel schema $(*)$ in the $\in$-language as in **ZFC**ϑ (see § 1.5f).

Exercise 4.4.10. Show that **BST**ϑ implies **ZFC**ϑ in the $\in$-language (with the constant symbol ϑ allowed), and in addition:

(i) **BST**ϑ is a conservative extension of **ZFC**ϑ in the sense that if **BST**ϑ proves an $\in$-sentence Φ (perhaps including ϑ) then **ZFC**ϑ proves Φ ;

(ii) **BST**ϑ is a conservative extension of **BST** in the sense that if **BST**ϑ proves a st-$\in$-sentence Φ (not including ϑ) then **BST** proves Φ ;

(iii) **BST**ϑ proves $\Phi \iff \Phi^{\mathbf{V}_\vartheta} \iff \Phi^{\mathtt{st}} \iff (\Phi^{\mathbf{V}_\vartheta})^{\mathtt{st}}$ for any $\in$-formula Φ with parameters in $\mathbb{S} \cap \mathbf{V}_\vartheta$;

(iv) **BST**ϑ proves $\Phi \iff \Phi^{\mathbf{V}_\vartheta}$ for any st-$\in$-formula Φ with parameters in $\mathbb{S} \cap \mathbf{V}_\vartheta$.

Hints. (i) Show that the standard core interpretation of **BST** in **ZFC** in § 4.3c can be transformed to a standard core interpretation of **BST**ϑ in **ZFC**ϑ.

(ii) Suppose that **BST**ϑ proves a st-$\in$-sentence Φ. As Φ is equivalent to a $\in$-sentence in **BST** by Theorem 3.2.3(ii) we can assume that Φ is an $\in$-sentence. To accomplish the proof in this case argue as in Exercise 1.5.17(i).

(iii) Apply Inner Transfer and the schema $(*)$ of § 1.5f.

(iv) Apply (iii) and the reduction to $\in$-formulas as mentioned for (ii). □

Arguing in **BST**ϑ, define $\mathtt{st}_\vartheta\, x$ iff $\mathtt{st}\, x$ and $x \in \mathbf{V}_\vartheta$.

Theorem 4.4.11. *The structure* $\mathbf{v} = \langle \mathbf{V}\, ; \in, \mathtt{st}_\vartheta \rangle$ *is an interpretation of* **IST** *in* **BST**ϑ*. In addition the schema of* $\mathbb{S}$*-Size Choice, and hence also* $\mathbb{S}$*-Separation, Inner S. S. Choice, Inner Extension, hold in* $\mathbf{v}$.

Thus to cook up an interpretation of **IST** we keep the internal universe of **BST**ϑ intact but drastically reduce the amount of standard sets!

Proof (Sketch). **ZFC**$^{\mathtt{st}}$. If Φ is an axiom of **ZFC** with *st-standard parameters then by definition $\Phi^{\mathtt{st}_\vartheta}$ is $(\Phi^{\mathbf{V}_\vartheta})^{\mathtt{st}}$. However $\Phi^{\mathbf{V}_\vartheta}$ is a theorem of **BST**ϑ by 4.4.10(iii), and hence the result follows by the **BST**ϑ Inner Transfer.

Inner Transfer: a similar argument.

Inner Standardization follows from the **BST**ϑ Inner Standardization in the whole universe, based on the fact that $Y \subseteq X \in \mathbf{V}_\vartheta$ implies $Y \in \mathbf{V}_\vartheta$.

The schema of Idealization for $\mathtt{st}_\vartheta$ is easily equivalent to the schema of Inner Saturation for $A_0 = \mathbf{V}_\vartheta$ (see § 3.2a). Thus the result follows from Lemma 3.2.2. Similarly the schema of $\mathbb{S}$-Size Choice for $\mathbf{v}$ is simply Inner S. S. Choice (with the domain $\mathbf{V}_\vartheta \cap \mathbb{S}$), true in **BST** by Theorem 3.2.11. □

Exercise 4.4.12. (1) Prove Corollary 4.4.7 using 4.4.10 and Theorem 4.4.11.

(2) Prove in **BST**ϑ that the structure $\mathbf{v}$ as in Theorem 4.4.11 satisfies Map-Standardization iff ϑ is a strongly inaccessible cardinal.

(3) Arguing in **BST**ϑ, let $V = \bigcup_{n \in \mathbb{N} \cap \mathbb{S}} \mathbf{V}_{\vartheta+n}$. Prove that the structure $\langle V\, ; \in \restriction V, \mathtt{st} \restriction V \rangle$ interprets ***ZCN**. (See Theorem 3.6.4.) □

4.5 Non-reducibility of IST

Our plan to prove the non-reducibility of **IST** in Theorem 4.1.10(iii) is as follows. Our ground standard theory will be **ZFC**ϑ strengthened by two hypotheses, the axiom of constructibility "$\mathbf{V} = \mathbf{L}$" and a hypothesis saying that ϑ is *minimal* in a certain sense. Recall that the axiom "$\mathbf{V} = \mathbf{L}$" says that every set is *constructible*, that is, it can be obtained after a transfinite number of steps of a certain construction starting from the empty set.

Under these assumptions we define a pair of standard core extensions of $\mathbf{V}_\vartheta$, each of them satisfying **IST**, such that Inner Extension holds in one of the extensions (essentially by Theorem 4.4.11) but a certain example of the Inner Extension schema fails in the other extension. The latter belongs to a special type of quotient powers: unlike the quotient power constructions considered earlier, only definable functions will be involved.

4.5a The minimality axiom

Recall that **ZFC**ϑ does not prove that $\mathbf{V}_\vartheta$ is a *model* of **ZFC**, see discussion in § 1.5f. In other words, arguing in **ZFC**ϑ we don't know whether $\mathbf{V}_\vartheta$ is an $\in$-model of **ZFC** or not. In the latter case Σ_n-Collection (Collection for Σ_n formulas) fails in $\mathbf{V}_\vartheta$ for some natural n, and we let $\mathbf{n} = \mathbf{n}(\vartheta)$ be the largest number such that $\Sigma_{\mathbf{n}}$-Collection still holds in $\mathbf{V}_\vartheta$. If $\mathbf{V}_\vartheta$ is indeed a model of **ZFC** then put $\mathbf{n}(\vartheta) = \infty$. Consider the following axiom:

Minimality: There is no ordinal $\kappa < \vartheta$ such that $\mathbf{V}_\kappa$ is an elementary submodel of $\mathbf{V}_\vartheta$ with respect to all $\Sigma_{\mathbf{n}(\vartheta)}$ formulas (with parameters in $\mathbf{V}_\kappa$) — that means all $\in$-formulas in the case $\mathbf{n}(\vartheta) = \infty$.

To show that this axiom is consistent with **ZFC**ϑ let $\kappa = \kappa(\vartheta)$ be the least ordinal $\le \vartheta$ such that $\mathbf{V}_\kappa$ is an elementary submodel of $\mathbf{V}_\vartheta$ with respect to all $\Sigma_{\mathbf{n}(\vartheta)}$ formulas — meaning all $\in$-formulas in the case $\mathbf{n}(\vartheta) = \infty$.

Lemma 4.5.1. (i) *For any natural number m,* **ZFC**ϑ *proves $m < \mathbf{n}(\vartheta)$.*

(ii) *For any $\in$-formula $\Phi(x_1, ..., x_n)$,* **ZFC**ϑ *proves*

$$\forall x_1, \ldots, x_n \in \mathbf{V}_\kappa \left(\Phi(x_1, \ldots, x_n) \Longleftrightarrow \Phi(x_1, \ldots, x_n)^{\mathbf{V}_\kappa}\right).$$

(iii) *$\langle \mathbf{V}_\kappa ; \in \restriction \mathbf{V}_\kappa \rangle$ models $\Sigma_{\mathbf{n}}$-Collection.*

(iv) *$\langle \mathbf{V} ; \in, \kappa \rangle$ is an interpretation of* **ZFC**ϑ + *Minimality in* **ZFC**ϑ.

Proof. (i) For any m **ZFC** proves Σ_m-Collection, and hence **ZFC**ϑ proves that Σ_m-Collection holds in $\mathbf{V}_\vartheta$, thus $m < \mathbf{n} = \mathbf{n}(\vartheta)$.

(ii) It follows from (i) by the definition of $\kappa = \kappa(\vartheta)$ that for any $\in$-formula Φ **ZFC**ϑ proves $\Phi^{\mathbf{V}_\vartheta} \Longleftrightarrow \Phi^{\mathbf{V}_\kappa}$. Yet $\Phi^{\mathbf{V}_\vartheta} \Longleftrightarrow \Phi$ is an axiom of **ZFC**ϑ.

(iii) By definition, $\mathbf{V}_\kappa$ is an elementary submodel of $\mathbf{V}_\vartheta$ with respect to all $\Sigma_{\mathbf{n}}$ formulas, while $\mathbf{V}_\vartheta$ is a model of **ZC** plus $\Sigma_{\mathbf{n}}$-Collection. In this case the proof of Theorem 1.5.4(ii) works.

(iv) With $\boldsymbol{\kappa}$ playing the role of $\boldsymbol{\vartheta}$, **ZFC**$\boldsymbol{\vartheta}$ is satisfied by (ii), thus it remains to prove Minimality w. r. t. $\boldsymbol{\kappa}$. Suppose towards the contrary that $\boldsymbol{\kappa}' < \boldsymbol{\kappa}$ and $\mathbf{V}_{\boldsymbol{\kappa}'}$ is an elementary submodel of $\mathbf{V}_{\boldsymbol{\kappa}}$ with respect to all $\Sigma_{\mathbf{n}(\boldsymbol{\kappa})}$ formulas (with parameters in $\mathbf{V}_{\boldsymbol{\kappa}'}$). Note that $\mathbf{n}(\boldsymbol{\kappa}) \geq \mathbf{n}(\boldsymbol{\vartheta})$ by (iii). Thus $\mathbf{V}_{\boldsymbol{\kappa}'}$ is an elementary submodel of $\mathbf{V}_{\boldsymbol{\vartheta}}$ with respect to all $\Sigma_{\mathbf{n}(\boldsymbol{\vartheta})}$ formulas, in contradiction to the choice of $\boldsymbol{\kappa}$. □

Definition 4.5.2. **ZFC**$\boldsymbol{\vartheta}^+$ is **ZFC**$\boldsymbol{\vartheta}$ plus "$\mathbf{V} = \mathbf{L}$" plus Minimality. □

Corollary 4.5.3. *The theory* **ZFC**$\boldsymbol{\vartheta}^+$ *is equiconsistent with* **ZFC**$\boldsymbol{\vartheta}$ *and hence with* **ZFC**.

Proof. It is known that **ZFC** + "$\mathbf{V} = \mathbf{L}$" is equiconsistent with **ZFC**. It follows by 1.5.17 that **ZFC**$\boldsymbol{\vartheta}$ + "$\mathbf{V} = \mathbf{L}$" is equiconsistent with **ZFC**. It remains to apply Lemma 4.5.1(iv). (Note that the interpretation by $\boldsymbol{\kappa}$ does not change the set universe, thus does not violate "$\mathbf{V} = \mathbf{L}$".) □

Blanket agreement 4.5.4. We argue in **BST**$\boldsymbol{\vartheta}^+$ in the remainder of this section unless explicitly specified otherwise. □

4.5b The source of counterexamples

An important consequence of Minimality is the existence of a sequence of sets $\mathbf{y}_n \in \mathbf{V}_{\boldsymbol{\vartheta}}$, "almost" definable, but not really definable in $\mathbf{V}_{\boldsymbol{\vartheta}}$. Let $\{\psi_k(v)\}_{k \in \mathbb{N}}$ be a recursive enumeration of all parameter-free $\in$-formulas with the only free variable v. Let $\varphi_k(v)$ say that either v is the only set satisfying $\psi_k(v)$, or $v = \varnothing \wedge \neg \exists ! x \, \psi_k(x)$, or, more formally,

$$\varphi_k(v) \quad \text{is} \quad \big(\psi_k(v) \wedge \exists ! x \, \psi_k(x)\big) \vee \big(v = \varnothing \wedge \neg \exists ! x \, \psi_k(x)\big).$$

For each $k \in \mathbb{N}$, let $\mathbf{y}_k$ be the unique set in $\mathbf{V}_{\boldsymbol{\vartheta}}$ satisfying $\mathbf{V}_{\boldsymbol{\vartheta}} \models \varphi_k(\mathbf{y}_k)$, and $\boldsymbol{\alpha}_k$ be the least infinite ordinal $< \boldsymbol{\kappa}$ such that $\mathbf{y}_k \subseteq \mathbf{V}_{\boldsymbol{\alpha}_k}$ and $\boldsymbol{\alpha}_k > \boldsymbol{\alpha}_{k-1}$.

Lemma 4.5.5. $\boldsymbol{\vartheta} = \sup_{k \in \mathbb{N}} \boldsymbol{\alpha}_k$.

Proof. Let, on the contrary, $\boldsymbol{\vartheta} > \alpha = \sup_{k \in \mathbb{N}} \boldsymbol{\alpha}_k$. We claim that $\mathbf{V}_\alpha$ is an elementary submodel of $\mathbf{V}_{\boldsymbol{\vartheta}}$ w. r. t. all $\Sigma_{\mathbf{n}(\boldsymbol{\vartheta})}$ formulas with parameters in $\mathbf{V}_\alpha$, in contradiction to Minimality. It suffices to prove that, for any m,

$$\exists x \in \mathbf{V}_{\boldsymbol{\vartheta}} \; (\mathbf{V}_{\boldsymbol{\vartheta}} \models \Phi(x)) \Longrightarrow \exists k \geq m \; \exists x \in \mathbf{V}_{\boldsymbol{\alpha}_k} \; (\mathbf{V}_{\boldsymbol{\vartheta}} \models \Phi(x)),$$

where $\Phi(x)$ is a $\Sigma_{\mathbf{n}(\boldsymbol{\vartheta})}$ formula with x as the only free variable and parameters in $\mathbf{V}_{\boldsymbol{\alpha}_m}$. We can assume that there is only one parameter p_0 ($p_0 \in \mathbf{V}_{\boldsymbol{\alpha}_m}$), so that Φ is $\Phi(p_0, x)$. As $\mathbf{V}_{\boldsymbol{\vartheta}}$ models $\Sigma_{\mathbf{n}(\boldsymbol{\vartheta})}$-Collection by Lemma 4.5.1, there is an ordinal $\nu < \boldsymbol{\kappa}$ such that $\forall p \in \mathbf{V}_{\boldsymbol{\alpha}_m} \big(\exists x \, \Phi(p, x) \Longrightarrow \exists x \in \mathbf{V}_\nu \; \Phi(p, x)\big)$ is true in $\mathbf{V}_{\boldsymbol{\vartheta}}$. The least ordinal ν of this kind is definable in $\mathbf{V}_{\boldsymbol{\vartheta}}$, and hence ν is equal to $\boldsymbol{\alpha}_k$ for some k. We have then

$$\forall p \in \mathbf{V}_{\boldsymbol{\alpha}_m} \big(\exists x \, \Phi(p, x) \Longrightarrow \exists x \in \mathbf{V}_{\boldsymbol{\alpha}_k} \; \Phi(p, x)\big)$$

in $\mathbf{V}_{\boldsymbol{\vartheta}}$, as required. □

Lemma 4.5.6. *The sequence* $\{\mathbf{y}_k\}_{k\in\mathbb{N}}$ *is not definable in* $\mathbf{V}_\vartheta$ *by an* $\in$-*formula (with parameters in* $\mathbf{V}_\vartheta$ *).*

Proof. Let, on the contrary, $a_0 \in \mathbf{V}_\vartheta$, $\Phi(a,k,x)$ be an $\in$-formula, and

$$(\mathbf{V}_\vartheta \models \Phi(a_0,k,x)) \iff x = \mathbf{y}_k \quad - \quad \text{for all } k \in \mathbb{N} \text{ and } x \in \mathbf{V}_\vartheta .$$

Then $\{a_0\} \times \mathbb{N}$ also belongs to $\mathbf{V}_\vartheta$. Therefore, by Lemma 4.5.5, there is $m \in \mathbb{N}$ such that $a_0 \in \mathbf{V}_{\alpha_m}$ and moreover, all pairs $\langle a_0, k\rangle$, $k \in \mathbb{N}$, belong to $\mathbf{V}_{\alpha_m}$. For any $p = \langle a, k\rangle \in \mathbf{V}_{\alpha_m}$, if there is a unique set $x \in \mathbf{V}_\vartheta$ satisfying $\mathbf{V}_\vartheta \models \Phi(a,k,x)$, then this x is denoted by $x(p)$; otherwise we put $x(p) = 0$. In particular every $\mathbf{y}_k$ belongs to the set $\{x(p) : p \in \mathbf{V}_{\alpha_m}\}$.

Note that the set $Z = \{p \in \mathbf{V}_{\alpha_m} : p \notin x(p)\}$ belongs to $\mathbf{V}_\vartheta$ and is $\in$-definable in $\mathbf{V}_\vartheta$, and hence Z is equal to a set $\mathbf{y}_k$, $k \in \mathbb{N}$, therefore, equal to $x(p_0)$, where $p_0 = \langle a_0, k\rangle \in \mathbf{V}_{\alpha_m}$. This leads to a contradiction by the diagonal argument: $p_0 \in Z \iff p_0 \notin x(p_0) = Z$. □

4.5c The ultrafilter

Our priority in the construction of a standard core extension of $\mathbf{V}_\vartheta$ will be to ensure that the map $k \mapsto \mathbf{y}_k$ does not penetrate into the extension. Lemma 4.5.6 suggests the method: since no function definable in $\mathbf{V}_\vartheta$ can realize such a map, we have to define the extension in a form essentially definable in $\mathbf{V}_\vartheta$. To achieve this goal, maps $f : {\mathbf{V}_\vartheta}^{\mathtt{fin}} \to \mathbf{V}_\vartheta$ *definable in* $\mathbf{V}_\vartheta$ *with parameters* will be taken to form a quotient power. Accordingly we employ a $\mathbf{V}_\vartheta$-adequate ultrafilter with a very special property: the corresponding quantifier preserves the $\in$-definability in $\mathbf{V}_\vartheta$.

Now let us consider details. For any transitive set V, we use $\mathtt{Def}(V)$ to denote the set of all sets $X \subseteq V$, which are $\in$-definable in V with parameters. More exactly, a set $X \subseteq V$ belongs to $\mathtt{Def}(V)$ iff there exists an $\in$-formula φ with parameters in V and a single free variable x, such that

$$X = \{x \in V : V \models \varphi(x)\} = \{x \in V : \varphi(x) \text{ is true in } V\}.$$

Let $J = {\mathbf{V}_\vartheta}^{\mathtt{fin}}$. Recall that, for an ultrafilter $U \subseteq \mathscr{P}(J)$, $Ui\,P(i,x)$ means that the set $\{i \in J : \langle i, x\rangle \in P\}$ belongs to U.

Theorem 4.5.7. *There is a* $\mathbf{V}_\vartheta$-*adequate ultrafilter* $U \subseteq \mathscr{P}(J)$ *satisfying the following: if* $P \subseteq J \times \mathbf{V}_\vartheta$, $P \in \mathtt{Def}(\mathbf{V}_\vartheta)$, *then the set* $\{z \in \mathbf{V}_\vartheta : Ui\,P(i,z)\}$ *also belongs to* $\mathtt{Def}(\mathbf{V}_\vartheta)$.

Proof. One of the most important consequences of the axiom of constructibility "$\mathbf{V} = \mathbf{L}$" is the existence of a well-ordering $<_{\mathbf{L}}$ of the universe $\mathbf{V}$ of all sets definable by a parameter-free $\in$-formula. (See [Jech 78, Kun 80].) Let x_α be α-th element of $\mathbf{V}_\vartheta$ w. r. t. $<_{\mathbf{L}}$ for every $\alpha < \vartheta$. It follows from Lemma 4.5.1 that the $\in$-formula that defines $<_{\mathbf{L}}$ is absolute for $\mathbf{V}_\vartheta$, hence $<_{\mathbf{L}} \restriction \mathbf{V}_\vartheta$ belongs to $\mathtt{Def}(\mathbf{V}_\vartheta)$ and so does the sequence $\{x_\alpha\}_{\alpha<\vartheta}$.

We claim that $\mathbf{V}_\vartheta = \{x_\alpha : \alpha < \vartheta\}$. Indeed otherwise $\mathbf{V}_\vartheta$ has the $<_{\mathbf{L}}$-order type bigger than ϑ. Then there is $x \in \mathbf{V}_\vartheta$ such that the set $X = \{y : y <_{\mathbf{L}} x\}$ admits a definable map onto ϑ. All definitions involved are absolute for $\mathbf{V}_\vartheta$ by Lemma 4.5.1, hence $X \in \mathbf{V}_\vartheta$. Thus there is a map $X \stackrel{\text{onto}}{\longrightarrow} \vartheta$ $\in$-definable in $\mathbf{V}_\vartheta$, contrary to the fact that $\mathbf{V}_\vartheta$ interprets **ZFC** still by 4.5.1.

Let $\mathscr{D} = \mathscr{P}(J) \cap \mathtt{Def}(\mathbf{V}_\vartheta)$ be the algebra of all subsets of J $\in$-definable (with parameters allowed) in $\mathbf{V}_\vartheta$. To enumerate $\mathscr{D}$ by ordinals fix once and for all a recursive enumeration $\{\chi_n(x,i)\}_{1 \le n \in \mathbb{N}}$ of all parameter-free $\in$-formulas with i and x as the only free variables. Put

$$A_n(\alpha) = \{i \in J : \mathbf{V}_\vartheta \models \chi_n(x_\alpha, i)\}, \ \ A_n^1(\alpha) = A_n(\alpha), \ \ A_n^0(\alpha) = J \smallsetminus A_n(\alpha).$$

Exercise 4.5.8. Prove that $\mathscr{D} = \{A_n(\alpha) : n \ge 1 \wedge \alpha < \vartheta\}$.

Hint: if $X \in \mathscr{D}$ is defined by an $\in$-formula with a single parameter $p \in \mathbf{V}_\vartheta$ then $p = x_\alpha$ for to some $\alpha < \vartheta$ and the formula is χ_n for some $n \ge 1$. □

Definition 4.5.9. We define a set $B_n(\alpha)$, equal to either $A_n^1(\alpha)$ or $A_n^0(\alpha)$, by induction on $n \in \mathbb{N}$ and $\alpha < \vartheta$ as follows.

(i) $B_0(\alpha) = \{i \in J : x_\alpha \in i\}$ whenever $x_\alpha \in J$, and $B_0(\alpha) = J$ otherwise.

(ii) Suppose that $n \ge 1$ and all sets $B_k(\gamma) \in \mathscr{D}$ with $k < n$ and $\gamma < \vartheta$ have been defined. Then, by induction on $\alpha < \vartheta$, define $B_n(\alpha) = A_n^1(\alpha)$ or $B_n(\alpha) = A_n^0(\alpha)$ whenever the family

$$\{B_k(\xi) : k < n \wedge \xi \in \mathtt{Ord}\} \cup \{B_n(\xi) : \xi < \alpha\} \cup \{A_n^1(\alpha)\}$$

does, or resp., does not satisfy f. i. p., the finite intersection property.

Finally put $D = \{B_n(\alpha) : n \in \mathbb{N} \wedge \alpha < \vartheta\}$. □

Exercise 4.5.10. Prove that the set $W_n = \{\alpha < \vartheta : B_n(\alpha) = A_n(\alpha)\}$ belongs to $\mathtt{Def}(\mathbf{V}_\vartheta)$, by induction on $n \ge 1$.

Hint. If $W_1, ..., W_{n-1}$ belong to $\mathtt{Def}(\mathbf{V}_\vartheta)$ then (ii) determines, by induction on α, whether $\alpha \in W_n$. The construction is relativizable to $\mathbf{V}_\vartheta$. □

Clearly D is f. i. p. (of two complementary sets at least one can be added to a f. i. p. family with the f. i. p. preserved). It follows that D is an ultrafilter in $\mathscr{D}$ containing all sets of the form $\{i \in J : x \subseteq i\}$, $x \in J$, by (i). Let U be any ultrafilter over J with $D \subseteq U$. To verify the required property of U, consider a set $P \subseteq J \times \mathbf{V}_\vartheta$, $P \in \mathtt{Def}(\mathbf{V}_\vartheta)$. Then P has the form

$$P = \{\langle i, z\rangle \in J \times \mathbf{V}_\vartheta : \mathbf{V}_\vartheta \models \psi(i, z, p)\},$$

where ψ is an $\in$-formula and $p \in \mathbf{V}_\vartheta$ (a parameter). Let $\varphi(x, i)$ be the formula $\exists q\, \exists z\, (x = \langle q, z\rangle \wedge \psi(i, z, q))$. Thus $\varphi(x, i)$ is $\chi_n(x, i)$ for some $n \ge 1$. Let $\alpha(q, z)$ denote the unique ordinal $\alpha < \vartheta$ satisfying $x_\alpha = \langle q, z\rangle$. We observe that the map $\langle p, z\rangle \mapsto \alpha(p, z)$ belongs to $\mathtt{Def}(\mathbf{V}_\vartheta)$ because so does the sequence $\{x_\alpha\}_{\alpha<\vartheta}$. On the other hand, by definition, $U i\, P(i, z)$ iff $\alpha(p, z) \in W_n$. Thus the result follows from 4.5.10. □ (*Theorem 4.5.7*)

4.5d "Definable" adequate quotient power

Still arguing in $\mathbf{ZFC}\vartheta^+$, we let U be the $<_{\mathbf{L}}$-minimal $\mathbf{V}_{\boldsymbol{\vartheta}}$-adequate ultrafilter satisfying the requirements of Theorem 4.5.7.

The U-part of the "definable" quotient power construction is the same as in § 4.4b (with $\vartheta = \boldsymbol{\vartheta}$): $C = \mathbb{N} \times \mathbf{V}_{\boldsymbol{\vartheta}}$, $I = C^{\mathtt{fin}}$, $U = \bigtimes_{p \in \mathbb{N}} U_p$, and so on. But the $\mathscr{F}$-part drastically changes. For any finite $s \subseteq \mathbb{N}$ we let $\mathscr{F}^s$ be the set of all functions $f : I \to \mathbf{V}_{\boldsymbol{\vartheta}}$ with finite support $\|f\| \subseteq s$, which belong to $\mathtt{Def}(\mathbf{V}_{\boldsymbol{\vartheta}})$. Put $\mathscr{F} = \bigcup_{s \subseteq \mathbb{N} \text{ finite}} \mathscr{F}^s$. Thus only functions definable in $\mathbf{V}_{\boldsymbol{\vartheta}}$ (with parameters) are now admitted to $\mathscr{F}$; this is why the quotient power obtained this way is called "*definable*".

Lemma 4.5.11. *Conditions 1° – 4° of § 4.2b are satisfied for* $\mathbf{U}$, $\mathscr{F}$, *and the standard structure* $\langle V ; \varepsilon \rangle = \langle \mathbf{V}_{\boldsymbol{\vartheta}} ; \in \restriction \mathbf{V}_{\boldsymbol{\vartheta}} \rangle$.

Proof. The only claim that is not trivial and does not immediately follow from 4.3.8, 4.3.10 is 4° of § 4.2b. The new detail is that the function f in 4° now has to be definable in $\mathbf{V}_{\boldsymbol{\vartheta}}$, and hence a plain reference to Choice is not sufficient any more. However we can define $f(i)$ $(i \in I)$ to be the $<_{\mathbf{L}}$-least $x \in \mathbf{V}_{\boldsymbol{\vartheta}}$ satisfying $\mathbf{V}_{\boldsymbol{\vartheta}} \models \varphi(x, f_1(i), \ldots, f_n(i))$. As $<_{\mathbf{L}}$ is $\in$-definable in $\mathbf{V}_{\boldsymbol{\vartheta}}$, such a function f belongs to $\mathtt{Def}(\mathbf{V}_{\boldsymbol{\vartheta}})$ provided so do all functions f_k, and moreover if all f_k are of finite support then so is f, and $\|f\| \subseteq \bigcup_k \|f_k\|$. □

Thus we can define the quotient power ${}^*\mathbf{v} = \langle {}^*\mathbf{V}_{\boldsymbol{\vartheta}} ; {}^*\!\in, {}^*\mathtt{st} \rangle = \mathscr{F}/\mathbf{U}$ together with the natural elementary standard core embedding $* : \mathbf{V}_{\boldsymbol{\vartheta}} \to {}^*\mathbf{V}_{\boldsymbol{\vartheta}}$ (see Definition 4.2.8).

Theorem 4.5.12. (i) $(\mathbf{ZFC}\vartheta^+)$ *The structure* ${}^*\mathbf{v}$ *is an elementary standard core extension of* $\mathbf{v} = \langle \mathbf{V}_\vartheta ; \in \rangle$ *via* $*$ *and a model of* **IST**[**ZC**].

(ii) ${}^*\mathbf{v}$ *is an interpretation of* **IST** *in* $\mathbf{ZFC}\vartheta^+$.

(iii) $(\mathbf{ZFC}\vartheta^+)$ *Some concrete, well-defined cases of the schemata of* $\mathbb{S}$*-Size Choice*, $\mathbb{S}$*-Separation*, *Map-Standardization*, *Inner S. S. Choice*, *and Inner Extension fail in* ${}^*\mathbf{v}$.

Proof. (i) The following is the only amendment to the scheme of the proof of Basic Idealization in Theorem 4.3.13 and Idealization in Proposition 4.4.3: we define $f(i)$ to be the $<_{\mathbf{L}}$-least $x \in \mathbf{V}_{\boldsymbol{\vartheta}}$ satisfying $\forall\, a \in g(i)\, \Phi(a, x, h(i \restriction s))$ (see the proof of 4.3.13). Then f belongs to $\mathtt{Def}(\mathbf{V}_{\boldsymbol{\vartheta}})$ and hence to $\mathscr{F}$.

(ii) See the proof of Proposition 4.4.4(ii). However here the parameter-free character of the interpretation is restored: "$\mathbf{V} = \mathbf{L}$" enables us to choose U as the $<_{\mathbf{L}}$-least ultrafilter satisfying a certain property.

(iii) To prove that the schemata of $\mathbb{S}$-Size Choice *etc.* fail in ${}^*\mathbf{v}$ we make use of the following two results related to the sets $\mathbf{y}_k \in \mathbf{V}_{\boldsymbol{\vartheta}}$ defined in § 4.5b:

(A) the map ${}^*k \mapsto {}^*\mathbf{y}_k$ is $\mathtt{st}$-$\in$-definable in ${}^*\mathbf{v}$;

(B) there is no $F \in {}^*\mathbf{v}$ such that "F is a function, ${}^*k \in \mathtt{dom}\, F$, and $F({}^*k) = {}^*\mathbf{y}_k$" is true in ${}^*\mathbf{v}$ for any $k \in \mathbb{N}$.

(A) We employ the st-$\in$-formula "φ is f. true in $\mathbb{S}$" (φ a finite sequence of symbols coding an $\in$-formula with parameters in $\mathbb{S}$), defined in §3.5a, which expresses in **IST** the truth of φ in $\mathbb{S}$. Let $\Xi(k,x)$ be the st-$\in$-formula saying that: "*Both* $k \in \mathbb{N}$ *and* x *are standard and* $\ulcorner\varphi_k(x)\urcorner$ *is f. true in* $\mathbb{S}$".

We claim that

$$\{\langle K,X\rangle \in {}^*\mathbf{V}_\vartheta : {}^*\mathbf{v} \models \Xi(K,X)\} = \{\langle {}^*k, {}^*\mathbf{y}_k\rangle : k \in \mathbb{N}\}.$$

Indeed by definition $\varphi_k(\mathbf{y}_k)$ holds in $\mathbf{V}_\vartheta$ for any $k \in \mathbb{N}$. It follows that $\varphi_k({}^*\mathbf{y}_k)^{\mathtt{st}}$ holds in ${}^*\mathbf{v}$ because * is a standard core embedding. Thus, by Theorem 3.5.5, we have in ${}^*\mathbf{v}$: "$\ulcorner\varphi_{{}^*k}({}^*\mathbf{y}_k)\urcorner$ is f. true in $\mathbb{S}$", that is, $\Xi({}^*k, {}^*\mathbf{y})$, as required. [17] To prove the converse suppose that $\Xi(K,X)$ holds in ${}^*\mathbf{v}$. Then we have ${}^*\mathtt{st}\,K$ and ${}^*\mathtt{st}\,X$, and hence $K = {}^*k$ and $X = {}^*x$ for some $k \in \mathbb{N}$ and $x \in \mathbf{V}_\vartheta$, and $\Xi({}^*k, {}^*x)$ holds in ${}^*\mathbf{v}$. The same arguments as above, but in the opposite direction, yield $\varphi_k(x)$ in $\mathbf{V}_\vartheta$, so that $x = \mathbf{y}_k$, as required. $\dashv$

(B) Suppose that $F = [f] \in \mathbf{V}_\vartheta$ is a counterexample; $f \in \mathscr{F}$. We claim that the set $X = \{x \in \mathbf{V}_\vartheta : {}^*x \;{}^*\!\!\in F\}$ belongs to $\mathtt{Def}(\mathbf{V}_\vartheta)$. It follows from Lemma 4.2.10 that ${}^*x \;{}^*\!\!\in F$ iff $\mathbf{U}\,i\,(x \in f(i))$, that is, $U^s\,i\,(x \in f(i))$ by 4.3.8, where $s = \|f\| = \{p_1 < \dots < p_n\} \subseteq \mathbb{N}$. This is equivalent to

$$U_{p_n}\,i_n \dots U_{p_1}\,i_1\;(x \in f(i_1 \cup \dots \cup i_n)).$$

However $f \in \mathtt{Def}(\mathbf{V}_\vartheta)$ and the ultrafilters U_p preserve the definability in $\mathbf{V}_\vartheta$ (as their prototype U does). Thus $X \in \mathtt{Def}(\mathbf{V}_\vartheta)$. But then the set $\{\langle k, \mathbf{y}_k\rangle : k \in \mathbb{N}\}$ also belongs to $\mathtt{Def}(\mathbf{V}_\vartheta)$ by the choice of F, a contradiction to Lemma 4.5.6. $\dashv$

Let $\mathtt{Fun}\,f$ mean: "f is a function". It follows from (A) and (B) that

$$\forall^{\mathtt{st}} k \in \mathbb{N}\;\exists!^{\mathtt{st}} x\;\Xi(k,x) \implies \exists f\,\big(\mathtt{Fun}\,f \wedge \forall^{\mathtt{st}} k \in \mathbb{N}\;\Xi(k, f(k))\big) \qquad (\dagger)$$

is false in ${}^*\mathbf{v}$. This immediately witnesses that $\mathbb{S}$-Size Choice, $\mathbb{S}$-Separation, Map-Standardization, Inner S. S. Choice, Inner Extension fail in ${}^*\mathbf{v}$. □

Exercise 4.5.13. Suppose that $X \subseteq \mathbf{V}_\vartheta$, $X \in \mathtt{Def}(\mathbf{V}_\vartheta)$. Prove that the function $f(i) = \{x \in X : \langle 0, x\rangle \in i\}$, $\forall i \in I$, belongs to $\mathscr{F}$ and satisfies $X = \{x \in \mathbf{V}_\vartheta : {}^*x \;{}^*\!\!\in [f]\}$. (*Hint.* Show that $\mathbf{U}\,i\,(x \in f(i))$ for any $x \in X$ and $\mathbf{U}\,i\,(x \notin f(i))$ for any $x \in \mathbf{V}_\vartheta \smallsetminus X$. Then apply the Łoś theorem.) □

4.5e Corollaries and remarks

Corollary 4.5.14. (i) *Each of the principles of* $\mathbb{S}$-*Size Choice*, $\mathbb{S}$-*Separation, Map-Standardization, Inner S. S. Choice, Inner Extension can be neither proved nor falsified in* **IST**.

(ii) **IST** *is not a reducible theory in the sense of Definition 4.1.3.*

[17] Theorem 3.5.5 is applicable in ${}^*\mathbf{v}$ by 3.5.7, since ${}^*\mathbf{v}$ is a **model** of **IST**[**ZC**] by (i).

Proof. (i) follows from Corollary 4.4.7 and the result of Exercise 4.4.8(i) – from the one hand, and Theorem 4.5.12 – from the other hand.

(ii) Let Ψ be an example of Inner Extension false in $^*\mathbf{v}$, *e.g.* (†) in the proof of Theorem 4.5.12. Suppose on the contrary that **IST** proves $\Psi \Longleftrightarrow \psi^{\mathtt{st}}$, where ψ is an $\in$-sentence. Arguing in $\mathbf{ZFC}\vartheta^+$ note that according to theorems 4.4.6 and 4.5.12, the model $\mathbf{V}_\vartheta$ admits two different standard core extensions interpreting **IST**, such that Ψ holds in one of them but fails in the other one. Thus $\psi^{\mathtt{st}}$ also is true in one of the extensions and false in the other one, so that, by 4.1.2, ψ is both true and false in $\mathbf{V}_\vartheta$, contradiction. □

□ (*Non-reducibility of* **IST** *in Theorem 4.1.10*(iii))

Exercise 4.5.15. Prove that the restricted Inner S. S. Choice schema (as in Exercise 3.6.10) also fails in the structure $^*\mathbf{v}$ considered in Theorem 4.5.12.

Hint. Let C, $H \in {}^*\mathbf{V}_\vartheta$ satisfy, in $^*\mathbf{v}$, the following:

$$C \text{ is finite} \quad \wedge \quad \mathbb{S} \subseteq C \quad \wedge \quad H : C \to \mathbb{N} \text{ is a } 1-1 \text{ map.}$$

Take $X = Y = \mathbb{N}$ and $\Phi(k,m)$ to be the formula $k \in \mathbb{N} \wedge m \in \mathtt{ran}\, H \wedge \Xi(k, H^{-1}(m))$ in 3.6.10.

Note that the counterexample contains a nonstandard parameter H. It is an interesting ***problem*** to obtain a parameter-free counterexample to the restricted Inner S. S. Choice in **IST**. □

Exercise 4.5.16. Use the following argument [18] to prove the irreducibility of **IST** without "definable" ultralimits. Let $\kappa < \vartheta$ be a pair of cardinals such that κ has countable cofinality while $\mathtt{cof}\, \vartheta = \omega_1$, $\mathbf{V}_\kappa$ is an elementary submodel of $\mathbf{V}_\vartheta$, and both $\mathbf{V}_\vartheta$ and $\mathbf{V}_\kappa$ are models of **ZFC**. (The existence of such a pair follows, for instance, from a strongly inaccessible cardinal.) Fix an increasing sequence of ordinals α_n cofinal in κ.

Consider a standard core extension $^*\mathbf{v} = \langle {}^*\mathbf{V}_\kappa \,;\, {}^*\!\in, {}^*\mathtt{st} \rangle$ of $\mathbf{V}_\kappa$ as in § 4.4b. We easily define $H \in {}^*\mathbf{V}_\kappa$ such that $^*\alpha_n \;{}^*\!\in H$ for all n and, for any $x \in \mathbf{V}_\kappa$, if $^*x \;{}^*\!\in H$ then $x = \alpha_n$ for some n. Let $\Phi(n,x)$ formally say: x is the n-th element of H (or nothing if meaningless). Prove that Map-Standardization fails in $^*\mathbf{v}$ for this Φ and $X = \mathbb{N}$. Prove that, on the contrary, this case of Map-Standardization is true in any standard core extension of $\mathbf{V}_\vartheta$ of the same kind just because $\mathtt{cof}\, \vartheta > \omega$. Thus if **IST** were a reducible theory then the same $\in$-sentence would be true in $\mathbf{V}_\vartheta$ and false in $\mathbf{V}_\kappa$, in contradiction to the choice of κ, ϑ. □

Exercise 4.5.17. Is there any reasonable version of "definable" ultrapowers or quotient powers based on good rather than adequate ultrafilters?

There can be obstacles related to the fact that both the construction of good ultrafilters and the techniques involved in the derivation of Saturation are of somewhat different and generally more complicated nature than those for adequate ultrafilters. □

[18] Communicated to one of the authors (V. Kanovei) by R. Solovay in 1993.

4.6 Interpretability of IST in a standard theory

This section is devoted to claim (iv) of Theorem 4.1.10 the interpretability of **IST** in a standard theory. We also consider the related question of **IST**-extendibility of transitive models of **ZFC**. It will be demonstrated that while all models of **ZFC** are **BST**-extendible by Corollary 4.3.14, not every countable transitive model of **ZFC** is **IST**-extendible, for instance minimal models of **ZFC** are not extendible. The question of exactly which models are extendible to models of **IST** remains open, yet we shall find a convenient characterization of extendibility to models of a close version of **IST**, which in the same time will be a sufficient condition of **IST**-extendibility. To achieve this goal, we employ a "standard" extension **ZFGT** of **ZFC** in which **IST** is standard core interpretable.

4.6a Standard theory with a global choice and a truth predicate

First of all let us extend the "gödelization" of the $\in$-language in § 3.5a to the language $\mathcal{L}_{\in,\mathbf{G}}$, where $\mathbf{G}$ is a symbol for a global choice function. To simplify technicalities, *we allow* $\mathbf{G}$ *to occur, in* $\mathcal{L}_{\in,\mathbf{G}}$, *only in expressions of the form* $\mathbf{G}(x) = y$ (essentially equivalent to $\langle x, y\rangle \in \mathbf{G}$). Other occurrences of $\mathbf{G}$ in formulas are considered as shortcuts, for instance, $\mathbf{G}(x) \in z$ is a shortcut for the formula $\exists\, y\, (y = \mathbf{G}(x) \land y \in z)$. Define $\mathtt{Form}_\mathbf{G}$ similarly to $\mathtt{Form}$ in § 3.5a, with the following addition:

- integer 7 identified with $\mathbf{G}$.

Define $\mathtt{Form}_\mathbf{G}$, $\mathtt{ClForm}_\mathbf{G}$, $\mathtt{Form}_\mathbf{G}[\varphi]$, *etc.* as in § 3.5a. Elements of $\mathtt{Form}_\mathbf{G}$ (that is, finite sequences of certain kind) will be referred to as *coded formulas*.

Definition 4.6.1. **ZFGT** is a theory in the language $\mathcal{L}_{\in,\mathbf{G},\mathbf{T}}$ which extends the language $\mathcal{L}_{\in,\mathbf{G}}$ by a unary predicate symbol $\mathbf{T}$ used mostly as $\varphi \in \mathbf{T}$ rather than $\mathbf{T}(\varphi)$. The axioms of **ZFGT** include **ZFC** in the $\in$-language, the Global Choice axiom (saying that $\mathbf{G}$ is a global choice function, that is $\mathbf{G}(x) \in x$ for any $x \neq \varnothing$), and in addition

(A) the Separation schema in the language $\mathcal{L}_{\in,\mathbf{G},\mathbf{T}}$;

(B) an axiom saying that $\mathbf{T}$ is a truth predicate for $\mathcal{L}_{\in,\mathbf{G}}$, in the sense that it satisfies appropriate Tarski conditions for a truth definition, that is,

(1′) $\forall \varphi\, (\varphi \in \mathbf{T} \Longrightarrow \varphi \in \mathtt{ClForm}_\mathbf{G})$;

(2′) $\forall p\, \forall q\, \big((\ulcorner p = q \urcorner \in \mathbf{T} \iff p = q) \land (\ulcorner p \in q \urcorner \in \mathbf{T} \iff p \in q)\big)$;

(3′) $\forall \varphi\, \forall \psi\, (\ulcorner \varphi \land \psi \urcorner \in \mathbf{T} \iff \varphi \in \mathbf{T} \land \psi \in \mathbf{T})$;

(4′) $\forall \psi \in \mathtt{ClForm}_\mathbf{G}\, (\ulcorner \neg\, \psi \urcorner \in \mathbf{T} \iff \psi \notin \mathbf{T})$;

(5′) $\forall \varphi(v_i) \in \mathtt{Form}_\mathbf{G}\, \big(\ulcorner \exists\, v_i\, \varphi(v_i) \urcorner \in \mathbf{T} \iff \exists\, x\, (\ulcorner \varphi(x) \urcorner \in \mathbf{T})\big)$;

(6′) $\forall p\, \forall q\, \big(\ulcorner p = \mathbf{G}(q) \urcorner \in \mathbf{T} \iff p = \mathbf{G}(q)\big)$. □

Thus the universe of **ZFGT** is the universe of **ZFC** equipped with a global choice function **G** and a truth predicate **T** for formulas of $\mathcal{L}_{\in,\mathbf{G}}$. The following result shows that **T** includes all sentences true in the ordinary sense.

Exercise 4.6.2 (Compare with Theorem 3.5.5). Prove by induction on the complexity of φ that for any formula $\varphi(x_1, ..., x_n)$ of $\mathcal{L}_{\in,\mathbf{G}}$ it is a theorem of **ZFGT** that $\forall x_1 ... \forall x_n \left(\varphi(x_1, ..., x_n) \iff (\ulcorner\varphi(x_1, ..., x_n)\urcorner \in \mathbf{T})\right)$. □

Theorem 4.6.3. **ZFGT** *is a conservative* (in the sense of Definition 1.5.16) *and hence equiconsistent extension of* **ZFC**: *any theorem of* **ZFGT** *in the* $\in$*-language is provable in* **ZFC**.

Proof. Suppose that Φ, an $\in$-sentence, is a theorem of **ZFGT** but, on the contrary, $\neg\Phi$ is consistent with **ZFC**. Then Φ is provable in a finite fragment H of **ZFGT** that consists of a finite fragment ζ of **ZFC**, a finite fragment of (A)+(B) above, and the Global Choice axiom for **G**. Then, by Theorem 1.5.4, the consistent theory $\mathbf{ZFC} + \neg\Phi$ proves that there is an ordinal κ such that $\langle \mathbf{V}_\kappa ; \in\rangle$ is a model of ζ and Φ fails in $\mathbf{V}_\kappa$. Let $G : \mathbf{V}_\kappa \to \mathbf{V}_\kappa$ be any choice function for $\mathbf{V}_\kappa$ and T be the set of all formulas of $\mathcal{L}_{\in,\mathbf{G}}$ with parameters in $\mathbf{V}_\kappa$, true in $\langle \mathbf{V}_\kappa ; \in, G\rangle$. Then $\langle \mathbf{V}_\kappa ; \in, G, T\rangle$ is a model of ζ + (A) + (B) (indeed every set of the form $\mathbf{V}_\kappa$ models Separation), hence of H, but Φ fails in $\langle \mathbf{V}_\kappa ; \in, G, T\rangle$, contradiction. □

Note that unlike Separation the schemata of Collection and Replacement are included in **ZFGT** in the $\in$-language only.

Exercise 4.6.4. (1) Prove that if Replacement (or Collection) in $\mathcal{L}_{\in,\mathbf{G},\mathbf{T}}$ is added to (A) then such a stronger version of **ZFGT** will not be conservative, in fact it is strong enough to prove the consistency of **ZFC**.

(2) Prove that if Collection is added only in $\mathcal{L}_{\in,\mathbf{G}}$ then Theorem 4.6.3 remains true (with a less elementary proof based on Felgner's [Fel 71] method of adjoining a generic global choice function which preserves Collection). □

In spite of the absence of Collection and Replacement, **ZFGT** still allows some transfinite constructions that involve **G**. The following is an example.

Exercise 4.6.5. (i) In **ZFGT**, define a set x_α by induction on $\alpha \in \mathtt{Ord}$ as follows: $x_0 = \varnothing$, and, for any $\alpha > 0$,

(∗) $x_\alpha = \mathbf{G}(\mathbf{V}_{\mu(\alpha)} \smallsetminus \{x_\gamma : \gamma < \alpha\})$, where $\mu(\alpha)$ is the least ordinal μ such that $\mathbf{V}_\mu \not\subseteq \{x_\gamma : \gamma < \alpha\}$.

(ii) Prove that this is a legitimate definition in the sense that **ZFGT** proves that for any $\lambda \in \mathtt{Ord}$ there is a set of the form $\{x_\alpha\}_{\alpha<\lambda}$ such that $x_0 = \varnothing$ and (∗) holds for all $0 < \alpha < \lambda$.
(*Hint*. First prove that $\mathbf{G} \restriction X$ is a set for any set X — this is because $\mathbf{G}(x) \in x$ for any x, thus the result required needs only Separation in the language with **G**. Then note that to define $\{x_\alpha\}_{\alpha<\lambda}$ we need only $\mathbf{G} \restriction \lambda$, and hence this is a **ZFC** construction with $\mathbf{G} \restriction \lambda$ as a parameter.)

(iii) Prove in **ZFGT** that $\alpha \mapsto x_\alpha$ is a bijection $\mathtt{Ord} \xrightarrow{\text{onto}} \mathbf{V}$. Prove that the relation $x <_\mathbf{G} y$ iff $x = x_\alpha$ and $y = x_\beta$ for $\alpha < \beta$ well-orders the universe so that each $<_\mathbf{G}$-initial segment is a set. □

4.6b Formally definable classes

The following notion of formal $\mathcal{L}_{\in,\mathbf{G}}$-definability is a reasonable approximation of the metamathematical notion of definability, available due to the properties of $\mathbf{T}$ postulated by Definition 4.6.1(B)

Definition 4.6.6. In **ZFGT**, a *formally* $\mathcal{L}_{\in,\mathbf{G}}$*-definable class* is any class of the form $\mathscr{C}_{\varphi(x)} = \{x : \ulcorner\varphi(x)\urcorner \in \mathbf{T}\}$, where $\varphi(x) \in \mathtt{Form}_\mathbf{G}$ contains x as the only free variable x and may contain any sets as parameters. □

Remark 4.6.7. The parametrization $\mathscr{C}_{\varphi(x)}$ of formally $\mathcal{L}_{\in,\mathbf{G}}$-definable classes provided by Definition 4.6.6 is itself definable in $\mathcal{L}_{\in,\mathbf{G},\mathbf{T}}$. This enables us to incorporate definable classes in usual arguments with sets, consistently consider collections of $\mathcal{L}_{\in,\mathbf{G}}$-definable classes (viewed as collections of coded formulas in $\mathtt{Form}_\mathbf{G}$ which define those classes as in Definition 4.6.6), legitimately quantify over $\mathcal{L}_{\in,\mathbf{G}}$-definable classes, and so on.

On the other hand, the results of 4.6.8 and 4.6.9 just below show that the family of all definable classes looks very much like a "power class" of the set universe $\mathbf{V}$ of **ZFGT**. □

Proposition 4.6.8. (i) *Given any formula $\psi(x, v_1, ..., v_n)$ of $\mathcal{L}_{\in,\mathbf{G}}$ it is a theorem of* **ZFGT** *that for any parameters $p_1, ..., p_n$, the class $\{x : \psi(x, p_1, ..., p_n)\}$ is formally $\mathcal{L}_{\in,\mathbf{G}}$-definable. In addition,* **ZFGT** *proves:*

(ii) *if X is a formally $\mathcal{L}_{\in,\mathbf{G}}$-definable class then so are the complement $X^\mathtt{C} = \mathbf{V} \smallsetminus X$ and the projection $Y = \operatorname{dom} X = \{y : \exists\, z\, (\langle y, z\rangle \in X)\}$;*

(iii) *if $X_1, ..., X_n$ are formally $\mathcal{L}_{\in,\mathbf{G}}$-definable classes then so are $X_1 \cup ... \cup X_n$, $X_1 \cap ... \cap X_n$, $X_1 \times ... \times X_n$.*

Proof. (i) $\{x : \psi(x, p_1, ..., p_n)\} = \{x : \ulcorner\psi(x, p_1, ..., p_n)\urcorner \in \mathbf{T}\}$ by 4.6.2.

(ii) Suppose that $X = \{x : \ulcorner\varphi(x)\urcorner \in \mathbf{T}\}$, $\varphi(x) \in \mathtt{Form}_\mathbf{G}$. It immediately follows from 4.6.1(B)(4′) that $X^\mathtt{C} = \{x : \ulcorner\varphi'(x)\urcorner \in \mathbf{T}\}$, where φ' is $\ulcorner\neg\,\varphi\urcorner$. Now let $\psi(y)$ be $\ulcorner\exists\, x\, \exists\, z\, (x = \langle y, z\rangle \wedge \varphi(x))\urcorner$. Then

$$\begin{aligned} \ulcorner\psi(y)\urcorner \in \mathbf{T} &\iff \exists\, x\, \exists\, z\, (\ulcorner x = \langle y, z\rangle\urcorner \in \mathbf{T} \wedge \ulcorner\varphi(x)\urcorner \in \mathbf{T}) \\ &\iff \exists\, x\, \exists\, z\, (x = \langle y, z\rangle \wedge x \in X) \qquad \iff y \in Y \end{aligned}$$

by (B)(5′), (B)(3′) of 4.6.1 and Exercise 4.6.2 for the formula $x = \langle y, z\rangle$.

(iii) Suppose that $X_i = \{x : \ulcorner\varphi_i(x)\urcorner \in \mathbf{T}\}$, $i = 1, ..., n$. Let $\varphi(x)$ be $\ulcorner\varphi_1(x) \wedge ... \wedge \varphi_n(x)\urcorner$. Then $X_1 \cap ... \cap X_n = \{x : \ulcorner\varphi(x)\urcorner \in \mathbf{T}\}$. Indeed it follows from 4.6.1(B)(3′) that $\ulcorner\varphi\urcorner \in \mathbf{T}$ iff $\ulcorner\varphi_i\urcorner \in \mathbf{T}$ for all $i \le n$. Finally, let $\psi(x)$ be $\ulcorner\exists\, x_1 \dots \exists\, x_n\, (x = \langle x_1, ..., x_n\rangle \wedge \varphi_1(x_1) \wedge ... \wedge \varphi_n(x_n))\urcorner$. We have $X_1 \times ... \times X_n = \{x : \ulcorner\psi(x)\urcorner \in \mathbf{T}\}$ by (B)(5′), (B)(3′) of 4.6.1. □

Corollary 4.6.9. *Suppose that $\Phi(x, V_1, ..., V_n)$ is a formula of $\mathcal{L}_{\in,\mathbf{G}}$ in which all variables V_i occur only through expressions of the form $v \in V_i$* (that is, to the right of $\in$). *Then* **ZFGT** *proves that for any formally $\mathcal{L}_{\in,\mathbf{G}}$-definable classes $X_1, ..., X_n$ the class $X = \{x : \Phi(x, X_1, ..., X_n)\}$ is a formally $\mathcal{L}_{\in,\mathbf{G}}$-definable class, too.*

Proof. Argue by induction on the number of symbols in Φ, with (i) of Proposition 4.6.8 used for elementary formulas Φ and (ii), (iii) for the inductive steps for $\neg$, $\wedge$, $\exists$. We leave the proof as an easy **exercise** for the reader. □

4.6c A nonstandard theory extending IST

Our goal is to define a standard core interpretation of **IST**, even of a somewhat stronger nonstandard theory, in **ZFGT**. To introduce the stronger theory, define $\mathtt{Truth}_G\,T$, a modified truth predicate, as the conjunction of the following $\mathtt{st}$-$\in$-formulas (compare with Definition 3.5.2 !) with free variables T and G:

(1) $T \subseteq \mathtt{ClForm}_{\mathbf{G}}$, G is a function, $\mathbb{S} \subseteq \operatorname{dom} G$ and $G(x) \in x$ for all $x \in \mathbb{S}$;

(2) $\forall^{\mathtt{st}} p\, \forall^{\mathtt{st}} q\, \big((\ulcorner p = q \urcorner \in T \iff p = q) \wedge (\ulcorner p \in q \urcorner \in T \iff p \in q)\big)$;

(3) for any standard φ, ψ: $\ulcorner \varphi \wedge \psi \urcorner \in T \Longrightarrow \varphi \in T \wedge \psi \in T$, and
$\ulcorner \neg\, (\varphi \wedge \psi) \urcorner \in T \Longrightarrow \ulcorner \neg\, \varphi \urcorner \in T \vee \ulcorner \neg\, \psi \urcorner \in T$.

(4) for any standard φ : $\ulcorner \neg\, \varphi \urcorner \in T \Longrightarrow \varphi \notin T$ and
$\ulcorner \neg \neg\, \varphi \urcorner \in T \Longrightarrow \varphi \in T$;

(5) for any standard $\varphi(v_i)$: $\ulcorner \exists\, v_i\, \varphi(v_i) \urcorner \in T \Longrightarrow \exists^{\mathtt{st}} x\, (\ulcorner \varphi(x) \urcorner \in T)$ and
$\ulcorner \neg \exists\, v_i\, \varphi(v_i) \urcorner \in T \Longrightarrow \forall^{\mathtt{st}} x\, (\ulcorner \neg\, \varphi(x) \urcorner \in T)$;

(6) $\forall^{\mathtt{st}} p\, \forall^{\mathtt{st}} q\, \big(\ulcorner \mathbf{G}(p) = q \urcorner \in T \iff G(p) = q\big)$.

Accordingly, sets T satisfying $\mathtt{Truth}_G\,T$ are called *truth sets for* $\langle \mathbb{S}; G \rangle$.[19]

A formula $\varphi \in \mathtt{ClForm}_{\mathbf{G}}$ is *formally true* (f. true) in $\langle \mathbb{S}; G \rangle$ if there is a truth set T for $\langle \mathbb{S}; G \rangle$ containing φ. A formula φ is *formally false* (f. false) in $\langle \mathbb{S}; G \rangle$ iff $\neg\varphi$ is f. true. Thus "φ is f. true (false) in $\mathbb{S}$" are $\mathtt{st}$-$\in$-formulas with φ as the unique free variable. Similarly to Theorem 3.5.4(ii), no standard $\varphi \in \mathtt{ClForm}_{\mathbf{G}}$ can be both f. true and f. false in $\langle \mathbb{S}; G \rangle$.

Definition 4.6.10. ISTGT is **IST** plus the following axiom:

(†) There is a function G such that $\mathbb{S} \subseteq \operatorname{dom} G$, $G(x) \in x \cap \mathbb{S}$ for any standard $x \neq \varnothing$, and any standard $\varphi \in \mathtt{ClForm}_{\mathbf{G}}$ is either f. true or f. false in $\langle \mathbb{S}; G \rangle$.

In **ISTGT**, if G satisfies (†) then let $\mathbf{T}_G$ be the collection (not necessarily a set) of all standard formulas $\varphi \in \mathtt{ClForm}_{\mathbf{G}}$ which are f. true in $\langle \mathbb{S}; G \rangle$. $G \restriction \mathbb{S}$ is intended to interpret the formal symbol $\mathbf{G}$ of the language $\mathcal{L}_{\in,\mathbf{G}}$. □

[19] *Partial* truth sets, as in § 3.5a, see footnote 23 on page 119.

Exercise 4.6.11 (ISTGT). Prove, using (†), that if G is as indicated then the structure $\langle \mathbb{S}; \in, G \restriction \mathbb{S}, \mathbf{T}_G\rangle$ interprets **ZFGT**. □

Thus **ZFGT** is interpretable in **ISTGT**, in a sense slightly different from § 1.5d as now the interpretation depends on a parameter G. Note that the basic universe of the interpretation is the class $\mathbb{S}$ of all standard sets. The result of 4.6.11 is the easier part of the following theorem:

Theorem 4.6.12. *The theories* **ZFGT**, **ISTGT** *are interpretable in each other, in particular, the interpretation of* **ZFGT** *in* **ISTGT** *is given in 4.6.11 and it has* $\mathbb{S}$ *as the set universe, while* **ISTGT** *is standard core interpretable in* **ZFGT** *by Theorem 4.6.19 below.*

Remark 4.6.13. It takes some effort to derive, or even properly formulate, a reasonable conservativity result from Theorem 4.6.12. We conjecture that a sentence Φ of $\mathcal{L}_{\in,\mathbf{G},\mathbf{T}}$ is a theorem of **ZFGT** iff **ISTGT** proves

$$\forall G\,(G \text{ satisfies } 4.6.10(\dagger) \Longrightarrow \Phi^{\langle \mathbb{S};\in,G\restriction\mathbb{S},\mathbf{T}_G\rangle}). \qquad \square$$

4.6d The ultrafilter

To prove the nontrivial part of Theorem 4.6.12 we define an interpretation of the nonstandard theory **ISTGT** in **ZFGT**.

We argue in **ZFGT** *in this subsection.*

As usual, $\mathbf{V} = \bigcup_{\xi\in\mathtt{Ord}} \mathbf{V}_\xi$ is the set universe of **ZFGT**.

Put $J = \mathbf{V}^{\mathtt{fin}} = \{x : x \text{ is finite}\}$ (a proper class, of course). Let $\mathscr{D}$ consist of all formally $\mathcal{L}_{\in,\mathbf{G}}$-definable classes $X \subseteq J$. Our first goal is to define, in **ZFGT**, a $\mathbf{V}$-adequate ultrafilter $U \subseteq \mathscr{D}$ preserving the formal $\mathcal{L}_{\in,\mathbf{G}}$-definability as a quantifier.

Fix a recursive enumeration $\{\chi_n(v)\}_{1\le n\in\mathbb{N}}$ of all parameter-free coded formulas in $\mathtt{Form}_{\mathbf{G}}$ with one free variable. Put $\mathbf{A}_n = \{z : \ulcorner\chi_n(z)\urcorner \in \mathbf{T}\}$ and $A_n(\alpha) = \{i \in J : \langle i, x_\alpha\rangle \in \mathbf{A}_n\}$ for $n \in \mathbb{N}$, $\alpha \in \mathtt{Ord}$. (See 4.6.5 on x_α.)

Exercise 4.6.14. Prove that $\mathscr{D}$ is equal to the family of all classes $A_n(\alpha)$.

Hint. The sequence $\{x_\alpha\}_{\alpha\in\mathtt{Ord}}$ is definable by a formula of $\mathcal{L}_{\in,\mathbf{G}}$ and hence formally $\mathcal{L}_{\in,\mathbf{G}}$-definable by Proposition 4.6.8, thus so is each class $A_n(\alpha)$. For the other direction, if $X \subseteq J$ is formally $\mathcal{L}_{\in,\mathbf{G}}$-definable then $X = \{i \in J : \ulcorner\varphi(i, x_\alpha)\urcorner \in \mathbf{T}\}$ where $\alpha \in \mathtt{Ord}$ and $\varphi(\cdot,\cdot) \in \mathtt{Form}_{\mathbf{G}}$. Easily $X = A_n(\alpha)$, where n is such that $\chi_n(v)$ is $\exists i\, \exists p\, (v = \langle p, i\rangle \wedge \varphi(i,p))$. □

Put $A_n^1(\alpha) = A_n(\alpha)$ and $A_n^0(\alpha) = J \smallsetminus A_n(\alpha)$.

The following definition is obviously analogous to Definition 4.5.9. However we have to be careful in details because now $\mathbf{V}$ and J are proper classes and the usual $\in$-definability is replaced by the formal $\mathcal{L}_{\in,\mathbf{G}}$-definability in the sense of 4.6.6.

Definition 4.6.15. We define a set $B_n(\alpha)$, equal to either $A^1_k(\alpha)$ or $A^0_k(\alpha)$, by induction on $n \in \mathbb{N}$ and $\alpha \in \mathtt{Ord}$, as follows.

(i) $B_0(\alpha) = \{i \in J : x_\alpha \in i\}$ whenever $x_\alpha \in J$, and $B_0(\alpha) = J$ otherwise.

(ii) Suppose that $n \geq 1$ and all classes $B_k(\gamma) \in \{A^1_k(\gamma), A^0_k(\gamma)\}$ with $k < n$ and $\gamma \in \mathtt{Ord}$ have been defined. Then, by induction on $\alpha \in \mathtt{Ord}$, define $B_n(\alpha) = A^1_n(\alpha)$ or $B_n(\alpha) = A^0_n(\alpha)$ whenever the family

$$\{B_k(\xi) : k < n \wedge \xi \in \mathtt{Ord}\} \cup \{B_n(\xi) : \xi < \alpha\} \cup \{A^1_n(\alpha)\}$$

resp. satisfies or does not satisfy f. i. p., the finite intersection property.

Finally, set $U = \{B_n(\alpha) : n \in \mathbb{N} \wedge \alpha \in \mathtt{Ord}\}$. □

The most essential issue concerning this definition is that the objects defined by induction are proper classes. Therefore, according to the well-known formalization of transfinite induction, the definition involves transfinite sequences of proper classes, which makes it look suspicious. To circumpass this problem, we put

$$W_n = \{\alpha \in \mathtt{Ord} : B_n(\alpha) = A_n(\alpha)\}$$

for any $n \geq 1$. The following lemma shows that at least these classes are fully legitimate objects!

Lemma 4.6.16 (ZFGT). *For any $n \geq 1$, W_n is a formally $\mathcal{L}_{\in,\mathbf{G}}$-definable class: there exists $\varphi_n(\alpha) \in \mathtt{Form}_{\mathbf{G}}$ such that $W_n = \{\alpha : \ulcorner\varphi_n(\alpha)\urcorner \in \mathbf{T}\}$.*

Proof. We argue by induction on $n \geq 1$. For any $k < n$, let $\mathscr{B}_k$ denote the collection $\{B_k(\xi) : \xi \in \mathtt{Ord}\}$ of classes. Then $\alpha \in W_n$ is equivalent to the following long formula:

$\exists\, h : [0, \alpha] \to \{0, 1\}$:

$h(\alpha) = 1 \;\wedge\;$ the family $\bigcup_{k<n} \mathscr{B}_k \cup \{A^{h(\xi)}_n(\xi) : \xi \leq \alpha\}$ is f. i. p. $\wedge$

$\wedge\, \forall\, \gamma \leq \alpha\, (h(\gamma) = 1 \Longleftrightarrow \bigcup_{k<n} \mathscr{B}_k \cup \{A^{h(\xi)}_n(\xi) : \xi < \gamma\} \cup \{A^1_n(\gamma)\}$ is f. i. p.).

The meaning of h is as follows: $h(\gamma) = 1$ iff $\gamma \in W_n$. The displayed formula defines W_n in terms of the classes W_k, $1 \leq k < n$ (to properly substitute $A^1_k(\xi)$ or $A^0_k(\xi)$ for $B_k(\xi)$ in $\mathscr{B}_k$), the classes $\mathbf{A}_k$, $1 \leq k < n$, and the sequence $\{x_\alpha\}_{\alpha \in \mathtt{Ord}}$ considered as a class. In addition $\mathscr{B}_0$ participates in the displayed formula in directly $\mathcal{L}_{\in,\mathbf{G}}$-definable fascion. All these classes are formally $\mathcal{L}_{\in,\mathbf{G}}$-definable (the classes W_k by the inductive hypothesis) and clearly occur only to the right side of the membership, while the f. i. p. can be adequately expressed in the form that for any relevant finite set of indices the corresponding intersection is non-empty. It follows that W_n is formally $\mathcal{L}_{\in,\mathbf{G}}$-definable as well by Corollary 4.6.9. □

The proof of the lemma and also of Corollary 4.6.9 can be carried out in such a form that all steps of transformation of the coded formulas involved are explicitly defined. This leads to a recursive sequence of coded formulas $\varphi_n(v) \in \mathtt{Form}_{\mathbf{G}}$ such that $W_n = \{\alpha \in \mathtt{Ord} : \ulcorner\varphi_n(\alpha)\urcorner \in \mathbf{T}\}$. Let $\tau(n,\alpha)$ say: $\ulcorner\varphi_n(\alpha)\urcorner \in \mathbf{T}$; this is a formula of $\mathcal{L}_{\in,\mathbf{G},\mathbf{T}}$ with n, α as free variables. In other words we have $\tau(n,\alpha)$ iff $\alpha \in W_n$ for all n, α. [20]

We observe that the $\mathcal{L}_{\in,\mathbf{G},\mathbf{T}}$-definition of U in 4.6.15 can be presented in the following compact form involving τ :

- $U = \{B_n(\alpha) : n \in \mathbb{N} \wedge \alpha \in \mathtt{Ord}\}$, where the classes $B_0(\alpha)$ are defined by (i) of Definition 4.6.15, while, for $n \ge 1$, $B_n(\alpha)$ is either the class $A_n(\alpha) = \{i \in J : \ulcorner\chi_n(x_\alpha, i)\urcorner \in \mathbf{T}\}$ or the complementary class $J \smallsetminus A_n(\alpha)$ in the cases, resp., $\tau(n,\alpha)$ and $\neg\,\tau(n,\alpha)$.

Clearly U is a collection of proper classes, a suspicious-looking object in **ZFGT**, yet any occurrence of U in our arguments can be considered as just a figure of speech intended to present reasoning with relevant coded formulas (like χ_n, φ_n, τ) in a manner easier to grasp. See Remark 4.6.7.

Exercise 4.6.17. Show that U is an ultrafilter in $\mathscr{D}$, in the sense that U is f. i. p. and, for any formally $\mathcal{L}_{\in,\mathbf{G}}$-definable class $X \subseteq J$, exactly one of the classes X, $J \smallsetminus X$ belongs to U. (*Hint*. Classes are adjoined to U one by one, and if adding some $A_n(\alpha)$ kills f. i. p. then the complement is added, so that f. i. p. is obviously preserved in the course of the construction.) □

We further define the quantifier $\mathsf{U}i$ as follows: for any $\mathcal{L}_{\in,\mathbf{G},\mathbf{T}}$-formula $\varphi(i)$, $\mathsf{U}i\,\varphi(i)$ is the $\mathcal{L}_{\in,\mathbf{G},\mathbf{T}}$-formula

$$\exists n \ge 1\, \exists \alpha \in \mathtt{Ord}\, \forall i \in J\, \big(\varphi(i) \iff (\ulcorner\chi_n(x_\alpha, i)\urcorner \in \mathbf{T} \iff \tau(n,\alpha))\big),$$

whose meaning is that $\{i \in J : \varphi(i)\}$ coincides with a set of the form $B_n(\alpha)$.

Lemma 4.6.18 (ZFGT). *If a relation $P \subseteq J \times \mathbf{V}$ is formally $\mathcal{L}_{\in,\mathbf{G}}$-definable then so is the relation $\mathsf{U}i\, P(i,x)$.*

Proof. Suppose that $P(i,x)$ iff $\ulcorner\psi(\langle i,x\rangle, p')\urcorner \in \mathbf{T}$, where $\psi \in \mathtt{Form}_{\mathbf{G}}$ and $p' \in \mathbf{V}$ (a parameter). Let $\varphi(z)$ say

$$\exists i\, \exists x\, \exists p\, \exists y\, (z = \langle i, y\rangle \wedge y = \langle x, p\rangle \wedge \psi(\langle i,x\rangle, p)),$$

so that $\ulcorner\psi(\langle i,x\rangle,p)\urcorner \in \mathbf{T} \iff \ulcorner\varphi(\langle i,\langle x,p\rangle\rangle)\urcorner \in \mathbf{T}$. Then φ is χ_n for some $n \ge 1$. Let $\alpha(x,p)$ be the only $\alpha \in \mathtt{Ord}$ such that $x_\alpha = \langle x,p\rangle$. Then by definition $\mathsf{U}i\, P(i,x)$ iff $\alpha(x,p') \in W_n$, and the result follows from Corollary 4.6.9 because the class W_n is formally $\mathcal{L}_{\in,\mathbf{G}}$-definable by Lemma 4.6.16, the map $\langle x,p\rangle \mapsto \alpha(x,p)$ is just definable by a formula of $\mathcal{L}_{\in,\mathbf{G}}$, and so is the sequence $\{x_\alpha\}_{\alpha \in \mathtt{Ord}}$. □

---

[20] This does not imply the existence of a single $\varphi(n,\alpha) \in \mathtt{Form}_{\mathbf{G}}$ such that $\alpha \in W_n$ iff $\ulcorner\varphi(n,\alpha)\urcorner \in \mathbf{T}$. Actually, it is a good **exercise** to show that the binary relation $\alpha \in W_n$ is not formally $\mathcal{L}_{\in,\mathbf{G}}$-definable in the sense of Definition 4.6.6.

4.6e The interpretation

The ground standard structure of the interpretation will be the standard universe $\mathbf{V}$ of all sets in **ZFGT** as in §4.3c, but the construction of the interpretation will rather resemble the definable quotient power of §4.5d.

Following the definitions of §4.5d, but with the universe $\mathbf{V}$ instead of the set $\mathbf{V}_\vartheta$, we define $C = \mathbb{N} \times \mathbf{V}$, $I = C^{\mathtt{fin}}$, $C_p = \{p\} \times \mathbf{V}$, $C^s = s \times \mathbf{V}$, $I_p = {C_p}^{\mathtt{fin}}$, $I^s = (C^s)^{\mathtt{fin}}$ (here $p \in \mathbb{N}$ while $s \subseteq \mathbb{N}$ is finite). Note that these objects are $\in$-definable, and hence formally $\mathcal{L}_{\in,\mathbf{G}}$-definable proper classes.

For any $p \in \mathbb{N}$ let U_p be the image of the ultrafilter U defined in §4.6d under the map $x \mapsto \langle p, x\rangle$ ($p \in \mathbf{V}$); this is a I_p-adequate ultrafilter in the algebra of all formally $\mathcal{L}_{\in,\mathbf{G}}$-definable subclasses of I_p, and U_p preserves the formal $\mathcal{L}_{\in,\mathbf{G}}$-definability in $\mathbf{V}$ the same way as U itself does by Lemma 4.6.18. The notion of a (formally $\mathcal{L}_{\in,\mathbf{G}}$-definable) class $X \subseteq I$ of finite support, $\|X\|$ for any such class $X \subseteq I$, the Fubini product $U^s = U_{p_1} \times \cdots \times U_{p_n}$ for any finite set $s = \{p_1 < ... < p_n\} \subseteq \mathbb{N}$, and the product $\mathbf{U} = \bigtimes_{p \in \mathbb{N}} U_p$ are introduced as in §4.3b (with $P = \mathbb{N}$).

Finally, if $s \subseteq \mathbb{N}$ is finite then let $\mathscr{F}^s$ be the family of all formally $\mathcal{L}_{\in,\mathbf{G}}$-definable mappings (that is classes satisfying the definition of being a function) $f : I \to \mathbf{V}$ with finite support $\|f\| \subseteq s$ (this means $f(i) = f(i \restriction s)$ for all $i \in I$). Put $\mathscr{F} = \bigcup_{n \in \mathbb{N}} \mathscr{F}^n$.

Principal results of §4.3b remain true in this case, with obvious amendments, of course. In particular any U^s is a C^s-adequate ultrafilter in the algebra of all formally $\mathcal{L}_{\in,\mathbf{G}}$-definable classes $X \subseteq I^s$, $\mathbf{U}$ is a C-adequate filter (it consists of $\mathcal{L}_{\in,\mathbf{G}}$-definable subclasses of I), and the requirements $1^\circ - 4^\circ$ of §4.2b are formally satisfied with $\langle V ; \varepsilon\rangle = \langle \mathbf{V} ; \in\rangle$, $\mathbf{U}$, $\mathscr{F}$. (However differently from the proof of Lemma 4.5.11 the validation of 4° here involves the choice function $\mathbf{G}$, so that $f(i)$ is defined as the $\mathbf{G}$-choice in an appropriate set rather than the choice of a $<_{\mathbf{L}}$-minimal element.)

It follows that we can define the quotient power ${}^*\mathbf{v} = \langle {}^*\mathbf{V} ; {}^*\!\in, {}^*\mathtt{st}\rangle = \mathscr{F}/\mathbf{U}$ together with the natural embedding ${}^* : \mathbf{V} \to {}^*\mathbf{V}$ (see Definition 4.2.8).

Theorem 4.6.19. *The structure ${}^*\mathbf{v}$ is a standard core extension of $\mathbf{V}$ and a standard core interpretation of* **ISTGT** *in* **ZFGT**. *The map * is an elementary standard core $\in$-embedding of $\mathbf{V}$ into ${}^*\mathbf{V}$.*

Note that even the underlying domain $\mathscr{F}$ of this structure consists of proper (formally $\mathcal{L}_{\in,\mathbf{G}}$-definable) classes, and hence the quotient domain $\mathscr{F}/{}^*\!\!=$ consists of collections of classes. However there is an isomorphic structure with its domain consisting of sets. Indeed, for any $f \in \mathscr{F}$ we let $\varphi_f(v)$ be the $<_{\mathbf{G}}$-least coded formula in $\mathtt{Form}_{\mathbf{G}}$ such that the equivalence $f(i) = x \iff \ulcorner \varphi_f(\langle i, x\rangle)\urcorner \in \mathbf{T}$ holds for all $i \in I$ and $x \in \mathbf{V}$, and let $[f]'$ be the $<_{\mathbf{G}}$-least amongst all $\varphi_h \in \mathtt{Form}_{\mathbf{G}}$ where $h \in \mathscr{F}$ and $h \,{}^*\!\!= f$. We have $[f] = [g]$ iff $[f]' = [g]'$, but $[f]'$ is a set !

Exercise 4.6.20. Prove, following an argument in (B) in the proof of Theorem 4.5.12, that if $f \in \mathscr{F}$ then the class $\{x \in \mathbf{V} : {}^*x \; {}^*\!\in [f]\}$ is formally $\mathcal{L}_{\in,\mathbf{G}}$-definable. *Hint*: apply Lemma 4.6.18. □

Proof (Theorem 4.6.19). We concentrate on the verification of Inner Standardization and (†) of Definition 4.6.10 in ${}^*\mathbf{v}$, the rest of the proof is just the same as in Theorems 4.5.12 or 4.3.13 above.

Inner Standardization. It suffices to show that, for any set x, if a collection $Y \subseteq x$ is definable by a st-$\in$-formula relativized to ${}^*\mathbf{v}$ then Y is a set. Note that, by the construction of ${}^*\mathbf{v}$, any such a set Y is also definable by a formula of $\mathcal{L}_{\in,\mathbf{G},\mathbf{T}}$. However **ZFGT** contains Separation in $\mathcal{L}_{\in,\mathbf{G},\mathbf{T}}$.

Axiom 4.6.10(†). For every $i \in I$ define

$$\mathbf{g}(i) = \{\langle x, y\rangle : \langle 0, \langle x, y\rangle\rangle \in i \wedge \mathbf{G}(x) = y\}.$$

Then $\mathbf{g}$ is is formally $\mathcal{L}_{\in,\mathbf{G}}$-definable by Proposition 4.6.8(i) and $\|g\| = \{0\}$, therefore $g \in \mathscr{F}$. Thus ${}^*\mathbf{G} = [\mathbf{g}] \in {}^*\mathbf{V}$, and moreover, by the Łoś theorem, the following is true in ${}^*\mathbf{v}$: " ${}^*\mathbf{G}$ is a function, ${}^*x \in \mathtt{dom}\,{}^*\mathbf{G}$, and ${}^*\mathbf{G}({}^*x) = {}^*y$" whenever $x, y \in \mathbf{V}$ satisfy $y = \mathbf{G}(x)$. (Compare with Exercise 4.6.20.) In other words, it is true in ${}^*\mathbf{v}$ that $\mathbb{S} \subseteq \mathtt{dom}\,{}^*\mathbf{G}$ and ${}^*\mathbf{G}(x) \in x \cap \mathbb{S}$ for any standard $x \neq \varnothing$, as required by 4.6.10(†).

Check the second part of (†). Since $*$ is an isomorphism between $\mathbf{V}$ and the standard core of ${}^*\mathbf{v}$, it suffices to prove that for any $\varphi \in \mathtt{ClForm}_{\mathbf{G}}$ in $\mathbf{V}$ it is true in ${}^*\mathbf{v}$ that there is a truth set τ for $\langle \mathbb{S}; {}^*\mathbf{G}\rangle$ containing one of the coded formulas ${}^*\varphi$, $\neg\,{}^*\varphi$. Let φ be $\varphi(x_1, ..., x_n)$, all parameters $x_i \in \mathbf{V}$ indicated. Consider the parameter-free "matrix" $\varphi(v_1, ..., v_n)$ (all v_i being free variables). Let $\{\varphi_k(v_1, ..., v_{a(k)})\}_{k \leq K}$ be the list of all subformulas of φ (including φ itself) and their negations. We claim that

$$T = \{\ulcorner \varphi_k(x_1, ..., x_{a(k)})\urcorner \in \mathbf{T} : k \leq K \wedge x_1, ..., x_{a(k)} \text{ arbitrary sets}\}$$

is a formally $\mathcal{L}_{\in,\mathbf{G}}$-definable class.

Indeed by Proposition 4.6.8(iii) it suffices to show that any class

$$T' = \{\ulcorner \vartheta(x_1, ..., x_m)\urcorner \in \mathbf{T} : x_1, ..., x_m \text{ arbitrary sets}\}, \quad \vartheta \in \mathtt{Form}_{\mathbf{G}},$$

is formally $\mathcal{L}_{\in,\mathbf{G}}$-definable. However any $\psi \in \mathtt{ClForm}_{\mathbf{G}}$ belongs to T' iff

$$\exists x_1 \ldots \exists x_m \left(\psi \doteq \ulcorner \vartheta(x_1, ..., x_m)\urcorner \wedge \ulcorner \vartheta(x_1, ..., x_{m)})\urcorner \in \mathbf{T}\right), \qquad (\dagger)$$

where $\doteq$ is the graphical equality of coded formulas (actually the same as the equality of sets). We observe that the first term $\psi \doteq \ulcorner \vartheta(x_1, ..., x_m)\urcorner$ in the brackets can be considered as an $\in$-formula with free variables ψ, ϑ, and $\mathbf{x} = \langle x_1, ..., x_m\rangle$, therefore by the result of Exercise 4.6.2

$$\psi \doteq \ulcorner \vartheta(x_1, ..., x_m)\urcorner \quad \text{iff} \quad \ulcorner \psi \doteq \ulcorner \vartheta(x_1, ..., x_m)\urcorner\urcorner \in \mathbf{T}.$$

Thus, by the properties of $\mathbf{T}$ postulated, (†) is equivalent to

$$\ulcorner \exists x_1 ... \exists x_m \left(\psi \doteq \ulcorner \vartheta(x_1, ..., x_m)\urcorner \wedge \vartheta(x_1, ..., x_m)\right)\urcorner \in \mathbf{T}$$

— and hence this is equivalent to $\psi \in T'$. Let $\Phi(\psi)$ indicate the coded formula $\exists x_1 ... \exists x_m (\psi \doteq \ulcorner \vartheta(x_1, ..., x_m)\urcorner \wedge \vartheta(x_1, ..., x_m))$; it belongs to $\mathtt{Form}_{\mathbf{G}}$. Thus $T' = \{\psi \in \mathtt{Form}_{\mathbf{G}} : \ulcorner \Phi(\psi)\urcorner \in \mathbf{T}\}$, so that by definition T' is formally $\mathcal{L}_{\in,\mathbf{G}}$-definable, and hence so is T by the above.

It follows (see the construction of $\mathbf{g}$ and ${}^*\mathbf{G}$ above) that there is a map $\tau \in \mathscr{F}$ such that the element ${}^*T = [\tau] \in {}^*\mathbf{V}$ satisfies $T = \{\varphi : {}^*\varphi \;{}^*{\in}\; {}^*T\}$.

On the other hand, one easily proves that T satisfies in $\mathbf{V}$ conditions (1) – (6) of § 4.6c with $\forall^{\mathtt{st}}$ changed to $\forall$, G changed to $\mathbf{G}$, and $\mathbb{S} \subseteq \mathtt{dom}\, G$ dropped. It follows, since $*$ is the natural embedding, that ${}^*T$ satisfies $\mathtt{Truth}_G\, {}^*T$ in ${}^*\mathbf{v}$. Finally, by definition, one of φ, $\neg\,\varphi$ belongs to T, hence, one of the coded formulas ${}^*\varphi$, $\neg\, {}^*\varphi$ ${}^*{\in}$-belongs to ${}^*T$.

□ (*Theorems 4.6.19 and 4.6.12*)

□ (*The interpretability of* **IST** *in Theorem 4.1.10*)

4.6f Extendibility of standard models

We argue in **ZFC**. Recall the notion of extendibility (Definition 4.1.6).

Corollary 4.6.21 (ZFC). *Any transitive set M such that, for some $G : M \to M$ and $T \subseteq M$, $\langle M ; \in, G, T\rangle$ is a model of* **ZFGT**, *is* **IST**-*extendible.*

Proof. Apply Theorem 4.6.19 in $\langle M ; \in, G, T\rangle$. □

Unlike **BST**-extendibility, it is not the case that any countable transitive model of **ZFC** is **IST**-extendible. Indeed, if there exist transitive models of **ZFC** then there is a unique *minimal* transitive model M of **ZFC** : it has the form $M = \mathbf{L}_\mu$, where μ is the least ordinal such that $\mathbf{L}_\mu$ (the set of all Gödel-constructible sets that appear at a level earlier than μ in the construction of $\mathbf{L}$, the class of all constructible sets) is a model of **ZFC**.

Exercise 4.6.22. Prove that μ and M are countable, the axiom of constructibility "$\mathbf{V} = \mathbf{L}$" holds in M, and any $x \in M$ is $\in$-definable in M, that is there is an $\in$-formula $\varphi(\cdot)$ with one free variable such that x is the only element of M satisfying $M \models \varphi(x)$.

Hint. Show that the set M' of all $x \in M$ which are $\in$-definable in M coincides with M. It follows from "$\mathbf{V} = \mathbf{L}$" in M that there is a well-ordering $<_{\mathbf{L}}$ of M, $\in$-definable in M, and hence M' is an elementary submodel of M and a model of **ZFC** by the Löwenheim – Skolem argument. There is a transitive set T and an $\in$-isomorphism $\pi : M' \xrightarrow{\text{onto}} T$. Then T is also a model of **ZFC**, therefore $T = M$ by the minimality of M. It follows by $\in$-induction that π is the identity and $M' = M$. □

Theorem 4.6.23 (ZFC). *The minimal transitive model M of* **ZFC** *is not* **IST**-*extendible.*

It follows that if transitive models of **ZFC** exist then not all of them are **IST**-extendible. In fact an appropriate modification of the argument proves that if models of **ZFC** of any kind (not necessarily transitive) exist then not all of them are **IST**-extendible.

Proof. Let $\{\varphi_n(\cdot)\}_{n\in\mathbb{N}}$ be a recursive enumeration of all $\in$-formulas with one free variable and, for any n, ξ_n be the least ordinal $\xi < \omega_1^M$ satisfying $\varphi_n(\xi)$ in M — if such ordinals exist, otherwise $\xi_n = 0$. Then $\omega_1^M = \{\xi_n : n \in \mathbb{N}\}$ by 4.6.22. Thus the map $n \mapsto \xi_n$ does not belong to M. On the other hand, if $^*M$ is a model of **IST** whose standard core is M then, by Theorem 3.5.5, the map $n \to \xi_n$ is st-$\in$-definable in $^*M$, and hence by Inner Standardization in $^*M$ it is true in $^*M$ that there is a standard function f such that $f(n) = \xi_n$ for all $n \in \mathbb{N}$. But standard elements of $^*M$ belong to M, which yields a contradiction to the above. □

Corollary 4.6.24. **IST** *is not standard core interpretable in* **ZFC**.

Proof. Otherwise the minimal model of **ZFC** would be **IST**-extendible, contrary to the theorem. (This argument explicitly assumes the consistency of **ZFC** + the existence of transitive models of **ZFC**, but actually `Consis` **ZFC** is enough.) □

Problem 4.6.25. Which transitive models V of **ZFC** are **IST**-extendible ? Note that **ZFGT**-extendibility, *i.e.*, the existence of a map $G : V \to V$ and a set $T \subseteq V$ such that $\langle V ; \in, G, T\rangle$ is a model of **ZFGT**, is a sufficient condition of **IST**-extendibility by Theorem 4.6.19 (as well as a necessary and sufficient condition of **ISTGT**-extendibility). □

Historical and other notes to Chapter 4

Section 4.1. Our principal metamathematical definitions in §§ 4.1a–4.1c are generally speaking compatible with the common practice in this area, but details can be different of course.

Section 4.2. See [Hen 97, § 7] and [CK 92, 4.4] on enlargements and saturated structures in model theoretic nonstandard analysis and [CK 92, 3.1] on elementary chains of structures. The construction called quotient power in § 4.2b is based on model theoretic folklore, yet we were not able to find an explicit definition at such a level of generality.

See [CK 92] on the history of good ultrafilters.

We employ adequate ultrafilters to obtain nonstandard extensions of standard structures mainly because of their extremely transparent connection with the Idealization (or Saturation) property in ultrapowers that does not need any special knowledge of ultrafilters (see the proof of Theorem 4.2.17(i)). It is worth to remember that adequate ultrafilters in our sense form just a subpopulation within regular ultrafilters. (See 4.2.14, and also [CK 92, 4.3.4] on the equivalence: countably incomplete = $\aleph_0$-regular.)

Sections 4.3, 4.4. The proof of conservativity of **IST** was given in [Nel 77] ([Theorem 8.8) with a reference to William C. Powell. This proof utilizes an $\mathbb{N}$-long chain of consecutive ultrapower extensions ${}^0V \xrightarrow{e_0} {}^1V \xrightarrow{e_1} {}^2V \xrightarrow{e_2} \dots$, such that $V = {}^0V$ is a transitive set satisfying a given finite list of axioms of **ZFC** while each model ${}^{n+1}V$ is an ultrapower ${}^nV^{I_n}/U_n$ *adequate* in the sense that for any ultrafilter U on V there is an element $x \in {}^nV^{I_n}$ such that $U = \{X \subseteq V : [x] \in {}^*X\}$, where $[x] \in {}^*X$ means that the set $\{i \in I_n : x(i) \in X\}$ belongs to U_n. It is observed in [Nel 77] that to fulfill this property it suffices to define $I_n = \mathscr{P}_{\mathtt{fin}}(\mathscr{P}({}^nV))$ and take any ultrafilter U_n over I_n $\mathscr{P}({}^nV)$-adequate in the sense of our Definition 4.2.13.

It was observed in [Kan 91] that ${}^nV$-adequate ultrafilters lead to the same goal, that is, Idealization in the final structure at the step ω — this is essentially the chain construction outlined in § 4.3a, Example 2. The definition of adequate ultrafilter as in 4.2.13 was explicitly given in [KanR 95, Part 1].

Meanwhile Chang and Keisler [CK 92, §4.4] proved the conservativity of **IST** by means of sufficiently saturated extensions of transitive models V of finite fragments of **ZFC**, getting saturated extensions themselves as ultrapowers modulo good ultrafilters as in § 4.4a.

On iterated ultrapowers as in § 4.3b see [CK 92, 6.5].

The conservativity (hence, equiconsistency) and reducibility of **BST** with respect to **ZFC** was established in [Kan 91] (however the conservativity result can be abstracted from Hrbaček [Hr 78]). The interpretability of **BST** in **ZFGC**^{weak}, a conservative Global Choice extension of **ZFC**, was proved in our paper [KanR 95]. The construction of §§ 4.3c, 4.3d leading to the interpretation of **BST** in **ZFC** is a modification of a method introduced in [KanS 04] in order to obtain a definable saturated elementary extension of $\mathbb{R}$ (in **ZFC**).

The idea to use theories like **ZFC**ϑ instead of finite fragments of **ZFC** was first applied in the context of nonstandard set theories in [Hr 78] (see also the paper [Kan 91]).

Section 4.5. The non-reducibility of **IST** was established in [Kan 94a, Kan 94b]. In the context of models of **IST** the method of "definable" ultralimits was introduced in [Kan 91], but it had already been known for a while in studies of nonstandard models of Peano arithmetic. In a very general form see [CK 92, 6.4.30ff].

Section 4.6. That a minimal transitive model of **ZFC** is not **IST**-extendible, was demonstrated in [KanR 95, Part 1]. Exercise 4.6.22, the key part of the argument, is taken from Cohen [Coh 66, § III.6]. Standard theories with a truth predicate were employed in proofs of **IST**-extendibility (as in second part of Corollary 4.6.21) in our paper [KanR 99c]. The proof of Theorem 4.6.12 is a modification of an argument in [KanR 99c, KanR 00a].

See [Sh 94] on different aspects of the truth predicate in proof theory, and [FS 87] on the truth predicate for Peano arithmetic.

5 Definable external sets and metamathematics of HST

Metamathematical studies of nonstandard theories continue in this Chapter with the aim to prove the main metamathematical properties of **HST** including its standard core interpretability in **ZFC** (Theorem 1.1.14) and internal core interpretability in **BST** (essentially, Theorem 3.1.10) and consequences related to conservativity *etc.* Section 5.1 introduces all necessary notation and presents the main results (Theorem 5.1.4 and Corollary 5.1.5).

The interpretation of **HST** in **BST** involves an intermediate theory **EEST**, elementary external set theory, essentially a theory of "external sets" of § 3.2f. Accordingly "external sets" (more exactly, their codes) form an interpretation of **EEST** in **BST** defined in 5.2. Unlike **BST** such nonstandard set theories as **EEST** and **HST** do provide the existence of "external sets" as sets — then we call them *elementary external sets* — moreover this category $\mathbb{E}$ turns out to be the same as the class $\boldsymbol{\Delta}_2^{\mathtt{ss}}$ briefly considered in § 1.4a.

A natural way to obtain more complicated external sets in **HST** is to apply more complicated operations to internal sets. Pursuing this plan, we define in Section 5.5 a much bigger class $\mathbb{L}[\mathbb{I}]$ of sets *constructible over* $\mathbb{I}$. This is based on another idea: iterated assemblage of sets, beginning with internal sets. The construction (considered in Section 5.3), known as *decoration* in graph theory, applies to a well-founded tree decorated by internal sets, such that both the tree and the decoration are elementary external sets, that is, sets $\mathtt{st}$-$\in$-definable in $\mathbb{I}$. For instance, elementary external sets can be obtained, in the framework of this construction, with trees only of height 1 (with a single branching point). Due to some implicit affinities with Gödel's constructibility in **ZFC**, $\mathbb{L}[\mathbb{I}]$ will be the smallest transitive class containing all internal sets and satisfying all axioms of **HST**. As in the case of constructible sets in **ZFC**, the class $\mathbb{L}[\mathbb{I}]$ appears to be essentially more definite than a general **HST** universe: for instance internal sets of different internal cardinalities remain non-equinumerous in the external sense in $\mathbb{L}[\mathbb{I}]$.

The ideas connected with $\mathbb{L}[\mathbb{I}]$ play the key role in Section 5.4 in the construction of an interpretation of **HST** in **EEST**. This interpretation, obviously the most complicated part of our metamathematical program in this Chapter, will be formed essentially by codes of sets in $\mathbb{L}[\mathbb{I}]$ in the **EEST** universe.

5.1 Introduction to metamathematics of HST

To properly formulate some results related to the metamathematics of **HST** in § 5.1b, we begin in § 5.1a with a convenient notation that describes relations between nonstandard structures and theories in a way similar to §§ 4.1a, 4.1b. A theory **EEST**, *elementary external set theory*, intermediate between **BST** and **HST**, will be introduced in § 5.1b.

5.1a Internal core embeddings and interpretability

The reader is recommended to refresh the content of §§ 4.1a, 4.1b. The interpretations of external set theories in **BST** considered below will be mainly invariant $\mathtt{st}$-$\in$-structures that likely do not admit reductions to structures with true equality. This is why the following definitions are mostly oriented to invariant structures rather than to those with true equality as in Chapter 4.

For any invariant $\mathtt{st}$-$\in$-structure $\mathbf{v} = \langle V\,;\,\varepsilon, \mathtt{st}\,;\,\equiv\rangle$ we define

$$\mathbb{I}^{(\mathbf{v})} = \{x \in V : (\mathtt{int}\,x)^{\mathbf{v}}\} = \{x \in V : \exists y \in V\,(\mathtt{st}\,y \wedge x \,\varepsilon\, y)\}\,,$$

the *internal core* of $\mathbf{v}$. (This is applicable for structures with true equality as well.) Recall that $\mathtt{int}\,x$ is the $\mathtt{st}$-$\in$-formula $\exists^{\mathtt{st}} y\,(x \in y)$. For instance if a structure $\mathbf{v} = \langle V\,;\,\varepsilon, \mathtt{st}\,;\,\equiv\rangle$ interprets **BST** then obviously $\mathbb{I}^{(\mathbf{v})} = V$ but if $\mathbf{v}$ interprets **HST** then $\mathbb{I}^{(\mathbf{v})} \subsetneqq V$.

Suppose that $\mathbf{v}_1 = \langle V_1\,;\,\varepsilon_1, \mathtt{st}_1\rangle$ and $\mathbf{v}_2 = \langle V_2\,;\,\varepsilon_2, \mathtt{st}_2\,;\,\equiv_2\rangle$ are $\mathtt{st}$-$\in$-structures whose universes V_1 and V_2 can be both sets and proper classes. (The first one is assumed to be a structure with true equality.) By an *internal core embedding* of $\mathbb{I}^{(\mathbf{v}_1)}$ in $\mathbf{v}_2$ we understand any map $\star : \mathbb{I}^{(\mathbf{v}_1)} \to V_2$ satisfying the two following properties: [1]

(1) $\star$ is a $\mathtt{st}$-$\in$-*embedding*, i.e. for all $x,\, y \in \mathbb{I}^{(\mathbf{v}_1)}$:

$$x = y \quad \text{iff} \quad {}^{\star}x \equiv_2 {}^{\star}y\,;$$
$$\mathtt{st}_1\, x \quad \text{iff} \quad \mathtt{st}_2\, {}^{\star}x\,;$$
$$x \,\varepsilon_1\, y \quad \text{iff} \quad {}^{\star}x \,\varepsilon_2\, {}^{\star}y\,.$$

(2) $\mathbb{I}^{(\mathbf{v}_2)} = \{z \in V_2 : \exists x \in \mathbb{I}^{(\mathbf{v}_1)}\,(z \equiv_2 {}^{\star}x)\}$.

In this case, $\mathbf{v}_2$ is said to be an *internal core extension of* $\mathbf{v}_1$ *via* $\star$.

It will be a typical case below that the $\equiv_2$-class $[x]_{\equiv_2} = \{y \in V_2 : x \equiv_2 y\}$ of any $x \in \mathbb{I}^{(\mathbf{v}_2)}$ is equal to $\{x\}$ — then the restriction $\equiv_2 \restriction \mathbb{I}^{(\mathbf{v}_2)}$ is the equality on $\mathbb{I}^{(\mathbf{v}_2)}$ and hence any internal core embedding is just a $\mathtt{st}$-$\in$-isomorphism of the internal core $\langle \mathbb{I}^{(\mathbf{v}_1)}\,;\,\varepsilon_1, \mathtt{st}_1\rangle$ of $\mathbf{v}_1$ onto the internal core $\langle \mathbb{I}^{(\mathbf{v}_2)}\,;\,\varepsilon_2, \mathtt{st}_2\rangle$ of $\mathbf{v}_2$. But even in the general case the map $r : \mathbb{I}^{(\mathbf{v}_2)} \to \mathbb{I}^{(\mathbf{v}_1)}$ sending any $z \in \mathbb{I}^{(\mathbf{v}_2)}$ to the only $r(z) \in \mathbb{I}^{(\mathbf{v}_1)}$ such that $z \equiv_2 {}^{\star}(r(z))$ is

[1] $\star$ here is a generic symbol, chosen to make a destinction from the sterisk $*$ typically used to denote embeddings of standard structures into nonstandard ones as in the last Chapter.

obviously a reduction of the invariant structure $\langle \mathbb{I}^{(\mathbf{v}_2)} ; \varepsilon_2, \mathtt{st}_2 ; \equiv_2 \rangle$ to the structure $\langle \mathbb{I}^{(\mathbf{v}_1)} ; \varepsilon_1, \mathtt{st}_1 \rangle$ with true equality (in the sense of Definition 1.5.9). Therefore Proposition 1.5.10 implies:

Proposition 5.1.1. *If* $\star$ *is an internal core embedding of* $\mathbb{I}^{(\mathbf{v}_1)}$ *in* $\mathbf{v}_2$ *then for any* st-$\in$-*formula* $\Phi(x_1, ..., x_n)$ *and any* $x_1, ..., x_n \in \mathbb{I}^{(\mathbf{v}_1)}$ *we have*

$$\Phi(x_1, ..., x_n)^{\langle \mathbb{I}^{(\mathbf{v}_1)} ; \varepsilon_1, \mathtt{st}_1 \rangle} \iff \Phi({}^\star x_1, ..., {}^\star x_n)^{\langle \mathbb{I}^{(\mathbf{v}_2)} ; \varepsilon_2, \mathtt{st}_2 ; \equiv_2 \rangle}. \qquad \square$$

Thus being an internal core extension simply means that the internal core of the extended st-$\in$-structure admits a reduction onto the the internal core of the original st-$\in$-structure, and in fact in the applications the internal cores mostly will be just st-$\in$-isomorphic. Nothing is said regarding the non-internal parts. In fact the word "extension" is not really suitable here, but we would like to preserve some analogy with standard core extensions.

Definition 5.1.2. Suppose that $\mathfrak{T}$, $\mathfrak{U}$ are theories in the st-$\in$-language.

(1) $\mathfrak{T}$ is an *internal core extension* of $\mathfrak{U}$ if for any st-$\in$-sentence Φ, if $\mathfrak{U}$ proves the relativized sentence $\Phi^{\mathtt{int}}$ then $\mathfrak{T}$ proves $\Phi^{\mathtt{int}}$,
$\mathfrak{T}$ is a *conservative internal core extension* of $\mathfrak{U}$ if for any st-$\in$-sentence Φ, $\mathfrak{U}$ proves $\Phi^{\mathtt{int}}$ iff $\mathfrak{T}$ proves $\Phi^{\mathtt{int}}$. Note that if $\mathfrak{U}$ is **BST**, where all sets are internal, then the condition can be re-written as follows: $\mathfrak{U}$ proves Φ iff $\mathfrak{T}$ proves $\Phi^{\mathtt{int}}$.

(2) $\mathfrak{T}$ is *internal core interpretable* in $\mathfrak{U}$ if there exist:
   a) an interpretation ${}^\star\mathbf{v} = \langle {}^\star\mathbf{V} ; {}^\star\!\in, {}^\star\mathtt{st} ; \equiv \rangle$ of $\mathfrak{T}$ in $\mathfrak{U}$, where ${}^\star\!\in$ and ${}^\star\mathtt{st}$ interpret the atomic predicates $\in$ and st, and
   b) an internal core embedding $\star : \mathbb{I}^{(\mathbf{v})} \to {}^\star\mathbf{v}$, where $\mathbf{v} = \langle \mathbf{V} ; \in, \mathtt{st} \rangle$ is the set universe of the theory $\mathfrak{U}$. [2]

Such an interpretation is called: an *internal core* interpretation. $\square$

5.1b Metamathematics of HST : an overview

The main goal of our study of metamathematical properties of **HST** is the standard core interpretability of **HST** in **ZFC** (the key part of Theorem 1.1.14). The task is divided in three parts, that is, three interpretability claims, and accordingly with two intermediate theories involved.

One of the intermediate theories is **BST**, and one of the interpretability claims is the standard core interpretability of **BST** in **ZFC** already established by Theorem 4.1.10(i). The other intermediate theory is **EEST**, essentially a theory of "external sets" in the sense of § 3.2f, introduced by the following definition:

---

[2] As usual, both the interpretation and the embedding must be defined by formulas of the language of $\mathfrak{U}$, and their indicated properties provable in $\mathfrak{U}$.

Definition 5.1.3. *Elementary external set theory* **EEST** is a theory in the st-$\in$-language containing the axioms of Extensionality, Union, Infinity in their **HST** forms, Separation in the st-$\in$-language, $\mathbf{BST}^{\mathtt{int}}$, *i. e.* all axioms of **BST** formally relativized to the class $\mathbb{I} = \{x : \mathtt{int}\, x\}$ of all internal sets, Transitivity of $\mathbb{I}$, and finally

Parametrization: $\forall C\, \exists^{\mathtt{int}} p\, (C = \mathsf{E}_p)$. (See Definition 3.2.14 on E_p.) □

Internal sets are understood in **EEST** exactly as in **HST**, that is as elements of standard sets, accordingly, $\mathtt{int}\, x$ is still the formula $\exists^{\mathtt{st}} y\, (x \in y)$.

Parametrization may be seen as a rather artificial statement saying that all sets can be presented in a certain special form. In particular, since we also have Separation in the st-$\in$-language, Parametrization implies that every st-$\in$-definable part of any internal set has the form E_p, that is, it is st-$\in$-definable in a very special way. This is a rare property, indeed theories of set theoretic type usually satisfy some form of hierarchy theorem, that is, no single formula can replace all formulas in matters of definability. It is a very special property of **BST** that such a single formula based on the the parametrization E_p does exist by Theorem 3.2.16.

It follows from the axioms of Parametrization and Transitivity of $\mathbb{I}$ that the universe of **EEST** consists of internal sets and non-internal sets of the form E_p, $p \in \mathbb{I}$, in particular any set in **EEST** contains only internal elements. See §§ 5.2c, 5.2d on more theorems on the structure of the **EEST** universe.

The diagram **ZFC** $\to$ **BST** $\to$ **EEST** $\to$ **HST** describes the relation of natural interpretability between these four theories, in the sense that by trivial reasons the class $\mathbb{S}$ (standard sets) interprets **ZFC** in **BST**, the class $\mathbb{I}$ (internal sets) interprets **BST** in **EEST**, and the class $\mathbb{E} = \mathbf{\Delta}_2^{\mathtt{ss}}$ (elementary external sets, see § 5.2b) interprets **EEST** in **HST**.

The next theorem presents interpretability in the opposite direction.

Theorem 5.1.4. 1°. **BST** *is standard core interpretable in* **ZFC**.

2°. **EEST** *is internal core interpretable in* **BST**.

3°. **HST** *is internal core interpretable in* **EEST**.

Assertion 1° of this theorem has been proved: see Theorem 4.1.10(i). We formulate it here once again only to present the picture in full integrity.

We prove in § 5.2a that **EEST** admits an internal core interpretation in **BST** by means of "external sets" of § 3.2f, and this will give us 2°.

Assertion 3° of the theorem is the most complicated one. An internal core interpretation of **HST** in **EEST**, based on an assembling construction of external sets, will be defined in Section 5.4. A preliminary study of the assembling construction is carried out in Section 5.3. Another application of this construction, leading to the class $\mathbb{L}[\mathbb{I}]$ of all sets constructible from internal sets, will be given in Section 5.5.

The following corollary contains additional related results.

Corollary 5.1.5. 4°. **HST** *is internal core interpretable in* **BST**.

5°. **EEST** *and* **HST** *are standard core interpretable in* **ZFC** *and hence are "realistic" theories* (in the sense of Definition 4.1.8).

6°. **EEST** *is reducible to* **BST** *and to* **ZFC** *in the sense that for any* st-$\in$-*sentence* Φ *there exist a* st-$\in$-*sentence* φ *and an* $\in$-*sentence* ψ *such that* **EEST** *proves* $\Phi \Longleftrightarrow \varphi^{\mathtt{int}} \Longleftrightarrow \psi^{\mathtt{st}}$.

7°. **HST** *and* **EEST** *are conservative internal core extensions of* **BST**, *that is for any* st-$\in$-*formula* Φ, **BST** *proves* Φ *iff* **EEST** *proves* $\Phi^{\mathtt{int}}$ *iff* **HST** *proves* $\Phi^{\mathtt{int}}$.

8°. **HST** *and* **EEST** *are conservative standard core extensions of* **ZFC**, *that is for any* $\in$-*formula* Φ, **ZFC** *proves* Φ *iff* **EEST** *proves* $\Phi^{\mathtt{st}}$ *iff* **HST** *proves* $\Phi^{\mathtt{wf}}$ *iff* **HST** *proves* $\Phi^{\mathtt{st}}$.

The proof of the corollary is organized as follows.

Claims 4° and 5° will be proved in § 5.4f by means of suitable superpositions of the interpretations involved in 1°, 2°, 3° of Theorem 5.1.4.

An easy proof of 6° is given in § 5.2b.

The conservativity over **BST** in 7° follows from the interpretability given by 2° of Theorem 5.1.4 (for **EEST**) and 4° of Corollary 5.1.5 (for **HST**) (see for instance the proof of Proposition 4.1.9).

Finally, 8°. The last "iff" here follows because the classes $\mathbb{S}$ and $\mathbb{WF}$ are $\in$-isomorphic in **HST**. As for the first "iff", given an $\in$-formula Φ, **ZFC** proves Φ iff **BST** proves $\Phi^{\mathtt{st}}$ (because **BST** is conservative over **ZFC** by Theorem 4.1.10) iff **HST** proves $(\Phi^{\mathtt{st}})^{\mathtt{int}}$ – by 7°. Note, finally, that the formula $(\Phi^{\mathtt{st}})^{\mathtt{int}}$ coincides with $\Phi^{\mathtt{st}}$. ⊣

The **HST** parts of Theorem 5.1.4 and Corollary 5.1.5 are obviously sufficient to imply Theorems 1.1.14 and 3.1.10.

□ (*Thms 1.1.14, 3.1.10 modulo Thm 5.1.4 and Corollary 5.1.5*)

It remains to comment upon the reducibility issue with respect to **HST**. We know that **BST** is reducible to **ZFC** by Theorem 4.1.10 while **EEST** is reducible to **BST** and hence to **ZFC** by 6° of Corollary 5.1.5. A natural question whether **HST** is reducible in a similar sense is answered in the negative: the axiomatics of **HST** does not imply that strict control over the structure of its whole universe from the side of its standard (or even internal or elementary external) core as the axiomatics of **BST** and **EEST** do. With respect to this property **HST** is similar to **IST** (see the non-reducibility claim in Theorem 4.1.10) but the non-reducibility of **HST** will be established by different methods in Chapter 7 (Corollary 7.2.3).

However a certain reasonable extension of **HST** is reducible, see Exercise 5.5.10.

5.2 From internal to elementary external sets

In this section we concentrate on the relationships between the theories **BST** and **EEST**. To prove 2° of Theorem 5.1.4 we show in § 5.2a that "external sets" (introduced in § 3.2f) form an internal core interpretation of **EEST** in **BST**. Then we study "external sets" in § 5.2b from the point of view of the theories **EEST** and **HST** where they turn out to be legitimate sets — this is why "external sets" will be renamed to *elementary external* sets. The section ends with some basic theorems of **EEST** in §§ 5.2c, 5.2d.

5.2a Interpretation of EEST in BST

By definition "external sets" in **BST** are st-∈-definable subclasses of sets, that is, collections of the form $\{x \in Z : \varphi(x)\}$, where Z is a set (internal) while φ is a st-∈-formula with arbitrary (internal) parameters. "External sets" are generally speaking not sets in **BST** (Exercise 3.2.15). However they admit a common description in the form E_p by Theorem 3.2.16, and this enables us to legitimately study them in **BST**. In particular we prove here that in a certain exact sense "external sets" satisfy the axioms of **EEST**.

We argue in **BST**.

The reader should recall basic definitions related to coding in § 3.2f.

Definition 5.2.1. Let **E** be the class of all E-codes $p \in \mathbb{D}$ such that either E_p is not an (internal) set or $p = {}^{\mathsf{e}}x$ for some (internal) x. □

Thus by definition $\mathbf{E} \subseteq \mathbb{D}$. The reservation that E_p is not internal unless $p = {}^{\mathsf{e}}x$ for some x provides uniqueness of coding of internal sets.

Definition 5.2.2. We define the following st-∈-formulas, which express the basic relations of the st-∈-language, *i.e.*, =, ∈, and st, between coded "external sets" in $\mathbb{E}$ in terms of codes; $p \in \mathbf{E}$ is to be understood as the st-∈-formula saying "p is a function and $\exists^{\mathtt{st}} A\, \exists^{\mathtt{st}} B\, (\operatorname{dom} p = A \times B)$".

$$\begin{array}{lll} x \,\varepsilon\, p & \text{is the formula} & p \in \mathbf{E} \wedge \exists^{\mathtt{st}} a \in A\, \forall^{\mathtt{st}} b \in B\, (x \in p(a,b)), \\ & & \quad\text{where } A, B \in \mathbb{S} \text{ satisfy } \operatorname{dom} p = A \times B; \\ p \;{}^{\mathsf{e}}\!\!= q & \text{is the formula} & p, q \in \mathbf{E} \wedge \forall x\, (x \,\varepsilon\, p \Longleftrightarrow x \,\varepsilon\, q)\,; \\ p \;{}^{\mathsf{e}}\!\!\in q & \text{is the formula} & p, q \in \mathbf{E} \wedge \exists x\, (p = {}^{\mathsf{e}}x \wedge x \,\varepsilon\, q)\,; \\ {}^{\mathsf{e}}\mathtt{st}\, p & \text{is the formula} & \exists^{\mathtt{st}} x\, (p = {}^{\mathsf{e}}x) \quad (\text{this implies } p \in \mathbf{E})\,; \\ {}^{\mathsf{e}}\mathtt{int}\, p & \text{is the formula} & \exists x\, (p = {}^{\mathsf{e}}x) \quad (\text{this implies } p \in \mathbf{E})\,. \end{array}$$ □

Definition 5.2.3. Define $\mathbf{e} = \langle \mathbf{E}; {}^{\mathsf{e}}\!\!\in, {}^{\mathsf{e}}\mathtt{st}; {}^{\mathsf{e}}\!\!= \rangle$. The invariance claim in the next lemma implies that $\mathbf{e}$ is an invariant structure. □

Lemma 5.2.4. ${}^{\mathsf{e}}\!\!=$ *is an equivalence on* **E** *while* ${}^{\mathsf{e}}\!\!\in$, ${}^{\mathsf{e}}\mathtt{st}$, ${}^{\mathsf{e}}\mathtt{int}$ *are* ${}^{\mathsf{e}}\!\!=$*-invariant relations. In addition for any (internal)* x *and* $p, q \in \mathbf{E}$ *we have:*

(i) $\mathsf{E}_{{}^{e}x} = x$ and $x \mathrel{\varepsilon} p \iff x \in \mathsf{E}_p \iff {}^{e}x \mathrel{{}^{e}\!\in} p$.

(ii) $p \mathrel{{}^{e}\!=} q \iff \mathsf{E}_p = \mathsf{E}_q$ and ${}^{e}x \mathrel{{}^{e}\!=} p \iff x = \mathsf{E}_p \iff {}^{e}x = p$.

(iii) $p \mathrel{{}^{e}\!\in} q \iff \mathsf{E}_p \in \mathsf{E}_q \iff \exists x\,({}^{e}x = p \wedge x \in \mathsf{E}_q)$.

(iv) ${}^{e}\mathtt{st}\, p \iff \exists^{\mathtt{st}} x\,(x = \mathsf{E}_p)$ *and* ${}^{e}\mathtt{int}\, p \iff \exists x\,(x = \mathsf{E}_p)$.

Proof. (i), (ii), (iii), (iv) immediately follow from definitions. To prove the invariance of ${}^{e}\!\in$ suppose that $p \mathrel{{}^{e}\!\in} q$ and $p \mathrel{{}^{e}\!=} p'$, $q \mathrel{{}^{e}\!=} q'$. Then $p = {}^{e}x$ for some (internal) x by (iii). Thus $p' \mathrel{{}^{e}\!=} {}^{e}x$, and hence $p' = {}^{e}x$ as well by (ii). Now $p = {}^{e}x \mathrel{{}^{e}\!\in} q$ implies $x \in \mathsf{E}_q$ by (i). Yet $\mathsf{E}_q = \mathsf{E}_{q'}$ follows from $q \mathrel{{}^{e}\!=} q'$ by (ii). Thus $x \in \mathsf{E}_{q'}$, and hence $p' = {}^{e}x \mathrel{{}^{e}\!\in} q'$ still by (i). □

Note that the universe $\mathbf{E}/{}^{e}\!=$ of the *quotient structure* $\overline{\mathbf{e}} = \mathbf{e}/({}^{e}\!=) = \langle \mathbf{E}/{}^{e}\!= ; {}^{e}\!\in, {}^{e}\mathtt{st}\rangle$ (see § 1.5c) consists of ${}^{e}\!=$*-classes* $[p]^{e} = \{q \in \mathbf{E} : p \mathrel{{}^{e}\!=} q\}$ of elements $p \in \mathbf{E}$, that is, generally speaking, proper classes. Theorem 1.5.11 does not help to reduce $\overline{\mathbf{e}}$ to a structure with true equality because **BST**, our ground set theory here, does not have Separation in the $\mathtt{st}$-$\in$-language.

Problem 5.2.5. Prove that $\mathbf{e}$ does not admit reduction in **BST** to a structure with true equality having (internal) sets as its elements. □

For any $\mathtt{st}$-$\in$-formula Φ let ${}^{e}\Phi$ be its *relativization* to $\mathbf{e}$ ($\Phi^{\mathbf{e}}$ in the notation of § 1.5c), obtained as follows:

(A) change all occurrences $p = q$, $p \in q$, $\mathtt{st}\, p$ to resp. $p \mathrel{{}^{e}\!=} q$, $p \mathrel{{}^{e}\!\in} q$, ${}^{e}\mathtt{st}\, p$;

(B) relativize all quantifiers to $\mathbf{E}$.

Theorem 5.2.6. (i) *The structure* $\mathbf{e} = \langle \mathbf{E}; {}^{e}\!\in, {}^{e}\mathtt{st}; {}^{e}\!=\rangle$ *is an internal core interpretation of* **EEST** *in* **BST**.

(ii) (**BST**) *The mapping* $x \mapsto {}^{e}x$ *is an internal core embedding of the universe* $\langle \mathbb{I}; \in, \mathtt{st}\rangle$ *into* $\mathbf{e}$, *moreover, a* $\mathtt{st}$-$\in$*-isomorphism of* $\langle \mathbb{I}; \in, \mathtt{st}\rangle$ *onto the internal core* $\langle \mathbb{I}^{(\mathbf{e})}; {}^{e}\!\in, {}^{e}\mathtt{st}\rangle$ *of the structure* $\mathbf{e}$.

(iii) (**BST**) *We have* $\Phi(x_1, ..., x_n) \iff {}^{e}(\Phi({}^{e}x_1, ..., {}^{e}x_n)^{\mathtt{int}})$ *for any* $\mathtt{st}$-$\in$-*formula* Φ *and any (internal) sets* $x_1, ..., x_n$.

Proof. (ii) First of all we claim that

$$\mathbb{S}^{(\mathbf{e})} = \{{}^{e}x : x \in \mathbb{S}\} \quad \text{and} \quad \mathbb{I}^{(\mathbf{e})} = \{p \in \mathbf{E} : {}^{e}\mathtt{int}\, p\} = \{{}^{e}x : x \in \mathbb{I}\}$$

(the standard core of $\mathbf{e}$, § 4.1a and the internal core of $\mathbf{e}$, § 5.1a; $\mathbb{I}$ is the universe of all sets in **BST**). Indeed by definition $\mathbb{S}^{(\mathbf{e})} = \{p \in \mathbf{E} : {}^{e}(\mathtt{st}\, p)\}$ and still by definition ${}^{e}(\mathtt{st}\, p)$ is ${}^{e}\mathtt{st}\, p$, and this is equivalent to $\exists^{\mathtt{st}} x\,(p = {}^{e}x)$, as required. The proof for $\mathbb{I}^{(\mathbf{e})}$ is equally simple.

The equivalences $x = y \iff {}^{e}x \mathrel{{}^{e}\!=} {}^{e}y$ and $x \in y \iff {}^{e}x \mathrel{{}^{e}\!\in} {}^{e}y$ for all $x, y \in \mathbb{I}$ immediately follow from Lemma 5.2.4. We conclude that the map $x \mapsto {}^{e}x$ is an internal core embedding of $\langle \mathbb{I}; \in, \mathtt{st}\rangle$ in $\mathbf{e}$, in this case actually a $\mathtt{st}$-$\in$-isomorphism of $\langle \mathbb{I}; \in, \mathtt{st}\rangle$ onto $\langle \mathbb{I}^{(\mathbf{e})}; {}^{e}\!\in, {}^{e}\mathtt{st}\rangle$.

(iii) This follows from (ii) and Proposition 5.1.1 because in the case considered the formula $\Phi(x_1, ..., x_n)$ is the same as $\Phi(x_1, ..., x_n)^{\langle \mathbb{I}; \in, \mathtt{st}\rangle}$ while ${}^{\mathsf{e}}(\Phi({}^{\mathsf{e}}x_1, ..., {}^{\mathsf{e}}x_n)^{\mathtt{int}})$ is the same as $\Phi({}^{\mathsf{e}}x_1, ..., {}^{\mathsf{e}}x_n)^{\langle \mathbb{I}^{(e)}; {}^{\mathsf{e}}\in, {}^{\mathsf{e}}\mathtt{st}\rangle}$.

(i) We have to prove ${}^{\mathsf{e}}\Phi$ in **BST** for any axiom Φ of **EEST**.

Extensionality. Suppose $\forall r \in \mathbf{E}\,(r \;{}^{\mathsf{e}}\!\!\in p \iff r \;{}^{\mathsf{e}}\!\!\in q)$ where $p, q \in \mathbf{E}$. In particular ${}^{\mathsf{e}}x \;{}^{\mathsf{e}}\!\!\in p \iff {}^{\mathsf{e}}x \;{}^{\mathsf{e}}\!\!\in q$ for any x. However ${}^{\mathsf{e}}x \;{}^{\mathsf{e}}\!\!\in p$ means $x \,\varepsilon\, p$ by Lemma 5.2.4(i), and hence we have $\forall x\,(x \,\varepsilon\, p \iff x \,\varepsilon\, q)$, that is $p \;{}^{\mathsf{e}}\!\!= q$.

Transitivity of $\mathbb{I}$. It suffices to show that, for $p, q \in \mathbf{E}$, if $p \;{}^{\mathsf{e}}\!\!\in q$ then ${}^{\mathsf{e}}\mathtt{int}\,p$. However $p \;{}^{\mathsf{e}}\!\!\in q$ by definition implies that $p \;{}^{\mathsf{e}}\!\!= {}^{\mathsf{e}}x$ for some $x \,\varepsilon\, q$. Now $p = {}^{\mathsf{e}}x$ and ${}^{\mathsf{e}}\mathtt{int}\,p$ follow from Lemma 5.2.4(ii).

To prove Separation in $\mathbf{e}$ suppose that $P \in \mathbf{E}$ and $\Phi(x)$ is a st-$\in$-formula with parameters in $\mathbf{E}$; we have to find $Q \in \mathbf{E}$ such that

$$\forall x\,\big(x \;{}^{\mathsf{e}}\!\!\in Q \iff (x \;{}^{\mathsf{e}}\!\!\in P \wedge {}^{\mathsf{e}}\Phi(x))\big).$$

By Theorem 3.2.16 it suffices to show that $X = \{x : x \;{}^{\mathsf{e}}\!\!\in P \wedge {}^{\mathsf{e}}\Phi(x)\}$ is an "external set." The st-$\in$-definability of X (with the parameter P and the parameters involved in Φ) is obvious, thus it remains to find a set Y (internal) with $X \subseteq Y$. Clearly $Y = \mathtt{ran}\,P$ is as required because $\mathsf{E}_P \subseteq Y$.

Union easily follows from Separation. To prove ${}^{\mathsf{e}}(\mathbf{BST}^{\mathtt{int}})$ apply (iii).

Parametrization holds by the construction of $\mathbf{e}$. Indeed consider any $p \in \mathbf{E}$. Let $q = {}^{\mathsf{e}}p$, then $q \in \mathbf{E}$ and $p = \mathsf{E}_q$. Formally, $q \in \mathbf{E} \wedge p = \mathsf{E}_q$. Thus we have ${}^{\mathsf{e}}((q \in \mathbf{E} \wedge p = \mathsf{E}_q)^{\mathtt{int}})$ by (iii). However the formula $q \in \mathbf{E} \wedge p = \mathsf{E}_q$ is absolute for the internal universe: apply the Transitivity of $\mathbb{I}$ and ${}^{\mathsf{e}}(\mathbf{BST}^{\mathtt{int}})$. Thus finally ${}^{\mathsf{e}}(q \in \mathbf{E} \wedge p = \mathsf{E}_q)$ as required. □

□ (*2° of Theorem 5.1.4*)

5.2b Elementary external sets in external theories

By definition, both in **HST** and in **EEST**, the class $\mathbb{I}$ of all internal sets (that is elements of standard sets) or, to be more exact, the structure $\langle \mathbb{I}; \in, \mathtt{st}\rangle$, satisfies **BST** (for **HST** by Theorem 3.1.8). In addition $\mathbb{I}$ is transitive by the axiom of Transitivity of $\mathbb{I}$.

Definition 5.2.7 (HST or EEST). Define E_p for any internal p as in Definition 3.2.14, and $\mathbf{E}$ as in Definition 5.2.1.

Define the formulas ε, ${}^{\mathsf{e}}\!\!=$, ${}^{\mathsf{e}}\!\!\in$, ${}^{\mathsf{e}}\mathtt{st}$, ${}^{\mathsf{e}}\mathtt{int}$ as in Definition 5.2.2.

An *elementary external* set is any set of the form E_p, $p \in \mathbf{E}$.

$\mathbb{E} = \{\mathsf{E}_p : p \in \mathbf{E}\}$, the class of all elementary external sets. □

Lemma 5.2.8 (HST or EEST). *If* $p \in \mathbf{E}$ *then* E_p *is a set.*

Proof. Apply Separation in the st-$\in$-language. □

Thus differently from **BST** all "external sets" are true sets in **HST** and in **EEST**! This enables us to change notation in external theories from "external sets" to elementary external sets as in Definition 5.2.7. Note also that the definitions of $\mathbf{E}$ and $x \in \mathsf{E}_p$ are absolute for $\mathbb{I}$ because $\mathbb{I}$ is a transitive class and an interpretation of **BST** both in **HST** and in **EEST**. In other words, it does not matter whether we define $\mathbf{E}$ or E_p for some $p \in \mathbf{E}$ in $\mathbb{I}$ or in the whole external set universe of **HST** or **EEST**. This allows us to use all related theorems in §§ 3.2f, 5.2a in **HST** and **EEST**.

Exercise 5.2.9 (EEST). Prove using the Parametrization axiom that $\mathbb{E}$ contains all sets! Why is this not the case in **HST**? □

Thus elementary external sets in external theories are the same as "external sets" in **BST**. The word *elementary* refers to the fact that, first, all sets in $\mathbb{E}$ are subsets of $\mathbb{I}$, and second, all of them admit a direct coding by means of internal sets. Hence $\mathbb{E}$ is arguably the family of simplest possible external sets (some of them are internal, of course) — this explains why we call sets in $\mathbb{E}$ **elementary** external. The next theorem shows that $\mathbb{E}$ coincides with another family of sets considered in § 1.4a.

Theorem 5.2.10 (HST). *The following classes coincide:* $\mathbf{\Delta}_2^{\mathbf{ss}}$, $\mathbb{E}$, *and the class of all "external sets" in the sense of the universe* $\mathbb{I}$, *that is, the class of all sets* $X \subseteq \mathbb{I}$ $\mathtt{st}$-$\in$-*definable in* $\mathbb{I}$ *(with parameters in* $\mathbb{I}$*).*

Proof. That "external sets" = $\mathbb{E}$ follows from Theorem 3.2.16. [3]

To show that $\mathbf{\Delta}_2^{\mathbf{ss}} \subseteq \mathbb{E}$, consider a $\mathbf{\Delta}_2^{\mathbf{ss}}$ set $X = \bigcup_{a \in U} \bigcap_{b \in V} X_{ab}$, where $U, V \in \mathbb{WF}$ and all sets X_{ab} are internal. By Corollary 1.3.13(ii) (Extension) there exists an internal function p defined on ${}^*U \times {}^*V$ so that $p({}^*a, {}^*b) = X_{ab}$ for all $a \in U$, $b \in B$. Now we have $p \in \mathbf{E}$ and $X = \mathsf{E}_p$.

To prove the converse, let $X = \mathsf{E}_p$, where $\mathtt{dom}\, p = {}^*U \times {}^*V$; U, V being well-founded sets. We put $X_{ab} = p({}^*a, {}^*b)$ for all $a \in U$ and $b \in V$. Then $X = \bigcup_{a \in U} \bigcap_{b \in V} X_{ab}$. To get a $\bigcap\bigcup$-presentation of X, take a standard set S such that $X \subseteq S$ and consider the complement $X' = S \smallsetminus X$. □

In the remainder, it will be more important that the class $\mathbb{E}$ contains all sets $\mathtt{st}$-$\in$-definable in $\mathbb{I}$, while the presentation implied by $\mathbf{\Delta}_2^{\mathbf{ss}}$ will have some technical applications.

Theorem 5.2.11 (HST or EEST). $\mathbb{E}$ *is a transitive class containing all internal (hence all standard) sets and satisfying* **EEST**. *In particular the structure* $\mathbb{e} = \langle \mathbb{E}; \in, \mathtt{st} \rangle$ *is an interpretation of* **EEST** *in* **HST**.

Proof. It follows from Lemma 5.2.4 that the map $p \mapsto \mathsf{E}_p$ is a reduction of the invariant structure $\boldsymbol{e} = \langle \mathbf{E}; {}^{\mathsf{e}}\!\in, {}^{\mathsf{e}}\mathtt{st}; {}^{\mathsf{e}}\!\!= \rangle$ to the structure $\mathbb{e} = \langle \mathbb{E}; \in, \mathtt{st} \rangle$ with true equality in the sense of Definition 1.5.9, and hence we have

$$\left.\begin{array}{ll} \mathbf{HST}: & \Phi(\mathsf{E}_{p_1}, ..., \mathsf{E}_{p_n})^{\mathbb{E}} \iff {}^{\mathsf{e}}\Phi(p_1, ..., p_n) \iff ({}^{\mathsf{e}}\Phi(p_1, ..., p_n))^{\mathtt{int}} \\ \mathbf{EEST}: & \Phi(\mathsf{E}_{p_1}, ..., \mathsf{E}_{p_n}) \iff {}^{\mathsf{e}}\Phi(p_1, ..., p_n) \iff ({}^{\mathsf{e}}\Phi(p_1, ..., p_n))^{\mathtt{int}} \end{array}\right\} \quad (*)$$

[3] Give a precise formulation of this fact, as in Theorem 3.2.16(i).

for any st-∈-formula $\Phi(x_1, ..., x_n)$ and any $p_1, ..., p_n \in \mathbf{E}$ by Proposition 1.5.10. The rightmost equivalence in both lines holds since the domain $\mathbf{E}$ of the structure $\boldsymbol{e}$ is a subclass of $\mathbb{I}$ anyway. The superscript $^{\mathbb{E}}$ is omitted in the first term of the **EEST** line because $\mathbb{E}$ contains all sets in **EEST**.

Now the theorem immediately follows from Theorem 5.2.6(i). □

Exercise 5.2.12. Prove the equivalences (∗) directly by induction on the complexity of Φ using Lemma 5.2.4. Also prove the following in **HST**:

(1) $\mathbb{E}$ contains all sets $X \subseteq \mathbb{I}$ of standard size.

(2) $\mathbb{N}$ and $\mathbb{HF}$ (both non-internal sets; see § 1.2e on $\mathbb{HF}$) belong to $\mathbb{E}$.

(3) $\mathbb{I} \subsetneqq \mathbb{E} \subsetneqq \mathbb{H}$.

(4) The $\mathbb{E}$-"power set" $\mathscr{P}(X) \cap \mathbb{E}$ is not a set for any infinite internal set X.

Hints. (1) By Theorem 1.3.12 there is a standard set S and an internal function f defined on S such that $X = \{f(y) : y \in S \cap \mathbb{S}\}$. Then X is st-∈-definable in $\mathbb{I}$ with parameters f, S, and hence $X \in \mathbb{E}$.

(2) Both sets are standard size subsets of $\mathbb{I}$.

(3) The axiom of Transitivity of $\mathbb{I}$ proves $X \subseteq \mathbb{I}$ for any $X \in \mathbb{E}$. Codes of the form ${}^{\mathbf{e}}x$ witness that $\mathbb{I} \subseteq \mathbb{E}$. Finally, $\mathbb{N} \in \mathbb{E} \setminus \mathbb{I}$ while $\{\mathbb{N}\} \notin \mathbb{E}$.

(4) Apply Theorem 1.3.9. □

Proof of part 6° of Corollary 5.1.5

Given a st-∈-sentence Φ, let φ be the sentence ${}^{\mathbf{e}}\Phi$. Then **EEST** proves $\Phi \iff \varphi^{\mathrm{int}}$ by the **EEST**-equivalence in the proof of Theorem 5.2.11. To get an ∈-sentence ψ satisfying $\Phi \iff \psi^{\mathrm{st}}$ in **EEST** apply the reducibility of **BST** by Theorem 4.1.10(ii) to φ, together with the fact that $\mathbb{I}$ interprets **BST** in **EEST** because the latter includes $\mathbf{BST}^{\mathrm{int}}$.

5.2c Some basic theorems of EEST

Here several important theorems of **EEST** are presented. All of them capitalize on basic theorems of **BST** (§§ 3.2d, 3.2e). In general the axiom of Parametrization effectively reduces properties of the **EEST** set universe $\mathbb{E}$ to its internal universe $\mathbb{I}$ which, as we know, satisfies **BST**.

Note that by Theorem 5.2.11 formal deduction in **EEST** can be employed to study elementary external sets in **HST**.

Lemma 5.2.13 (EEST). *Every set C is a subset of a standard set.*

Proof. By Parametrization, $C = \mathsf{E}_p$, where $p \in \mathbf{E}$, thus, $p \in \mathbb{I}$ and $C \subseteq Y = \bigcup \operatorname{ran} p$. Yet the set Y is internal (define $\bigcup \operatorname{ran} p$ in $\mathbb{I}$ and prove, using Transitivity of $\mathbb{I}$, that this is Y). Now apply 3.1.7(iii) in $\mathbb{I}$. □

Theorem 5.2.14 (EEST). *Let $\Phi(x, y)$ be a st-∈-formula with arbitrary sets as parameters. For any set X there exist standard sets S, Y and an internal function F such that the following holds:*

Standardization: $S \cap \mathbb{S} = X \cap \mathbb{S}$;

Collection: $\forall x \in X \left(\exists^{\mathtt{int}} y \, \Phi(x,y) \implies \exists y \in Y \, \Phi(x,y)\right)$;

Standard Size Choice: $\forall^{\mathtt{st}} x \in X \left(\exists^{\mathtt{int}} y \, \Phi(x,y) \implies \Phi(x, F(x))\right)$.

Proof. By Parametrization any parameter in Φ has the form E_q ; let, for brevity, Φ be $\Phi(x, y, \mathsf{E}_q)$, $q \in \mathbf{E}$. Let $\Psi(x, y, q)$ be the formula ${}^{\mathsf{e}}\Phi({}^{\mathsf{e}}x, {}^{\mathsf{e}}y, q)$. Then $\Phi(x, y, \mathsf{E}_q)$ iff $\Psi(x, y, q)$ holds in $\mathbb{I}$ (by $(*)$ of Theorem 5.2.6). Now to obtain Y apply Theorem 3.2.8 (**BST** Collection) in $\mathbb{I}$ to the formula Ψ. To obtain F and S use resp. the **BST** Inner S. S. Choice (Theorem 3.2.11) and the **BST** axiom of Inner Standardization the same way. □

Definition 5.2.15. In **EEST** a *set-like collection* is any collection of the form $\{\mathsf{E}_p : p \in P\}$, where $P \subseteq \mathbf{E}$ is a set. □

This informal definition is convenient to meaningfully consider in **EEST** collections that are not sets. For instance, if at least one of sets x, y is not internal then $\{x, y\}$ is not a set (see Exercise 5.2.17(1) below) but clearly a set-like collection. The following theorem ensures a rather good behaviour of these objects.

Theorem 5.2.16 (EEST). *Let $\varphi(x)$ and $\Phi(x,y)$ be* $\mathtt{st}$-$\in$-*formulas with arbitrary sets as parameters. For any set-like collection X there exist set-like collections X', Y such that*

Separation: $\forall x \in X \left(x \in X' \iff \varphi(x)\right)$;

Collection: $\forall x \in X \left(\exists y \, \Phi(x,y) \implies \exists y \in Y \, \Phi(x,y)\right)$.

Proof. Let $X = \{\mathsf{E}_p : p \in P\}$, where $P \subseteq \mathbf{E}$ is a set. Then, by the Separation axiom, $P' = \{p \in P : \varphi(\mathsf{E}_p)\}$ is a set. Yet $X' = \{\mathsf{E}_p : p \in P'\}$. To prove Collection apply Collection of Theorem 5.2.14 to the formula $\Psi(x, p)$ saying $p \in \mathbf{E} \wedge \Phi(x, \mathsf{E}_p)$. □

Exercise 5.2.17. Rewrite the statement of Theorem 5.2.16 in the ordinary language of **EEST**, in terms of E-codes. Also, prove the following in **EEST**:

(1) If $x \in y$ then x is necessarily internal (by Lemma 5.2.13, $y \subseteq S$ for a standard set S), but y may be non-internal.

(2) Any set-like collection that consists only of internal sets is a set.

(3) (Difficult !) If X is an infinite standard set then $\mathscr{P}(X)$ is <u>not</u> a set-like collection. (*Hint*. See Hrbaček paradox, Theorem 1.3.9.) □

5.2d Standard size, natural numbers, finiteness in EEST

We accept, for **EEST**, the same definition of sets of standard size (Definition 1.1.12) as in **HST**. Note that not all of Theorem 1.3.1 remains true in **EEST** because the well-founded universe is too small, see below.

Theorem 5.2.18 (EEST). *The Saturation axiom, as in* § 1.1f, *holds.*

Proof. Let $\mathscr{X} \subseteq \mathbb{I}$ be a $\cap$-closed set of standard size consisting of nonempty sets; we prove that $\bigcap \mathscr{X} \neq \varnothing$. By definition there is a set $S \subseteq \mathbb{S}$ and a function f with $S \subseteq \operatorname{dom} f$ such that $\mathscr{X} = \{f(s) : s \in S\}$. We can assume that f is internal, by Theorem 5.2.14. Now apply Theorem 3.1.23 in $\mathbb{I}$. □

Exercise 5.2.19 (EEST). Prove that the Dependent Choice axiom, as in § 1.1f, holds. *Hint*: apply, in $\mathbb{I}$, Theorem 3.2.11 (Inner Dependent Choice). □

Exercise 5.2.20 (EEST). Prove that the set $\mathbb{N} = \omega$ of all standard $\mathbb{I}$-natural numbers is the largest ordinal. (*Hint.* To prove that $\mathbb{N}$ is well-ordered argue as in the proof of Lemma 3.1.18(i). Then note that in **EEST** a non-internal set like $\mathbb{N}$ cannot be a member of a set.) □

In **EEST**, by *ordinals* we still mean transitive sets well-ordered by $\in$. Thus by 5.2.20 the class `Ord` of all ordinals is too miserable, just $\omega \cup \{\omega\} = \omega + 1$. Fortunately there is a good replacement. Let `SOrd` be the class of all $\mathbb{S}$-*ordinals*, *i.e.* standard sets that are ordinals in the sense of $\mathbb{S}$.

Lemma 5.2.21 (EEST). *The class* `SOrd` *is well-ordered by* $\in$. *Moreover for any set* $X \subseteq$ `SOrd` *there exists a least* $\mathbb{S}$-*ordinal* $\alpha \notin X$. *This ordinal will be denoted by* $\alpha = \sup^{\mathbb{S}} X$.

Proof. By Theorem 5.2.14 there is a standard set $Y \subseteq$ `SOrd` such that $X \cap$ `SOrd` $= Y \cap$ `SOrd`. As $\mathbb{S}$ satisfies **ZFC** there exists a least $\mathbb{S}$-ordinal $\alpha \notin Y$. By the choice of Y this α is as required. □

Elements of the set $\mathbb{N}$ are called *natural numbers*. A *finite set* is a set equinumerous to $\{1, 2, \dots, n\} = \{k : 1 \le k \le n\}$, where $n \in \mathbb{N}$. Thus, natural numbers in **EEST** are $\mathbb{S}$-natural numbers, *i.e.* standard sets n such that it is true in $\mathbb{S}$ (or, equivalently, in $\mathbb{I}$) that n is a natural number.

Similarly to § 1.2e, define $\mathbb{HF} = \bigcup_{n \in \mathbb{N}} \mathscr{P}^n(\varnothing)$ (all hereditarily finite sets). The next exercise shows that the domain of $*$-methods in **EEST** is restricted to elements and subsets of $\mathbb{HF}$.

Exercise 5.2.22 (EEST). Prove the following:

(1) $\mathbb{HF} = {}^*\mathbb{HF} \cap \mathbb{S}$, where ${}^*\mathbb{HF}$ is the internal set of all internal sets hereditarily finite in $\mathbb{I}$, hence, $\mathbb{HF}$ is a set;

(2) a set x is well-founded iff $x \subseteq \mathbb{HF}$, *i.e.*, $\mathbb{WF} = \mathscr{P}(\mathbb{HF})$ (compare with Exercise 3.2.20), in particular, $\mathbb{N} \in \mathbb{WF}$ and $\mathbb{N} \subseteq \mathbb{WF}$;

(3) ${}^*x \in \mathbb{I}$ can be defined, as in 1.1.6, for any $x \subseteq \mathbb{HF}$, and we have ${}^*x = x$ for any $x \in \mathbb{HF}$ but $x \subsetneqq {}^*x$ for any infinite $x \subseteq \mathbb{HF}$;

(4) $\mathbb{R} \subseteq \mathbb{WF}$, therefore, ${}^*r \in \mathbb{I}$ is defined for any real r, but $\mathbb{R}$ itself is not a set in **EEST**. □

5.3 Assembling of external sets in HST

Elementary external sets (the class $\mathbb{E} = \mathbf{\Delta}_2^{\mathtt{ss}}$) are characterized, within all external sets, by two properties: 1st, their definability, and 2nd, the fact that they contain only internal elements, that is, they are sets of the 1st von Neumann level over $\mathbb{I}$ (see § 1.5b). Obviously there are many definable external sets of higher levels. For instance, any monad of a standard real (§ 2.1a) is an elementary external (and non-internal) set, thus the set of all monads is a definable external set of second level over $\mathbb{I}$. Sets of higher levels can also be defined.

There is a universal method to present this multitude of external sets. Recall that in set theories containing Regularity, like **ZFC**, the construction of an arbitrary set x can be presented as a well-founded tree T, with the empty set assigned to every endpoint of T, such that at any preceding point we assemble all sets already assigned to its immediate successors, and x, the given set, comes out at the root. In **HST**, the axiom of Regularity over $\mathbb{I}$ allows to define sets in a similar manner, but endpoints of trees have to be assigned, or "decorated" with arbitrary internal sets, not necessarily the empty set. Graph theory calls such a construction a *decoration* of a tree.

This section presents the construction itself. It will have two major applications: an interpretation of **HST** in **EEST** in Section 5.4, and a class $\mathbb{L}[\mathbb{I}]$ of all sets obtained by assembling beginning with internal sets in Section 5.5. (Sets in $\mathbb{L}[\mathbb{I}]$ will be called sets constructible from internal sets.) In those applications, a particular form of the construction will be used, such that the trees and assigments to endpoints belong to $\mathbb{E}$, in order to obtain all external sets $\mathtt{st}$-$\in$-definable in the broadest sense.

5.3a Well-founded trees

Let $\mathtt{Seq}$ denote the class of all sequences $\langle a_1, ..., a_n\rangle$ (of arbitrary sets a_i, but mostly only internal a_i will be considered) of finite length. For $t \in \mathtt{Seq}$ and every set a, $t^\wedge a$ is the sequence in $\mathtt{Seq}$ obtained by adjoining a as the rightmost additional term to t. The notation $a^\wedge t$ is understood correspondingly. Generally, $s^\wedge t \in \mathtt{Seq}$ is the concatenation of two sequences $s, t \in \mathtt{Seq}$. The formula $t' \subseteq t$ means that the sequence $t \in \mathtt{Seq}$ *extends* $t' \in \mathtt{Seq}$ (perhaps $t' = t$), while $t' \subset t$ will mean that t is a proper extension of t' (so that $t' \neq t$). $\langle a\rangle$ is a sequence with the only term a.

- A *tree* is a nonempty set $T \subseteq \mathtt{Seq}$ such that, for any pair of sequences $t', t \in \mathtt{Seq}$ satisfying $t' \subseteq t$, we have $t \in T \Longrightarrow t' \in T$.
  Note that every tree contains Λ, *the empty sequence.*
- Define $\mathtt{Max}\, T$ to be the set of all $\subseteq$-*maximal* elements $r \in T$.
- If $t \in T$ then let $\mathtt{Succ}_T(t) = \{a : t^\wedge a \in T\}$.
- Define $\mathtt{Min}\, T = \mathtt{Succ}_T(\Lambda) = \{a : \langle a\rangle \in T\}$; then $\mathtt{Min}\, T = \varnothing$ iff $T = \{\Lambda\}$.

– A tree T is *well-founded* (*wf tree*, in brief) if every nonempty set $T' \subseteq T$ contains an element $\subseteq$-maximal in T'.

Thus, a tree T is well-founded if the inverse relation $\prec$ ($t \prec t'$ iff $t' \subset t$) is a well-founded relation on T in the sense of Definition 1.1.4. Therefore, the next definition is a legitimate definition by well-founded induction (see Remark 1.1.7), as so is any definition of the following kind: a function f is defined on a given wf tree T so that each value $f(t)$ depends only on values $f(t^\wedge a)$, where $t^\wedge a \in T$.

Definition 5.3.1. Let T be a wf tree. Define an ordinal $|t|_T$ (the *rank* of t in T) for each $t \in T$ so that $|t|_T = \sup_{t^\wedge a \in T} |t^\wedge a|_T$.

In particular, $|t|_T = 0$ for $t \in \mathtt{Max}\, T$ (since $\sup \varnothing = 0$).

Put $|T| = |\Lambda|_T$ (the *height* of T). □

Exercise 5.3.2. Suppose that T is a wf tree. Prove that if a set $X \subseteq T$ satisfies $\mathtt{Max}\, T \subseteq X$ and is *inductive* in T (that is $t \in X$ whenever $t \in T$ is such that $t^\wedge a$ belongs to X for all $a \in \mathtt{Succ}_T(t)$) then $X = T$.

Hint. Assume otherwise and consider a $\subseteq$-maximal element $t \in T \setminus X$.)

Prove that for any $t \in T$ then there is $t' \in \mathtt{Max}\, T$ with $t \subseteq t'$. □

5.3b Coding of the assembling construction

The following definition formalizes the idea of assembling construction.

Definition 5.3.3.[4] An *A-code* (or: *assembling* code) is any function $\mathbf{x} : D \to \mathbb{I}$ defined on a set $D \subseteq \mathtt{Seq}$ consisting of pairwise $\subseteq$-incomparable sequences, such that $T_{\mathbf{x}} = \{t \in \mathtt{Seq} : \exists t' \in \mathtt{dom}\, \mathbf{x} : t \subseteq t'\}$ is a wf tree. (Note that then $\mathtt{Max}\, T_{\mathbf{x}} = D = \mathtt{dom}\, \mathbf{x}$, therefore $\mathbf{x}$ is a map $\mathtt{Max}\, T_{\mathbf{x}} \to \mathbb{I}$.)

In this case, a function $\mathsf{F}_{\mathbf{x}}(\cdot)$ can be defined on $T_{\mathbf{x}}$, by the same kind of well-founded induction as above:

1) if $t \in \mathtt{Max}\, T_{\mathbf{x}}$ then $\mathsf{F}_{\mathbf{x}}(t) = \mathbf{x}(t)$;

2) if $t \notin \mathtt{Max}\, T_{\mathbf{x}}$ then $\mathsf{F}_{\mathbf{x}}(t) = \{\mathsf{F}_{\mathbf{x}}(t^\wedge a) : t^\wedge a \in T\}$.

We define $\mathsf{A}_{\mathbf{x}} = \mathsf{F}_{\mathbf{x}}(\Lambda)$ (the set coded by $\mathbf{x}$). □

It is, perhaps, more natural to define an A-code to consist of a well-founded tree $T \subseteq \mathtt{Seq}$ and a function $\mathbf{x} : \mathtt{Max}\, T \to \mathbb{I}$. Then the function $\mathsf{F}_{\mathbf{x}}(\cdot)$ defined on T in accordance with 1) and 2) is a *decoration* of T (relative to $\mathbf{x}$) in the notational system of graph theory (see, e.g., Devlin [Dev 98]). However this would lead to a certain technical inconvenience. Indeed, we shall be mainly interested in those A-codes which belong to $\mathbb{E}$. As the class $\mathbb{E}$ is, generally speaking, not closed under pairing, a pair consisting of a wf tree T and a function $\mathbf{x}$ is not necessarily a member of $\mathbb{E}$ even if both T and

[4] In this definition, A in all shapes refers to "assembling".

$\mathbf{x}$ separately belong to $\mathbb{E}$. This technical nuisance can be fixed by different means; our solution is based on the fact that T, a wf tree, is obviously a function of any suitable $\mathbf{x}$: as above just take the transitive $\subset$-closure $T_{\mathbf{x}}$ of $\mathtt{dom}\,\mathbf{x}$ downwards.

5.3c Examples of codes

We introduce here several useful types of A-codes. To begin with, consider codes of intermediate sets $\mathsf{F}_{\mathbf{x}}(t)$.

Exercise 5.3.4. Suppose that $\mathbf{x}$ is an A-code. For any element $t \in T = T_{\mathbf{x}}$, we put $T|_t = \{s : t^\wedge s \in T\}$. Further, define $\mathbf{x}|_t(s) = \mathbf{x}(t^\wedge s)$ for any $s \in \mathtt{Max}\,T|_t = \{s : t^\wedge s \in \mathtt{Max}\,T\}$. In particular, if $a \in \mathtt{Min}\,T$ then $\langle a \rangle$ (a one-term sequence) belongs to T — thus we can define $T|_a = \{s : a^\wedge s \in T\}$ and $\mathbf{x}|_a(s) = \mathbf{x}(a^\wedge s)$ for any $s \in \mathtt{Max}\,T|_a = \{s : a^\wedge s \in T\}$.

Exercise: prove that $\mathbf{x}|_t$ is an A-code and $\mathsf{A}_{\mathbf{x}|_t} = \mathsf{F}_{\mathbf{x}}(t)$ for any $t \in T = T_{\mathbf{x}}$, in particular, $\mathbf{x}|_a$ is an A-code and $\mathsf{A}_{\mathbf{x}|_a} = \mathsf{F}_{\mathbf{x}}(\langle a \rangle)$ for any $a \in \mathtt{Min}\,T$, moreover, if $|T| \geq 1$ (so that $T \neq \{\Lambda\}$) then $\mathsf{A}_{\mathbf{x}} = \{\mathsf{A}_{\mathbf{x}|_a} : a \in \mathtt{Min}\,T\}$. □

It occurs that any set $x \in \mathbb{H}$ is equal to $\mathsf{A}_{\mathbf{x}}$ for a suitable A-code $\mathbf{x}$. This will be another useful family of codes.

Definition 5.3.5. Let x be any set. Define an A-code ${}^{\mathsf{a}}x$ with $x = \mathsf{A}_{{}^{\mathsf{a}}x}$ as follows. If x is internal let $T_{{}^{\mathsf{a}}x} = \{\Lambda\}$ and ${}^{\mathsf{a}}x(\Lambda) = x$. In other words, in this case ${}^{\mathsf{a}}x = \{\langle \Lambda, x \rangle\}$. If $x \notin \mathbb{I}$ then let $T_{{}^{\mathsf{a}}x}$ be the set of all finite sequences of the form $t = \langle y_0, y_1, ..., y_n \rangle$, where $n \in \mathbb{N}$, y_i are arbitrary sets all of which except possibly y_n are non-internal, and $x \ni y_0 \ni y_1 \ni ... \ni y_n$, together with the empty sequence Λ. ($T_{{}^{\mathsf{a}}x}$ is a set, for instance, because any sequence $t \in T_{{}^{\mathsf{a}}x}$ consists of sets which belong to the transitive closure of x.) Then $T_{{}^{\mathsf{a}}x} \subseteq \mathsf{Seq}$ is a tree and $\mathtt{Max}\,T_{{}^{\mathsf{a}}x}$ consists of all sequences $t = \langle y_0, ..., y_n \rangle$ with $y_n \in \mathbb{I}$. Put ${}^{\mathsf{a}}x(t) = y_n$ for any such t. □

Lemma 5.3.6. *$T_{{}^{\mathsf{a}}x} \subseteq \mathsf{Seq}$ is a wf tree, ${}^{\mathsf{a}}x$ is an A-code, and $\mathsf{A}_{{}^{\mathsf{a}}x} = x$. Moreover, $\mathsf{F}_{{}^{\mathsf{a}}x}(t) = y_n$ and $({}^{\mathsf{a}}x)|_t = {}^{\mathsf{a}}y_n$ for any $t = \langle y_0, ..., y_n \rangle \in T_{{}^{\mathsf{a}}x}$.*

Proof. Suppose towards the contrary that $T_{{}^{\mathsf{a}}x}$ is not a wf tree. Then, by Dependent Choice, there exists an infinite sequence $y \ni b_0 \ni b_1 \ni ...$ of non-internal sets, clearly a contradiction to Regularity over $\mathbb{I}$. Therefore, we can prove the first equality of the "moreover" statement by well-founded induction, on the base of Exercise 5.3.2, *i.e.*, prove that it holds for all $t \in \mathtt{Max}\,T_{{}^{\mathsf{a}}x}$, and also holds for any $t \in T_{{}^{\mathsf{a}}x}$ provided it holds for all immediate successors $t^\wedge a \in T_{{}^{\mathsf{a}}x}$. If $t = \langle a_0, ..., a_n \rangle \in \mathtt{Max}\,T_{{}^{\mathsf{a}}x}$, so that $a_n \in \mathbb{I}$, then by definition $\mathsf{F}_{{}^{\mathsf{a}}x}(t) = {}^{\mathsf{a}}x(t) = a_n$. Suppose that $t \notin \mathtt{Max}\,T_{{}^{\mathsf{a}}x}$. All immediate successors of t in $T_{{}^{\mathsf{a}}x}$ are of the form $t^\wedge a = \langle a_0, ..., a_n, a \rangle$, where $a \in a_n$. If $\mathsf{F}_{{}^{\mathsf{a}}x}(t^\wedge a) = a$ for all $a \in a_n$ then

$$\mathsf{F}_{^{\mathsf{a}}x}(t) = \{\mathsf{F}_{^{\mathsf{a}}x}(t^\wedge a) : a \in a_n\} = a_n \,,$$

as required. In particular, $\mathsf{A}_{^{\mathsf{a}}x} = \mathsf{F}_{^{\mathsf{a}}x}(\Lambda) = \{\mathsf{F}_{^{\mathsf{a}}x}(\langle a \rangle) : a \in x\} = x$. □

Let us mention a related problem in passing by.

Problem 5.3.7. Is it true that every set x is equal to $\mathsf{A}_{\mathbf{x}}$ for an A-code $\mathbf{x}$ such that $T_{\mathbf{x}} \subseteq \mathbb{I}$ — that is, the tree $T_{\mathbf{x}}$ consists only of (finite) sequences of internal terms? A positive answer would follow from the hypothesis that every set is a functional image of an internal set. This hypothesis is consistent with **HST** (claim 8° of Theorem 5.5.4 below), but most likely not provable in **HST**. □

Example 5.3.8. The codes $^{\mathsf{a}}x$ will be especially important for sets $x \subseteq \mathbb{I}$. If x is internal then, by definition, $T_{^{\mathsf{a}}x} = \{\Lambda\}$ and $^{\mathsf{a}}x(\Lambda) = x$. If $x \subseteq \mathbb{I}$ is non-internal then $T_{^{\mathsf{a}}x} = \{\Lambda\} \cup \{\langle a \rangle : a \in x\}$ (hence, $\mathtt{Max}\, T_{^{\mathsf{a}}x} = \{\langle a \rangle : a \in x\}$ and $\mathtt{Min}\, T_{^{\mathsf{a}}x} = \{a : a \in x\} = x$), and $^{\mathsf{a}}x(\langle a \rangle) = a$ for any $a \in x$. □

Now let us consider A-coding of well-founded sets. We leave it as an **exercise** for the reader to show that if $v \in \mathbb{WF}$ then $^{\mathsf{a}}x \in \mathbb{WF}$ as well. In particular, any $t = \langle a_0, ..., a_n \rangle \in T_{^{\mathsf{a}}x}$ is a (finite) sequence of well-founded sets a_i, which, except possibly a_n, are non-internal. Recall that

$$\mathbb{WF} \cap \mathbb{I} = \mathbb{WF} \cap \mathbb{S} = \mathbb{HF} = \{v \in \mathbb{WF} : {}^*v = v\}$$

is the (well-founded, non-internal) set of all hereditarily finite sets (Exercise 1.2.17). thus the requirement of non-internality can be reformulated as follows: none of the (well-founded) sets a_i, except possibly for a_n, belongs to $\mathbb{HF}$. The next definition introduces an isomorphic copy of $^{\mathsf{a}}v$, whose advantage is that the associated wf tree consists of internal (moreover, standard) sequences.

Definition 5.3.9. Suppose that $v \in \mathbb{WF}$ and $y = {}^*v$ (a standard set).

If $v \in \mathbb{HF}$, and hence $y = v$, then put $\mathbf{c}[y] = {}^{\mathsf{a}}y$.

Suppose that $v \notin \mathbb{HF}$. Let $T[y]$ be the set of all finite sequences of the form $t = \langle b_0, b_1, ..., b_n \rangle$, where $n \in \mathbb{N}$, $v \ni b_0 \ni b_1 \ni ... \ni b_n$, and b_i are arbitrary standard sets (thus, $b_i = {}^*a_i$ for a well-founded set a_i) — with the restriction that all of them, except possibly for b_n, do <u>not</u> belong to $\mathbb{HF}$, together with the empty sequence Λ. Thus $\mathtt{Max}\, T[y]$ consists of all sequences $t = \langle b_0, ..., b_n \rangle \in T[y]$ with $b_n \in \mathbb{HF}$. Put $\mathbf{c}[y](t) = b_n$ (then $= {}^*b_n$ as $x = {}^*x$ for $x \in \mathbb{HF}$) for any such t. □

Exercise 5.3.10. Prove, using Lemma 5.3.6, that, in both cases, $T[y] \subseteq \mathtt{Seq} \cap \mathbb{S}$ is a wf tree, $\mathbf{c}[y]$ is an A-code, and $\mathsf{A}_{\mathbf{c}[y]} = v$. Finally, if $v \notin \mathbb{HF}$ then $\mathsf{F}_{\mathbf{c}[y]}(t) = a_n$ and $\mathbf{c}[y]\|_t = \mathbf{c}[a_n]$ for any $t = \langle {}^*a_0, ..., {}^*a_n \rangle \in T[y]$. □

5.3d Regular codes

Any internal set x admits not only the "natural" code ${}^{\mathsf{a}}x$, but many other codes, for instance, a code that assembles x in one step from its elements. To inhibit such a non-uniqueness, consider the following special class of A-codes that produce internal sets only through "natural" codes of the form ${}^{\mathsf{a}}x$.

Definition 5.3.11. An A-code $\mathbf{x}$ is *regular* if for each $t \in T = T_{\mathbf{x}}$ satisfying $|t|_T = 1$ the set $\mathsf{F}_{\mathbf{x}}(t) = \{\mathbf{x}(t^\wedge a) : t^\wedge a \in \mathtt{Max}\,T\}$ is not internal. □

Thus regularity requires that internal sets do not appear at the first assembling level. The next lemma shows that this requirement is sufficient to forbid internal sets to appear at all higher levels. Let $\mathtt{DI}_{\mathbf{x}} = \{t \in T_{\mathbf{x}} : \mathsf{F}_{\mathbf{x}}(t) \in \mathbb{I}\}$, the *domain of internality*. Note that any code $\mathbf{x}$ with $T_{\mathbf{x}} = \{\Lambda\}$ is regular, and $\mathtt{DI}_{\mathbf{x}} = T_{\mathbf{x}} = \{\Lambda\}$.

Lemma 5.3.12. *An A-code* $\mathbf{x}$ *is regular iff* $\mathtt{DI}_{\mathbf{x}} = \mathtt{Max}\,T_{\mathbf{x}}$.

Proof. If $\mathtt{DI}_{\mathbf{x}} = \mathtt{Max}\,T_{\mathbf{x}}$ then $\mathbf{x}$ is regular by definition. To prove the converse suppose that $\mathbf{x}$ is regular. Since $\mathtt{Max}\,T_{\mathbf{x}} \subseteq \mathtt{DI}_{\mathbf{x}}$ for any A-code, it remains to check the opposite inclusion. Let $X = \{t \in T_{\mathbf{x}} : t \notin \mathtt{DI}_{\mathbf{x}} \vee t \in \mathtt{Max}\,T_{\mathbf{x}}\}$. We have to show that $X = T_{\mathbf{x}}$. According to Exercise 5.3.2, it suffices to prove $t \in X$, assuming that $t \in T_{\mathbf{x}}$ and every extension $t^\wedge a \in T_{\mathbf{x}}$ belongs to X. Let, on the contrary, $t \notin X$. Then $t \in T_{\mathbf{x}} \smallsetminus \mathtt{Max}\,T_{\mathbf{x}}$ and $x = \mathsf{F}_{\mathbf{x}}(t) \in \mathbb{I}$. We have $|t|_{T_{\mathbf{x}}} \geq 2$ because of regularity, hence there is $t^\wedge a \in T_{\mathbf{x}}$ such that $|t^\wedge a|_{T_{\mathbf{x}}} \geq 1$. Thus $t^\wedge a \notin \mathtt{Max}\,T_{\mathbf{x}}$. Yet $t^\wedge a \in X$, and hence $t^\wedge a \notin \mathtt{DI}_{\mathbf{x}}$ and $y = \mathsf{F}_{\mathbf{x}}(t^\wedge a) \notin \mathbb{I}$. However $y \in x$, a contradiction to Transitivity of $\mathbb{I}$. □

Exercise 5.3.13. Prove that if an A-code $\mathbf{x}$ is regular then $\mathtt{irk}\,\mathsf{F}_{\mathbf{x}}(t) = |t|_T$ for all $t \in T = T_{\mathbf{x}}$. (Recall that $\mathtt{irk}$ is the rank over $\mathbb{I}$, see § 1.5b.) □

Yet our coding potential does not really suffer, because any A-code $\mathbf{x}$ can be reduced to a regular A-code $\mathbf{x}^{\mathrm{R}}$ such that $\mathsf{A}_{\mathbf{x}} = \mathsf{A}_{\mathbf{x}^{\mathrm{R}}}$. To define $\mathbf{x}^{\mathrm{R}}$ note that the set D of all $\subseteq$-minimal elements of $\mathtt{DI}_{\mathbf{x}}$ is obviously pairwise $\subseteq$-incomparable, that is $s \not\subseteq t$ for all $s \neq t$ in D. Put $\mathbf{x}^{\mathrm{R}}(t) = \mathsf{F}_{\mathbf{x}}(t)$ for $t \in D$; note that $\mathbf{x}^{\mathrm{R}}(t) \in \mathbb{I}$ because $D \subseteq \mathtt{DI}_{\mathbf{x}}$.

Exercise 5.3.14. Prove that then $\mathbf{x}^{\mathrm{R}}$ is a regular A-code satisfying $\mathsf{A}_{\mathbf{x}} = \mathsf{A}_{\mathbf{x}^{\mathrm{R}}}$, $T_{\mathbf{x}^{\mathrm{R}}} = (T_{\mathbf{x}} \smallsetminus \mathtt{DI}_{\mathbf{x}}) \cup D$, $\mathtt{Max}\,T_{\mathbf{x}^{\mathrm{R}}} = D$. In addition, prove the following:

(i) All codes ${}^{\mathsf{a}}x$ and $\mathbf{c}[y]$, $y \in \mathbb{S}$ (Definitions 5.3.5, 5.3.9) are regular. (*Hint*: regarding $\mathbf{c}[y]$, apply the result of 1.3.8(2) that a standard size set of internal sets is internal iff it is finite.)

(ii) An A-code $\mathbf{x}$ with $|T_{\mathbf{x}}| \geq 1$ is regular iff all codes $\mathbf{x}|_a$, $a \in \mathtt{Min}\,T_{\mathbf{x}}$ (Exercise 5.3.4) are regular and either $|T_{\mathbf{x}}| \geq 2$ or $|T_{\mathbf{x}}| = 1$ and $\mathsf{A}_{\mathbf{x}} \notin \mathbb{I}$.

(iii) If $\mathbf{x}$ is a regular A-code and $\mathsf{A}_{\mathbf{x}} = x \in \mathbb{I}$ then $\mathbf{x} = {}^{\mathsf{a}}x$. □

5.4 From elementary external to all external sets

The main goal of this section is to prove 3° of Theorem 5.1.4. In the course of the proof we define an interpretation of **HST** in **EEST** based on the assembling construction outlined in Section 5.3.

Blanket agreement 5.4.1. In the arguments below $\mathbb{E}$ will denote:

– *either* the set universe of **EEST** – and then we study it by means of **EEST** (the **EEST** *case*);

– *or* the class of all elementary external sets in **HST** – and then we study it by means of **HST** (the **HST** *case*).

Note that in the **HST** case $\mathbb{E}$ still satisfies **EEST** by Theorem 5.2.11. We shall make clear distinctions whenever it is necessary to avoid ambiguity.

As usual $\mathbb{S} \subsetneqq \mathbb{I} \subsetneqq \mathbb{E}$ are classes of resp. standard and internal sets in $\mathbb{E}$. Thus $\mathbb{S} = \{x \in \mathbb{E} : \mathtt{st}\, x\}$ and $\mathbb{I} = \{x \in \mathbb{E} : \mathtt{int}\, x\}$. □

The **HST** case will lead to the class $\mathbb{L}[\mathbb{I}]$ of all sets constructible from internal sets in Section 5.5 while the **EEST** case is directly connected with the interpretation of **HST** in **EEST** defined below in this section.

The domain of the interpretation will consist of all regular A-codes $\mathbf{x} \in \mathbb{E}$. The intended meaning of the basic relations ${}^{\mathrm{a}}\!\in$, ${}^{\mathrm{a}}\!=$, ${}^{\mathrm{a}}\mathtt{st}$ is connected with the coded sets $\mathsf{A}_{\mathbf{x}}$, for instance $\mathbf{x} \;{}^{\mathrm{a}}\!\in \mathbf{y}$ iff $\mathsf{A}_{\mathbf{x}} \in \mathsf{A}_{\mathbf{y}}$. The main difficulty here is that the sets $\mathsf{A}_{\mathbf{x}}$ themselves generally speaking do not belong to $\mathbb{E}$, and hence we cannot explicitly appeal to any relation between them. To solve the problem we shall find adequate definitions of basic relations within $\mathbb{E}$.

Proofs of items 4° and 5° of Corollary 5.1.5 follow in § 5.4f. This section ends with a continuation of our discussion of external sets in **BST** which began in § 3.2f.

5.4a The domain of the interpretation

First of all let us have another look at different notions introduced in Section 5.3 from the point of view of $\mathbb{E}$ as the principal domain.

In $\mathbb{E}$ only internal sets can be elements of other sets, and hence $\mathtt{Seq}$ consists of finite internal sequences of internal sets. The method of definition by well-founded induction on a wf tree has to be somewhat changed in $\mathbb{E}$. Indeed it follows from 5.2.20 that the class $\mathtt{Ord}$ of all ordinals is equal to $\omega \cup \{\omega\}$ in **EEST**, and hence is too small to support transfinite induction of any bigger length. Thus the rank function $|t|_T : T \to \mathtt{Ord}$ generally does not exist in $\mathbb{E}$ for a wf tree $T \subseteq \mathtt{Seq}$. Yet Lemma 5.2.21 provides us with an equivalent substitution in the class of $\mathbb{S}$-ordinals.

Say that $T \in \mathbb{E}$, $T \subseteq \mathtt{Seq}$ is a $\mathbb{E}$-*wf* tree if every nonempty set $T' \in \mathbb{E}$, $T' \subseteq T$ contains an element $\subseteq$-maximal in T'. This is the same as just being wf in the **EEST** case (see 5.4.1 on the cases). Let us show that this is also the same in the **HST** case.

Definition 5.4.2. Let $T \in \mathbb{E}$ be a wf tree. Define an $\mathbb{S}$-ordinal $|t|^*_T$ for each $t \in T$ so that $|t|^*_T = \sup^{\mathbb{S}}{}_{t^\wedge a \in T} |t^\wedge a|^*_T$ for any $t \in T$ (the least $\mathbb{S}$-ordinal strictly bigger than all $\mathbb{S}$-ordinals $|t^\wedge a|^*_T$, $t^\wedge a \in T$). In particular, $|t|^*_T = 0$ for $t \in \mathtt{Max}\,T$. Put $|T|^* = |\Lambda|^*_T$. □

Lemma 5.4.3. (i) *For any $\mathbb{E}$-wf tree $T \in \mathbb{E}$ there is a unique map $t \mapsto |t|^*_T$ from T to* $\mathtt{SOrd}$ *that belongs to $\mathbb{E}$ and satisfies $|t|^*_T = \sup^{\mathbb{S}}{}_{t^\wedge a \in T} |t^\wedge a|^*_T$ for all $t \in T$;*

(ii) (**HST**) $|T|^*_t = {}^*(|T|_t)$ *for any $\mathbb{E}$-wf tree $T \in \mathbb{E}$;*

(iii) (**HST**) *any tree $T \in \mathbb{E}$ is wf iff it is $\mathbb{E}$-wf.*

Proof. (i) Let, for any $t \in T$, a *t-function* be any map $f \in \mathbb{E}$ defined on the set $\{t' \in T : t \subseteq t'\}$ and satisfying $|t|^*_T = \sup^{\mathbb{S}}{}_{t^\wedge a \in T} |t^\wedge a|^*_T$ on its domain. It suffices to prove that for any $t \in T$ there is a unique t-function f_t.

Note that "being a t-function" and the existence of a t-function are $\mathtt{st}$-$\in$-formulas relativized to $\mathbb{E}$. It follows that if $f \neq g$ are two t-functions fot some $t \in T$ then the set $X_{tfg} = \{t' \in T : t \subseteq t' \wedge f(t') \neq g(t')\}$ is $\mathtt{st}$-$\in$-definable in $\mathbb{E}$. We conclude that $X_{tfg} \in \mathbb{E}$ because $\mathbb{E}$ satisfies **EEST** by Theorem 5.2.11. If $f \neq g$ then $X_{tfg} \neq \varnothing$, and hence X_{tfg} contains a $\subseteq$-maximal element t' (inteed T is $\mathbb{E}$-wf). Thus $f(t') \neq g(t')$, but $f(\tau) = g(\tau)$ for any $\tau \in T$ with $t' \subset \tau$, easily leading to contradiction. This proves the uniqueness of f_t.

To prove the existence suppose towards the contrary that $X = \{t \in T : \text{there is no } t\text{-function}\} \neq \varnothing$. Still X belongs to $\mathbb{E}$ and hence it contains a $\subseteq$-maximal element t. In other words a unique $f_{t^\wedge a} \in \mathbb{E}$ does exist for any $a \in \mathtt{Succ}_T(t)$, but f_t does not exist. Define $f = \bigcup_{a \in \mathtt{Succ}_T(t)} f_{t^\wedge a}$. In addition, let $f(t)$ be the least $\mathbb{S}$-ordinal bigger than all $\mathbb{S}$-ordinals $f_{t^\wedge a}(t^\wedge a)$, $a \in \mathtt{Succ}_T(t)$. (To see that this is well-defined use Lemma 5.2.21.) Clearly f is $\mathtt{st}$-$\in$-definable in $\mathbb{E}$, and hence it belongs to $\mathbb{E}$ by Theorem 5.2.11. It follows that f is a t-function, contradiction.

(ii) A routine proof with the help of usual **HST** methods including $*$-Transfer is left for the reader.

(iii) As $\mathbb{S}$-ordinals are isomorphic to the true ordinals (those in $\mathbb{WF}$) in **HST** via $*$, and hence are well-ordered, the map $t \mapsto |t|^*_T$ proves that any $\mathbb{E}$-wf tree $T \in \mathbb{E}$ is wf in the universe of the **HST** universe as well. □

We keep the definition of A-code and $T_{\mathbf{x}}$ as in 5.3.3.

Definition 5.4.4 (EEST). $\underline{\mathbf{A}}$ is the class of all A-codes $\mathbf{x} \in \mathbb{E}$.

A is the class of all *regular* A-codes in $\underline{\mathbf{A}}$, where the regularity means that $\{\mathbf{x}(t^\wedge a) : t^\wedge a \in T_{\mathbf{x}}\} \notin \mathbb{I}$ whenever $t \in T_{\mathbf{x}}$ satisfies $|T|^*_t = 1$ (*i.e.* $t \notin \mathtt{Max}\,T_{\mathbf{x}}$ but any $t' \in T_{\mathbf{x}}$ with $t \subset t'$ belongs to $\mathtt{Max}\,T_{\mathbf{x}}$). □

Due to the restrictive character of the **EEST** set universe the functions $\mathsf{F}_{\mathbf{x}}(\cdot)$ generally speaking do not exist for $\mathbf{x} \in \underline{\mathbf{A}}$, accordingly, $\mathsf{A}_{\mathbf{x}}$, generally speaking, cannot be defined as in Definition 5.3.3.

Theorem 5.4.5. *The class* $\mathbf{A}$ *is* st-$\in$-*definable in* $\mathbb{E}$.

Proof. In the **EEST** case (see 5.4.1) the result is obvious. In the **HST** case we have to prove that the definition of $\mathbf{A}$ is absolute for $\mathbb{E}$ in **HST**. The absoluteness easily follows from Lemma 5.4.3(iii), because all other elements of the definition of $\mathbf{A}$ (that is except for the well-foundedness of the tree) are absolute by rather obvious reasons. □

Lemma 5.4.6. *If* $\mathbf{x} \in \mathbf{A}$ *then* $T_{\mathbf{x}}$, $\mathtt{Max}\,T_{\mathbf{x}}$, $\mathtt{Min}\,T_{\mathbf{x}}$ *are sets in* $\mathbb{E}$ *while* $\mathbf{x}|_t$, $\mathbf{x}|_a$ *for all* $t \in T_{\mathbf{x}}$, $a \in \mathtt{Min}\,T_{\mathbf{x}}$ *are sets in* $\mathbb{E}$ *and codes in* $\mathbf{A}$. *In addition,* ${}^{\mathbf{a}}z$ *for any* $z \in \mathbb{E}$ *and* $\mathbf{c}[x]$ *for any standard* x *are sets in* $\mathbb{E}$ *and codes in* $\mathbf{A}$.

Proof. Routine verification based on some results of § 5.2c, most notably, Lemma 5.2.13 and **EEST** Separation. For instance, if $\mathbf{x} \in \mathbf{A}$ then formally $\mathbf{x}$ is a function with $D = \mathtt{dom}\,\mathbf{x} \subseteq \mathsf{Seq}$. On the other hand, by Lemma 5.2.13 there is a standard, hence, internal set P with $\mathbf{x} \subseteq P$. Then $\mathtt{dom}\,P = \{x : \exists y\, (\langle x, y\rangle \in P)\}$ is still an internal set (define it in $\mathbb{I}$, which is an $\in$-interpretation of **ZFC**). It follows that D is a set by Separation. Now, still by Lemma 5.2.13, $D \subseteq T$, where T is internal, and we can assume that $T \subseteq \mathsf{Seq}$. Moreover, $T' = \{t' \in \mathsf{Seq} : \exists t \in T\, (t' \subseteq t)\}$ is an internal set, hence, $T_{\mathbf{x}}$ (a definable subset of T') is a set by Separation.

If x is internal then by definition ${}^{\mathbf{a}}x = \{\langle \Lambda, x\rangle\} \in \mathbb{I}$. If $x \in \mathbb{E} \setminus \mathbb{I}$ then ${}^{\mathbf{a}}x = \{\langle\langle a\rangle, a\rangle : a \in x\}$ (see Example 5.3.8). Choose an internal set S with $x \subseteq S$ (Lemma 5.2.13). Then ${}^{\mathbf{a}}x$ is a subset of the internal set $X = \{\langle\langle a\rangle, a\rangle : a \in S\}$ st-$\in$-definable in $\mathbb{E}$, anf hence ${}^{\mathbf{a}}x \in \mathbb{E}$ by Theorem 5.2.16. It follows that ${}^{\mathbf{a}}x \in \underline{\mathbf{A}}$ for any $x \in \mathbb{E}$. As the regularity is obvious we have ${}^{\mathbf{a}}x \in \mathbf{A}$.

We leave the rest of the lemma as an **exercise** for the reader. □

The class $\mathbf{A}$ will be the domain of the interpretation.

5.4b Basic relations between codes

We continue to argue under the assumptions of 5.4.1.

Suppose that $\mathbf{x}, \mathbf{y} \in \mathbb{E}$ are A-codes in $\mathbf{A}$. In principle, to figure out whether, say, $A_{\mathbf{x}} = A_{\mathbf{y}}$, we have to compute both sets and check whether they are equal — but this is impossible within $\mathbb{E}$ because the coded sets do not necessarily belong to $\mathbb{E}$. Yet there is a way to avoid the actual computation of coded sets, based on the following definition taken from graph theory.

Definition 5.4.7. A map $j : T_{\mathbf{x}} \times T_{\mathbf{y}} \to \{0, 1\}$ is *a bisimulation* for A-codes $\mathbf{x}$ and $\mathbf{y}$ if it satisfies the following requirements:

1*. If $t \in \mathtt{Max}\,T_{\mathbf{x}}$ and $r \in \mathtt{Max}\,T_{\mathbf{y}}$ then $j(t, r) = 1$ iff $\mathbf{x}(t) = \mathbf{y}(r)$.

2*. If $t \in \mathtt{Max}\,T_{\mathbf{x}}$ but $r \notin \mathtt{Max}\,T_{\mathbf{y}}$, or conversely, $t \notin \mathtt{Max}\,T_{\mathbf{x}}$ but $r \in \mathtt{Max}\,T_{\mathbf{y}}$, then $j(t, r) = 0$.

3*. Suppose that $t \notin \mathtt{Max}\,T_{\mathbf{x}}$ and $r \notin \mathtt{Max}\,T_{\mathbf{y}}$. Then $j(r, t) = 1$ iff

(a) $\forall r^\wedge b \in T_{\mathbf{y}}\ \exists t^\wedge a \in T_{\mathbf{x}}\ (\jmath(t^\wedge a, r^\wedge b) = 1)$, and
(b) $\forall t^\wedge a \in T_{\mathbf{x}}\ \exists r^\wedge b \in T_{\mathbf{y}}\ (\jmath(t^\wedge a, r^\wedge b) = 1)$. □

Since we consider only well-founded trees, in **HST** for any two A-codes $\mathbf{x}, \mathbf{y} \in \mathbf{A}$ there exists a bisimulation $\jmath : T_{\mathbf{x}} \times T_{\mathbf{y}} \to \{0,1\}$ defined so that $\jmath(t,r) = 1$ whenever $\mathsf{F}_{\mathbf{x}}(t) = \mathsf{F}_{\mathbf{y}}(r)$ and $\jmath(t,r) = 0$ otherwise. (The requirement of regularity validates 2*; in the non-regular case that would be more cumbersome.) Now, under the assumptions of 5.4.1, we prove

Lemma 5.4.8. *For any two codes* $\mathbf{x}, \mathbf{y} \in \mathbf{A}$ *there exists a unique bisimulation* $\jmath$. *This unique bisimulation belongs to* $\mathbb{E}$. *It will be denoted by* $\mathbf{j}_{\mathbf{xy}}$.

Proof. We argue as in the proof of Lemma 5.4.3(i). Let, for $t \in T_{\mathbf{x}}$ and $r \in T_{\mathbf{y}}$, a (t,r)*-function* be any function $\jmath \in \mathbb{E}$ defined on the set $\{\langle t', r' \rangle \in T_{\mathbf{x}} \times T_{\mathbf{y}} : t \subseteq t' \wedge r \subseteq r'\}$ and satisfying 1*, 2*, 3* of Definition 5.4.7 on this domain. Let $P(t)$ say: "for any $r \in T_{\mathbf{y}}$, there is a unique (t,r)-function". To prove $P(\Lambda)$, the desired result, it suffices to show that the set of all $t \in T_{\mathbf{x}}$ with $P(t)$ is inductive (see Exercise 5.3.2). Take any $t \in T_{\mathbf{x}}$, suppose $P(t^\wedge a)$ for all extensions $t^\wedge a \in T_{\mathbf{x}}$ (for instance, this holds for $t \in \mathtt{Max}\, T_{\mathbf{x}}$), and derive $P(t)$. Let $r \in T_{\mathbf{y}}$. By the inductive hypothesis, for any $t^\wedge a \in T_{\mathbf{x}}$ and $r^\wedge b \in T_{\mathbf{y}}$ there exists a unique $(t^\wedge a, r^\wedge b)$-function, say, $\jmath_{ab} \in \mathbb{E}$. Moreover, it follows from the uniqueness that these functions are pairwise compatible on intersections of the domains, and hence the union $\jmath = \bigcup_{ab} \jmath_{ab}$ is a function. Since "to be a (t,r)-function" is a notion absolute for $\mathbb{E}$, the formula which defines $\jmath$ witnesses that $\jmath$ is st-∈-definable in $\mathbb{E}$, therefore $\jmath \in \mathbb{E}$ because $\mathbb{E}$ satisfies **EEST** by Theorem 5.2.11. It remains to define, additionally, values $\jmath(t, r')$ and $\jmath(t', r)$ for all $r' \in T_{\mathbf{y}}$ with $r \subseteq r'$ and all $t' \in T_{\mathbf{x}}$ with $t \subseteq t'$ (in particular, $\jmath(t,r)$) applying Definition 5.4.7: the result belongs to $\mathbb{E}$ still by Theorem 5.2.11 and is a unique (t,r)-function. □

Exercise 5.4.9. Prove that if $\mathbf{x} \in \mathbf{A}$ and $a \in \mathtt{Min}\, T_{\mathbf{x}}$ then $\mathbf{j}_{\mathbf{x}\mathbf{x}|_a}(\langle a \rangle, \Lambda) = 1$. Describe the whole structure of the bisimulation $\mathbf{j}_{\mathbf{x}\mathbf{x}|_a}$. □

The notion of bisimulation allows us to introduce st-∈-formulas which define, in $\mathbb{E}$, the basic relations between coded sets in terms of A-codes. Index ^{a} still refers to "assembling".

Definition 5.4.10. $\mathbf{x} \ {}^{\mathsf{a}}\!\!= \mathbf{y}$ is the st-∈-formula "$\mathbf{x}, \mathbf{y} \in \mathbf{A} \wedge \mathbf{j}_{\mathbf{xy}}(\Lambda, \Lambda) = 1$".
$\mathbf{x} \ {}^{\mathsf{a}}\!\!\in \mathbf{y}$ is the st-∈-formula "$\mathbf{x}, \mathbf{y} \in \mathbf{A} \wedge ((1) \vee (2))$", where

(1) $\mathbf{x} = {}^{\mathsf{a}}x$ and $\mathbf{y} = {}^{\mathsf{a}}y$ for some internal $x \in y$;

(2) $T_{\mathbf{y}} \neq \{\Lambda\}$ and there is $b \in \mathtt{Min}\, T_{\mathbf{y}}$ such that $\mathbf{j}_{\mathbf{xy}}(\Lambda, \langle b \rangle) = 1$.

(Note that these two cases are incompatible.) Finally,

${}^{\mathsf{a}}\mathtt{st}\, \mathbf{x}$ is the st-∈-formula $\exists^{\mathtt{st}} y\, (\mathbf{x} = {}^{\mathsf{a}}y)$ (this implies $\mathbf{x} \in \mathbf{A}$);

${}^{\mathsf{a}}\mathtt{int}\, \mathbf{x}$ is the st-∈-formula $\exists^{\mathtt{int}} y\, (\mathbf{x} = {}^{\mathsf{a}}y)$ (this implies $\mathbf{x} \in \mathbf{A}$). □

The relations $\stackrel{\mathtt{a}}{=}$, ${}^{\mathtt{a}}\!\in$, ${}^{\mathtt{a}}\mathtt{st}$, ${}^{\mathtt{a}}\mathtt{int}$ have a pretty clear meaning in **HST** where the coded sets $\mathsf{A}_{\mathbf{x}}$ do exist:

Theorem 5.4.11 (HST). *Suppose that* $\mathbf{x}$ *and* $\mathbf{y}$ *belong to* $\mathbf{A}$. *Then*

(i) $\mathbf{x} \stackrel{\mathtt{a}}{=} \mathbf{y}$ *iff* $\mathsf{A}_{\mathbf{x}} = \mathsf{A}_{\mathbf{y}}$;

(ii) $\mathbf{x} \;{}^{\mathtt{a}}\!\!\in \mathbf{y}$ *iff* $\mathsf{A}_{\mathbf{x}} \in \mathsf{A}_{\mathbf{y}}$;

(iii) ${}^{\mathtt{a}}\mathtt{st}\,\mathbf{x}$ *iff* $\mathsf{A}_{\mathbf{x}} \in \mathbb{S}$ *and* ${}^{\mathtt{a}}\mathtt{int}\,\mathbf{x}$ *iff* $\mathsf{A}_{\mathbf{x}} \in \mathbb{I}$.

Proof. (i) As the map $j : T_{\mathbf{x}} \times T_{\mathbf{y}} \to \{0,1\}$, defined so that $j(t,r) = 1$ whenever $\mathsf{F}_{\mathbf{x}}(t) = \mathsf{F}_{\mathbf{y}}(r)$ and $j(t,r) = 0$ otherwise, is clearly a bisimulation, it coincides with $\mathbf{j}_{\mathbf{xy}}$ by Lemma 5.4.10. This implies the required result.

(ii) Suppose that $\mathsf{A}_{\mathbf{x}} \in \mathsf{A}_{\mathbf{y}}$. If $T_{\mathbf{y}} = \{\Lambda\}$ then $\mathbf{y} = {}^{\mathtt{a}}y$, where $y = \mathbf{y}(\Lambda) = \mathsf{A}_{\mathbf{y}}$ is internal, and hence $\mathsf{A}_{\mathbf{x}}$ is internal as well (because $\mathbb{I}$ is transitive), so that $T_{\mathbf{x}} = \{\Lambda\}$ because the code $\mathbf{x}$ is regular, therefore, $\mathbf{x} = {}^{\mathtt{a}}x$, where $x = \mathbf{x}(\Lambda) = \mathsf{A}_{\mathbf{x}} \in y$. If $T_{\mathbf{y}} \neq \{\Lambda\}$ then clearly $\mathsf{A}_{\mathbf{x}} = \mathsf{F}_{\mathbf{y}}(\langle b\rangle)$ for some $b \in \mathtt{Min}\,T_{\mathbf{y}}$, thus we have $\mathbf{j}_{\mathbf{xy}}(\Lambda, \langle b\rangle) = 1$. This implies $\mathbf{x} \;{}^{\mathtt{a}}\!\!\in \mathbf{y}$ by (2) of Definition 5.4.10. The converse can be proved the same way.

(iii) Assume that $\mathsf{A}_{\mathbf{x}} = y \in \mathbb{S}$. Codes ${}^{\mathtt{a}}y$ belong to $\mathbf{A}$ (Lemma 5.4.6), and satisfy $\mathsf{A}_{{}^{\mathtt{a}}y} = y$ (Lemma 5.3.6). It follows, by (i), that $\mathbf{x} \stackrel{\mathtt{a}}{=} {}^{\mathtt{a}}y$, hence, ${}^{\mathtt{a}}\mathtt{st}\,\mathbf{x}$. The converse can be proved similarly. □

5.4c The structure of basic relations

We continue to argue under the assumptions of 5.4.1.

We are going to consider $\mathfrak{a} = \langle \mathbf{A}; {}^{\mathtt{a}}\!\!\in, {}^{\mathtt{a}}\mathtt{st}; \stackrel{\mathtt{a}}{=}\rangle$ as an invariant st-$\in$-structure. Then in particular we have to show that $\stackrel{\mathtt{a}}{=}$ is an equivalence on the domain $\mathbf{A}$ while ${}^{\mathtt{a}}\!\!\in$ and ${}^{\mathtt{a}}\mathtt{st}$ are $\stackrel{\mathtt{a}}{=}$-invariant relations. In **HST** this task is pretty easy on the grounds of Theorem 5.4.11. But the **EEST** case needs more work with codes and bisimulations.

Lemma 5.4.12. *The bisimulations* $\mathbf{j}_{\mathbf{xy}}$ $(\mathbf{x}, \mathbf{y} \in \mathbf{A})$ *satisfy the following:*

(i) $\mathbf{j}_{\mathbf{xx}}(t,t) = 1$ *for all* $t \in T_{\mathbf{x}}$;

(ii) *if* τ, t, ρ, r *are finite sequences such that* $\tau^\wedge t \in T_{\mathbf{x}}$ *and* $\rho^\wedge r \in T_{\mathbf{y}}$ (then, of course, $\tau \in T_{\mathbf{x}}$ and $\rho \in T_{\mathbf{y}}$), *then* $\mathbf{j}_{\mathbf{x}|_\tau\, \mathbf{y}|_\rho}(t,r) = \mathbf{j}_{\mathbf{xy}}(\tau^\wedge t, \rho^\wedge r)$;

(iii) *if* $t \in T_{\mathbf{x}}$, $r \in T_{\mathbf{y}}$, $u \in T_{\mathbf{z}}$, *then* $\mathbf{j}_{\mathbf{xy}}(t,r) = \mathbf{j}_{\mathbf{yz}}(r,u) = 1 \Longrightarrow \mathbf{j}_{\mathbf{xz}}(t,u) = 1$.

Proof. (i) According to Definition 5.4.7, the set T of all $t \in T_{\mathbf{x}}$ such that $\mathbf{j}_{\mathbf{xx}}(t,t) = 1$ is inductive, that is, $\mathtt{Max}\,T_{\mathbf{x}} \subseteq T$ and $t \in T$ provided any $t^\wedge a \in T_{\mathbf{x}}$ belongs to T. Thus, $T = T_{\mathbf{x}}$ by the result of Exercise 5.3.2.

(ii) The map $j(t,r) = \mathbf{j}_{\mathbf{xy}}(\tau^\wedge t, \rho^\wedge r)$ is clearly a bisimulation for the pair of codes $\mathbf{x}|_\tau$, $\mathbf{y}|_\rho$, and hence it coincides with $\mathbf{j}_{\mathbf{x}|_\tau\, \mathbf{y}|_\rho}$ by the uniqueness.

(iii) Consider this as a property of $t \in T_{\mathbf{x}}$ (beginning with $\forall r\, \forall u$), say, $P(t)$. As T is a wf tree, it suffices to prove $P(t)$ for any $t \in \mathtt{Max}\,T_{\mathbf{x}}$ and, if $t \notin \mathtt{Max}\,T_{\mathbf{x}}$, prove $P(t)$ assuming $P(t^\wedge a)$ for any $t^\wedge a \in T_{\mathbf{x}}$.

First prove $P(t)$ for $t \in \mathtt{Max}\, T_{\mathbf{x}}$. In this assumption, $\mathbf{j}_{\mathbf{xy}}(t,r) = 1$ implies $r \in \mathtt{Max}\, T_{\mathbf{y}}$ and $\mathbf{x}(t) = \mathbf{y}(r)$, similarly, we have $u \in \mathtt{Max}\, T_{\mathbf{z}}$ and $\mathbf{y}(r) = \mathbf{z}(u)$, hence, $\mathbf{x}(t) = \mathbf{z}(u)$, now $\mathbf{j}_{\mathbf{xz}}(t,u) = 1$ holds by Definition 5.4.7.

Now suppose that $t \notin \mathtt{Max}\, T_{\mathbf{x}}$. It follows from $\mathbf{j}_{\mathbf{xy}}(t,r) = \mathbf{j}_{\mathbf{yz}}(r,u) = 1$ that $r \notin \mathtt{Max}\, T_{\mathbf{y}}$ and $u \notin \mathtt{Max}\, T_{\mathbf{z}}$. Consider any $t{}^\wedge a \in T_{\mathbf{x}}$. By definition there exist $r{}^\wedge b \in T_{\mathbf{y}}$ and then $u{}^\wedge c \in T_{\mathbf{z}}$ such that $\mathbf{j}_{\mathbf{xy}}(t{}^\wedge a, r{}^\wedge b) = \mathbf{j}_{\mathbf{yz}}(r{}^\wedge b, u{}^\wedge c) = 1$. We conclude that $\mathbf{j}_{\mathbf{xz}}(t{}^\wedge a, u{}^\wedge c) = 1$, by the assumption of $P(t{}^\wedge a)$. Thus,

$$\forall t{}^\wedge a \in T_{\mathbf{x}}\, \exists u{}^\wedge c \in T_{\mathbf{z}}\, (\mathbf{j}_{\mathbf{xz}}(t{}^\wedge a, u{}^\wedge c) = 1)\,.$$

The same argument, in the opposite direction, shows that

$$\forall u{}^\wedge c \in T_{\mathbf{z}}\, \exists t{}^\wedge a \in T_{\mathbf{x}}\, (\mathbf{j}_{\mathbf{xz}}(t{}^\wedge a, u{}^\wedge c) = 1)\,.$$

It follows, by definition, that $\mathbf{j}_{\mathbf{xz}}(t,u) = 1$, as required. □

The proof of the lemma demonstrates the inconvenience of the absence of coded sets for arguments with A-codes in **EEST** : for instance, to prove (iii) in **HST** we can simply note that, say, $\mathbf{j}_{\mathbf{xy}}(t,r) = 1$ is equivalent to $\mathsf{F}_{\mathbf{x}}(t) = \mathsf{F}_{\mathbf{y}}(r)$. (This is similar to Theorem 5.4.11(i).) The equality $\mathsf{A}_{\mathbf{x}} = \{\mathsf{A}_{\mathbf{x}|_a} : a \in \mathtt{Min}\, T_{\mathbf{x}}\}$ (Exercise 5.3.4) is also meaningless in $\mathbb{E}$ in any direct sense, yet we can attach an adequate meaning: ${}^{\mathsf{a}}\!\in$-elements of any $\mathbf{y} \in \mathbf{A}$ are, modulo ${}^{\mathsf{a}}\!\!=$, codes of the form $\mathbf{y}|_b$, $b \in \mathtt{Min}\, T_{\mathbf{y}}$, and only those codes:

Lemma 5.4.13. *Suppose that* $\mathbf{x}$, $\mathbf{y} \in \mathbf{A}$. *Then*

(i) $\mathbf{x} \,{}^{\mathsf{a}}\!\!= \mathbf{y}$ *iff <u>either</u>* $\mathbf{x} = \mathbf{y} = {}^{\mathsf{a}}x$ *for some internal* x *<u>or</u>* $T_{\mathbf{x}} \neq \{\Lambda\} \neq T_{\mathbf{y}}$ *and*
 (a) $\forall b \in \mathtt{Min}\, T_{\mathbf{y}}\, \exists a \in \mathtt{Min}\, T_{\mathbf{x}}\, (\mathbf{x}|_a \,{}^{\mathsf{a}}\!\!= \mathbf{y}|_b)$, *and*
 (b) $\forall a \in \mathtt{Min}\, T_{\mathbf{x}}\, \exists b \in \mathtt{Min}\, T_{\mathbf{y}}\, (\mathbf{x}|_a \,{}^{\mathsf{a}}\!\!= \mathbf{y}|_b)$.

(ii) $\mathbf{x} \,{}^{\mathsf{a}}\!\in \mathbf{y}$ *iff <u>either</u>* $\mathbf{x} = {}^{\mathsf{a}}x$ *and* $\mathbf{y} = {}^{\mathsf{a}}y$ *for some internal sets* $x \in y$ *<u>or</u>* $T_{\mathbf{y}} \neq \{\Lambda\}$ *and there is* $b \in \mathtt{Min}\, T_{\mathbf{y}}$ *such that* $\mathbf{x} \,{}^{\mathsf{a}}\!\!= \mathbf{y}|_b$.

Proof. (i) Suppose that $\mathbf{x} \,{}^{\mathsf{a}}\!\!= \mathbf{y}$, so that $\mathbf{j}_{\mathbf{xy}}(\Lambda,\Lambda) = 1$. If at least one of $T_{\mathbf{x}}$, $T_{\mathbf{y}}$ is $\{\Lambda\}$ then $T_{\mathbf{x}} = T_{\mathbf{y}} = \{\Lambda\}$ and $\mathbf{x}(\Lambda) = \mathbf{y}(\Lambda)$ (Definition 5.4.7). Suppose that $T_{\mathbf{x}} \neq \{\Lambda\} \neq T_{\mathbf{y}}$, so that $\Lambda \notin \mathtt{Max}\, T_{\mathbf{x}} \cup \mathtt{Max}\, T_{\mathbf{y}}$. Prove, for instance, (i)(a). Let $b \in \mathtt{Min}\, T_{\mathbf{y}}$. It follows from 3*(a) of Definition 5.4.7 (with $t = r = \Lambda$) that there is $a \in \mathtt{Min}\, T_{\mathbf{x}}$ with $\mathbf{j}_{\mathbf{xy}}(\langle a\rangle, \langle b\rangle) = 1$. Yet $\mathbf{j}_{\mathbf{x}|_a \mathbf{y}|_b}(\Lambda,\Lambda) = \mathbf{j}_{\mathbf{xy}}(\langle a\rangle, \langle b\rangle)$ by Lemma 5.4.12(ii), hence, $\mathbf{x}|_a \,{}^{\mathsf{a}}\!\!= \mathbf{y}|_b$.

As for the converse, if $T_{\mathbf{x}} = T_{\mathbf{y}} = \{\Lambda\}$ and $\mathbf{x}(\Lambda) = \mathbf{y}(\Lambda)$ then $\mathbf{j}_{\mathbf{xy}}(\Lambda,\Lambda) = 1$ by 1* of Definition 5.4.7, hence, $\mathbf{x} \,{}^{\mathsf{a}}\!\!= \mathbf{y}$. It remains to consider the "or" hypothesis of (i). We are going to prove $\mathbf{j}_{\mathbf{xy}}(\Lambda,\Lambda) = 1$ applying 3* of Definition 5.4.7. Let us check, say, 3*(a) of Definition 5.4.7 ($t = r = \Lambda$). Let $b \in \mathtt{Min}\, T_{\mathbf{y}}$. Then, by (i)(a), there is $a \in \mathtt{Min}\, T_{\mathbf{x}}$ with $\mathbf{j}_{\mathbf{x}|_a\, \mathbf{y}|_b}(\Lambda,\Lambda) = 1$. As above, this implies $\mathbf{j}_{\mathbf{xy}}(\langle a\rangle, \langle b\rangle) = 1$, as required for 3*(a).

(ii) By definition, it suffices to show that, for any $b \in \mathtt{Min}\, T_{\mathbf{y}}$, $x \,{}^{\mathsf{a}}\!\!= \mathbf{y}|_b$ is equivalent to $\mathbf{j}_{\mathbf{xy}}(\Lambda, \langle b\rangle) = 1$. But the latter formula implies $\mathbf{j}_{\mathbf{x}\, \mathbf{y}|_b}(\Lambda,\Lambda) = 1$ by Lemma 5.4.12(ii), as required. □

Corollary 5.4.14. $\overset{a}{=}$ *is an equivalence on* $\mathbf{A}$ *while the relations* ${}^{a}\!\in$, ${}^{a}\mathtt{st}$, ${}^{a}\mathtt{int}$ *are* $\overset{a}{=}$*-invariant.*

Proof. $\overset{a}{=}$ is an equivalence relation on $\mathbf{A}$ by Lemma 5.4.12(iii). That the relation ${}^{a}\!\in$ is $\overset{a}{=}$-invariant follows from Lemma 5.4.13: for instance, if $\mathbf{x} \,{}^{a}\!\in\, \mathbf{y} \overset{a}{=} \mathbf{y}'$ then (in the nontrivial case $T_{\mathbf{y}} \neq \{\Lambda\}$) we have $x \overset{a}{=} \mathbf{y}|_b$ for some $b \in \mathtt{Min}\, T_{\mathbf{y}}$ by 5.4.13(ii), on the other hand, $\mathbf{y}|_b \overset{a}{=} \mathbf{y}'|_{b'}$ for some $b' \in \mathtt{Min}\, T_{\mathbf{y}'}$ by 5.4.13(i), and thus $x \overset{a}{=} \mathbf{y}'|_{b'}$ because $\overset{a}{=}$ is an equivalence relation, thus $\mathbf{x} \,{}^{a}\!\in\, \mathbf{y}'$ still by 5.4.13(ii). To see that the relations ${}^{a}\mathtt{st}$, ${}^{a}\mathtt{int}$ are $\overset{a}{=}$-invariant apply Lemma 5.4.13(i). □

5.4d The interpretation and the embedding

We continue to argue under the assumptions of 5.4.1.

Definition 5.4.15. The results of Corollary 5.4.14 allow us to define an invariant $\mathtt{st}$-$\in$-structure $\mathfrak{a} = \langle \mathbf{A}\,;\, {}^{a}\!\in, {}^{a}\mathtt{st}\,;\, \overset{a}{=} \rangle$. □

Let ${}^{a}\Phi$ be the relativization of any $\mathtt{st}$-$\in$-formula Φ to $\mathbf{e}$ ($\Phi^{\mathfrak{a}}$ in the notation of § 1.5c). Thus ${}^{a}\Phi$ is the formula obtained from Φ as follows:

(A) change all occurrences $p = q$, $p \in q$, $\mathtt{st}\, p$ to resp. $p \overset{a}{=} q$, $p \,{}^{a}\!\in\, q$, ${}^{a}\mathtt{st}\, p$;

(B) relativize all quantifiers to $\mathbf{A}$.

Lemma 5.4.16 (HST). *The map* $\mathbf{x} \mapsto \mathsf{A}_{\mathbf{x}}$ *is a reduction of* $\mathfrak{a}$ *to the structure* $\langle \mathbb{L}[\mathbb{I}]; \in, \mathtt{st}\rangle$, *where* $\mathbb{L}[\mathbb{I}] = \{\mathsf{A}_{\mathbf{x}} : \mathbf{x} \in \mathbf{A}\}$. [5] *Therefore we have*

$$\Phi(\mathsf{A}_{\mathbf{x}_1}, ..., \mathsf{A}_{\mathbf{x}_n})^{\mathbb{L}[\mathbb{I}]} \iff {}^{a}\Phi(\mathbf{x}_1, ..., \mathbf{x}_n) \iff ({}^{a}\Phi(\mathbf{x}_1, ..., \mathbf{x}_n))^{\mathbb{E}} \qquad (*)$$

for any $\mathtt{st}$-$\in$*-formula* $\Phi(x_1, ..., x_n)$ *and any* $\mathbf{x}_1, ..., \mathbf{x}_n \in \mathbf{A}$.

Proof. The reduction claim follows from Theorem 5.4.11. The consequence is a particular case of Proposition 1.5.10. The rightmost equivalence follows from the fact that the domain $\mathbf{A}$ of the structure $\mathfrak{a}$ is a subclass of $\mathbb{E}$. □

Theorem 5.4.17. (i) *The structure* $\mathfrak{a} = \langle \mathbf{A}\,;\, {}^{a}\!\in, {}^{a}\mathtt{st}\,;\, \overset{a}{=} \rangle$ *is an internal core interpretation of* **HST** *in* **EEST**.

(ii) (**EEST**) *The map* $x \mapsto {}^{a}x$ *restricted to* $\mathbb{I}$ *is an internal core embedding of* $\langle \mathbb{I}; \in, \mathtt{st}\rangle$ (the internal core of the **EEST** set universe) *into* $\mathfrak{a}$, *moreover, a* $\mathtt{st}$-$\in$*-isomorphism of* $\langle \mathbb{I}; \in, \mathtt{st}\rangle$ *onto* $\langle \mathbb{I}^{(\mathfrak{a})}; {}^{a}\!\in, {}^{a}\mathtt{st}\rangle$.

(iii) (**EEST**) *We have* $\Phi(x_1, ..., x_n)^{\mathtt{int}} \iff {}^{a}(\Phi({}^{a}x_1, ..., {}^{a}x_n)^{\mathtt{int}})$ *for any* $\mathtt{st}$-$\in$*-formula* Φ *and any (internal) sets* $x_1, ..., x_n$.

The proof of the theorem will continue until the end of § 5.4e.

First of all let us study properties of the map $x \mapsto {}^{a}x : \mathbb{E} \to \mathbf{A}$. We are going to prove slightly more than asserted by (ii) of the theorem, namely that the map $\mathtt{st}$-$\in$-isomorphically embeds $\mathbb{E}$ onto a meaningful part of $\mathfrak{a}$.

[5] The class $\mathbb{L}[\mathbb{I}]$ will be considered in detail in Section 5.5.

Definition 5.4.18. A set x is *sub-internal* if it consists of internal elements. Accordingly $\mathtt{subint}\,x$ is the $\mathtt{st}$-$\in$-formula $\forall y\,(y \in x \Longrightarrow \mathtt{int}\,y)$.
$\mathbb{P} = \{x : \mathtt{subint}\,x\}$ is the class of all sub-internal sets. We define

$$\mathbb{P}^{(\mathfrak a)} = \{\mathbf{x} \in \mathbf{A} : (\mathtt{subint}\,x)^{\mathfrak a}\} = \{\mathbf{x} \in \mathbf{A} : \forall \mathbf{y} \in \mathbf{A}\,(\mathbf{y} \,{}^{\mathtt a}{\in}\, \mathbf{x} \Longrightarrow \mathbf{y} \in \mathbb{I}^{(\mathfrak a)})\}\,,$$

the *sub-internal core* of $\mathfrak a$. □

Obviously $\mathbb{E} \subseteq \mathbb{P}$ in **HST** and $\mathbb{E} = \mathbb{P} =$ all sets in **EEST**.

Lemma 5.4.19. *The map $x \mapsto {}^{\mathtt a}x : \mathbb{E} \to \mathbf{A}$ satisfies the following:*

(a) *For any $\mathbf{x} \in \mathbf{A}$, $z \in \mathbb{E}$ we have:*

(1) *if $\mathbf{x} \,{}^{\mathtt a}{=}\, {}^{\mathtt a}z$ and z is internal then simply $\mathbf{x} = {}^{\mathtt a}z$;*

(2) *if $\mathbf{x} \,{}^{\mathtt a}{\in}\, {}^{\mathtt a}z$ then there is $x \in z$ (necessarily internal) with $\mathbf{x} = {}^{\mathtt a}x$.*

(b) *For all $x, y \in \mathbb{E}$:* $x = y$ *iff* ${}^{\mathtt a}x \,{}^{\mathtt a}{=}\, {}^{\mathtt a}y$; $\mathtt{st}\,x$ *iff* ${}^{\mathtt a}\mathtt{st}\,{}^{\mathtt a}x$;
$x \in y$ *iff* ${}^{\mathtt a}x \,{}^{\mathtt a}{\in}\, {}^{\mathtt a}y$; $\mathtt{int}\,x$ *iff* ${}^{\mathtt a}\mathtt{int}\,{}^{\mathtt a}x$.

(c) *We have:*

$$\mathbb{S}^{(\mathfrak a)} = \{\mathbf{x} \in \mathbf{A} : {}^{\mathtt a}\mathtt{st}\,\mathbf{x}\} = \{{}^{\mathtt a}x : x \in \mathbb{S}\}\,;$$
$$\mathbb{I}^{(\mathfrak a)} = \{\mathbf{x} \in \mathbf{A} : {}^{\mathtt a}\mathtt{int}\,\mathbf{x}\} = \{{}^{\mathtt a}x : x \in \mathbb{I}\}\,;$$
$$\mathbb{P}^{(\mathfrak a)} = \{\mathbf{x} \in \mathbf{A} : \exists x \in \mathbb{E}\,({}^{\mathtt a}x \,{}^{\mathtt a}{=}\, \mathbf{x})\}\,;$$

and the classes $\mathbb{I}^{(\mathfrak a)}$ and $\mathbb{P}^{(\mathfrak a)}$ are ${}^{\mathtt a}{\in}$-transitive in $\mathfrak a$.

Proof. (a) Both (a)(1) and the case of internal z in (a)(2) immediately follow from Lemma 5.4.13. If z is not internal in (a)(2) then $T_{{}^*z} = \langle\Lambda\rangle \cup \{\langle c\rangle : c \in z\}$, $\mathtt{Min}\,T_{{}^*z} = z$ and ${}^*z|_y = {}^{\mathtt a}y$ for any $c \in z$. Thus, if $\mathbf{x} \,{}^{\mathtt a}{\in}\, {}^{\mathtt a}z$ then $\mathbf{x} \,{}^{\mathtt a}{=}\, {}^{\mathtt a}y$ for some (internal) $y \in z$ by Lemma 5.4.13(ii), and thus $\mathbf{x} = {}^{\mathtt a}y$ by (a)(1).

(b) Suppose that ${}^{\mathtt a}x \,{}^{\mathtt a}{=}\, {}^{\mathtt a}y$, and hence $\mathbf{j}_{{}^{\mathtt a}x\,{}^{\mathtt a}y}(\Lambda,\Lambda) = 1$. We can apply Lemma 5.4.13(i): the "either" case is easy, thus, we can assume that $T_{{}^{\mathtt a}x} \neq \{\Lambda\} \neq T_{{}^{\mathtt a}y}$. Then by definition (Definition 5.3.5) sets x, y are non-internal, in addition, $T_{{}^{\mathtt a}x} = \{\Lambda\} \cup \{\langle a\rangle : a \in x\}$ and $T_{{}^{\mathtt a}y} = \{\Lambda\} \cup \{\langle b\rangle : b \in y\}$, accordingly $\mathtt{Max}\,T_{{}^{\mathtt a}x} = \{\langle a\rangle : a \in x\}$, $\mathtt{Max}\,T_{{}^{\mathtt a}y} = \{\langle b\rangle : b \in y\}$, and finally, ${}^{\mathtt a}x(\langle a\rangle) = a$ and ${}^{\mathtt a}y(\langle b\rangle) = b$ for all $a \in x$, $b \in y$. In this case, it follows from Definition 5.4.7 that the only chance for $\mathbf{j}_{{}^{\mathtt a}x\,{}^{\mathtt a}y}(\Lambda,\Lambda) = 1$ is that $x = y$.

Suppose that ${}^{\mathtt a}x \,{}^{\mathtt a}{\in}\, {}^{\mathtt a}y$. If y is internal then $T_{{}^{\mathtt a}y} = \{\Lambda\}$, and hence by Definition 5.4.10 ${}^{\mathtt a}x \,{}^{\mathtt a}{\in}\, {}^{\mathtt a}y$ holds iff x is also internal and $x \in y$. Now assume that $y \notin \mathbb{I}$. Then $T_{{}^{\mathtt a}y} = \{\Lambda\} \cup \{\langle b\rangle : b \in y\}$, $\mathtt{Min}\,T_{{}^{\mathtt a}y} = y$, and clearly $({}^{\mathtt a}y)|_b = {}^{\mathtt a}b$ for any $b \in y$. Now we have ${}^{\mathtt a}x \,{}^{\mathtt a}{\in}\, {}^{\mathtt a}y$ iff ${}^{\mathtt a}x \,{}^{\mathtt a}{=}\, {}^{\mathtt a}b$ for some $b \in y$ (by Definition 5.4.10) iff $x = b$ by the above.

The equivalences for $\mathtt{st}$ and $\mathtt{int}$ hold by Definition 5.4.10.

(c). The equivalence for $\mathbb{S}^{(\mathfrak a)}$ holds by the definition of ${}^{\mathtt a}\mathtt{st}$. Prove the equivalence for $\mathbb{I}^{(\mathfrak a)}$. If $y \in \mathbb{I}$ then there is a standard x with $y \in x$. We have ${}^{\mathtt a}y \,{}^{\mathtt a}{\in}\, {}^{\mathtt a}x$ by (b) and ${}^{\mathtt a}x \in \mathbb{S}^{(\mathfrak a)}$ by the equivalence for $\mathbb{S}^{(\mathfrak a)}$, therefore ${}^{\mathtt a}y \in \mathbb{I}^{(\mathfrak a)}$. Conversely if $\mathbf{x} \in \mathbb{I}^{(\mathfrak a)}$ then, by definition, there exists $\mathbf{y} \in \mathbb{S}^{(\mathfrak a)}$ with $\mathbf{x} \,{}^{\mathtt a}{\in}\, \mathbf{y}$.

It follows from the equivalence for $\mathbb{S}^{(\mathfrak{a})}$ that $\mathbf{y} = {}^{\mathfrak{a}}y$ for a standard y, and hence $\mathbf{x} \mathrel{{}^{\mathfrak{a}}\!\!\in} {}^{\mathfrak{a}}y$. It remains to apply (a)(2).

Prove the equivalence for $\mathbb{P}^{(\mathfrak{a})}$. Suppose that $x \in \mathbb{E}$, $\mathbf{x} \in \mathbf{A}$, $\mathbf{x} \mathrel{{}^{\mathfrak{a}}\!\!=} {}^{\mathfrak{a}}x$. To prove that $\mathbf{x} \in \mathbb{P}^{(\mathfrak{a})}$ take any $\mathbf{y} \in \mathbf{A}$ such that $\mathbf{y} \mathrel{{}^{\mathfrak{a}}\!\!\in} \mathbf{x}$ and prove that $\mathbf{y} \in \mathbb{I}^{(\mathfrak{a})}$. We have $\mathbf{y} \mathrel{{}^{\mathfrak{a}}\!\!\in} {}^{\mathfrak{a}}x$ by the ${}^{\mathfrak{a}}\!\!=$-invariance of ${}^{\mathfrak{a}}\!\!\in$, and hence there is a (necessarily internal) $y \in x$ such that $\mathbf{y} = {}^{\mathfrak{a}}y$ by (a)(2). Thus $\mathbf{y} \in \mathbb{I}^{(\mathfrak{a})}$ as required. Conversely, suppose that $\mathbf{x} \in \mathbb{P}^{(\mathfrak{a})}$. To avoid trivialities, we may assume that $\mathbf{x}$ is not ${}^{\mathfrak{a}}y$ for some $y \in \mathbb{I}$, thus $T_{\mathbf{x}} \neq \{\Lambda\}$. Let $A = \mathtt{Min}\, T_{\mathbf{x}}$ (a set by Lemma 5.4.6). If $a \in A$ then $\mathbf{x}|_a \mathrel{{}^{\mathfrak{a}}\!\!\in} \mathbf{x}$ by Lemma 5.4.13, and hence $\mathbf{x}|_a \in \mathbb{I}^{(\mathfrak{a})}$, that is, $\mathbf{x}|_a = {}^{\mathfrak{a}}y$ for some (obviously unique) internal $y = y_a$ by the above. It follows from the **EEST** Collection Theorem 5.2.14 and Separation that $x = \{y_a : a \in A\}$ is a set in $\mathbb{E}$, furthermore, ${}^{\mathfrak{a}}x \mathrel{{}^{\mathfrak{a}}\!\!=} \mathbf{x}$ by Lemma 5.4.13.

The ${}^{\mathfrak{a}}\!\!\in$-transitivity of the classes follows from (a). □

We observe that (ii) of Theorem 5.4.17 follows from Lemma 5.4.19 and implies (iii) of Theorem 5.4.17 by the same reasons as in Theorem 5.2.6.

5.4e Verification of the HST axioms

We finally prove the principal claim (i) of Theorem 5.4.17: the structure $\mathfrak{a}$ is an interpretation of **HST**.

We continue to argue under the assumptions of 5.4.1.

Our verification of the axioms of **HST** in $\mathfrak{a}$ is divided into three parts, in accordance with the structure of the axiomatics of **HST** in § 1.1. Our main technical tool will be the following lemma:

Lemma 5.4.20 (EEST). *For any set-like collection $Z \subseteq \mathbf{A}$ there is a code $\mathbf{X} \in \mathbf{A}$ such that codes $\mathbf{x} \in Z$ are the only ${}^{\mathfrak{a}}\!\!\in$-elements of $\mathbf{X}$ modulo ${}^{\mathfrak{a}}\!\!=$, in other words, we have $\mathbf{x} \mathrel{{}^{\mathfrak{a}}\!\!\in} \mathbf{X}$ for any $\mathbf{x} \in Z$, and conversely, if $\mathbf{y} \in \mathbf{A}$ satisfies $\mathbf{y} \mathrel{{}^{\mathfrak{a}}\!\!\in} \mathbf{X}$ then there is a code $\mathbf{x} \in Z$ such that $\mathbf{y} \mathrel{{}^{\mathfrak{a}}\!\!=} \mathbf{x}$.* □

The lemma is applicable under the assumptions 5.4.1. Note that in **HST** set-like collections $Z \subseteq \mathbb{E}$ (see Definition 5.2.15) are sets while in **EEST** no set can contain non-internal elements.

Proof. By definition there is a set $C \in \mathbb{E}$ such that $Z \subseteq \{\mathsf{E}_p : p \in C\}$. By Lemma 5.2.13 we have $C \subseteq S$ for a standard set S. The set $P = \{a \in S : \mathsf{E}_a \in Z\}$ is $\mathtt{st}$-$\in$-definable in $\mathbb{E}$, and hence it is a set in $\mathbb{E}$ because $\mathbb{E}$ satisfies **EEST** by Theorem 5.2.11. Define $\mathbf{X}(a{}^\wedge t) = \mathsf{E}_a(t)$ whenever $a \in P$ and $t \in \mathtt{Max}\, T_{\mathsf{E}_a}$, then $\mathbf{X}$ is an A-code, and $\mathsf{A}_{\mathbf{X}} = \{\mathsf{A}_{\mathbf{x}} : \mathbf{x} \in Z\}$. Note that $\mathbf{X}$ is $\mathtt{st}$-$\in$-definable in $\mathbb{E}$ using only $P \in \mathbb{E}$ as a parameter, hence $\mathbf{X}$ is a set in $\mathbb{E}$ (Theorem 5.2.11 again) and thus $\mathbf{X} \in \underline{\mathbf{A}}$. Now (see Exercise 5.3.14) either $\mathbf{X}$ is a regular code, thus $\mathbf{X} \in \mathbf{A}$ as required, or $y = \mathsf{A}_{\mathbf{X}} \in \mathbb{I}$, and then the code ${}^{\mathfrak{a}}y \in \mathbf{A}$ proves the result. □

Part 1: axioms for the external universe

We prove that $\mathfrak{a}$ satisfies all **HST** axioms of the first group (see § 1.1b).

Extensionality. We have to prove that if codes $\mathbf{x}$, $\mathbf{y} \in \mathbf{A}$ satisfy

($\dagger$) $\mathbf{z} \,{}^{\mathsf{a}}\!\!\in \mathbf{x} \iff \mathbf{z} \,{}^{\mathsf{a}}\!\!\in \mathbf{y}$ for any $\mathbf{z} \in \mathbf{A}$

then $\mathbf{x} \,{}^{\mathsf{a}}\!\!= \mathbf{y}$. If both $\mathbf{x}$, $\mathbf{y}$ have the form resp. ${}^{\mathsf{a}}x$, ${}^{\mathsf{a}}y$ for $x, y \in \mathbb{I}$ then ($\dagger$) means, by Lemma 5.4.13(ii), that $z \in x \iff z \in y$ for any internal z, hence $x = y$ and $\mathbf{x} = \mathbf{y}$. If $\mathbf{x}$, $\mathbf{y}$ satisfy $T_{\mathbf{x}} \neq \{\Lambda\} \neq T_{\mathbf{y}}$ then it follows from Lemma 5.4.13(ii) and the transitivity of ${}^{\mathsf{a}}\!\!=$ that the "or" case of Lemma 5.4.13(i) holds, thus $\mathbf{x} \,{}^{\mathsf{a}}\!\!= \mathbf{y}$. Consider the mixed case, *e.g.*, $\mathbf{x} = {}^{\mathsf{a}}x$ for some $x \in \mathbb{I}$ but $T_{\mathbf{y}} \neq \{\Lambda\}$. According to Lemma 5.4.13(ii), ${}^{\mathsf{a}}\!\!\in$-elements of $\mathbf{x} = {}^{\mathsf{a}}x$ are codes of the form ${}^{\mathsf{a}}z$, $z \in x$, hence, by ($\dagger$), they are the only ${}^{\mathsf{a}}\!\!\in$-elements of $\mathbf{y}$, which implies, still by Lemma 5.4.13(ii), that $|T_{\mathbf{y}}| = 1$. Then $T_{\mathbf{y}} = \{\Lambda\} \cup \{\langle a \rangle : a \in A = \mathtt{Min}\, T_{\mathbf{y}}\}$, and in addition $\{\mathbf{y}(a) : a \in A\} = x$. But this contradicts the regularity of $\mathbf{y}$ because the set x is internal.

Pair. Apply Lemma 5.4.20 for $Z = \{\mathbf{x}, \mathbf{y}\}$, a given pair of codes in $\mathbf{A}$.

Union. Let $\mathbf{x} \in \mathbf{A}$. We have to find a code $\mathbf{U} \in \mathbf{A}$ such that $\mathbf{z} \,{}^{\mathsf{a}}\!\!\in \mathbf{U}$ iff $\mathbf{z} \,{}^{\mathsf{a}}\!\!\in \mathbf{y} \,{}^{\mathsf{a}}\!\!\in \mathbf{x}$ for some $\mathbf{y}$. If $\mathbf{x} = {}^{\mathsf{a}}x$, $x \in \mathbb{I}$ then let $U = \bigcup x$ (an internal set) and $\mathbf{U} = {}^{\mathsf{a}}U$. If $T_{\mathbf{x}} \neq \{\Lambda\}$ then let Z be the collection of all codes of the form $\mathbf{x}|_{\langle a,b \rangle}$, where $\langle a, b \rangle$ (a 2-term sequence) belongs to $T_{\mathbf{x}}$, and all codes ${}^{\mathsf{a}}u$ such that $u \in \mathbf{x}(\langle a \rangle)$ and $\langle a \rangle \in \mathtt{Max}\, T_{\mathbf{y}}$ (then u is internal). It follows from Theorem 5.2.16 that Z is a set-like collection. Let, by Lemma 5.4.20, $\mathbf{U} \in \mathbf{A}$ ${}^{\mathsf{a}}\!\!\in$-contain all codes in Z and nothing more (modulo ${}^{\mathsf{a}}\!\!=$). Lemma 5.4.13 implies that in both cases $\mathbf{U}$ is as required.

Infinity. This axiom in fact follows from the other axioms of **HST**, for instance any set $x \in \mathbb{S}$ infinite in the sense of $\mathbb{S}$ will be infinite in the **HST** universe as well. Yet it is a useful **exercise** to verify that it is true in $\mathfrak{a}$ that, say, the code ${}^{\mathsf{a}}\mathbb{N} \in \mathbf{A}$ is an infinite set in $\mathfrak{a}$.

Separation. Let $\mathbf{X}$ be an A-code in $\mathbf{A}$ and $\Phi(x)$ be a $\mathtt{st}$-$\in$-formula with codes in $\mathbf{A}$ as parameters. We have to find a code $\mathbf{Y} \in \mathbf{A}$ which ${}^{\mathsf{a}}\!\!\in$-contains any code $\mathbf{x} \,{}^{\mathsf{a}}\!\!\in \mathbf{X}$ with ${}^{\mathsf{a}}\Phi(\mathbf{x})$, and (modulo ${}^{\mathsf{a}}\!\!=$) nothing more. If $\mathbf{X} = {}^{\mathsf{a}}X$, $X \in \mathbb{I}$, then let Z be the collection of all codes ${}^{\mathsf{a}}x$, $x \in X$ satisfying ${}^{\mathsf{a}}\Phi({}^{\mathsf{a}}x)$. If $T_{\mathbf{X}} \neq \{\Lambda\}$ then let Z be the collection of all codes of the form $\mathbf{X}|_a$, $a \in \mathtt{Min}\, T_{\mathbf{X}}$, still satisfying ${}^{\mathsf{a}}\Phi(\mathbf{X}|_a)$. Applying Lemma 5.4.20 to Z (which is a set-like collection by Theorem 5.2.16) we obtain a code $\mathbf{Y} \in \mathbf{A}$ as required.

Collection. Let $\mathbf{X}$ be a code in $\mathbf{A}$ and $\Phi(x, y)$ be a $\mathtt{st}$-$\in$-formula with codes in $\mathbf{A}$ as parameters. We have to find a code $\mathbf{Y} \in \mathbf{A}$ such that

$$\exists\, \mathbf{y}\; {}^{\mathsf{a}}\Phi(\mathbf{x}, \mathbf{y}) \implies \exists\, \mathbf{y}\, \big(\mathbf{y} \,{}^{\mathsf{a}}\!\!\in \mathbf{Y} \wedge {}^{\mathsf{a}}\Phi(\mathbf{x}, \mathbf{y})\big)$$

for any A-code $\mathbf{x}$ with $\mathbf{x} \,{}^{\mathsf{a}}\!\!\in \mathbf{X}$. According to Lemma 5.4.20, it suffices to find a set-like collection $Z \subseteq \mathbf{A}$ such that (assuming that $T_{\mathbf{X}} \neq \{\Lambda\}$)

$$(\forall\, a \in \mathtt{Min}\, T_{\mathbf{X}})\, \big(\exists\, \mathbf{y}\; {}^{\mathsf{a}}\Phi(\mathbf{X}|_a, \mathbf{y}) \implies \exists\, \mathbf{y} \in Z\; {}^{\mathsf{a}}\Phi(\mathbf{X}|_a, \mathbf{y})\big)\,.$$

To get such a Z, just apply Theorem 5.2.16.

Part 2: axioms for standard and internal sets

It follows from Lemma 5.4.19 that the $\mathtt{st}$-$\in$-structure of the internal core $\mathbb{I}^{(\mathfrak{a})}$ of $\mathfrak{a}$ is identical to the $\mathtt{st}$-$\in$-structure of the internal core $\mathbb{I}$ of $\mathbb{E}$. Therefore **ZFC**$^{\mathbf{st}}$ and Transfer of **HST** for $\mathfrak{a}$ are immediate corollaries of the corresponding axioms of **EEST**.

Transitivity of $\mathbb{I}$: follows from Lemma 5.4.19(c).

Standardization. Let $\mathbf{X}$ be a code in $\mathbf{A}$. Assume that $T_{\mathbf{X}} \neq \{\Lambda\}$. Let $A = \mathtt{Min}\, T$. Then $D = \{x \in \mathbb{S} : \exists\, a \in A\, (\mathbf{X}|_a \overset{\mathsf{a}}{=} {}^{\mathsf{a}}x)\}$ is a set in $\mathbb{E}$ (by Theorem 5.2.14). Moreover there is a standard set S with $D = S \cap \mathbb{S}$. If $T_{\mathbf{X}} = \{\Lambda\}$, that is $\mathbf{X} = {}^{\mathsf{a}}X$ for some $X \in \mathbb{I}$, then we let S be a standard set with $X \cap \mathbb{S} = S \cap \mathbb{S}$. In both cases, we have ${}^{\mathsf{a}}\mathtt{st}\, {}^{\mathsf{a}}S$, and moreover, the equivalence ${}^{\mathsf{a}}x \in {}^{\mathsf{a}}S \Longleftrightarrow {}^{\mathsf{a}}x \in \mathbf{X}$ holds for any standard x.

Regularity over $\mathbb{I}$. Let $\mathbf{X}$ be a code in $\mathbf{A}$, nonempty in $\mathfrak{a}$ in the sense that there is at least one code in $\mathbf{A}$ which ${}^{\mathsf{a}}\!\in$-belongs to $\mathbf{X}$. We have to find another code $\mathbf{x}$ with $\mathbf{x} \;{}^{\mathsf{a}}\!\!\in \mathbf{X}$ such that any $\mathbf{y} \in \mathbf{A}$ which is an ${}^{\mathsf{a}}\!\in$-element of both $\mathbf{x}$ and $\mathbf{X}$ satisfies ${}^{\mathsf{a}}\mathtt{int}\,\mathbf{y}$.

We leave it as an **exercise** to show using Lemma 5.4.19 that if $\mathbf{X} = {}^{\mathsf{a}}X$, X internal then a code $\mathbf{x} = {}^{\mathsf{a}}x$, where x is any element of X, is as required.

Now consider the case when $T_{\mathbf{X}} \neq \{\Lambda\}$. Then the set

$$T = \{t \in T_{\mathbf{X}} : \exists\, a \in \mathtt{Min}\, T_{\mathbf{X}}\, (\mathbf{X}|_t \overset{\mathsf{a}}{=} \mathbf{X}|_a)\}$$

is nonempty as well, for instance, $\mathtt{Min}\, T_{\mathbf{X}} \subseteq T$. As $T_{\mathbf{X}}$ is well-founded, there exists $t \in T$ such that none among the extensions $t^\wedge b \in T_{\mathbf{X}}$ belongs to T. Let $a \in \mathtt{Min}\, T_{\mathbf{X}}$ witness that $t \in T$. Then $\mathbf{x} = \mathbf{X}|_a$ (Example 5.3.4) is a code in $\mathbf{A}$ and $\mathbf{x} \;{}^{\mathsf{a}}\!\!\in \mathbf{X}$ by Lemma 5.4.13. If now $\langle a \rangle \in \mathtt{Max}\, T_{\mathbf{X}}$ then $\mathbf{x} = {}^{\mathsf{a}}x$, where $x = \mathbf{X}(\langle a \rangle)$, so that $\mathbf{x} \in \mathbb{I}^{(\mathfrak{a})}$. Thus in this case $\mathbf{X}$ contains a ${}^{\mathsf{a}}\mathtt{int}$-internal ${}^{\mathsf{a}}\!\in$-element $\mathbf{x}$. It remains to apply Transitivity of $\mathbb{I}$ in $\mathfrak{a}$.

Suppose that $\langle a \rangle \notin \mathtt{Max}\, T_{\mathbf{X}}$. We claim that $\mathbf{x} \cap \mathbf{X} = \varnothing$ in $\mathfrak{a}$. Let, on the contrary, a code $\mathbf{y} \in \mathbf{A}$ satisfy $\mathbf{y} \;{}^{\mathsf{a}}\!\!\in \mathbf{x}$ and $\mathbf{y} \;{}^{\mathsf{a}}\!\!\in \mathbf{X}$. By Lemma 5.4.13 there is $a' \in \mathtt{Min}\, T_{\mathbf{X}}$ such that $\mathbf{y} \overset{\mathsf{a}}{=} \mathbf{X}|_{a'}$, and there is $b \in \mathtt{Min}\, T_{\mathbf{x}}$ such that $\mathbf{y} \overset{\mathsf{a}}{=} \mathbf{x}|_b$, which implies $\mathbf{y} \overset{\mathsf{a}}{=} \mathbf{X}|_{\langle a,b \rangle}$. We conclude that $\mathbf{X}|_{a'} \overset{\mathsf{a}}{=} \mathbf{X}|_{\langle a,b \rangle}$. Since $\mathbf{X}|_t \overset{\mathsf{a}}{=} \mathbf{X}|_a$ and $\langle a \rangle \notin \mathtt{Max}\, T_{\mathbf{X}}$, there exists b' such that $t^\wedge b' \in T_{\mathbf{X}}$ and $\mathbf{X}|_{t^\wedge b'} \overset{\mathsf{a}}{=} \mathbf{X}|_{\langle a,b \rangle}$. Then $\mathbf{X}|_{t^\wedge b'} \overset{\mathsf{a}}{=} \mathbf{X}|_{a'}$, therefore $t^\wedge b' \in T$, contradiction.

Part 3: axioms for sets of standard size

Note that Saturation (as defined in § 1.1f) is obviously relativized to the class $\mathbb{P} = \{x : x \subseteq \mathbb{I}\}$ (of sets which contain only internal elements). However it follows from Lemma 5.4.19 that the map sending every $\mathbf{x} \in \mathbb{P}^{(\mathfrak{a})}$ to the unique $x \in \mathbb{E}$ with $\mathbf{x} \overset{\mathsf{a}}{=} {}^{\mathsf{a}}x$ is a reduction of the invariant structure $\langle \mathbb{P}^{(\mathfrak{a})}; {}^{\mathsf{a}}\!\in, {}^{\mathsf{a}}\mathtt{st}; \overset{\mathsf{a}}{=} \rangle$ to $\langle \mathbb{E}; \in, \mathtt{st} \rangle$ (in the sense of Definition 1.5.9), and hence both structures have the same true $\mathtt{st}$-$\in$-statements. But Saturation holds in $\mathbb{E}$ by Theorem 5.2.18.

The following lemma demonstrates that the other two axioms of this group, Standard Size Choice and Dependent Choice, also are essentially relativized to the same class, although this is not immediately clear.

Lemma 5.4.21. *For any code* $\mathbf{x} \in \mathbf{A}$ *there is a set* D *and a code* $\mathbf{f} \in \mathbf{A}$ *such that the following is true in* $\mathfrak{a}$: "$\mathbf{f}$ *is a function mapping* $\mathbf{D}$ *onto* $\mathbf{x}$", *where* $\mathbf{D} = {}^{\mathtt{a}}D \in \mathbf{A}$.

This lemma, together with the already verified axioms, shows that both Standard Size Choice and Dependent Choice follow from the instances where the domain of choices consists of internal sets. Thus the same argument as for Saturation above derives Standard Size Choice and Dependent Choice in $\mathfrak{a}$ from the relevant results in $\mathbb{E}$ (Theorem 5.2.14 and Exercise 5.2.19).

Proof (Lemma). We can assume that $T_{\mathbf{x}} \neq \{\Lambda\}$ (the case $\mathbf{x} = {}^{\mathtt{a}}x$ for some $x \in \mathbb{I}$ is rather elementary). Informally, as $\mathbf{x}$ is assumed to ${}^{\mathtt{a}}{\in}$-contain $\mathbf{x}|_a$, $a \in D = \mathtt{Min}\, T_{\mathbf{x}}$, as elements, we can map D onto $\mathbf{x}$ sending every $a \in D$ to $\mathbf{x}|_a$. To be more accurate, let, for any codes $\mathbf{u}$, $\mathbf{v}$ in $\mathbf{A}$, $[\mathbf{u}, \mathbf{v}]$ denote a code $\mathbf{p} \in \mathbf{A}$ such that "$\mathbf{p}$ is a set containing $\mathbf{u}$, $\mathbf{v}$ and nothing more" holds in the structure $\mathfrak{a} = \langle \mathbf{A}\,;\, {}^{\mathtt{a}}{\in}, {}^{\mathtt{a}}\mathtt{st}\,;\, {}^{\mathtt{a}}{=} \rangle$. We put $(\mathbf{u}, \mathbf{v}) = [[\mathbf{u}, \mathbf{u}], [\mathbf{u}, \mathbf{v}]]$, a code, in $\mathbf{A}$, for the ordered pair $\langle \mathbf{u}, \mathbf{v} \rangle = \{\{\mathbf{u}\}, \{\mathbf{u}, \mathbf{v}\}\}$. Let finally $\mathbf{f}$ be an A-code defined to ${}^{\mathtt{a}}{\in}$-contain codes $({}^{\mathtt{a}}a, \mathbf{x}|_a)$, where $a \in D$, and only them (as in the proofs of Separation and Collection). □

□ (*Theorems 5.4.17 and 5.1.4*)

5.4f Superposition of interpretations

To accomplish the proof of Corollary 5.1.5, we now prove its claims 4°, 5° by a rather straightforward superposition of the interpretations involved in the proofs of items 1°, 2°, 3° of Theorem 5.1.4.

Part 4° of Corollary 5.1.5

Recall that $\mathfrak{a} = \langle \mathbf{A}\,;\, {}^{\mathtt{a}}{\in}, {}^{\mathtt{a}}\mathtt{st}\,;\, {}^{\mathtt{a}}{=} \rangle$ is an invariant internal core interpretation of **HST** in **EEST** defined in § 5.4d (Theorem 5.4.17). Thus each of $\mathbf{A}$, ${}^{\mathtt{a}}{\in}$, ${}^{\mathtt{a}}\mathtt{st}$, ${}^{\mathtt{a}}{=}$ is $\mathtt{st}$-$\in$-definable in the **EEST** universe, **EEST** proves that $\mathfrak{a}$ is an invariant structure and proves $\Phi^{\mathfrak{a}}$ for any axiom Φ of **HST**. Finally there is a map $x \mapsto {}^{\mathtt{a}}x$, provably in **EEST** a $\mathtt{st}$-$\in$-isomorphism of $\mathbb{I}$, the internal universe of the **EEST** set universe, onto the internal core $\mathbb{I}^{(\mathfrak{a})}$ of $\mathfrak{a}$.

Recall that $\mathfrak{e} = \langle \mathbf{E}\,;\, {}^{\mathtt{e}}{\in}, {}^{\mathtt{e}}\mathtt{st}\,;\, {}^{\mathtt{e}}{=} \rangle$ is an invariant internal core interpretation of **EEST** in **BST**, § 5.2a. Each of $\mathbf{E}$, ${}^{\mathtt{e}}{\in}$, ${}^{\mathtt{e}}\mathtt{st}$, ${}^{\mathtt{e}}{=}$ is $\mathtt{st}$-$\in$-definable in the **EEST** universe, **BST** proves that $\mathfrak{e}$ is an invariant structure, and proves $\Phi^{\mathfrak{e}}$ for any axiom Φ of **HST**, and there is a $\mathtt{st}$-$\in$-isomorphism $x \mapsto {}^{\mathtt{e}}x$ of the (internal) universe $\mathbb{I}$ of **BST** onto the internal core $\mathbb{I}^{(\mathfrak{e})}$ of $\mathfrak{e}$. (Theorem 5.2.6.)

Arguing in **BST**, consider the *superposition* $\mathfrak{u}$ of $\mathfrak{a}$ and $\mathfrak{e}$. Thus $\mathfrak{u} = \langle \mathbf{U}\,;\, {}^{\mathtt{u}}{\in}, {}^{\mathtt{u}}\mathtt{st}\,;\, {}^{\mathtt{u}}{=} \rangle$, where $\mathbf{U} = \{p \in \mathbf{E} : {}^{\mathtt{e}}(p \in \mathbf{A})\}$ and for $u, v \in \mathbf{U}$:

$$u \stackrel{\mathsf{u}}{=} v \text{ iff } {}^{\mathsf{e}}(u \stackrel{\mathsf{a}}{=} v), \quad u \,{}^{\mathsf{u}}\!\!\in v \text{ iff } {}^{\mathsf{e}}(u \,{}^{\mathsf{a}}\!\!\in v), \quad {}^{\mathsf{u}}\mathtt{st}\, v \text{ iff } {}^{\mathsf{e}}({}^{\mathsf{a}}\mathtt{st}\, v).$$

Thus $\mathbf{U}$, the domain of $\mathfrak{u}$, consists of those elements $u \in \mathbf{E}$ which belong to the domain $\mathbf{A}$ of the structure $\mathfrak{a}$ *defined within* $\mathfrak{e}$. The relations ${}^{\mathsf{u}}\!\!\in$, ${}^{\mathsf{u}}\mathtt{st}$, $\stackrel{\mathsf{u}}{=}$ have a similar meaning, and hence in general $\mathfrak{u}$ is $\mathfrak{a}$ defined in $\mathfrak{e}$.

Proposition 5.4.22. $\mathfrak{u}$ *is an interpretation of* **HST** *in* **BST**.

Proof. This is based on the following claim: for any $\mathtt{st}$-$\in$-formula Φ with parameters in $\mathbf{U}$, $\Phi^{\mathfrak{u}}$ is equivalent to ${}^{\mathsf{e}}({}^{\mathsf{a}}\Phi)$ in **BST**. This can be proved by induction on the syntactical structure of Φ. For Φ an elementary formula this follows immediately from the definition of $\stackrel{\mathsf{u}}{=}$, ${}^{\mathsf{u}}\!\!\in$, ${}^{\mathsf{u}}\mathtt{st}$. To carry out the nontrivial step for $\exists$ let Φ be $\exists\, x\, \varphi(x)$. Then $\Phi^{\mathfrak{u}}$ is $\exists\, x \in \mathbf{U}\, \varphi^{\mathbf{U}}(x)$. This can be converted, by the definition of $\mathbf{U}$ and the inductive hypothesis, to

$$\exists\, x \in \mathbf{E}\, ({}^{\mathsf{e}}(x \in \mathbf{A}) \wedge {}^{\mathsf{e}}({}^{\mathsf{a}}\varphi(x))), \quad \text{that is, to} \quad {}^{\mathsf{e}}(\exists\, x\, (x \in \mathbf{A} \wedge {}^{\mathsf{a}}\varphi(x))).$$

However the subformula in brackets in the right-hand formula is ${}^{\mathsf{a}}(\exists\, x\, \varphi(x))$.

Now let Φ be any axiom of **HST**; we have to prove $\Phi^{\mathfrak{u}}$ in **BST**. By the above, it suffices to prove ${}^{\mathsf{e}}({}^{\mathsf{a}}\Phi)$. Since $\mathfrak{e}$ is an interpretation of **EEST** in **BST** it remains to show that **EEST** proves ${}^{\mathsf{a}}\Phi$. Yet this holds because $\mathfrak{a}$ is an interpretation of **HST** in **EEST**. □

Still arguing in **BST**, put ${}^{\mathsf{u}}x = {}^{\mathsf{e}}({}^{\mathsf{a}}x)$ for any x. In reality this means that ${}^{\mathsf{u}}x$ is a function defined on the singleton $\{\varnothing\} \times \{\varnothing\}$ by ${}^{\mathsf{u}}x(\varnothing, \varnothing) = \{\langle \Lambda, x\rangle\}$, where Λ, the empty sequence, is equal to $\varnothing$. (Recall that ${}^{\mathsf{a}}x = \{\langle \Lambda, x\rangle\}$ for any internal x by Definition 5.3.5.)

Let $\Psi(x, y)$ be the formula $y = \{\langle \Lambda, x\rangle\}$. Thus if x, y are internal then $\Psi(x, y)$ expresses the equality $y = {}^{\mathsf{a}}x$ according to Definition 5.3.5.

Lemma 5.4.23 (BST). *Let x be any (internal) set, $p = {}^{\mathsf{e}}x$, $u = {}^{\mathsf{u}}x$. Then we have* ${}^{\mathsf{e}}(\mathtt{int}\, p \wedge \mathtt{int}\, u \wedge \Psi(p, u))$. *Less formally, u is equal to ${}^{\mathsf{a}}p$ in $\mathfrak{e}$.*

Proof. Note that $p \in \mathbb{I}^{(\mathfrak{e})}$ (see the second line in the proof of Theorem 5.2.6), in other words ${}^{\mathsf{e}}(\mathtt{int}\, p)$. By the same reasons $u = {}^{\mathsf{u}}x = {}^{\mathsf{e}}({}^{\mathsf{a}}x)$ satisfies ${}^{\mathsf{e}}(\mathtt{int}\, u)$. We observe that the formula $\Psi(p, u)$ is equivalent to $\Psi(p, u)^{\mathtt{int}}$ in **EEST** provided p, u are internal. Thus it remains to show ${}^{\mathsf{e}}\Psi(p, u)^{\mathtt{int}}$. But this is equivalent to $\Psi(x, {}^{\mathsf{a}}x)$ by Theoorem 5.2.6(iii) since $p = {}^{\mathsf{e}}x$ and $u = {}^{\mathsf{e}}({}^{\mathsf{a}}x)$. Finally $\Psi(x, {}^{\mathsf{a}}x)$ holds by definition. □

Note a remarkable inversion: ${}^{\mathsf{u}}x$ defined in the **BST** universe as ${}^{\mathsf{e}}({}^{\mathsf{a}}x)$ turns out to be rather ${}^{\mathsf{a}}({}^{\mathsf{e}}x)$ in $\mathfrak{e}$ by the lemma. (In one and the same universe ${}^{\mathsf{e}}({}^{\mathsf{a}}x) = \{\langle\langle 0, 0\rangle, \{\langle \Lambda, x\rangle\}\rangle\}$ and ${}^{\mathsf{a}}({}^{\mathsf{e}}x) = \{\langle \Lambda, \{\langle\langle 0, 0\rangle, x\rangle\}\rangle\}$ are obviously different.) But this enables us to prove:

Proposition 5.4.24 (BST). *The map $x \mapsto {}^{\mathsf{u}}x$ is an internal core embedding of the set universe $\langle \mathbb{I}; \in, \mathtt{st}\rangle$ into the structure $\mathfrak{u} = \langle \mathbf{U}; {}^{\mathsf{u}}\!\!\in, {}^{\mathsf{u}}\mathtt{st}; \stackrel{\mathsf{u}}{=}\rangle$, and moreover a $\mathtt{st}$-$\in$-isomorphism of $\langle \mathbb{I}; \in, \mathtt{st}\rangle$ onto $\langle \mathbb{I}^{(\mathfrak{u})}; {}^{\mathsf{u}}\!\!\in, {}^{\mathsf{u}}\mathtt{st}\rangle$.*

Proof. The map $x \mapsto {}^{\mathbf{e}}x$ is a $\mathtt{st}$-$\in$-isomorphism of $\langle \mathbb{I}; \in, \mathtt{st}\rangle$ onto the internal core $\langle \mathbb{I}^{(\mathbf{e})}; {}^{\mathbf{e}}\!\in, {}^{\mathbf{e}}\mathtt{st}\rangle$ of $\mathbf{e}$ (Theorem 5.2.6). On the other hand $\mathbf{e}$ is an interpretation of **EEST** (still by Theorem 5.2.6), and hence by Theorem 5.4.17 the map $y \mapsto {}^{\mathbf{a}}y$ <u>defined in</u> $\mathbf{e}$ is a $\mathtt{st}$-$\in$-isomorphism of $\langle \mathbb{I}^{(\mathbf{e})}; {}^{\mathbf{e}}\!\in, {}^{\mathbf{e}}\mathtt{st}\rangle$ onto the internal core of the structure $\mathbf{a}$ <u>defined in</u> $\mathbf{e}$. However the map $y \mapsto {}^{\mathbf{a}}y$ defined in $\mathbf{e}$ is just the map ${}^{\mathbf{e}}x \mapsto {}^{\mathbf{u}}x$ by Lemma 5.4.23, while the structure $\mathbf{a}$ defined in $\mathbf{e}$ is by definition just the structure $\mathbf{u} = \langle \mathbf{U}; {}^{\mathbf{u}}\!\in, {}^{\mathbf{u}}\mathtt{st}; {}^{\mathbf{u}}\!\!=\rangle$. It follows that the superposition $x \mapsto {}^{\mathbf{e}}x \mapsto {}^{\mathbf{u}}x$ of the two maps is a $\mathtt{st}$-$\in$-isomorphism of $\langle \mathbb{I}; \in, \mathtt{st}\rangle$ onto the internal core $\langle \mathbb{I}^{(\mathbf{u})}; {}^{\mathbf{u}}\!\in, {}^{\mathbf{u}}\mathtt{st}\rangle$ of $\mathbf{u}$ as required. □

Part 5° of Corollary 5.1.5

To show that **HST** is standard core interpretable in **ZFC** we take the superposition of the internal core interpretation $\mathbf{u}$ of **HST** in **BST** defined just above and the standard core interpretation ${}^{*}\mathbf{v}$ of **BST** in **ZFC** defined in § 4.3c. The only notable extra issue is related to the fact that we require any standard core interpretation to be a structure with true equality in § 4.1b while both $\mathbf{u}$ and hence the superposition are invariant structures. But this discrepancy is immediately fixed by Theorem 1.5.11.

□ (*Corollary 5.1.5*)

5.4g The problem of external sets revisited

Here we come back to the problem of external sets briefly considered in § 3.2f. The results of our study of metamathematics of **HST** (Theorem 5.1.4, Corollary 5.1.5) enable us to give a satisfactory solution to the problem of external sets in **BST**. Recall that the problem appears because many useful objects of study, for instance, $\mathtt{st}$-$\in$-definable parts (subclasses) of sets turn out to be not sets in **BST**, see § 3.2f.

The solution given by Theorem 5.2.6 (= 2° of Theorem 5.1.4) incorporates only those external "non-sets" which themselves consist of internal sets: recall that they were called *"external sets"*, § 3.2f. The theorem asserts that **EEST** is internal core interpretable in **BST**. Less formally, this means that **BST** is strong enough to build up a kind of external "envelope" or "hull" over its internal set universe $\mathbb{I}$, which turns out to be a much more complete universe of the elementary external set theory **EEST**! In other words, **BST** contains full information regarding a large universe of external "non-sets", including the opportunity to quantify over them. This advantage of **BST** (by the way, so far unknown for Nelson's internal set theory **IST**) is based on the parametrization theorem (Theorem 3.2.16). The existence of such an "envelope" explains why somewhat naïve "internal" considerations of external sets by **IST** practitioners are in fact consistent: those sets, mostly non-existing as internal sets in theories like **IST** or **BST**, are elements of a correctly defined envelope of "external sets" over the universe of all bounded sets in **IST** or the full universe of **BST**.

The treatment of the external extension $\mathfrak{e}$ in $\mathbb{I}$, the internal universe of **BST**, is in principle analogous to the treatment of complex numbers as pairs of real numbers. In other words, assuming that $\mathbb{I}$ is extended to $\mathfrak{e}$, the "universe" of all elementary external sets, the **BST** mathematician does not face any problem with uncertainty or illegality. Similar to the case of complex numbers, there is no need to translate everything back into the ground universe all the time, however the interpretation ${}^{\mathfrak{e}}\Phi$ (defined in § 5.2a) can be employed.

Claim 4° of Corollary 5.1.5 provides us with a much more comprehensive solution. Not only external subsets of the internal universe of **BST** but all reasonable external sets of any kind (in particular those which contain other external sets as elements) can be consistently adjoined to the internal **BST** universe $\mathbb{I}$ in the form of an "envelope" or "hull" which satisfies axioms of **HST**! In other words, a mathematician working in **BST** can legitimately assume that the universe $\mathbb{I}$ of **BST** is the internal universe of an external universe which satisfies the axioms of **HST**. This fact practically equalizes the bounded set theory **BST** with such an advanced theory as **HST** in the capability of treatment of external sets. Elements of the **HST** "envelope" can be visualized in the ground **BST** universe $\mathbb{I}$ by means of A-codes which are "external sets" from the $\mathbb{I}$-point of view.

To explain what we have in mind in detail consider a couple of examples.

Example 5.4.25 (a monad). Let $\mathbb{R}$ denote the set of real numbers in $\mathbb{I}$, the universe of **BST**. A *monad* of a standard $x \in \mathbb{R}$ is the "external set" $\mu_x = \{y \in {}^*\mathbb{R} : x \simeq y\}$ (not a set in $\mathbb{I}$), where $x \simeq y$ means $\forall^{\text{st}} \varepsilon > 0\,(|x-y| < \varepsilon)$. We put $T_{\mathbf{x}} = \{\Lambda\} \cup \{\langle y\rangle : y \simeq x\}$ and $\mathsf{F}_{\mathbf{x}}(\langle y\rangle) = y$ for any $y \simeq x$. Then $\mathbf{x}$ is an "external set" (**exercise**: prove that it is non-internal). Moreover, $\mathbf{x}$ is a code in $\mathbf{A}$ and $\mathsf{A}_{\mathbf{x}} = \mu_x$. □

Example 5.4.26 (the set of all monads). Every monad is a bounded definable class, so that this is still in the framework of "external sets". However the collection of all monads is not a bounded definable class (of internal sets), therefore this is the point where the A-coding construction seriously enters the reasoning. We put $\mathsf{F}_{\mathbf{x}}(\langle x, y\rangle) = y$ for any $x \in \mathbb{R} \cap \mathbb{S}$ and $y \simeq x$, and

$$T_{\mathbf{x}} = \{\Lambda\} \cup \{\langle x\rangle : x \in \mathbb{R} \cap \mathbb{S}\} \cup \{\langle x, y\rangle : x \in \mathbb{R} \cap \mathbb{S} \wedge y \simeq x\},$$

so that still $\mathbf{x} \in \mathbf{A}$ and $\mathsf{A}_{\mathbf{x}}$ is the collection of all monads of standard reals. Saying it differently, $\mathbf{x}$ is the set of all monads of standard $\mathbb{I}$--reals in $\mathfrak{a}$. □

One can develop in this manner in **BST** most of typical external constructions of nonstandard mathematics. This is restricted only by properties of the theory **HST** itself, of course. Of those restrictions, the most notable is the fact that **HST** contradicts the Power Set axiom (see § 1.3b). It will be shown in Chapter 6 how to define external universes which do satisfy Power Set, at the cost of the full Saturation axiom (which is replaced by Saturation restricted to a fixed cardinal).

5.5 The class $\mathbb{L}[\mathbb{I}]$: sets constructible from internal sets

Here we are going to pursue the **HST** case (see 5.4.1) for the considerations in Section 5.4. In other words, the structure $\mathfrak{a} = \langle \mathbf{A}\,;\, {}^{\mathsf{a}}\!\!\in, {}^{\mathsf{a}}\mathtt{st}\,;\, {}^{\mathsf{a}}\!\!= \rangle$ (defined in § 5.4d) will be considered in the **HST** universe.

Note that a set $A_{\mathbf{x}}$ does exist for any code $\mathbf{x} \in \mathbf{A}$ in **HST**. The class $\mathbb{L}[\mathbb{I}] = \{A_{\mathbf{x}} : \mathbf{x} \in \mathbf{A}\}$ is studied in this section: by some reasons given below we call sets in $\mathbb{L}[\mathbb{I}]$ *sets constructible from internal sets*. The class $\mathbb{L}[\mathbb{I}]$ will be shown to be a transitive interpretation of **HST** having several additional properties unavailable in **HST**. For instance we prove that different infinite internal cardinalities remain externally different in $\mathbb{L}[\mathbb{I}]$ — a result are not provable in **HST**. Another statement true in $\mathbb{L}[\mathbb{I}]$: for any cardinal κ all κ-complete partially ordered sets are κ-distributive in $\mathbb{L}[\mathbb{I}]$ (in the absence of the axiom of Choice!).

Blanket agreement 5.5.1. We argue in **HST** in this section. Accordingly $\mathbb{S}$, $\mathbb{I}$, $\mathbb{E}$ indicate the classes of all resp. standard, internal, and elementary external (as in Definition 5.2.7) sets.

Recall §§ 5.4a–5.4d on the class $\mathbf{A}$ of all regular A-codes $\mathbf{x} \in \mathbb{E}$, the relations ${}^{\mathsf{a}}\!\!\in$, ${}^{\mathsf{a}}\mathtt{st}$, ${}^{\mathsf{a}}\!\!=$ on $\mathbf{A}$ and the $\mathfrak{a} = \langle \mathbf{A}\,;\, {}^{\mathsf{a}}\!\!\in, {}^{\mathsf{a}}\mathtt{st}\,;\, {}^{\mathsf{a}}\!\!= \rangle$. □

5.5a Sets constructible from internal sets

The idea of relative constructibility is well known: following the **ZFC** patterns, we should define as $\mathbb{L}[\mathbb{I}]$, the class of all sets constructible from internal sets, something like $\bigcup_{\xi \in \mathtt{Ord}} \mathbb{L}_\xi[\mathbb{I}]$, where the initial level $\mathbb{L}_0[\mathbb{I}] = \mathbb{I}$ consists of all internal sets, the union is taken at all limit steps, and any $\mathbb{L}_{\xi+1}[\mathbb{I}]$ consists of all sets $\mathtt{st}$-$\in$-definable in $\mathbb{L}_\xi[\mathbb{I}]$ — in particular, $\mathbb{L}_1[\mathbb{I}] = \mathbb{E}$. But in this case such an $\mathtt{Ord}$-long inductive definition can be avoided: the following definition yields the same result (see Exercise 5.5.6).

Definition 5.5.2 (HST). We define $\mathbb{L}[\mathbb{I}] = \{A_{\mathbf{x}} : \mathbf{x} \in \mathbf{A}\}$, the collection of all sets which admit regular A-codes $\mathbf{x} \in \mathbb{E}$.

Sets in $\mathbb{L}[\mathbb{I}]$ are called *sets constructible from internal sets*. □

It follows from Lemma 5.4.16 that the structures $\mathbb{L}[\mathbb{I}]$ and $\mathfrak{a}$ have essentially the same properties!

Exercise 5.5.3. Prove that the domain $\mathbf{A}/{}^{\mathsf{a}}\!\!=$ of the quotient structure $\mathfrak{a}/{}^{\mathsf{a}}\!\!=$ consists of ${}^{\mathsf{a}}\!\!=$*-classes* $[\mathbf{x}]^{\mathsf{a}} = \{\mathbf{y} \in \mathbf{A} : \mathbf{x} \;{}^{\mathsf{a}}\!\!= \mathbf{y}\}$ of codes $\mathbf{x} \in \mathbf{A}$, generally speaking, proper classes. Apply (1) of Exercise 1.5.7 to reduce the equivalence classes to sets as in the proof of Theorem 1.5.11. Prove that after such a reduction $\mathfrak{a}/{}^{\mathsf{a}}\!\!=$ will be $\mathtt{st}$-$\in$-isomorphic to $\langle \mathbb{L}[\mathbb{I}]; \in, \mathtt{st} \rangle$. □

Theorem 5.5.4 (HST). *The class* $\mathbb{L}[\mathbb{I}]$ (that is, to be more precise, the structure $\langle \mathbb{L}[\mathbb{I}]; \in, \mathtt{st} \rangle$) *is an interpretation of* **HST**. *In addition,*

1°. $\mathbb{E} \cup \mathbb{WF} \subseteq \mathbb{L}[\mathbb{I}]$, *in other words, all elementary external, internal, standard, and well-founded sets belong to* $\mathbb{L}[\mathbb{I}]$.

2°. $\mathbb{L}[\mathbb{I}]$ *is a transitive subclass of* $\mathbb{H}$.

3°. *Every set* $X \in \mathbb{L}[\mathbb{I}]$ *satisfying* $X \subseteq \mathbb{I}$ *belongs to* $\mathbb{E}$.

4°. *If a set* $X \subseteq \mathbb{L}[\mathbb{I}]$ *is definable in* $\mathbb{L}[\mathbb{I}]$ *by a* st-$\in$-*formula with parameters in* $\mathbb{L}[\mathbb{I}]$ *then* $X \in \mathbb{L}[\mathbb{I}]$.

5°. *If a set* $X \subseteq \mathbb{I}$ *is definable in* $\mathbb{L}[\mathbb{I}]$ *by a* st-$\in$-*formula with parameters in* $\mathbb{I}$ *then* X *is definable in* $\mathbb{I}$ *by a* st-$\in$-*formula with the same parameters.*

6°. *Every set* $X \subseteq \mathbb{L}[\mathbb{I}]$ *of standard size belongs to* $\mathbb{L}[\mathbb{I}]$.

7°. $\mathbb{WF}$ *is still the class of all well-founded sets in the sense of* $\mathbb{L}[\mathbb{I}]$.

8°. *In* $\mathbb{L}[\mathbb{I}]$, *every set is a functional image of a standard set.*

The primary goal of this section is to prove the theorem. In addition, we prove in § 5.5d that $\mathbb{L}[\mathbb{I}]$ satisfies a useful transfinite form of Dependent Choice, most likely not available on the base of the axioms of **HST**.

Technical arrangements in the proof of Theorem 5.5.4 will consist of transformations of codes in **A**, mainly on the base of the following lemma.

Lemma 5.5.5 (HST). *If a set* $Z \subseteq \mathbf{A}$ *is* st-$\in$-*definable in* $\mathbb{E}$ *(parameters in* $\mathbb{E}$ *allowed) then there is a code* $\mathbf{X} \in \mathbf{A}$ *such that* $\mathsf{A}_{\mathbf{X}} = \{\mathsf{A}_{\mathbf{x}} : \mathbf{x} \in Z\}$.

Proof. The argument is pretty analogous to the proof of Lemma 5.4.20; we leave it as an **exercise** for the reader. □

5.5b Proof of the theorem on I-constructible sets

We begin with claims 1° – 8° of Theorem 5.5.4.

1°. This is a consequence of Lemma 5.4.6: indeed the codes ${}^{\mathsf{a}}x$, $\mathbf{c}[{}^*v]$ (Definitions 5.3.5, 5.3.9) satisfy $\mathsf{A}_{{}^{\mathsf{a}}x} = x$, $\mathsf{A}_{\mathbf{c}[{}^*v]} = v$ (Exercises 5.3.6, 5.3.10).

2°. Suppose that $x \in X = \mathsf{A}_{\mathbf{x}} \in \mathbb{L}[\mathbb{I}]$, where $\mathbf{x} \in \mathbf{A}$. If $T_{\mathbf{x}} = \{\Lambda\}$ then $X = \mathbf{x}(\Lambda) \in \mathbb{I}$ by definition. Thus $X \subseteq \mathbb{I}$, and it remains to apply 1°. If $T_{\mathbf{x}} \neq \{\Lambda\}$ then $x = \mathsf{A}_{\mathbf{x}|_a}$ for some $a \in \mathtt{Min}\, T_{\mathbf{x}}$, where $\mathbf{x}|_a$ is a regular A-code (Exercise 5.3.14). Moreover, $\mathbf{x}|_a$ is st-$\in$-definable in $\mathbb{E}$ by a formula containing only $\mathbf{x}$ and a as parameters, where $\mathbf{x}$ and a belong to $\mathbb{E}$, in fact a is even internal. Thus $\mathbf{x}|_a \in \mathbf{A}$ because $\mathbb{E}$ interprets **EEST** by Theorem 5.2.11. It follows that $x \in \mathbb{L}[\mathbb{I}]$.

3°. Assume that $\mathbf{y} \in \mathbf{A}$ and $X = \mathsf{A}_{\mathbf{y}} \subseteq \mathbb{I}$. By 1.1.11(3) there is a standard set S such that $X \subseteq S$. Then $X = \{x \in S : {}^{\mathsf{a}}x \mathrel{{}^{\mathsf{a}}\!\!\in} \mathbf{y}\}$ by Theorem 5.4.11 (because $\mathsf{A}_{{}^{\mathsf{a}}x} = x$), and hence X is definable in $\mathbb{E}$ by a st-$\in$-formula with only $\mathbf{y}$, $S \in \mathbb{E}$ as parameters. It follows that $X \in \mathbb{E}$ by Theorem 5.2.11.

4°. Any $x \in X$ has the form $x = \mathsf{A}_{\mathbf{x}}$ where $\mathbf{x} \in \mathbf{A}$. According to the **HST** Collection, there is a set $Z' \subseteq \mathbf{A}$ such that such a code $\mathbf{x}$ can be chosen in Z' for any $x \in X$. As in the proof of 3° above, the set $Z = \{\mathbf{c} \in Z' : \mathsf{A}_{\mathbf{c}} \in X\}$ is $\mathtt{st}$-$\in$-definable in $\mathbb{E}$ (using only some elements of $\mathbf{A}$ as parameters), thus, $Z \in \mathbb{E}$ by Theorem 5.2.11. Now apply Lemma 5.5.5.

5°. If $X = \{x \in \mathbb{I} : \Phi^{\mathbb{L}[\mathbb{I}]}(x, y)\}$, where $y \in \mathbb{I}$ (a parameter) then by $(*)$ in the proof of Theorem 5.2.11 and in Lemma 5.4.16 we have

$$x \in X \iff ({}^{\mathsf{a}}\Phi({}^{\mathsf{a}}x, {}^{\mathsf{a}}y))^{\mathbb{E}} \iff ({}^{\mathsf{e}}({}^{\mathsf{a}}\Phi({}^{\mathsf{e}}({}^{\mathsf{a}}x), {}^{\mathsf{e}}({}^{\mathsf{a}}y))))^{\mathtt{int}}.$$

6°. By definition there exist: a set $S \subseteq \mathbb{S}$ and a map $g : S$ onto X. Using Extension (Theorem 1.3.12) we obtain an internal function f with $S \subseteq \operatorname{\mathtt{dom}} f$ such that, for each standard $s \in S$, the set $f(s)$ is a code in $\mathbf{E}$ satisfying $\mathsf{E}_{f(s)} \in \mathbf{A}$ and $g(s) = \mathsf{A}_{\mathsf{E}_{f(s)}}$. Then the set $X = \{\mathsf{E}_{f(s)} : s \in S\}$ is $\mathtt{st}$-$\in$-definable in $\mathbb{L}[\mathbb{I}]$ with sets S, f as parameters. It remains to apply 4°.

7°. That $\mathbb{WF} \subseteq \mathbb{L}[\mathbb{I}]$ follows from 1°. Further, a well-founded set obviously remains such in $\mathbb{L}[\mathbb{I}]$. If $X \in \mathbb{L}[\mathbb{I}]$ is not well-founded then, by Dependent Choice, there is an infinite $\in$-decreasing chain $X \ni x_0 \ni x_1 \ni x_2 \ni \ldots$. As $\mathbb{L}[\mathbb{I}]$ is transitive, this chain belongs to $\mathbb{L}[\mathbb{I}]$ by 6°, where it still witnesses that X is not a well-founded set in $\mathbb{L}[\mathbb{I}]$, as required.

8°. Consider a set $X = \mathsf{A}_{\mathbf{x}} \in \mathbb{L}[\mathbb{I}]$; $\mathbf{x} \in \mathbf{A}$. The set $P = \mathtt{Min}\, T_{\mathbf{x}} \subseteq \mathbb{I}$ belongs to $\mathbb{E}$ because $T_{\mathbf{x}} \in \mathbb{E}$. In particular, $P \in$ and $\subseteq \mathbb{L}[\mathbb{I}]$. Note that $X = \{\mathsf{A}_{\mathbf{x}|_a} : a \in P\}$, and the map $a \longmapsto \mathsf{A}_{\mathbf{x}|_a}$ is $\mathtt{st}$-$\in$-definable in $\mathbb{L}[\mathbb{I}]$, and hence it belongs to $\mathbb{L}[\mathbb{I}]$ by 4°. Thus X is an image of a set $P \subseteq \mathbb{I}$ in $\mathbb{L}[\mathbb{I}]$. It remains to cover P by a standard set, using (3) of Exercise 1.1.11.

As for the **HST** axioms in $\mathbb{L}[\mathbb{I}]$, the result in principle follows from Theorem 5.4.17 (see Lemma 5.4.16). However an independent proof on the base of 1° – 8° is very simple. The axioms of § 1.1c, Regularity over $\mathbb{I}$, and Saturation are inherited from $\mathbb{H}$ because $\mathbb{WF} \cup \mathbb{I} \subseteq \mathbb{E} \subseteq \mathbb{L}[\mathbb{I}]$. To prove Standard Size Choice or Dependent Choice in $\mathbb{L}[\mathbb{I}]$, we first get a choice function in $\mathbb{H}$. The function is a standard size subset of $\mathbb{L}[\mathbb{I}]$, so it belongs to $\mathbb{L}[\mathbb{I}]$ by 6°. As for the axioms of § 1.1b, all of them except for Collection are easy consequences of 4°, and we leave this an an **exercise** for the reader.

Collection. Since we have Collection in $\mathbb{H}$, it suffices to check the following: for any set $X \subseteq \mathbb{L}[\mathbb{I}]$ there is a set $X' \in \mathbb{L}[\mathbb{I}]$ such that $X \subseteq X'$. Using Collection in $\mathbb{H}$ and (3) of Exercise 1.1.11, we obtain a standard P with

$$\forall x \in X \, \exists a \in P \, (\mathsf{E}_a \in \mathbf{A} \wedge x = \mathsf{A}_{\mathsf{E}_a}).$$

The set $P' = \{a \in P : \mathsf{E}_a \in \mathbf{A}\}$ belongs to $\mathbb{E}$. Applying 4° as above, we easily prove that $X \subseteq X' = \{\mathsf{A}_{\mathsf{E}_a} : a \in P'\} \in \mathbb{L}[\mathbb{I}]$.

□ (*Theorem 5.5.4*)

Exercise 5.5.6 (Difficult !). Prove that the class $\mathbb{L}[\mathbb{I}]$ as defined by 5.5.2 is equal to $\bigcup_{\xi \in \mathtt{Ord}} \mathbb{L}_\xi[\mathbb{I}]$ defined as in the beginning of § 5.5, as well as to the least transitive class which contains all internal sets and satisfies **HST**. □

5.5c The axiom of $\mathbb{I}$-constructibility

Following patterns known from **ZFC**, we introduce the axiom of $\mathbb{I}$-constructibility: let "$\mathbb{H} = \mathbb{L}[\mathbb{I}]$" be the statement: *all sets belong to* $\mathbb{L}[\mathbb{I}]$.

Corollary 5.5.7. "$\mathbb{H} = \mathbb{L}[\mathbb{I}]$" *is consistent with* **HST**. □

It is known from numerous set theoretic studies that Gödel's axiom of constructibility "$\mathbf{V} = \mathbf{L}$" allows to prove many results which **ZFC** alone does not prove, in particular, it greatly simplifies the structure of cardinals *etc.* The axiom "$\mathbb{H} = \mathbb{L}[\mathbb{I}]$" plays a similar role in **HST**. The applications are mainly based on assertion 3° of Theorem 5.5.4 which allows us to extend properties of elementary external sets to all sets $X \subseteq \mathbb{I}$. The next theorem gives some examples (see also Theorem 5.5.12):

Theorem 5.5.8. *The following statements are consequences of* "$\mathbb{H} = \mathbb{L}[\mathbb{I}]$", *therefore they are consistent with* **HST** :

(i) *every cut (initial segment)* $U \subseteq {}^*\mathtt{Ord}$ *is standard size cofinal or standard size coinitial*;

(ii) *the axiom of Choice, in the form of* § 1.1h, *fails*;

(iii) *if* $X, Y \in \mathbb{I}$ *and* $f : X \xrightarrow{\text{onto}} Y$ *be any, possibly non-internal, function then* ${}^*\mathtt{card}\, Y \le n\, {}^*\mathtt{card}\, X$ *for some* $n \in \mathbb{N}$, *in particular, if* X *is* $*$*-infinite then* ${}^*\mathtt{card}\, Y \le {}^*\mathtt{card}\, X$. (Recall that ${}^*\mathtt{card}$ is the cardinality in $\mathbb{I}$.);

(iv) *there exist infinite* $*$*-finite non-equinumerous sets*;

(v) *every set* X *is either "large" or of standard size.* (Recall that a set is "large" if it contains a subset equinumerous to an infinite internal set.)

Proof. (i) The set U belongs to $\mathbb{E}$ by 3° of Theorem 5.5.4, hence, to $\mathbf{\Delta}_2^{\mathrm{ss}}$ by Theorem 5.2.10. It remains to apply Theorem 1.4.6(i).

(ii) It follows from Theorem 1.4.7 that there is a partition of ${}^*\mathbb{N}$ which does not admit a $\mathbf{\Delta}_2^{\mathrm{ss}}$ transversal, and hence does not admit a transversal of any kind under the assumption of "$\mathbb{H} = \mathbb{L}[\mathbb{I}]$".

(iii) As in (i), f belongs to $\mathbf{\Delta}_2^{\mathrm{ss}}$. Apply Theorem 1.4.9.

(iv) Let $h \in {}^*\mathbb{N} \setminus \mathbb{N}$. Take internal sets X, Y with ${}^*\mathtt{card}\, Y = 2^h$ and ${}^*\mathtt{card}\, X = h$ and apply (iii).

(v) We have $X = \mathsf{A}_{\mathbf{x}}$ for a code $\mathbf{x} \in \mathbf{A}$. Let $T = T_{\mathbf{x}}$ and $A = \mathtt{Min}\, T_{\mathbf{x}}$, so that $X = \{\mathsf{A}_{\mathbf{x}|_a} : a \in A\}$ (Exercise 5.3.4). It follows that X is equinumerous to the quotient A/E, where E is an equivalence relation on A defined so that $a \,\mathsf{E}\, b$ iff $\mathbf{x}|_a \overset{\mathtt{a}}{=} \mathbf{x}|_b$. Yet both A and E belong to $\mathbb{E}$ (because so does $\mathbf{x}$), hence to $\mathbf{\Delta}_2^{\mathrm{ss}}$, so that the result follows from Theorem 1.4.11. □

Problem 5.5.9. It follows from (ii) of the theorem that the negation of Choice is compatible with **HST**. Does **HST** prove the negation of Choice? Is the negation of Choice given by the proof of (ii) the strongest possible? □

Exercise 5.5.10. Prove that the theory **HST** + "$\mathbb{H} = \mathbb{L}[\mathbb{I}]$" is reducible to **EEST** in the sense that for any st-$\in$-sentence Φ there is a st-$\in$-sentence φ such that **HST** proves $\Phi \iff \varphi^{\mathbb{E}}$. Argue as in the proof of 6° of Theorem 5.1.4 in the end of § 5.2b using Lemma 5.4.16. □

5.5d Transfinite constructions in $\mathbb{L}[\mathbb{I}]$

It was announced in the preamble to this section that $\mathbb{L}[\mathbb{I}]$ models an additional Choice–like property. The property we shall prove is, perhaps, not everything one can obtain in $\mathbb{L}[\mathbb{I}]$; one should try to prove for instance the existence of a maximal chain in each p. o. set. Nevertheless the one we prove will be of extreme importance in the development of forcing over $\mathbb{L}[\mathbb{I}]$ below.

Let us recall some notation related to ordered sets.

Definition 5.5.11. Let κ be a cardinal. A *transitive relation* is any structure $P = \langle P\,; \lhd \rangle$ such that $x \lhd y \lhd z$ implies $x \lhd z$ (but $x \lhd x$ is, generally, not assumed). A subset $Q \subseteq P$ is *open dense* in P iff

1) $\forall p \in P \, \exists q \in Q \, (q \lhd p)$, and
2) $\forall p \in P \, \forall q \in Q \, (p \lhd q \Longrightarrow p \in Q)$.

Sets Q satisfying only 1) are called *dense*.

A transitive relation P is *κ-closed* if every decreasing chain $\{p_\alpha\}_{\alpha<\lambda}$ (that is, $p_\alpha \lhd p_\beta$ whenever $\beta < \alpha < \kappa$) of length $\lambda \le \kappa$ in P has a lower $\lhd$-bound in P, and $(<\kappa)$-*closed* if this holds for all decreasing chains of length $< \kappa$. A transitive relation P is *κ-distributive* if any intersection of κ-many open dense subsets of P is dense. In addition, a transitive relation P is *κ-specially distributive* [6] if for any $p \in P$ and any family $\{D_\alpha\}_{\alpha<\kappa}$ of open dense subsets of P there is a $\lhd$-decreasing chain $\{x_\alpha\}_{\alpha<\kappa}$ of elements $x_\alpha \in P$ such that $x_0 \lhd p$ and $x_\alpha \in D_\alpha$ for all $\alpha < \kappa$. □

Examples of transitive relations include both strict ($<$) and non-strict ($\le$) order relations. In **ZFC** a closed transitive relation is specially distributive, hence, distrubutive in the same cardinality, but this simple observation is based on Choice in a form which **HST** does not provide.

Theorem 5.5.12 (HST + "$\mathbb{H} = \mathbb{L}[\mathbb{I}]$"). *For each cardinal κ, every $(<\kappa)$-closed transitive relation is κ-specially distributive, therefore, every κ-closed transitive relation is κ-distributive.*

Proof (*We argue in* **HST** + "$\mathbb{H} = \mathbb{L}[\mathbb{I}]$"). Let $P = \langle P\,; \lhd\rangle \in \mathbb{L}[\mathbb{I}]$ be a $(<\kappa)$-closed transitive relation on a *standard* (we refer to item 8° of Theorem 5.5.4) set P. Let $\{D_\alpha\}_{\alpha<\kappa}$ be a family of open sets in $\langle P\,; \lhd\rangle$, and $\hat{x} \in P$. For

[6] Equivalent to the ordinary distributivity provided $\lhd$ is a non-strict partial order. This version is introduced to provide an application below (Theorem 6.2.9) where the relation is a strict order.

$x \in P$, let $\alpha(x)$ denote the largest ordinal $\alpha \le \kappa$ such that $x \in D_\beta$ for all $\beta < \alpha$. For $x, y \in P$, we let $x \prec y$ mean: $x \lhd y$, and either $\alpha(y) < \alpha(x)$ or $\alpha(x) = \alpha(y) = \kappa$, so that $\prec$ is still a transitive relation. To prove the theorem, it suffices to obtain a $\prec$-decreasing κ-sequence $\mathbf{x} = \{x_\alpha\}_{\alpha<\kappa}$ of elements $x_\alpha \in P$, satisfying $x_0 \lhd \hat{x}$.

The relation $\prec$ belongs to $\mathbb{E}$ by 3° of Theorem 5.5.4, hence, to $\mathbf{\Delta}_2^{\mathtt{ss}}$ by Theorem 5.2.10, so that $x \prec y$ iff $\exists\, a \in A\, \forall\, b \in B\, Q_{ab}(x, y)$, where $A, B \in \mathbb{WF}$ and $\{Q_{ab}\}_{a\in A,\, b\in B}$ is and a family of internal subsets of P^2.

The principal idea of the following argument is to divide the problem into a Choice argument in the $\in$-setting and a saturation argument. It can be traced to the proof of a choice theorem in Nelson [Nel 88]

Let us say that $a \in A$ *witnesses* $x \prec y$ iff $\forall\, b \in B\, Q_{ab}(x, y)$.

For any $\alpha \le \kappa$, let M_α be the family of all maps $\mathbf{a} : \alpha \times \alpha \to A$ such that there is a function $\mathbf{x} : \alpha \to P$ satisfying $\mathbf{x}(0) = \hat{x}$ and the requirement that $\mathbf{a}(\delta, \gamma) \in A$ witnesses $\mathbf{x}(\gamma) \prec \mathbf{x}(\delta)$ whenever $\delta < \gamma < \alpha$. Then, by Theorem 1.1.9(ii), each function $\mathbf{a} \in \bigcup_{\alpha\le\kappa} M_\alpha$, every set M_α, and the sequence $\langle M_\alpha : \alpha \le \kappa \rangle$ belong to $\mathbb{WF}$. To prove the theorem we have to check that $M_\kappa \ne \varnothing$. Since the sequence of sets M_α, $\alpha \le \kappa$, belongs to $\mathbb{WF}$, a **ZFC** universe, the following two claims immediately imply $M_\kappa \ne \varnothing$ as required.

Claim 1. *If $\alpha < \kappa$ and $\mathbf{a} \in M_\alpha$ then there is $\mathbf{a}' \in M_{\alpha+1}$ extending $\mathbf{a}$.*

Claim 2. *If $\alpha \le \kappa$ is a limit ordinal and a function $\mathbf{a} : \alpha \times \alpha \to A$ satisfies $\mathbf{a} \restriction (\beta \times \beta) \in M_\beta$ for all $\beta < \alpha$ then $\mathbf{a} \in M_\alpha$.*

Proof (Claim 1). By definition there exists a function $\mathbf{x} : \alpha \to P$ such that $\mathbf{x}(0) = \hat{x}$ and $\mathbf{a}(\delta, \gamma)$ witnesses $\mathbf{x}(\gamma) \prec \mathbf{x}(\delta)$ whenever $\delta < \gamma < \alpha$. Since P is $(< \kappa)$-closed, some $x \in P$ is $\le \mathbf{x}(\delta)$ for each $\delta < \alpha$. By the density of the sets D_β, we can assume that in fact $x \prec \mathbf{x}(\delta)$ for all $\delta < \alpha$. Using Standard Size Choice, we obtain a function $f : \alpha \to A$ such that $f(\delta)$ witnesses $x \prec \mathbf{x}(\delta)$ for each $\delta < \alpha$. We define $\mathbf{a}' \in M_{\alpha+1}$ by $\mathbf{a}'(\delta, \gamma) = \mathbf{a}(\delta, \gamma)$ whenever $\delta < \gamma < \alpha$, and $\mathbf{a}'(\delta, \alpha) = f(\delta)$ for $\delta < \alpha$. □ *(Claim)*

Proof (Claim 2). Suppose that $\delta < \gamma < \alpha$ and $b \in B$. We let $\Xi_{b\delta\gamma}$ be the set of all internal functions $\xi : {}^*\alpha \to P$ such that $\xi({}^*0) = \hat{x}$ and $Q_{\mathbf{a}(\delta,\gamma)\, b}(\xi({}^*\gamma), \xi({}^*\delta))$. The sets $\Xi_{b\delta\gamma}$ are internal because so are all Q_{ab}.

We assert that the intersection $\Xi_\beta = \bigcap_{b\in B;\ \delta<\gamma<\beta} \Xi_{b\delta\gamma}$ is non-empty for any $\beta < \alpha$. Indeed, since $\mathbf{a} \restriction (\beta \times \beta) \in M_\beta$, there exists a function $\mathbf{x} : \beta \to P$ such that $\mathbf{a}(\delta, \gamma)$ witnesses $\mathbf{x}(\gamma) \prec \mathbf{x}(\delta)$ whenever $\delta < \gamma < \beta$. Corollary 1.3.13 yields an internal function ξ defined on ${}^*\alpha$ and satisfying $\xi({}^*\gamma) = \mathbf{x}(\gamma)$ for all $\gamma < \alpha$. Then $\xi \in \Xi_\beta$.

It follows that the total intersection $\Xi = \bigcap_{b\in B;\ \delta<\gamma<\alpha} \Xi_{b\delta\gamma}$ is also non-empty, by the Saturation Theorem 1.3.5. Let $\xi \in \Xi$. By definition, we have $Q_{\mathbf{a}(\delta,\gamma)\, b}(\xi({}^*\gamma), \xi({}^*\delta))$ whenever $\delta < \gamma < \alpha$ and $b \in B$. To see that $\mathbf{a} \in M_\alpha$, we let $\mathbf{x}(\delta) = \xi({}^*\delta)$ for all $\delta < \alpha$. □ *(Claim)*

□ *(Theorem 5.5.12)*

Historical and other notes to Chapter 5

Section 5.1. Elementary external set theory **EEST** introduced in our paper [KanR 95, part 2] is not to be confused with Ballard's *enlargement set theory* **EST** (see [Bal 94] and a brief discussion in § 8.4b below). Claims 3° and 2° of Theorem 5.1.4, as well as claim 1° in the weaker form of interpretability of **BST** in **ZFGC**^{weak} were established in our paper [KanR 95].

Section 5.2. Elementary external sets, in their generality, were introduced in [Kan 91], but, of course, particular species of them (sets ${}^{\sigma}X = X \cap \mathbb{S}$, halos, galaxies, monads) were typical objects of study by **IST** practitioners since Nelson [Nel 77]. Theorem 5.2.11: [KanR 95, Part 2].

Section 5.3. The method of coding of sets by well-founded trees is borrowed from works on 2nd order Peano arithmetic, where the method (with countable trees) has been used since the 1960s to define interpretations of fragments of **ZFC** (like **ZFC** minus the Power Set axiom) in 2nd order PA.

Section 5.4. The strange word *bisimulation* (Definition 5.4.7) is used, in this context, in modern works on non-well-founded set theories, see, *e.g.*, [A 88, HrJ 98, Dev 98], where it is applied in a much more general case, in particular, for non-well-founded trees and even (directed) graphs which are not trees, where the uniqueness as in Lemma 5.4.8, generally speaking, fails. Originally it came from studies in computer science.

The conservativity of **HST** over **ZFC** (8° of our Corollary 5.1.5) was established by Hrbaček [Hr 78] (modulo minor details in the axiomatical system of **HST**) by a complicated argument which in a sense combined three distinct arguments of a very different nature: the conservativity of **BST** over **ZFC** by means of appropriate ultrapowers, as in Section 4.3, the conservativity of **EEST** over **BST** by means of the E-coding and the parametrization theorem, as in Section 5.2, and the conservativity of **HST** over **EEST** by means of the A-coding, as in Section 5.4. The idea to split the argument into these three rather transparent partial arguments with the same final goal was realized in our paper [KanR 95].

Section 5.5. The content of this section, including Theorems 5.5.4 and 5.5.12, first appeared in our papers [KanR 95, KanR 97]. Note that κ-closed and κ-distributive p. o. sets are frequently used in set theory, in particular in the practice of forcing, see, *e.g.*, Kunen [Kun 80].

6 Partially saturated universes and the Power Set problem

Unlike the model theoretic version of nonstandard analysis, which offers a multiplicity of nonstandard structures with various properties, **HST** directly provides us with a unique universe $\mathbb{I}$ of all internal sets, saturated in a certain maximally possible way, and embedded in the external universe $\mathbb{H}$ of all sets. This may appear too boring for a specialist accustomed to deal with peculiar nonstandard models with sometimes hardly achievable properties.

Fortunately, **HST** still has opportunities for this style of research. We introduce the notions of *internal subuniverse*, which is just a subclass $\mathscr{I} \subseteq \mathbb{I}$ closed under applications of standard functions (which is enough to interpret a big fragment of **BST**), and *external subuniverse*, *i.e.*, a class $\mathscr{H} \subseteq \mathbb{H}$ which interprets a certain big fragment of **HST**. Those subuniverses allow to achieve various effects, for instance, various amounts of Saturation and Choice available, and, most notably, the Power Set axiom!

Recall that the Power Set axiom contradicts **HST**, by Theorem 1.3.9. The proof shows the origin of the problem: Saturation implies that any infinite internal set has a proper class of (external) subsets. To fix the problem, we show that **HST** admits a system of partially saturated external subuniverses which model **HST** with Saturation restricted by a fixed cardinal, together with the Power Set axiom. Their definition is based on the assembling method used in Section 5.5 in the construction of $\mathbb{L}[\mathbb{I}]$, yet now the initial level of the assembling will be an internal subuniverse $\mathscr{I} \subseteq \mathbb{I}$ rather than the whole $\mathbb{I}$.

Internal subuniverses $\mathscr{I}$ taking part in this construction will generally be of the form $\mathbb{S}(X)$, the class of all sets standard relative to elements of a given set $X \subseteq \mathbb{I}$ (Section 6.1). Most important are classes $\mathbb{I}_\kappa = \mathbb{S}({}^*\kappa)$ (Section 6.2), κ being an infinite well-founded cardinal: we show that each $\mathbb{I}_\kappa$ can be extended to an external subuniverse $\mathbb{L}[\mathbb{I}_\kappa]$ satisfying a κ-version of **HST** and the Power Set axiom. Moreover, there exist sets $X \subseteq {}^*\kappa$, such that $\mathbb{S}(X)$ admits an internal core extension which models an appropriate κ-version of **HST** with the Power Set and the full Choice axioms. This is summarized in Corollaries 6.4.4 and 6.4.14. A scheme of applications is outlined in § 6.4h.

We also present a more complicated construction of external classes which are elementary submodels of $\mathbb{L}[\mathbb{I}_\kappa]$ and each other in the $\mathtt{st}$-$\in$-language.

Blanket agreement 6.0.1. We argue in **HST** in this Chapter. □

6.1 Internal subuniverses

The goal of this section is to define and study, in **HST**, subclasses of $\mathbb{I}$ which satisfy the closure property of the next definition. Among them, there are structures which interpret partially saturated internal theories $\mathbf{BST}_\kappa$ and $\mathbf{BST}'_\kappa$ introduced in 3.3 (see Section 6.2).

Definition 6.1.1. A class $\mathscr{I} \subseteq \mathbb{I}$ is an *internal subuniverse* if ${}^*\!f(x) \in \mathscr{I}$ whenever $f \in \mathbb{WF}$ is a function and $x \in \mathscr{I}^{<\omega} \cap \operatorname{dom} {}^*\!f$, or equivalently, if $f(x) \in \mathscr{I}$ whenever $f \in \mathbb{S}$ is a function and $x \in \mathscr{I}^{<\omega} \cap \operatorname{dom} f$. [1] □

Definability in $\mathbb{I}$ will be an important issue in some results below.

Definition 6.1.2. Suppose that $P \subseteq \mathbb{I}$ is a set or class. By $\mathrm{Def}^{\mathbb{I}}_{\in,\mathtt{st}}(P)$ we denote the collection of all sets and classes $X \subseteq \mathbb{I}$ definable in the structure $\mathbb{I} = \langle \mathbb{I}; \in, \mathtt{st} \rangle$ by a $\mathtt{st}$-$\in$-formula with sets in P as parameters. □

6.1a Some basic definitions and results

We begin with a theorem on general properties of internal subuniverses.

Theorem 6.1.3. *Let $\mathscr{I} \subseteq \mathbb{I}$ be an internal subuniverse. Then $\mathbb{S} \subseteq \mathscr{I}$, and*

(i) *$\mathscr{I}$ is an elementary substructure of $\mathbb{I}$ in the $\in$-language;*

(ii) *$\mathscr{I}$ is extensional, i.e., $x \cap \mathscr{I} \neq y \cap \mathscr{I}$ whenever $x \neq y$ belong to $\mathscr{I}$;*

(iii) *$\mathscr{I}$, or, more precisely, the structure $\langle \mathscr{I}; \in, \mathtt{st} \rangle$, is an interpretation of **BST** minus the axiom of Basic Idealization;*

(iv) *the equivalence $x \subseteq \mathscr{I} \iff x \in \mathscr{I}$ holds for any finite x ;*

(v) *any set $x \in \mathbb{I}$, $x \in \mathrm{Def}^{\mathbb{I}}_{\in,\mathtt{st}}(\mathscr{I})$, belongs to $\mathscr{I}$;*

(vi) *if $\mathscr{I}$ contains a number in ${}^*\mathbb{N} \smallsetminus \mathbb{N}$, $z \in \mathscr{I}$, and $z \cap \mathscr{I}$ is an at most countable subset of $\mathbb{S}$, then z is finite and $z \subseteq \mathscr{I}$.*

Proof. To see that $x = {}^*w \in \mathbb{S}$ belongs to $\mathscr{I}$ note that $\Lambda \in \mathscr{I}^{<\omega}$ by definition and $x = {}^*\!f(\Lambda)$, where $f \in \mathbb{WF}$ is a function satisfying $f(\Lambda) = w$.

(i) To prove the result by induction on the complexity of $\in$-formulas, it suffices to show, for any $\in$-formula $\varphi(x)$ with parameters in $\mathscr{I}$, that if there is an internal x satisfying $\varphi(x)$ in $\mathbb{I}$ then such an x exists in $\mathscr{I}$. We can assume that φ has only one parameter $p_0 \in \mathscr{I}$, so that $\varphi(x)$ is $\varphi(p_0, x)$. There is a set $P \in \mathbb{WF}$ such that $p_0 \in {}^*P$, and a function $F \in \mathbb{WF}$ defined on P and satisfying the following in $\mathbb{WF}$: if $p \in P$ and $\exists x\, \varphi(p,x)$ then $\varphi(p, F(p))$. Then, by $*$-Transfer, we have $\varphi(p_0, x_0)$ in $\mathbb{I}$, where $x_0 = {}^*F(p_0) \in \mathscr{I}$.

[1] Recall that for any set or class X, $X^{<\omega}$ is the set or class of all finite sequences of elements of X, containing, by definition, the empty sequence Λ even if $X = \varnothing$. Note that $X^{<\omega}$ is a set in **HST** for any set X by (5) of Exercise 1.2.15.

(ii) Follows from (i) because $\mathbb{I}$, as any transitive class, is extensional.

(iii) Follows from (i) and (ii).

(iv) Argue by induction on $n = \#x \in \mathbb{N}$. If $x \subseteq \mathscr{I}$ then $y = x \smallsetminus \{a\}$, where a is any element of x, belongs to $\mathscr{I}$ by the inductive hypothesis, and hence $x = y \cup \{a\} \in \mathscr{I}$. (The map $f(y,a) = y \cup \{a\}$ is obviously standard.) Conversely, if $x \in \mathscr{I}$ then by (i) there is an element $a \in x \cap \mathscr{I}$, and $y = x \smallsetminus \{a\}$ still belongs to $\mathscr{I}$. (The map $g(x,a) = x \smallsetminus \{a\}$ is standard.) We have $y \subseteq \mathscr{I}$ by the inductive hypothesis, thus $x \subseteq \mathscr{I}$ as well.

(v) follows from Theorem 3.2.11 (Uniqueness in **BST**, recall that $\mathbb{I}$ is a model of **BST** by Theorem 3.1.8).

(vi) If z is infinite then the set $U = \{u \in \mathbb{WF} : {}^*u \in z \cap \mathscr{I}\}$ is countable, hence, $X = {}^*U$ is a $*$-countable standard set, satisfying $X \cap \mathbb{S} = z \cap \mathscr{I}$. We can w. l. o. g. assume that $X = {}^*\mathbb{N}$ (otherwise apply a standard bijection of ${}^*\mathbb{N}$ onto X). This observation reduces the problem to the case when $z \subseteq {}^*\mathbb{N}$.

Let, in $\mathbb{I}$, for any $n < \nu$, $f(n)$ be the n-th element of X in the increasing order, where $\nu = \#z$ in $\mathbb{I}$ provided z is $*$-finite, or ν is any fixed number in $\mathscr{I} \cap ({}^*\mathbb{N} \smallsetminus \mathbb{N})$ otherwise. Use Theorem 6.1.3(i) to show that if z is infinite then $\nu \notin \mathbb{N}$ and $f(\nu - 1) \in \mathscr{I} \cap ({}^*\mathbb{N} \smallsetminus \mathbb{N})$, contradiction with $z \cap \mathscr{I} \subseteq \mathbb{S}$. □

6.1b Relative standardness

The following definition introduces a natural method to obtain internal subuniverses: the application of all standard functions to finite sequences of elements of a given set or class of internal sets.

Definition 6.1.4. If $X \subseteq \mathbb{I}$ then $\mathbb{S}(X)$ is the class of all sets of the form ${}^*\!f(x)$ where $f \in \mathbb{WF}$ is a function and $x \in X^{<\omega} \cap \mathtt{dom}\,{}^*\!f$, or equivalently, of the form $f(x)$ where f is a standard function and $x \in X^{<\omega} \cap \mathtt{dom}\, f$.

Sets in $\mathbb{S}(X)$ are called *standard relative to elements of* X.

Particular cases: $\mathbb{S}[w] = \mathbb{S}(\{w\})$ (simple relative standardness, see § 6.1c), and a slightly more general $\mathbb{S}[c_1, ..., c_n] = \mathbb{S}(\{c_1, ..., c_n\})$. □

Exercise 6.1.5. (1) Prove that $\mathbb{S}(X)$ is an internal subuniverse and the least internal subuniverse with $X \subseteq \mathbb{S}(X)$. Moreover, if a set $X \subseteq \mathbb{I}$ satisfies:

(†) for any natural n there is a function $h \in \mathbb{WF}$ such that $X \subseteq \mathtt{dom}\,{}^*h$ and $X^n \subseteq \{{}^*h(x) : x \in X\}$ (this will be the typical case below)

then $\mathbb{S}(X) = \{{}^*\!f(x) : f \in \mathbb{WF}$ is a function and $x \in X \cap \mathtt{dom}\,{}^*\!f\}$.

(2) Prove that $\mathbb{S}({}^*X) = \mathbb{S}({}^*Y)$ provided sets $X, Y \in \mathbb{WF}$ have the same cardinality, but $\mathbb{S}({}^*X) \subsetneqq \mathbb{S}({}^*Y)$, in particular, ${}^*Y \not\subseteq \mathbb{S}({}^*X)$, whenever X, Y are sets of infinite cardinalities $\kappa < \lambda$.

Hint: otherwise ${}^*Y$ is covered by the union of all standard subsets of ${}^*Y$ of $*$-cardinality $\leq {}^*\kappa$, in contradiction to Corollary 1.3.6. □

Exercise 6.1.6. (1) Find an internal set $A \subseteq {}^*\mathbb{N}$ such that $A \notin \mathbb{S}(A)$.

(2) A class $\mathscr{H}$ is *thin* if any set $X \subseteq \mathscr{H}$ is a set of standard size. Prove that if $K \subseteq \mathbb{I}$ is a set of standard size then $\mathbb{S}(K)$ is a thin internal subuniverse.

Hints. (1) Applying 6.1.5(2) for $X = \mathbb{N}$ and $Y = \mathscr{P}(\mathbb{N})$ we obtain an internal set $A \subseteq {}^*\mathbb{N}$ such that even $A \notin \mathbb{S}({}^*\mathbb{N})$.

(2) $\mathbb{S}(K) = \bigcup_{z \in Z} \mathbb{S}[z]$, where $\mathbb{S}[z] = \{{}^*\!f(z) : f \in \mathbb{WF}\}$ and $Z = K^{<\omega}$. Let $X \subseteq \mathbb{S}(K)$. There are well-founded sets B, W with $X \subseteq {}^*B$ and $Z \subseteq {}^*W$. Let $F = B^W$, then $X \subseteq \{{}^*\!f(w) : f \in F \wedge w \in W\}$. Apply 1.3.3. □

Proposition 6.1.7. *Suppose that $\mathscr{I} \subseteq \mathbb{I}$ is an internal subuniverse. Then*

(i) *if $X \subseteq \mathbb{I}$, $X \in \mathrm{Def}^{\mathbb{I}}_{\in,\mathtt{st}}(\mathscr{I})$ is a set of standard size then $X \subseteq \mathscr{I}$;* [2]

(ii) *if $\mathscr{I} \in \mathrm{Def}^{\mathbb{I}}_{\in,\mathtt{st}}(\mathbb{I})$ and $\mathscr{I} \subsetneqq \mathbb{I}$ then there exists an ordinal γ such that $\mathscr{I} \cap {}^*\gamma$ is not internal.*

Proof. (i) Thus X is st-$\in$-definable in $\mathbb{I}$, a **BST** universe. Applying Theorem 3.2.16(ii) in $\mathbb{I}$, we obtain $X = \mathsf{E}_p = \bigcup_{a \in A \cap \mathbb{S}} X_a$, where $X_a = \bigcap_{b \in B \cap \mathbb{S}} p(a,b)$, A, B are standard sets, p is an internal function defined on $A \times B$, and moreover $p \in \mathbb{S}[c_1, ..., c_n]$ for some $n \in \mathbb{N}$ and $c_1, ..., c_n \in \mathscr{I}$. Then p itself belongs to $\mathscr{I}$ since this class is an internal subuniverse.

We observe that each X_a is a $\mathbf{\Pi}^{\mathtt{ss}}_1$ set of standard size, hence an internal set by Theorem 1.4.2(ii) because all sets of standard size are $\mathbf{\Sigma}^{\mathtt{ss}}_1$. It follows that X_a is a finite set by 1.3.8(2). However $X_a \in \mathrm{Def}^{\mathbb{I}}_{\in,\mathtt{st}}(\mathscr{I})$ (because $p \in \mathscr{I}$, thus $X_a \in \mathscr{I}$ by Theorem 6.1.3(v). It follows that $X_a \subseteq \mathscr{I}$ by (iv) of Theorem 6.1.3.

(ii) Suppose that $\mathscr{I}$ is in $\mathrm{Def}^{\mathbb{I}}_{\in,\mathtt{st}}(\{p\})$, where $p \in \mathbb{I}$, so that there is a set $P \in \mathbb{WF}$ such that $p \in {}^*P$. Let $\kappa = \mathtt{card}\, P$ in $\mathbb{WF}$. Take $\gamma \in \mathtt{Ord}$ with $\gamma > 2^\kappa$ in $\mathbb{WF}$. We claim that $X = \mathscr{I} \cap {}^*\gamma \notin \mathbb{I}$. Indeed, otherwise $X \in \mathbb{S}(\{p\})$ by Theorem 6.1.3(v) (for $\mathscr{I} = \mathbb{S}(\{p\})$) because X is in $\mathrm{Def}^{\mathbb{I}}_{\in,\mathtt{st}}(\{p, {}^*\gamma\})$. We have $X = {}^*\!f(p)$, for a function $f \in \mathbb{WF}$, $f : P \to \mathscr{P}(\gamma)$.

Put $P_\xi = \{q \in P : \xi \in f(q)\}$ for $\xi < \gamma$. As $\gamma > 2^\kappa$, there are sets $Q \subseteq P$ and $G \subseteq \gamma$ with $\mathtt{card}\, G = \gamma$ such that $P_\xi = Q$ for all $\xi \in G$, therefore, $G \subseteq f(q)$ for any $q \in Q$, so that ${}^*G \subseteq {}^*\!f(q)$ for any $q \in {}^*Q$ by $*$-Transfer.

Coming back to the parameter p, note that ${}^*\xi \in X = {}^*\!f(p)$ for any $\xi < \gamma$, hence $p \in {}^*(P_\xi)$ for any $\xi < \gamma$, and finally $p \in {}^*Q$. This implies ${}^*G \subseteq X = {}^*\!f(p)$ by the above, in particular, ${}^*G \subseteq \mathscr{I}$. It follows that ${}^*\gamma \subseteq \mathscr{I}$ as well because $G \subseteq \gamma$ are sets of the same cardinality (see Exercise 6.1.5(2)), which is a contradiction. □

6.1c Simple relative standardness

We consider here a rather special type of internal subuniverses: classes of the form $\mathbb{S}[w] = \mathbb{S}(\{w\})$, where w is an internal set. By definition $\mathbb{S}[w]$ consists of all sets of the form ${}^*\!f(w)$, where $f \in \mathbb{WF}$ is a function and $w \in \mathtt{dom}\, {}^*\!f$.

[2] Note that this is a generalization of (v) of Theorem 6.1.3.

Let w-$\mathtt{st}\,x$ mean that $x \in \mathbb{S}[w]$. Sets in $\mathbb{S}[w]$ are called *standard relative to* w, or w*-standard*. Internal subuniverses of this type are interesting, in particular, because some of them can be interpretations of $\mathbf{BST}'_\kappa$.

The following is a list of basic facts related to classes of the form $\mathbb{S}[w]$.

Exercise 6.1.8. Suppose that $w \in \mathbb{I}$. Prove the following:

(1) $\mathbb{S}[w]$ coincides with the class $\mathbb{S}[w]$ defined earlier in $\mathbb{I}$ as a **BST** universe (Definition 3.1.13);

(2) $\mathbb{S}[w]$ is a thin class (apply 6.1.6(2));

(3) if $X \subseteq \mathbb{S}[w]$ then there exist a standard set Z and a map $g \in \mathbb{S}[w]$ defined on Z such that $X = \{g(z) : z \in Z \cap \mathbb{S}\}$;

(4) the set $\mathbb{N}[w] = {}^*\mathbb{N} \cap \mathbb{S}[w]$ is neither cofinal in ${}^*\mathbb{N}$ nor coinitial in ${}^*\mathbb{N} \smallsetminus \mathbb{N}$;

(5) if $w \notin \mathbb{S}$ then the set $\mathbb{N}[w]$ is uncountable, moreover, if $x \in \mathbb{N}[w] \smallsetminus \mathbb{N}$ then $\mathbb{N}[w] \cap [0, x)$ is still uncountable.

Hints. (3) By Standard Size Choice and Standardization, there is a standard set F such that any $f \in F$ is a function with $w \in \mathtt{dom}\, f$, and we have $X = \{f(w) : f \in F \cap \mathbb{S}\}$. Put $Z = F$ and $g(f) = f(w)$ for $f \in F$.

(4) Apply Exercise 6.1.6(2) and (3) of Exercice 1.3.8.

(5) Prove that $\mathbb{N}[w]$ is uncountable. Otherwise there is a function $f : \mathbb{N}^2 \to \mathbb{N}$ such that $\mathbb{N}[w] = \{{}^*\!f(n, w) : n \in \mathbb{N}\}$. Put $g(n) = \mathtt{max}_{n' \leq n} f(n', n)$. Then, for any n, we have $g(k) > f(n, k)$ for all $k \geq n$, therefore ${}^*g(w) > {}^*\!f(n, w)$ for any $n \in \mathbb{N}$ by Transfer. However ${}^*g(w) \in \mathbb{N}[w]$, contradiction. □

Classes $\mathbb{S}[w]$ admit a characterization in terms of ultrapowers. Recall that if U is an ultrafilter over a set I then, for any set or class M, to obtain the *ultrapower* M^I/U we define $f \approx_U g$ iff the set $\{i \in I : f(i) = g(i)\}$ belongs to U — for any $f, g \in M^I$ (M^I is the set or class of all functions $f : I \to M$), then put $[f] = \{g : f \approx_U g\}$ and define $[f] \;{}^*\!\!\in [g]$ iff the set $\{i \in I : f(i) \in g(i)\}$ belongs to U. Finally, $M^I/U = \langle \{[f] : f \in M^W\}; {}^*\!\!\in \rangle$ is the ultrapower. (See § 4.2b for details.)

Note that if $I \in \mathbb{WF}$ then any ultrafilter over I still belongs to $\mathbb{WF}$.

Proposition 6.1.9. *If $w \in \mathbb{I}$ then there exists an ultrafilter U over a set $I \in \mathbb{WF}$ such that $\langle \mathbb{S}[w]; \in \rangle$ is isomorphic to $\mathbb{WF}^I/U$.*

Conversely, if U is an ultrafilter over a set $I \in \mathbb{WF}$ then there is $w \in {}^*I$ such that $\langle \mathbb{S}[w]; \in \rangle$ is isomorphic to $\mathbb{WF}^I/U$.

Proof. If $w \in \mathbb{I}$ then there is a set $I \in \mathbb{WF}$ such that $w \in {}^*I$. Consider the set $U = U_w = \{X \subseteq I : w \in {}^*X\}$, the *associated ultrafilter*: one easily shows that U is an ultrafilter over I in $\mathbb{WF}$, and that $\langle \mathbb{S}[w]; \in \rangle$ and $\langle \mathbb{WF}^I/U; {}^*\!\!\in \rangle$ are isomorphic via the map sending any ${}^*\!f(w) \in \mathbb{S}[w]$ (where $f : I \to \mathbb{WF}$, so that ${}^*\!f : {}^*I \to \mathbb{S}$) to $[f] \in \mathbb{WF}^I/U$. Conversely, given an ultrafilter U over a set $I \in \mathbb{WF}$, we obtain, using Saturation, a set $w \in \mathbb{I}$ such that $U = U_w$. (**Exercise**: fill in the details of the proof.) □

Exercise 6.1.10. Consider a class $\mathbb{S}(R)$, where $R = \{x_\xi : \xi \in \kappa\} \subseteq \mathbb{I}$ is a set of standard size, $\kappa \in \mathbb{WF}$ being a (well-founded) cardinal.

By Extension there is an internal map $\mathbf{x}$, $\operatorname{dom}\mathbf{x} = {}^*\kappa$, with $x_\xi = \mathbf{x}({}^*\xi)$, $\forall \xi < \kappa$. There is an indexed family $\{D_\xi\}_{\xi<\kappa} \in \mathbb{WF}$ of sets $D_\xi \in \mathbb{WF}$ with $x_\xi \in {}^*D_\xi$, $\forall \xi$. Let $U \in \mathbb{WF}$ be the ultrafilter of all sets $X \subseteq I = \{\langle \xi, d\rangle : \xi < \kappa \wedge d \in D_\xi\}$ such that $\mathbf{x} \in {}^*X$. Let $\mathscr{F}$ be the class of all $f : I \to \mathbb{WF}$ of finite support, *i.e.* there exists a finite set $s \subseteq D$ such that $f(y) = f(z)$ whenever $y, z \in I$ satisfy $y(\xi, d) = z(\xi, d)$ for all $\xi \in s$, $d \in D_\xi$.

Prove the following generalization of 6.1.9: *$\langle \mathbb{S}(R); \in\rangle$ is isomorphic to the quotient power $\mathscr{F}/U$.* (See § 4.2b on quotient powers.) Why would the modification with $\mathscr{F} = \mathbb{WF}^I$ not be satisfactory in this case? □

6.1d Gordon classes

The notion of simple relative standardness admits an interesting modification which yields less sparse internal subuniverses:

Definition 6.1.11. Let $w \in \mathbb{I}$. Define $\mathbb{N}_{\mathrm{M}}[w] = \bigcup \mathbb{N}[w]$ (the least initial segment of ${}^*\mathbb{N}$ containing $\mathbb{N}[w] = \mathbb{S}[w] \cap {}^*\mathbb{N}$).

Put $\mathbb{S}_{\mathrm{M}}[w] = \mathbb{S}(\{w\} \cup \mathbb{N}_{\mathrm{M}}[w])$ [3] (*Gordon's classes*). Say that a set x is *w-standard in the modified sense*, in brief $w\text{-}\mathtt{st}_{\mathrm{M}}\, x$, if it belongs to $\mathbb{S}_{\mathrm{M}}[w]$. □

Lemma 6.1.12. *$x \in \mathbb{S}_{\mathrm{M}}[w]$ iff there is a $*$-finite set $y \in \mathbb{S}[w]$ containing x.*

Proof. Since $x, w \in \mathbb{I}$, there are well-founded sets X, W with $x \in {}^*X$, $w \in {}^*W$. The set $K = \mathbb{N}_{\mathrm{M}}[w]$ satisfies (†) of 6.1.5(1). (For instance, if $i, j \in K$ then easily $k = 2^i 3^j \in K$ and i, j belong to $\mathbb{S}[k]$. This argument holds for any finite number of elements of K.) It follows from 6.1.5(1) that any $x \in \mathbb{S}_{\mathrm{M}}[w]$ has the form $x = {}^*\!f(w, n)$, where $n \in \mathbb{N}_{\mathrm{M}}[w]$ and $f \in \mathbb{WF}$ is a function satisfying $\langle w, n\rangle \in \operatorname{dom}{}^*\!f$. We can assume that simply $\operatorname{dom} f = W \times \mathbb{N}$. Further, $n \le m$ for some $m \in \mathbb{N}[w]$. Take $y = \{{}^*\!f(w, n') : n' \le m\}$, then $x \in y$, y is $*$-finite, and $y \in \mathbb{S}[w]$ by Theorem 6.1.3(i).

Conversely, assume that $x \in y \in \mathbb{S}[w]$, and y is $*$-finite. There is a function $g \in \mathbb{WF}$ with $\operatorname{dom} g = W$ with $y = {}^*g(w)$. Further there are functions $h, f \in \mathbb{WF}$, defined, resp., on W and $W \times \mathbb{N}$, such that, for any $w' \in W$, if $g(w')$ is finite then $h(w') = \#g(w')$ and $g(w') = \{f(w', n) : n < h(w')\}$. Then, by $*$-Transfer, $x \in y = {}^*g(w) = \{{}^*\!f(w, n) : n < {}^*h(w)\} \subseteq \mathbb{S}_{\mathrm{M}}[w]$. □

Corollary 6.1.13. (1) $\mathbb{N}_{\mathrm{M}}[w] = \mathbb{S}_{\mathrm{M}}[w] \cap {}^*\mathbb{N}$;

(2) *if* $\mathbb{N} \subsetneqq \mathbb{N}[w]$ *then* $\mathbb{N} \subsetneqq \mathbb{N}[w] \subsetneqq \mathbb{N}_{\mathrm{M}}[w] = \mathbb{S}_{\mathrm{M}}[w] \cap {}^*\mathbb{N} \subsetneqq {}^*\mathbb{N}$.

Proof. If $k \in \mathbb{S}_{\mathrm{M}}[w] \cap {}^*\mathbb{N}$ then we have $k \in y \in \mathbb{S}[w]$ for a $*$-finite $y \subseteq {}^*\mathbb{N}$. However $n = \sup y$ belongs to $\mathbb{S}[w]$, hence, to $\mathbb{N}[w]$, and $k \le n$. To see that the rightmost $\subseteq$ in (2) is $\subsetneqq$ apply (1) and Exercise 6.1.8(4) (non-cofinality). To see that the middle $\subseteq$ is $\subsetneqq$ use Exercise 6.1.8(4) (non-coinitiality). □

[3] The subscript $_{\mathrm{M}}$ will indicate "the modified sense" of simple relative standardness.

Lemma 6.1.14. *Let $w \in \mathbb{I}$. Any set $X \subseteq \mathbb{S}_{\mathrm{M}}[w]$ can be covered by a union of standard size many $*$-finite sets $y \in \mathbb{S}[w]$.*

Proof. Note that by (3) of Exercise 1.1.11, there is a set $C \in \mathbb{WF}$ such that $X \subseteq {}^*C$. The power set $P = \mathscr{P}(C)$ is well-founded, too. Let Y be the set of all $*$-finite sets $y \in \mathbb{S}[w]$, $y \in {}^*P$, then Y is a set of standard size by the first part of the lemma. Finally $X \subseteq \bigcup Y$ by Lemma 6.1.12: any $*$-finite set $y \subseteq {}^*C$ is internal, hence, belongs to ${}^*P$ by $*$-Transfer. □

6.1e Associated structures

Given an internal subuniverse $\mathscr{I}$, the following two related structures can be considered:

type I : inflated standardness: $\langle \mathbb{I}; \in, \mathtt{st}_{\mathscr{I}} \rangle$, where $\mathtt{st}_{\mathscr{I}}\, x$ means $x \in \mathscr{I}$;

type II: deflated internal domain: $\langle \mathscr{I}; \in, \mathtt{st} \rangle$.

Either of them can be naturally viewed as a $\mathtt{st}$-$\in$-structure. (The atomic predicate $\mathtt{st}$ is interpreted as the unary relation $\mathtt{st}_{\mathscr{I}}$ in $\langle \mathbb{I}; \in, \mathtt{st}_{\mathscr{I}} \rangle$, of course.) Since the internal universe $\mathbb{I} = \langle \mathbb{I}; \in, \mathtt{st} \rangle$ satisfies bounded set theory **BST** (Theorem 3.1.8), one can ask whether these structures still interpret any significant part of **BST**.

We observe that structures of both types satisfy $\mathbf{ZFC}^{\mathtt{st}}$ and Inner Transfer by Theorem 6.1.3(i). Structures of type II obviously satisfy Inner Standardization, and some of them satisfy Basic Idealization restricted by a certain (well-founded) cardinal — their study follows in Section 6.2. On the contrary, structures of type I fail to satisfy Inner Standardization (except for two trivial cases), but those associated with Gordon classes do satisfy Basic Idealization. The proof of these two results follows in this subsection.

Theorem 6.1.15. *Suppose that $\mathscr{I}$ is an internal subuniverse that belongs to $\mathrm{Def}^{\mathbb{I}}_{\in,\mathtt{st}}(\mathbb{I})$, $\mathbb{S} \subsetneqq \mathscr{I} \subsetneqq \mathbb{I}$. Then Inner Standardization fails in $\langle \mathbb{I}; \in, \mathtt{st}_{\mathscr{I}} \rangle$.*

Proof. *Case 1*: $\mathbb{N} \subsetneqq \mathscr{I} \cap {}^*\mathbb{N}$. It follows from Proposition 6.1.7(ii) that the class $X = \{\xi \in {}^*\mathtt{Ord} \cap \mathscr{I} : \xi \cap \mathscr{I} \in \mathbb{I}\}$ is bounded in ${}^*\mathtt{Ord}$, therefore, X is a set. Accordingly, the least initial segment $C \subseteq {}^*\mathtt{Ord}$ with $X \subseteq C$ is a set. As $\mathscr{I}$ belongs to $\mathrm{Def}^{\mathbb{I}}_{\in,\mathtt{st}}(\mathbb{I})$, so do X and C. It follows that C is a $\mathbf{\Delta}^{\mathtt{ss}}_2$ set (in the **HST** universe of all sets) by Theorem 5.2.10.

Suppose towards the contrary that Inner Standardization holds in the structure $\langle \mathbb{I}; \in, \mathtt{st}_{\mathscr{I}} \rangle$. We claim that *under this assumption the cut C is internal.* Indeed otherwise the cut C is a gap by Theorem 1.4.6(i), that is C has no maximal element while the complementary class ${}^*\mathtt{Ord} \smallsetminus C$ has no minimal element, and either C is standard size cofinal or else ${}^*\mathtt{Ord} \smallsetminus C$ is standard size coinitial.

Let $Y \subseteq {}^*\mathtt{Ord}$ be a set of standard size, cofinal in C in the first case, or coinitial in ${}^*\mathtt{Ord} \smallsetminus C$ in the second case. Choose any $h \in ({}^*\mathbb{N} \smallsetminus \mathbb{N}) \cap$

$\mathscr{I}$. There is an internal set $S \subseteq {}^*\mathtt{Ord}$ with $\#S \le h$, such that $Y \subseteq S$ (Exercise 1.3.8). Thus S has $\mathtt{st}_{\mathscr{I}}$-standard number of elements in $\mathbb{I}$. Then, applying Lemma 3.1.18 in the structure $\langle \mathbb{I}; \in, \mathtt{st}_{\mathscr{I}} \rangle$, we conclude that $S \cap C$ is not cofinal in C and $S \smallsetminus C$ is not coinitial in ${}^*\mathtt{Ord} \smallsetminus C$ since C is a gap. This is a contradiction as $Y \subseteq S$.

Thus C is internal, so that obviously $C \in {}^*\mathtt{Ord}$. By definition X is a cofinal subset of C, satisfying $X \cap \xi = \mathscr{I} \cap \xi \in \mathbb{I}$ for any $\xi \in X$. Theorem 1.4.6(ii) implies that X itself is internal. Yet $X = C \cap \mathscr{I}$ — it follows that $C \in X$, contradiction. [4]

Case 2: $\mathscr{I} \cap {}^*\mathbb{N} = \mathbb{N}$. [5] Suppose, towards the contrary, that Inner Standardization holds in $\langle \mathbb{I}; \in, \mathtt{st}_{\mathscr{I}} \rangle$. Then $\mathscr{I}$ is a well-founded class: any non-empty set $X \subseteq \mathscr{I}$ contains an $\in$-minimal element [6]. (Otherwise by a simple argument there is an $\in$-descending sequence $\{x_n\}_{n \in \mathbb{N}}$ of elements of $\mathscr{I}$. By Theorem 1.3.12 (Extension) there is an internal function f, $\mathtt{dom}\, f = {}^*\mathbb{N}$, with $f(n) = x_n$ for all standard n. By Inner Standardization in $\langle \mathbb{I}; \in, \mathtt{st}_{\mathscr{I}} \rangle$, there is a function $g \in \mathscr{I}$ such that $g \cap \mathscr{I} = f \cap \mathscr{I}$, in particular, $g(n) = f(n) = x_n$ at least for all $n \in \mathbb{N}$. By the **ZFC** Regularity axiom in $\mathscr{I}$, there is a number $n \in {}^*\mathbb{N} \cap \mathscr{I}$ such that $x_{n+1} \notin x_n$. But ${}^*\mathbb{N} \cap \mathscr{I} = \mathbb{N}$, contradiction.)

The well-foundedness allows us to define, by $\in$-induction, a set $F(x) \in \mathbb{S}$ for any $x \in \mathscr{I}$ so that $F(x) = {}^{\mathbb{S}}\{F(y) : y \in x \cap \mathscr{I}\}$ (a standard set satisfying $F(x) \cap \mathbb{S} = \{F(y) : y \in x \cap \mathscr{I}\}$). Then F is a $1-1$ map (a simple argument by $\in$-induction is left as an **exercise**), and $x \in y \iff F(x) \in F(y)$. In addition, F is definable in $\mathbb{I}$, by a parameter-free $\mathtt{st}$-$\in$-formula.

We claim that $\mathtt{ran}\, F = \mathbb{S}$. Indeed, as $\mathbb{S}$ is well-founded, it suffices to prove that $s \in \mathbb{S}$ belongs to $\mathtt{ran}\, F$ assuming that $s \subseteq \mathtt{ran}\, F$. Then the set $X = \{x \in \mathscr{I} : F(x) \in s\}$ (not necessarily internal) is $\mathtt{st}$-$\in$-definable in $\mathbb{I}$ (since so is F), therefore, by Inner Standardization in $\langle \mathbb{I}; \in, \mathtt{st}_{\mathscr{I}} \rangle$, there is a set $x \in \mathscr{I}$ with $X = x \cap \mathscr{I}$. Obviously $s = F(x)$.

Thus $F : \mathscr{I} \xrightarrow{\text{onto}} \mathbb{S}$ is an $\in$-isomorphism. It follows that $G = F \restriction \mathbb{S}$ is an elementary embedding $\mathbb{S} \to \mathbb{S}$, that is, for any $\in$-formula $\vartheta(v_1, ..., v_n)$ and any $x_1, ..., x_n \in \mathbb{S}$, we have $\vartheta(x_1, ..., x_n) \iff \vartheta(F(x_1), ..., F(x_n))$ in $\mathbb{S}$. (This is because $\mathscr{I}$ is an internal subuniverse.) Moreover, G is not the identity because $\mathbb{S} \subsetneqq \mathscr{I}$. Finally, by (ii) of Exercise 3.2.4(3), G is $\in$-definable in $\mathbb{S}$. (Recall that F is $\mathtt{st}$-$\in$-definable in $\mathbb{I}$.) But this contradicts Kunen's well-known theorem (Theorem 6.1.21 below). □

[4] Andreev [An 99] gave a somewhat simpler argument for the subcase $\mathbb{N} \subsetneqq \mathscr{I} \cap {}^*\mathbb{N} \subsetneqq {}^*\mathbb{N}$. Then $C = {}^*\mathbb{N} \cap \mathscr{I}$ is a proper initial segment of ${}^*\mathbb{N}$ by Lemma 3.1.18(iv) (applied in the structure $\langle \mathbb{I}; \in, \mathtt{st}_{\mathscr{I}} \rangle$). As in the main construction, C must be internal, thus it has, as any bounded subset of ${}^*\mathbb{N}$, a maximal element, easily leading to contradiction.

[5] This case is rather peculiar: it leads to measurable cardinals by Theorem 3.1.25.

[6] The well-foundedness of $\mathscr{I}$ cannot be established here without the assumption of Inner Standardization: a counterexample, not to be considered here, was communicated to the authors by Hrbaček.

Let us study what happens with Idealization in structures of the form $\langle \mathbb{I}; \in, w\text{-}\mathtt{st}\rangle$ and $\langle \mathbb{I}; \in, w\text{-}\mathtt{st}_{\mathrm{M}}\rangle$.

Theorem 6.1.16. *Let w be an internal set. Then*

(i) *the structure $\langle \mathbb{I}; \in, w\text{-}\mathtt{st}_{\mathrm{M}}\rangle$ satisfies Basic Idealization;*

(ii) *the structure $\langle \mathbb{I}; \in, w\text{-}\mathtt{st}\rangle$ satisfies $\Longrightarrow$ in Basic Idealization, but does not satisfy $\Longleftarrow$ in Basic Idealization at least in the case when $\mathbb{N} \subsetneqq \mathbb{N}[w]$.*

The following proof takes advantage of the **HST** world of external sets, but in fact all arguments can be maintained internally, *i.e.*, entirely on the base of the **BST** axioms.

Proof. Prove $\Longleftarrow$ in Basic Idealization for $w\text{-}\mathtt{st}_{\mathrm{M}}$. It suffices to show that $w\text{-}\mathtt{st}_{\mathrm{M}}\, x \Longrightarrow w\text{-}\mathtt{st}_{\mathrm{M}}\, z$ whenever $z \in x$ are internal sets and x is $*$-finite. By Lemma 6.1.12 there is a $*$-finite set $y \in \mathbb{S}[w]$ containing x. Let u be the union of all $*$-finite elements of y : then $u \in \mathbb{S}[w]$ and u is $*$-finite by Theorem 6.1.3(i). Yet $z \in u$, hence, $z \in \mathbb{S}_{\mathrm{M}}[w]$ by Lemma 6.1.12.

Prove the implication $\Longrightarrow$ in Basic Idealization.

The case of $w\text{-}\mathtt{st}$ is rather simple. Indeed suppose that $I \subseteq \mathbb{S}[w]$, $\mathscr{X} = \{X_i : i \in I\}$ is a family of internal sets, and $\bigcap_{i \in I'} X_i \neq \varnothing$ whenever a set $I' \subseteq I$ is w-standard and $*$-finite. It follows from claim (2) of Exercise 6.1.8 that I is a set of standard size, so that $\bigcap \mathscr{X} \neq \varnothing$ by Theorem 1.3.5 (Saturation in **HST**), as required.

Consider now the the predicate $w\text{-}\mathtt{st}_{\mathrm{M}}$. Suppose that $A \in \mathbb{S}_{\mathrm{M}}[w]$, $I = A \cap \mathbb{S}_{\mathrm{M}}[w]$, $\mathscr{X} = \{X_i : i \in I\}$ is a family of internal sets, and $\bigcap_{i \in I'} X_i \neq \varnothing$ whenever a set $I' \subseteq I$ is w-standard in the modified sense and $*$-finite; prove that $\bigcap \mathscr{X} \neq \varnothing$. According to Lemma 6.1.14, there is a standard size set $Y \subseteq \mathbb{S}[w]$, hence $\subseteq \mathbb{S}_{\mathrm{M}}[w]$, such that every $y \in Y$ is $*$-finite and $I \subseteq \bigcup Y$.

The sets $y' = y \cap A$, where $y \in Y$, are still $*$-finite sets in $\mathbb{S}_{\mathrm{M}}[w]$. Moreover, it follows from the above that $y' \subseteq \mathbb{S}_{\mathrm{M}}[w]$, hence, $I = \bigcup Y'$, where $Y' = \{y' : y \in Y\}$. We observe that if $y' \in Y'$ then the intersection $X_{y'} = \bigcap_{i \in y'} X_i$ is non-empty by the choice of sets X_i, moreover, if $n \in \mathbb{N}$ and $y'_1, \dots, y'_n \in Y'$ then $\bigcap_{k=1}^{n} X_{y'_k} \neq \varnothing$ by the same reason. It remains to apply Theorem 1.3.5: the set Y' is of standard size and all $X_{y'}$ are internal, and hence we have $\bigcap \mathscr{X} = \bigcap_{y' \in Y'} X_{y'} \neq \varnothing$.

To see that $\Longleftarrow$ in Basic Idealization fails in the structure $\langle \mathbb{I}; \in, w\text{-}\mathtt{st}\rangle$ $\mathbb{N}[w] \neq \mathbb{N}$, note that it would imply that the w-standard numbers form an initial segment of ${}^*\mathbb{N}$. (Indeed, let us argue in $\langle \mathbb{I}; \in, w\text{-}\mathtt{st}\rangle$. Assuming $\Longleftarrow$ in Basic Idealization, any w-standard finite set contains only w-standard elements. Apply this result to w-standard finite sets of the form $\{k : k \leq n\}$, where n is w-standard.) On the other hand, $\mathbb{N}[w]$ is not an initial segment by (4) of Exercise 6.1.8. $\square$

6.1f More on internal subuniverses

The structure of internal subuniverses remains largely unclear except for some particular cases like $\mathbb{S}[x]$ or $\mathbb{I}_\kappa$ in the next section.

Problem 6.1.17. Study the structure of the internal subuniverses $\mathscr{I} \subseteq \mathbb{I}$ $\mathtt{st}$-$\in$-definable in $\mathbb{I}$ (with parameters in $\mathbb{I}$ allowed). Once it was thought that any such $\mathscr{I}$ has the form $\mathscr{I} = \mathbb{S}(C \cup R)$ where $R \subseteq \mathbb{I}$ is a set of standard size and $C \subseteq {}^*\mathtt{Ord}$ is an initial segment but [Hr **] gives a counterexample. □

There is a wide spectrum of interesting related questions, sometimes quite difficult to answer.

Exercise 6.1.18. Let $\mathscr{I} \subseteq \mathbb{I}$ be a class in $\mathrm{Def}^{\mathbb{I}}_{\in,\mathtt{st}}(\{p\})$, $p \in \mathbb{I}$, that is only p is admitted in the $\mathtt{st}$-$\in$-definition of $\mathscr{I}$. Prove the following.

(1) $\mathscr{I}$ is $\Sigma_2^{\mathtt{st}}(\{p\})$, *i.e.* $\mathscr{I} = \{x \in \mathbb{I} : \exists^{\mathtt{st}} a\, \forall^{\mathtt{st}} b\, \varphi(x, a, b, p)^{\mathtt{int}}\}$ where φ is a $\in$-formula with p as the only parameter. (Use Theorem 3.2.3.)

(2) If $\mathscr{I}$ has no infinite internal subsets then $\mathscr{I} \subseteq \mathbb{S}[p]$ and $\mathscr{I}$ is thin. [7]
(Any set $X_{ab} = \{x \in \mathbb{I} : \varphi(x, {}^*a,, p)^{\mathtt{int}}\}$, a, $b \in \mathbb{WF}$, is internal. For any $a \in \mathbb{WF}$ there is a finite $B \in \mathbb{WF}$ such that $X_a^B = \bigcap_{b \in B} X_{ab}$ is finite as otherwise $X_a = \bigcap_{b \in \mathbb{WF}} X_{ab}$ contains an infinite internal subset by an appropriate Collection – Saturation argument. Then $X_a \subseteq X_a^B \in \mathbb{S}[p]$ and $\subseteq \mathbb{S}[p]$ by Theorem 6.1.3 for the class $\mathbb{S}[p]$.) □

It follows from (2) that any **non**-thin internal subuniverse $\mathscr{I}$ contains an infinite internal subset. Does it necessarily contain an infinite subset X which itself belongs to $\mathscr{I}$? [8]

We add a couple more problems of different sort.

Problem 6.1.19. Characterize Gordon's classes $\mathbb{S}_{\mathrm{M}}[w]$ in terms of ultraproducts. This cannot be exactly the same as in Proposition 6.1.9. □

Problem 6.1.20. Study structures of the form $\langle \mathscr{K}\,; \in, \mathtt{st}_{\mathscr{I}} \rangle$ where $\mathscr{I} \subsetneqq \mathscr{K}$ are arbitrary internal subuniverses, for instance of the form $\mathbb{S}[x]$. Hrbaček [Hr **] presented a structure of this type, with $\mathbb{S} \subsetneqq \mathscr{I} \subsetneqq \mathscr{K} \subsetneqq \mathbb{I}$, satisfying Inner Standardization in spite of Theorem 6.1.15. □

[7] Hrbaček informed us that if, in addition to (2), $\mathscr{I}$ is an internal subuniverse then it has the form $\mathscr{I} = \mathbb{S}(R)$ for a suitable set $R \subseteq \mathbb{S}[p]$. This follows from some results that will appear in Hrbaček [Hr **].

[8] Hrbaček informed us that this question answers in the negative assuming the existence of selective ultrafilters on $\mathbb{N}$.

6.1g Appendix: Kunen's theorem

To make the exposition self-contained, we outline the proof of a sufficient fragment of Kunen's result cited in the proof of Theorem 6.1.15.

Theorem 6.1.21 (ZFC). *If $G : \mathbf{V} \to \mathbf{V}$ is a definable elementary embedding of the universe $\mathbf{V}$ of* **ZFC** *in itself, then it is the identity.*

Proof. We argue in **ZFC**. Suppose that $G = \{\langle x, y\rangle : \varphi(p_0, x, y)\}$, φ is an $\in$-formula and p_0 any set. Let $G_p = \{\langle x, y\rangle : \varphi(p, x, y)\}$ for any p, so that $G = G_{p_0}$. Let $\Phi(p)$ be the conjunction of the following formulas:

- G_p is a $1-1$ map, $\mathtt{dom}\, G_p = \mathbf{V}$, and $x \in y \iff F(x) \in F(y)$ for all x, y;
- the class $R_p = \mathtt{ran}\, G_p$ contains arbitrarily large von Neumann sets $\mathbf{V}_\xi$;
- for any $\mathbf{V}_\xi \in R_p$, the structure $\langle \mathbf{V}_\xi \cap R_p ; \in\rangle$ is an elementary submodel of $\langle \mathbf{V}_\xi ; \in\rangle$.

Then $\Phi(p_0)$ holds by the choice of $G = G_{p_0}$. We claim that, conversely, if p satisfies $\Phi(p)$ then G_p is an elementary embedding, in other words,

(*) $\varphi \iff \varphi^{R_p}$ for any $\in$-formula φ with sets in R_p as parameters.

To prove (*) assume that φ has the form $\mathsf{Q}_1 y_1 \mathsf{Q}_2 y_2 \dots \mathsf{Q}_n y_n \psi(y_1, ..., y_n)$, where Q_k are quantifiers $\exists$, $\forall$ while ψ is a bounded formula with parameters in R_p. We argue by induction on n. The case $n = 0$ (of bounded formulas φ) easily follows from the assumption $\Phi(p)$. To carry out the step note that, by Collection, there exists a sequence $Y_1 \subseteq \dots \subseteq Y_n$ of sets $Y_k = \mathbf{V}_{\xi_k}$, $\xi \in \mathtt{Ord}$, such that for all $k \ge 0$ and $y_1 \in Y_1, \dots, y_{k-1} \in Y_{k-1}$ satisfying (†) on page 97. Moreover, we can assume, by $\Phi(p)$, that $Y_1, ..., Y_n \in R_p$. Then the arguments in the proof of Theorem 3.2.6 can be carried out for R_p instead of $\mathbb{S}$. We leave it to the reader to accomplish the induction step.

Coming back to the proof of the theorem, we claim that $G(\xi) = \xi$ for any ordinal ξ. Otherwise consider the least ordinal ξ such that there exists a set p such that $\Phi(p)$ holds, all parameters of φ (hence, of Φ as well) belong to R_p, and $G_p(\xi) \ne \xi$. Then $G_p : \mathbf{V} \xrightarrow{\text{onto}} \mathbf{V}$ is an elementary embedding by (*). It follows that $\xi \in R_p = \mathtt{ran}\, G_p$, because ξ is definable in $\mathbf{V}$ by an $\in$-formula with parameters in R_p. Thus $\xi = G_p(\eta)$ for some $\eta \in \mathtt{Ord}$. Note that $\eta > \xi$ because $G_p(\eta) = \eta$ for $\eta < \xi$ by the choice of ξ. On the other hand, $G_p(\eta) \ge \eta$ for any ordinal η (by induction on η), contradiction.

We finally prove that $G(x) = x$ for any set x, not necessarily an ordinal. Note that G preserves the von Neumann rank: $\mathtt{rank}\, G(x) = G(\mathtt{rank}\, x) = \mathtt{rank}\, x$ by the above. Take, towards the contrary, a set x with $G(x) \ne x$, of the least possible $\mathtt{rank}\, x$. Then $G(y) = y$ for all $y \in x$, hence $x \subseteq G(x)$. Suppose that $z \in G(x) \smallsetminus x$. Then $\mathtt{rank}\, z < \mathtt{rank}\, G(x) = \mathtt{rank}\, x$, so that $G(z) = z \in G(x)$, which implies $z \in x$, contradiction. □

6.2 Partially saturated internal universes

Here we study internal subuniverses satisfying certain forms of Saturation. Let κ be a (well-founded) cardinal; then ${}^*\kappa$ is a standard $*$-cardinal by Inner Transfer. Say that an internal subuniverse $\mathscr{I}$ is:

κ-deep saturated: if for any set $\mathscr{X} \subseteq \mathscr{I}$ of standard size and consisting of subsets of a given set of the form ${}^*W$, where $W \in \mathbb{WF}$, $\operatorname{card} W \le \kappa$, it follows from $\bigcap \mathscr{X} \neq \varnothing$ that $\mathscr{I} \cap \bigcap \mathscr{X} \neq \varnothing$;

κ-size saturated: if for any set $\mathscr{X} \subseteq \mathscr{I}$ of cardinality $\le \kappa$, it follows from $\bigcap \mathscr{X} \neq \varnothing$ that $\mathscr{I} \cap \bigcap \mathscr{X} \neq \varnothing$. [9]

The goal of this section is to study internal subuniverses of these types along with associated structures $\langle \mathscr{I} ; \in, \mathtt{st} \rangle$ in **HST**. We also show how partially saturated internal subuniverses can be defined using relative standardness.

6.2a Partially saturated classes $\mathbb{I}_\kappa$

According to Exercise 6.1.5(2), the class $\mathbb{S}({}^*X)$, $X \in \mathbb{WF}$, depends only on the cardinality of X. This leads us to the following definition. Note that ${}^*\kappa$, a $*$-cardinal, is equal to the set of all $*$-ordinals smaller than ${}^*\kappa$.

Definition 6.2.1. $\mathbb{I}_\kappa = \mathbb{I}_{{}^*\kappa} = \mathbb{S}({}^*\kappa)$ for any (well-founded) cardinal κ. □

Each $\mathbb{I}_\kappa$ is an internal subuniverse by 6.1.5(1), and $\mathbb{I}_\kappa \subsetneqq \mathbb{I}_{\kappa^+}$ by 6.1.5(2). Classes of the form $\mathbb{I}_\kappa$ will be extensively used below in this Chapter.

Exercise 6.2.2. Prove the following:

$$\begin{aligned}
\mathbb{I}_\kappa &= \{{}^*\!f(\xi) : \xi \in {}^*\kappa \wedge f \in \mathbb{WF} \text{ is a function} \wedge \operatorname{dom} f = \kappa\} = \\
&= \{f(\xi) : \xi \in {}^*\kappa \wedge \ f \in \mathbb{S} \ \text{ is a function} \wedge \operatorname{dom} f = {}^*\kappa\} = \\
&= \{x : \exists^{\mathtt{wf}} W \, (x \in {}^*W \wedge \operatorname{card} W \le \kappa)\} = \\
&= \{x : \exists^{\mathtt{st}} X \, (x \in X \wedge \operatorname{card} X \le {}^*\kappa \text{ in } \mathbb{S} \text{ or in } \mathbb{I})\}.
\end{aligned}$$ [10]

Hint. The first equality: the set ${}^*\kappa$ satisfies (†) of 6.1.5(1). Further if $x \in {}^*W$ and $\operatorname{card} W \le \kappa$ then consider any map $f \in \mathbb{WF} : \kappa$ onto W. By $*$-Transfer, ${}^*\!f : {}^*\kappa \xrightarrow{\text{onto}} {}^*W$, so $x = {}^*\!f(\alpha) \in \mathbb{I}_\kappa$ for some $\alpha < {}^*\kappa$. To prove the converse apply the result of Exercise 6.1.5(2). □

The next theorem proves, among other things, that the internal subuniverses of the form $\mathbb{I}_\kappa$ can be employed to interpret the partially saturated internal theories $\mathbf{BST}_\kappa$ and $\mathbf{BST}'_\kappa$ defined in § 3.3a.

[9] This is a form of κ^+-Saturation rather than κ-Saturation, of course, but we would like to keep uniformity with the κ-deep version.

[10] Thus $\mathbb{I}_\kappa$ is equal to the class $\mathbb{I}_{{}^*\kappa}$ defined internally in $\mathbb{I}$ as in § 3.4a.

Theorem 6.2.3. *Let κ be an infinite (well-founded) cardinal. Then $\mathbb{I}_\kappa$ is an internal subuniverse* (which implies properties mentioned in Theorem 6.1.3). *In addition $\mathbb{I}_\kappa$ is κ-deep saturated and κ-size saturated.*

Suppose that $\mathscr{I} \subseteq \mathbb{I}_\kappa$ is an internal subuniverse. Then

(i) *If $\mathscr{I} \subseteq \mathbb{I}_\kappa$ is κ-deep saturated then $\mathscr{I}$ is κ-size saturated, is an interpretation of* **BST**$_{^*\kappa}$ [11] *and satisfies the following form of 2^κ-size Choice:*

  (∗) *if $W \in \mathbb{WF}$ is a set of cardinality $\leq 2^\kappa$ and $\varnothing \neq X_w \subseteq \mathscr{I}$ for any $w \in W$ then there is a function $f \in \mathscr{I}$ defined on $^*W$ such that $f(^*w) \in X_w$ for all $w \in W$.*

(ii) *If $\mathscr{I} \subseteq \mathbb{I}_\kappa$ is κ-size saturated then it satisfies the axioms of $^*\kappa$-size BI and Inner $^*\kappa$-Boundedness of* § 3.3a; *if, in addition, $\mathscr{I} = \mathbb{S}[w]$ for some $w \in \mathscr{I}$* (see Definition 3.1.13) *then $\mathscr{I}$ is an interpretation of* **BST**$'_{^*\kappa}$.

(iii) *For each $X \in \mathbb{WF}$ there is a set $P \in \mathbb{WF}$ such that for any $q \in \mathscr{I}$ satisfying $\mathsf{E}_q \subseteq {}^*X$ there exists $p \in {}^*P \cap \mathscr{I}$ with $\mathsf{E}_q = \mathsf{E}_p$.*

Proof. $\mathbb{I}_\kappa$ is κ-deep saturated because $^*\kappa$ retains its elements in $\mathbb{I}_\kappa$.

(i) To verify κ-size Saturation, suppose that $\mathscr{X} \subseteq \mathscr{I}$, $\mathtt{card}\,\mathscr{X} \leq \kappa$, and $\bigcap \mathscr{X} \neq \varnothing$. The set $\mathscr{F}$ of all finite intersections of sets in $\mathscr{X}$ has cardinality $\leq \kappa$. Since $\mathscr{I} \subseteq \mathbb{I}_\kappa$, for any $X \in \mathscr{F}$ there is a set $u_X \in \mathbb{WF}$ of cardinality $\leq \kappa$ with $X \cap {}^*u_X \neq \varnothing$. (Standard Size Choice enables us to choose the sets u_X simultaneously.) Then $u = \bigcup_{X \in \mathscr{F}} u_X \in \mathbb{WF}$ is a set of cardinality $\leq \kappa$, moreover, $X' = X \cap {}^*u \neq \varnothing$ for all $X \in \mathscr{F}$. Then $\bigcap_{X \in \mathscr{F}} X' \neq \varnothing$ by the **HST** Saturation. We have $\mathscr{I} \cap \bigcap_{X \in \mathscr{F}} X' \neq \varnothing$, because $\mathscr{I}$ is κ-deep saturated.

(∗) (Compare with the proof of a 2^κ-size Choice in Theorem 3.3.5.) For any $w \in W$ we have $X_w \cap \mathscr{I} \neq \varnothing$ by Theorem 6.1.3(i), hence, as $\mathscr{I} \subseteq \mathbb{I}_\kappa$, there exist: a set $R_w \in \mathbb{WF}$ of cardinality κ, a bijection $h_w : \kappa \xrightarrow{\text{onto}} R_w$, an element $x_w \in {}^*R_w \cap X_w$, and a ∗-ordinal $\xi_w = {}^*h_w{}^{-1}(x_w) \in {}^*\kappa \cap \mathscr{I}$ such that $x_w = {}^*h_w(\xi_w)$. (The axiom of Standard Size Choice in the entire universe is applied here.) Define, in $\mathbb{WF}$, $h(w,\xi) = h_w(\xi)$, thus, $\mathtt{dom}\,h = W \times \kappa$.

We can w. l. o. g. assume that $W = 2^\kappa$, the set of all maps $\kappa \to 2$.

The set $P = \mathscr{P}_{\mathtt{fin}}(\kappa)$ still has cardinality κ, hence applying κ-deep Saturation to the family of internal sets $P_\xi = \{Z \in {}^*P : {}^*\xi \in Z\}$, we obtain a ∗-finite set $\Omega \in \mathscr{I}$, $\Omega \subseteq {}^*\kappa$ with ${}^*\kappa \cap \mathbb{S} \subseteq \Omega$. Then $\eta(w) = {}^*w \restriction \Omega$ is an injection $W \to {}^*P$. Let $G \in \mathbb{WF}$ be the set of all functions g with $\mathtt{dom}\,g \in P$ and $\mathtt{ran}\,g \subseteq \kappa$, still a set of cardinality κ. Put, for each $w \in W$, $G_w = \{g \in {}^*G : g({}^*w \restriction \Omega) = \xi_w\}$. The family of all sets G_w is easily f. i. p. (and we leave it to the reader to show that $G_w \in \mathscr{I}$), hence, as $\mathscr{I}$ is κ-deep saturated, there is $g \in {}^*G \cap \mathscr{I}$ which belongs to each G_w, in other words, $g({}^*w \restriction \Omega) = \xi_w$ for all $w \in W$. Define $f(w) = {}^*h(w, g(w \restriction \Omega))$ for any $w \in {}^*W$; this is a function in $\mathscr{I}$. We have, for any $w \in W$, $f({}^*w) = {}^*h({}^*w, g({}^*w \restriction \Omega)) = {}^*h({}^*w, \xi_w) = {}^*h_w(\xi_w) = x_w$, as required.

[11] This means that the structure $\langle \mathscr{I}\,; \in, \mathtt{st}, {}^*\kappa\rangle$ is an interpretation of **BST**$_\kappa$ such that the constant κ of the theory is interpreted as $^*\kappa$.

(i), (ii): interpretations. First of all recall Theorem 6.1.3(iii). Inner ${}^*\kappa$-Boundedness holds in (i) because $\mathscr{I} \subseteq \mathbb{I}_\kappa$. Inner Strong ${}^*\kappa$-Boundedness holds in (ii) by Lemma 3.3.2. κ-size BI, resp., κ-deep BI follow from the assumption that $\mathscr{I}$ is resp. κ-size, κ-deep saturated, as in the proof of Theorem 3.1.8.

(iii) The relation $p \doteq q$ between E-codes defined in § 3.3b is absolute for $\mathscr{I}$: indeed, it can be defined by a formula of the form $Q\,\Phi$, where Q is a prefix containing quantifiers $\exists^{\text{st}}$ and $\forall^{\text{st}}$ while Φ is an $\in$-formula — yet any formula of this type is absolute for $\mathscr{I}$ by Theorem 6.1.3(i). [12] It remains to apply (∗) of Theorem 3.3.5 because $\mathscr{I}$ satisfies $\mathbf{BST}_\kappa$ by (i). □

The following lemma will be quite useful below.

Lemma 6.2.4. *For any internal subuniverse $\mathscr{I}$, each of the following two conditions is sufficient for a set $z \in \mathscr{I}$ to be finite and satisfy $z \subseteq \mathscr{I}$:*

(1) *$\mathscr{I}$ is κ-size saturated and $\mathtt{card}(z \cap \mathscr{I}) \le \kappa$,*

(2) *$\mathscr{I}$ is $\aleph_0$-deep saturated and $z \cap \mathscr{I}$ is a set of standard size.*

Proof. $z \subseteq \mathscr{I}$ follows from the finiteness and Theorem 6.1.3(iv). If z is infinite then, since $\mathscr{I}$ is an $\in$-interpretation of **ZFC** by Theorem 6.1.3(iii), there is infinite and ∗-countable (countable in $\mathbb{I}$) set $y \in \mathscr{I}$, $y \subseteq x$. Thus, we can assume that x itself is ∗-countable. Then there is a bijection $\pi \in \mathscr{I}$, $\pi : {}^*\mathbb{N} \xrightarrow{\text{onto}} z$. This allows us to w. l. o. g. suppose that $z = {}^*\mathbb{N}$. Applying either κ-size Saturation or $\aleph_0$-deep Saturation to the family of sets $Z_a = z \smallsetminus \{a\}$, $a \in z \cap \mathscr{I}$, we get an $x \in \mathscr{I} \cap z$, $x \notin \bigcup_{a \in z \cap \mathscr{I}} Z_a$, contradiction. □

Exercise 6.2.5 (Hrbaček). Prove that $\mathscr{I} = \bigcup_{n \in \mathbb{N}} \mathbb{I}_{\aleph_n}$ is a $\aleph_0$-deep saturated but not $\aleph_0$-size saturated internal subuniverse. (This does not contradict (i) of Theorem 6.2.3 because $\mathscr{I} \not\subseteq \mathbb{I}_{\aleph_0}$.) □

6.2b Good internal subuniverses

Our next goal is to show that κ-size saturated classes can be found among the internal subuniverses of the form $\mathbb{S}[w]$.

Let us fix an infinite (well-founded) cardinal κ.

Any ∗-ordinal $\zeta < {}^*\kappa$ defines an ultrafilter $U_\zeta = \{X \subseteq \kappa : \zeta \in {}^*X\} \in \mathbb{WF}$ on κ. One can reasonably expect that properties of the ultrafilter U_ζ are somehow reflected in the properties of the internal subuniverse $\mathbb{S}[\zeta]$. We are going to show that if U_ζ is a good ultrafilter then $\mathbb{S}[\zeta]$ is a κ-size saturated internal subuniverse. Say that a ∗-ordinal $\zeta < {}^*\kappa$ is *κ-good* if U_ζ is a κ^+-good countably incomplete ultrafilter. (See Definition 4.2.15 on good ultrafilters.)

Theorem 6.2.6 (HST). *If κ is an infinite cardinal then κ-good ∗-ordinals $\zeta < {}^*\kappa$ do exist, and for any κ-good ∗-ordinal $\zeta < {}^*\kappa$ the class $\mathbb{S}[\zeta]$ is a thin κ-size saturated internal subuniverse and an interpretation of $\mathbf{BST}'_{{}^*\kappa}$.* [13]

[12] Recall that $p \doteq q$ implies $\mathsf{E}_p = \mathsf{E}_q$. The equality $\mathsf{E}_p = \mathsf{E}_q$ is, generally speaking, not absolute; this is essentially why we introduced the relation $\doteq$ in § 3.3b.

This theorem is equivalent to Theorem 4.2.17(ii): the equivalence is ensured by Proposition 6.1.9. Yet it is worthwhile to outline a proof.

Proof. Let $U \subseteq \mathscr{P}(\kappa)$ be a κ^+-good ultrafilter (Proposition 4.2.16 is applied). By Saturation, there exists a $*$-ordinal $\zeta < {}^*\kappa$ such that $U = U_\zeta$.

Suppose that $\zeta < {}^*\kappa$ is a κ-good $*$-ordinal, $X_\alpha \in \mathbb{S}[\zeta]$ for any $\alpha < {}^*\kappa$, and the family $\{X_\alpha\}_{\alpha<\kappa}$ satisfies f. i. p.; we have to prove that $\bigcap_{\alpha<\kappa} X_\alpha \cap \mathbb{S}[\zeta] \neq \varnothing$. For any $\alpha < \kappa$ there is a function $H_\alpha \in \mathbb{WF}$ defined on κ such that $X_\alpha = {}^*H_\alpha(\zeta) = {}^*H({}^*\alpha, \zeta)$, where $H(\gamma, \zeta) = H_\gamma(\zeta)$ in $\mathbb{WF}$.

We argue in $\mathbb{WF}$. By definition $U = U_\zeta \in \mathbb{WF}$ is a κ^+-good countably incomplete ultrafilter on κ, in particular there is a decreasing chain $\kappa = I_0 \supseteq I_1 \supseteq I_2 \supseteq \ldots$ of sets $I_n \in U$ with $\bigcap_{n\in\mathbb{N}} I_n = \varnothing$. For any finite $s \subseteq \kappa$ with n elements and any ordinal $\xi < \kappa$ define $H_s(\xi) = \bigcap_{\alpha\in s} H(\alpha, \xi)$ and $f(s) = I_n \cap \{\xi < \kappa : H_s(\xi) \neq \varnothing\}$. Note that by definition ${}^*H_s(\zeta) = \bigcap_{\alpha\in s} X_\alpha \neq \varnothing$, therefore $\zeta \in {}^*f(s)$ by $*$-Transfer and hence $f(s) \in U$. The map $f : \kappa^{\text{fin}} \to U$ is obviously monotone, thus there is an additive map $g : \kappa^{\text{fin}} \to U$, $g \leq f$.

Arguing in $\mathbb{WF}$ as in the proof of Theorem 4.2.17(ii), we get a function $x \in \mathbb{WF}$ defined on κ such that $x(\xi) \in H_\alpha(\xi)$ whenever $\alpha, \xi < \kappa$ and $\xi \in g(\{\alpha\})$. However for any $\alpha < \kappa$ the set $Z = g(\{\alpha\})$ belongs to U by the choice of g, and hence $\zeta \in {}^*Z$ and ${}^*x(\zeta) \in X_\alpha = {}^*H(\zeta)$. Finally ${}^*h(\zeta) \in \mathbb{S}[\zeta]$.

Now to show that $\mathbb{S}[\zeta]$ interprets $\mathbf{BST}'_{*\kappa}$ use Theorem 6.2.3(ii). □

Arguing similarly to the proof of Corollary 4.1.12(i), we can infer that $\mathbf{BST}'_\kappa$ is a "realistic" theory (in a modified sense mentioned in footnote 15 on page 146), and hence it is an equiconsistent and conservative standard core extension of **ZFC** — but this has been proved earlier (Proposition 4.3.2).

Problem 6.2.7. Does Corollary 4.1.12(iii) hold for $\mathbf{BST}'_\kappa$ with classes $\mathbb{I}_\kappa$ replaced by classes of the form $\mathbb{S}[\zeta]$, ζ being a κ-good ordinal? □

6.2c Internal universes over complete sets

By Theorem 6.1.3(i), any internal subuniverse is an elementary substructure of $\mathbb{I}$ in the $\in$-language. Here we introduce a family of thin κ-size saturated internal subuniverses $\mathscr{I} \subseteq \mathbb{I}_\kappa$ which are elementary substructures of $\mathbb{I}_\kappa$ even in the $\mathtt{st}$-$\in$-language, and hence interpretations of $\mathbf{BST}_\kappa$.

Definition 6.2.8. Let κ, λ be infinite (well-founded) cardinals. A set $R \subseteq {}^*\kappa$ is λ-*complete* if for each family $\mathscr{X} \subseteq \mathbb{S}(R)$ with $\mathtt{card}\, \mathscr{X} \leq \lambda$ and consisting of subsets of ${}^*\kappa$, it follows from $\bigcap \mathscr{X} \neq \varnothing$ that $R \cap \bigcap \mathscr{X} \neq \varnothing$. □

Any such set $R \subseteq {}^*\kappa$ (for any λ) obviously satisfies $R = \mathbb{S}(R) \cap {}^*\kappa$.

Thus the completeness is a special type of Saturation, in particular, by the next theorem, κ-completeness implies κ-size Saturation of $\mathbb{S}(R)$, while 2^κ-completeness leads to elementary submodels of $\mathbb{I}_\kappa$ in the $\mathtt{st}$-$\in$-language.

[13] In other words, the structure $\langle \mathbb{S}[\zeta]; \in, \mathtt{st}, {}^*\kappa \rangle$ is an interpretation of $\mathbf{BST}'_\kappa$.

Theorem 6.2.9. *Let κ be an infinite (well-founded) cardinal. If $\varnothing \neq R \subseteq {}^*\kappa$ then $\mathbb{S} \subseteq \mathbb{S}(R) \subseteq \mathbb{I}_\kappa$ and $\mathbb{S}(R)$ is an internal subuniverse. Moreover,*

(i) *if R is κ-complete then class $\mathbb{S}(R)$ is κ-size saturated and satisfies the axioms of ${}^*\kappa$-size BI and Inner ${}^*\kappa$-Boundedness of § 3.3a;*

(ii) *if R is 2^κ-complete then, in addition, $\mathbb{S}(R)$ is an elementary substructure of $\mathbb{I}_\kappa$ in the $\mathtt{st}$-$\in$-language, even in the following strong sense:*

(‡) *if $\Phi(x)$ is a $\mathtt{st}$-$\in$-formula with parameters in $\mathbb{S}(R)$, and there is $x \in \mathbb{I}_\kappa$ such that $\Phi(x)$ holds in $\mathbb{I}$ then such a set x exists in $\mathbb{S}(R)$;*

in particular, in this case $\mathbb{S}(R)$ is an interpretation of $\mathbf{BST}_{{}^*\kappa}$.

Remark 6.2.10. Condition (ii)(‡) implies that $\mathbb{S}(R)$ is an elementary substructure of $\mathbb{I}_\kappa$ in the sense of the $\mathtt{st}$-$\in$-language: if ψ is a $\mathtt{st}$-$\in$-formula with parameters in $\mathbb{S}(R)$ then $\psi^{\mathbb{S}(R)} \Longleftrightarrow \psi^{\mathbb{I}_\kappa}$. To prove this claim by induction on the complexity of ψ (the ordinary argument), note that, by (ii)(‡) (take $\Phi(x)$ to be $\Psi^{\mathbb{I}_\kappa}(x)$, the relativization of Ψ to $\mathbb{I}_\kappa$), for any $\mathtt{st}$-$\in$-formula $\Psi(x)$ with parameters in $\mathbb{S}(R)$ we have: $\exists\, x \in \mathbb{I}_\kappa\, \Psi^{\mathbb{I}_\kappa}(x) \Longrightarrow \exists\, x \in \mathbb{S}(R)\, \Psi^{\mathbb{I}_\kappa}(x)$. □

Proof (Theorem). (i) Consider a set $\mathscr{X} \subseteq \mathbb{S}(R)$ with $\mathtt{card}\, \mathscr{X} \leq \kappa$, $\bigcap \mathscr{X} \neq \varnothing$. As in the proof of Theorem 6.2.3(i), there exists a set $u \in \mathbb{WF}$ of cardinality $\leq \kappa$ such that ${}^*u \cap \bigcap \mathscr{X} \neq \varnothing$. Let h be any map from κ onto u. Applying the κ-completeness to the family of the ${}^*f$-preimages of sets ${}^*u \cap X$, $X \in \mathscr{X}$, we easily obtain $\mathbb{S}(R) \cap \bigcap \mathscr{X} \neq \varnothing$.

(ii) We can suppose, that $\Phi(\cdot)$ contains only standard sets and some $w_0 \in R^n$, $n \in \mathbb{N}$, as parameters, thus Φ will be written as $\Phi(x, w_0)$. We can further assume that $\Phi(x, \cdot)$ explicitly says that x is an $*$-ordinal and $x < {}^*\kappa$, simply because $\mathbb{S}(R) \subseteq \mathbb{I}_\kappa = \mathbb{S}({}^*\kappa)$. In addition, it can be assumed by Theorem 3.2.3 that $\Phi(x, \cdot)$ is a $\Sigma_2^{\mathtt{st}}$ formula. Since $\mathbb{S} \subseteq \mathbb{S}(R)$, the left-most quantifier $\exists^{\mathtt{st}}$ in this formula can be eliminated, thus, let $\Phi(x, \cdot)$ be $\forall^{\mathtt{st}} b\, \varphi(b, x, \cdot)$, where φ is an $\in$-formula with standard parameters.

After these simplifications, condition (ii)(‡) takes the form:

> if (1) there is an $\mathbb{I}$-ordinal $\xi < {}^*\kappa$ such that $\forall^{\mathtt{st}} b\, \varphi(b, \xi, w_0)$ in $\mathbb{I}$ then (2) such an ordinal ξ exists in R.

To prove this claim, we are going to restrict the variable b by a standard set of cardinality $\leq {}^*\lambda$ in $\mathbb{S}$, where $\lambda = 2^\kappa$. As $w_0 \in R^n \subseteq \mathbb{I}_\kappa$, there exists a set $W \in \mathbb{WF}$ such that $\mathtt{card}\, W \leq \kappa$ and $w_0 \in {}^*W$. Let, in $\mathbb{I}$,

$$\Xi_b = \{\langle \xi, w \rangle \in {}^*\kappa \times {}^*W : \neg\, \varphi(b, \xi, w)\}$$

for all internal b, so that $\Xi_b \in \mathbb{I}$ for all $b \in \mathbb{I}$ since φ is an $\in$-formula. Applying in $\mathbb{I}$ (which satisfies of **ZFC**) the **ZFC** Collection and Choice, we get a set B of cardinality $\leq {}^*\lambda$ in $\mathbb{I}$ such that $\forall\, b\, \exists\, b' \in B\, (\Xi_b = \Xi_{b'})$.

Such a set B can be chosen in $\mathbb{S}$ by Transfer. Then, by Transfer again, we have $\forall^{\mathtt{st}} b\, \exists^{\mathtt{st}} b' \in B\, (\Xi_b = \Xi_{b'})$. This implies, in $\mathbb{I}$,

$$\forall \xi < {}^*\kappa \, \forall w \in {}^*W \left(\exists^{\text{st}} b \neg \varphi(b, \xi, w) \implies \exists^{\text{st}} b \in B \neg \varphi(b, \xi, w) \right).$$

We observe that, since B is a standard set satisfying $\mathtt{card}\, B \leq {}^*\lambda$ in $\mathbb{S}$, there exists a surjection h mapping $\lambda = 2^\kappa$ (a well-founded cardinal) onto the set $B \cap \mathbb{S}$. Now the last displayed formula takes the form:

$$\forall \xi < {}^*\kappa \, \forall w \in {}^*W \left(\forall^{\text{st}} b \, \varphi(b, \xi, w) \iff \forall \nu < \lambda \, \varphi(h(\nu), \xi, w) \right). \qquad (*)$$

We define $X_\nu = \{\xi < {}^*\kappa : \varphi(h(\nu), \xi, w_0)\}$ for every $\nu < \lambda$. It follows from (1) that $\bigcap_{\nu<\lambda} X_\nu \neq \varnothing$. Then there exists $\xi \in R \cap \bigcap_{\nu<\lambda} X_\nu$ (because R is λ-complete). Now we have $\forall^{\text{st}} b \, \varphi(b, \xi, w_0)$ by $(*)$, *i.e.* (2), as required. □

Let us now prove the existence of complete sets.

Theorem 6.2.11. *Let $\kappa \leq \lambda < \Omega$ be (well-founded) infinite cardinals and $\Omega^\lambda = \Omega$. Then for any set $R_0 \subseteq {}^*\kappa$ of standard size with $\mathtt{card}\, R_0 \leq \Omega$ there is a λ-complete set $R \subseteq {}^*\kappa$ of standard size with $R_0 \subseteq R$ and $\mathtt{card}\, R \leq \Omega$.*

Proof. Suppose that $Q \subseteq R \subseteq {}^*\kappa$. We say that R *completes* Q iff $Q \subsetneqq R$ and for every set $\{X_\nu : \nu < \lambda\} \subseteq \mathbb{S}(Q)$ such that $X_\nu \subseteq {}^*\kappa$ for all ν and $X = \bigcap_{\nu<\lambda} X_\nu$ is non-empty, the intersection $X \cap R$ is non-empty, too.

Lemma 6.2.12. *Let $Q \subseteq {}^*\kappa$ be a set of standard size, $\mathtt{card}\, Q \leq \Omega$. Then there exists a set $R \subseteq {}^*\kappa$ with $\mathtt{card}\, R \leq \Omega$ which completes Q.*

Proof (Lemma)**.** To obtain a set R as required, let us first of all enumerate all relevant λ-sequences of sets $X \in \mathbb{S}(Q)$, $X \subseteq {}^*\kappa$, by ordinals $\delta < \Omega$.

Let $Q = \{\rho_\gamma : \gamma < \Omega\}$. Consider, in $\mathbb{WF}$, a pair of maps, $g : \lambda \longrightarrow \Omega^{<\omega}$ and $h : \lambda \times \kappa^{<\omega} \longrightarrow \mathscr{P}(\kappa)$. Let $\nu < \lambda$. Then $g(\nu) = \langle \gamma_1, \dots, \gamma_n \rangle \in \Omega^{<\omega}$. Define $w_\nu = \langle \rho_{\gamma_1}, \dots, \rho_{\gamma_n} \rangle \in Q^{<\omega}$, so that $w_\nu \in ({}^*\kappa)^{<\omega}$. We further put $h_\nu(w) = h(\nu, w)$, for any $w \in \kappa^{<\omega}$, thus $h_\nu \in \mathbb{WF}$ maps $\kappa^{<\omega}$ into $\mathscr{P}(\kappa)$. We finally put $X_\nu^{gh} = {}^*(h_\nu)(w_\nu)$. Thus $\{X_\nu^{gh} : \nu < \lambda\}$ is a set of subsets of ${}^*\kappa$, each X_ν^{gh} being a member of $\mathbb{S}(Q)$ since all functions ${}^*(h_\nu)$ are standard.

Clearly for *every* set $\{X_\nu : \nu < \lambda\} \subseteq \mathbb{S}(Q)$ such that $X_\nu \subseteq {}^*\kappa$ for all ν, there exists a pair of functions g, $h \in \mathbb{WF}$ of the type described just above, satisfying $X_\nu = X_\nu^{gh}$ for all $\nu < \lambda$. We observe that, since $\Omega^\lambda = \Omega$, the set of all such pairs of functions h, g has cardinality Ω in $\mathbb{WF}$. Standard Size Choice yields a set $Q' = \{\sigma_\delta : \delta < \Omega\} \subseteq {}^*\kappa$, such that for any such a pair of functions g, h, there is an index $\delta < \Omega$ such that if $\bigcap_{\nu<\lambda} X_\nu^{gh} \neq \varnothing$ then $\sigma_\delta \in \bigcap_{\nu<\lambda} X_\nu^{gh}$. Put $R = Q \cup Q'$. □ *(Lemma)*

Coming back to the theorem, let $\mathbf{P}$ be the set of all sets $Q \subseteq {}^*\kappa$ of cardinality $\leq \Omega$. For $Q, R \in \mathbf{P}$ let $R \lhd Q$ mean that R completes Q. Then $\lhd$ is a strict partial order with no minimal elements by Lemma 6.2.12. Moreover, $\mathbf{P}$ is Ω-closed. (Given a $\lhd$-decreasing, hence, $\subset$-increasing chain $\{Q_\gamma\}_{\gamma<\alpha}$, where $\alpha \leq \Omega$ is a limit ordinal, the union $Q = \bigcup_{\gamma<\alpha} Q_\gamma$ belongs to $\mathbf{P}$ and $Q \lhd Q_\gamma$ for all γ.) In addition, both $\mathbf{P}$ and the relation $\lhd$ belong to $\mathbb{L}[\mathbb{I}]$

because this class contains all subsets of $\mathbb{I}$ of standard size by Theorem 5.5.4. It follows from Theorem 5.5.12 that there is a $\triangleleft$-decreasing λ^+-sequence $\{R_\gamma\}_{\gamma<\lambda^+}$ of elements of $\mathbf{P}$, beginning with R_0. The set $R = \bigcup_{\gamma<\lambda^+} R_\gamma$ is as required; $\mathtt{card}\, R \le \Omega$ holds because $\lambda < \Omega$. □

Exercise 6.2.13. (1) Prove that if $R \subseteq {}^*\kappa$ is a κ-complete set then $\mathbb{S}(R)$ does not have the form $\mathbb{S}[w]$, and hence Theorem 6.2.3(ii) is not fully applicable.

(2) Prove that $\mathbb{I}_\kappa$ is not an elementary substructure of $\mathbb{I}$ in the $\mathtt{st}$-$\in$-sense (compare with (i) in Theorem 6.1.3).

(3) Prove that $\mathbb{S}(R)$ is not a κ-deep saturated internal subuniverse provided $R \subseteq \mathbb{I}$ is a set of standard size. □

Remark 6.2.14. Classes of the form $\mathbb{S}(R)$, $R \subseteq {}^*\kappa$ being a 2^κ-complete set, can be used as a convenient environment for iterated extensions.

Indeed consider any 2^κ-complete set $R \subseteq {}^*\kappa$ of standard size. Let $\mathscr{X} \subseteq \mathbb{S}(R)$ be either a f. i. p. family of cardinality $\le \kappa$ or a f. i. p. family of any cardinality that consists of subsets of a fixed standard set S of $*$-cardinality $\le {}^*\kappa$, for instance, $S = {}^*\kappa$. In the "or" case $\mathscr{X}$ is a set of standard size since $\mathbb{S}(R)$ is a thin class by 6.1.6(2). Then by Theorem 6.2.3 there exists an element $x \in \bigcap \mathscr{X} \cap \mathbb{I}_\kappa$. It follows from Theorem 6.2.11 that there is a 2^κ-complete set $R' \subseteq {}^*\kappa$ of standard size such that $R \cup \{x\} \subseteq R'$. Then $\mathbb{S}(R')$ is an elementary substructure of $\mathbb{I}_\kappa$ and hence an elementary extension of $\mathbb{S}(R)$ in the $\mathtt{st}$-$\in$-language, and $\mathbb{S}(R')$ contains an element in $\bigcap \mathscr{X}$. □

Exercise 6.2.15. Consider any (well-founded) infinite cardinal $\kappa \in \mathbb{WF}$ and a 2^κ-complete set $R \subseteq {}^*\kappa$ of standard size. Then $\mathbb{S}(R)$ is an interpretation of $\mathbf{BST}_{^*\kappa}$ by Theorem 6.2.13. On the other hand, $\langle \mathbb{S}(R); \in, \mathtt{st}\rangle$ is isomorphic to a certain set-indexed [14] quotient power of the well-founded universe $\mathbb{WF}$ (see Exercise 6.1.10.)

(1) Expand this argument to show, in **ZFC**, that for any infinite cardinal κ there exists a set-indexed quotient power of the **ZFC** set universe that interprets $\mathbf{BST}_{^*\kappa}$. (The conservativity of **HST** over **ZFC** by Corollary 5.1.5 must be employed.)

(2) Prove that no class of the form $\mathbb{S}[w]$, $w \in \mathbb{I}$ can be an interpretation of $\mathbf{BST}_\kappa$, and accordingly the **ZFC** set universe does not admit an ultrapower that interprets $\mathbf{BST}_\kappa$. (*Hint.* It is true in $\mathbb{S}[w]$ that any standard set X is a set of standard size because $X = \{f(w) : f \in X^W \cap \mathbb{S}\}$, where W is any standard set containing w. But ${}^*\kappa$ is not a set of standard size in $\mathbf{BST}_\kappa$ by a simple argument.)

(3) In spite of (1) it is an interesting **problem** to explicitly define a set-indexed quotient power of the **ZFC** universe satisfying $\mathbf{BST}_\kappa$. □

---

[14] In the sense that the index set is really a set rather than a proper class as e.g. in the constructions in §§ 4.3c, 4.3d.

6.3 External universes

In this section, we concentrate on *internal core extensions* of a given internal subuniverse $\mathscr{I} \subseteq \mathbb{I}$, *i.e.* classes $\mathscr{H} \subseteq \mathbb{H}$ satisfying $\mathscr{I} = \mathscr{H} \cap \mathbb{I}$ (that is, $\mathscr{H}$ contains no new internal sets) and a reasonably big fragment of **HST**.

6.3a External universes and internal core extensions

Definition 6.3.1. Let $\mathscr{I} \subseteq \mathbb{I}$ be an internal subuniverse.

A set Y is said to be $\mathscr{I}$*-wrong* if there is $z \in \mathscr{I}$ with $Y = z \cap \mathscr{I} \subsetneqq z$.

Suppose that $\mathscr{H}$ is any class satisfying $\mathscr{I} \subseteq \mathscr{H}$. We say that $\mathscr{H}$ is:

- *extensional* if $X \cap \mathscr{H} \neq Y \cap \mathscr{H}$ for any $X \neq Y \in \mathscr{H}$;
- an *internal core extension of* $\mathscr{I}$ if $\mathscr{I} = \mathscr{H} \cap \mathbb{I}$;
- *transitive over* $\mathscr{I}$, or a *transitive extension of* $\mathscr{I}$, if any set $X \in \mathscr{H} \setminus \mathscr{I}$ satisfies $X \subseteq \mathscr{H}$;
- *complete over* $\mathscr{I}$ if any set $Y \subseteq \mathscr{H}$, $Y \notin \mathscr{H}$ is $\mathscr{I}$-wrong.

Definition 6.3.2. Let ^{0}**HST** be a subtheory of **HST** containing only the axioms of the first and second group (§§ 1.1b, 1.1c), that is **HST** minus Saturation, Standard Size Choice, and Dependent Choice.

An *external subuniverse* is any class $\mathscr{H}$ satisfying ^{0}**HST** [15] and containing all standard sets. □

We observe that the theory ^{0}**HST** is strong enough to define ${}^*x \in \mathbb{S}$ for any $x \in \mathbb{WF}$ and to prove that $x \mapsto {}^*x$ is an $\in$-isomorphism of $\mathbb{WF}$ onto $\mathbb{S}$, because the content of § 1.1d does not at all depend on the 3rd group of the **HST** axioms. In addition ^{0}**HST** is strong enough to consistently define the class of all internal sets (*i.e.*, elements of standard sets) as in § 1.1a.

Note that ^{0}**HST** includes the axiom of Extensionality, and hence any external subuniverse, generally any class $\mathscr{H}$ satisfying ^{0}**HST**, is extensional.

Our goal will be the construction of extensional transitive internal core extensions of internal subuniverses, which sometimes will be complete. The next theorem contains some rather simple facts.

Theorem 6.3.3. *Let $\mathscr{I}$ be an internal subuniverse and $\mathscr{I} \subseteq \mathscr{H}$. Then:*

(i) *An internal set $A \subseteq \mathscr{I}$ cannot be $\mathscr{I}$-wrong, thus if $\mathscr{H}$ is a complete extension of $\mathscr{I}$ then any internal $A \subseteq \mathscr{I}$ belongs to $\mathscr{H}$.*

(ii) *If $\mathscr{H}$ is a transitive internal core extension of $\mathscr{I}$ then $\mathscr{H}$ is extensional if and only if $\mathscr{H}$ does not contain $\mathscr{I}$-wrong sets.*

Suppose, in addition to the above, that $\mathscr{H}$ is an external subuniverse and a transitive internal core extension of $\mathscr{I}$. Then:

(iii) *Any finite set $Y \subseteq \mathscr{H}$ belongs to $\mathscr{H}$.*

(iv) *If $\mathscr{I} \cap ({}^*\mathbb{N} \setminus \mathbb{N}) \neq \varnothing$ then $\mathbb{WF} \subseteq \mathscr{H}$, otherwise even $\mathtt{Ord} \cap \mathscr{H} = \mathbb{N}$.*

---

[15] That is the structure $\langle \mathscr{H}; \in, \mathtt{st} \rangle$ is an interpretation of ^{0}**HST**.

Note that the case $\mathscr{I} \cap ({}^*\mathbb{N} \smallsetminus \mathbb{N}) = \varnothing$ in (iv) is rather pathological: indeed it implies the existence of a measurable cardinal in $\mathbb{WF}$ unless $\mathscr{I} = \mathbb{S}$ (Theorem 3.1.25) and other peculiar properties of $\mathscr{I}$ which we are not interested in. In fact, it will alvays be the case below (see e.g. Theorem 6.4.3) that $\mathscr{I} \cap ({}^*\mathbb{N} \smallsetminus \mathbb{N}) \neq \varnothing$.

Proof. (i) Suppose towards the contrary that $X \in \mathscr{I}$ and $A = X \cap \mathscr{I} \subsetneqq X$; A is an internal set. Let $<$ be any well-ordering of X in $\mathscr{I}$ (formally, $<$ is a set in $\mathscr{I}$). As $A \subsetneqq X$, there is a $<$-least element x of $X \smallsetminus A$. Then $x \notin \mathscr{I}$, thus x is not the $<$-least element of X. (**Exercise**: why does the $<$-least element of X belong to $\mathscr{I}$?) Let y be the $<$-predecessor of x in X, so that $y \in A$ and $y \in \mathscr{I}$. Yet x is the $<$-least element of X above y, and hence x also belongs to $\mathscr{I}$ by Theorem 6.1.3(v), contradiction.

(ii) Suppose that $X \neq Y \in \mathscr{H}$. If both X and Y belong to $\mathscr{I}$ then we have already $X \cap \mathscr{I} \neq Y \cap \mathscr{I}$ because $\mathscr{I}$ satisfies Extensionality. If neither of X, Y belongs to $\mathscr{I}$ then $X \cup Y \subseteq \mathscr{H}$ by the transitivity over $\mathscr{I}$. Consider the mixed case: $X \in \mathscr{I}$ while $Y \subseteq \mathscr{H}$. Suppose towards the contrary that $X \cap \mathscr{H} = Y \cap \mathscr{H}$, that is, actually $X \cap \mathscr{H} = Y$, thus $X \cap \mathscr{I} = Y \subsetneqq X$ because $\mathscr{I} = \mathscr{H} \cap \mathbb{I}$. Thus, X is $\mathscr{I}$-wrong, hence $X \notin \mathscr{H}$, contradiction.

(iii) The class $\mathscr{H}$ is closed under finite set formation (because it satisfies ^{0}**HST**), at least in the sense that there is a set $z \in \mathscr{H}$ such that $Y = z \cap \mathscr{H}$. By the transitivity, either $z \subseteq \mathscr{H}$, and then $Y = z \in \mathscr{H}$, or $z \in \mathscr{I}$, and then $Y = z \cap \mathscr{I}$, which easily leads to $z = Y$ by the finiteness of Y.

(iv) To prove the first claim, let us show, by $\in$-induction, that any set $X \in \mathbb{WF}$ belongs to $\mathscr{H}$. The inductive hypothesis here implies that $X \subseteq \mathscr{H}$, hence if, on the contrary, $X \notin \mathscr{H}$ then $X = z \cap \mathscr{I} \subsetneqq z$ for some $z \in \mathscr{I}$, in particular $X \subseteq \mathbb{I}$, and hence $X \subseteq \mathbb{S}$ (as $\mathbb{I} \cap \mathbb{WF} \subseteq \mathbb{HF} \subseteq \mathbb{S}$, Exercise 1.2.17). It follows that z is finite and $z \subseteq \mathscr{I}$ (Theorem 6.1.3(vi); here the assumption that $\mathscr{I} \cap ({}^*\mathbb{N} \smallsetminus \mathbb{N}) \neq \varnothing$ is applied), hence, $X = z \in \mathscr{I}$, which is contradiction.

To prove the "otherwise" claim suppose on the contrary that $\mathscr{I} \cap {}^*\mathbb{N} = \mathbb{N}$ but there is an ordinal $\alpha \geq \omega$, $\alpha \in \mathscr{H}$. Then $\alpha \subseteq \mathscr{H}$ by the transitivity over $\mathscr{I}$, because if $\alpha \in \mathscr{I}$ then α is finite, contradiction. It follows that $\omega = \mathbb{N} \in \mathscr{H}$. However $\mathbb{N} = \mathscr{I} \cap {}^*\mathbb{N}$ by the choice of $\mathscr{I}$, in other words, $\mathbb{N}$ is $\mathscr{I}$-wrong and cannot belong to $\mathscr{H}$, contradiction. □

Exercise 6.3.4. (1) Prove that if an internal set $A \subseteq {}^*\mathbb{N}$ satisfies $A \notin \mathbb{S}(A)$ (see Exercise 6.1.6(1)) then the class $\mathscr{I} = \mathbb{S}(A)$ does not have *complete* internal core extensions by (i) of the theorem.

(2) If $\mathscr{I}$ is a thin internal subuniverse then any internal set $A \subseteq \mathscr{I}$ belongs to $\mathscr{I}$. (*Hint*: use Exercise 1.3.8(2) and Theorem 6.1.3(iv).) [16]

(3) Prove that $\mathscr{H} \cap \mathbb{I}$ is an internal subuniverse (in the sense of 6.1.1) whenever $\mathscr{H}$ is an external subuniverse (in the sense of 6.3.2). □

[16] We don't know any other population of internal subuniverses $\mathscr{I}$ which satisfy the requirement of the non-existence of internal sets $A \subseteq \mathscr{I}$, $A \notin \mathscr{I}$.

6.3b Von Neumann construction over non-transitive classes

The most natural idea of how to define a transitive internal core extension $\mathscr{H}$ of an internal subuniverse $\mathscr{I} \subseteq \mathbb{I}$ is to apply von Neumann's construction, as in § 1.5b, beginning with sets in $\mathscr{I}$. However suitable measures must be taken to cut off $\mathscr{I}$-wrong sets (see Theorem 6.3.3(ii)). This observation leads us to the following version of the von Neumann construction:

$$\begin{array}{rl} \text{for } \xi = 0: & \mathbf{V}^{\mathtt{E}}_0[\mathscr{I}] = \mathscr{I}\,, \\ \text{for each ordinal } \xi: & \mathbf{V}^{\mathtt{E}}_{\xi+1}[\mathscr{I}] = \mathscr{I} \cup \{Y : Y \subseteq \mathbf{V}^{\mathtt{E}}_{\xi}[\mathscr{I}] \text{ is not } \mathscr{I}\text{-wrong}\}\,, \\ \text{for limit ordinals } \lambda: & \mathbf{V}^{\mathtt{E}}_{\lambda}[\mathscr{I}] = \bigcup_{\xi<\lambda} \mathbf{V}^{\mathtt{E}}_{\xi}[\mathscr{I}]\,, \end{array}$$

where index $^{\mathtt{E}}$ refers to "extensional". Let finally $\mathbb{WF}[\mathscr{I}] = \bigcup_{\xi \in \mathtt{Ord}} \mathbf{V}^{\mathtt{E}}_{\xi}[\mathscr{I}]$, the class of all sets *well-founded over* $\mathscr{I}$.

Exercise 6.3.5. Prove that $\mathbf{V}^{\mathtt{E}}_{\eta}[\mathscr{I}] \subseteq \mathbf{V}^{\mathtt{E}}_{\xi}[\mathscr{I}]$ whenever $\eta < \xi$. □

Theorem 6.3.6. *If $\mathscr{I}$ is an internal subuniverse then $\mathscr{H} = \mathbb{WF}[\mathscr{I}]$ is transitive and complete over $\mathscr{I}$. If, in addition, every internal set $A \subseteq \mathscr{I}$ belongs to $\mathscr{I}$ then $\mathscr{H}$ is an external subuniverse and an internal core extension of $\mathscr{I}$.*

Proof. That $\mathscr{H}$ is transitive over $\mathscr{I}$ is clear. That $\mathscr{H}$ is complete over $\mathscr{I}$ follows from the construction: any $X \subseteq \mathscr{H}$ is a subset of some $\mathbf{V}^{\mathtt{E}}_{\xi}[\mathscr{I}]$, and hence if it is not $\mathscr{I}$-wrong then it belongs to $\mathscr{H}$.

As for the additional claim, prove first that $\mathscr{H} \cap \mathbb{I} = \mathscr{I}$. It suffices to show that any internal $A \in \mathbf{V}^{\mathtt{E}}_{\xi}[\mathscr{I}]$ belongs to $\mathscr{I}$, by induction on ξ. Note that by definition $A \subseteq \bigcup_{\eta<\xi} \mathbf{V}^{\mathtt{E}}_{\eta}[\mathscr{I}]$, hence, by the inductive hypothesis, $A \subseteq \mathscr{I}$. We have $A \in \mathscr{I}$ by the assumption of the corollary.

Thus, $\mathscr{H}$ is an internal core extension of $\mathscr{I}$. It remains to check axioms of ^{0}**HST** (this is included in the definition of external subuniverse). We consider several ^{0}**HST** axioms, leaving the rest of them as an easy **exercise** for the reader.

Pair. $\mathscr{H}$ is closed w. r. t. finite subsets (Theorem 6.3.3(iii)).

Separation. If $Y \subseteq \mathscr{H}$ then, by the above, either $Y \in \mathscr{H}$ or there is a set z in $\mathscr{I}$ (and hence in $\mathscr{H}$) with $z \cap \mathscr{I} = z \cap \mathscr{H} = Y$.

Union. Let $X \in \mathscr{H}$. Then $U = \bigcup X$ may not be a member of $\mathscr{H}$, but we can define $Y = U \cap \mathscr{H}$ and end the argument as in Separation above.

Transfer. Note that $\mathbb{S} \subseteq \mathscr{I} \subseteq \mathscr{H}$. On the other hand, as $\mathscr{I} = \mathscr{H} \cap \mathbb{I}$, the class $\mathscr{I}$ is the collection of all internal sets in the sense of $\langle \mathscr{H}; \in, \mathtt{st} \rangle$. It remains to apply Theorem 6.1.3(i). □

Exercise 6.3.7. Prove the following:

(1) $\mathbb{WF}[\mathbb{I}]$ contains all sets while $\mathbb{WF}[\mathbb{S}] = \mathbb{S}$ (use Standardization);

(2) if $\mathscr{I} \cap ({}^*\mathbb{N} \setminus \mathbb{N}) = \varnothing$ then $\mathbb{WF}[\mathscr{I}] \cap \mathbb{WF} = \mathbb{HF}$. □

6.3c Absoluteness for external subuniverses

It follows from Theorem 6.3.3 that under certain conditions even non-transitive classes $\mathscr{H}$ in the **HST** universe can be extensional. (All transitive classes are extensional, of course.) The extensionality is a necessary condition for a non-transitive class $\mathscr{H}$ to be considered a set theoretic universe in any reasonable sense. Another important characteristic is the degree of absoluteness of basic set theoretic notions for $\mathscr{H}$.

Many basic notions and relations are absolute for any *transitive* class satisfying rather mild requirements (see, for instance, § 1.2b). In principle, the non-transitivity of external subuniverses makes our life difficult because some fundamental notions can change their meaning in a nontransitive class. Fortunately, the property of transitivity of $\mathscr{H}$ over $\mathscr{I}$ divides $\mathscr{H}$ in two parts, one of which is the internal subuniverse $\mathscr{I}$ which, by definition, satisfies **ZFC** and is an elementary substructure of $\mathbb{I}$ in the $\in$-language, while the other one is closed under subsets. This helps to establish some absoluteness.

Lemma 6.3.8. *Suppose that $\mathscr{I}$ is an internal subuniverse and $\mathscr{H}$ is an external subuniverse and a transitive internal core extension of $\mathscr{I}$ (not necessarily complete). Then the properties "x is an unordered (ordered) pair", $z = \{x, y\}$, $z = \langle x, y\rangle$, "f is a function", "f is a function and $f(x) = y$", "n is a natural number", "x is a finite set" are absolute for $\mathscr{H}$.*

Proof. First of all note that, by Theorem 6.3.3, $\mathscr{H}$ is closed under finite subsets, hence, closed under the operations of unordered and ordered pairs. Moreover, for any $x, y \in \mathscr{H}$, $z = \{x, y\}$ is the unique set in $\mathscr{H}$ which satisfies the requirement that its only elements are x and y — this is because $\mathscr{H}$ is extensional. The same for $\langle x, y\rangle$. The absoluteness is now an easy corollary.

To see that "f is a function" is absolute for $\mathscr{H}$ consider two cases. If $f \in \mathscr{I}$ then $f \cap \mathscr{H} = f \cap \mathscr{I}$, thus by the above f is a function in $\mathscr{H}$ iff f a is function in $\mathscr{I}$. We continue: iff f is a function in $\mathbb{I}$, — by Theorem 6.1.3(i). But $\mathbb{I}$ is a transitive class. If $f \notin \mathscr{I}$ (the second case) then $f \subseteq \mathscr{H}$ by the transitivity of $\mathscr{H}$ over $\mathscr{I}$, which implies the absoluteness.

Further, since $\mathscr{H}$ is closed under finite subsets, we have $\mathbb{HF} \subseteq \mathscr{H}$, and hence $\mathbb{N} \subseteq \mathscr{H}$. It follows that the notions of being a finite set and being a natural number are absolute for $\mathscr{H}$. □

However, some relations between sets are non-absolute, especially when the sets belong to two different "populations" in $\mathscr{H}$, that is, $\mathscr{I}$ and subsets of $\mathscr{H}$. For instance, $X = {}^*\mathbb{N}$ belongs to $\mathbb{S}$, hence, to $\mathscr{I}$ and $\mathscr{H}$. If ${}^*\mathbb{N} \not\subseteq \mathscr{I}$ then the set $Z = X \cap \mathscr{I}$ is $\mathscr{I}$-wrong, but $Y = Z \cup \{y\}$, where $y \in \mathscr{H} \setminus \mathscr{I}$, belongs to $\mathscr{H}$ and we have $X \subseteq Y$ in $\mathscr{H}$, that is

$$X \cap \mathscr{H} = X \cap \mathscr{I} = Z \subseteq Z \cup \{y\} = Y = Y \cap \mathscr{H},$$

but clearly $X \not\subseteq Y$ in the entire universe. Yet $X \subseteq Y$ is obviously absolute provided X, Y belong to one and the same "population" of elements of $\mathscr{H}$.

6.4 Partially saturated external universes

In this section, we orient the general considerations of Section 6.3 towards structures that interpret the Power Set axiom and a certain amount of Saturation and Choice (restricted by a cardinal) in addition to the axioms of ^{0}**HST**. We begin with a definition of nonstandard set theories $\mathbf{HST}_\kappa$ and $\mathbf{HST}'_\kappa$ of "external" kind, designed as suitable external extensions of the theories $\mathbf{BST}_\kappa$ and $\mathbf{BST}'_\kappa$ (partially saturated internal theories introduced in § 3.3a), and containing the Power Set axiom. Our main goal will be to figure out when and how an internal subuniverse $\mathscr{I}$ satisfying $\mathbf{BST}_\kappa$ or $\mathbf{BST}'_\kappa$ can be extended to an external subuniverse satisfying resp. $\mathbf{HST}_\kappa$, $\mathbf{HST}'_\kappa$.

6.4a Partially saturated external theories

Let $\mathbf{HST}_\kappa$ and $\mathbf{HST}'_\kappa$ be theories, in the $\mathtt{st}$-$\in$-language with an additional constant κ, containing all of ^{0}**HST**, an axiom saying that κ is an infinite (well-founded) cardinal, the Power Set axiom, and the following axiom:

κ-Boundedness: every internal set belongs to $\mathbb{I}_{*\kappa}$,

and finally the following axioms:

for $\mathbf{HST}_\kappa$: 1) *κ-deep Saturation*, *i.e.* Saturation for the case when $\mathscr{X}$ is a set of standard size and $\mathscr{X}$ consists of internal subsets of a set of the form $^*X$, where X is a fixed well-founded set of cardinality $\leq \kappa$,

2) *2^κ-size Choice*, *i.e.* Standard Size Choice for the case when the domain of a choice function postulated to exist is a set of cardinality $\leq 2^\kappa$,

3) Dependent Choice;

for $\mathbf{HST}'_\kappa$: 1′) *κ-size Saturation*, *i.e.* Saturation for the case when $\mathscr{X}$ is a collection of internal sets, satisfying $\mathtt{card}\,\mathscr{X} \leq \kappa$ [17],

2′) the axiom: *every set is a set of standard size*,

3′) the axiom: $^*\kappa$ is a set of standard size via an internal map f, *i.e.* *there is an internal bijection f defined on $^*\kappa$ such that the f-image $f``{}^*\kappa = \{f(\xi) : \xi \in {}^*\kappa\}$ of $^*\kappa$ satisfies $f``{}^*\kappa \subseteq \mathbb{S}$.*

Let $\mathbf{HST}^-_\kappa$ be $\mathbf{HST}'_\kappa$ minus the axiom 3′); this is a somewhat weaker theory, accordingly, its internal universes form a much broader family.

Remark 6.4.1. The axioms κ-deep Saturation and κ-size Saturation say, in terms of § 6.2, that the class of internal sets is κ-deep, resp., κ-size saturated.

κ-deep Saturation implies of κ-size Saturation, as in Theorem 6.2.3(i).

The axiom of 2^κ-size Choice is somewhat stronger than $(*)$ of (ii) of Theorem 6.2.3 because it is not necessarily related to internal sets. Recall that 2^κ-size Choice and κ-deep Saturation correspond to each other, see Theorem 3.3.5 and Remark 3.3.6. □

[17] See footnote 9 on page 230.

As (3), (4) of the next Exercise show, the theories $\mathbf{HST}_\kappa$ and $\mathbf{HST}'_\kappa$ are approximately in the same relation to the internal theories $\mathbf{BST}_\kappa$ and $\mathbf{BST}'_\kappa$ as **HST** itself to **BST**.

Exercise 6.4.2. Prove the following.

(1) in $\mathbf{HST}_\kappa$, infinite internal sets are not sets of standard size, and hence $\mathbf{HST}_\kappa$ is incompatible with $\mathbf{HST}'_\kappa$ (even with $\mathbf{HST}'^{-}_\kappa$);

(2) $\mathbf{HST}_\kappa$ is incompatible with $\mathbf{HST}_\lambda$ provided κ, λ are different (well-founded) cardinals, the same for $\mathbf{HST}'$ (this needs some care to be accurately formulated, see Exercise 3.3.3);

(3) in $\mathbf{HST}_\kappa$, the class of internal sets satisfies $\mathbf{BST}_{{}^*\kappa}$ (why does κ change to ${}^*\kappa$ in the index ?);

(4) in $\mathbf{HST}'_\kappa$, the class of internal sets satisfies $\mathbf{BST}'_{{}^*\kappa}$, in particular by Lemma 3.3.2 the class of all internal sets has the form $\mathbb{S}[\xi]$, $\xi < {}^*\kappa$ and every internal set is a set of standard size via an internal map f;

(5) in $\mathbf{HST}^{-}_\kappa$, every set X can be well-ordered, therefore the axiom of Choice holds.

Hints. (5): this is the same as the implication (1) $\Longrightarrow$ (2) in Theorem 1.3.1. In brief, by definition, there is a standard set Z and a map f such that X is the f-image of $Z \cap \mathbb{S}$. There is a standard binary relation $\prec$ on Z such that it is true in $\mathbb{I}$ that $\prec$ is a well-ordering of Z. By Standardization, for $x \in X$ there is a (unique) standard set $Z_x \subseteq Z$ such that $f(z) = x \Longleftrightarrow z \in Z_x$ for any standard $z \in Z$. Put $x \prec' y$ iff the $\prec$-least element of Z_x $\prec$-precedes the $\prec$-least element of Z_y. Prove that $\prec'$ is a well-ordering of X. □

Claims (3) and (4) give a definite clue regarding the choice of internal subuniverses to be extended to external subuniverses satisfying $\mathbf{HST}_\kappa$ and $\mathbf{HST}'_\kappa$. As for the methods of extension, two of them will be considered. The first method, the more elementary of the two, works for internal core extensions of thin internal subuniverses $\mathscr{I}$, for instance those of the form $\mathbb{S}(X)$, where X is a set of standard size. This will lead us to external subuniverses $\mathscr{H}$ satisfying $\mathbf{HST}'_\kappa$ for a given well-founded cardinal κ [18]. The other, more complicated method, will work for internal subuniverses of the form $\mathscr{I} = \mathbb{I}_\kappa$ and their subclasses produced by complete sets, leading to external subuniverses satisfying $\mathbf{HST}_\kappa$.

The Power Set axiom is the most interesting issue here. Recall that it fails in **HST** (Theorem 1.3.9). Yet we include Power Set in the axiomatics of $\mathbf{HST}_\kappa$ and $\mathbf{HST}'_\kappa$. In the internal core extensions of the first kind, the Power Set axiom will be based on the fact that (in **HST**) any set X of standard size has a power set $\mathscr{P}(X)$ which is also a set of standard size (Exercise 1.3.3).

[18] Meaning that the structure $\langle \mathscr{H}; \in, \mathtt{st}; \kappa\rangle$ is an interpretation of $\mathbf{HST}'_\kappa$ under the map sending the constant $\boldsymbol{\kappa}$ to κ.

As for the extensions of the second type, the Power Set axiom will follow from Theorem 6.2.3(iii).

6.4b Extensions of thin classes

Recall that by Theorem 6.3.6 any internal subuniverse $\mathscr{I}$ such that every internal set $A \subseteq \mathscr{I}$ belongs to $\mathscr{I}$, admits a transitive and complete internal core extension $\mathscr{H}$ which is an external subuniverse. It follows from 6.3.4(2) that this is applicable for instance to thin classes $\mathscr{I}$. (Recall that $\mathscr{I}$ is thin iff all sets $X \subseteq \mathscr{I}$ are sets of standard size.)

Theorem 6.4.3. *Suppose that $\mathscr{I}$ is a thin internal subuniverse, and the intersection $\mathscr{I} \cap ({}^*\mathbb{N} \smallsetminus \mathbb{N})$ is non-empty. Then*

(i) *$\mathscr{H} = \mathbb{WF}[\mathscr{I}]$ is an external subuniverse (in particular it satisfies ^{0}**HST**) and internal core extension of $\mathscr{I}$, transitive and complete over $\mathscr{I}$;*

(ii) *$\mathscr{H}$ satisfies the Power Set and Choice axioms;*

(iii) *$\mathbb{WF} \subseteq \mathscr{H}$, and any set $X \in \mathscr{H} \smallsetminus \mathscr{I}$ is a set of standard size.*

(iv) *If in addition κ is an infinite (well-founded) cardinal, $\mathscr{I} \subseteq \mathbb{I}_\kappa$, and $\mathscr{I}$ is κ-size saturated then $\mathscr{H}$ satisfies $\mathbf{HST}_\kappa^-$.*

(v) *If in addition to (iv) $\mathscr{I}$ satisfies $\mathbf{BST}'_{*\kappa}$ then $\mathscr{H}$ satisfies $\mathbf{HST}'_\kappa$.*

Proof. (iii) $\mathbb{WF} \subseteq \mathscr{H}$ follows from Theorem 6.3.3(iv). Prove that every $X \in \mathscr{H} \smallsetminus \mathscr{I}$ is a set of standard size. It suffices to show by induction on $\xi \in \mathtt{Ord}$ that any $X \subseteq \mathbf{V}_\xi^{\mathrm{E}}[\mathscr{I}]$ is a set of standard size. That any $X \subseteq \mathbf{V}_0^{\mathrm{E}}[\mathscr{I}] = \mathscr{I}$ is a set of standard size holds by the choice of $\mathscr{I}$. The limit step follows because standard size unions of sets of standard size are sets of standard size (Exercise 1.3.3). As for the step $\xi \to \xi + 1$, if $X \subseteq \mathbf{V}_{\xi+1}^{\mathrm{E}}[\mathscr{I}]$ then, by the **HST** Collection, there is a set $Y \subseteq \mathbf{V}_\xi^{\mathrm{E}}[\mathscr{I}]$ such that $X \subseteq \mathscr{P}(Y)$, but the power set $\mathscr{P}(Y)$ of a set Y of standard size is also a set of standard size.

(i) Apply Theorem 6.3.6 and 6.3.4(2).

(ii) Choice. Let $X \in \mathscr{H}$; prove that X is well-orderable in $\mathscr{H}$. If $X \notin \mathscr{I}$ then $X \subseteq \mathscr{H}$ by the transitivity over $\mathscr{I}$, hence, X is a set of standard size by the above, and well-orderable by Theorem 1.3.1 – this is witnessed by a set $W \subseteq X \times X$, also a set of standard size and a subset of $\mathscr{H}$, of course. Then $W \in \mathscr{H}$ by the completeness over $\mathscr{I}$ (or there is an internal $z \in \mathscr{I}$ such that $W = z \cap \mathscr{I} = z \cap \mathscr{H}$), thus W (resp., z) witnesses the well-orderability of X in $\mathscr{H}$. If $X \in \mathscr{I}$ then we apply the same argument to the standard size set $Y = X \cap \mathscr{I}$ (not a member of $\mathscr{H}$), with suitable corrections.

Power Set. Let $X \in \mathscr{H}$. If $X \cap \mathscr{I} = \varnothing$ then, as above, X is a set of standard size, so that the power set $\mathscr{P}(X)$ is a set of standard size (Exercise 1.3.3). Yet under the assumption $X \cap \mathscr{I} = \varnothing$ all subsets of X belong to $\mathscr{H} = \mathbb{WF}[\mathscr{I}]$ and $\mathscr{P}(X) \in \mathscr{H}$ by the completeness over $\mathscr{I}$ (or otherwise $\mathscr{P}(X) \subseteq \mathscr{I}$ which contradicts the choice of X). Thus, Power Set holds in $\mathscr{H}$

for all sets $X \in \mathscr{H}$ satisfying $X \cap \mathscr{I} = \varnothing$. The general case can be reduced to the case $X \cap \mathscr{I} = \varnothing$ by a pure argument within ^{0}**HST**.

(iv) If $\mathscr{I} \subseteq \mathbb{I}_\kappa$ is a κ-size saturated class then κ-size Saturation holds in $\mathscr{H}$ because we have $\mathscr{I} = \mathscr{H} \cap \mathbb{I}$. To see that that "all sets are sets of standard size" is true in $\mathscr{H}$, note that this principle is equivalent, in ^{0}**HST**, to "all sets are well-orderable" (see hints to (5) of Exercise 6.4.2).

(v) If $\mathscr{I}$ satisfies $\mathbf{BST}'_{^*\kappa}$ then $^*\kappa$ is a set of standard size in $\mathscr{I}$, that is, there is a map $f \in \mathscr{I}$ defined on a standard set Z such that $^*\kappa \cap \mathscr{I}$ is the f-image of $Z' = Z \cap \mathbb{S}$. Since $^*\kappa \cap \mathscr{H} = {}^*\kappa \cap \mathscr{I}$, the same (internal) map f witnesses in $\mathscr{H}$ that $^*\kappa$ is a set of standard size. □

Corollary 6.4.4. *For any infinite cardinal κ there is an internal subuniverse $\mathscr{I} \subseteq \mathbb{I}_\kappa$ such that the external subuniverse $\mathbb{WF}[\mathscr{I}]$ satisfies $\mathbf{HST}'_\kappa$.*

Proof. Let, by Theorem 6.2.6, $\xi < {}^*\kappa$ is a κ-good $*$-ordinal. Then $\mathscr{I} = \mathbb{S}[\xi]$ is a thin κ-size saturated internal subuniverse satisfying $\mathbf{BST}'_{^*\kappa}$ (still by Theorem 6.2.6). It remains to apply Theorem 6.4.3. □

Exercise 6.4.5. Suppose that κ is an infinite cardinal and $R \subseteq {}^*\kappa$ is a κ-complete set of standard size (Theorem 6.2.11). Prove that $\mathscr{I} = \mathbb{S}(R)$ is a κ-size saturated thin class, $\mathscr{I} \subseteq \mathbb{I}_\kappa$, and $\mathbb{WF}[\mathscr{I}]$ is an interpretation of the theory $\mathbf{HST}^-_\kappa$.

Why does $\mathbb{WF}[\mathscr{I}]$ not satisfy $\mathbf{HST}'_\kappa$, a stronger theory, in this case? □

6.4c Constructible extensions

That the Power Set axiom holds in internal core extensions $\mathbb{WF}[\mathscr{I}]$ of thin internal subuniverses $\mathscr{I}$ (Theorem 6.4.3) is true only because the power set of any set of standard size is still a set, moreover a set of standard size in **HST**. The argument fails for non-thin internal subuniverses, for instance for the classes $\mathbb{I}_\kappa$. Yet there is a different construction of extensions, especially of classes $\mathscr{I} = \mathbb{I}_\kappa$, in which the Power Set axiom holds by reasons related rather to (iii) of Theorem 6.2.3 than to the thinness of $\mathscr{I}$.

To define external subuniverses of this kind, we modify the definition of $\mathbb{WF}[\mathscr{I}]$ in § 6.3b in the spirit of the construction of $\mathbb{L}[\mathbb{I}]$ in Section 5.5, so that sets in $\mathscr{I}$ are the only initial sets, and the A-codes admitted are E-coded in $\mathscr{I}$ themselves. With an appropriate choice of initial classes $\mathscr{I}$, this will result in external subuniverses which interpret $\mathbf{HST}_\kappa$.

This new family of external subuniverses will consist of *incomplete* (Definition 6.3.1) transitive internal core extensions of their internal universes.

The next definition leads us back to the notions considered earlier in Sections 5.3, 5.5. Recall (§ 5.4a) that $\underline{\mathbf{A}}$ is the class of all A-codes $\mathbf{x} \in \mathbb{E}$, and $\mathbf{A}$ is the class of all *regular* codes in $\underline{\mathbf{A}}$, where regularity means that intermediate sets in the assembling construction cannot be internal.

Definition 6.4.6. Let $\mathscr{I}$ be an internal subuniverse. Define

$$\mathbb{E}[\mathscr{I}] = \{\mathsf{E}_p : p \in \mathscr{I}\} ;$$

$$\underline{\mathbf{A}}(\mathscr{I}) = \text{ all A-codes } \mathbf{x} \in \mathbb{E}[\mathscr{I}] \text{ such that } T_{\mathbf{x}} \subseteq \mathscr{I} \text{ and } \mathbf{ran}\,\mathbf{x} \subseteq \mathscr{I}.$$

($\mathsf{E}_p \subseteq \mathbb{I}$ is an elementary external set coded by $p \in \mathbb{I}$, Definition 3.2.14.)

Say that a code $\mathbf{x} \in \underline{\mathbf{A}}(\mathscr{I})$ is *$\mathscr{I}$-regular* if for any $t \in T_{\mathbf{x}}$ such that $|t|_{T_{\mathbf{x}}} = 1$ the set $\mathsf{F}_{\mathbf{x}}(t) = \{\mathbf{x}(t^\wedge a) : t^\wedge a \in T_{\mathbf{x}}\}$ is not of the form $z \cap \mathscr{I}$, where $z \in \mathscr{I}$, in particular, this set is not $\mathscr{I}$-wrong.

Let $\mathbf{A}(\mathscr{I})$ be the class of all $\mathscr{I}$-regular codes in $\underline{\mathbf{A}}(\mathscr{I})$.

Finally, define $\mathbb{L}[\mathscr{I}] = \{\mathsf{A}_{\mathbf{x}} : \mathbf{x} \in \mathbf{A}(\mathscr{I})\}$. □

The definition of $\mathbb{L}[\mathscr{I}]$ applies, for instance, in the case when $\mathscr{I} = \mathbb{I}$, the class of all internal sets. As the $\mathbb{I}$-regularity in the sense of the last definition is obviously the same as regularity in the sense of Definition 5.3.11, definitions of $\mathbb{L}[\mathbb{I}]$ given by 5.5.2 and 6.4.6 (for $\mathscr{I} = \mathbb{I}$) are equivalent.

Exercise 6.4.7. Prove that $\mathscr{I} \subseteq \mathbb{L}[\mathscr{I}] \subseteq \mathbb{WF}[\mathscr{I}]$. □

We begin our study of these notions with the following technical result.

Proposition 6.4.8. *If $\mathscr{I}$ is an internal subuniverse then any set $X \subseteq \mathbb{I}$, $\mathtt{st}$-$\in$-definable in $\mathbb{L}[\mathbb{I}]$ with sets in $\mathbb{E}[\mathscr{I}] \cup \mathbb{L}[\mathscr{I}]$ as parameters, belongs to $\mathbb{E}[\mathscr{I}]$, moreover, if X is internal then $X \in \mathscr{I}$.*

Proof. The set X is $\mathtt{st}$-$\in$-definable in $\mathbb{L}[\mathbb{I}]$ with parameters in $\mathscr{I}$, because so are all sets in $\mathbb{E}[\mathscr{I}] \cup \mathbb{L}[\mathscr{I}]$. It follows by 5° of Theorem 5.5.4. that X is $\mathtt{st}$-$\in$-definable in $\mathbb{I}$ with parameters in $\mathscr{I}$. Now apply Theorem 3.2.16. The "moreover" statement now follows from Theorem 6.1.3(v). □

The strategy for the study of classes of the form $\mathbb{L}[\mathscr{I}]$ is somewhat different from that of classes of the form $\mathbb{WF}[\mathscr{I}]$. In particular, the key equality $\mathbb{L}[\mathscr{I}] \cap \mathbb{I} = \mathscr{I}$ ((iii) in the next lemma) is an immediate corollary of the definitions.

Lemma 6.4.9. *Let $\mathscr{I}$ be an internal subuniverse and $\mathscr{H} = \mathbb{L}[\mathscr{I}]$. Then*

(i) *if $\mathbf{x} \in \mathbf{A}(\mathscr{I})$ and $t \in T_{\mathbf{x}}$ then $\mathbf{x}|_t$* (Example 5.3.4) *belongs to $\mathbf{A}(\mathscr{I})$;*

(ii) *any code $\mathbf{x} \in \mathbf{A}(\mathscr{I})$ is regular, hence, belongs to $\mathbf{A}$;*

(iii) *if $X \in \mathscr{H}$ is internal then $X \in \mathscr{I}$;*

(iv) *if $X \in \mathscr{H}$ satisfies $X \subseteq \mathbb{I}$ then X is not $\mathscr{I}$-wrong.*

Proof. (i) The code $\mathbf{x}|_t$ is $\mathtt{st}$-$\in$-definable in $\mathbb{L}[\mathbb{I}]$ with parameters $\mathbf{x}, t$, hence, $\mathbf{x}|_t$ belongs to $\mathbb{E}[\mathscr{I}]$ by Proposition 6.4.8.

(ii) If $t \in T_{\mathbf{x}}$ and $|t|_{T_{\mathbf{x}}} = 1$ then $X = \mathsf{F}_{\mathbf{x}}(t) \subseteq \mathscr{I}$ and X is $\mathtt{st}$-$\in$-definable in $\mathbb{I}$ with parameters in $t, \mathbf{x} \in \mathscr{I}$, hence, if X is internal then $X \in \mathscr{I}$ by Proposition 6.4.8, which is a contradiction with the $\mathscr{I}$-regularity of $\mathbf{x}$.

(iii) Suppose that $X = \mathsf{A}_{\mathbf{x}}$, $\mathbf{x} \in \mathbf{A}(\mathscr{I})$. Obviously X is st-∈-definable in $\mathbb{L}[\mathbb{I}]$ with $\mathbf{x}$ as the only parameter. It remains to apply Proposition 6.4.8.

(iv) Suppose that $X = \mathsf{A}_{\mathbf{x}}$, $\mathbf{x} \in \mathbf{A}(\mathscr{I})$. The code $\mathbf{x}$ is regular by (ii). But $X \subseteq \mathbb{I}$; it follows from the regularity (for instance, by Lemma 5.3.12) that $|T| \leq 1$. If $|T| = 0$ then $T_{\mathbf{x}} = \{\Lambda\}$ and $X = \mathsf{A}_{\mathbf{x}} \in \mathscr{I}$, hence, X cannot be $\mathscr{I}$-wrong. If $|T| = 1$ then $X = \mathsf{F}_{\mathbf{x}}(\Lambda) = \{\mathsf{F}_{\mathbf{x}}(\langle a \rangle) : \langle a \rangle \in T_{\mathbf{x}}\}$ cannot be $\mathscr{I}$-wrong by the $\mathscr{I}$-regularity of $\mathbf{x}$. □

It follows from (ii) that $\mathbb{L}[\mathscr{I}] \subseteq \mathbb{L}[\mathbb{I}]$.

Theorem 6.4.10. *Let $\mathscr{I}$ be an internal subuniverse. Then $\mathscr{H} = \mathbb{L}[\mathscr{I}]$ is an extensional class, transitive over $\mathscr{I}$, and $\mathscr{H} \cap \mathbb{I} = \mathscr{I}$, thus $\mathscr{H}$ is an internal core extension of $\mathscr{I}$.*

If $Y \subseteq \mathscr{H}$ is finite then $Y \in \mathscr{H}$.

If, in addition, $\mathscr{I} \cap ({}^*\mathbb{N} \setminus \mathbb{N}) \neq \varnothing$ then $\mathbb{WF} \subseteq \mathbb{L}[\mathscr{I}]$.

Proof. The transitivity of $\mathscr{H}$ follows from (i) of Lemma 6.4.9 because $\mathsf{A}_{\mathbf{x}}$ consists of sets of the form $\mathsf{A}_{\mathbf{x}|_t}$ where $t \in T_{\mathbf{x}}$ and $|t|_{T_{\mathbf{x}}} = 1$. The equality $\mathscr{H} \cap \mathbb{I} = \mathscr{I}$ follows from Lemma 6.4.9(iii). Now extensionality follows from Theorem 6.3.3(ii) and Lemma 6.4.9(iv).

That $\mathscr{H}$ is closed under finite subsets is left as an **exercise** for the reader.

Prove the additional claim. Let $v \in \mathbb{WF}$. We make use of the A-code $\mathbf{c}[{}^*v]$ (Definition 5.3.9). To show that $\mathbf{c}[{}^*v] \in \mathbf{A}(\mathscr{I})$ note that $\mathbf{c}[{}^*v]$ is st-∈-definable in $\mathbb{I}$ with ${}^*v \in \mathbb{S}$ as the only parameter, hence, it belongs to $\mathbb{E}[\mathscr{I}]$ by Theorem 3.2.16. If $\mathbf{c}[{}^*v]$ is <u>not</u> $\mathscr{I}$-regular then there is $t \in T = T_{\mathbf{c}[{}^*v]}$ with $|t|_T = 1$ and $\mathsf{F}_{\mathbf{c}[{}^*v]}(t) = \{\mathbf{c}[{}^*v](t^\wedge a) : t^\wedge a \in T\}$ is equal to $z \cap \mathscr{I}$ for some $z \in \mathscr{I}$. However, by definition, we have $\mathsf{F}_{\mathbf{c}[{}^*v]}(t) \subseteq \mathbb{HF} \subseteq \mathbb{S}$, and hence $z \cap \mathscr{I} \subseteq \mathbb{S}$. Thus z is finite and $z \subseteq \mathscr{I}$ (Lemma 6.2.4). We have $\mathsf{F}_{\mathbf{c}[{}^*v]}(t) = z \in \mathbb{S}$, in contradiction to the regularity of $\mathbf{c}[{}^*v]$. □

6.4d Constructible extensions of self-definable classes

Due to obvious analogies between $\mathbb{L}[\mathscr{I}]$ and $\mathbb{L}[\mathbb{I}]$ (a particular case of $\mathbb{L}[\mathscr{I}]$), we could expect that classes of the form $\mathbb{L}[\mathscr{I}]$ satisfy suitable fragments of **HST**. Unfortunately, this plan does not seem to be realizable in full generality. The main obstacle is that the transformations of A-codes used in the proof of Theorem 5.5.4 (see, *e.g.*, Lemma 5.5.5) now will have to refer to classes $\mathscr{I}$ and $\mathbf{A}(\mathscr{I})$ in a manner which does not guarantee that the A-codes obtained belong to $\mathbb{E}[\mathscr{I}]$. To overcome this difficulty, we simply distinguish a family of internal subuniverses $\mathscr{I}$ which allow to solve the problem.

Definition 6.4.11. An internal subuniverse $\mathscr{I} \subseteq \mathbb{I}$ is *self-definable* if it is definable in $\mathbb{I}$ by a st-∈-formula with sets in $\mathscr{I}$ as parameters. □

For instance, classes $\mathbb{S}[w]$, $\mathbb{S}_{\mathrm{M}}[w]$ (§ 6.1c), $\mathbb{I}_\kappa = \mathbb{S}({}^*\kappa)$ are self-definable.

The proof of the next lemma (compare with Lemma 5.5.5 and 4° of Theorem 5.5.4) shows how the property of self-definability can be utilized.

Lemma 6.4.12. *Assume that $\mathscr{I}$ is a self-definable internal subuniverse, and a set $Y \subseteq \mathbb{L}[\mathscr{I}]$ is* $\mathtt{st}$-$\in$*-definable in $\mathbb{L}[\mathbb{I}]$ with parameters in $\mathbb{E}[\mathscr{I}] \cup \mathbb{L}[\mathscr{I}]$. Then $Y \in \mathbb{L}[\mathscr{I}]$* or *$Y = z \cap \mathscr{I}$ for some $z \in \mathscr{I}$.*

Proof. Using (3) of Exercise 1.1.11, we get a standard set P such that for any $y \in Y$ there is $p \in P \cap \mathscr{I}$ with $\mathsf{E}_p \in \mathbf{A}(\mathscr{I})$ and $\mathsf{A}_{\mathsf{E}_p} = y$. Then

$$Z = \{p \in P \cap \mathscr{I} : \mathsf{E}_p \in \mathbf{A}(\mathscr{I}) \wedge \mathsf{A}_{\mathsf{E}_p} \in X\}$$

is a subset of P, $\mathtt{st}$-$\in$-definable in $\mathbb{L}[\mathbb{I}]$ with parameters in $\mathscr{I}$, namely, S, X, and those sets in $\mathscr{I}$ which occur in a fixed $\mathtt{st}$-$\in$-formula which defines $\mathscr{I}$ in $\mathbb{I}$. (Here the self-definability of $\mathscr{I}$ is applied!) It follows that Z belongs to $\mathbb{E}[\mathscr{I}]$ (Proposition 6.4.8). Define an A-code $\mathbf{x}$ so that

$$T_{\mathbf{x}} = \{\Lambda\} \cup \{p^\wedge t : p \in Z \wedge t \in \mathsf{E}_p\}$$

and $\mathbf{x}(a^\wedge t) = \mathsf{E}_p(t)$ for all $p \in Z$ and $t \in \mathtt{Max}\,\mathsf{E}_p$, thus, $\mathsf{A}_{\mathbf{x}} = X$ (because $\mathsf{F}_{\mathbf{x}}(\langle p\rangle) = \mathsf{A}_{\mathsf{E}_p}$ whenever $p \in Z$), and $\mathbf{x} \in \underline{\mathbf{A}}(\mathscr{I})$. The only way for the $\mathscr{I}$-regularity of $\mathbf{x}$ to fail is that there is a set $z \in \mathscr{I}$ with $z \cap \mathscr{I} = \mathsf{A}_{\mathbf{x}} = X$. □

Theorem 6.4.13. *Suppose that $\mathscr{I} \subseteq \mathbb{I}$ is a self-definable internal subuniverse containing a number in ${}^*\mathbb{N} \smallsetminus \mathbb{N}$. Then*

(i) *$\mathscr{H} = \mathbb{L}[\mathscr{I}]$ is an external subuniverse (in particular, it satisfies* ^{0}**HST***) and a transitive internal core extension of $\mathscr{I}$;*

(ii) $\mathbb{WF} \subseteq \mathscr{H}$;

(iii) *if, in addition, κ is an infinite (well-founded) cardinal, $\mathscr{I} \subseteq \mathbb{I}_\kappa$, and $\mathscr{I}$ is a κ-deep saturated class, then $\mathscr{H}$ satisfies* $\mathbf{HST}_\kappa$ *and any set $X \subseteq \mathbb{L}[\mathscr{I}]$ of cardinality $\leq 2^\kappa$ (in the universe) belongs to $\mathbb{L}[\mathscr{I}]$.*

Proof. Theorem 6.4.10 implies (ii) and (i) except for the axioms of ^{0}**HST** in $\mathscr{H}$. Of the axioms, Extensionality holds in $\mathscr{H}$ because it is an extensional class still by Theorem 6.4.10.

Other axioms of ^{0}**HST** can be verified in $\mathbb{L}[\mathscr{I}]$ exactly as in the proof of Theorem 5.5.4, with the only essential difference that some items of Lemma 6.4.9 and Lemma 6.4.12 are used instead of Lemma 5.5.5.

(iii) Assume that $\mathscr{I} \subseteq \mathbb{I}_\kappa$ is a κ-deep saturated internal subuniverse.

The axiom of κ-deep Saturation holds in $\mathbb{L}[\mathscr{I}]$, because $\mathbb{L}[\mathscr{I}] \cap \mathbb{I} = \mathscr{I}$.

To verify the Power Set axiom in $\mathbb{L}[\mathscr{I}]$, note that every set is a functional image of a standard set in $\mathbb{L}[\mathscr{I}]$. (This can be proved similarly to the proof of claim 8° of Theorem 5.5.4 in § 5.5b, with a reference to Lemma 6.4.12 at certain point.) Thus it suffices to demonstrate that every *standard* set X has a power set in $\mathbb{L}[\mathscr{I}]$. Furthermore by Lemma 6.4.12 it suffices to prove that $\mathscr{P}(X) \cap \mathbb{L}[\mathscr{I}]$ is a set for any standard X. However $\mathscr{P}(X) \cap \mathbb{L}[\mathscr{I}] \subseteq \mathbb{E}[\mathscr{I}]$ (Proposition 6.4.8), thus, it suffices to show that $\mathscr{P}(X) \cap \mathbb{E}[\mathscr{I}]$ is a set for any $X \in \mathbb{S}$. But this follows from (iii) of Theorem 6.2.3 because $\mathscr{I} \subseteq \mathbb{I}_\kappa$.

Prove that any set $Y = \{y_\xi : \xi < 2^\kappa\} \subseteq \mathbb{L}[\mathscr{I}]$ of cardinality $\le 2^\kappa$ belongs to $\mathbb{L}[\mathscr{I}]$. By definition, for any $\xi < 2^\kappa$ there is a code $\mathbf{x}_\xi \in \mathbf{A}(\mathscr{I})$ such that $y_\xi = \mathsf{A}_{\mathbf{x}_\xi}$, therefore, as $\mathbf{A}(\mathscr{I}) \subseteq \mathbb{E}[\mathscr{I}]$, there is $p_\xi \in \mathscr{I}$ with $\mathbf{x}_\xi = \mathsf{E}_{p_\xi}$. Yet $\mathscr{I}$ satisfies 2^κ-size Choice by Theorem 6.2.3, so that there is a function $f \in \mathscr{I}$, defined on ${}^*\lambda$ so that $f({}^*\xi) = p_\xi$ for each ordinal $\xi < \lambda$. We observe that the set $Y = \{\mathsf{A}_{\mathsf{E}_{f({}^*\xi)}} : \xi < 2^\kappa\}$ is definable in $\mathbb{L}[\mathbb{I}]$ by a st-$\in$-formula with f, κ as the only parameters, therefore, by Lemma 6.4.12, if $Y \notin \mathbb{L}[\mathscr{I}]$ then $Y = z \cap \mathscr{I}$ for some $z \in \mathscr{I}$. It follows that $z \subseteq \mathscr{I}$ by Lemma 6.2.4 (condition (2); $\mathscr{I}$ is $\aleph_0$-deep saturated), thus, $Y \in \mathbb{L}[\mathscr{I}]$ anyway.

That Dependent Choice and the axiom of 2^κ-size Choice hold in $\mathbb{L}[\mathscr{I}]$ is now clear: 1st, both axioms (Standard Size Choice in the full form) hold in the entire universe of **HST**, 2nd, the axioms assert the existencc of sets of cardinality $\le 2^\kappa$ ($\le \aleph_0$ for Dependent Choice), 3rd, $\mathbb{L}[\mathscr{I}]$ is closed under subsets of cardinality $\le 2^\kappa$ by the above. □

6.4e The classes $\mathbb{L}[\mathbb{I}_\kappa]$

There is an important particular case: $\mathscr{I} = \mathbb{I}_\kappa$, where κ is any infinite (well-founded) cardinal. Note that $\mathbb{I}_\kappa$ is a κ-deep saturated internal subuniverse by Theorem 6.2.3, and obviously $\mathbb{I}_\kappa$ is self-definable (in fact definable with ${}^*\kappa \in \mathbb{S}$ as the only parameter). Thus the main part of the next result immediately follows from Theorem 6.4.13.

Corollary 6.4.14. *For any infinite cardinal κ, the class $\mathbb{L}[\mathbb{I}_\kappa]$ is an internal core extension of $\mathbb{I}_\kappa$, transitive over $\mathbb{I}_\kappa$, and an external subuniverse satisfying* $\mathbf{HST}_\kappa$. *Any set $X \subseteq \mathbb{L}[\mathbb{I}_\kappa]$ of cardinality $\le 2^\kappa$ belongs to $\mathbb{L}[\mathbb{I}_\kappa]$.*

In addition, it is true in $\mathbb{L}[\mathbb{I}_\kappa]$ that ${}^*\mathbb{R}$ contains a well-orderable set of cardinality strictly greater than 2^κ.

Proof. Let us prove the additional claim. Suppose towards the contrary that $\lambda > 2^\kappa$ is a cardinal in $\mathbb{WF}$ and $F \in \mathbb{L}[\mathbb{I}_\kappa]$, $F : \lambda \to {}^*\mathbb{R}$ is an injection. By definition there is $p \in \mathbb{I}_\kappa$ such that $\mathsf{E}_p \in \mathbf{A}(\mathbb{I}_\kappa)$ and $\mathsf{A}_{\mathsf{E}_p} = F$, in particular, F is st-$\in$-definable in $\mathbb{L}[\mathbb{I}]$ with p as the only parameter. On the other hand, p, as any other member of $\mathbb{I}_\kappa$, has the form $f(\xi)$, where $\xi < {}^*\kappa$ is a $*$-ordinal and f is a standard function with $\xi \in \operatorname{dom} f$, by 6.2.2.

In other words F is st-$\in$-definable in $\mathbb{L}[\mathbb{I}]$ with standard parameters and ξ as the only nonstandard parameter. Therefore so is every value $F(\alpha)$, $\alpha < \lambda$ (with the addition of one more standard parameter ${}^*\alpha$). It follows that $F(\alpha)$ belongs to the class $\mathbb{S}[\xi] = \mathbb{S}(\{\xi\})$, by Theorem 6.1.3(v) (for $\mathscr{I} = \mathbb{S}[\xi]$). Thus, by Standard Size Choice, for any $\alpha < \lambda$ there is a map $f_\alpha : \kappa \to \mathbb{R}$ ($f_\alpha \in \mathbb{WF}$, of course) such that $F(\alpha) = {}^*f_\alpha(\xi)$. And, as F is injective, we have $f_\alpha \ne f_\gamma$ whenever $\alpha \ne \gamma < \lambda$. Therefore $al \mapsto f_\alpha$ is an injection of λ into $\mathbb{R}^\kappa$, which contradicts the assumption that $\lambda > 2^\kappa$. □

6.4f External universes over complete sets

Let κ be an infinite (well-founded) cardinal.

We demonstrated in § 6.2c that there exist sets $R \subseteq {}^*\kappa$ of standard size, called 2^κ-*complete*, such that the corresponding internal subuniverses $\mathbb{S}(R)$ are elementary submodels of $\mathbb{I}_\kappa$ in the st-$\in$-language. Since the constructible internal core extensions $\mathbb{L}[\mathbb{I}_\kappa]$ and $\mathbb{L}[\mathbb{S}(R)]$ are obtained on the base of resp. $\mathbb{I}_\kappa$ and $\mathbb{S}(R)$, it is a pretty natural question whether the external subuniverse $\mathbb{L}[\mathbb{S}(R)]$ is a st-$\in$-elementary submodel of $\mathbb{L}[\mathbb{I}_\kappa]$ in this case.

The question should be answered in the negative in such a straight form, in particular, because the collection (of A-codes) $\mathbf{A}(\mathscr{I})$, which participates in the construction of $\mathbb{L}[\mathscr{I}]$, is not defined by a st-$\in$-formula relativized to $\mathscr{I}$ but rather by a st-$\in$-formula which refers to $\mathscr{I}$ in a more complicated manner. Let alone the fact that internal subuniverses of the form $\mathscr{I} = \mathbb{S}(R)$, where $\mathscr{I} \subseteq \mathbb{I}_\kappa$ is a 2^κ-complete set, are not necessarily self-definable classes, thus the results of § 6.4d are generally speaking not applicable. The goal of this subsection is to modify the definition of $\mathbb{L}[\mathscr{I}]$ for internal subuniverses $\mathscr{I} \subseteq \mathbb{I}_\kappa$ of this type so that the modified extensions of $\mathscr{I}$ will be elementary substructures of $\mathbb{L}[\mathbb{I}_\kappa]$ in the st-$\in$-language.

Recall that $\mathbb{E}[\mathbb{I}_\kappa] = \{\mathsf{E}_p : p \in \mathbb{I}_\kappa\}$ and $\mathbb{L}[\mathbb{I}_\kappa] = \{\mathsf{A}_\mathbf{x} : \mathbf{x} \in \mathbf{A}(\mathbb{I}_\kappa)\}$, where $\mathbf{A}(\mathbb{I}_\kappa)$ is, by definition, the class of all $\mathbb{I}_\kappa$-regular (in the sense of § 6.3b) A-codes $\mathbf{x} \in \mathbb{E}(\mathbb{I}_\kappa)$ such that $T_\mathbf{x} \subseteq \mathbb{I}_\kappa$, $\mathtt{ran}\,\mathbf{x} \subseteq \mathbb{I}_\kappa$.

Now, for any internal subuniverse $\mathscr{I} \subseteq \mathbb{I}_\kappa$, let

$$^\kappa\mathbf{A}(\mathscr{I}) = \mathbf{A}(\mathbb{I}_\kappa) \cap \mathbb{E}[\mathscr{I}] \quad \text{and} \quad ^\kappa\mathbb{L}[\mathscr{I}] = \{\mathsf{A}_\mathbf{x} : \mathbf{x} \in {}^\kappa\mathbf{A}(\mathscr{I})\}\,.$$

The class $^\kappa\mathbf{A}(\mathscr{I})$ is different from the abovedefined $\mathbf{A}(\mathscr{I})$, in particular, since now it is required that $T_\mathbf{x} \subseteq \mathbb{I}_\kappa$ and $\mathtt{ran}\,\mathbf{x} \subseteq \mathbb{I}_\kappa$ rather than $T_\mathbf{x} \subseteq \mathscr{I}$ or $\mathtt{ran}\,\mathbf{x} \subseteq \mathscr{I}$, yet we keep the condition of E-coding in $\mathscr{I}$.

Theorem 6.4.15. *Let $R \subseteq {}^*\kappa$ be a 2^κ-complete set. Then $^\kappa\mathbb{L}[\mathbb{S}(R)]$ is an elementary substructure of $\mathbb{L}[\mathbb{I}_\kappa]$ in the* st-$\in$-*language, and hence an external subuniverse and an interpretation of* $\mathbf{HST}_\kappa$. *In addition, $^\kappa\mathbb{L}[\mathbb{S}(R)]$ is an internal core extension of $\mathbb{S}(R)$, that is $^\kappa\mathbb{L}[\mathbb{S}(R)] \cap \mathbb{I} = \mathbb{S}(R)$.*

Proof. As $\mathbb{L}[\mathbb{I}_\kappa] \subseteq \mathbb{L}[\mathbb{I}]$ is st-$\in$-definable in $\mathbb{L}[\mathbb{I}]$ with only $^*\kappa \in \mathbb{S}$ as a parameter, it suffices to show that, for any st-$\in$-formula $\Phi(x)$ with parameters in $^\kappa\mathbb{L}[\mathbb{S}(R)]$, if there is $x \in \mathbb{L}[\mathbb{I}_\kappa]$ satisfying $\Phi(x)$ in $\mathbb{L}[\mathbb{I}]$ then such a set x exists in $^\kappa\mathbb{L}[\mathbb{S}(R)]$, or in a more formal manner

$$\exists x \in \mathbb{L}[\mathbb{I}_\kappa]\,(\Phi(x)^{\mathbb{L}[\mathbb{I}]}) \implies \exists x \in {}^\kappa\mathbb{L}[\mathbb{S}(R)]\,(\Phi(x)^{\mathbb{L}[\mathbb{I}]})\,. \tag{1}$$

(Compare with Remark 6.2.10 !) The key idea is to obtain the result required from Theorem 6.2.9(ii)(‡), but in order to carry out this argument we use a reduction from $\mathbb{L}[\mathbb{I}]$ to $\mathbb{I}$ (with $\mathbb{E}$ as an intermediate structure) provided by the coding systems studied in Chapter 5.

Suppose that $y_0 \in {}^{\kappa}\mathbb{L}[\mathbb{S}(R)]$ is the only parameter in Φ, thus, $\Phi(x)$ is $\Phi(x, y_0)$. By definition, we have $y_0 = \mathsf{A}_{\mathbf{y}_0}$ where $\mathbf{y}_0 \in {}^{\kappa}\mathbf{A}(\mathbb{S}(R))$, and hence $\mathbf{y}_0 = \mathsf{E}_{q_0}$ for some $q_0 \in \mathbb{S}(R) \cap \mathbf{E}$. Now let $\Psi(\vartheta, x, y)$ be the $\mathtt{st}$-$\in$-formula

$$\vartheta \in \mathbb{S} \text{ is a } *\text{-ordinal} \wedge x, y \in \mathbb{L}[\mathbb{I}_\vartheta] \wedge \Phi(x, y),$$ [19]

with the free variables ϑ, x, y. Let further $\psi(\vartheta, x, y)$ be the $\mathtt{st}$-$\in$-formula ${}^{\mathsf{e}}({}^{\mathsf{a}}(\Psi(\vartheta, x, y)))$. Then it follows from the equivalences $(*)$ in the proof of Theorem 5.2.11 (the second equivalence) and in Lemma 5.4.16 that

$$\Psi(\mathsf{A}_{\mathsf{E}_t}, \mathsf{A}_{\mathsf{E}_p}, \mathsf{A}_{\mathsf{E}_q})^{\mathbb{L}[\mathbb{I}]} \iff ({}^{\mathsf{a}}(\Psi(\mathsf{E}_t, \mathsf{E}_p, \mathsf{E}_q)))^{\mathbb{E}} \iff \psi(t, p, q)^{\mathtt{int}}, \qquad (2)$$

for all t, p, $q \in \mathbf{U} = \{u \in \mathbf{E} : \mathsf{E}_u \in \mathbf{A}\}$. Note that q_0 (see above) belongs to $\mathbf{U}$ and so does $t_0 = {}^{\mathsf{e}}({}^{\mathsf{a}}({}^{*}\kappa))$ (then ${}^{*}\kappa = \mathsf{A}_{\mathsf{E}_{t_0}}$).

Let us come back to our task. Suppose that $x \in \mathbb{L}[\mathbb{I}_\kappa]$ witnesses the left-hand side of (1), *i.e.* we have $\Psi({}^{*}\kappa, x, y_0)^{\mathbb{L}[\mathbb{I}]}$. Then $x = \mathsf{A}_{\mathbf{x}}$ where $\mathbf{x} \in \mathbf{A}(\mathbb{I}_\kappa)$, and hence $\mathbf{x} \in \mathbb{E}(\mathbb{I}_\kappa)$, thus $\mathbf{x} = \mathsf{E}_p$ for some $p \in \mathbb{I}_\kappa \cap \mathbf{E}$ – and then $p \in \mathbf{U}$. In other words we have $\Psi(\mathsf{A}_{\mathsf{E}_{t_0}}, \mathsf{A}_{\mathsf{E}_p}, \mathsf{A}_{\mathsf{E}_{q_0}})^{\mathbb{L}[\mathbb{I}]}$, and hence $\psi(t_0, p, q_0)^{\mathtt{int}}$ by (2). It follows from Theorem 6.2.9(ii)(‡) that the very same properties, that is $p \in \mathbb{I}_\kappa \cap \mathbf{U}$, $\mathsf{E}_p \in \mathbf{A}(\mathbb{I}_\kappa)$, and $\psi(t_0, p, q_0)^{\mathtt{int}}$ can be fulfilled by some $p \in \mathbb{S}(R)$. But in this case $\mathbf{x} = \mathsf{E}_p \in \mathbb{E}[\mathbb{S}(R)]$, therefore $\mathbf{x} \in {}^{\kappa}\mathbf{A}(\mathbb{S}(R))$, and then $x = \mathsf{A}_{\mathbf{x}} \in {}^{\kappa}\mathbb{L}[\mathbb{S}(R)]$. Finally $\Psi({}^{*}\kappa, x, y_0)^{\mathbb{L}[\mathbb{I}]}$ follows from (2), thus we have the right-hand side of (1) with this x, as required.

That ${}^{\kappa}\mathbb{L}[\mathbb{S}(R)] \cap \mathbb{I} = \mathbb{S}(R)$ is left as an **exercise** for the reader. □

Absoluteness

The major difference of the classes ${}^{\kappa}\mathbb{L}[\mathscr{I}]$ from those studied above (like $\mathbb{L}[\mathscr{I}]$) is that they are not transitive over the internal subclass $\mathscr{I} = {}^{\kappa}\mathbb{L}[\mathscr{I}] \cap \mathbb{I}$ any more: indeed, this is because codes $\mathbf{x} \in {}^{\kappa}\mathbf{A}(\mathscr{I})$ do not necessarily satisfy $\mathtt{ran}\,\mathbf{x} \subseteq \mathscr{I}$. This makes the absoluteness arguments outlined in § 6.3c not directly applicable for classes of this form.

Yet under the conditions of Theorem 6.4.15 the absoluteness for ${}^{\kappa}\mathbb{L}[\mathbb{S}(R)]$ holds to exactly the same degree as for the external subuniverse $\mathbb{L}[\mathbb{I}_\kappa]$, which is an internal core extension of $\mathbb{I}_\kappa$ transitive over $\mathbb{I}_\kappa$ by Corollary 6.4.14. For instance, the formula "f is a function" is absolute for $\mathbb{L}[\mathbb{I}_\kappa]$ by a general argument given in § 6.3c, but on the other hand ${}^{\kappa}\mathbb{L}[\mathbb{S}(R)]$ is an elementary substructure of $\mathbb{L}[\mathbb{I}_\kappa]$ w. r. t. this formula (and any other $\mathtt{st}$-$\in$-formula) by Theorem 6.4.15, and hence we can conclude that "f is a function" is absolute for ${}^{\kappa}\mathbb{L}[\mathbb{S}(R)]$ as well. The reader can easily verify that many other similar simple formulas like being a pair, being a union, *etc*, are absolute for ${}^{\kappa}\mathbb{L}[\mathbb{S}(R)]$ in virtue of the same argument.

[19] If $\vartheta = {}^{*}\kappa$ where $\kappa \in \mathbb{WF}$ is a (well-founded) cardinal then by definition $\mathbb{I}_\vartheta = \mathbb{I}_\kappa$ and accordingly $\mathbb{L}[\mathbb{I}_\vartheta] = \mathbb{L}[\mathbb{I}_\kappa]$. We prefer here to substitute the standard parameter ${}^{*}\kappa$ for the well-founded parameter $\kappa \in \mathbb{WF}$.

6.4g Collapse onto a transitive class

The transitivity of ${}^{\kappa}\mathbb{L}[\mathscr{I}]$ over $\mathscr{I}$ can be restored by a procedure that resembles Mostowski's collapse known from works on **ZFC**. Recall that in **ZFC** any extensional set or class X admits a unique $\in$-isomorphism f of X onto a (also unique) *transitive* set or class T. In the **HST** context, we prove

Theorem 6.4.16 (HST). *Suppose that $\mathscr{I} \subseteq \mathbb{I}$ is an internal subuniverse and any internal set $X \subseteq \mathscr{I}$ belongs to $\mathscr{I}$* [20]. *Then*

(i) *Any extensional class $\mathscr{H}$ satisfying $\mathscr{H} \cap \mathbb{I} = \mathscr{I}$ admits a unique $\in$-isomorphism $\phi = \phi_{\mathscr{H}}$ onto a class $\mathscr{H}^{\text{MC}}$ transitive over $\mathscr{I}$.* [21]

(ii) *$\mathscr{H}^{\text{MC}} \cap \mathbb{I} = \mathscr{I}$ and ϕ is the identity on $\mathscr{I}$.*

(iii) *$\mathscr{H}^{\text{MC}} \subseteq \mathbb{WF}[\mathscr{I}]$ and $\mathscr{H}^{\text{MC}}$ is a transitive internal core extension of $\mathscr{I}$.*

Proof. We define $\phi(x)$ for all $x \in \mathscr{H}$ by induction on $\mathtt{irk}\, x$, the rank over $\mathbb{I}$ (see § 1.5b). If $\mathtt{irk}\, x = 0$ then $x \in \mathscr{H}$ is internal, thus $x \in \mathscr{I}$. Put $\phi(x) = x$ in this case. Put $\phi(x) = \{\phi(y) : y \in x \cap \mathscr{H}\}$ provided $\mathtt{irk}\, x \geq 1$. In particular if $\mathtt{irk}\, x = 1$, so that $x \subseteq \mathbb{I}$ but $x \notin \mathbb{I}$, then $\phi(x) = x \cap \mathscr{I}$.

We claim that $\phi(x)$ is not internal for any $x \in \mathscr{H} \setminus \mathscr{I}$. Indeed let $x \in \mathscr{H} \setminus \mathscr{I}$ be a counterexample with the least possible $\mathtt{irk}\, x$. Thus $\phi(x) = \{\phi(y) : y \in x \cap \mathscr{H}\} \in \mathbb{I}$, and hence any $y \in x \cap \mathscr{H}$ is internal. Thus $\phi(x) = x \cap \mathscr{H} = x \cap \mathscr{I} \in \mathbb{I}$. It follows that $x \cap \mathscr{I} = x \cap \mathscr{H} \in \mathscr{I}$ under the assumptions of the theorem, hence $x = x \cap \mathscr{I} \in \mathscr{I}$, a contradiction.

We observe that ϕ is a bijection. Indeed if $x \in \mathscr{I}$ and $y \in \mathscr{H} \setminus \mathscr{I}$ then $\phi(x) = x$ is internal while $\phi(y)$ is not so by the above. Prove $\phi(x) = \phi(y) \Longrightarrow x = y$ for $x, y \in \mathscr{H} \setminus \mathscr{I}$ by induction on $\mathtt{irk}\, x$, $\mathtt{irk}\, y$. If

$$\phi(x) = \{\phi(x') : x' \in \mathscr{H} \cap x\} = \{\phi(y') : y' \in \mathscr{H} \cap y\} = \phi(y)$$

then $\mathscr{H} \cap x = \mathscr{H} \cap y$ by the inductive assumption and hence $x = y$ by the extensionality of $\mathscr{H}$.

Prove finally that $\phi(x) \in \phi(y) \Longrightarrow x \in y$. Suppose that $\phi(x) \in \phi(y)$. If $y \in \mathscr{I}$ then $\phi(y) = y$ is internal, thus so are both $\phi(x)$ and x by the above, and hence $x \in \mathscr{I}$ and $\phi(x) = x$. If $y \notin \mathscr{I}$ then $\phi(y) = \{\phi(y') : y' \in \mathscr{H} \cap y\}$, thus $\phi(x) = \phi(y')$ for some $y' \in \mathscr{H} \cap y$, and $x = y' \in y$ by the above.

We leave it as an **exercise** for the reader to prove the uniqueness of the collapse map $\phi_{\mathscr{H}}$ and the rest of the theorem. □

Now, let κ still be a fixed infinite (well-founded) cardinal.

Consider an internal subuniverse of the form $\mathscr{I} = \mathbb{S}(R)$, where $R \subseteq {}^*\kappa$ is a 2^κ-complete set of standard size. In this case the collapse map $\phi_{{}^{\kappa}\mathbb{L}[\mathbb{S}(R)]}$ provided by Theorem 6.4.16 for $\mathscr{H} = {}^{\kappa}\mathbb{L}[\mathbb{S}(R)]$ admits a pretty transparent

[20] This condition is satisfied for instance if $\mathscr{I}$ is a thin class, see 6.3.4(2).

[21] MC means: Mostowski collapse. Note that $\mathscr{H}^{\text{MC}}$ depends on $\mathscr{I}$ as well, not only on $\mathscr{H}$, but we suppress this dependence in the notation.

presentation in terms of A-codes involved. Indeed if $\mathbf{x} \in {}^\kappa\mathbf{A}(\mathbb{S}(R))$ then clearly $\mathbf{x}\restriction_{\mathbb{S}(R)} = \mathbf{x} \restriction (\mathbf{dom}\,\mathbf{x} \cap \mathbb{S}(R))$ is a code in $\underline{\mathbf{A}}$. But not only that. The following exercise contains a list of related results.

Exercise 6.4.17. Let $\mathbf{x} \in {}^\kappa\mathbf{A}(\mathbb{S}(R))$. Prove the following:

(1) $\mathbf{x}$ and $T_\mathbf{x}$ belong to ${}^\kappa\mathbb{L}[\mathbb{S}(R)]$.

(2) $T_{\mathbf{x}\restriction_{\mathbb{S}(R)}} = T_\mathbf{x} \cap \mathbb{S}(R)$.

(3) $\mathbf{x}\restriction_{\mathbb{S}(R)}$ is a regular code and hence belongs to $\mathbf{A}$.

(4) If $t \in T_\mathbf{x} \cap \mathbb{S}(R)$, $t^\wedge a \in T_\mathbf{x}$ ($t^\wedge a$ is an extension of t by a term a), and $\mathsf{F}_\mathbf{x}(t^\wedge a) \in {}^\kappa\mathbb{L}[\mathbb{S}(R)]$ then there is an extension $t^\wedge b \in T_\mathbf{x} \cap \mathbb{S}(R) = T_{\mathbf{x}\restriction_{\mathbb{S}(R)}}$ such that $\mathsf{F}_\mathbf{x}(t^\wedge a) = \mathsf{F}_\mathbf{x}(t^\wedge b)$.

(5) $\mathsf{A}_{\mathbf{x}\restriction_{\mathbb{S}(R)}} = \phi_{{}^\kappa\mathbb{L}[\mathbb{S}(R)]}(\mathsf{A}_\mathbf{x})$.

(6) The collapsed class $({}^\kappa\mathbb{L}[\mathbb{S}(R)])^{\mathrm{MC}}$ is equal to $\{\mathsf{A}_{\mathbf{x}\restriction_{\mathbb{S}(R)}} : \mathbf{x} \in {}^\kappa\mathbf{A}(\mathbb{S}(R))\}$.

(7) $({}^\kappa\mathbb{L}[\mathbb{S}(R)])^{\mathrm{MC}}$ satisfies $\mathbf{HST}_\kappa$ together with ${}^\kappa\mathbb{L}[\mathbb{S}(R)]$ itself.

Hints. (1) $\mathbf{x}$ belongs to $\mathbb{E}[\mathbb{S}(R)] = \{\mathsf{E}_p : p \in \mathbf{E} \cap \mathbb{S}(R)\}$, and hence is $\mathtt{st}$-$\in$-definable with only $p \in \mathbb{S}(R)$ as a parameter, and so is $T_\mathbf{x}$, thus $\mathbf{x}$ and $T_\mathbf{x}$ belong to ${}^\kappa\mathbb{L}[\mathbb{S}(R)]$ by Theorem 6.4.15.

(2) By a similar argument if $t \in T_\mathbf{x} \cap \mathbb{S}(R)$ then "t has an extension in $\mathtt{Max}\, T_\mathbf{x} \cap \mathbb{S}(R)$" is absolute for ${}^\kappa\mathbb{L}[\mathbb{S}(R)]$.

(3) As $\mathbb{S}(R)$ is thin any internal set $X \subseteq \mathbb{S}(R)$ is finite. Then apply absoluteness as in (1), (2).

(4): a similar absoluteness argument.

(5) follows from (4). □

6.4h Outline of applications: subuniverses satisfying Power Set

Theories $\mathbf{HST}_\kappa$ and $\mathbf{HST}'_\kappa$ contain less Saturation (and Standard Size Choice, for the first theory) than **HST** does. However as soon as a particular application is fixed, where all the cardinalities of sets involved are naturally bounded by a certain cardinal, the opportunitities offered by these partially saturated versions of **HST** are practically equal to those of **HST**; in addition, $\mathbf{HST}_\kappa$ and $\mathbf{HST}'_\kappa$ contain the Power Set axiom!

To see how this can be used in the practice of nonstandard analysis, recall that in **ZFC** any particular mathematical structure $\mathfrak{A}$ is a member of a certain transitive set W of the form $\mathbf{V}_\alpha$, $\alpha \in \mathtt{Ord}$. Usually we can take $W = \mathbf{V}_{\omega+\omega}$, the $(\omega+\omega)$-th level of the von Neumann hierarchy (see § 1.5a): indeed, all natural numbers belong to $\mathbf{V}_\omega = \mathbb{HF}$, so do all rationals, viewed as pairs of natural numbers, hence, all reals, defined by Dedekind, belong to $\mathbf{V}_{\omega+1}$, all sets of reals, including the set $\mathbb{R}$ of all reals itself, to $\mathbf{V}_{\omega+2}$, all real functions appear at appropriate higher level, *et cetera*.)

If, arguing in **HST**, we consider such a structure $\mathfrak{A}$ in the class $\mathbb{WF}$ of all well-founded sets, then, accordingly, $\mathfrak{A}$ belongs to a well-founded set of the form $W = \mathbf{V}_\alpha$, where α is a (well-founded) ordinal. Let us fix any (well-founded) cardinal $\kappa \geq \alpha$ as the amount of Saturation required to study $\mathfrak{A}$ and its $*$-extension ${}^*\mathfrak{A}$ in suitable nonstandard manner. Something like $\kappa = (\mathtt{card}\, W)^+$ will normally be sufficient.

First option

The class $\mathbb{L}[\mathbb{I}_\kappa]$ is a transitive internal core extension of $\mathbb{I}_\kappa$ satisfying $\mathbf{HST}_\kappa$ by Corollary 6.4.14, that is, all of **ZF** (minus Regularity), together with such tools of the "nonstandard" instrumentarium as κ-deep Saturation, hence also κ-size Saturation (that is, Saturation for families of cardinality $\leq \kappa$) and the 2^κ-version of Standard Size Choice [22] and finally the Power Set axiom. Recall that the latter is incompatible with our basic theory **HST** itself.

However any element of ${}^*\mathbf{V}_\alpha$, including the set ${}^*\mathfrak{A}$ itself and all its elements, belongs to $\mathbb{I}_\kappa$ (because $\mathfrak{A} \in \mathbf{V}_\alpha \subseteq \mathbf{V}_\kappa$), and hence to $\mathbb{L}[\mathbb{I}_\kappa]$. This allows us to carry out in $\mathbb{L}[\mathbb{I}_\kappa]$ any ordinary "nonstandard" argument related to $\mathfrak{A}$ and ${}^*\mathfrak{A}$ which requires not more than the mentioned κ-forms of Saturation and, possibly, uses the Power Set axiom. If the results, obtained in the course of this study conducted in $\mathbb{L}[\mathbb{I}_\kappa]$, are related only to $\mathfrak{A}$ and ${}^*\mathfrak{A}$ and their elements, then they retain their meaning in the whole **HST** universe because both $\mathbf{V}_\alpha$ and ${}^*\mathbf{V}_\alpha$ are transitive sets which belong to $\mathbb{L}[\mathbb{I}_\kappa]$ together with all their elements.

We call this approach *the scheme* "$\mathbb{WF} \xrightarrow{*} \mathbb{I}_\kappa$ [*in* $\mathbb{L}[\mathbb{I}_\kappa]$] ".

Classes of the form ${}^\kappa\mathbb{L}[\mathbb{S}(R)]$ where $R \subseteq {}^*\kappa$ is a 2^κ-complete set of standard size (see § 6.4f) offer an additional opportunity. They are still internal core (non-transitive) extensions of $\mathbb{I}_\kappa$ satisfying $\mathbf{HST}_\kappa$ and elementary substructures of $\mathbb{L}[\mathbb{I}_\kappa]$ in the $\mathtt{st}$-$\in$-language. In addition they can be used to define consecutive extensions, see Remark 6.2.14.

Second option

Classes satisfying $\mathbf{HST}'_\kappa$ (with full Choice) can also be used. According to Corollary 6.4.4, there is a κ-saturated thin internal subuniverse $\mathscr{I} \subseteq \mathbb{I}_\kappa$ such that its internal core extension $\mathbb{WF}[\mathscr{I}]$ satisfiess $\mathbf{HST}'_\kappa$ including even full Choice (and Power Set), together with κ-size Saturation. The class $\mathbb{WF}[\mathscr{I}]$ contains $\mathfrak{A}$ and ${}^*\mathfrak{A}$, of course, but it is not true any more that ${}^*\mathfrak{A} \subseteq \mathbb{WF}[\mathscr{I}]$, in fact, any set $X \subseteq \mathbb{WF}[\mathscr{I}]$ is a set of standard size. Yet one may employ the

[22] That we have Saturation for families of cardinality κ but Choice for domains of cardinality 2^κ in $\mathbb{L}[\mathbb{I}_\kappa]$ is remarkably in line with the practice of model-theoretic nonstandard analysis, where it is customary to assume countable Saturation (also called $\aleph_1$-Saturation) but sometimes to employ constructions which require continuum-many choices, see, *e.g.*, the choice of r_A in the proof of Theorem 9.7.10 below.

absoluteness between $\mathbb{WF}[\mathscr{I}]$ and the entire universe, as in §6.3c, to obtain, in the latter, an adequate meaning of facts established in $\mathbb{WF}[\mathscr{I}]$.

We can identify such a method as *the scheme* "$\mathbb{WF} \xrightarrow{*} \mathscr{I}$ [*in* $\mathbb{WF}[\mathscr{I}]$]".

An application

The following example, albeit rather elementary, shows how these schemes can be utilized.

Consider, in **HST**, an infinite $*$-finite set $H = [1, ..., h] \subseteq {}^*\mathbb{N}$, where $h \in {}^*\mathbb{N} \setminus \mathbb{N}$. By **Borel** (H) they denote the least σ-algebra of subsets of H containing all internal sets, which means, most naturally, the intersection of all σ-algebras of subsets of H that include $\mathscr{P}_{\mathtt{int}}(H)$. But how to get at least one such a σ-algebra? In **ZFC** there is no problem to take $\mathscr{P}(H)$, the power set. However $\mathscr{P}(H)$ is definitely not a set in **HST** for any infinite internal set H by Theorem 1.3.9, so that this argument does not work directly.

Let us show how partially saturated subuniverses can be employed to solve this problem. Fix any infinite (well-founded) cardinal κ, for instance, $\kappa = \aleph_0$. It follows from Corollary 6.4.14 that the subuniverse $\mathbb{L}[\mathbb{I}_\kappa]$ satisfies $\mathbf{HST}_\kappa$, a rather rich partially saturated version of **HST** which includes the Power Set axiom. On the other hand ${}^*\mathbb{N} \subseteq \mathbb{L}[\mathbb{I}_\kappa]$ still by 6.4.14, in particular $H \in \mathbb{L}[\mathbb{I}_\kappa]$ and $H \subseteq \mathbb{L}[\mathbb{I}_\kappa]$. It follows that the power set $P_\kappa = \mathscr{P}(H)^{\mathbb{L}[\mathbb{I}_\kappa]}$, equal to $\mathscr{P}(H) \cap \mathbb{L}[\mathbb{I}_\kappa]$ by the above, is really a set and belongs to $\mathbb{L}[\mathbb{I}_\kappa]$.

Finally, we claim that P_κ is a σ-algebra. Indeed recall that any set $Q \subseteq \mathbb{L}[\mathbb{I}_\kappa]$ of cardinality $\leq 2^\kappa$ in the **HST** universe belongs to $\mathbb{L}[\mathbb{I}_\kappa]$ by Corollary 6.4.14. It follows that P_κ is even $(2^\kappa)^+$-additive, that is, closed under unions and intersections of $\leq 2^\kappa$ sets. Thus we have defined a σ-algebra of subsets of H containing all internal subsets of H, and this is sufficient to consistently define **Borel** (H).

Historical and other notes to Chapter 6

Section 6.1. The notion of relative standardness (Definition 3.1.13) can be traced down to [CherH 70] (in the context of the model theoretic nonstandard analysis). Relative standardness, in the form of classes $\mathbb{S}[x]$ and $\mathbb{S}_{\mathrm{M}}[x]$, together with Theorem 6.1.16, is due to Gordon [Gor 89]. Lemma 6.1.12 presents the original definition in [Gor 89].

Thin classes: the definition is due to Andreev.

Proposition 6.1.7 and Theorem 6.1.15 are due to Andreev and Hrbaček [AnH 04]. The particular case $\mathbb{N} \subsetneqq {}^*\mathbb{N} \cap \mathscr{I} \subsetneqq {}^*\mathbb{N}$ in Theorem 6.1.15 is due to Andreev [An 99] (also [Gor 89] for classes $\mathscr{I}$ of the form $\mathbb{S}[x]$ and $\mathbb{S}_{\mathrm{M}}[x]$). Hrbaček [Hr 01] explores more in this direction.

The proof of Theorem 6.1.21 is based on ideas from [Suz 99]. Theorem 68 in [Jech 78] gives a more general result, essentially saying that **ZFCj**, an extension of **ZFC** by a symbol **j** for an elementary embedding of the set universe in itself, with appropriate axioms, proves that **j** is the identity.

Section 6.2. Classes $\mathbb{I}_\kappa$ (Definition 6.2.1) appeared in [Kan 91]. In the particular case $\kappa = \aleph_0$, sets that belong to countable standard sets were introduced by Luxemburg [Lux 62] under the name of σ-quasistandard objects. A general definition was first given in a nonpublished version of Hrbaček [Hr 79], which the author of [Kan 91] was not aware of.

The main parts of Theorem 6.2.3 appeared in our paper [KanR 95, Part 3]. The concept of λ-complete sets and Theorem 6.2.9 appeared in [KanR 98] (partially in [KanR 95, Part 3] where the corresponding classes were denoted by $\mathbb{I}'_\kappa$).

Sections 6.3, 6.4. The content of these sections is mainly due to [KanR 95, part 3] ([KanR 97] contains an updated version), in particular, internal core extensions $\mathbb{L}[\mathbb{I}_\kappa]$ and $\mathbb{WF}[\mathbb{I}'_\kappa]$, introduced in [KanR 95] under the names, resp., $\mathbb{H}_\kappa$ and $\mathbb{H}'_\kappa$, and their main properties as in Theorems 6.4.3 and 6.4.13. Theorem 6.2.11, and applications similar to § 6.4f, appeared in [KanR 98].

See [Kun 80, III.5] on the Mostowski collapse theorem in **ZFC**.

7 Forcing extensions of the nonstandard universe

Recall that the class $\mathbb{L}[\mathbb{I}]$ of sets constructible from internal sets was employed in Chapter 5 to obtain some consistency theorems. For instance Theorem 5.5.8 implies that it is consistent with **HST** that $\mathbb{I}$-infinite internal sets of different $\mathbb{I}$-cardinalities are necessarily non-equinumerous. It would be in the spirit of mathematical foundations to ask whether the negation of this sentence, that is the existence of equinumerous $\mathbb{I}$-infinite internal sets of different $\mathbb{I}$-cardinalities, is also consistent.

In **ZFC**, questions of this kind are often solved by forcing, [1] and it will be our goal in this Chapter to show how forcing works in **HST**.

There are remarkable differences from the **ZFC** setting.

First of all, the **HST** universe $\mathbb{H}$ is <u>not</u> well-founded inside. This makes it difficult to define the forcing relation for atomic sentences by induction on the ranks of involved "names", as in the **ZFC** case. We solve this problem using the well-foundedness of the universe $\mathbb{H}$ over the internal universe $\mathbb{I}$. This property allows us to treat $\mathbb{H}$ as a sort of **ZFC**-like model with urelements; internal sets playing the role of urelements. Of course internal sets do not behave completely like urelements; in particular they participate in the common membership relation. But this gives us the key idea: generic extensions should not introduce new internal (and thereby new standard) sets.

This leads us to another problem, connected with Standardization. Since new standard sets do not appear, a set of standard size cannot acquire new *subsets* in the extension. To obey this restriction, we apply a classical forcing argument: if the forcing notion is "standard size distributive" in the ground model then no new standard size subsets of $\mathbb{H}$ appear in the extension.

These ideas will be demonstrated on two examples. The first of them is a model of **HST** which "glues" $\mathbb{I}$-cardinalities of two given infinite internal sets having different cardinalities in the ground model. This example will be considered in Section 7.2 The other, a much more complicated example is a model of **HST** in Section 7.3, in which the isomorphism property (saying, in the context of **HST**, that any two elementarily equivalent structures of a language of standard size are isomorphic) holds.

[1] We assume that the reader is acquainted with elements of forcing and has some experience in it. Jech [Jech 78], Kunen [Kun 80], Shoenfield [Shoen 71] can be given as general references in this matter.

7.1 Generic extensions of models of HST

This section discusses three principal elements of forcing in **HST** : the ground model, the forcing notion, and generic extensions.

7.1a Ground model

In Chapter 7 we argue in the **ZFC** universe **V** unless clearly stated otherwise. $\mathbb{H} = \langle \mathbb{H}; \in_{\mathbb{H}}, \mathtt{st}_{\mathbb{H}} \rangle$ [2] is supposed to be a fixed model of **HST**, *the ground model*. We shall consider the well-founded, standard, and internal cores

$$\begin{aligned} \mathbb{WF} = \mathbb{WF}^{(\mathbb{H})} &= \{x \in \mathbb{H} : \mathbb{H} \models \mathtt{wf}\, x\} && \text{(all } \mathbb{H}\text{-well-founded sets)}, \\ \mathbb{S} = \mathbb{S}^{(\mathbb{H})} &= \{x \in \mathbb{H} : \mathtt{st}_{\mathbb{H}}\, x\} && \text{(all } \mathbb{H}\text{-standard sets)}, \quad \text{and} \\ \mathbb{I} = \mathbb{I}^{(\mathbb{H})} &= \{x \in \mathbb{H} : \mathtt{int}_{\mathbb{H}}\, x\} && \text{(all } \mathbb{H}\text{-internal sets)} \end{aligned}$$

of the model $\mathbb{H}$, where $\mathtt{int}_{\mathbb{H}}\, x$ is the formula $\exists y\, (\mathtt{st}_{\mathbb{H}}\, y \wedge x \in_{\mathbb{H}} y)$.

Unlike the case of models of **ZFC**, no model of **HST** can be an $\in$-model in the **ZFC** universe simply because **HST** implies infinite $\in$-decreasing chains of sets. Yet some regularity can be postulated.

Blanket agreement 7.1.1. (a) All sets $x \in \mathbb{I}_0 = \mathbb{I} \smallsetminus \mathbb{HF}^{(\mathbb{H})}$ (*i.e.* internal but **not** hereditarily finite in $\mathbb{H}$) have one and the same von Neumann rank κ in the **ZFC** universe **V**, where κ is a cardinal $> \mathtt{card}\, \mathbb{H}$.

(b) If $x \in \mathbb{H} \smallsetminus \mathbb{I}_0$ then the set $x^{(\mathbb{H})} = \{y \in \mathbb{H} : y \in_{\mathbb{H}} x\}$ is equal to x. Thus $\in_{\mathbb{H}}$-elements of any set $x \in \mathbb{H} \smallsetminus \mathbb{I}_0$ and $\in$-elements of x in the universe is one and the same. In particular $x \subseteq \mathbb{H}$ for any $x \in \mathbb{H} \smallsetminus \mathbb{I}_0$. □

Exercise 7.1.2. Prove the following, using 7.1.1:

(1) $x^{(\mathbb{H})} = x$ for any $x \in \mathbb{WF}^{(\mathbb{H})}$ (recall that $\mathbb{WF} \cap \mathbb{I} = \mathbb{HF}$ in **HST** by 1.2.17).

(2) $\mathbb{H}$ is *well-founded over* $\mathbb{I}$ in the sense that the set $\mathtt{Ord}^{(\mathbb{H})}$ of all $\mathbb{H}$-ordinals defined in $\mathbb{H}$ as in § 1.2c coincides with the initial segment $\{\xi : \xi < \mathtt{ht}_{\mathbb{H}}\}$ of true **ZFC** ordinals, with one and the same order, where $\mathtt{ht}_{\mathbb{H}} \in \mathtt{Ord}$, the *height* of $\mathbb{H}$, is the order type of $\langle \mathtt{Ord}^{(\mathbb{H})}; \in_{\mathbb{H}} \rangle$.

(3) $\mathbb{WF}^{(\mathbb{H})}$ is a transitive subset of $\mathbf{V}_{\mathtt{ht}_{\mathbb{H}}}$ in the **ZFC** universe, and $\in_{\mathbb{H}} \restriction \mathbb{WF}^{(\mathbb{H})}$ coincides with $\in \restriction \mathbb{WF}^{(\mathbb{H})}$.

(4) If $x \in \mathbb{H}$ and $y \in \mathbb{H} \smallsetminus \mathbb{I}$ then the ordered pair $p = \langle x, y \rangle = \{\{x,y\}\{y\}\}$ belongs to $\mathbb{H} \smallsetminus \mathbb{I}$ and $\mathbb{H} \models p = \langle x, y \rangle$. (However we have $p \notin \mathbb{H}$ by 7.1.1(a) provided x, $y \in \mathbb{I}$ and at least one of x, y belongs to $\mathbb{I}_0$.)

(5) $\mathbb{N} \in \mathbb{H} \smallsetminus \mathbb{I}$, and hence $\breve{x} = \langle x, \mathbb{N} \rangle$ belongs to $\mathbb{H} \smallsetminus \mathbb{I}$ and the equality $\breve{x} = \langle x, \mathbb{N} \rangle$ is true in $\mathbb{H}$ for any $x \in \mathbb{H}$. □

[2] All **HST**-based arguments below will be restricted to this model, and hence it is rather convenient to denote it by $\mathbb{H}$, normally our symbol of the **HST** universe. The same applies to $\mathbb{WF}$, $\mathbb{S}$, $\mathbb{I}$ just below in the text.

Requirement 7.1.1(b) can be interpreted as saying that any $x \in \mathbb{H} \setminus \mathbb{I}_0$ is a true set in the sense that it is in the universe what it seems to be in $\mathbb{H}$: the elements are the same. On the other hand sets in $\mathbb{I}_0$ are just $\mathbb{H}$-sets: their true elements may have nothing to do with $\in_{\mathbb{H}}$-elements.

It follows from (3) that natural numbers and hereditarily finite sets in $\mathbb{H}$ are equal to those in the **ZFC** universe. In particular $\varnothing \in \mathbb{H}$ and $\varnothing^{(\mathbb{H})} = \varnothing$, so that $\varnothing$ still is the $\mathbb{H}$-empty set. Moreover the set $\mathbb{N} = \mathbb{N}^{(\mathbb{H})} \in \mathbb{H}$ is equal to $\mathbb{N}^{(\mathbb{H})}$ (that is, $\mathbb{N}$ in the sense of $\mathbb{H}$). Saying it differently, $\varnothing$ and $\mathbb{N}$ are absolute for $\mathbb{H}$. Assertion (4), an easy consequence of (1), says that the operation of ordered pair is absolute for $\mathbb{H}$ as well, provided at least the second term of the pair considered is not $\mathbb{H}$-internal. There are other simple absoluteness results, for instance "being a subset of $X \times Y$" provided $X, Y \in \mathbb{H}$ and $Y \subseteq \mathbb{H} \setminus \mathbb{I}$, not mentioned in 7.1.2.

Requirement 7.1.1(a) looks rather artificial; but we make use of it in the proof of Lemma 7.1.10.

Exercise 7.1.3. Prove that any $x \in \mathbb{I}_0$ satisfies $x \not\subseteq \mathbb{H}$, even $x \cap \mathbb{H} = \varnothing$. □

Exercise 7.1.4. Prove that any model $\mathbb{H}$ of **HST** well-founded in the sense of 7.1.2(2) is isomorphic to a model satisfying 7.1.1 and then 7.1.2, 7.1.3.

Hint. 7.1.1(a) can be assumed immediately. To achieve 7.1.1(b) define $f(x) = x$ for $x \in \mathbb{I}_0$ and $f(x) = \{f(y) : y \in_{\mathbb{H}} x\}$ otherwise. This is sound because of 7.1.2(2) and since $\mathbb{H} \setminus \mathbb{I}_0 = (\mathbb{H} \setminus \mathbb{I}) \cup \mathbb{HF}$ is well-founded in **HST**. The f-image of $\mathbb{H}$ is as required. □

Thus the real content of 7.1.1 is the well-foundedness of $\mathbb{H}$ over its internal core while the rest is just simple cosmetical rearrangements.

7.1b Regular extensions

Forcing is a powerful method that allows to extend models of certain theories to models (of the same or a closely related theory) which have some desired additional properties. Let us formulate some basic requirements to be satisfied by such an extension in the case of models of **HST**.

Definition 7.1.5. A $\mathtt{st}$-$\in$-structure $\mathbb{H}' = \langle \mathbb{H}'; \in_{\mathbb{H}'}, \mathtt{st}_{\mathbb{H}'} \rangle$ is a *regular extension* of a model $\mathbb{H} = \langle \mathbb{H}; \in_{\mathbb{H}}, \mathtt{st}_{\mathbb{H}} \rangle$ satisfying 7.1.1 if

(1) $\mathbb{H} \subseteq \mathbb{H}'$, $\in_{\mathbb{H}}$ is equal to the restriction $\in_{\mathbb{H}'} \restriction \mathbb{H}$, and $\mathbb{H}$ is an $\in_{\mathbb{H}'}$-transitive subset of $\mathbb{H}'$,

(2) the classes $\mathbb{S}^{(\mathbb{H}')}$, $\mathbb{I}^{(\mathbb{H}')}$ (standard and internal sets in $\mathbb{H}'$) coincide with resp. $\mathbb{S} = \mathbb{S}^{(\mathbb{H})}$, $\mathbb{I} = \mathbb{I}^{(\mathbb{H})}$, and finally

(3) for any $x \in \mathbb{H}' \setminus \mathbb{H}$ we have:
   a) $x = x^{(\mathbb{H}')}$, where $x^{(\mathbb{H}')} = \{y \in \mathbb{H}' : y \in_{\mathbb{H}'} x\}$, and
   b) there is no $z \in \mathbb{H}$ such that $x = z^{(\mathbb{H})} = \{y \in \mathbb{H} : y \in_{\mathbb{H}} z\}$. □

Note that standard and internal sets are postulated to be preserved by this type of extensions — it is not clear whether there exist reasonably useful extensions **not** of this type. (3) implies that any "new" set $x \in \mathbb{H}' \smallsetminus \mathbb{H}$ is a true set, and none of them has the same $\in_{\mathbb{H}'}$-elements as some "old" set.

Lemma 7.1.6. *If $\mathbb{H}'$ is a regular extension of a model $\mathbb{H} \models$ **HST** satisfying 7.1.1 then Extensionality and Regularity over $\mathbb{I}$ are true in $\mathbb{H}'$.*

Proof. To prove Extensionality suppose that $x, x' \in \mathbb{H}'$ have the same $\in_{\mathbb{H}'}$-elements. If each of them belongs to $\mathbb{H}$ then $x = x'$ follows from Extensionality in $\mathbb{H}$. If each of them belongs to $\mathbb{H}' \smallsetminus \mathbb{H}$ then $x = x'$ by 7.1.5(3)a. The mixed case is impossible by 7.1.5(3)b.

If Regularity over $\mathbb{I}$ fails in $\mathbb{H}'$ then by 7.1.5(2) there exists (in the universe) an infinite chain $\{x_n\}_{n\in\mathbb{N}}$ of $x_n \in \mathbb{H}' \smallsetminus \mathbb{I}$ such that $x_{n+1} \in_{\mathbb{H}'} x_n, \forall n$. At least one term x_n belongs to $\mathbb{H}$ as otherwise we would have an $\in$-decreasing chain by 7.1.5(3)a. Then $x_m \in \mathbb{H}$ for any $m \ge n$ by 7.1.5(1). Thus we have (in the universe) a strictly decreasing sequence of $\mathbb{H}$-ordinals — namely the ranks $\mathtt{irk}^{(\mathbb{H})} x_n$ in the sense of $\mathbb{H}$, see § 1.5b, so that $\mathtt{Ord}^{(\mathbb{H})}$ is not well-founded in the universe. But this contradicts 7.1.2(2). □

7.1c Forcing notions and names

We argue under the assumptions of 7.1.1.

The following will be our blanket assumption regarding forcing notions used to obtain generic extensions of $\mathbb{H}$.

Blanket agreement 7.1.7. Let *the forcing notion* $\mathbf{P} \subseteq \mathbb{H} \smallsetminus \mathbb{I}$ be a partially ordered $\mathbb{H}$-*class*, that is both $\mathbf{P}$ and the order $\le_{\mathbf{P}}$ (as a subset of $\mathbf{P} \times \mathbf{P}$ in the universe) are definable in $\mathbb{H}$ by $\mathtt{st}$-$\in$-formulas with parameters in $\mathbb{H}$.

It will be an important special case when $\mathbf{P} \in \mathbb{H}$ – *set-size forcing.* □

Elements of $\mathbf{P}$ are called *(forcing) conditions*. The inequality $p \le_{\mathbf{P}} q$ means that p is a *stronger* condition. We shall systematically shorten $\le_{\mathbf{P}}$ to $\le$ as long as $\mathbf{P}$ remains fixed.

The assumption $\mathbf{P} \subseteq \mathbb{H} \smallsetminus \mathbb{I}$ is not really restrictive because the map $x \mapsto \breve{x} : \mathbb{H} \to \mathbb{H} \smallsetminus \mathbb{I}$ (see 7.1.2(5)) can be applied to define an order isomorphic p. o. set within $\mathbb{H} \smallsetminus \mathbb{I}$. On the other hand this assumption is extremely important since in connection with 7.1.1 it implies that in many issues it does not matter where $\mathbf{P}$ is considered, whether in $\mathbb{H}$ or in the **ZFC** universe, and this resembles, to some extent, the case of $\in$-models of **ZFC**:

Exercise 7.1.8. Prove the following, using 7.1.1, 7.1.2 (and $\mathbf{P} \subseteq \mathbb{H} \smallsetminus \mathbb{I}$):

(i) $\mathbb{H}$-pairs $\langle x, y\rangle^{\mathbb{H}}$ of elements of $\mathbf{P}$ are equal to their true pairs $\langle x, y\rangle$.

(ii) If $\mathbf{P} \in \mathbb{H}$ (and $\mathbf{P} \subseteq \mathbb{H} \smallsetminus \mathbb{I}$) then $\mathbf{P} = \mathbf{P}^{(\mathbb{H})}$, the order $\le_{\mathbf{P}}$ belongs to $\mathbb{H}$ as a set, $\mathbb{H} \models$ "$\le_{\mathbf{P}}$ is a partial order of $\mathbf{P}$", and, for any $p, q \in \mathbf{P}$, we have the equivalence: $\mathbb{H} \models p \le_{\mathbf{P}} q$ iff $p \le_{\mathbf{P}} q$ in the universe. □

Define, in the **ZFC** universe, a set $\mathtt{Nms}(\mathbf{P}) = \bigcup_{\xi<\mathtt{ht}_{\mathbb{H}}} \mathtt{Nms}_\xi(\mathbf{P})$ of *names* for sets in an intended $\mathbf{P}$-generic extension of $\mathbb{H}$ by induction on ξ:

$$\mathtt{Nms}_0 = \mathtt{Nms}_0(\mathbf{P}) = \{\breve{x} : x \in \mathbb{H}\}, \quad \text{where, we recall,} \quad \breve{x} = \langle x, \mathbb{N}\rangle \in \mathbb{H} \setminus \mathbb{I}\ ^3;$$
$$\mathtt{Nms}_\xi(\mathbf{P}) = \{a : \varnothing \neq a \subseteq \mathbf{P} \times \textstyle\bigcup_{\eta<\xi} \mathtt{Nms}_\eta(\mathbf{P})\} \text{ whenever } \xi > 0.$$

(Clearly $\mathtt{Nms}_0(\mathbf{P})$ does not depend on $\mathbf{P}$.)

For $a \in \mathtt{Nms}(\mathbf{P})$, we let $\mathtt{nrk}\, a$, the *name-rank* of a, indicate the least ordinal ξ such that $a \in \mathtt{Nms}_\xi(\mathbf{P})$.

Exercise 7.1.9. (1) Prove that $\mathtt{Nms}_0 \cap \mathtt{Nms}_\xi(\mathbf{P}) = \varnothing$ (any $a \in \mathtt{Nms}_0$ is a pair while any $a \in \mathtt{Nms}_\xi(\mathbf{P})$, $\xi > 0$, is a set of pairs).

(2) Prove, using 7.1.2 (in particular 7.1.2(5)), that the construction of $\mathtt{Nms}_\xi(\mathbf{P})$ is absolute for $\mathbb{H}$, and hence $\breve{x} \in \mathbb{H} \setminus \mathbb{I}$ for any $x \in \mathbb{H}$, $\mathtt{Nms}(\mathbf{P}) \subseteq \mathbb{H} \setminus \mathbb{I}$, $\mathtt{Nms}(\mathbf{P})$ and all $\mathtt{Nms}_\xi(\mathbf{P})$, $\xi < \mathtt{ht}_{\mathbb{H}}$, are classes in $\mathbb{H}$, while the maps $x \mapsto \breve{x}$ and $a \mapsto \mathtt{nrk}\, a$ are $\mathtt{st}$-$\in$-definable in $\mathbb{H}$.

(3) Prove that any $a \in \mathbb{H}$, $\varnothing \neq a \subseteq \mathbf{P} \times \mathtt{Nms}(\mathbf{P})$, belongs to $\mathtt{Nms}(\mathbf{P})$. □

7.1d Adding a set

We continue to argue under the assumptions 7.1.1 and 7.1.7.

The goal of this Subsection is to show how to extend $\mathbb{H}$ by adding a set $\mathbf{G}$ which, generally speaking, does not belong to $\mathbb{H}$. The main issue is that many other sets have to be adjoined to $\mathbb{H}$ together with $\mathbf{G}$ in order that the extension continues to be a reasonable structure (and a model of **HST** with an appropriate choice of $\mathbf{G}$). It turns out that suitable $\mathbf{G}$-interpretations of names in $\mathtt{Nms}(\mathbf{P})$ form a sufficient family of sets to be adjoined.

Suppose that $\mathbf{G} \subseteq \mathbf{P}$ (usually $\mathbf{G} \notin \mathbb{H}$). Define a set $a[\mathbf{G}]$ for each name $a \in \mathtt{Nms}(\mathbf{P})$ by induction on $\mathtt{nrk}\, a$:

1. $a[\mathbf{G}] = x$ for any $a = \breve{x} \in \mathtt{Nms}_0$.
2. Assume that $\mathtt{nrk}\, a > 0$. Then we have two subcases:
   a) if the set $a^{\mathtt{put}}[\mathbf{G}] = \{c[\mathbf{G}] : \exists p \in \mathbf{G}\ (\langle p, c\rangle \in a)\}$ (a *putative* $\mathbf{G}$-*interpretation* of a) is equal to $x^{(\mathbb{H})}$ for some (unique by Extensionality in $\mathbb{H}$) $x \in \mathbb{H}$ then put $a[\mathbf{G}] = x$;
   b) otherwise define $a[\mathbf{G}] = a^{\mathtt{put}}[\mathbf{G}]$ (the *true* $\mathbf{G}$-*interpretation* of a).

Note that even in case 2a $a[\mathbf{G}] = a^{\mathtt{put}}[\mathbf{G}]$ holds provided the unique set $x \in \mathbb{H}$ satisfying $a^{\mathtt{put}}[\mathbf{G}] = x^{(\mathbb{H})}$ belongs to $\mathbb{H} \setminus \mathbb{I}$ — because then $x^{(\mathbb{H})} = x$ by 7.1.1(b). A sufficient condition for this subcase is $a^{\mathtt{put}}[\mathbf{G}] \not\subseteq \mathbb{I}$.

We put $\mathbb{H}[\mathbf{G}] = \{a[\mathbf{G}] : a \in \mathtt{Nms}(\mathbf{P})\}$. Define the *membership* $\in_{\mathbf{G}}$ in $\mathbb{H}[\mathbf{G}]$ as follows: $y \in_{\mathbf{G}} x$ in either of the two following cases:

[3] $\mathbb{N}$ can be replaced by any fixed non-internal set in $\mathbb{H}$ in the defintion of $\breve{x}$; all we need is that $x \mapsto \breve{x}$ is an injection $\mathbb{H} \to \mathbb{H} \setminus \mathbb{I}$.

A) x, y belong to $\mathbb{H}$ and $y \in_{\mathbb{H}} x$,

B) $x \notin \mathbb{H}$ and $y \in x$ in the **ZFC** universe — thus $\in_{\mathbf{G}}$-elements of any $x \in \mathbb{H}[\mathbf{G}] \smallsetminus \mathbb{H}$ and $\in$-elements of x is one and the same.

Define the *standardness* $\mathtt{st}_{\mathbf{G}}$ in $\mathbb{H}[\mathbf{G}]$ so that $\mathtt{st}_{\mathbf{G}}\, x$ iff $x \in \mathbb{H}$ and x is standard in $\mathbb{H}$; thus $\mathtt{st}_{\mathbf{G}}$ coincides with $\mathtt{st}_{\mathbb{H}}$.

This completes the definition of the model $\mathbb{H}[\mathbf{G}] = \langle \mathbb{H}[\mathbf{G}]; \in_{\mathbf{G}}, \mathtt{st}_{\mathbf{G}} \rangle$.

Suppose that $a, b \in \mathtt{Nms}(\mathbf{P})$ and $p \in \mathbf{P}$. Define a preliminary forcing relation, only for atomic formulas of the form $b \in a$, as follows:

$$p \;\mathtt{forc}\; b \in a \quad \text{iff} \quad \begin{cases} \exists y \in x \, (b = \breve{y}) & \text{whenever } a = \breve{x} \in \mathtt{Nms}_0 \,; \\ \exists q \geq p \, (\langle q, b \rangle \in a) & \text{whenever } \mathtt{nrk}\, a > 0 \,. \end{cases}$$

The next lemma explains in more detail how the membership in the extension is organized in terms of $\mathtt{forc}$.

Lemma 7.1.10. *Assume that $a \in \mathtt{Nms}(\mathbf{P})$ while $\mathbf{G} \subseteq \mathbf{P}$ is a is a filter in the sense that $p \in \mathbf{G} \Longrightarrow q \in \mathbf{G}$ whenever $p, q \in \mathbf{P}$ and $q \geq p$. Then for any $y \in \mathbb{H}[\mathbf{G}]$ each of the following conditions* (i), (ii) *is equivalent to $y \in_{\mathbf{G}} a[\mathbf{G}]$:*

(i) a) $y \in x^{(\mathbb{H})} = \{y \in \mathbb{H} : y \in_{\mathbb{H}} x\}$, *provided* $a = \breve{x} \in \mathtt{Nms}_0$,
  b) $y \in a^{\mathtt{put}}[\mathbf{G}]$, *provided* $\mathtt{nrk}\, a > 0$;

(ii) $\exists b \in \mathtt{Nms}(\mathbf{P})\, \exists p \in \mathbf{G}\, (y = b[\mathbf{G}] \wedge p \;\mathtt{forc}\; b \in a)$.

Proof. Since (i) $\Longleftrightarrow$ (ii) is an immediate corollary of the assumption that $\mathbf{G}$ is a filter, we can concentrate on the equivalence $y \in_{\mathbf{G}} a[\mathbf{G}] \Longleftrightarrow$ (i). The only possible counterexample to this equivalence is a name $a \in \mathtt{Nms}(\mathbf{P})$ with $\mathtt{nrk}\, a > 0$ such that the set $x = a^{\mathtt{put}}[\mathbf{G}]$ belongs to $\mathbb{H}$ but $x \neq x^{(\mathbb{H})}$, and hence $x \in \mathbb{I}_0$ by 7.1.1.

It follows from our definitions that $x = a^{\mathtt{put}}[\mathbf{G}]$ is the result of an assembling construction, of the type considered in Section 5.3, which begins with sets in $\mathbb{H}$ and contains at least one step (since $\mathtt{nrk}\, a > 0$) but has a total height $\mathtt{nrk}\, a < \mathtt{ht}_{\mathbb{H}}$. If the initial sets of the construction all belong to $\mathbb{H} \smallsetminus \mathbb{I}_0$ then $x = a(G)$ is a set of the von Neumann rank $\leq \mathtt{ht}_{\mathbb{H}} + \mathtt{ht}_{\mathbb{H}} < \kappa$ by 7.1.1(b), thus $x \notin \mathbb{I}_0$ by 7.1.1, contrary to the above. Thus at least one of the initial sets belongs to $\mathbb{I}_0$. But then the result $x = a(G)$ is a set of the von Neumann rank $> \kappa$ in the universe by 7.1.1(a), and hence $x \notin \mathbb{H}$ by 7.1.1, still a contradiction. □

Corollary 7.1.11. *Suppose that $\mathbf{G} \subseteq \mathbf{P}$. Then $\mathbb{H}[\mathbf{G}]$ is a regular extension of $\mathbb{H}$ satisfying 7.1.2(2) and $\mathtt{ht}_{\mathbb{H}[\mathbf{G}]} = \mathtt{ht}_{\mathbb{H}}$.*

If in addition $\mathbf{P} \in \mathbb{H}$ then $\mathbf{G} \in \mathbb{H}[\mathbf{G}]$. [4]

---

[4] Theorem 7.1.20 below shows that $\mathbb{H}[\mathbf{G}]$ satisfies **HST** for a wide category of sets $\mathbf{G} \subseteq \mathbf{P}$ (generic sets) provided $\mathbf{P}$ itself satisfies certain requirements.

Proof. To prove $\mathbb{H} \subseteq \mathbb{H}[\mathbf{G}]$ note that $\breve{x}[\mathbf{G}] = x$ for any $x \in \mathbb{H}$ and $\mathbf{G} \neq \varnothing$.

To verify the transitivity of $\mathbb{H}$ in $\mathbb{H}[\mathbf{G}]$, assume that $x \in \mathbb{H}$, $y \in \mathbb{H}[\mathbf{G}]$, $y \in_{\mathbf{G}} x$. Then by definition $x = \breve{x}[\mathbf{G}]$ where $\breve{x} \in \mathtt{Nms}_0$, and hence y belongs to the set $x^{(\mathbb{H})}$ by Lemma 7.1.10, that is $y \in \mathbb{H}$ and $y \in_{\mathbb{H}} x$.

If $\mathbf{P} \in \mathbb{H}$ then $\underline{G} = \{\langle p, \breve{p}\rangle : p \in \mathbf{P}\}$ is still a set in $\mathbb{H}$, and moreover $\underline{G} \in \mathtt{Nms}(\mathbf{P})$ by 7.1.9, while on the other hand $\underline{G}[\mathbf{G}] = \mathbf{G}$ for any $\varnothing \neq \mathbf{G} \subseteq \mathbf{P}$! (This argument obviously fails if $\mathbf{P}$ is a proper class in $\mathbb{H}$.)

Prove 7.1.5(3)b. Suppose that $a \in \mathtt{Nms}(\mathbf{P})$ and $a[\mathbf{G}] \in \mathbb{H}[\mathbf{G}] \setminus \mathbb{H}$, and hence $a \notin \mathtt{Nms}_0$. Then $a[\mathbf{G}] \neq z^{(\mathbb{H})}$ for any $z \in \mathbb{H}$ because otherwise we would have $a[\mathbf{G}] = z \in \mathbb{H}$ by definition 2a.

The rest of the corollary is left as a simple **exercise** for the reader. □

We finish with two boundedness-type results. Define, for any $a \in \mathtt{Nms}(\mathbf{P})$,

$$\Delta_a = \begin{cases} \{\breve{y} : y \in x\} & \text{whenever } a = \breve{x} \in \mathtt{Nms}_0\,, \\ \mathtt{ran}\, a = \{b : \exists\, q\, (\langle q, b\rangle \in a)\} & \text{whenever } a \in \mathtt{Nms}(\mathbf{P}) \setminus \mathtt{Nms}_0\,. \end{cases}$$

Exercise 7.1.12. (1) Prove that $\Delta_a \in \mathbb{H}$, $\Delta_a \subseteq \mathtt{Nms}(\mathbf{P})$, and $p \;\mathtt{forc}\; b \in a$ implies that $b \in \Delta_a$ and either $a, b \in \mathtt{Nms}_0$ or $\mathtt{nrk}\, b < \mathtt{nrk}\, a$.

(2) Prove that if $a \in \mathtt{Nms}(\mathbf{P})$, $\mathbf{G} \subseteq \mathbf{P}$, and $x \in_{\mathbf{G}} a[\mathbf{G}]$ then $x = b[\mathbf{G}]$ for some $b \in \mathtt{Nms}(\mathbf{P})$ satisfying $b \in_{\mathbb{H}} \Delta_a$. □

Lemma 7.1.13. *For any name $a \in \mathtt{Nms}(\mathbf{P})$ there is a set $s(a) \in \mathbb{S}$ such that we have $y \in_{\mathbb{H}} s(a)$ whenever $y \in \mathbb{I}$, $\mathbf{G} \subseteq \mathbf{P}$, and $y \in_{\mathbf{G}} a[\mathbf{G}]$.*

Proof. Define $s(a)$ in $\mathbb{H}$ by induction on $\mathtt{nrk}\, a$. If $a = \breve{x} \in \mathtt{Nms}_0$ then $x \cap \mathbb{I}$ can be covered by a standard set by the axiom of Boundedness in $\mathbb{H}$, and hence there is a standard $*$-ordinal ξ such that $x \cap \mathbb{I} \subseteq \mathbf{V}_\xi$ (the ξ-th level of the von Neumann hierarchy in $\mathbb{I}$). Put $s(a) = \mathbf{V}_\xi$, where ξ is the least standard $*$-ordinal of this sort. By definition any $y \in \mathbb{I}$ with $y \in_{\mathbf{G}} a[\mathbf{G}] = x$ satisfies $y \in s(a)$ in $\mathbb{H}$.

If $\mathtt{nrk}\, a > 0$ and $s(b) \in \mathbb{S}$ is defined for all $b \in \Delta_a$ then put, in $\mathbb{H}$, $s(a) = \mathscr{P}_{\mathtt{int}}(\mathbf{V}_\xi)$ (the power set in $\mathbb{I}$; $s(a)$ is standard together with ξ and the set $\mathbf{V}_\xi$ by Transfer) where ξ the least *standard* $*$-ordinal with $\bigcup_{b \in \Delta_a} s(b) \subseteq \mathbf{V}_\xi$. Suppose that $y \in \mathbb{I}$ and $y \in_{\mathbf{G}} a[\mathbf{G}]$. Then, in $\mathbb{H}$, $y = b[\mathbf{G}]$ for some $b \in \Delta_a$ by 7.1.12(2), and hence $s(b) \subseteq \mathbf{V}_\xi$, then $y \cap \mathbb{I} \subseteq \mathbf{V}_\xi$ by the inductive assumption, $y \subseteq \mathbf{V}_\xi$ as $\mathbb{I}$ is transitive in **HST**, and finally $y \in s(a) = \mathscr{P}_{\mathtt{int}}(\mathbf{V}_\xi)$. □

7.1e Forcing relation

We continue to argue under the assumptions of 7.1.1, 7.1.7.

Definition 7.1.14. A $\mathbf{P}$-*forcing relation* is any relation $p \Vdash \Phi$ whose arguments are conditions $p \in \mathbf{P}$ and closed $\mathtt{st}$-$\in$-formulas Φ with parameters in $\mathtt{Nms}(\mathbf{P})$, satisfying the following requirements F1 – F7:

F1: For any $x, y \in \mathbb{H}$: $p \Vdash \breve{y} = \breve{x}$ iff $y = x$, and
$p \Vdash \breve{y} \in \breve{x}$ iff $y \in x$.

F2: $p \Vdash a = b$ iff for each condition $q \le p$ and every name $c \in \mathtt{Nms}(\mathbf{P})$:
1) $q \;\mathtt{forc}\; c \in a$ implies $q \Vdash c \in b$, and
2) $q \;\mathtt{forc}\; c \in b$ implies $q \Vdash c \in a$.

F3: $p \Vdash b \in a$ iff for each condition $q \le p$ there exist $r \le q$ and a name $c \in \mathtt{Nms}(\mathbf{P})$ such that $r \;\mathtt{forc}\; c \in a$ and $r \Vdash b = c$.

F4: $p \Vdash \mathtt{st}\, a$ iff $\forall q \le p\, \exists r \le q\, \exists^{\mathtt{st}} s\, (r \Vdash a = \breve{s})$.

F5: $p \Vdash \neg \Phi$ iff none of stronger forcing conditions $q \le p$ forces Φ.

F6: $p \Vdash (\Phi \wedge \Psi)$ iff $p \Vdash \Phi$ and $p \Vdash \Psi$.

F7: $p \Vdash \forall x\, \Phi(x)$ iff $p \Vdash \Phi(a)$ for every name $a \in \mathtt{Nms}(\mathbf{P})$.

In the whole scheme F1 – F7 p, q, r are forcing conditions in $\mathbf{P}$. □

F1 obviously implies both F2 and F3, and hence there is no need to stress that at least one of the names a, b does not belong to $\mathtt{Nms}_0$ in F2, F3.

Items F4 – F7 handle st and non–atomic formulas. It is assumed that other logic connectives are combinations of $\neg$, $\wedge$, $\forall$.

Theorem 7.1.15. *Under the assumptions of 7.1.1, 7.1.7, there exists a unique* $\mathbf{P}$*-forcing notion, denoted by* $\Vdash_{\mathbf{P}}$ *henceforth. This forcing notion satisfies the following definability requirements, in which* $\varphi(x_1, ..., x_n)$ *is an arbitrary* st-$\in$*-formula:*

(i) *If* $\mathbf{P} \in \mathbb{H}$ (set-size forcing) *then the relation* $p \Vdash_{\mathbf{P}} \varphi(a_1, ..., a_n)$, *with* $p, a_1, ..., a_n$ *as arguments, is* st-$\in$*-definable* (with parameters in $\mathbb{H}$ allowed, including $\mathbf{P}$ as a parameter, of course) *in* $\mathbb{H}$, *i.e. the set*

$$F_{\mathbf{P}} = \{\langle p, a_1, ..., a_n\rangle : p \in \mathbf{P} \wedge a_1, ..., a_n \in \mathtt{Nms}(\mathbf{P}) \wedge p \Vdash_{\mathbf{P}} \varphi(a_1, ..., a_n)\}$$

is is st-$\in$*-definable in* $\mathbb{H}$ (with parameters in $\mathbb{H}$ allowed).

(ii) *Moreover, the relation* $p \Vdash_{\mathbf{P}} \varphi(a_1, ..., a_n)$ *with the arguments* p, $\mathbf{P}$, *and* $a_1, ..., a_n$, *is also* st-$\in$*-definable in* $\mathbb{H}$ (parameters allowed), *i.e. the set*

$$\{\langle \mathbf{P}, p, a_1, ..., a_n\rangle : \mathbf{P} \in \mathbb{H} \text{ is a p. o. set satisfying 7.1.7} \wedge \\ p \in \mathbf{P} \wedge a_1, ..., a_n \in \mathtt{Nms}(\mathbf{P}) \wedge p \Vdash_{\mathbf{P}} \varphi(a_1, ..., a_n)\}$$

is st-$\in$*-definable in* $\mathbb{H}$ (parameters allowed).

The definability of the forcing relation $\le_{\mathbf{P}}$ in the case when the forcing notion $\mathbf{P}$ is a proper class in the ground model is too complicated an issue to be considered here. In the only example of such a "class" forcing, studied below in Section 7.3, the definability will be obtained by reduction to set-size subforcings.

Proof. To prove the existence and uniqueness we have to show that F1, F2, F3 form a legitimate scheme of well-founded induction. Let Φ be the collection of all formulas of the form $b = a$ and $b \in a$, where $a, b \in \mathtt{Nms}(\mathbf{P})$. For any $\varphi \in \Phi$ let R_φ indicate the collection of all formulas $\psi \in \Phi$ to which the definition of $p \Vdash \varphi$ can directly refer according to F2 and F3 for different $p \in \mathbf{P}$. To be more precise,

1) if φ is $\breve{y} = \breve{x}$ or $\breve{y} \in \breve{x}$, where $x, y \in \mathbb{H}$, then $R_\varphi = \varnothing$;
2) if φ is $a = b$ where $a, b \in \mathtt{Nms}(\mathbf{P})$ and at least one of a, b does not belong to $\mathtt{Nms}_0$ then R_φ consists of all formulas of the form $c \in a$ where $c \in \Delta_a$ and all formulas $c \in b$ where $c \in \Delta_b$ (see 7.1.12 on Δ_a, Δ_b);
3) if φ is $b \in a$ where $a, b \in \mathtt{Nms}(\mathbf{P})$ and at least one of a, b does not belong to $\mathtt{Nms}_0$ then R_φ consists of all formulas $b = c$ where $c \in \Delta_a$.

Define a partial order $\prec$ on Φ as follows: $\varphi \prec \psi$ iff there exists a finite sequence $\varphi = \varphi_0, \varphi_1, ..., \varphi_n = \psi$ $(n \geq 1)$ such that $\varphi_k \in R_{\varphi_{k+1}}$ for all $k < n$.

Lemma 7.1.16. *$\prec$ is a well-founded partial order on Φ.*

Proof. Let, on the contrary, $\varphi_0 \succ \varphi_1 \succ \varphi_2 \succ \dots$ be an infinite decreasing chain in Φ, so that $\varphi_{n+1} \in R_{\varphi_n}$, $\forall n$. Assume that φ_0 has the form $a = b$. (Otherwise φ_1 has such a form.) Then φ_{2n} is $a_n = b_n$ for all n, and moreover, by definition, either $a_{n+1}, b_{n+1} \in \Delta_{a_n}$ or $a_{n+1}, b_{n+1} \in \Delta_{b_n}$ for any n. Thus either some φ_{2n} is of the form $\breve{x} = \breve{y}$ – and then the chain breaks because formulas of the form $\breve{x} = \breve{y}$ and $\breve{x} \in \breve{y}$ are $\prec$-minimal in Φ, or we have an infinite $\mathtt{nrk}$-decreasing chain of names by 7.1.12, also a contradiction. □

Thus F1, F2, F3 give a legitimate definition by well-founded induction in the **ZFC** universe (as explained in Remark 1.1.7): that is sets of the form $\mathbf{P}_\varphi = \{p \in \mathbf{P} : p \Vdash \varphi\}$ are defined by well-founded induction on $\varphi \in \Phi$. We shall call this argument *induction on the $\prec$-rank*. Induction on the $\prec$-rank means that to prove something for all formulas of the form $a = b$ and $b \in a$, where $a, b \in \mathtt{Ind}$, we first establish the result in case F1 and then show that the steps F2 and F3 preserve the result.

Now to prove (ii) note that the same argument based on Lemma 7.1.16 enables us to define $\Vdash_{\mathbf{P}}$ for formulas of the form $a \in b$ and $a = b$ inside $\mathbb{H}$, using $\mathbf{P}$ as a parameter of the definition. The extension to more complicated formulas is obvious. Claim (i) is an easy corollary.

□ (*Theorem 7.1.15*)

We finish here with a simple technical lemma. It is still assumed that $\mathbf{P} \subseteq \mathbb{H}$ satisfies 7.1.7. Let $\Vdash$ denote $\Vdash_{\mathbf{P}}$.

Lemma 7.1.17. (i) *If $a, b \in \mathtt{Nms}(\mathbf{P})$ and p $\mathtt{forc}$ $b \in a$ then $p \Vdash b \in a$.*

(ii) *If $a, b \in \mathtt{Nms}(\mathbf{P})$, p $\mathtt{forc}$ $b \in a$, and $q \leq p$ then q $\mathtt{forc}$ $b \in a$.*

(iii) *If Φ is a closed* st-$\in$*-formula with parameters in* $\mathtt{Nms}(\mathbf{P})$, *and* $p, q \in \mathbf{P}$ *satisfy* $q \le p$ *then* $p \Vdash \Phi$ *implies* $q \Vdash \Phi$.

(iv) *If Φ is a closed* st-$\in$*-formula with parameters in* $\mathtt{Nms}(\mathbf{P})$, *and* $p \in \mathbf{P}$, $\neg\, p \Vdash \Phi$, *then there is a stronger condition* $q \le p$ *such that* $q \Vdash \neg\, \Phi$.

Proof. (i), (ii) are quite obvious. Claims (iii) and (iv) can be easily proved by induction on the complexity of Φ. Consider (iv), a more complicated example. The induction steps are entirely similar to the well-known case of **ZFC** models, and hence we concentrate on atomic formulas.

We argue by induction on the $\prec$-rank.

Assume $\neg\, p \Vdash b \in a$. By F3 there is a condition $q \le p$ such that:

$$\neg\, \exists r \le q\, \exists c\, (r \;\mathtt{forc}\; c \in a \wedge r \Vdash b = c).$$

To see that $q \Vdash \neg\, b \in a$, suppose towards the contrary that a condition $q' \le q$ satisfies $q' \Vdash b \in a$. Then, by F3, $r \;\mathtt{forc}\; c \in a$ and $r \Vdash b = c$ for a condition $r \le q'$ and a name c, contradiction.

Assume that $\neg\, p \Vdash a = b$. Then, by F2, there exist a condition $q' \le p$ and a name c such that for instance $q' \;\mathtt{forc}\; c \in a$ but $\neg\, q' \Vdash c \in b$. It follows by the inductive hypothesis that there is a condition $q \le q'$ with $q \Vdash \neg\, c \in b$. We prove that $q \Vdash a \ne b$. Let on the contrary a condition $r \le q$ force $a = b$. As $r \;\mathtt{forc}\; c \in a$ by (ii), we have $r \Vdash c \in b$, contradiction.

A similar reasoning proves the result for formulas of the form $\mathtt{st}\, a$. □

7.1f Generic extensions and the truth lemma

We continue to consider a p. o. $\mathbb{H}$-class $\mathbf{P} \subseteq \mathbb{H} \setminus \mathbb{I}$ under the assumptions 7.1.1.

It is not too difficult to define sets $\mathbf{G}$ that code something explicitly incompatible with $\mathbb{H}[\mathbf{G}]$ being a model of **HST**. On the other hand there is a special type of extensions $\mathbb{H}[\mathbf{G}]$ in which **HST** still holds — *generic models*.

Recall that a set $\mathbf{G} \subseteq \mathbf{P}$ is $\mathbf{P}$*-generic over* $\mathbb{H}$ if we have

i) for any $p, q \in \mathbf{G}$ there is $r \in \mathbf{G}$ with $r \le p$ and $r \le q$,

ii) for any $p \in \mathbf{G}$ and $q \in \mathbf{P}$, $q \ge p$ we have $q \in \mathbf{G}$, and

iii) $\mathbf{G} \cap D \ne \varnothing$ for any open dense [5] set $D \subseteq \mathbf{P}$ (D can be a class in $\mathbb{H}$) st-$\in$-definable in $\mathbb{H}$ (with parameters from $\mathbb{H}$).

Theorem 7.1.18 just below ties the forcing relation $\Vdash \;=\; \Vdash_{\mathbf{P}}$ with the truth in generic extensions of $\mathbb{H}$.

If Φ is a st-$\in$-formula with parameters in $\mathtt{Nms}(\mathbf{P})$ and $\mathbf{G} \subseteq \mathbf{P}$ then $\Phi[\mathbf{G}]$ indicates the formula obtained by the substitution of $a[\mathbf{G}]$ for any name $a \in \mathtt{Nms}(\mathbf{P})$ occurring in Φ. Obviously $\Phi[\mathbf{G}]$ is a st-$\in$-formula with parameters in $\mathbb{H}[\mathbf{G}]$.

[5] See § 5.5d on dense and open dense sets.

Theorem 7.1.18 (Truth lemma). *Suppose that $\Vdash_{\mathbf{P}}$ is definable in* $\mathbb{H}$ *in the sense of Theorem 7.1.15*(i). [6] *Let* $\mathbf{G} \subseteq \mathbf{P}$ *be* $\mathbf{P}$*-generic over* $\mathbb{H}$ *and* Φ *a formula as above. Then* $\Phi[\mathbf{G}]$ *is true in* $\mathbb{H}[\mathbf{G}]$ *iff* $\exists p \in \mathbf{G}\,(p \Vdash_{\mathbf{P}} \Phi)$.

Proof. Let us write $\Vdash$ instead of $\Vdash_{\mathbf{P}}$. Prove the result for atomic formulas $a = b$ and $b \in a$ by induction on the $\prec$-rank. If $a, b \in \mathtt{Nms}_0$ (item F1) then the result is obvious.

Step F2. Suppose that none of $p \in \mathbf{G}$ forces $a = b$. Note that the sets

$$D_+ = \{q \in \mathbf{P} : q \Vdash a = b\} \quad \text{and} \quad D_- = \{q \in \mathbf{P} : q \Vdash \neg\, a = b\}$$

are $\mathbb{H}$-classes by the assumption of definability of $\Vdash$, and hence so is $D = D_+ \cup D_-$. Moreover D is dense in $\mathbf{P}$ by Lemma 7.1.17(iv). Thus by the genericity of $\mathbf{G}$ there exists a condition $q \in \mathbf{G} \cap D$. We observe that $q \in D_-$ since $D_+ \cap \mathbf{G} = \varnothing$. By F2 this means that, say, $q\,\mathtt{forc}\,c \in a$ but $q \Vdash c \notin b$ for some $c \in \mathtt{Nms}(\mathbf{P})$. Then $c[\mathbf{G}] \in_{\mathbf{G}} a[\mathbf{G}]$ by Lemma 7.1.10 but $c[\mathbf{G}] \notin_G b[\mathbf{G}]$ by the inductive hypothesis.

Suppose now that $a[\mathbf{G}] \neq b[\mathbf{G}]$. Then, since $\mathbb{H}[\mathbf{G}]$ satisfies Extensionality, there is a name $c \in \mathtt{Nms}(\mathbf{P})$ such that, say, $c[\mathbf{G}] \in_{\mathbf{G}} a[\mathbf{G}]$ but $c[\mathbf{G}] \notin_G b[\mathbf{G}]$. By the inductive hypothesis and Lemma 7.1.10 there exist: a condition $p \in \mathbf{G}$ and a name c' such that $p\ \mathtt{forc}\ c' \in a$ but $p \Vdash c' \notin b$. Then $p \Vdash a \neq b$ as otherwise we would have a condition $q \leq p$ which forces $a = b$, immediately getting a contradiction.

Step F3. Let a condition $p \in \mathbf{G}$ force $b \in a$. As above, it follows from the genericity of $\mathbf{G}$ there is a condition $r \in \mathbf{G}$ such that $r\ \mathtt{forc}\ c \in a$ and $r \Vdash c = b$ for a name $c \in \mathtt{Nms}(\mathbf{P})$. Then $c[\mathbf{G}] \in_{\mathbf{G}} a[\mathbf{G}]$ by definition and $c[\mathbf{G}] = b[\mathbf{G}]$ by the inductive hypothesis. To prove the converse assume that $b[\mathbf{G}] \in_{\mathbf{G}} a[\mathbf{G}]$. Lemma 7.1.10 yields a condition $p \in \mathbf{G}$ and a name $b' \in \mathtt{Nms}(\mathbf{P})$ such that $b'[\mathbf{G}] = b[\mathbf{G}]$ and $p\ \mathtt{forc}\ b' \in a$. We can assume, by the inductive hypothesis, that $p \Vdash b = b'$. Then $p \Vdash b \in a$ by definition.

Formulas of the form $\mathtt{st}\,a$ are treated similarly.

The proof for non-atomic formulas proceeds by induction on the complexity of the formula involved, entirely as in the **ZFC** case. □

7.1g The extension models HST

In some dissimilarity with the **ZFC** case, it is perhaps not true that a generic extension $\mathbb{H}[\mathbf{G}]$ models **HST** independently of the choice of the notion of forcing $\mathbf{P}$. To guarantee Standardization in the extension, "old" sets of standard size cannot contain "new" subsets. Standard size distributivity provides a sufficient condition.

Definition 7.1.19 (HST). A p. o. set $\mathbf{P}$ is *standard size closed* iff it is κ-closed for every cardinal κ, that is any decreasing κ-sequence of elements of

[6] This holds for instance in the case when $\mathbf{P} \in \mathbb{H}$ by Theorem 7.1.15(i).

$\mathbf{P}$ has a lower bound in $\mathbf{P}$. (Compare with Definition 5.5.11.) A p. o. set $\mathbf{P}$ is *standard size distributive* iff it is κ-distributive for any cardinal κ. □

The next theorem says that set-generic extensions preserve **HST** assuming the forcing is standard size distributive. Similarly to the **ZFC** case, class-generic extensions generally speaking don't preserve **HST** unless the forcing notion of proper class size has a special structure. An example of this kind will be defined and studied in Section 7.3.

Theorem 7.1.20. *Under the assumptions 7.1.1, let $\mathbf{P} \in \mathbb{H}$, $\mathbf{P} \subseteq \mathbb{H} \setminus \mathbb{I}$ be a p. o. set standard size distributive in $\mathbb{H}$, and $\mathbf{G} \subseteq \mathbf{P}$ be $\mathbf{P}$-generic over $\mathbb{H}$. Then $\mathbb{H}[\mathbf{G}]$ is a model of* **HST**, *a regular extension of $\mathbb{H}$ containing $\mathbf{G}$, and*

(i) *it is true in $\mathbb{H}[\mathbf{G}]$ that $\mathbf{G}$ is $\mathbf{P}$-generic over $\mathbb{H}$;*

(ii) *if $X \in \mathbb{H}[\mathbf{G}] \setminus \mathbb{H}$, $X \subseteq \mathbb{H}$, then $\mathbb{H}[\mathbf{G}] \models X$ is not a set of standard size.* [7]

Proof. The regularity and $\mathbf{G} \in \mathbb{H}[\mathbf{G}]$ hold by Corollary 7.1.11. Moreover, since $\mathbb{H}$ is an $\in_{\mathbf{G}}$-transitive subset of $\mathbb{H}[\mathbf{G}]$ and $\in_{\mathbb{H}}$ coincides with $\in_{\mathbf{G}} \restriction \mathbb{H}$ by the regularity, $\mathbf{G}$ remains $\mathbf{P}$-generic over $\mathbb{H}$ in $\mathbb{H}[\mathbf{G}]$, that is (i).

Let us check **HST** axioms in $\mathbb{H}[\mathbf{G}]$.

Separation. Suppose that $X \in \mathtt{Nms}(\mathbf{P})$ and $\Phi(x)$ is a $\mathtt{st}$-$\in$-formula with parameters in $\mathtt{Nms}(\mathbf{P})$. We have to find a name $Y \in \mathtt{Nms}(\mathbf{P})$ satisfying $Y[\mathbf{G}] = \{x \in X[\mathbf{G}] : \Phi[\mathbf{G}](x)\}$ in $\mathbb{H}[\mathbf{G}]$. It follows from 7.1.12 that all elements of $X[\mathbf{G}]$ in $\mathbb{H}[\mathbf{G}]$ are of the form $a[\mathbf{G}]$ where a belongs to the set of names $A = \Delta_X \in \mathbb{H}$. We claim that

$$Y = \{\langle p, a\rangle \in \mathbf{P} \times A : p \Vdash a \in X \wedge \Phi(a)\}$$

is a name required. ($A \in \mathtt{Nms}(\mathbf{P})$ by 7.1.9.) Indeed let $x \in_{\mathbf{G}} Y[\mathbf{G}]$. By Lemma 7.1.10 there is a name $a \in A$ and a condition $p \in \mathbf{G}$ such that $x = a[\mathbf{G}]$ and $\langle p, a\rangle \in Y$. Theorem 7.1.18 implies that both $x \in X[\mathbf{G}]$ and $\Phi[\mathbf{G}](x)$ hold in $\mathbb{H}[\mathbf{G}]$ as required. Conversely if $x \in \mathbb{H}[\mathbf{G}]$ and $x \in X[\mathbf{G}] \wedge \Phi[\mathbf{G}](x)$ in $\mathbb{H}[\mathbf{G}]$ then $x = a[\mathbf{G}]$ for some $a \in A$, and hence by Theorem 7.1.18 there is a condition $p \in \mathbf{G}$ forcing $a \in X \wedge \Phi(a)$. Thus $\langle p, a\rangle \in Y$ and hence $x = a[\mathbf{G}] \in Y[\mathbf{G}]$.

Collection. Suppose that $X \in \mathtt{Nms}(\mathbf{P})$ and $\Phi(x, y)$ is a formula with parameters in $\mathtt{Nms}(\mathbf{P})$. We have to find a name $Y \in \mathtt{Nms}(\mathbf{P})$ such that

$$\text{in } \mathbb{H}[\mathbf{G}]: \quad \forall x \in X[\mathbf{G}]\, (\exists y\, \Phi[\mathbf{G}](x, y) \Longrightarrow \exists y \in Y[\mathbf{G}]\, \Phi[\mathbf{G}](x, y)) . \qquad (*)$$

Put $A = \Delta_X$ as above. Applying Collection, we obtain a set of names $B \subseteq \mathtt{Nms}(\mathbf{P})$ such that for any $a \in A$ and any $p \in \mathbf{P}$, if $p \Vdash \exists y\, \Phi(a, y)$ then there is a name $b \in B$ and a stronger condition $q \leq p$ with $q \Vdash \Phi(a, b)$. In other words the set D_a of all $p \in \mathbf{P}$ such that either $p \Vdash \neg \exists y\, \Phi(a, y)$ or

[7] That standard size distributive forcings do not add new sets of standard size has well known analogies in **ZFC** forcing.

$\exists\, b \in B\, (p \Vdash \Phi(a,b))$ is dense in $\mathbf{P}$. Define $Y = \mathbf{P} \times B$; then $Y \in \mathtt{Nms}(\mathbf{P})$ and $b[\mathbf{G}] \in_{\mathbf{G}} Y[\mathbf{G}]$ for any $b \in B$.

We claim that the name Y satisfies $(*)$. Indeed consider any $x \in \mathbb{H}[\mathbf{G}]$ such that $x \in_{\mathbf{G}} X[\mathbf{G}]$. Then $x = a[\mathbf{G}]$ for some $a \in A$ (see above). Suppose that $\exists\, y\, \Phi[\mathbf{G}](x,y)$ is true in $\mathbb{H}[\mathbf{G}]$. Then by Theorem 7.1.18 there is a condition $p \in D_a$ forcing $a \in X \wedge \exists\, y\, \Phi(a,y)$. Then $p \Vdash \Phi(a,b)$ for some $b \in B$ by the definition of D_a, and hence $\Phi[\mathbf{G}](x,y)$ holds in $\mathbb{H}[\mathbf{G}]$, where $y = b[\mathbf{G}]$. Finally $y = b[\mathbf{G}] \in_{\mathbf{G}} Y[\mathbf{G}]$, see above.

The remaining **HST** axioms in § 1.1b are verified the same way as for generic extensions of **ZFC** models. (Extensionality and Regularity over $\mathbb{I}$ hold in $\mathbb{H}[\mathbf{G}]$ by Corollary 7.1.11 ang Lemma 7.1.6.) The **HST** axioms in § 1.1c, except for Standardization, are automatically inherited by $\mathbb{H}[\mathbf{G}]$ because the standard and internal sets in $\mathbb{H}[\mathbf{G}]$ are the same as in $\mathbb{H}$ and internal sets do not acquire new elements by Corollary 7.1.11.

Now we prove claim (ii) of the theorem. First of all note that by the definition of sets of standard size it suffices to consider in (ii) only sets X which themselves are functions with $\mathtt{dom}\, X \subseteq \mathbb{S}$ (and $\mathtt{ran}\, X \subseteq \mathbb{H}$). Thus assume that $f \in \mathtt{Nms}(\mathbf{P})$ and $X = f[\mathbf{G}]$ is such a function. Lemma 7.1.13 implies the existence of a set $U = \mathtt{dom}\, s(f) \in \mathbb{S}$ such that $\mathtt{dom}\, f[\mathbf{G}] \cap \mathbb{S} \subseteq U$ in $\mathbb{H}[\mathbf{G}]$. Since $\mathbf{P}$ is standard size distributive in $\mathbb{H}$, we got density of the set D of all conditions that decide the value $f(\breve{x})$ for any standard $x \in U$, i.e. force $f(\breve{x}) = \breve{y}$ for some $y = y_x \in \mathbb{H}$ or force that $f(\breve{x})$ is meaningless or does not belong to $\mathbb{H}$, in $\mathbf{P}$. Thus $\mathbf{G}$ contains a condition $p \in D$. Now $f[\mathbf{G}] = \{\langle x, y_x\rangle : x \in U \wedge y_x \text{ is defined}\}$ is a set in $\mathbb{H}$, a contradiction to the assumption that $X = f[\mathbf{G}] \notin \mathbb{H}$.

Standardization immediately follows from (ii) and Standardization in $\mathbb{H}$.

It remains to consider the axioms of § 1.1f on sets of standard size.

Standard Size Choice. The problem can be reduced to the following form. Suppose that $U \in \mathbb{S}$, $P \in \mathtt{Nms}(\mathbf{P})$, and $P[\mathbf{G}]$ is a set of pairs in $\mathbb{H}[\mathbf{G}]$. Find a name $F \in \mathtt{Nms}(\mathbf{P})$ such that the following is true in $\mathbb{H}[\mathbf{G}]$:

"$F[\mathbf{G}]$ is a function defined on $U \cap \mathbb{S}$ and satisfying

$$\exists\, y\, P[\mathbf{G}](x,y) \Longrightarrow P[\mathbf{G}](x, F[\mathbf{G}](x)) \text{ for each standard } x \in U\text{"}.$$

Arguing as above in the proof of (ii) and using Standard Size Choice in $\mathbb{H}$, we obtain a condition $p \in \mathbf{G}$ and a function $f \in \mathbb{H}$, $f : U \cap \mathbb{S} \to \mathtt{Nms}(\mathbf{P})$, such that, for every standard $x \in U$, either p forces $\neg \exists\, y\, P(\breve{x}, y)$ or p forces $P(\breve{x}, y_x)$ where $y_x = f(x) \in \mathtt{Nms}(\mathbf{P})$. One easily converts f to a name F as required.

Dependent Choice can be verified by a similar reduction to DC in $\mathbb{H}$.

Saturation. Using the same argument, one can prove that each family of internal sets in $\mathbb{H}[\mathbf{G}]$, of standard size in $\mathbb{H}[\mathbf{G}]$, already belongs to $\mathbb{H}$. □

7.2 Applications: collapse maps and isomorphisms

We consider here two applications of forcing for **HST** models. The first and easier of them is the extension of a given model $\mathbb{H}$ of **HST** by a generic bijection between two fixed infinite internal sets in $\mathbb{H}$. As a consequence, we prove that the existence of a pair of internal $*$-infinite sets of different $*$-cardinality, equinumerous [8] in the external universe, is consistent with **HST**. (This existence statement fails in $\mathbb{L}[\mathbb{I}]$, the universe of all sets constructible over $\mathbb{I}$, by Theorem 5.5.8.) The other application is a generic isomorphism between a pair of elementarily equivalent internally presented structures.

7.2a Making two internal sets equinumerous

The following theorem is the key ingredient.

Theorem 7.2.1. *Let $\mathbb{H}$ be a countable model of* **HST** *satisfying 7.1.1 and "$\mathbb{H} = \mathbb{L}[\mathbb{I}]$" [9] and let A, B be infinite [10] internal sets in $\mathbb{H}$. There is a generic extension $\mathbb{H}[\mathbf{G}] \models$* **HST** *of $\mathbb{H}$ such that A, B are equinumerous in $\mathbb{H}[\mathbf{G}]$.*

Proof. *Arguing in* $\mathbb{H}$, we define $\mathbb{P} = \mathbb{P}_{AB}$ to be the set of all internal partial bijections $A \to B$ with co-infinite $\mathtt{dom}$ and $\mathtt{ran}$. In other words $p \in \mathbb{P}$ iff p is an internal bijection, $\mathtt{dom}\, p \subseteq A$, $\mathtt{ran}\, p \subseteq B$ and the sets $A \setminus \mathtt{dom}\, p$, $B \setminus \mathtt{ran}\, p$ are infinite (perhaps $*$-finite). We put $q \le p$ (q is stronger than p) iff $p \subseteq q$. Note that $\mathbb{P}$ is $\mathbf{\Pi}^{\mathrm{ss}}_1$ (in $\mathbb{H}$): indeed we have $\mathbb{P} = \bigcup_{n \in \mathbb{N}} Q_n$, where each Q_n is the internal set of all internal partial bijections $A \to B$ such that either of the sets $A \setminus \mathtt{dom}\, p$, $B \setminus \mathtt{ran}\, p$ contains at least n elements.

We claim that $\mathbb{P}$ is standard size closed (in $\mathbb{H}$). Indeed if p_ξ, $\xi < \lambda$, is a $\subseteq$-decreasing sequence of conditions, $\lambda < \mathtt{ht}_{\mathbb{H}}$ being a limit ordinal, then apply Theorem 1.4.2(i) to the standard size family of nested $\mathbf{\Pi}^{\mathrm{ss}}_1$ sets $P_\xi = \{p \in \mathbb{P} : p \subseteq p_\xi\}$, $\xi < \lambda$. It follows that $\mathbb{P}$ is standard size distributive (in $\mathbb{H}$) by Theorem 5.5.12 — this is the only point where "$\mathbb{H} = \mathbb{L}[\mathbb{I}]$" is used.

The set $\mathbb{P}$ cannot be used as a forcing notion because it does not satisfy our blanket assumption 7.1.7 to consist of non-internal elements in $\mathbb{H}$; in fact on the contrary all $\in_{\mathbb{H}}$-elements of $\mathbb{P}$ belong to $\mathbb{I}$. But this is easy to fix.

Let, in $\mathbb{H}$, $\mathbf{P} = \{\breve{p} : p \in \mathbb{P}\}$. Recall that $\breve{x} = \langle x, \mathbb{N}\rangle \in \mathbb{H} \setminus \mathbb{I}$ for any $x \in \mathbb{H}$ by 7.1.2(5), and hence $\mathbf{P} \subseteq \mathbb{H} \setminus \mathbb{I}$. Order $\mathbf{P}$ accordingly: $\breve{p} \le_{\mathbf{P}} \breve{q}$ iff $p \le_{\mathbb{P}} q$. This is a definition in $\mathbb{H}$, but it easily follows from 7.1.8 that, in the **ZFC** universe, still $\mathbf{P} = \{\breve{p} : p \in_{\mathbb{H}} \mathbb{P}\}$ and $\breve{p} \le_{\mathbf{P}} \breve{q}$ iff $\mathbb{H} \models p \le_{\mathbb{P}} q$. However $\mathbf{P}$ as a forcing notion satisfies 7.1.7 because $\mathbf{P} \subseteq \mathbb{H} \setminus \mathbb{I}$. In addition, $\mathbf{P}$ is order isomorphic to $\mathbb{P}$ in $\mathbb{H}$, thus $\mathbf{P}$ is standard size distributive in $\mathbb{H}$.

---

[8] Sets X, Y are called equinumerous if there exists a bijection $f : X \xrightarrow{\text{onto}} Y$.

[9] "$\mathbb{H} = \mathbb{L}[\mathbb{I}]$" says that all sets belong to $\mathbb{L}[\mathbb{I}]$, see § 5.5c.

[10] Note that an infinite internal set can be $*$-finite, *i.e.* hyperfinite. We recall that $*$-notions are those related to the internal universe $\mathbb{I}$ in **HST**. Thus a $*$-infinite set is any set $X \in \mathbb{I}$ such that it is true in $\mathbb{I}$ that X is infinite.

By the countability of $\mathbb{H}$ there is a $\mathbf{P}$-generic, over $\mathbb{H}$, set $\mathbf{G} \subseteq \mathbf{P}$. We obtain a generic extension $\mathbb{H}[\mathbf{G}] \models \mathbf{HST}$ of $\mathbb{H}$ containing $\mathbf{G}$ by Theorem 7.1.20. Prove that A and B are equinumerous in $\mathbb{H}[\mathbf{G}]$.

Note that $\mathbb{P}$, $\mathbf{P}$ remain order isomorphic in $\mathbb{H}[\mathbf{G}]$, by means of a map that belongs to $\mathbb{H}$, because $\mathbb{H}[\mathbf{G}]$ is a regular extension of $\mathbb{H}$ by Corollary 7.1.11. It follows that, in $\mathbb{H}[\mathbf{G}]$, $\mathbb{G} = \{p \in \mathbb{P} : \breve{p} \in \mathbf{G}\}$ is a set $\mathbb{P}$-generic over $\mathbb{H}$, in particular a pairwise compatible subset of $\mathbb{P}$, therefore $F = \bigcup \mathbb{G}$ is a bijection from some $A' \subseteq A$ onto $B' \subseteq B$ in $\mathbb{H}[\mathbf{G}]$. To show that $A' = A$ note that, in $\mathbb{H}$, given a condition $p \in \mathbb{P}$ and some $x \in A$, there is a stronger condition $q \in \mathbb{P}$ with $x \in \mathtt{dom}\, q$. (Indeed $A \smallsetminus \mathtt{dom}\, p$ and $B \smallsetminus \mathtt{ran}\, p$ are infinite sets.) Thus $D_x = \{p \in \mathbb{P} : x \in \mathtt{dom}\, p\}$ is a dense subset of $\mathbb{P}$ definable in $\mathbb{H}$, therefore $D_x \cap \mathbb{G} \neq \varnothing$ by the genericity of $\mathbb{G}$, and $x \in A' = \mathtt{dom}\, F$. Thus $A' = A$, and similarly $B' = B$, and hence F is a bijection $A \overset{\text{onto}}{\longrightarrow} B$ in $\mathbb{H}[\mathbf{G}]$. □

Corollary 7.2.2. *It is consistent with* **HST** *that there are internal $*$-infinite sets of different $*$-cardinality, equinumerous in the external universe.*

Proof. Arguing in **ZFC**, assume that $\mathbb{H}$ is a countable model of **HST** satisfying 7.1.1. The class $\mathbb{L}[\mathbb{I}]$ of all sets constructible from internal sets defined in $\mathbb{H}$ is then also a model of **HST** satisfying 7.1.1, thus we can simply assume that "$\mathbb{H} = \mathbb{L}[\mathbb{I}]$" is true in $\mathbb{H}$. Theorem 7.2.1 (applied to an arbitrary pair of internal $*$-infinite sets of different $*$-cardinality in $\mathbb{H}$, say $A = {}^*\mathbb{N}$ and $B = {}^*\mathbb{R}$) leads to a model that proves the required consistency.

There is another, purely syntactical method to prove consistency results, also well-known from studies of forcing in **ZFC**. This method does not involve any models. We argue in **HST**, take a pair of $*$-infinite internal sets A, B of differend $*$-cardinality, define the forcing notion $\mathbb{P} = \mathbb{P}_{AB}$ as in the proof of Theorem 7.2.1, then define, in the **HST** universe $\mathbb{H}$, the relation $\Vdash\, = \,\Vdash_{\mathbb{P}}$ satisfying F1 – F7 of § 7.1e. Then we can prove in **HST**, following the proof of Theorem 7.1.20 with some amendments, that 1) $p \Vdash \Phi$ for any $p \in \mathbb{P}$ and any axiom Φ of **HST**, and, following the proof of Theorem 7.2.1, that 2) any $p \in \mathbb{P}$ forces the statement: "$\breve{A}$ and $\breve{B}$ are $*$-infinite internal sets of different $*$-cardinality but equinumerous in the whole universe". On the other hand, it follows from very general properties of forcing that 1) and 2) imply that the existence statement contained in 2) is consistent with **HST**.

The amount of technicalities involved in such a scheme does not allow us to present it in this book in full detail. □

Corollary 7.2.3. **HST** *is not a reducible theory (unless it is inconsistent).*

Proof. Let $\mathbb{H}_0$ be any model of **HST** satisfying 7.1.1. Taking $\mathbb{H} = \mathbb{L}[\mathbb{I}]$ in $\mathbb{H}_0$ we obtain a model for **HST** where "$\mathbb{H} = \mathbb{L}[\mathbb{I}]$" holds, and hence the statement Φ saying that "there exist infinite internal sets not equinumerous in the whole universe" **holds** in $\mathbb{H}$ by Theorem 5.5.8(iv). On the other hand, it follows from Theorem 7.2.1 that there exists a generic extension $\mathbb{H}[\mathbf{G}] \models \mathbf{HST}$ in which the same statement Φ **fails**. Note that $\mathbb{H}$ and $\mathbb{H}[\mathbf{G}]$ have the same standard core by Corollary 7.1.11. It follows that there is no $\in$-sentence φ such that $\Phi \iff \varphi^{\mathrm{st}}$ is a theorem of **HST**. □

7.2b Internal preserving bijections

We still suppose that $\mathbb{H}$ is a countable model of **HST** satisfying 7.1.1.

Generic bijections F obtained by the method of Theorem 7.2.1 are *locally internal* in the sense that for any $a \in A$ there is an infinite (possibly $*$-finite) internal set $A' \subseteq A$ containing a such that $F \restriction A'$ is internal, and for any $b \in B$ there is an infinite internal set $B' \subseteq B$ containing b such that $F^{-1} \restriction B'$ is internal. Another forcing can be used to obtain bijections $F : A \xrightarrow{\text{onto}} B$ *internal-preserving* in the sense that for any $X \subseteq A$ the F-image $F"X = \{F(x) : x \in X\}$ is internal iff X is internal.

Theorem 7.2.4. *In the conditions of Theorem 7.2.1 there is a generic extension* $\mathbb{H}[\mathbf{G}] \models$ **HST** *of* $\mathbb{H}$ *in which it is true that there exists an internal-preserving bijection* $F : A \xrightarrow{\text{onto}} B$.

Proof. A *partition* of a given set C is any set $\mathscr{C}$ of pairwise disjoint subsets of C such that $\bigcup \mathscr{C} = C$. A partition $\mathscr{C}'$ (of the same set C) is *finer* than $\mathscr{C}$ if for any $X' \in \mathscr{C}'$ there exists $X \in \mathscr{C}$ with $X' \subseteq X$.

Arguing in $\mathbb{H}$, consider the set $\mathbb{P}$ of all triples of the form $p = \langle \mathscr{A}, \mathscr{B}, f \rangle$ such that $\mathscr{A}$ is an internal partition of A, $\mathscr{B}$ is an internal partition of B, either of $\mathscr{A}, \mathscr{B}$ contains at least one infinite set, $f : \mathscr{A} \xrightarrow{\text{onto}} \mathscr{B}$ satisfies $X' \subseteq X \iff f(X') \subseteq f(X)$, and for any $X \in \mathscr{A}$ either both X and $f(X) \in \mathscr{B}$ are infinite or both are finite with the same number of elements.

We define $\langle \mathscr{A}', \mathscr{B}', f' \rangle \leq \langle \mathscr{A}, \mathscr{B}, f \rangle$ iff $\mathscr{A}'$ is finer than $\mathscr{A}$, $\mathscr{B}'$ is finer than $\mathscr{B}$, and $f'(X') \subseteq f(X)$ holds whenever $X \in \mathscr{A}$, $X' \in \mathscr{A}'$, $X' \subseteq X$.

As in the proof of Theorem 7.2.1, there is a regular generic extension $\mathbb{H}[\mathbf{G}]$ of $\mathbb{H}$ containing a set $\mathbb{G} \subseteq \mathbb{P}$ $\mathbb{P}$-generic over $\mathbb{H}$ in $\mathbb{H}[\mathbf{G}]$. To extract a bijection $F : A \xrightarrow{\text{onto}} B$ note that for any $x \in A$ the set D_x of all $p = \langle \mathscr{A}, \mathscr{B}, f \rangle \in \mathbb{P}$ such that the singleton $\{x\}$ belongs to $\mathscr{A}$ is dense in $\mathbb{P}$. (Indeed, given a condition $p = \langle \mathscr{A}, \mathscr{B}, f \rangle \in \mathbb{P}$, x belongs to a unique $X \in \mathscr{A}$. Take an arbitrary $y \in Y = f(X)$. Put

$$\mathscr{A}' = (\mathscr{A} \smallsetminus \{X\}) \cup \{\{x\}, X \smallsetminus \{x\}\}, \quad \mathscr{B}' = (\mathscr{B} \smallsetminus \{Y\}) \cup \{\{y\}, Y \smallsetminus \{y\}\},$$

$f'(\{x\}) = \{y\}$, $f'(X \smallsetminus \{x\}) = Y \smallsetminus \{y\}$, and define f' to be equal to f on $\mathscr{A} \smallsetminus \{X\}$. Then $p' = \langle \mathscr{A}', \mathscr{B}', f' \rangle$ belongs to $\mathbb{P}$ and $p' \leq p$.) It follows by the genericity of $\mathbb{G}$ that there exists a condition $p = \langle \mathscr{A}, \mathscr{B}, f \rangle \in \mathbb{G} \cap D_x$. Put $F(x) = y$, where y is a unique element of B satisfying $f(\{x\}) = \{y\}$.

A similar argument proves that F is a bijection from A onto B.

Prove that F is internal-preserving.

Consider any internal set $U \subseteq A$. We claim that the set D of all conditions $p = \langle \mathscr{A}, \mathscr{B}, f \rangle \in \mathbb{P}$, such that every $X \in \mathscr{A}$ satisfies $X \subseteq U \vee X \cap U = \varnothing$, is dense in $\mathbb{P}$. Indeed given $p = \langle \mathscr{A}, \mathscr{B}, f \rangle \in \mathbb{P}$, define

$$\mathscr{A}' = \{X \cap U : X \in \mathscr{A}\} \cup \{X \smallsetminus U : X \in \mathscr{A}\}.$$

For any $X \in \mathscr{A}$, we can find a partition $f(X) = Y' \cup Y''$ such that either both $X \cap U$ and Y' are infinite or both are finite with the same number of elements, and the same for the other pair $X \smallsetminus U$ and Y''. This allows us to define $f'(X \cap U) = Y'$ and $f'(X \smallsetminus U) = Y''$. Finally let $\mathscr{B}'$ contain all sets Y', Y'' obtained this way from all $X \in \mathscr{A}$. This definition can be carried out inside $\mathbb{I}$, and hence the result can be assumed to be internal. Then easily $p' = \langle \mathscr{A}', \mathscr{B}', f' \rangle$ belongs to D and $p' \leq p$.

Thus there exists a condition $p = \langle \mathscr{A}, \mathscr{B}, f \rangle \in \mathbb{G} \cap D$. We leave it as an **exercise** for the reader to check that $F"A'$ is equal to the internal set $B' = \bigcup_{X \in \mathscr{A},\, X \subseteq U} f(X)$ and fill in other details in the argument. □

The next corollary will imply the independence of the choice of an internal domain in "hyperfinite" descriptive set theory in Chapter 9.

Corollary 7.2.5 (HST). *If A, B are infinite internal sets then the two-sort structures $\mathfrak{A} = \langle A; \mathscr{P}_{\mathtt{int}}(A); \in \rangle$ and $\mathfrak{B} = \langle B; \mathscr{P}_{\mathtt{int}}(B); \in \rangle$* ($\in$ acts only between the first and the second domain) *are elementarily equivalent.*

Proof. The result is obviously absolute for the class $\mathbb{L}[\mathbb{I}]$, and therefore we can assume "$\mathbb{H} = \mathbb{L}[\mathbb{I}]$" in the **HST** universe. Then $\mathfrak{A}$ and $\mathfrak{B}$ are isomorphic in a generic extension of $\mathbb{L}[\mathbb{I}]$ defined as in the proof of Theorem 7.2.4 because an internal-preserving bijection immediately induces an isomorphism between $\mathfrak{A}$ and $\mathfrak{B}$. Thus $\mathfrak{A}$, $\mathfrak{B}$ are elementarily equivalent in the extension, and thereby in the ground universe as well. [11] □

7.2c Making elementarily equivalent structures isomorphic

Isomorphic structures of the same language are obviously elementarily equivalent w. r. t. formulas of that language. It is known from model theory that under suitable cardinality and saturation assumptions the converse is true: elementarily equivalent structures turn out to be isomorphic. We have plenty of saturation in **HST** for internal structures by Theorem 1.5.20. Yet the back-and-forth argument typically used to establish an isomorphism, badly fails because infinite internal sets are not well-orderable in **HST**. On the other hand, *generic* isomorphisms can be obtained!

Definition 7.2.6 (HST). A *language of standard size* is any (first order) language $\mathcal{L}$ having standard size many atomic symbols. [12] In this case, an $\mathcal{L}$-structure $\mathfrak{A} = \langle A; ... \rangle$ is *internally presented* if both the underlying set A and every $\mathfrak{A}$-interpretation of a symbol in $\mathcal{L}$ are internal sets. □

[11] We don't know whether there is a comparably short but more elementary proof of this corollary.

[12] Recall that, in **HST**, sets of standard size are those equinumerous to a set $S \subseteq {}^*\mathbb{S}$, which turns out to be equivalent to any of the following: *well-orderable*, *equinumerous to a well-founded set*, *equinumerous to an ordinal*, see § 1.3a.

Theorem 7.2.7. *Let* $\mathbb{H}$ *be a countable model of* **HST** *satisfying 7.1.1. Suppose that* $\mathcal{L}$, $\mathfrak{A}$, $\mathfrak{B} \in \mathbb{H}$ *and the following is true in* $\mathbb{H}$:

(†) *"*$\mathbb{H} = \mathbb{L}[\mathbb{I}]$*" holds,* $\mathcal{L}$ *is a language of standard size,* $\mathfrak{A} = \langle A; ...\rangle$ *and* $\mathfrak{B} = \langle B; ...\rangle$ *are internally presented elementarily equivalent* $\mathcal{L}$*-structures.*

Then there is a generic extension $\mathbb{H}[\mathbf{G}] \models$ **HST** *where* $\mathfrak{A}$, $\mathfrak{B}$ *are isomorphic.*

A natural idea as how to force an isomorphism of $\mathfrak{A}$ onto $\mathfrak{B}$ is to use partial bijections p from A into B, $\mathcal{L}$*-preserving* in the sense that if $a \in \mathtt{dom}\, p$ satisfies an $\mathcal{L}$-formula $\Phi(a)$ in $\mathfrak{A}$ then $b = p(a)$ has to satisfy $\Phi(b)$ in $\mathfrak{B}$. The following property has to be achieved: given a condition p and some $\mathbf{a} \in A \setminus \mathtt{dom}\, p$, there exists $\mathbf{b} \in B \setminus \mathtt{ran}\, p$ being in the same relations with every $p(a)$, $a \in \mathtt{dom}\, p$, in $\mathfrak{B}$ as $\mathbf{a}$ is with a in $\mathfrak{A}$, and *vice versa*.

This requirement does not cause problems if only partial maps p with domains of standard size are considered as forcing conditions: then there exist only standard size many properties to be satisfied by $\mathbf{b}$, so Saturation yields the result. But such a forcing can be a proper class because in general the collection of standard size subsets of an internal infinite set is a proper class in **HST**. This would lead to great complications in the development of forcing.

The alternative is to use *internal* partial maps. Then we definitely have a set rather than a proper class as the forcing notion. But another problem arizes: since $\mathtt{dom}\, p$ may well be not a set of standard size, we have too many properties to be satisfied by $\mathbf{b}$ in $\mathfrak{B}$, and hence it is not clear how Saturation can be successfully applied.

To explain how we overcome this obstacle, let $\Phi(x, y)$ be an $\mathcal{L}$-formula. We are going to find $\mathbf{b} \in B$ so that $\Phi(\mathbf{a}, a)$ holds in $\mathfrak{A}$ iff $\Phi(\mathbf{b}, p(a))$ holds in $\mathfrak{B}$, for all $a \in \mathtt{dom}\, p$. The sets $u = \{a \in \mathtt{dom}\, p : \mathfrak{A} \models \Phi(\mathbf{a}, a)\}$ and $v = \mathtt{dom}\, p \setminus u$ are internal by the choice of $\mathfrak{A}$. We observe that the chosen element $\mathbf{a}$ satisfies $\forall\, a \in u\, \Phi(\mathbf{a}, a)$ and $\forall\, a \in v\, \neg\, \Phi(\mathbf{a}, a)$ in $\mathfrak{A}$, and thus the sentence $\exists\, x\, (\forall\, a \in u\, \Phi(x, a) \wedge \forall\, a \in v\, \neg\, \Phi(x, a))$ is true in $\mathfrak{A}$. Suppose that p preserves sentences of this form. Then $\exists\, y\, (\forall\, b \in p"u\, \Phi(y, b) \wedge \forall\, b \in p"v\, \neg\, \Phi(y, b))$ is true in $\mathfrak{B}$. ($p"u = \{p(a) : a \in u\}$ is the p-image of u.) This yields an element $\mathbf{b} \in B$ which may be put in correspondence with $\mathbf{a}$.

It follows that we have to assume the preservation of $\mathcal{L}$-formulas with some internal subsets of A as parameters under the action of p. In other words a stronger preservation hypothesis is involved than the result achieved. This leads to the next level *etc.*, and finally to a kind of type-theoretic extension of the language $\mathcal{L}$. This plan is unfolded in the remainder of this section.

7.2d The forcing notion

We argue in $\mathbb{H}$, *that is in* **HST**. In addition, (†) of Theorem 7.2.7 is assumed.

Let us define *types* as follows: 0 is a type, and $t = \diamond(t_1, ..., t_k)$ is a type provided $t_1, ..., t_k$ are types. (Here $\diamond$ is a formal sign and $k \in \mathbb{N}$.) Every type belongs to $\mathbb{HF}$ (hereditarily finite sets, § 1.2e), and hence to $\mathbb{S} \cap \mathbb{WF}$, while the set $\mathbf{T} \subseteq \mathbb{HF}$ of all types belongs to $\mathbb{WF}$.

Let D be an internal set. We put $D^0 = D$, and $D^t = \mathscr{P}_{\mathtt{int}}(D^{t_1} \times ... \times D^{t_n})$ whenever $t = \diamond(t_1, ..., t_k)$. Here $\mathscr{P}_{\mathtt{int}}(X)$ means the $\mathbb{I}$-power set of a set $X \in \mathbb{I}$. For instance $D^{\diamond(0,0)} = \mathscr{P}(D \times D)$ in $\mathbb{I}$. Note that $D = D^0 \subseteq C$. Easily D^t is internal for any type t by Lemma 1.2.16. Let $D^\infty = \bigcup_t D^t$. [13]

Recall that $\mathcal{L}$ is a standard size language by 7.2.7(†). Let $\mathcal{L}^\infty$ be the extension of $\mathcal{L}$ by variables $x^t, y^t, ...$ for each type t, which can be used only through expressions of the form $x^t(x^{t_1}, ..., x^{t_k})$ (may be written as $\langle x^{t_1}, ..., x^{t_k} \rangle \in x^t$), provided $t = \diamond(t_1, ..., t_k)$, and also $x = x^0$, where x is an $\mathcal{L}$-variable. $\mathcal{L}$-variables do not have an upper index and are formally distinguished from variables of type 0.

Definition 7.2.8. Suppose that $\mathfrak{C} = \langle C; ... \rangle$ is an internally presented $\mathcal{L}$-structure and $D \subseteq C$ is an internal set. Define a structure $\mathfrak{C}[D]$ containing the same underlying domain C with all related $\mathfrak{C}$-interpretations of $\mathcal{L}$-symbols, and the domain D^t for each type t. Every $\mathcal{L}^\infty$-formula (perhaps, containing parameters in $\mathfrak{C}[D]$) can be interpreted in $\mathfrak{C}[D]$ so that variables of any type t are interpreted in D^t while $\mathcal{L}$-variables are interpreted still in C. This converts $\mathfrak{C}[D]$ to an internally presented $\mathcal{L}^\infty$-structure. □

Note that *formulas*, with respect to the languages like $\mathcal{L}$ and $\mathcal{L}^\infty$, are finite sequences satisfying certain obvious requirements similar to those given in § 3.5a with respect to the $\in$-language. Since any finite tuple of internal sets is internal, a formula with internal parameters is an internal object.

Exercise 7.2.9. Under the assumptions of 7.2.8 prove that

$$X_\varphi = \{\langle x_1, ..., x_m, y^{t_1}, ..., y^{t_n} \rangle \in C^m \times D^{t_1} \times \ldots \times D^{t_n} : \\ \mathfrak{C}[D] \models \varphi(x_1, ..., x_m, y^{t_1}, ..., y^{t_n})\}$$

is an internal set for any $\mathcal{L}^\infty$-formula $\varphi(x_1, ..., x_m, y^{t_1}, ..., y^{t_n})$ with parameters in $\mathfrak{C}[D]$. (*Hint.* Argue by induction on the complexity of φ. The corresponding transformations of sets X_φ are absolute for $\mathbb{I}$ by 1.2.5.) □

Suppose that p is an internal $1-1$ map from an internal set $D \subseteq A$ onto a set $E \subseteq B$ (also internal). We expand p to all types t by induction:

$$p^t(x^t) = \{\langle p^{t_1}(x^{t_1}), ..., p^{t_k}(x^{t_k}) \rangle : \langle x^{t_1}, ..., x^{t_k} \rangle \in x^t\} \quad - \quad \text{for all } x^t \in D^t,$$

whenever $t = \diamond(t_1, ..., t_k)$. Then p^t internally $1-1$ maps D^t onto E^t.

If Φ is an $\mathcal{L}^\infty$-formula containing parameters in D^∞ then let $p\Phi$ be the formula obtained by changing each parameter $x \in D^t$ in Φ to $p^t(x) \in E^t$.

[13] Clearly $D^{t_1} \cap D^{t_2} \neq \varnothing$ for different types t_1, t_2; for instance $\varnothing$ belongs to any D^t, $t \neq 0$. However it is supposed that appropriate provisions are taken to distinguish equal sets which appear in different types.

Definition 7.2.10. Under the assumption (†) of Theorem 7.2.7, let $\mathbb{P}_{\mathcal{L}}(\mathfrak{A}, \mathfrak{B})$ be the set of all internal $1-1$ maps p such that $D = \mathtt{dom}\, p$ is an (internal) subset of A, $E = \mathtt{ran}\, p \subseteq B$ (also internal), and, for each closed $\mathcal{L}^\infty$-formula Φ with parameters in $D \cup D^\infty$, we have: $\mathfrak{A}[D] \models \Phi$ iff $\mathfrak{B}[E] \models p\Phi$.

We define $p \le q$ (p is stronger than q) iff $q \subseteq p$. □

For instance the empty map $\varnothing$ belongs to $\mathbb{P}_{\mathcal{L}}(\mathfrak{A}, \mathfrak{B})$ because $\mathfrak{A}$ and $\mathfrak{B}$ are elementarily equivalent. (Proper $\mathcal{L}^\infty$-variables can be eliminated in this case because the domains $\varnothing^t$ are finite.)

Lemma 7.2.11. *Under the assumption 7.2.7(†), the p. o. set $\mathbb{P} = \mathbb{P}_{\mathcal{L}}(\mathfrak{A}, \mathfrak{B})$ is standard size closed and hence standard size distributive by "$\mathbb{H} = \mathbb{L}[\mathbb{I}]$".*

Proof. Let λ be a limit ordinal. Suppose that p_α, $\alpha < \lambda$, are conditions in $\mathbb{P}$, and $p_\beta \le p_\alpha$ whenever $\alpha < \beta < \lambda$. We claim that $\mathbb{P}$ is $\mathbf{\Pi}_1^{\mathtt{ss}}$, *i.e.* a standard size intersection of internal sets. Indeed by definition $\mathbb{P} = \bigcap_\varphi P_\varphi$, where φ runs over all parameter-free $\mathcal{L}^\infty$-formulas having no proper $\mathcal{L}$-variable free, and, for any such formula $\varphi(y^{t_1}, ..., y^{t_n})$, P_φ consists of all internal partial $1-1$ maps $p : A \to B$ satisfying, for all $y^{t_j} \in D^{t_j}$, $1 \le j \le n$,

$$\mathfrak{A}[D] \models \varphi(y^{t_1}, ..., y^{t_n}) \quad \text{iff} \quad \mathfrak{B}[E] \models \varphi(p^{t_1}(y^{t_1}), ..., p^{t_n}(y^{t_n}))$$

where $D = \mathtt{dom}\, p$ and $E = \mathtt{ran}\, p$ (internal subsets of resp. A, B). However there exist only standard size many parameter-free $\mathcal{L}^\infty$-formulas φ, and on the other hand all sets P_φ are internal by 7.2.9 (and because all maps p^t are internal together with p). Therefore all sets $W_\alpha = \{p \in \mathbb{P} : p \le p_\alpha\}$ are $\mathbf{\Pi}_1^{\mathtt{ss}}$ as well. Furthermore $W_\alpha \ne \varnothing$ and we have $W_\beta \subseteq P_\alpha$ whenever $\alpha < \beta < \lambda$. Finally λ (as every set in $\mathbb{WF}$, the well-founded universe) is a set of standard size by Lemma 1.3.1, thus $\bigcap_{\alpha < \kappa} W_\alpha \ne \varnothing$ by Theorem 1.4.2(i). □

7.2e Key lemma

The next lemma is of key importance since it shows that $\mathbb{P} = \mathbb{P}_{\mathcal{L}}(\mathfrak{A}, \mathfrak{B})$ behaves more or less like a collapse forcing, in particular, a condition $p \in \mathbb{P}$ cannot satisfy both $\mathtt{dom}\, p \subsetneqq A$ and $\mathtt{ran}\, p = B$ (or *vice versa*). The difference from typical collapse forcing notions is that $\mathbb{P}_{\mathcal{L}}(\mathfrak{A}, \mathfrak{B})$ still can contain minimal (that is, strongest and $\subseteq$-maximal) conditions p, but any such condition p satisfies $\mathtt{dom}\, p = A$ and $\mathtt{ran}\, p = B$ by the lemma, and hence immediately exhibits an isomorphism between $\mathfrak{A}$ and $\mathfrak{B}$ already in M. It follows that there is no need to explicitly require that $\mathtt{dom}\, p$ has $*$-cardinality less than A and B, as in § 7.2a with respect to collapse forcing notions.

Theorem 7.2.12 (the key lemma). *Still under the assumption 7.2.7(†), suppose that $p \in \mathbb{P} = \mathbb{P}_{\mathcal{L}}(\mathfrak{A}, \mathfrak{B})$, $D = \mathtt{dom}\, p$, $E = \mathtt{ran}\, p$. If $\mathbf{a} \in A \smallsetminus D$ then there exists $\mathbf{b} \in B \smallsetminus E$ such that $p_+ = p \cup \{\langle \mathbf{a}, \mathbf{b} \rangle\} \in \mathbb{P}$. Conversely, if $\mathbf{b} \in B \smallsetminus E$ then there exists $\mathbf{a} \in A \smallsetminus D$ such that $p_+ = p \cup \{\langle \mathbf{a}, \mathbf{b} \rangle\} \in \mathbb{P}$.*

Proof. By symmetry, we concentrate on the first part. Let us fix a condition $p \in \mathbb{P}$. Let $D = \mathtt{dom}\, p$, $E = \mathtt{ran}\, p$. (For instance we may have $p = D = E = \varnothing$.) Consider an arbitrary $\mathbf{a} \in A \smallsetminus D$; we have to find a counterpart $\mathbf{b} \in B \smallsetminus E$ such that $p_+ = p \cup \{\langle \mathbf{a}, \mathbf{b} \rangle\}$ still belongs to $\mathbb{P}$.

Let $\kappa = \mathtt{card}\, \mathcal{L}$ (or $\kappa = \aleph_0$ provided $\mathcal{L}$ is finite). Let us enumerate by $\varphi_\alpha(x, x^{t_\alpha})$ $(\alpha < \kappa)$ all parameter–free $\mathcal{L}^\infty$-formulas containing only one free $\mathcal{L}$-variable x and only one free $\mathcal{L}^\infty$-variable x^{t_α} of a type $t_\alpha \neq 0$.

We define $X_\alpha = \{x^{t_\alpha} \in D^{t_\alpha} : \mathfrak{A}[D] \models \varphi_\alpha(\mathbf{a}, x^{t_\alpha})\}$ for all $\alpha < \kappa$; thus $X_\alpha \in D^{\diamond(t_\alpha)}$. (Note that X_α is internal by 7.2.9.) Let $\Psi_\alpha(X, x)$ be the $\mathcal{L}^\infty$-formula $\forall x^{t_\alpha}\, (X(x^{t_\alpha}) \iff \varphi_\alpha(x, x^{t_\alpha}))$, thus $\mathfrak{A}[D] \models \Psi_\alpha(X_\alpha, \mathbf{a})$ by definition. We have κ-many formulas $\Psi_\alpha(X_\alpha, x)$ realized in $\mathfrak{A}[D]$ by one and the same element $x = \mathbf{a} \in A$. We put $Y_\alpha = p^{\diamond(t_\alpha)}(X_\alpha)$; thus $Y_\alpha \in E^{\diamond(t_\alpha)}$.

Lemma 7.2.13. *There exists* $\mathbf{b} \in B$ *such that* $\mathfrak{B}[E] \models \Psi_\alpha(Y_\alpha, \mathbf{b})$ *for all* α.

Proof. It suffices to prove that any finite conjunction $\Psi_{\alpha_1}(Y_{\alpha_1}, y) \wedge \cdots \wedge \Psi_{\alpha_m}(Y_{\alpha_m}, y)$ can be realized in $\mathfrak{B}[E]$. By definition $\mathbf{a}$ witnesses that

$$\mathfrak{A}[D] \models \exists x\, (\Psi_{\alpha_1}(X_{\alpha_1}, x) \wedge \cdots \wedge \Psi_{\alpha_m}(X_{\alpha_m}, x)) .$$

Therefore $\mathfrak{B}[E] \models \exists y\, (\Psi_{\alpha_1}(Y_{\alpha_1}, y) \wedge \cdots \wedge \Psi_{\alpha_m}(Y_{\alpha_m}, y))$, since $p \in \mathbb{P}$. □

Let us fix an element $\mathbf{b} \in B$ satisfying $\Psi_\alpha(Y_\alpha, \mathbf{b})$ in $\mathfrak{B}[E]$ for all $\alpha < \kappa$. We set $p_+ = p \cup \{\langle \mathbf{a}, \mathbf{b} \rangle\}$, $D_+ = D \cup \{\mathbf{a}\}$, $E_+ = E \cup \{\mathbf{b}\}$.

We claim that $p_+ \in \mathbb{P}$, *i.e.* the equivalence $\mathfrak{A}[D_+] \models \Phi$ iff $\mathfrak{B}[E_+] \models p_+ \Phi$ holds for each closed $\mathcal{L}^\infty$-formula Φ with parameters in $D_+ \cup {D_+}^\infty$. The next lemma gives a partial result.

Lemma 7.2.14. *Let* $\varphi(x)$ *be an* $\mathcal{L}^\infty$*-formula possibly with sets in* D^∞ *as parameters. Then* $\varphi(\mathbf{a})$ *is true in* $\mathfrak{A}[D]$ *iff* $(p\varphi)(\mathbf{b})$ *is true in* $\mathfrak{B}[E]$.

Proof. Assume that φ contains only one parameter $\xi \in D^t$ and $t \neq 0$ (otherwise use a tuple coding). Then $\varphi(x)$ is $\varphi_\alpha(x, \xi)$ for some $\alpha < \kappa$ such that $t = t_\alpha$. Since $\Psi_\alpha(X_\alpha, \mathbf{a})$ is true in $\mathfrak{A}[D]$, we have: $X_\alpha(\xi)$ iff $\mathfrak{A}[D] \models \varphi_\alpha(\mathbf{a}, \xi)$. Note that $X_\alpha(\xi) \iff Y_\alpha(\eta)$ where $\eta = p^t(\xi) \in E^t$ because $p \in \mathbb{P}$. On the other hand we have $Y_\alpha(\eta)$ iff $\mathfrak{B}[E] \models \varphi_\alpha(\mathbf{b}, \eta)$, because $\Psi_\alpha(Y_\alpha, \mathbf{b})$ is true in $\mathfrak{B}[E]$ by the choice of $\mathbf{b}$. But $\varphi_\alpha(\mathbf{b}, \eta)$ coincides with $(p\varphi)(\mathbf{b})$. □

However there is a more serious problem: we have to check that p_+ transforms true $\mathcal{L}^\infty$-formulas with parameters in ${D_+}^\infty$ into true $\mathcal{L}^\infty$-formulas with parameters in ${E_+}^\infty$. The idea is to convert formulas with parameters in ${D_+}^\infty$ (not necessarily equal to $\mathbf{a}$) into formulas with parameters in D^∞ plus $\mathbf{a}$ as an extra parameter, and use Lemma 7.2.14.

Fortunately the structure of types over an internal set C depends only on the internal cardinality of C but does not depend on the exact choice of C. This allows to "model" ${D_+}^\infty$ in D^∞ identifying the $\mathbf{a}$ with $\varnothing$ and any $a \in D$ with $\{a\}$. To realize this plan, let us define $U = \{\varnothing\} \cup \{\{a\} : a \in D\}$, so that $U \subseteq D^\tau$, where $\tau = \diamond(0)$ (the type of subsets of D). Furthermore we have $U \in D^{\diamond(\tau)}$ because U is internal. Accordingly, on the other side, we put $V = \{\varnothing\} \cup \{\{b\} : b \in E\}$; then $V \subseteq E^\tau$, $V \in E^{\diamond(\tau)}$, and $V = p^{\diamond(\tau)}(U)$.

For each type t, we define a type $\delta(t)$ by $\delta(0) = \tau$ and $\delta(t) = \diamond(\delta(t_1), ..., \delta(t_n))$ provided $t = \diamond(t_1, ..., t_n)$. Put $\delta(\mathbf{a}) = \varnothing$, and $\delta(a) = \{a\}$ for all $a \in D$, so that δ is an internal $1-1$ map from D_+ onto U. The transform δ expands on higher types by $\delta(x) = \{\langle \delta(x_1), ..., \delta(x_n) \rangle : \langle x_1, ..., x_n \rangle \in x\}$; thus $\delta(x) \in U^t \subseteq D^{\delta(t)}$ whenever $x \in {D_+}^t$. Take notice that $\delta({D_+}^t) = U^t$. Thus $\delta = \delta_{D\mathbf{a}}$ defines a $1-1$ correspondence between ${D_+}^\infty$ and U^∞. Similarly the map $\varepsilon = \delta_{E\mathbf{b}}$ defined on E by $\varepsilon(\mathbf{b}) = \varnothing$ and $\varepsilon(b) = \{b\}$ for all $b \in E$ is an internal bijection from E_+ onto V expanding on higher types t as above, so that we get a $1-1$ correspondence between ${E_+}^\infty$ and V^∞.

Now, given a parameter-free $\mathcal{L}^\infty$-formula $\psi(x^t)$ with x^t as the only free variable, one easily defines another $\mathcal{L}^\infty$-formula, denoted by $\psi_D(x, \xi^{\delta(t)})$, containing D and some sets U^t as parameters — this is symbolized by the subscript D since the sets U^t involved are derivates of D — so that $\mathfrak{A}[D_+] \models \psi(x^t)$ iff $\mathfrak{A}[D] \models \psi_D(\mathbf{a}, \delta(x^t))$, for any $x^t \in {D_+}^t$. (For instance every quantifier $\exists\, y^t \dots y^t \dots$ is changed to $\exists\, \eta^{\delta(t)} \in U^t \dots \eta^{\delta(t)} \dots$ — this shows how the sets U^t appear as parameters.)

Then we have $\mathfrak{A}[D] \models \psi_D(\mathbf{a}, \delta(x^t))$ iff $\mathfrak{B}[E] \models \psi_E(\mathbf{b}, p^{\delta(t)}(\delta(x^t)))$ for any type t and every $x^t \in {D_+}^t$ by Lemma 7.2.14. In the last formula, one can easily verify that $p^{\delta(t)}(\delta(x^t)) = \varepsilon(y^t)$ where $y^t = {p_+}^t(x^t)$. We conclude that the final statement, $\mathfrak{B}[E] \models \psi_E(\mathbf{b}, \varepsilon(y^t))$, is equivalent to $\mathfrak{B}[E_+] \models \psi(y^t)$, similarly to the first step of this argument in the preceding paragraph.

Thus the equivalence $\mathfrak{A}[D_+] \models \psi(x^t)$ iff $\mathfrak{B}[E_+] \models \psi({p_+}^t(x^t))$ holds for any type t and any $x^t \in {D_+}^t$. The case of formulas with more than one variable does not differ much. $\square$ *(Theorem 7.2.12)*

7.2f Generic isomorphisms

To prove Theorem 7.2.7, define, in $\mathbb{H}$, the p. o. set $\mathbb{P} = \mathbb{P}_\mathcal{L}(\mathfrak{A}, \mathfrak{B})$ according to Definition 7.2.10. Then define the set and $\mathbf{P} = \mathbf{P}_\mathcal{L}(\mathfrak{A}, \mathfrak{B}) = \{\breve{p} : p \in \mathbb{P}\}$, where $\breve{p} = \langle p, \mathbb{N} \rangle \in \mathbb{H} \smallsetminus \mathbb{I}$, as in the proof of Theorem 7.2.1. Note that $\mathbf{P}$ belongs to $\mathbb{H}$, satisfies $\mathbf{P} \subseteq \mathbb{H} \smallsetminus \mathbb{I}$, and is standard size distributive in $\mathbb{H}$ together with $\mathbb{P}$ by Lemma 7.2.11, and hence can be used as a forcing notion.

It follows from the countability of $\mathbb{H}$ that there exists a set $\mathbf{G} \subseteq \mathbf{P}$, $\mathbf{P}$-generic over $\mathbb{H}$. Then $\mathbb{H}[\mathbf{G}] \models \mathbf{HST}$ by Theorem 7.1.20 and it is true in $\mathbb{H}[\mathbf{G}]$ that there exists a set $\mathbb{G} \subseteq \mathbb{P}$, $\mathbb{P}$-generic over $\mathbb{H}$. Define, in $\mathbb{H}[\mathbf{G}]$, $F = \bigcup \mathbb{G}$. Then F is a bijection of A onto B by Theorem 7.2.12 and ordinary forcing arguments, as in the proof of Theorem 7.2.1. Then, in $\mathbb{H}[\mathbf{G}]$, F turns out to be a union of compatible conditions in $\mathbb{P}$, thus it preserves the truth of $\mathcal{L}$-sentences, in particular, all atomic sentences. We conclude that the map F is an isomorphism of $\mathfrak{A}$ onto $\mathfrak{B}$ in $\mathbb{H}[\mathbf{G}]$.

$\square$ *(Theorem 7.2.7)*

7.3 Consistency of the isomorphism property

Let κ be a cardinal in the **ZFC** universe. In model theoretic nonstandard analysis, a nonstandard model is said to satisfy the *κ-isomorphism property*, $\mathbf{IP}_\kappa$ in brief, iff whenever $\mathcal{L}$ is a first-order language containing $< \kappa$ symbols, any two internally presented elementarily equivalent $\mathcal{L}$-structures are isomorphic.

It is known that even with $\kappa = \aleph_1$ $\mathbf{IP}_\kappa$ implies several strong consequences inavailable in the frameworks of ordinary postulates of nonstandard analysis, for instance the existence of a set of infinite Loeb outer measure which intersects every set of finite Loeb measure by a set of Loeb measure 0, the theorem that any two infinite internal sets have the same external cardinality, *etc.* (See some references in comments to this Chapter.)

HST admits the following general cardinal–free formulation of **IP**:

Isomorphism Property: If $\mathcal{L}$ is a first-order language of standard size then any two internally presented elementarily equivalent $\mathcal{L}$-structures are isomorphic.

In particular Isomorphism Property implies in **HST** that any two infinite internal sets are externally equinumerous. (Indeed take the empty language as $\mathcal{L}$. Any two infinite sets are elementarily equivalent if equality is the only atomic symbol.) It follows by Theorem 5.5.8 that Isomorphism Property fails in $\mathbb{L}[\mathbb{I}]$, and hence its negation is consistent with **HST**.

The aim of this section is to prove the following:

Theorem 7.3.1. *Isomorphism Property itself is consistent with* **HST**.

It follows that such an important technical tool of "nonstandard" mathematics as the isomorphism property can be adequately developed in the context of the nonstandard set theory **HST**.

Theorem 7.3.1 is a consequence of the following more concrete theorem; the derivation of the former from the latter is the same as in § 7.2a (the derivation of Corollary 7.2.2 from Theorem 7.2.1).

Theorem 7.3.2. *Let $\mathbb{H}$ be a countable model of* **HST** *satisfying 7.1.1. Suppose that "$\mathbb{H} = \mathbb{L}[\mathbb{I}]$" is true in $\mathbb{H}$. There exists a generic extension $\mathbb{H}[\mathbf{G}] \models$* **HST** *where Isomorphism Property holds.*

The forcing notion we employ to prove Theorem 7.3.2 will be a product, with $*$-finite internal support, of more elementary forcing notions, of the kind introduced by Definition 7.2.10, each of which forces a generic isomorphism between a pair of internally presented elementarily equivalent structures of a language of standard size. It is extremely important that the extension will not contain pairs of this form other than those which already exist in $\mathbb{H}$, the ground model — this enables us to use product rather than iterated forcing.

7.3a The product forcing notion

We argue in $\mathbb{H}$, *that is essentially in* **HST**.

($\mathbb{H}$ is a model satisfying the conditions of Theorem 7.3.2.)

Definition 7.3.3. For any $*$-cardinal κ, let $\mathbf{L}(\kappa) \in \mathbb{I}$ be the first-order language $\{s_\alpha\}_{\alpha<\kappa}$ where s_α is an n-ary relational symbol whenever $\alpha = \lambda+n < \kappa$ and λ is a limit ordinal or 0. We shall consider a *truncated* standard size language $\mathcal{L}(\kappa) = \{s_\alpha\}_{\alpha<\kappa \wedge \mathtt{st}\,\alpha}$.

Let **Ind** be the class of all tuples of the form $i = \langle w, \kappa, \mathbb{A}, \mathbb{B}\rangle$, called *indices*, such that $w \in {}^*\mathbb{N}$, κ is a $*$-cardinal, and $\mathbb{A}$, $\mathbb{B}$ are internal $\mathbf{L}(\kappa)$-structures. Obviously $\mathbf{Ind} \subseteq \mathbb{I}$.

Suppose that $i = \langle w, \kappa, \mathbb{A}, \mathbb{B}\rangle \in \mathbf{Ind}$.

We set $w_i = w$, $\kappa_i = \kappa$, $\mathbb{A}_i = \mathbb{A}$, $\mathbb{B}_i = \mathbb{B}$, and $\mathbf{L}_i = \mathbf{L}(\kappa_i)$.

Then $\mathbf{L}_i = \mathbf{L}(\kappa_i) = \{s_\alpha\}_{\alpha<\kappa_i}$ is an internal language and $\mathcal{L}_i = \mathcal{L}(\kappa_i) = \{s_\alpha\}_{\alpha<\kappa_i \wedge \mathtt{st}\,\alpha}$ is a standard size language. Let $\mathfrak{A}_i$ and $\mathfrak{B}_i$ denote the corresponding truncated forms of $\mathbb{A}_i$ and $\mathbb{B}_i$; they are internally presented $\mathcal{L}_i$-structures. Put $\mathbb{P}_i = \mathbb{P}_{\mathcal{L}_i}(\mathfrak{A}_i, \mathfrak{B}_i)$: a p. o. set ordered as in 7.2.10. □

The next definition introduces the forcing $\mathbb{P}$ in the form of a product of all p. o. sets $\mathbb{P}_i$, $i \in \mathbf{Ind}$, with $*$-finite support. Thus $\mathbb{P}$ will consist of internal functions π such that $\mathtt{dom}\,\pi$ is a $*$-finite (internal) subset of **Ind**. In this case we define $|\pi| = \mathtt{dom}\,\pi$.

Definition 7.3.4. $\mathbb{P}$ is the collection (clearly a proper class) of all internal functions π such that $|\pi| \subseteq \mathbf{Ind}$ is an $*$-finite (internal) set, $|\pi| \neq \varnothing$, and $\pi(i) \in \mathbb{P}_i$ for each $i \in |\pi|$. Define $\pi \leq \rho$ (π is stronger than ρ) iff $|\rho| \subseteq |\pi|$ and $\pi(i) \leq \rho(i)$ (in $\mathbb{P}_i$) for all $i \in |\rho|$.

We set $\mathbb{P}_C = \{\pi \in \mathbb{P} : |\pi| \subseteq C\}$ for any $C \subseteq \mathbf{Ind}$. □

Obviously the classes $\mathbf{Ind} \subseteq \mathbb{I}$ and $\mathbb{P} \subseteq \mathbb{I}$ are $\in$-definable in $\mathbb{I}$.

We observe that if the $\mathcal{L}_i$-structures $\mathfrak{A}_i$ and $\mathfrak{B}_i$ are not elementarily equivalent then $\mathbb{P}_i$ is empty; thus in this case $i \notin |\pi|$ for all $\pi \in \mathbb{P}$. The parameter $w = w_i$ does not actively participate in the definition of $\mathbb{P}$; its role will be to make $\mathbb{P}$ homogeneous enough to admit a restriction theorem.

Lemma 7.3.5. *If* $C \subseteq \mathbf{Ind}$ *is an internal set then* $\mathbb{P}_C$ *is a standard size closed, and hence, assuming "*$\mathbb{H} = \mathbb{L}[\mathbb{I}]$*", standard size distributive p. o. set.*

In fact the whole p. o. class $\mathbb{P}$ itself is standard size distributive in some sense, but this needs a more complicated proof, which we leave as an **exercise** for the reader. Anyway only the result for $\mathbb{P}_C$ will be used.

Proof. It suffices to prove that $\mathbb{P}_C$ is a $\mathbf{\Pi}_1^{\mathrm{ss}}$ set. (We refer to the proof of Lemma 7.2.11.) By Collection, there is a cardinal κ (in $\mathbb{WF}$) such that $\kappa_i \leq {}^*\kappa$ in $\mathbb{I}$ whenever $i \in C$. Then each internal language $\mathbf{L}_i = \mathbf{L}(\kappa_i)$, $i \in C$, is a sublanguage of the language $\mathbf{L} = \mathbf{L}({}^*\kappa) = \{s_\alpha\}_{\alpha<{}^*\kappa}$ (see Definition 7.3.3).

Accordingly every language $\mathcal{L}_i$, $i \in C$, is a sublanguage of the language $\mathcal{L} = \{s_\alpha\}_{\alpha < {}^*\kappa \wedge \mathtt{st}\,\alpha}$ of standard size. Let $\mathcal{L}^\infty$ and ${\mathcal{L}_i}^\infty$ be the type theoretic extensions of resp. $\mathcal{L}$, $\mathcal{L}_i$ as in § 7.2d. If φ is an $\mathcal{L}^\infty$-formula then let C_φ be the set of all $i \in C$ such that φ is a formula of ${\mathcal{L}_i}^\infty$ (that is, it does not contain symbols s_α, $\kappa_i \le \alpha < {}^*\kappa$). Then C_φ is an internal subset of C equal to the set of all $i \in C$ such that $\kappa_i > \alpha_\varphi$, where α_φ is the largest $*$-ordinal α such that s_α occurs in φ.

By definition we have $\mathbb{P}_C = \bigcap_\varphi P_\varphi$, where the intersection is taken over all parameter-free $\mathcal{L}^\infty$-formulas having no proper $\mathcal{L}$-variable free, and, for any such formula $\varphi(y^{t_1}, ..., y^{t_n})$, P_φ consists of of all internal functions π such that $|\pi| \subseteq C$ is an $*$-finite (internal) set and, for any $i \in C_\varphi \cap |\pi|$, $\pi(i)$ is an internal partial $1-1$ map $p : A_i \to B_i$ (where A_i, B_i are the underlying sets of resp. $\mathfrak{A}_i$, $\mathfrak{B}_i$) satisfying, for all $y^{t_k} \in {D_i}^{t_k}$, $1 \le k \le n$,

$$\mathfrak{A}[D_i] \models \varphi(y^{t_1}, ..., y^{t_n}) \quad \text{iff} \quad \mathfrak{B}[E_i] \models \varphi(\pi(i)^{t_1}(y^{t_1}), ..., \pi(i)^{t_n}(y^{t_n}))$$

where $D_i = \mathtt{dom}\,\pi_i$ and $E_i = \mathtt{ran}\,\pi_i$ (internal subsets of resp. A_i, B_i). We leave it as an **exercise** for the reader to prove that all P_φ are internal sets, with the help of an appropriate uniform version of 7.2.9. □

7.3b Externalization

We still argue in the model $\mathbb{H}$.

By definition $\mathbb{P}$ consists of internal sets, and hence cannot serve as a forcing notion in our system. We have to replace it by an isomorphic p. o. class containing only external elements. We follow the definitions in § 7.2f.

For any $i = \langle w, \kappa, \mathbb{A}, \mathbb{B} \rangle \in \mathbf{Ind}$ define $\mathbf{P}_i = \mathbf{P}_{\mathcal{L}_i}(\mathfrak{A}_i, \mathfrak{B}_i) = \{\breve{p} : p \in \mathbb{P}_i\}$, a p. o. set with the induced order $\breve{p} \le \breve{q}$ iff $p \le q$ in $\mathbb{P}_i$ iff $q \subseteq p$ (in $\mathbb{H}$).

Now to define $\mathbf{P}$, a product of all sets $\mathbf{P}_i$, note the following. By definition if $\pi \in \mathbb{P}$ then π is an internal function defined on a set $|\pi| = \mathtt{dom}\,\pi \subseteq \mathbf{Ind}$ such that $\pi(i) \in \mathbb{P}_i$ for all $i \in |\pi|$. Define π°, a function with $|\pi^\circ| = \mathtt{dom}\,\pi^\circ = |\pi|$, by the equality $\pi^\circ(i) = \langle \pi(i), \mathbb{N} \rangle = \breve{p}$, where $p = \pi(i)$, for all $i \in |\pi^\circ|$.

Finally define $\mathbf{P} = \{\pi^\circ : \pi \in \mathbb{P}\}$, with the order

$$\pi^\circ \le_{\mathbf{P}} \rho^\circ \quad \text{iff} \quad \pi \le_{\mathbb{P}} \rho \quad \text{iff} \quad \forall i \in |\pi| = |\pi^\circ| \, (\pi^\circ(i) \le_{\mathbf{P}_i} \rho^\circ(i)),$$

and $\mathbf{P}_C = \{\pi \in \mathbf{P} : |\pi| \subseteq C\}$ for any $C \subseteq \mathbf{Ind}$. Note that $\mathbf{P} \subseteq \mathbb{H} \setminus \mathbb{I}$ (together with each factor $\mathbf{P}_i$), and hence $\mathbf{P}_C \subseteq \mathbb{H} \setminus \mathbb{I}$ for any $C \subseteq \mathbf{Ind}$.

Since $\mathbb{P}$ basically will not be involved any more, the same Greek characters π, ρ, ϑ will be used to denote forcing conditions in $\mathbf{P}$.

Lemma 7.3.5 implies:

Corollary 7.3.6. *If* $C \subseteq \mathbf{Ind}$ *is an internal set then* $\mathbf{P}_C$ *is a standard size distributive p. o. set (assuming "*$\mathbb{H} = \mathbb{L}[\mathbb{I}]$*").* □

Our plan is to prove that $\mathbf{P}$-generic extensions of $\mathbb{H}$ are models of **HST** in which Isomorphism Property holds. This will be done in § 7.3e, after a preliminary study of the product forcing notion $\mathbf{P}$ inside $\mathbb{H}$.

7.3c Restricted forcing relations

We come back to the ground model $\mathbb{H}$ of Theorem 7.3.2.

The fact that $\mathbf{P}$, the forcing notion defined in $\mathbb{H}$ as in § 7.3b, is a proper class in $\mathbb{H}$ causes a problem: Theorem 7.1.20 is not immediately applicable while Theorem 7.1.18 needs to prove the following:

Theorem 7.3.7. *The forcing relation* $\Vdash = \Vdash_{\mathbf{P}}$ *is definable in* $\mathbb{H}$ *in the sense of Theorem 7.1.15*(i).

Proof. It suffices to prove the result for formulas of the form $x = y$ and $x \in y$; indeed, F4 – F7 of Definition 7.1.14 easily extend the result to all $\mathtt{st}$-$\in$-formulas. Our idea is to show that the forcing $\Vdash$ can be suitably approximated by set-size forcings.

Recall that $\mathbf{P}_C$ is a p. o. set in $\mathbb{H}$ for any internal set $C \subseteq \mathbf{Ind}$, and hence the forcing relation $\Vdash_C = \Vdash_{\mathbf{P}_C}$ is definable in $\mathbb{H}$ by Theorem 7.1.15(i).

Define a set $\|a\| \subseteq \mathbf{Ind}$ for each name $a \in \mathtt{Nms}(\mathbf{P})$ by induction on $\mathtt{nrk}\, a$ as follows: if $a \in \mathtt{Nms}_0$ then put $\|a\| = \varnothing$, otherwise put $\|a\| = \bigcup_{\langle \pi, b\rangle \in a}(\|b\| \cup |\pi|)$. Thus $\|a\|$ is the set of all indices in $\mathbf{Ind}$ on which a essentially depends.

Exercise 7.3.8 (HST). Prove that $a \in \mathtt{Nms}(\mathbf{P}_C) \iff \|a\| \subseteq C$ for any internal $C \subseteq \mathbf{Ind}$ and any $a \in \mathtt{Nms}(\mathbf{P})$. □

The next lemma is of crucial importance.

Lemma 7.3.9. *Suppose that* Φ *is a formula of the form* $a = b$ *or* $b \in a$, *where* $a, b \in \mathtt{Nms}(\mathbf{P})$, $\pi' \in \mathbf{P}$, $C \subseteq \mathbf{Ind}$ *is an internal set,* $\|a\| \cup \|b\| \subseteq C \subseteq |\pi'|$, *and* $\pi = \pi' \restriction C$. *Then* $\pi' \Vdash \Phi$ *iff* $\pi \Vdash_C \Phi$.

Proof. We proceed by induction on the $\prec$-rank (see the proof of Theorem 7.1.15). The result is obvious provided $\mathtt{nrk}\, a = \mathtt{nrk}\, b = 0$, that is when $a, b \in \mathtt{Nms}_0$, because in this case definition F1 in § 7.1e does not depend on the forcing notion.

Step F2. Let Φ be a formula $b = a$. Suppose that $\pi' \Vdash b = a$ and prove $\pi \Vdash_C b = a$. Consider a condition $\rho \in \mathbf{P}_C$, $\rho \leq \pi$ and a name $c \in \mathtt{Nms}(\mathbf{P}_C)$ such that ρ $\mathtt{forc}$ $c \in a$. Put $\rho' = \rho \cup (\pi' \restriction (|\pi'| \smallsetminus C)) \in \mathbf{P}$; thus $\rho' \restriction C = \rho$ and $\rho' \leq \pi'$ because $\rho \leq \pi$. It follows that $\rho' \Vdash c \in b$: indeed we assume $\pi' \Vdash b = a$. Thus $\rho \Vdash_C c \in b$ by the inductive hypothesis.

Conversely, suppose that $\pi \Vdash_C b = a$ and prove $\pi' \Vdash b = a$. Consider a condition $\rho' \in \mathbf{P}$, $\rho' \leq \pi'$, and a name $c \in \mathtt{Nms}(\mathbf{P})$ with ρ' $\mathtt{forc}$ $c \in a$. Then $\|c\| \subseteq \|a\| \subseteq C$, so that $c \in \mathtt{Nms}(\mathbf{P}_C)$. It easily follows that $\rho = \rho' \restriction C$ satisfies $\rho \leq \pi$ and ρ $\mathtt{forc}$ $c \in a$. Now $\pi \Vdash_C b = a$ implies $\rho \Vdash_C c \in b$, and we have $\rho' \Vdash c \in b$ by the inductive hypothesis.

Step F3. Let Φ be a formula of the form $b \in a$. Suppose that $\pi' \Vdash b \in a$ and prove $\pi \Vdash_C b \in a$. Consider a condition $\rho \in \mathbf{P}_C$, $\rho \leq \pi$. Then $\rho' = \rho \cup (\pi' \restriction (|\pi'| \smallsetminus C))$ belongs to $\mathbf{P}$ and $\rho' \leq \pi'$, thus there exist a condition

$\vartheta' \in \mathbf{P}$, $\vartheta' \le \rho'$ and a name $c \in \mathtt{Nms}(\mathbf{P})$ such that ϑ' $\mathtt{forc}$ $c \in a$ and $\vartheta' \Vdash b = c$. Then $\vartheta = \vartheta' \restriction C \in \mathbf{P}_C$ and $\vartheta \le \rho$. On the other hand, $\|c\| \subseteq \|a\| \subseteq C$, and hence ϑ $\mathtt{forc}$ $c \in a$ and $\vartheta \Vdash_C b = c$, as required.

Conversely, suppose that $\pi \Vdash_C b \in a$ and prove $\pi' \Vdash b \in a$. Assume that $\rho' \in \mathbf{P}$ and $\rho' \le \pi'$. Then $\rho = \rho' \restriction C \in \mathbf{P}_C$ and $\rho \le \pi$, and hence $\vartheta\,\mathtt{forc}\,c \in a$ and $\vartheta \Vdash_C b = c$ for a condition $\vartheta \in \mathbf{P}_C$, $\vartheta \le \rho$, and a name $c \in \mathtt{Nms}(\mathbf{P}_C)$. We put $\vartheta' = \vartheta \cup (\rho' \restriction (|\rho'| \smallsetminus C))$. Then $\vartheta' \in \mathbf{P}$, $\vartheta' \le \rho'$, $\vartheta = \vartheta' \restriction C$, and ϑ' $\mathtt{forc}$ $c \in a$. Finally $\vartheta' \Vdash b = c$ by the inductive hypothesis. □

Now we easily accomplish the proof of the theorem. Indeed,

$$p \Vdash a = b \iff \exists^{\mathrm{int}} C \subseteq \mathbf{Ind}\,(|p| \cup \|a\| \cup \|b\| \subseteq C \wedge p \Vdash_C a = b)$$

by the lemma, and similarly for $a \in b$, and hence the result follows from the definablity of $\Vdash_C$ by Theorem 7.1.15(i).

□ (*Theorem 7.3.7*)

It follows that the forcing notion $\mathbf{P}$ and the relation $\Vdash\, = \,\Vdash_{\mathbf{P}}$ satisfy Lemma 7.1.17 and Theorem 7.1.18.

7.3d Automorphisms and the restriction property

We still argue in $\mathbb{H}$.

Put $\|\Phi\| = \bigcup_{a \in \mathtt{Nms}(\mathbf{P}) \text{ occurs in } \Phi} \|a\|$ for any $\mathtt{st}$-$\in$-formula Φ with parameters in $\mathtt{Nms}(\mathbf{P})$.

Theorem 7.3.10 (restriction). *Suppose that Φ is a closed formula with parameters in $\mathtt{Nms}(\mathbf{P})$, $\pi \in \mathbf{P}$, and $\pi \Vdash \Phi$. If $C \subseteq \mathbf{Ind}$ is an internal set and $\|\Phi\| \subseteq C$ then $\pi \restriction C \Vdash \Phi$.*

Proof. The proof involves automorphisms of $\mathbf{P}$. Let $D \subseteq \mathbf{Ind}$ be an internal set. An internal bijection $h : D \xrightarrow{\text{onto}} D$ satisfying the requirement

(∗) If $i = \langle w, \kappa, \mathbb{A}, \mathbb{B} \rangle \in D$ then $h(i) = \langle w', \kappa, \mathbb{A}, \mathbb{B} \rangle$ for some (internal) w' and the same κ, $\mathbb{A}$, $\mathbb{B}$,

will be called a *correct* bijection. In this case we define $H(i) = h(i)$ for $i \in D$, and $H(i) = i$ for $i \in \mathbf{Ind} \smallsetminus D$, so that $H = H_h$ is a bijection of $\mathbf{Ind}$ onto $\mathbf{Ind}$. The extension H obviously inherits property (∗).

Let us expand the action of H onto forcing conditions and names.

1) Let $\pi \in \mathbf{P}$. Define $H\pi \in \mathbf{P}$ so that $|H\pi| = \{H(i) : i \in |\pi|\}$ and $(H\pi)(H(i)) = \pi(i)$ for each $i \in |\pi|$. Then the map $\pi \longmapsto H\pi$ is an order automorphism of the forcing notion $\mathbf{P}$.
2) Define $H[a]$ for each $a \in \mathtt{Nms}(\mathbf{P})$ by induction on $\mathtt{nrk}\,a$. If $a = \breve{x} \in \mathtt{Nms}_0$ then put $H[a] = a$. If $\mathtt{nrk}\,a > 0$ then let $H[a] = \{\langle H\pi, H[b] \rangle : \langle \pi, b \rangle \in a\}$. Obviously $H[a] \in \mathtt{Nms}(\mathbf{P})$ and $\mathtt{nrk}\,a = \mathtt{nrk}\,H[a]$.

3) For a st-∈-formula Φ containing names in $\mathtt{Nms}(\mathbf{P})$, we let $H\Phi$ denote the formula obtained by changing each name a in Φ to $H[a]$.

Proposition 7.3.11. *Let h be a correct bijection, and $H = H_h$. For any $\pi \in \mathbf{P}$ and any formula Φ with parameters in $\mathtt{Nms}(\mathbf{P})$, $\pi \Vdash \Phi$ iff $H\pi \Vdash H\Phi$.*

Proof. Routine verification carried out by induction on the complexity of the formulas involved, following known forcing patterns. □

Now come back to Theorem 7.3.10. Otherwise by Lemma 7.1.17 there exists a pair of conditions $\pi, \rho \in \mathbf{P}$ such that $\pi \restriction C = \rho \restriction C$, $\pi \Vdash \Phi$, but $\rho \Vdash \neg \Phi$. Let $D = |\pi|$, $E = |\rho|$. There exist: an internal set W satisfying $C \cup D \cup E \subseteq W$ and an internal *correct* bijection $h : W \xrightarrow{\text{onto}} W$ equal to the identity on C and satisfying $E \cap (h\text{"}D) \subseteq C$. Define $H = H_h$ as above. Let $\pi' = H\pi$. Then $\pi' \restriction C = \pi \restriction C = \rho \restriction C$ because $h \restriction C$ is the identity. Furthermore $|\pi'| = h\text{"}D$, so that $|\pi'| \cap |\rho| \subseteq C$. We conclude that the conditions π' and ρ are compatible in $\mathbf{P}$.

On the other hand, $\pi' \Vdash H\Phi$ by Proposition 7.3.11. Thus it suffices to demonstrate that Φ coincides with $H\Phi$. We recall that $\|\Phi\| \subseteq C$, so that each name a which occurs in Φ satisfies $\|a\| \subseteq C$. However one can easily prove, by induction on $\mathtt{nrk}\, a$ that $H[a] = a$ whenever $\|a\| \subseteq C$, still using the fact that $h \restriction C$ is the identity. We conclude that $H\Phi$ is Φ, as required. □

7.3e The product generic extension

Recall that, in Theorem 7.3.2, $\mathbb{H}$ is a countable model of **HST** in the **ZFC** universe satisfying 7.1.1. Define $\mathbf{P}$ in $\mathbb{H}$ as in Definition 7.3.4, thus $\mathbf{P}$ is a st-∈-definable p. o. class in $\mathbb{H}$ satisfying 7.1.7, and a p. o. set in the **ZFC** universe.

Theorem 7.3.12. *Under these assumptions, if $\mathbf{G} \subseteq \mathbf{P}$ is $\mathbf{P}$-generic over $\mathbb{H}$ then $\mathbb{H}[\mathbf{G}]$ is a model of* **HST** *satisfying* (ii) *of Theorem 7.1.20, and Isomorphism Property holds in $\mathbb{H}[\mathbf{G}]$.*

Note that in addition $\mathbb{H}[\mathbf{G}]$ is a regular extension of $\mathbb{H}$ by Corollary 7.1.11, although the last assertion of 7.1.11 (that is $\mathbf{G} \in \mathbb{H}[\mathbf{G}]$) is not applicable because $\mathbf{P}$ is a proper class in $\mathbb{H}$.

Proof. The verification of **HST** in $\mathbb{H}[\mathbf{G}]$ is carried out in principle similarly to the arguments in § 7.1g, yet we have to account for two major differences: first, $\mathbf{P}$ is a proper class in $\mathbb{H}$, second, the standard size distributivity is now available for restricted forcing notions of the form $\mathbf{P}_C$ (Corollary 7.3.6) rather than $\mathbf{P}$ itself. Theorem 7.3.10 will help us to overcome these difficulties.

Separation. In terms of the proof of Separation in the proof of of Theorem 7.1.20, Y can now be a proper class together with $\mathbf{P}$, and hence not a name in $\mathtt{Nms}(\mathbf{P})$. To fix the problem take an internal set $C \subseteq \mathbf{Ind}$ such that $\|\Phi\| \cup \|X\| \subseteq C$ and define

$$Y = \{\langle \pi, a\rangle \in \mathbf{P}_C \times A : \pi \Vdash a \in X \wedge \Phi(x)\}.$$

This is a set together with $\mathbf{P}_C$ and A, thus a name in $\mathtt{Nms}(\mathbf{P})$. Theorem 7.3.10 (the restriction theorem) makes it possible to prove that $Y[\mathbf{G}] = \{x \in X[\mathbf{G}] : \Phi(x)[\mathbf{G}]\}$ in $\mathbb{H}[\mathbf{G}]$ following usual patterns.

To check (ii) of Theorem 7.1.20, consider, as in the proof of Theorem 7.1.20 any name $f \in \mathtt{Nms}(\mathbf{P})$ such that $X = f[\mathbf{G}]$ is a function with $\mathtt{dom}\, X \subseteq \mathbb{S}$ and $\mathtt{ran}\, X \subseteq \mathbb{H}$. Take, in $\mathbb{H}$, an internal set $C \subseteq \mathbf{Ind}$ satisfying $||f|| \subseteq C$. It follows from Theorem 7.3.10 that for any $x,\, y \in \mathbb{H}$ and any condition $\pi \in \mathbf{P}$, $\pi \Vdash \langle \breve{x}, \breve{y}\rangle \in f$ implies $\pi \upharpoonright C \Vdash \langle \breve{x}, \breve{y}\rangle \in f$. We conclude that $f[\mathbf{G}] = f[\mathbf{G}_C]$, where $\mathbf{G}_C = \mathbf{G} \cap \mathbf{P}_C$. An ordinary product forcing argument shows that $\mathbf{G}_C$ is $\mathbf{P}_C$-generic over $\mathbb{H}$. However $\mathbf{P}_C$ (unlike $\mathbf{P}$) is a set in $\mathbb{H}$, therefore the extension $\mathbb{H}[\mathbf{G}_C]$ satisfies Theorem 7.1.20, and hence $X = f[\mathbf{G}_C] \in \mathbb{H}$.

Standardization follows from (ii) (see the proof of Theorem 7.1.20).

Collection. We suppose that $X \in \mathtt{Nms}(\mathbf{P})$ and $\Phi(x, y)$ is a formula with parameters in $\mathtt{Nms}(\mathbf{P})$. Let $A = \Delta_X \subseteq \mathtt{Nms}(\mathbf{P})$ be defined in $\mathbb{H}$ as in the proof of Separation. It suffices to find a set of names $B \in \mathbb{H},\ B \subseteq \mathtt{Nms}(\mathbf{P})$, such that for every $a \in A$ and every condition $\vartheta \in \mathbf{P}$, if $\vartheta \Vdash \exists y\, \Phi(x, y)$ then a stronger condition $\rho \leq \vartheta$ forces $\Phi(a, b)$ for a name $b \in B$.

We argue in $\mathbb{H}$.

Choose an internal set $C_0 \subseteq \mathbf{Ind}$ such that $||\Phi|| \cup ||X|| \subseteq C_0$. Then by definition we have $||a|| \subseteq C_0$ as well for any $a \in A$. As $\mathbf{P}_{C_0}$ is a set (in $\mathbb{H}$), there is a set $P \subseteq \mathbf{P}$ of forcing conditions and a set $B_0 \subseteq \mathtt{Nms}(\mathbf{P})$ such that for any $a \in A$ if $\pi_0 \in \mathbf{P}_{C_0}$ forces $\exists y\, \Phi(a, y)$ then there exist a condition $\pi \in P,\ \pi \leq \pi_0$ and a name $b \in B_0$ such that $\pi \Vdash \Phi(a, b)$.

The set B_0 is not yet the B we are looking for. To define B, we first of all choose an internal set C such that $C_0 \subseteq C$, $|\pi| \subseteq C$ for all $\pi \in P$, and for any $i = \langle w, \kappa, \mathbb{A}, \mathbb{B}\rangle \in C$ we have $\langle w', \kappa, \mathbb{A}, \mathbb{B}\rangle \in C$ for all $w' \in {}^*\mathbb{N}$. Each internal *correct* bijection $h : C \xrightarrow{\text{onto}} C$ generates an automorphism H_h of $\mathbf{P}$, see § 7.3d. Let us prove that

$$B = \{H_h[b] : b \in B_0 \text{ and } h \in \mathbb{I} \text{ is a } \textit{correct} \text{ bijection } C \xrightarrow{\text{onto}} C\}$$

is a set of names satisfying the required property. (To see that B is a set note that the collection of all internal correct bijections $C \xrightarrow{\text{onto}} C$ is an internal set simply because it can be $\in$-defined in $\mathbb{I}$.)

Suppose that $a \in A,\ \vartheta \in \mathbf{P}$, and $\vartheta \Vdash \exists y\, \Phi(a, y)$. Then the condition $\pi_0 = \vartheta \upharpoonright C_0$ also forces $\exists y\, \Phi(a, y)$ by Theorem 7.3.10. ($||\exists y\, \Phi(a, y)|| \subseteq C_0$ by the choice of C_0.) Then by the choice of P and B_0 there exist a condition $\pi \in P,\ \pi \leq \pi_0$, and a name $b \in B_0$, such that $\pi \Vdash \Phi(a, b)$.

Let $\vartheta' = \vartheta \upharpoonright C$. Then $E = |\pi|$ and $D' = |\vartheta'|$ are $*$-finite internal subsets of C. Therefore [14] by the choice of C there is an internal correct bijection

[14] This is the only point where the $*$-finiteness of the domains $|\pi|,\ \pi \in \mathbf{P}$, see Definition 7.3.4, is used. In fact the proof does not change much if the $*$-cardinals of the domains $|\pi|$ are restricted to be less than a fixed $*$-cardinal.

$h : C \xrightarrow{\text{onto}} C$ such that $h \restriction C_0$ is the identity and $(h''E) \cap D' \subseteq C_0$. Let $H = H_h$. Then $\pi' = H\pi$ belongs to $\mathbf{P}_C$, $\pi' \restriction C_0 = \pi \restriction C_0 \leq \pi_0$, and $|\pi'| \cap |\vartheta'| \subseteq C_0$, and hence ϑ' and π' are compatible. Therefore π' is also compatible with ϑ because $\pi' \in \mathbf{P}_C$ and $\vartheta' = \vartheta \restriction C$. Let $\rho \in \mathbf{P}$ be stronger than both π' and ϑ.

We observe that $\pi' \Vdash H\Phi(H[a], H[b])$ by Proposition 7.3.11. However $||\Phi|| \subseteq C_0$ and $||a|| \subseteq C_0$ by the choice of C_0, and hence $H\Phi$ coincides with Φ and $H[a] = a$ because $H \restriction C_0$ is the identity. It follows that $\rho \Vdash \Phi(a, b')$, where $b' = H[b] \in B$, as required.

We finally verify Isomorphism Property in the extension.

Since the models $\mathbb{H} \subseteq \mathbb{H}[\mathbf{G}]$ contain the same standard sets, the well-founded universe $\mathbb{WF}$ is also one and the same in the two models. Therefore $\mathbb{H}[\mathbf{G}]$ contains the same ordinals and cardinals as $\mathbb{H}$. Furthermore all triples of the form: *language – structure – structure*, to be considered in the scope of Isomorphism Property in $\mathbb{H}[\mathbf{G}]$, are already in $\mathbb{H}$. Thus let, in $\mathbb{H}$, $\mathcal{L}$ be a standard size first-order language containing κ symbols in $\mathbb{H}$ (κ being a cardinal in $\mathbb{H}$), and $\mathfrak{A}$, $\mathfrak{B}$ be a pair of internally presented elementarily equivalent $\mathcal{L}$-structures in $\mathbb{H}$.

Finally, let us prove that $\mathfrak{A}$ is isomorphic to $\mathfrak{B}$ in $\mathbb{H}[\mathbf{G}]$.

We argue in $\mathbb{H}$.

It can be assumed that $\mathcal{L} = \{s_\alpha\}_{\alpha<\kappa}$, where s_α is an n-ary relational symbol whenever $\alpha = \lambda + n < \kappa$ and λ is a limit ordinal or 0. Then $\mathbf{L} = {}^*\mathcal{L} \in \mathbb{S}$ is an internal (even standard) language equal to $\mathbf{L}({}^*\kappa)$ in the sense of Definition 7.3.3. Moreover we can identify s_α (a sybmol in $\mathcal{L}$) with $s_{{}^*\alpha}$ (a symbol in $\mathbf{L}({}^*\kappa)$) for any ordinal $\alpha < \kappa$, and hence identify $\mathcal{L}$ with the truncated language $\mathcal{L}({}^*\kappa)$ (see Definition 7.3.3). Accordingly we can consider $\mathfrak{A}$, $\mathfrak{B}$ as $\mathcal{L}({}^*\kappa)$-structures.

Now it follows from Corollary 1.3.13 (ii) (in $\mathbb{H}$) that there exist internal $\mathbf{L}({}^*\kappa)$-structures $\mathbb{A}$ and $\mathbb{B}$ such that $\mathfrak{A}$, $\mathfrak{B}$ are equal to the corresponding truncated substructures of $\mathbb{A}$, $\mathbb{B}$. Thus $i = \langle 0, {}^*\kappa, \mathbb{A}, \mathbb{B}\rangle$ belongs to $\mathbf{Ind}$ and $\mathcal{L} = \mathcal{L}_i$, $\mathfrak{A} = \mathfrak{A}_i$, $\mathfrak{B} = \mathfrak{B}_i$, and finally $\mathbf{P}_{\mathcal{L}}(\mathfrak{A}, \mathfrak{B}) = \mathbf{P}_{\mathcal{L}_i}(\mathfrak{A}_i, \mathfrak{B}_i) = \mathbf{P}_i$.

We argue in the **ZFC** *universe.*

Note that the set $\mathbf{G}_i = \{\pi_i : \pi \in \mathbf{G} \wedge i \in |\pi|\}$ belongs to $\mathbb{H}[\mathbf{G}]$. (Indeed, since $\mathbf{P}_i$ is a set in $\mathbb{H}$, a name for $\mathbf{G}_i$ can be defined in $\mathbb{H}$ as the set of all pairs of the form $\langle \pi, p\rangle$, where $\pi = \{\langle i, p\rangle\} \in \mathbf{P}$.) An ordinary product forcing argument shows that $\mathbf{G}_i$ is $\mathbf{P}_i$-generic over $\mathbb{H}$ in $\mathbb{H}[\mathbf{G}]$. But then the structures $\mathfrak{A}$ and $\mathfrak{B}$ are isomorphic in $\mathbb{H}[\mathbf{G}_i]$ (see the proof of Theorem 7.2.7 in §7.2f). It follows that $\mathfrak{A}$, $\mathfrak{B}$ are isomorphic in $\mathbb{H}[\mathbf{G}]$, a larger model. □

□ (*Theorems 7.3.2, 7.3.1*)

Problem 7.3.13. Suppose that $a \in {}^*\mathbb{N} \setminus \mathbb{N}$. Does there exist a generic extension of $\mathbb{L}[\mathbb{I}]$ in which all nonstandard $*$-integers in $U = \bigcup_{n\in\mathbb{N}}[0, na)$ are equinumerous to each other and all $*$-integers in ${}^*\mathbb{N} \setminus U$ are equinumerous to each other but not equinumerous to those in the first class? □

Historical and other notes to Chapter 7

Section 7.1. Forcing in a non-well-founded environment has been occasionally studied in several papers, for instance [Bof 72, Tz **, Mat 01]. Yet the version applied in this book is close to the ordinary **ZFC** forcing because the ill-founded "kernel" $\mathbb{I}$ of the **HST** universe does not change in generic extensions. Some details (including Standardization) need some effort to be settled, of course.

Section 7.2. See [CK 92, Theorem 5.1.13] on isomorphism between saturated elementarily equivalent structures in model theory and [CK 92, 5.1.11] on a back-and-forth argument in this context. The forcing $\mathbb{P}_{\mathcal{L}}(\mathfrak{A}, \mathfrak{B})$ (Definition 7.2.10 above), which induces a generic isomorphism between elementarily equivalent structures, was introduced in [KanR 97, KanR 97a].

Section 7.3. Isomorphism property $\mathbf{IP}_\kappa$ was introduced by Henson [Hen 74]. Studies carried out in the 1990s (see, for instance, [Jin 92, Jin 92a, Jin 96, Jin 99, Jin **, JinS 94, JinK 93, Schm 95], also Ross [Ross 90]) demonstrate that **IP** implies several strong consequences inavailable in the framework of ordinary postulates of nonstandard analysis, for instance the existence of a set of infinite Loeb outer measure which intersects every set of finite Loeb measure by a set of Loeb measure zero, the theorem that any two infinite internal sets have the same external cardinality, and many more. Typical consequences of $\mathbf{IP}_\kappa$ for different cardinals κ can be easily converted to consequences of Isomorphism Property in **HST**.

Isomorphism Property as a hypothesis in **HST** and the proof of its consistency with **HST** (Theorem 7.3.1 above) appeared in [KanR 97a]. [15]

Problem 7.3.13: compare with Theorem 8 in [Mil 90].

We close this Chapter with two more problems.

Problem 7.3.14. Find other significant properties on external universe whose consistency or independence can be proved by forcing. Possible candidates are several hypotheses of the existence of generic sets studied by Di Nasso and Hrbaček [DiNH 03] (see also [Jin 99, Jin **]) in the frameworks of model-theoretic nonstandard analysis. Another possible group consists of questions of the type considered by Miller [Mil 90]. □

Problem 7.3.15. Our forcing set-up includes the principle that no new internal sets can be added, and for good reasons. Nevertheless, if $\mathbb{S}[G]$ is a usual, **ZFC**-like generic extension of $\mathbb{S}$, a standard core of a model $\mathbb{H} \models$ **HST**, is there any way to naturally define a model $\mathbb{H}[\mathbf{G}] \models$ **HST** which is a standard core extension of $\mathbb{S}[G]$ and simultaneously a generic extension of $\mathbb{H}$? □

[15] The question how to accomodate advanced nonstandard tools like the isomorphism property in a reasonable nonstandard set theory was discussed in the course of a meeting between H. J. Keisler and one of the authors (V. Kanovei, during his visit to Madison in December 1994).

8 Other nonstandard theories

The "Hrbaček paradox" (Theorem 1.3.9) can be viewed as the statement of inconsistency of the conjunction of the four following axioms, over a weak nonstandard theory:

– Collection;

– either of the axioms of Choice and Power Set;

– standard size Saturation;

– Standardization.

Any solution of the paradox means that (at least) one of the axioms has to be abandoned or essentially weakened to a form compatible with the other ones. The theory **HST**, the main topic of this book, sacrifices both Choice and Power Set (keeping either of them in a standard size form, and fully in suitable partially saturated universes). Other solutions are possible: we can keep any three of the four axioms and a partial form of the fourth one, which leads to theories based on different views of the nonstandard universe.

This Chapter contains a brief exposition of the theories **NST**, **KST**, ***ZFC** obtained this way. We begin in § 8.1 with Kawaï's theory **KST** which keeps Collection, Power Set, and Choice but reduces Standardization to a form compatible with the assumption that $\mathbb{S}$ and $\mathbb{I}$ are sets rather than proper classes, and **NST**, another of Hrbaček's theories (§ 8.2) which abandons Collection in favour of Power Set and Choice. Di Nasso's theory ***ZFC**, designed to avoid the Hrbaček paradox by reducing the amount of Saturation available to cardinals $\in$-definable in $\mathbb{WF}$, is considered in § 8.4a. These theories offer adequate tools to develop nonstandard mathematics, and (especially, **KST**) have advantages relative to **HST** in some details. All of them are conservative, but, unlike **HST**, not "realistic" (in the sense of Definition 4.1.8) extensions of **ZFC**, hence, they hardly can be anything more than syntactical deduction schemes with respect to the "standard" universe of **ZFC**.

Some other nonstandard theories will be considered, most notably, the Ballard – Hrbaček system based on Boffa's non-well-founded set theory.

The connection between well-founded and standard sets, on which the scheme "$\mathbb{WF} \xrightarrow{*} \mathbb{I}$ [in $\mathbb{H}$]" is based in **HST** (see § 1.2a), will not be valid any longer for most of the theories considered in this Chapter, although partial schemes of this kind will usually work.

8.1 Nonstandard set theory of Kawaï

Unlike **HST**, Kawaï's nonstandard set theory **KST** describes the class $\mathbb{I}$ of internal sets as a universe satisfying internal set theory **IST** rather than bounded set theory **BST**. This does not allow to use Hrbaček's definition of internal sets as elements of standard sets, therefore, in Kawaï's system $\mathtt{int}\, x$ ("x is internal") is an atomic predicate.

The theory **KST** contains both Power Set, Choice, and Collection, actually, all of **ZFC** except for Regularity, so that it is well equipped technically as a nonstandard theory. This is a solution of the "Hrbaček paradox" at the cost of Standardization: this axiom is weakened to a form compatible with the assumption that $\mathbb{S}$ is a set rather than a proper class (as in most of the nonstandard set theories). Metamathematically, **KST** is still a conservative extension of **ZFC**, but not a "realistic" nonstandard theory.

Theories **NST** and ***ZFC**, which solve the paradox by weakening, resp., Collection and Saturation, will be presented below.

8.1a The axioms of Kawaï's theory

Thus, **KST** is a theory in $\mathcal{L}_{\in,\,\mathtt{st},\,\mathtt{int}}$, the language containing the membership $\in$ and the unary predicates , $\mathtt{st}$, $\mathtt{int}$ as atomic predicates. The list of axioms of **KST** includes:

1) all axioms of § 1.1b (the first group of **HST** axioms), with the schemata of Separation, Collection, Replacement in the language $\mathcal{L}_{\in,\,\mathtt{st},\,\mathtt{int}}$;
2) Power Set and Choice in their ordinary **ZFC** forms in the $\in$-language, as in § 1.1h;
3) Transfer, Transitivity of $\mathbb{I}$, Regularity over $\mathbb{I}$, $\mathbf{ZFC}^{\mathtt{st}}$ of § 1.1c (the second group of **HST** axioms), however, $\mathtt{int}$ is now an atomic predicate of the language rather than the formula $\exists^{\mathtt{st}} y\, (x \in y)$ as in **HST**;

and three more axioms:

Set–existence of $\mathbb{I}$: $\mathbb{I} = \{x : \mathtt{int}\, x\}$ is a set and $\mathbb{S} \subseteq \mathbb{I}$;

Restricted Standardization: $\forall^{\mathtt{st}} S\, \forall X \subseteq S\, \exists^{\mathtt{st}} Y\, (X \cap \mathbb{S} = Y \cap \mathbb{S})$;

Strong Saturation: if $\mathscr{X} \subseteq \mathbb{I}$ is a $\cap$-closed set of $\mathbb{S}$-*size*, *i.e.*, a set of the form $Y = \{f(x) : x \in \mathbb{S}\}$, then $\bigcap \mathscr{X} \neq \varnothing$.

The first axiom implies that $\mathbb{S}$ is also a set by Separation. This immediately makes the **HST** Standardization inconsistent with **KST**, yet essentially Restricted Standardization expresses the same property because in **HST** anyway every set $X \subseteq \mathbb{S}$ satisfies $X \subseteq S$ for a standard S by (3) of Exercise 1.1.11. On the other hand, Kawaï's theory **KST** admits a bigger amount of Saturation than **HST**: indeed, any set of standard size is obviously a set of $\mathbb{S}$-size as well, but not *vice versa*.

Exercise 8.1.1. Show that if Φ is an axiom of **IST** then $\Phi^{\mathtt{int}}$ is a theorem of **KST**, thus $\langle \mathbb{I}; \in, \mathtt{st}\rangle$ is an interpretation of **IST** in **KST**.

Yet **KST** does not prove that $\mathbb{I}$ is formally a model of **IST**: this fact follows from Theorem 8.1.5 by the same argument as in Exercise 1.5.17. □

Define, in **KST**, $\mathfrak{s} = \mathtt{card}\,\mathbb{S}$ and $\mathfrak{i} = \mathtt{card}\,\mathbb{I}$; both sets (as well as any other set) are well-orderable, hence, have cardinals in **KST**.

Similarly to **HST**, the whole universe of sets is postulated in **KST** to be a **ZFC**-like world over $\mathbb{I}$ as the collection of "atoms", but, unlike Hrbaček's theory, **KST** sees $\mathbb{I}$ (as well as $\mathbb{S} \subseteq \mathbb{I}$) as sets rather than proper classes. This property, and the weakened Standardization, is why the Hrbaček paradox does not work in **KST** despite the presence of Collection.

Exercise 8.1.2 (KST). Show that the class $\mathbb{WF}$ of well-founded sets is transitive and $\subseteq$-complete (that is, $y \subseteq x \in \mathbb{WF} \Longrightarrow y \in \mathbb{WF}$), and an interpretation of **ZFC**. (Compare with Theorem 1.1.9 in **HST**!) □

In **KST**, $*$-methods can be developed to a great extent.

Definition 8.1.3. A set $x \in \mathbb{S}$ is *condensable* if there exists a transitive set $X \in \mathbb{S}$ containing x and a map $y \to \hat{y}$ defined on $X \cap \mathbb{S}$ such that $\hat{y} = \{\hat{z} : z \in y \cap \mathbb{S}\}$ for all $y \in X \cap \mathbb{S}$. In particular, $\hat{x}$ is defined in this case, and, as $\in \restriction \mathbb{S}$ is a well-founded relation (Theorem 1.1.9(i) remains valid in **KST**), $\hat{x}$ is a well-founded set independent of the choice of X.

Put $\mathbb{WF}^{\mathtt{feas}} = \{\hat{x} : x \in \mathbb{S}\}$ (*feasible* well-founded sets). □

Collection makes all standard sets condensable in **KST** (as well as in **HST**, where $x = {}^*(\hat{x})$ for any $x \in \mathbb{WF}$, Exercise 1.1.8, and $\mathbb{WF}^{\mathtt{feas}} = \mathbb{WF}$).

Exercise 8.1.4 (KST). (1) Prove that $\mathbb{WF}^{\mathtt{feas}}$ is a transitive and $\subseteq$-complete subclass of $\mathbb{WF}$, and the map $x \mapsto \hat{x}$ is $1-1$, and hence for any $u \in \mathbb{WF}^{\mathtt{feas}}$ there is a unique set $x \in \mathbb{S}$, denoted by ${}^*u$, of course, such that $u = \hat{x}$. Thus $u \mapsto {}^*u$ is an $\in$-isomorphism of $\mathbb{WF}^{\mathtt{feas}}$ onto $\mathbb{S}$, so that $\mathbb{WF}^{\mathtt{feas}}$ is an interpretation of **ZFC**.

(2) Prove that $\mathbb{WF}^{\mathtt{feas}}$ is a set, moreover, $\mathbb{WF}^{\mathtt{feas}} = \mathbf{V}_{\mathfrak{s}}$, where $\mathfrak{s} = \mathtt{card}\,\mathbb{S}$, yet **KST** does not prove that $\mathbb{WF}^{\mathtt{feas}}$ is a *model* of **ZFC**, for otherwise **KST** proves $\mathtt{Consis}$ **ZFC**, which is impossible by Theorem 8.1.5 below.

(3) Prove that $\mathbf{V}_{\xi}[\mathbb{I}]$ is a transitive set in **KST** for any (well-founded) ordinal ξ, and every set x belongs to $\bigcup_{\xi \in \mathtt{Ord}} \mathbf{V}_{\xi}[\mathbb{I}]$. (Compare with § 1.5b.) □

It follows that the theory **KST** is strong enough to develop the $*$-approach to nonstandard mathematics practically to the same extent and in the same way as in **HST**, with the only difference that the domain $\mathbb{WF}^{\mathtt{feas}}$ of $*$ is a part (possibly proper) of the whole class of well-founded sets. This approach can be called *the scheme* "$\mathbb{WF}^{\mathtt{feas}} \xrightarrow{*} \mathbb{I}$ [*in* $\mathbb{H}$]".

8.1b Metamathematical properties

Recall that a nonstandard theory $\mathfrak{T}$ in the st-$\in$-language is a conservative standard core extension of **ZFC** in the case when any $\in$-sentence Φ is a theorem of **ZFC** iff the relativization $\Phi^{\mathtt{st}}$ is a theorem of $\mathfrak{T}$.

Theorem 8.1.5. *The theory* **KST** *is a conservative, and hence equiconsistent standard core extension of* **ZFC**.

Proof. Our plan is to define an interpretation of **KST** in **ZFC**ϑ, a standard theory studied in §§ 1.5f, 4.4c. The interpretation will be of a kind sufficient to derive the theorem.

We argue in **ZFC**ϑ.

Let $V = \mathbf{V}_\vartheta$. Let ${}^*\mathbf{v} = \langle {}^*V\,; {}^*\!\in, {}^*\mathtt{st}\rangle$ be a ϑ^+-saturated interpretation of **IST** plus S-Size Choice in **ZFC**ϑ, defined in § 4.4d (for $\gamma = \vartheta^+$), with an elementary standard core embedding $* : V \to {}^*V$. (Note that ${}^*V = V_\gamma$, $*$ is $e_{0\gamma}$, and ${}^*\!\in$ is ε_γ in the notation of § 4.4d.) To prove the theorem we define a "superstructure" over ${}^*V$ that interprets **KST**. To avoid unnecessary complications, assume that all elements of ${}^*V$ have one and the same von Neumann rank in the **ZFC**ϑ universe. (Otherwise choose an ordinal κ with ${}^*V \subseteq \mathbf{V}_\kappa$, replace each $x \in {}^*V$ by $\langle x, \kappa\rangle$, and change ${}^*\!\in$ and ${}^*\mathtt{st}$ accordingly.)

Define, by induction on $\xi < \gamma$, a set $\mathbf{P}_\xi$, so that $\mathbf{P}_0 = {}^*V$ and

(I) $\mathbf{P}_1$ is the set of all sets $X \subseteq \mathbf{P}_0$ different from any set of the form $x_{{}^*\in} = \{y \in {}^*V : y \;\varepsilon_\gamma\; x\}$, where $x \in {}^*V$;

(II) if $\xi \geq 2$ then $\mathbf{P}_\xi$ is the set of all sets $X \subseteq \mathbf{P}_{<\xi} = \bigcup_{\eta<\xi} \mathbf{P}_\eta$ such that $X \not\subseteq \mathbf{P}_{<\eta}$ for any $\eta < \xi$.

Put $\mathbf{P} = \bigcup_{\xi<\gamma} \mathbf{P}_\xi$. For $p,\, q \in \mathbf{P}$ define $p \;{}^\bullet\!\in\; q$ if either $p,\, q \in {}^*V$ and $x \;{}^*\!\in\; y$ or just $p \in q$. For $p \in \mathbf{P}$ put ${}^\bullet\mathtt{st}\, p$ if $p \in {}^*V$ and ${}^*\mathtt{st}\, p$ in ${}^*V$.

Exercise 8.1.6. Prove that $\xi \neq \eta \Longrightarrow \mathbf{P}_\xi \cap \mathbf{P}_\eta = \varnothing$.

Prove that the classes of all standard and internal elements of ${}^\bullet V$ satisfy $\mathbb{S}^{({}^\bullet V)} = \{{}^*u : u \in \mathbf{V}_\vartheta\}$ and $\mathbb{I}^{({}^\bullet V)} = {}^*V$, and ${}^\bullet\!\in \restriction {}^*V$ coincides with ${}^*\!\in$. □

We claim that the structure ${}^\bullet V = \langle \mathbf{P}\,; {}^\bullet\!\in, {}^*\mathtt{st}\rangle$ interprets **KST**.

Justification of the axioms of **ZFC** (minus Regularity) is a rather routine exercise. We consider four key axioms.

Extensionality. By definition, ${}^\bullet\!\in$-elements of any $X \in \mathbf{P}$ coincide with its $\in$-elements whenever $X \notin \mathbf{P}_0$, and its ${}^*\!\in$-elements whenever $X \in \mathbf{P}_0 = {}^*V$. As ${}^*V$ satisfies Extensionality, the only opportunity to violate Extensionality in ${}^\bullet V$ is to find a pair of sets $X \in \mathbf{P}_1$ and $x \in \mathbf{P}_0 = {}^*V$ with $x = y_{{}^*\in}$, which is indeed impossible by (I) of the definition.

Separation. If $X \in \mathbf{P} \smallsetminus \mathbf{P}_0$ then all ${}^\bullet\!\in$-elements of X are its $\in$-elements, moreover, any set $Y \subseteq X$ either belongs to $\mathbf{P} \smallsetminus \mathbf{P}_0$ or $Y = y_{{}^*\in}$ for some $y \in {}^*V = \mathbf{P}_0$. If $X = x \in P_0$ then the set of all ${}^\bullet\!\in$-elements of X is $x_{{}^*\in}$, thus, any $Y \subseteq x_{{}^*\in}$ either belongs to $\mathbf{P}_1$ or $Y = y_{{}^*\in}$ for some $y \in {}^*V = \mathbf{P}_0$.

Power Set. Let $X \in \mathbf{P}$. If $X \notin \mathbf{P}_0$ then the ${}^{\bullet}V$-power set Z of X consists, first, of all sets $Y \subseteq X$ different from any $y_{{}^{\bullet}\!\in}$, $y \in {}^{*}V$, and all $y \in {}^{*}V$ such that $y_{{}^{\bullet}\!\in} \subseteq X$. Clearly Z belongs to $\mathbf{P}$. Similarly, if $X \in \mathbf{P}_0$ then the ${}^{\bullet}V$-power set Z of X consists, first, of all sets $Y \subseteq X_{{}^{\bullet}\!\in}$ different from any set $y_{{}^{\bullet}\!\in}$, $y \in {}^{*}V$, and all $y \in {}^{*}V$ such that $y_{{}^{\bullet}\!\in} \subseteq X_{{}^{\bullet}\!\in}$.

Collection follows from the fact that the construction is $\mathtt{Ord}$-long, and hence for any set $X \subseteq \mathbf{P}$ there is an ordinal ξ such that $X \subseteq \bigcup_{\eta<\xi} \mathbf{P}_\eta$.

Consider now the axioms of **KST** related to standard and internal sets. Recall that $\mathbb{I}^{({}^{\bullet}V)} = {}^{*}V$ is an interpretation of **IST**. This reduces our task to verification of Saturation and Standardization in ${}^{\bullet}V$: the rest of the axioms easily follows from the construction of $\mathbf{P}$, and we leave their verification as an **exercise** for the reader.

Restricted Standardization. Let $S \in \mathbb{S}^{({}^{\bullet}V)}$. According to Exercise 8.1.6, $S = {}^{*}U$ for some $U \in \mathbf{V}_\vartheta$, moreover, any standard $x \in S$ is equal to ${}^{*}u$, $u \in U$, therefore, the task is to find, for any $X \subseteq U$, a set $Y \in \mathbb{S}^{({}^{\bullet}V)}$ whose ${}^{\bullet}\!\in$-elements are ${}^{*}y$, $y \in Y$. Clearly $Y = {}^{*}X$ is as required.

Strong Saturation. Recall that ${}^{*}V$ was chosen to be ϑ^{+}-saturated, and $\vartheta = \operatorname{card} \mathbf{V}_\vartheta$ by 1.5.18. Then $\mathbb{S}^{({}^{\bullet}V)} = \{{}^{*}u : u \in \mathbf{V}_\vartheta\}$ is a set of cardinality ϑ. This suffices to derive Strong Saturation.

□ (*Theorem 8.1.5*)

Exercise 8.1.7. Show (using Exercise 8.1.1 and the non-interpretability in Theorem 4.1.10) that **KST** is not standard core interpretable in **ZFC** and hence not a "realistic" theory. □

Problem 8.1.8. (1) Is there a reasonable standard extension of **ZFC** in which **KST** is standard core interpretable ?

(2) Is **KST** a conservative internal core extension of **IST**, in the sense of § 5.1a? (See also 8.2.17 below.)

(3) Is **KST** internal core interpretable in **IST**, in the sense of § 5.1a? □

A complete characterization of **KST**-extendible standard models of **ZFC** (see Definition 4.1.6) was obtained by Kanovei and Reeken [KanR 00a].

Exercise 8.1.9. Arguing as in Exercise 8.2.7 below, define an interpretation of **HST** in **KST**. □

8.1c Special model axiom

It follows from 8.1.7 that the theory **KST** badly lacks that type of proximity to **ZFC** or **BST** which **HST** has and which is connected with the notions of standard core and internal core interpretability; in brief it is not a "realistic" theory as we defined this concept in § 4.1c. On the other hand, **KST** is better equipped technically than **HST**, let alone internal theories. In particular, we have all of **ZFC**, with the mandatory exception of Regularity, but with

Choice and Power Set, and the schemata of Separation and Collection in the language $\mathcal{L}_{\in,\,\mathtt{st},\,\mathtt{int}}$.

The additional feature of $\mathbb{S}$ and $\mathbb{I}$ being sets rather than proper classes allows to consistently use such a tool as the special model axiom, known from the model theoretic version of nonstandard analysis.

Special Model Axiom, SMA: Any internally presented (see § 7.3) structure $\mathfrak{A}$ of a language $\mathcal{L}_{\mathfrak{A}}$ with $\mathbb{S}$-size many (*i.e.*, $\leq \mathfrak{s}$) symbols, admits a presentation of the form $\mathfrak{A} = \bigcup_{\xi<\lambda} \mathfrak{A}_\xi$, where $\lambda = \mathtt{card}\, A$, $A = |\mathfrak{A}|$ is the underlying set of $\mathfrak{A}$, and each $\mathfrak{A}_\xi$ is a $(\mathtt{card}\,\xi)^+$-saturated elementary submodel of $\mathfrak{A}$.

In the model theoretic version of nonstandard analysis, they define κ-SMA, where κ is a cardinal, meaning that $\mathcal{L}_{\mathfrak{A}}$ has $< \kappa$ symbols; in these terms, SMA in our sense is $\mathfrak{s}^+$-SMA. Neither SMA nor κ-SMA can be even formulated in **HST**, because infinite internal sets are not well-orderable in this theory.

Exercise 8.1.10 (KST). (1) Prove in **KST** that SMA implies Isomorphism Property (as in § 7.3), even in a stronger version for languages of $\mathbb{S}$-size. (*Hint*. Follow [Ross 90], where it is shown that κ-SMA implies κ-IP.)

(2) Infer that SMA implies, in **KST**, that all infinite internal sets are (externally) equinumerous. □

Exercise 8.1.11 (KST). Prove that $\mathfrak{i} \geq \mathfrak{s}^+$, and if $\mathfrak{i} = \mathfrak{s}^+$ then all infinite internal sets have one and the same cardinality $\mathfrak{i}$ and SMA trivially holds by taking $\mathfrak{A}_\xi = \mathfrak{A}$ for all $\xi < \mathfrak{i}$. □

Exercise 8.1.12. Prove that $\mathfrak{i} = \mathfrak{s}^+$ is consistent with **KST**.

Hint. Argue as in the proof of Theorem 8.1.5, but assuming that the generalized continuum-hypothesis **GCH** holds in the **ZFC**ϑ universe. Note that $\mathtt{card}\,{}^*V = (\mathtt{card}\,V)^+ = \vartheta^+$. □

It follows that SMA is consistent with **KST**, too, in other words, **KST** + SMA is equiconsistent with **ZFC**. This leaves open the following question (we conjecture a positive answer).

Problem 8.1.13. Is SMA consistent with **KST** $+\, \mathfrak{i} > \mathfrak{s}^+$? □

A rather complicated model of **KST** where SMA fails was constructed by Kanovei and Reeken [KanR 99b]. The idea is to employ a model of **KST** where there is a number $n \in {}^*\mathbb{N} \smallsetminus \mathbb{N}$ such that the set $\{1, 2, ..., n\}$ has a strictly smaller cardinality than the whole set ${}^*\mathbb{N}$, which is incompatible with SMA by the result of Exercise 8.1.10. Such a model was constructed using methods similar to the forcing notions considered in Chapter 7.

8.2 "Nonstandard set theory" of Hrbaček

One more solution of the "Hrbaček paradox", suggested by Hrbaček, is to keep the axioms of Standardization, Power Set, and Choice but abandon Collection. (Still some useful forms of Collection can be saved.) This leads to **NST**, a nonstandard theory which is, similarly to **HST**, a conservative standard core extension of **ZFC** (Theorem 8.2.10) but, unlike **HST** (and similarly to **IST**) does not admit a standard core interpretation in **ZFC** (Exercise 8.2.16), that is, it is not a "realistic" nonstandard theory in the sense of our Definition 4.1.8.

8.2a Axioms

The "nonstandard set theory" **NST** (or $\mathbf{NS}_2(\mathbf{ZFC})$ in [Hr 78]) is a theory in the $\mathtt{st}$-$\in$-language, a modification of **HST** obtained by removing the schemata of Replacement and Collection (but keeping Separation and Saturation) and adding the axioms of Power Set and Choice (in the form: any set is well-orderable). We also adjoin the following two additional axioms absent in Hrbaček's prototype version:

Transitive Hulls: any set belongs to a transitive set;

Full Boundedness: any set $X \subseteq \mathbb{I}$ is covered by an internal set.

The reason to adjoin these two axioms is that Regularity over $\mathbb{I}$ does not seem to exhibit its full capacity in the absence of Collection (or Replacement). In particular, simple **HST** proofs of Transitive Hulls and Full Boundedness (as in § 1.5b and Exercise 1.1.11) do not work in **NST** without the additional axioms.

Exercise 8.2.1. (1) Show that the definition of $\mathbf{V}_\xi[U]$ in § 1.5b is legitimate in **NST**, *i.e.*, $\mathbf{V}_\xi[U]$ (not asserted to be a set !) is well-defined in **NST** for any $U \in \mathbb{I}$ and any (well-founded) ordinal ξ.

(2) Prove, following arguments in § 1.5b and using Full Boundedness, that any set x belongs to some $\mathbf{V}_\xi[U]$. □

The classes $\mathbb{S} \subseteq \mathbb{I}$ of standard and internal sets, and $\mathbb{WF}$ of well-founded sets are defined in **NST** in the same way as in **HST**.

All axioms of **HST** related to standard and internal sets are still present in **NST**. Thus, similarly to **HST** (we refer to Theorem 3.1.8), the theory **NST** allows to prove that the internal universe $\mathbb{I} = \langle \mathbb{I}; \in, \mathtt{st} \rangle$ is an interpretation of **BST**. However the structure of the whole (external) universe appears to be rather different from that of **HST**. Presence of Power Set and Choice is an advantage of **NST**, but the absence of Collection/Replacement abolishes some simple theorems of **HST**, for instance, the connection between $\mathbb{S}$ and the class $\mathbb{WF}$ does not exist any more.

We shall see that the class $\mathbb{WF}$ can be, in **NST**, as small as the set $\mathbf{V}_{\omega+\omega} = \bigcup_{n\in\mathbb{N}} \mathbf{V}_{\omega+n}$, a union of rather low von Neumann levels (and then $\mathbf{V}_{\omega+\omega}$ is not a set), but also can be as large as $\mathbb{S}$ and even strictly bigger! In view of this, the scheme "$\mathbb{WF} \xrightarrow{*} \mathbb{I}$ [in $\mathbb{H}$]" of § 1.2a, generally speaking, fails, but a "safe" development of $*$-methods in **NST** can be maintained in reduced form.

Exercise 8.2.2 (NST). Prove that a set ${}^*x \in \mathbb{S}$ can be defined for each $x \in \mathbf{V}_{\omega+\omega}$ so that ${}^*x$ is the only standard set satisfying ${}^*x \cap \mathbb{S} = \{{}^*y : y \in x\}$. Prove that $*$ is an elementary embedding of $\langle \mathbf{V}_{\omega+\omega}\,; \in\rangle$ in $\langle (\mathbf{V}_{{}^*\omega+{}^*\omega})^{\mathbb{I}}\,; \in\rangle$, where $(\mathbf{V}_{{}^*\omega+{}^*\omega})^{\mathbb{I}}$ is the $({}^*\omega + {}^*\omega)$-th von Neumann level in $\mathbb{I}$. □

Still $\mathbf{V}_{\omega+\omega}$ suffices to define natural and real numbers, real functions *etc.*, and related concepts like finiteness, which is enough to develop, in the nonstandard $*$-manner, all mathematics based on **ZC** rather than full **ZFC**.

The following example presents a demonstrable failure of Collection in **NST**: there is a surjection from a set onto $\mathbb{S}$, a proper class.

Recall the definitions of A-codes, $A_{\mathbf{x}}$, $\stackrel{a}{=}$, ${}^a\!\in$ *etc.*, from § 5.3b, 5.3c, 5.4b.

Let, in **NST**, $\mathbf{S}$ be the set of all A-codes $\mathbf{x}$ such that $T_{\mathbf{x}} \subseteq ({}^*\mathbb{N})^{<\omega}$ is a set of standard size and $\mathbf{x}(s) = \varnothing$ for any $s \in \mathtt{Max}\,T_{\mathbf{x}}$. (That X is a set follows from the axioms of Power Set and Separation; X is not a set in **HST**.) For $\mathbf{x} \in \mathbf{S}$ and $t \in T_x$, define $\mathsf{F}^{\mathbb{S}}_{\mathbf{x}}(t) \in \mathbb{S}$ and $|t|^{\mathbb{S}}_{\mathbf{x}} \in \mathtt{Ord}^{(\mathbb{S})}$ by induction:

1) $\mathsf{F}^{\mathbb{S}}_{\mathbf{x}}(t) = \mathbf{x}(t) = \varnothing$ and $|t|^{\mathbb{S}}_{\mathbf{x}} = 0$ whenever $t \in \mathtt{Max}\,T$.
2) otherwise, $|t|^{\mathbb{S}}_{\mathbf{x}}$ is the least standard ordinal strictly bigger than all $|t^\wedge k|^{\mathbb{S}}_{\mathbf{x}}$, $t^\wedge k \in T_{\mathbf{x}}$, and $\mathsf{F}^{\mathbb{S}}_{\mathbf{x}}(t)$ is the only (by Standardization) standard set satisfying $\mathsf{F}^{\mathbb{S}}_{\mathbf{x}}(t) \cap \mathbb{S} = \{\mathsf{F}^{\mathbb{S}}_{\mathbf{x}}(t^\wedge k) : t^\wedge k \in T_{\mathbf{x}}\}$.

Exercise 8.2.3. Prove that this is well-defined in **NST**, that is, the collections $\{|t^\wedge k|^{\mathbb{S}}_{\mathbf{x}} : t^\wedge k \in T_{\mathbf{x}}\}$ and $\{\mathsf{F}^{\mathbb{S}}_{\mathbf{x}}(t^\wedge k) : t^\wedge k \in T_{\mathbf{x}}\}$ in 2) are sets.

Hint. The main problem is that **NST** does not contain the Collection (or Replacement) axiom to show this by a straightforward argument. Yet if $\{|t^\wedge k|^{\mathbb{S}}_{\mathbf{x}} : t^\wedge k \in T_{\mathbf{x}}\}$ is <u>not</u> a set then it is unbounded in $\mathtt{Ord}^{(\mathbb{S})}$, moreover, $\{|s|^{\mathbb{S}}_{\mathbf{x}} : s \in T_{\mathbf{x}} \wedge t \subset s\}$ is just equal to $\mathtt{Ord}^{(\mathbb{S})}$. Therefore, as $T_{\mathbf{x}}$ is of standard size, we can produce a map from $X \cap \mathbb{S}$ onto $A \cap \mathbb{S}$ for a pair of standard sets X, A of different $\mathbb{S}$-cardinalities (A is just a big enough $\mathbb{S}$-cardinal), which leads to contradiction, see Exercise 1.1.11(4).

Thus $|t|^{\mathbb{S}}_{\mathbf{x}}$ is well-defined. Yet the collection $\{\mathsf{F}^{\mathbb{S}}_{\mathbf{x}}(t^\wedge k) : t^\wedge k \in T_{\mathbf{x}}\}$ is bounded in $\mathbb{S}$ by the von Neumann rank $\le |t|_{\mathbf{x}}$, and hence Separation can be used instead of Collection. □

Now put $\mathsf{A}^{\mathbb{S}}_{\mathbf{x}} = \mathsf{F}^{\mathbb{S}}_{\mathbf{x}}(\Lambda)$ for any $\mathbf{x} \in \mathbf{S}$. ($\Lambda$ is the empty sequence.)

Exercise 8.2.4 (NST). (1) Prove that $\mathbf{x} \mapsto \mathsf{A}^{\mathbb{S}}_{\mathbf{x}}$ maps $\mathbf{S}$, a set, <u>onto</u> $\mathbb{S}$, a proper class. (*Hint*: define, using Choice and Saturation, a code $\mathbf{x} \in \mathbf{S}$ with $\mathsf{A}^{\mathbb{S}}_{\mathbf{x}} = s$ by $\in$-induction for any $s \in \mathbb{S}$.)

(2) Prove that $\mathbf{x} \overset{a}{=} \mathbf{y}$ iff $\mathsf{A}^{\mathbb{S}}_{\mathbf{x}} = \mathsf{A}^{\mathbb{S}}_{\mathbf{y}}$ and $\mathbf{x} \,{}^{a}\!\in \mathbf{y}$ iff $\mathsf{A}^{\mathbb{S}}_{\mathbf{x}} \in \mathsf{A}^{\mathbb{S}}_{\mathbf{y}}$. Thus $\mathbf{x} \mapsto \mathsf{A}^{\mathbb{S}}_{\mathbf{x}}$ is an isomorphism of $\langle \mathbf{S}/\overset{a}{=}\,; {}^{a}\!\in\rangle$ onto $\langle \mathbb{S}; \in\rangle$, so that $\langle \mathbf{S}/\overset{a}{=}\,; {}^{a}\!\in\rangle$, a structure whose domain $\mathbf{S}/\overset{a}{=}$ is a set, is an *interpretation* of **ZFC** in **NST**. (We'll see what it takes to upgrade $\langle \mathbf{S}/\overset{a}{=}\,; {}^{a}\!\in\rangle$ to a *model* of **ZFC**!)

(3) Prove, using the result of Exercise 1.1.11(4) as above, that $\mathbf{S}/\overset{a}{=}$ is <u>not</u> a set of standard size. □

8.2b Additional axioms of Collection

It is shown in [Hr 78] that even Σ_1-Collection contradicts **NST**. It is of interest to add yet weaker forms of Collection to **NST** in order to improve the structure of the universe, in particular, to increase the domain of applicability of $*$-methods. We begin with the following principles:

Standard Size Collection: Collection (§ 1.1h), in st-$\in$-language, in the case when X is any set of standard size (a functional image of a set $S \subseteq \mathbb{S}$).

Exercise 8.2.5 (**NST** + Standard Size Collection). Prove that $\langle \mathbf{S}/\overset{a}{=}\,; {}^{a}\!\in\rangle$ is a *model* of **ZFC**. (*Hint*. As any $c \in \mathbf{S}/\overset{a}{=}$ obviously has only standard size many ${}^{a}\!\in$-elements in $\mathbf{S}/\overset{a}{=}$, and we know already that $\langle \mathbf{S}/\overset{a}{=}\,; {}^{a}\!\in\rangle$ is an *interpretation* of **ZFC** (Exercise 8.2.4 (**NST**)), it suffices to show that for any set $M \subseteq \mathbf{S}$ of standard size there is $\mathbf{x} \in \mathbf{S}$ such that M is the set of all ${}^{a}\!\in$-elements of $\mathbf{x}$, which is easy by Standard Size Collection.)

Thus, **NST** + Standard Size Collection proves Consis **ZFC** (Hrbaček). Models of stronger theories, like the Kelley–Morse theory of classes, can also be obtained: add the power set $\mathscr{P}(\mathbf{S}/\overset{a}{=})$ as the next type level, *etc.* □

In the presence of Collection, as in **HST** or **KST**, all standard sets are condensable (Definition 8.1.3) and $x = {}^{*}(\hat{x})$ for any $x \in \mathbb{WF}$ (Exercise 1.1.8), similarly, in **NST**, Standard Size Collection suffices to prove:

Standard Condensation: All sets in $\mathbb{S}$ are condensable.

Exercise 8.2.6 (**NST** + Standard Condensation). Prove that $\mathbb{WF}^{\mathbf{feas}} = \{\hat{x} : x \in \mathbb{S}\}$ is a transitive and $\subseteq$-complete subclass of $\mathbb{WF}$ [1], and for any $u \in \mathbb{WF}^{\mathbf{feas}}$ there is a unique $x = {}^{*}u \in \mathbb{S}$ such that $u = \hat{x}$, and $u \mapsto {}^{*}u$ is an $\in$-isomorphism $\mathbb{WF}^{\mathbf{feas}}$ onto $\mathbb{S}$, hence, $\mathbb{WF}^{\mathbf{feas}}$ is an interpretation of **ZFC**.

Prove that $\mathsf{A}_{\mathbf{x}}$ is well-defined and $\mathsf{A}^{\mathbb{S}}_{\mathbf{x}} = {}^{*}(\mathsf{A}_{\mathbf{x}})$ for each $\mathbf{x} \in \mathbf{S}$. □

It follows that the scheme "$\mathbb{WF}^{\mathbf{feas}} \xrightarrow{*} \mathbb{I}$ [in $\mathbb{H}$]" (see § 8.1a) can be developed in **NST** + Standard Condensation.

The theory **NST** + Standard Size Collection is essentially stronger meta-mathematically than **NST** + Standard Condensation: while the latter is equiconsistent with **ZFC**, see below, **NST** + Standard Size Collection proves

[1] Both $\mathbb{WF}^{\mathbf{feas}} = \mathbb{WF}$ and $\mathbb{WF}^{\mathbf{feas}} \subsetneqq \mathbb{WF}$ is compatible with **NST**, see below.

$\mathtt{Consis}$ **ZFC**, on the other hand, the consistency of **NST** + Standard Size Collection follows from the assumption of an inaccessible cardinal.

In presence of the Power Set axiom, the following principle (unlike Standard Condensation, incompatible with **HST** since it implies Power Set) is a consequence of Standard Size Collection because $\mathbf{V}_{\xi+1}[U] = \mathscr{P}(\mathbf{V}_\xi[U]) \cup U$.

Hierarchy Existence: If U is internal and $\xi \in \mathtt{Ord}$ then $\mathbf{V}_\xi[U]$ is a set.

Let $\mathbf{NST}^+$ be **NST** + Standard Condensation + Hierarchy Existence.

Exercise 8.2.7. Prove that the class $\mathbb{L}[\mathbb{I}]$ can be defined in $\mathbf{NST}^+$.

Hint. We can define $\mathbb{E}$ and $\mathbf{A}$ as in Chapter 5 and prove, following § 5.4, that $\langle \mathbf{A}/{}^{\mathtt{a}}{=}\,; {}^{\mathtt{a}}{\in}, {}^{\mathtt{a}}\mathtt{st} \rangle$ is an interpretation of **HST** in **NST** simply because $\mathbb{I}$ interprets **BST** in **NST**. Thus, the problem comes down to the following: prove, in $\mathbf{NST}^+$, that a set $\mathsf{A}_{\mathbf{x}}$ can be defined for any $\mathbf{x} \in \mathbf{A}$ in accordance with Definition 5.3.3.

That $\mathsf{A}_{\mathbf{x}}$ is well-defined in **HST** is due to Collection. Straightforwardly, even Standard Size Collection does not seem to fulfill this goal in **NST** : we would need a stronger axiom of Internal Size Collection, see below. Yet this obstacle can be circumvented. Indeed, if $\mathbf{x} \in \mathbf{A}$ then $T_{\mathbf{x}}$ is a well-founded tree in $\mathbb{E}$, hence, we can define the rank function $\rho : T \longrightarrow \mathbb{S}$-ordinals as in the proof of Lemma 5.4.3, and $\mathtt{ran}\,\rho$ is bounded by a $\mathbb{S}$-ordinal ξ. In the assumption of Standard Condensation, $\widehat{\rho}(t) = \widehat{\rho(t)}$ is the rank function into well-founded ordinals, and $\mathtt{ran}\,\rho$ is bounded by an ordinal $\alpha = \widehat{\xi}$. It follows from Hierarchy Existence that $\mathbf{V}_{\alpha+1}[X]$ is a set, where X is any transitive internal set such that $\mathtt{ran}\,\mathbf{x} \subseteq X$ (X exists because $\mathbf{x} \in \mathbb{E}$). However $\mathsf{F}_{\mathbf{x}}(t)$ (see Definition 5.3.3) belongs to $\mathbf{V}_{\alpha+1}[X]$ for any $t \in T_{\mathbf{x}}$, actually, $\mathsf{F}_{\mathbf{x}}(t) \in \mathbf{V}_{|t|_T+1}[X]$, where $T = T_{\mathbf{x}}$, hence, Separation can be used instead of Collection to prove that $\mathsf{A}_{\mathbf{x}}$ is well-defined in this case !

It follows that **HST** admits a "standard" interpretation $\langle \mathbb{L}[\mathbb{I}]; \in, \mathtt{st} \rangle$ in $\mathbf{NST}^+$, *i.e.*, with the same atomic predicates $\in$ and $\mathtt{st}$. □

Internal Size Collection: Collection, in $\mathtt{st}$-$\in$-language, in the case when X is any set "of internal size", *i.e.*, a functional image of a set $I \subseteq \mathbb{I}$.

Exercise 8.2.8. Argue in **NST** plus Internal Size Collection. Let Y be the class of all A-codes $\mathbf{x}$ such that $T_{\mathbf{x}} \subseteq \mathbb{I}$. Prove that a set $\mathsf{A}_{\mathbf{x}}$ can be defined accordingly to Definition 5.3.3 for any $\mathbf{x} \in Y$. Prove that $\mathbb{H} = \{\mathsf{A}_{\mathbf{x}} : \mathbf{x} \in Y\}$ is a transitive class and $\langle \mathbb{H}; \in, \mathtt{st} \rangle$ is an interpretation of **HST**. □

Problem 8.2.9. Is Internal Size Collection compatible with **NST**? □

8.2c Conservativity and consistency

Consider the following theories:

(A) **NST** + "$\mathbb{WF} = \mathbf{V}_{\omega+\omega}$";

(B) $\mathbf{NST}^+$ + "$\mathbb{WF}^{\mathtt{feas}} = \mathbb{WF}$" (quotes are just brackets);

(C) $\mathbf{NST}^+$ + "$\mathbb{WF}^{\mathtt{feas}} \subsetneqq \mathbb{WF}$".

(D) **NST** + Standard Size Collection

Theorem 8.2.10. *The theories* **NST**, $\mathbf{NST}^+$, *and* (A), (B), (C) *are conservative and equiconsistent standard core extensions of* **ZFC**. *In addition, the consistency of* (D) *can be proved in* **ZFCI** = **ZFC** + *"there is a strongly inaccessible cardinal"*.

Proof. Our plan is to define interpretations of theories (A), (B), (C) in Kawaï's theory **KST** and then apply Theorem 8.1.5. Theory (D) will require a bit more work.

We argue in **KST**. $\mathbb{S}$, $\mathbb{I}$, $\mathbb{WF}$ denote classes of resp. standard, internal, and well-founded sets. ($\mathbb{S}$ and $\mathbb{I}$ are in fact sets, $\mathbb{WF}$ a proper class.) Then $\langle \mathbb{I}; \in, \mathtt{st}\rangle$ interprets **IST** (Exercise 8.1.1), therefore, the set of all bounded sets $\mathbb{B} = \{y : \exists^{\mathtt{st}} x\,(y \in x)\}$ is an interpretation of **BST** (Theorem 3.4.5(i)).

Exercise 8.2.11. Prove that $\mathbb{B} = \bigcup_{v \in \mathbb{WF}^{\mathtt{feas}}} {}^*v$ and $\mathbb{B}$ is a transitive set. □

Define $\mathbb{H}_\xi = \bigcup_{U \in \mathbb{B} \text{ is transitive}} \mathbf{V}_\xi[U]$ for any (well-founded) ordinal ξ, where $\mathbf{V}_\xi[U]$ is the ξ-th level of the von Neumann hierarchy over U, see § 1.5b. Note that, unlike **HST** or **NST**, Kawaï's theory proves that $\mathbf{V}_\xi[U]$ is really a set for any set $U \in \mathbb{I}$ and (well-founded) ordinal ξ, because it contains both Collection and the Power Set axiom.

Our interprtetations of theories (A), (B), (C) will be of this form.

Lemma 8.2.12 (KST). *If* λ *is a limit ordinal then* $\mathbb{H}_\lambda$ *is a transitive and* $\subseteq$*-complete set, satisfying* $\mathbb{H}_\lambda \cap \mathbb{WF} = \mathbb{WF}_{\omega+\lambda}$ *and* $\mathbb{H}_\lambda \cap \mathbb{I} = \mathbb{B}$, *and an interpretation of* **NST**. *If* $\lambda \geq \omega \times \omega$ *then* Hierarchy Existence *is true in* $\mathbb{H}_\lambda$.

Once again, by the interpretability claim we here mean that for any axiom Φ of **NST**, it is a theorem of **KST** that, for any limit ordinal λ, Φ is true in $\mathbb{H}_\lambda$. This is weaker than to claim that $\mathbb{H}_\lambda$ is a model of **NST**.

Proof (*Lemma*)**.** Recall that $\mathbb{B} \cap \mathbb{WF} = \mathbb{HF}$ (Exercise 1.2.17 remains true in **KST**), which is the same as $\mathbf{V}_0$. As each next level $\mathbf{V}_{\xi+1}[U]$ adds subsets of $\mathbf{V}_\xi[U]$, we have the equality $\mathbb{H}_\xi \cap \mathbb{WF} = \mathbb{WF}_{\omega+\xi}$ by induction on ξ.

In particular, we have $\mathbb{H}_\lambda \cap \mathtt{Ord} = \omega + \lambda$.

The transitivity and $\subseteq$-completeness easily follows from the definition.

To prove $\mathbb{I} \cap \mathbb{H}_\lambda \subseteq \mathbb{B}$ note that, similarly to Lemma 1.5.8, if $U \in \mathbb{B}$ and $\xi \in \mathtt{Ord}$ then any internal $x \in \mathbf{V}_\xi[U]$ belongs to $\mathscr{P}_{\mathtt{int}}{}^n(U)$ for some $n \in \mathbb{N}$, hence, belongs to $\mathbb{B}$. This argument also justifies Full Boundedness in $\mathbb{H}_\lambda$. (We leave details here as an **exercise** for the reader.)

Validation of axioms of **ZC** (minus Regularity) and the axiom of Transitive Hulls is a rather routine exercise, for instance, Extensionality holds in any transitive class, Separation and Choice hold in any $\subseteq$-complete class of the

KST universe, and the axiom of Power Set holds in $\mathbb{H}_\lambda$ because if $x \in \mathbf{V}_\xi[U]$ (U transitive) then $\mathscr{P}(x) \in \mathbf{V}_{\xi+1}[U]$. To verify Hierarchy Existence in $\mathbb{H}_\lambda$ under the assumption $\lambda \geq \omega \times \omega$, note that any ordinal $\xi \in \mathbb{H}_\lambda$ satisfies $\xi < \omega + \lambda$ by the above, hence, if $\lambda \geq \omega \times \omega$ then actually $\xi < \lambda$.

Consider the axioms of **NST** related to standard and internal sets. Axioms of **ZFC**st, Transfer, Transitivity of $\mathbb{I}$, Regularity over $\mathbb{I}$ easily follow from the construction and the fact that $\mathbb{H}_\lambda \cap \mathbb{I} = \mathbb{B}$, an interpretation of **BST**.

Standardization. Suppose that $X \in \mathbb{H}_\lambda$, $X \subseteq \mathbb{S}$. We have got Full Boundedness already, hence, there is a set $B \in \mathbb{B}$ with $X \subseteq B$. There is a standard set s with $B \in s$. Then, as usual, $S = \bigcup s \in \mathbb{S}$ and $X \subseteq S$, hence, by Restricted Standardization of **KST**, there is a standard set Y with $X = Y \cap \mathbb{S}$.

Saturation: follows from Strong Saturation of **KST** because any set internal in the $\mathbb{H}_\lambda$-sense belongs to $\mathbb{B}$, and, on the other hand, as $\mathbb{B}$ is transitive, all elements of sets in $\mathbb{B}$ belong to $\mathbb{B}$.

Standard Size Choice and Dependent Choice: this is entailed by the full Choice axiom of **NST**. □ *(Lemma)*

In view of the lemma, it now suffices to provide additional sentences of the theories (A), (B), (C) to be true in $\mathbb{H}_\lambda$, by an appropriate choice of λ.

Version 1: $\gamma = \omega$. Then, by the lemma, the universe of well-founded sets in $\mathbb{H}_\lambda$ is $\mathbf{V}_{\omega+\omega}$, and hence $\mathbb{H}_\lambda$ is a model of the theory (A) by the above.

Version 2: $\gamma = \mathfrak{s}$, where $\mathfrak{s} = \operatorname{card} \mathbb{WF}^{\mathbf{feas}}$, see § 8.1a. The universe of well-founded sets in $\mathbb{H}_\mathfrak{s}$ is $\mathbf{V}_\mathfrak{s} = \mathbb{WF}^{\mathbf{feas}}$, which easily implies that $\mathbb{H}_\mathfrak{s}$ is a model of the theory (B).

Version 2: $\lambda = \mathfrak{s}^+$. Such an assignment leads to $\mathbb{WF} \cap \mathbb{H}_{\mathfrak{s}^+} = \mathbf{V}_{\mathfrak{s}^+}$, which is strictly bigger than $\mathbb{WF}^{\mathbf{feas}} = \mathbf{V}_\mathfrak{s}$, hence, $\mathbb{H}_{\mathfrak{s}^+}$ is a model of (C).

Finally consider the theory (D). First of all, as **KST** is a conservative extension of **ZFC**, the theory **KSTI** = **KST** + "it is true in $\mathbb{S}$ that there is a strongly inaccessible cardinal" is a conservative extension of **ZFCI**, hence, it suffices to prove the consistency of (D) in **KSTI**.

Arguing in **KSTI**, let $\kappa \in \mathbb{WF}^{\mathbf{feas}}$ be such that ${}^*\kappa$ is a strongly inaccessible cardinal in $\mathbb{S}$. Then κ itself is an inaccessible (well-founded) cardinal in $\mathbb{WF}$, hence, $V = \mathbf{V}_\kappa$ is a (transitive and $\subseteq$-complete) $\in$-model (not merely an interpretation, as $\mathbb{I}$) of **ZFC**.

Exercise 8.2.13. Prove that ${}^*V$ (*i.e.*, the structure $\langle {}^*V ; \in, \mathtt{st} \rangle$) is a model of **IST**, while ${}^*V^{\mathrm{B}} = \bigcup_{v \in V} {}^*v$ (the *bounded part* of ${}^*V$) is a model of **BST** and a transitive subset of ${}^*V$. (*Hint.* To prove Transfer for ${}^*V^{\mathrm{B}}$ argue as in the proof of Transfer for $\mathbb{B}$ in Theorem 3.4.5(i).) □

Similarly to the above, define $\mathbb{H}_\kappa[{}^*V^{\mathrm{B}}] = \bigcup_{U \in {}^*V^{\mathrm{B}}} \mathbf{V}_\kappa[V]$.

Exercise 8.2.14. Prove that $\mathbb{H}_\kappa[{}^*V^{\mathrm{B}}]$ is a *model* of **NST**. (*Hint*: accomodate the arguments used above w. r. t. theories (A) – (B).) □

This reduces our task to the axiom of Standard Size Collection: this is where the inaccessibility of κ will be most essential. As we have $\mathbb{S} \cap \mathbb{H}_\kappa[{}^*V^{\mathrm{B}}] = \mathbb{S} \cap {}^*V^{\mathrm{B}} = \{{}^*v : v \in V\}$, it suffices to check that any set $X \subseteq \mathbb{H}_\kappa[{}^*V^{\mathrm{B}}]$ of cardinality $\mathtt{card}\, X < \kappa$ (strictly) can be covered by a set in $\mathbb{H}_\kappa[{}^*V^{\mathrm{B}}]$. For any $x \in X$ there is a set $U_x \in {}^*V^{\mathrm{B}}$ and an ordinal $\xi_x < \kappa$ such that $x \in \mathbf{V}_{\xi_x}[U_x]$. Moreover, by definition, there is a set $v_x \in V$ such that $U_x \subseteq {}^*v_x$, hence, $x \in \mathbf{V}_{\xi_x}[{}^*v_x]$. As κ is inaccessible, $\xi = \mathtt{sup}_{x \in X}\, \xi_x$ is an ordinal $< \kappa$, and similarly, $v = \bigcup_{x \in X} v_x \in V$. It follows that $X \subseteq \mathbf{V}_\xi[{}^*v]$, but the set $\mathbf{V}_\xi[{}^*v]$ easily belongs to $\mathbb{H}_\kappa[{}^*V^{\mathrm{B}}]$.

□ (*Theorem 8.2.10*)

8.2d Remarks and exercises

Our strategy was to derive metamathematical properties of **NST** and its versions by interpretations in Kawaï's theory **KST**. Yet the result can be obtained directly.

Exercise 8.2.15. Arguing in **ZFC**$\boldsymbol{\vartheta}$, consider the following amendments in the proof of Theorem 8.1.5. Let ${}^*\mathbf{V}_{\boldsymbol{\vartheta}} = \langle {}^*V\,; {}^*{\in}, {}^*\mathtt{st}\rangle$ now be a $\boldsymbol{\vartheta}$-saturated standard core extension of $V = \mathbf{V}_{\boldsymbol{\vartheta}}$. (To obtain such a model apply the quotient power chain construction of § 4.3d with $\gamma = \boldsymbol{\vartheta}$ (the length of the chain) and $\langle V\,; \in\rangle$ as the initial structure.) Define $\mathbf{P}_\xi$ as in the proof of Theorem 8.1.5. Choose a limit ordinal γ and put $\mathbf{P} = \bigcup_{\xi<\gamma} \mathbf{P}_\xi$. For any $x \in \mathbf{P}$ define $\|x\|_{{}^*V} \subseteq {}^*V$ by induction on ξ, where $x \in \mathbf{P}_\xi$, as follows: $\|x\|_{{}^*V} = \{x\}$ for $x \in {}^*V = \mathbf{P}_0$, otherwise $\|x\|_{{}^*V} = \bigcup_{y \in x} \|y\|_{{}^*V}$. Let $\mathbf{P}^{\mathrm{R}}_\xi$ be the set of all $x \in \mathbf{P}_\xi$ such that $\|x\|_{{}^*V}$ can be covered by a set of the form $x_{{}^*\in} = \{y \in {}^*V : y \,{}^*{\in}\, x\}$, where $x \in {}^*V$. (^{R} stand for: "restricted".)

Show that $\mathbf{P}^{\mathrm{R}} = \bigcup_{\xi<\gamma} \mathbf{P}^{\mathrm{R}}_\xi$ is then a model of **NST**, and an appropriate choice of $\boldsymbol{\lambda}$ yields models of other theories mentioned in Theorem 8.2.10. □

The minimal model argument of § 4.6f can be used to show that, similarly to **IST** and unlike **HST**, the theory **NST** does <u>not</u> admit a standard core interpretation in **ZFC**. Recall that **ZC**$\boldsymbol{\vartheta}$, a theory in the $\in$-language with a constant symbol $\boldsymbol{\vartheta}$, was defined in § 3.6a. In **ZFC**, a *transitive model* of **ZC**$\boldsymbol{\vartheta}$ has the form $\langle \mathfrak{M}; \in, \vartheta\rangle$, or simply $\langle \mathfrak{M}; \vartheta\rangle$, where $\mathfrak{M}$ is a transitive model of **ZC** and $\vartheta \in \mathfrak{M}$ is an ordinal, the intended interpretation of $\boldsymbol{\vartheta}$, hence by necessity $\mathbf{V}_\vartheta \cap \mathfrak{M}$ is a transitive model of **ZFC**.

Exercise 8.2.16. Let M be a minimal transitive model of **ZFC** (see § 4.6f). Prove that M is not **NST**-extendible, *i.e.* not isomorphic to the standard core of a model of **NST**.

Hint. Assume that $M = \mathbb{S}^{(H)}$, where $H = \langle H\,; {}^*{\in}, \mathtt{st}\rangle$ is a model of **NST**. Recall (Exercise 8.2.4 (**NST**)) that there is a set, say $X \in H$ (*i.e.*, $\mathbf{S}/ \overset{\mathrm{a}}{=}$) and a map $f : X \xrightarrow{\text{onto}} M$ definable in H. But X has a power set $\mathscr{P}(X)$ in H, as well as the next power $\mathscr{P}(\mathscr{P}(X))$, *etc.*, and f maps this sequence of powers onto a transitive set $\mathfrak{M} \subseteq \bigcup_n \mathscr{P}^n(M)$, which is a model of **ZC**. Moreover, there is an ordinal $\vartheta \in \mathfrak{M}$ such that $M = \mathbf{V}_\vartheta \cap \mathfrak{M}$, therefore, $\langle \mathfrak{M}; \vartheta \rangle$ is a transitive model of **ZC**$\boldsymbol{\vartheta}$. In addition, it is true in $\mathfrak{M}$ that M is uncountable, even not a set of standard size (Exercise 8.2.4 (**NST**)). Now, the Löwenheim – Skolem argument within $\mathfrak{M}$ yields a transitive model of **ZFC** which is *countable*, hence, smaller than M, contradiction.

Infer that **NST** is not standard core interpretable in **ZFC**. Infer that **NST** is not internal core interpretable in **BST** (in the sense of § 5.1a). □

Exercise 8.2.17. Prove, using Theorem 8.2.10 and reducibility of **BST** (Theorem 4.1.10) that **NST** is a conservative external extension of **BST** (in the sense of § 5.1a). □

Exercise 8.2.18. It follows from arguments in 8.2.16 that for a transitive set M to be **NST**-extendible it is necessary that M is the $\mathbf{V}_\vartheta$-part of a transitive model of **ZC**$\boldsymbol{\vartheta}$. Prove that this is also a sufficient condition.

Hint. Let $\langle \mathfrak{M}; \in, \vartheta \rangle$ be a transitive model of **ZC**$\boldsymbol{\vartheta}$. Carry out the arguments, outlined in 8.2.15 above, in $\mathfrak{M}$, with $V = \mathbf{V}_\vartheta$ and $\gamma = \omega$. The result will be a model $\mathbf{P}^{\mathrm{R}}$ of theory (A) of Theorem 8.2.10. □

Exercise 8.2.19. Prove that **NST** admits an interpretation in **ZC**$\boldsymbol{\vartheta}$ such that the standard universe of the interpretation is (isomorphic to) $\mathbf{V}_{\boldsymbol{\vartheta}}$.

Hint. Carry out the arguments, outlined in 8.2.15 above, in **ZC**$\boldsymbol{\vartheta}$, with $V = \mathbf{V}_\vartheta$ and $\gamma = \omega$. The result will be an interpretation $\mathbf{P}^{\mathrm{R}}$ of **NST** (actually, of theory (A) in Theorem 8.2.10) in **ZC**$\boldsymbol{\vartheta}$. □

Interpretations of other theories mentioned in Theorem 8.2.10 can also be obtained this way, but the theory **ZC**$\boldsymbol{\vartheta}$ has to be strengthened, for instance, by the "axiom" that any well-ordered set is isomorphic to an ordinal and for any ordinal ξ there exists a set equal to $\mathbf{V}_\xi$.

Problem 8.2.20. Find standard extensions of **ZFC** in which the theories **NST** and **NST**$^+$ are standard core interpretable. (Interpretations outlined in 8.2.19 are not standard core interpretations in the sense of § 4.1b, of course.)

Find a standard set theory $\mathfrak{U}$ in which **NST** + Standard Size Collection is standard core interpretable, or at least such that the theory **NST** + Standard Size Collection is a conservative extension of $\mathfrak{U}$. (As for the "at least" part, the theory $\mathfrak{U}$ which consists of all $\in$-sentences Φ such that **NST** + Standard Size Collection proves $\Phi^{\mathbf{st}}$, is as required, but the real problem is to find a reasonably simple, at least recursive, axiomatization for $\mathfrak{U}$.) □

The interrelation between Special Model Axiom and **NST** is not clear: all sets are well-orderable, but, on the other hand, the absence of Collection may lead to rather destructive effects.

8.3 Non-well-founded set theories

This set theoretic scheme in foundations is explicitly based on a principle to which virtually all nonstandard systems can be reduced.

Definition 8.3.1. A *graph*, also, a *graph on* X, is any structure of the form $\mathsf{G} = \langle X ; \mathsf{G}\rangle$, where $\mathsf{G} \subseteq X \times X$. A graph G on X is *extensional* iff we have $x = y$ whenever $x, y \in X$ satisfy $\forall z \in X\ (z\,\mathsf{G}\,x \Longleftrightarrow z\,\mathsf{G}\,y)$. □

Universality: Any extensional graph G on a set X is isomorphic to $\in$ restricted to a transitive set T. [2]

This principle can also be called *Ill-founded Mostowski Collapse*. Mostowski's original collapse theorem says that in **ZFC** any extensional *well-founded* relation is isomorphic to $\in$ on a unique transitive set T. (There is no uniqueness requirement in Universality.)

Seen from this angle, we can view, say, **HST** as a scheme which has a "standard" **ZFC** universe (realized as $\mathbb{WF}$), together with a transitive copy (that is, $\mathbb{I}$) of a certain ultrapower (or ultralimit) of it, in a common set universe. In other words, the existence of transitive saturated extensions of "standard" structures is just one of possible applications of the ill-founded collapse.

Universality is obviously incompatible with the axiom of Regularity of **ZFC** (for instance, define $x\,\mathsf{G}\,x$ on a singleton x). Various versions of Universality lead to several set theoretic schemes known as *ill-founded* or *non-well-founded* theories (see Aczel [A 88]), one of which, Boffa's theory, turns out to be a reasonable foundational scheme for nonstandard analysis.

8.3a Boffa's non-well-founded set theory

Define **ZFBC**$^-$ [3] to be the theory **ZFC** without the Regularity axiom but with the following axiom that strengthens Universality:

SuperUniversality: Suppose that $\mathsf{F} \subseteq \mathsf{G}$ are extensional graphs on resp. sets $Y \subseteq X$, and F is *transitive* in G, *i.e.*, $\{y \in Y : y\,\mathsf{F}\,y'\} = \{x \in X : x\,\mathsf{G}\,y'\}$ for any $y' \in Y$. Then any isomorphism of $\langle Y ; \mathsf{F}\rangle$ onto a transitive set A (more exactly, onto $\langle A; \in \restriction A\rangle$) can be extended to an isomorphism of $\langle X ; \mathsf{G}\rangle$ onto a transitive set $B \supseteq A$.

(To get Universality take $Y = \varnothing$.)

Recall that **ZFGC** is the "global choice" extension of **ZFC**, see Definition 4.3.4. The axioms of **ZFGC** include in particular the Global Choice axiom saying that **G** is a global choice function. **ZFGC** is a conservative, hence equiconsistent extension of **ZFC** (Felgner [Fel 71]).

[2] Or, in terms of graph theory, *any extensional graph has an injective decoration.*

[3] The letter **B** is from Boffa who was perhaps the first who systematically considered theories of this sort, in [Bof 72].

Define **ZFBC** to be **ZFBC**$^-$ + Global Choice (the symbol **G** of a global choice function is allowed to occur in the schemata of Separation and Collection). Arguing in **ZFBC**$^-$, let $\mathbb{WF}$ be the class of all well-founded sets (Definition 1.1.4), while the whole universe will be $\mathbb{V}$, the *universal class*.

Exercise 8.3.2. Show that $\langle \mathbb{WF}\,; \in \restriction \mathbb{WF} \rangle$ is an interpretation of **ZFC** in **ZFBC**$^-$ while $\langle \mathbb{WF}\,; \in \restriction \mathbb{WF}, \mathbf{G} \restriction \mathbb{WF} \rangle$ is an interpretation of **ZFGC** in **ZFBC**. (*Hint*: $\mathbb{WF}$ is transitive and $x \subseteq \mathbb{WF} \Longrightarrow x \in \mathbb{WF}$.) □

Thus, we can use $\mathbb{WF}$ in **ZFBC** as the basic **ZFC** universe for notions like natural numbers, $\mathbb{N}$, finiteness, ordinals, *etc*., as in **HST** in § 1.2.

The next theorem (the proof follows in § 8.3e) describes the main metamathematical properties of **ZFBC**. Recall that $\Phi^{\tt wf}$ denotes the relativization of an $\in$-formula Φ to $\mathbb{WF}$, and the notions of `wf`-core interpretability and `wf`-conservativity were introduced earlier, see Remark 4.1.5.

Aczel [A 88] gives a rather lengthy proof of the next theorem:

Theorem 8.3.3 (originally, Boffa [Bof 72]). **ZFBC** *is* `wf`*-core interpretable*[4] *in* **ZFGC**, *therefore* **ZFBC** *is a conservative* `wf`*-core extension of* **ZFC**, *so that for any* $\in$*-sentence* Φ, **ZFC** *proves* Φ *iff* **ZFBC** *proves* $\Phi^{\tt wf}$. □

On the other hand **ZFBC** is **not** `wf`-core interpretable in **ZFC**. Indeed such an interpretation would imply that a global choice function is definable in **ZFC**, which is known to be not the case. Thus **ZFBC** is **not** a realistic theory even if Definition 4.1.8 is understood in the sense of wf core rather than standard core interpretability.

To demonstrate how SuperUniversality works, note that, in **ZFBC**$^-$, there exist sets a satisfying $a = \{a\}$: indeed, take $Y = \mathsf{F} = \varnothing$ and $X = \{\varnothing\}$, with $\mathsf{G} = \langle \varnothing, \varnothing \rangle$, and apply SuperUniversality for the empty map as the given isomorphism. Sets x satisfying $x = \{x\}$ can be called *atoms*, because, although it is not entirely true that they lie outside of the domain of the membership relation, each of them is at least rather isolated in the $\in$-structure. Needless to say that there are no atoms in this sense in **ZFC**.

Exercise 8.3.4. Modify the argument above to prove that the class $\mathbb{A} = \{a : a = \{a\}\}$ of all atoms is a proper class in **ZFBC**$^-$. □

The following is a more meaningful example.

Example 8.3.5. Arguing in **ZFBC**$^-$, let $U \in \mathbb{WF}$ be an ultrafilter on $\mathbb{N}$, and X a transitive set. Consider the ultrapower $\mathtt{Ult}_U(\langle X\,; \in \rangle) = \langle {}^\bullet X\,; {}^\bullet\!\in \rangle$, with an associated elementary embedding $x \mapsto {}^\bullet x$. The relation ${}^\bullet\!\in$ may have nothing to do with the true membership $\in$, however, by SuperUniversality, there is a <u>transitive</u> set ${}^*X$ and an isomorphism $\pi : \langle {}^\bullet X\,; {}^\bullet\!\in \rangle \xrightarrow{\text{onto}} \langle {}^*X\,; \in \rangle$.

[4] We cannot consider standard core interpretability because **ZFBC** does not introduce standard sets.

Accordingly, we have an elementary embedding ${}^*x = \pi({}^\bullet x)$ from $\langle X\,;\in\rangle$ to $\langle {}^*X\,;\in\rangle$. Thus, **ZFBC**$^-$ provides us with "standard" (referring to the true membership relation) elementary extensions of "standard" structures.

An interesting related issue is: for which elements $x \in X$ can it be arranged that ${}^*x = x$. Let F be the set of all $x \in X$ such that any $y' \in {}^*X$ with $y' \;{}^*\!\!\in {}^*x$ satisfies $y' = {}^*y$ for some $y \in X$ (necessarily $y \in x$). Then $\{{}^*x : x \in F\}$ is ${}^*\!\!\in$-transitive in ${}^*X$, hence, we can define π in two steps, where the first step is just $\pi({}^*x) = x$ for any $x \in F$. Thus, ${}^*x = x$ can be provided for all $x \in F$. We leave it as an **exercise** for the reader to prove that $x \in F$ iff $\mathtt{TC}(x)$ is finite, *i.e.*, x has only finitely many hereditary $\in$-predecessors, yet now this is not, as in **ZFC**, only $\mathbb{HF}$ over $\varnothing$; for instance, the transitive closure $\mathtt{TC}(x) = \{x\}$ of an arbitrary set $x \in \mathbb{A}$ (see 8.3.4) is finite. Compare this with Exercise 1.2.17(1) ! □

8.3b Extensions of proper classes

Suppose now that $\langle X\,;\mathsf{G}\rangle$ is an extensional graph whose domain X is a proper class [5]. Does G admit an isomorphism onto $\langle T\,;\in\rangle$, where T is a transitive (also proper) class ? Even SuperUniversality apparently does not answer this always in the positive, yet there is a useful family of classes which admit a positive answer. Say that a graph G is *locally of set size* if $\{x \in X : x\,\mathsf{G}\,y\}$ is a set (that is, not a proper class) for any y in the domain X of G.

Lemma 8.3.6 (ZFBC). *Suppose that* G *is an extensional graph locally of set size on a class* X. *Then there exists an isomorphism of* $\langle X\,;\mathsf{G}\rangle$ *onto* $\langle T\,;\in\rangle$, *where* T *is a transitive (also proper) class.*

Proof. Say that $S \subseteq X$ is G-*transitive* if $s \in S$ implies $x \in S$ whenever $x\mathsf{G}s$. Using the Global Choice axiom of **ZFBC**, we easily define a decomposition $X = \bigcup_{\xi\in\mathtt{Ord}} X_\xi$, where all X_ξ are <u>sets</u> G-transitive in X, and $X_\xi \subseteq X_\eta$ whenever $\xi < \eta$. The Global Choice and SuperUniversality axioms of **ZFBC** can be used to obtain T and the isomorphism required as the limit of an increasing $\mathtt{Ord}$-long sequence of isomorphisms defined on sets X_ξ. □

The lemma fails, of course, for graphs not locally of set size.

There are plenty of extensional graphs locally of set size, for example the membership $\in$ restricted onto any transitive class is such, and it turns out that ultrapowers preserve this property !

Exercise 8.3.7. Let V be a transitive class. Prove that if U is an ultrafilter on a set I, then the ultrapower $\langle {}^*V\,;{}^*\!\!\in\rangle = V^I/U$ of $\langle V\,;\in\rangle$ is a structure locally of set size. (*Hint*. Any element $\mathbf{x} \in {}^*V$ is an equivalence class $[f]$ of a function $f : I \to V$. Moreover, all ${}^*\!\!\in$-elements of such an $\mathbf{x}$ are classes $[g]$ of functions $g : I \to V$ such that $g(i) \in f(i)$ for all i.) □

[5] By *classes* we understand collections of sets defined by formulas, see § 1.5c.

A class V is *almost universal* if any set $X \subseteq V$ is a subset of some $Y \in V$.

Theorem 8.3.8 (ZFBC). *Let V be a transitive almost universal class, and κ a cardinal. Then there is a transitive, almost universal, and κ^+-saturated class ${}^*V$ and an elementary embedding $* : V \to {}^*V$.*

Proof. Using appropriately the chain quotient power construction of § 4.3d, we can obtain a κ^+-saturated standard core extension of $\langle V ; \in \rangle$ together with the natural embedding $x \mapsto {}^\bullet x$. Then $\langle {}^\bullet V ; {}^\bullet\!\in \rangle$ is locally of set size: this is true by the same reasons as in Exercise 8.3.7. Finally, $\langle {}^\bullet V ; {}^\bullet\!\in \rangle$ is extensional as an elementary extension of an extensional (since it is transitive) class $\langle V ; \in \rangle$. It follows, by Lemma 8.3.6, that there is an isomorphism π : $\langle {}^\bullet V ; {}^\bullet\!\in \rangle \xrightarrow{\text{onto}} \langle {}^*V ; \in \rangle$, where ${}^*V$ is a transitive class. The superposition ${}^*x = \pi({}^\bullet x)$ is then an elementary embedding $\langle V ; \in \rangle \to \langle {}^*V ; \in \rangle$. The class ${}^*V$ is κ^+-saturated together with ${}^\bullet V$. It remains to show that ${}^*V$ is almost universal, *i.e.*, for any set $X \subseteq {}^*V$ there exists $Y \in {}^*V$ with $X \subseteq Y$. Any $x \in X$ is equal to $\pi([f_x])$, where $f_x \in \mathscr{F}$ and $[f]$ is the element of ${}^*V$ generated by f as usual. But any $f_x \in \mathscr{F}$ has $\operatorname{ran} f \subseteq Z_x$ for a set $Z_x \in V$ because V is almost universal. Also, there is $Z \in V$ with $Z_x \subseteq Z$ for any $x \in X$. Thus, every $x \in X$ belongs to $Y = {}^*X$, as required. □

8.3c Applications to nonstandard analysis

Theorem 8.3.8 enables us to use the following scheme of applications.

Starting with a transitive and almost universal class V, for instance, the class $\mathbb{WF}$ of all well-founded sets, and a (well-founded) cardinal κ, we apply Theorem 8.3.8 to obtain a transitive, κ^+-saturated, and almost universal class ${}^*V$, together with an elementary embedding $* : V \to {}^*V$. If V satisfies **ZFC** then so does ${}^*V$. This allows to smoothly carry out "nonstandard" arguments which require not more than κ^+-Saturation, taking the universe of all sets as the source of external sets. In principle, a suitable transitive class $H = H(V)$ can be defined, so that $V \cup {}^*V \subseteq H$ and $\langle H ; \in, {}^*\mathtt{st} \rangle$ (where, for $x \in H$, ${}^*\mathtt{st}\, x$ means that $x = {}^*v$ for some $v \in V$) is an interpretation of $\mathbf{HST}^-_\kappa$, similarly to some subuniverses in **HST**, see § 6.4b, but this does not make much sense as the whole universe of **ZFBC** satisfies all of **ZFC** except for Regularity.

Can we take the "universal class", *i.e.*, the whole **ZFBC** universe $\mathbb{V}$, rather than a class like $\mathbb{WF}$, as the initial class? See below!

Generally, **ZFBC** provides a degree of flexibility of nonstandard methods not available in theories like **HST**, including repetitive saturated extensions of transitive classes.

On the negative side, the main flaw is the absence of a unique standard size saturated universe of internal sets. Accordingly, the unique existence of a set of all hyperreals, even all hyperintegers, which is, perhaps, one of the major goals of axiomatic foundations of nonstandard analysis (see Preface) is not achieved. The amount of Saturation available is always restricted by a (well-founded) cardinal, standard size Saturation is out of question in **ZFBC** as it yields structures not locally of set size. Another negative factor is that the Global Choice axiom has to be systematically used for the construction of saturated extensions of proper classes, something a **ZFC**-oriented mathematician would try to avoid. (In theories like **HST** applications of Global Choice are made once and for all at the metamathematical level, so that all useful subuniverses, like $\mathbb{I}_\kappa$ or $\mathbb{L}[\mathbb{I}_\kappa]$, admit a definition by concrete $\in$-formulas or $\mathtt{st}$-$\in$-formulas not containing such a deep abstraction as a Global Choice function.)

8.3d Alpha theory

This is a peculiar derivative of Boffa's theory, whose main feature is a uniform definition of ${}^*x$ for *any* set x in the set universe. According to Benci and di Nasso [BenDN 03], the system consists of two theories, one of which is formulated, strictly speaking, not in accordance with modern standards, but the other one, with which we begin, is unobjectionable.

Let **ZFC[J]** be a theory in the $\in$-language extended by a binary predicate symbol **J**. The axioms include all of **ZFC** minus Regularity (and **J** is allowed to occur in the schemata), together with the following special axioms:

Alpha-1: **J** is a function defined on all $\mathbb{N}$-sequences of arbitrary sets, *i.e.*,

$$\forall \varphi \left(\mathsf{Seq}_{\mathbb{N}}(\varphi) \Longrightarrow \exists!\, x\, \mathbf{J}(\varphi, x)\right) \wedge \forall \varphi \,\forall x \left(\mathbf{J}(\varphi, x) \Longrightarrow \mathsf{Seq}_{\mathbb{N}}(\varphi)\right),$$

where $\mathsf{Seq}_{\mathbb{N}}(\varphi)$ means that φ is an $\mathbb{N}$-*sequence*, *i.e.*, a function with $\operatorname{dom}\varphi = \mathbb{N}$. Assuming this, let, for any φ with $\mathsf{Seq}_{\mathbb{N}}(\varphi)$, $\mathbf{J}(\varphi)$ be that unique x which satisfies $\mathbf{J}(\varphi, x)$.

Alpha-2: If f is a function defined on a set A and $\varphi, \psi : \mathbb{N} \to A$ then $\mathbf{J}(\varphi) = \mathbf{J}(\psi)$ implies $\mathbf{J}(f \circ \varphi) = \mathbf{J}(f \circ \psi)$.

Alpha-3: $\mathbf{J}(c_m) = m$ for any m, where $c_m(k) = m$ for all $k \in \mathbb{N}$.

Alpha-4: If $\vartheta(n) = \{\varphi(k), \psi(k)\}$ for all k then $\mathbf{J}(\vartheta) = \{\mathbf{J}(\varphi), \mathbf{J}(\psi)\}$.

Alpha-5: Generally, $\mathbf{J}(\varphi) = \{\mathbf{J}(\psi) : \psi(k) \in \varphi(k) \text{ for all } n\}$ for any φ.

Alpha-6: $\mathbf{J}(\mathtt{id}) \notin \mathbb{N}$, where $\mathtt{id}(k) = k$ for all k.

Define ${}^*x = \mathbf{J}(c_x)$ for any set x (where $c_x(k) = x$ for all k in $\mathbb{N}$). Sets of the form ${}^*x$ can be called "standard" while sets of the form $\mathbf{J}(\varphi)$, φ being an $\mathbb{N}$-sequence, "internal". Let $\mathbb{S}$, $\mathbb{I}$ be the classes of all "standard", resp., "internal" sets. Finally, define $\boldsymbol{\alpha} = \mathbf{J}(\mathtt{id})$: a very important set, see below.

Direct arguments in **ZFC[J]** are quite special and do not follow any ordinary "nonstandard" intuition; even typical basic facts like $\mathbf{J}(\varphi) = \mathbf{J}(\psi)$ whenever $\varphi(k) = \psi(k)$ for almost all k need some tricks. Still there is a sequence of rather simple claims:

Exercise 8.3.9. Prove, in **ZFC[J]**, the following:

(1) If $\vartheta(k) = \langle \varphi(k), \psi(k) \rangle$, $\forall k$, then $\mathbf{J}(\vartheta) = \langle \mathbf{J}(\varphi), \mathbf{J}(\psi) \rangle$. (Recall that ordered pairs are formally defined by $\langle a, b \rangle = \{\{a\}, \{a, b\}\}$.)

(2) If $\varphi : \mathbb{N} \to X$ then $\mathbf{J}(\varphi) \in {}^*X$, conversely, any $x \in {}^*X$ has the form $\mathbf{J}(\varphi)$ for some $\varphi : \mathbb{N} \to X$.

(3) If $X \subseteq Y$ then ${}^*X \subseteq {}^*Y$, further, ${}^*(X \cup Y) = {}^*X \cup {}^*Y$, the same for $\cap$, $\setminus$, Δ (symmetric difference), $\times$ (Cartesian product).

(4) If X is finite then ${}^*X = \{{}^*x : x \in X\}$.

(5) The class $\mathbb{I}$ is transitive and equal to $\{y : \exists^{\text{st}} x\, (y \in x)\}$.

(6) If $f : A \to B$ then ${}^*f$ is a function ${}^*A \to {}^*B$ and ${}^*f(\mathbf{J}(\varphi)) = \mathbf{J}(f \circ \varphi)$ for any $\varphi : \mathbb{N} \to A$. In addition, if $C \subseteq A$ then ${}^*f \upharpoonright {}^*C = {}^*(f \upharpoonright C)$.

(7) If $\varphi(k) \notin \psi(k)$ for all k then $\mathbf{J}(\varphi) \notin \mathbf{J}(\psi)$. Similarly, $\varphi(k) \neq \psi(k)$ for all k then $\mathbf{J}(\varphi) \neq \mathbf{J}(\psi)$ — therefore, if $x \neq y$ then ${}^*x \neq {}^*y$.

Hints. (1) We have $\mathbf{J}(\vartheta) = \{\mathbf{J}(\varphi'), \mathbf{J}(\psi')\}$ by Alpha-4, where $\varphi'(k) = \{\varphi(k)\}$ and $\psi'(k) = \{\psi(k), \psi(k)\}$. Apply Alpha-4 for φ' and ψ'.

(2) Both assertions follow immediately from Alpha-5 because ${}^*X = \mathbf{J}(c_X)$, where $c_X(k) = X$ for all k.

(3) By (2), a typical element of ${}^*X$ is $\mathbf{J}(\varphi)$, where $\varphi : \mathbb{N} \to X$, but then $\mathbf{J}(\varphi) \in {}^*Y$ as well. Similar arguments validate the rest of the claim.

(4) Note that $\{{}^*x\} = {}^*\{x\}$ (generally, $\{{}^*x, {}^*y\} = {}^*\{x, y\}$) by Alpha-4. Then apply (3) for $\cup$ by induction on $\mathtt{card}\, X$.

(5) Suppose that $y = \mathbf{J}(\varphi)$. Let $x = \mathtt{ran}\, \varphi = \{\varphi(k) : k \in \mathbb{N}\}$, so that $\varphi : \mathbb{N} \to x$. Then $y \in {}^*x$ by (2). The converse is similar. The transitivity immediately follows from Alpha-5.

(6) We have ${}^*f \subseteq {}^*A \times {}^*B$ by (3). If $a \in {}^*A$, then, by Alpha-5, $a = \mathbf{J}(\varphi)$, where $\varphi : \mathbb{N} \to A$. Put $\vartheta(k) = \langle \varphi(k), f(\varphi(k)) \rangle$, then $\mathbf{J}(\vartheta) = \langle \mathbf{J}(\varphi), \mathbf{J}(f \circ \varphi) \rangle \in {}^*f$ by (1), (2), hence, $\mathtt{dom}\, {}^*f = {}^*A$. Now, let $\langle a, b \rangle$ and $\langle a', b' \rangle$ are two typical elements of ${}^*f$, so that, as above, $a = \mathbf{J}(\varphi)$, $a' = \mathbf{J}(\varphi')$, $b = \mathbf{J}(\psi)$, $b' = \mathbf{J}(\psi')$, where $\varphi, \varphi' : \mathbb{N} \to A$ and $\psi = f \circ \varphi$, $\psi' = f \circ \varphi'$. If $a = a'$ then we have $b = b'$ by Alpha-2. Thus ${}^*f$ is a map ${}^*A \to {}^*B$. The equality ${}^*f(\mathbf{J}(\varphi)) = \mathbf{J}(f \circ \varphi)$ has actually been established. The additional claim: both ${}^*f$ and ${}^*(f \upharpoonright C)$ are functions with domains ${}^*A$ and ${}^*C \subseteq {}^*A$, and ${}^*(f \upharpoonright C) \subseteq {}^*f$.

(7) Let $\vartheta(k) = \{\varphi(k)\}$. Then we have $\vartheta(k) \smallsetminus \psi(k) = \vartheta(k)$, $\forall k$, hence, by (3), $\mathbf{J}(\vartheta) \smallsetminus \mathbf{J}(\psi) = \mathbf{J}(\vartheta)$, in other words, $\mathbf{J}(\vartheta) \smallsetminus \mathbf{J}(\psi) = \varnothing$. However $\mathbf{J}(\vartheta) = \{\mathbf{J}(\varphi)\}$ by Alpha-4. The other claim is analogous. □

Taking $\varphi = \mathtt{id}$ and $f = \varphi$ in (6), we obtain ${}^*\varphi(\boldsymbol{\alpha}) = \mathbf{J}(\varphi)$ for any $\mathbb{N}$-sequence φ. This gives a much more meaningful form to the whole structure of the universe of **ZFC[J]**: **J**-extensions turn out to be just values of the $*$-extended functions on a nonstandard number $\boldsymbol{\alpha}$ ($\boldsymbol{\alpha} \in {}^*\mathbb{N}$ by (2) and $\notin \mathbb{N}$ by Alpha-6). We can now reformulate all axioms of **ZFC[J]**, for instance, Alpha-3 takes the form:

– if $\vartheta(n) = \{\varphi(k), \psi(k)\}$ for all k then ${}^*\vartheta(\boldsymbol{\alpha}) = \{{}^*\varphi(\boldsymbol{\alpha}), {}^*\psi(\boldsymbol{\alpha})\}$.

But to understand what ${}^*\vartheta$ is, we have to return to **J**-formulations. With heavy abuse of notation, [BenDN 03] gave all axioms *prima facie* in terms of $\boldsymbol{\alpha}$, simply dropping stars, *e.g.*, Alpha-3 takes the form:

– if $\vartheta(n) = \{\varphi(k), \psi(k)\}$ for all k then $\vartheta(\boldsymbol{\alpha}) = \{\varphi(\boldsymbol{\alpha}), \psi(\boldsymbol{\alpha})\}$,

followed by a comment that $\varphi(\boldsymbol{\alpha})$ should be understood as the value, on $\boldsymbol{\alpha}$, of an *extended* function φ. This version, called **ZFC**[α], can be rigorously understood only on the base of **ZFC**[**J**] or something like that.

To gain even more clarity, let $\mathscr{U} = \{X \subseteq \mathbb{N} : \boldsymbol{\alpha} \in {}^*X\}$.

Exercise 8.3.10. Prove that $\mathscr{U}$ is an ultrafilter on $\mathbb{N}$, containing all cofinite subsets of $\mathbb{N}$. (*Hint*. That $\mathscr{U}$ is an ultrafilter easily follows from (3). If $X = \{0, 1, 2, ..., n\}$ then $X = {}^*X$ by (4) of Exercise 8.3.9 and Alpha-3, hence, if $X \in \mathscr{U}$ then $\boldsymbol{\alpha} = n$ for some $n \in X$, contradiction with Alpha-6.) □

Theorem 8.3.11 (Łoś Theorem, **ZFC**[**J**])**.** *If $\sigma(x_1, ..., x_n)$ is an $\in$-formula and $\varphi_1, ..., \varphi_n$ are $\mathbb{N}$-sequences then*

(†) $\sigma({}^*\varphi_1(\boldsymbol{\alpha}), ..., {}^*\varphi_n(\boldsymbol{\alpha}))$ *is true in* $\mathbb{I}$ *iff* $\{k : \sigma(\varphi_1(k), ..., \varphi_n(k))\} \in \mathscr{U}$.

If, moreover, σ is a bounded formula (see § 1.5a) *then*

(‡) $\sigma({}^*\varphi_1(\boldsymbol{\alpha}), ..., {}^*\varphi_n(\boldsymbol{\alpha}))$ *iff* $\{k : \sigma(\varphi_1(k), ..., \varphi_n(k))\} \in \mathscr{U}$.

Proof. The "moreover" part follows because bounded formulas are absolute for any transitive class. ($\mathbb{I}$ is transitive by (5) of Exercise 8.3.9.)

The main part is proved by induction on the complexity of σ. Let σ be $x \in y$ and φ, ψ be $\mathbb{N}$-sequences. Consider the set $U = \{k : \varphi(k) \in \psi(k)\}$. Suppose that $\boldsymbol{\alpha} \in {}^*U$. Define φ', ψ' so that they coincide with resp. φ, ψ on U while $\varphi'(k) = 0$ and $\psi'(k) = 1 = \{0\}$ on $\mathbb{N} \setminus U$, so that $\varphi'(k) \in \psi'(k)$ for all k. Then ${}^*(\varphi')(\boldsymbol{\alpha}) \in {}^*(\psi')(\boldsymbol{\alpha})$ by Alpha-5. On the other hand, as $\boldsymbol{\alpha} \in {}^*U$, we have ${}^*(\varphi')(\boldsymbol{\alpha}) = {}^*(\varphi' \restriction U)(\boldsymbol{\alpha}) = {}^*(\varphi \restriction U)(\boldsymbol{\alpha}) = {}^*\varphi(\boldsymbol{\alpha})$ by the additional claim in (6) of Exercise 8.3.9, and similarly ${}^*(\psi')(\boldsymbol{\alpha}) = {}^*\psi(\boldsymbol{\alpha})$, hence, ${}^*\varphi(\boldsymbol{\alpha}) \in {}^*\psi(\boldsymbol{\alpha})$. If $\boldsymbol{\alpha} \notin {}^*U$ then ${}^*\varphi(\boldsymbol{\alpha}) \notin {}^*\psi(\boldsymbol{\alpha})$ by analogous arguments, but (7) of Exercise 8.3.9 is used instead of Alpha-5.

Formula $x = y$ is treated similarly.

As usual, the inductive steps for $\wedge$ and $\neg$ are rather easy, thus, we can concentrate on $\exists$. Suppose that $U = \{k : \exists\, x\, \sigma(x, \varphi_1(k), ..., \varphi_n(k))\}$ belongs to $\mathscr{U}$. By Choice, we obtain an $\mathbb{N}$-sequence φ such that $\sigma(\varphi(k), \varphi_1(k), ..., \varphi_n(k))$ for all $k \in U$. Then $\sigma({}^*\varphi(\boldsymbol{\alpha}), {}^*\varphi_1(\boldsymbol{\alpha}), ..., {}^*\varphi_n(\boldsymbol{\alpha}))$ is true in $\mathbb{I}$ by the inductive hypothesis, therefore, as $x = {}^*\varphi(\boldsymbol{\alpha}) = \mathbf{J}(\varphi) \in \mathbb{I}$, the formula $\exists\, x\, \sigma(x, {}^*\varphi_1(\boldsymbol{\alpha}), ..., {}^*\varphi_n(\boldsymbol{\alpha}))$ is true in $\mathbb{I}$, as required.

Conversely, suppose that $\sigma(x, {}^*\varphi_1(\boldsymbol{\alpha}), ..., {}^*\varphi_n(\boldsymbol{\alpha}))$ is true in $\mathbb{I}$ for some $x = {}^*\varphi(\boldsymbol{\alpha}) \in \mathbb{I}$. Then the set $W = \{k : \sigma(\varphi(k), \varphi_1(k), ..., \varphi_n(k))\}$ belongs to $\mathscr{U}$ by the inductive hypothesis. However clearly $W \subseteq U$. □

Corollary 8.3.12. *$^*\varphi(\alpha) = {}^*\psi(\alpha)$ iff the set $U = \{k : \varphi(k) = \psi(k)\}$ belongs to $\mathscr{U}$, in particular, it suffices that U is cofinite. The same for $\in$.* □

Corollary 8.3.13 ($*$-*Transfer*, ZFC[J]). *If $\sigma(x_1, ..., x_n)$, is an $\in$-formula and $x_1, ..., x_n$ are any sets then*

(†) $\sigma({}^*x_1, ..., {}^*x_n)$ is true in $\mathbb{I}$ iff $\sigma(x_1, ..., x_n)$.

If, moreover, σ is a bounded formula then

(‡) $\sigma({}^*x_1, ..., {}^*x_n) \Longleftrightarrow \sigma(x_1, ..., x_n)$. □

It follows that $\mathbb{I}$ is just (isomorphic to) the ultrapower $\mathtt{Ult}_{\mathscr{U}}(\mathbb{V})$ of the whole set universe $\mathbb{V}$ of **ZFC[J]**! This brings **ZFC[J]** back on the track of ordinary nonstandard methods, with the following special features:

1°. $*$ is an elementary embedding (in the sense of the $\in$-language) of the whole set universe $\mathbb{V}$ into the class $\mathbb{I}$ of all "internal" sets, and which is the transitive closure of the range $\mathtt{ran}\,*$ in the same universe $\mathbb{V}$.

2°. $\mathbb{I}$ is a Gordon class, in the sense that there is $\alpha \in {}^*\mathbb{N} \setminus \mathbb{N}$ such that $\mathbb{I}$ consists of all sets of the form ${}^*f(\alpha)$, where f is a function defined on $\mathbb{N}$.

Countable Saturation comes for free:

Proposition 8.3.14 (ZFC[J]). *The class $\mathbb{I}$ is countably saturated.*

Proof. Suppose that $X_n = {}^*\varphi_n(\alpha) \in \mathbb{I}$ are nonempty sets which form a f. i. p. family. We can assume that $X_{n+1} \subseteq X_n$ for all n. For any n, the set $U_n = \{k : \varphi_n(k) \neq \varnothing\}$ belongs to $\mathscr{U}$ by Theorem 8.3.11, hence, we can assume that $\varphi_n(k) \neq \varnothing$ for *all* k, for if not redefine φ_n outside of U_n and use Corollary 8.3.12. Similarly, it can be assumed that $\varphi_{n+1}(k) \subseteq \varphi_n(k)$ for all n, k. Choose any $\vartheta(k) \in \varphi_k(k)$. In our assumptions, $\vartheta(k) \in \varphi_n(k)$ for all $k \geq n$, thus, $x = {}^*\vartheta(\alpha)$ belongs to any $X_n = {}^*\varphi_n(\alpha)$ by Corollary 8.3.12. □

8.3e Interpretation of Alpha theory in ZFBC

Theorem 8.3.15. *There is an interpretation of **ZFC[J]** in **ZFBC** with the same set universe. Therefore **ZFC[J]** is $\mathtt{wf}$-core interpretable in **ZFGC** by Theorem 8.3.3, and hence **ZFC[J]** is a conservative $\mathtt{wf}$-core extension of **ZFC**.*

Proof. Arguing in **ZFBC**, fix a nonprincipal (*i.e.*, containing all cofinite subsets of $\mathbb{N}$) ultrafilter $\mathscr{U}$ on $\mathbb{N}$. The ultrapower $\mathtt{Ult}_{\mathscr{U}}(\mathbb{V}) = \langle {}^\bullet\mathbb{V}; {}^\bullet\!\in \rangle$ of the whole universe $\mathbb{V}$ is then an extensional structure, hence, by Lemma 8.3.6, there is a transitive class $\mathbb{I}$ and an isomorphism $\pi : \langle {}^\bullet\mathbb{V}; {}^\bullet\!\in \rangle \xrightarrow{\text{onto}} \langle \mathbb{I}; \in \rangle$. The superposition $x \mapsto {}^*x = \pi({}^\bullet x)$ of π and the canonical embedding $x \mapsto {}^\bullet x$ of $\mathbb{V}$ into ${}^\bullet\mathbb{V}$ is the an elementary embedding of $\langle \mathbb{V}; \in \rangle$ into $\langle \mathbb{I}; \in \rangle$.

To define **J**, let $c \in {}^\bullet\mathbb{V}$ be the $\mathscr{U}$-class of $\mathtt{id}$ (recall that $\mathtt{id}(k) = k$ for all k). Let $\alpha = \pi(c)$ and put $\mathbf{J}(\varphi) = {}^*\varphi(\alpha)$ for any map φ defined on $\mathbb{N}$. We leave it as a (difficult!) **exercise** for the reader to prove that $\langle \mathbb{V}; \in, \mathbf{J} \rangle$ is an interpretation of **ZFC[J]** (in **ZFBC**). □

Problem 8.3.16. Is **ZFC**[**J**] `wf`-core interpretable in **ZFC**? □

Coming back to principles 1° and 2° in § 8.3d which, in a sense, characterize the theory **ZFC**[**J**], it is quite clear that while the former is really important for development of nonstandard analysis in **ZFC**[**J**], the latter is rather special, moreover, an easy argument shows that 2° is incompatible with Saturation for families of cardinality $\geq 2^{\aleph_0}$. Therefore, it looks natural to drop 2° but add to 1° more Saturation. This leads to a theory (let us denote it by $\mathbf{ZFC}_{\boldsymbol{\kappa}}[*]$) in the $\in$-language enriched by two additional symbols, $*$ and κ, with the following axioms:

1) all of **ZFC** without Regularity ($*$ and κ can occur in the schemata),
2) axioms saying that $\boldsymbol{\kappa}$ is an infinite (well-founded) cardinal while $*$ is a map (a proper class) defined on the whole set universe $\mathbb{V}$,
3) $*$-Transfer for $*$ as a map $\mathbb{V} \to \mathbb{I}$, where $\mathbb{I} = \{y : \exists x\, (y \in {}^*x)\}$,
4) Transitivity of $\mathbb{I}$, and
5) Saturation for families $\mathscr{X} \subseteq \mathbb{I}$ of cardinality $\leq \boldsymbol{\kappa}$.

Exercise 8.3.17. Replace 5) by a stronger requirement: Saturation for well-orderable families $\mathscr{X} \subseteq \mathbb{I}$. Why is this inconsistent ? □

Exercise 8.3.18. Arguing in **ZFBC**, let κ be an infinite (well-founded) cardinal. Take $V = \mathbb{V}$ in Theorem 8.3.8, and let $*$ be an elementary embedding of $\mathbb{V}$ into a transitive and κ^+-saturated class ${}^*\mathbb{V}$. Show that then $\langle {}^*\mathbb{V}; \in, \kappa, * \rangle$ is an interpretation of $\mathbf{ZFC}_{\boldsymbol{\kappa}}[*]$. □

8.4 Miscellanea: some other theories

We begin this section with a nonstandard set theory, due to Di Nasso, which circumvents the Hrbaček paradox by reducing Saturation to a form that still incorporates all definable cardinals. Then three "stratified" nonstandard theories are considered: their common property is that a single "universe of discourse" is replaced by a conglomerate of universes related with each other in a certain way. Finally, a nonstandard class theory will be considered.

8.4a A theory with "definable" Saturation

The fourth, and last solution of the "Hrbaček paradox" (see the beginning of § 8.2) is to reduce Saturation to a form compatible with Power Set + Choice + Collection + Standardization. At first glance the task does not seem to have an adequate solution. In particular, because no cardinal, chosen as the amount of Saturation postulated, can be consistently argued to fulfill all needs of nonstandard mathematics once and for all. Moreover, fixing any cardinal for this purpose is neither esthetically nor philosophically acceptable.

A modification not connected with any particular cardinal was suggested by Di Nasso [DiN 99]. Let **DNST** be **HST** amended as follows [6]: Power Set and Choice are added, but Saturation is postulated for families whose cardinality is a cardinal $\in$-definable in $\mathbb{WF}$. Thus, Saturation in **DNST** is an axiom schema, which we call Definable Saturation, containing, for any $\in$-formula $\varphi(x)$, an axiom, say, SAT_φ, saying that Saturation holds for all $\cap$-closed families (of internal sets) of cardinality $\le \kappa$, where $\kappa = \kappa_\varphi$ is the least infinite cardinal satisfying $\varphi^{\mathtt{wf}}(\kappa)$, or $\aleph_0$ if no such cardinals exist. This does not imply Saturation for all standard size families (of any cardinality), moreover, we can consistently define the least cardinal κ for which Saturation fails, but this is not an $\in$-definition in $\mathbb{WF}$!

DNST is still a conservative standard core extension of **ZFC**. Indeed, it suffices to prove that each subtheory $\mathbf{DNST}_\varphi$, where we have only SAT_φ instead of the whole Definable Saturation, is a conservative standard core extension of **ZFC**. Arguing in **ZFC**, define a cardinal $\hat{\kappa} = \kappa_\varphi$ as above. As **HST** is a "realistic" theory, there is a standard core interpretation $\mathbb{H} = \langle \mathbb{H}; {}^*\!\in, {}^*\mathtt{st} \rangle$ of **HST** in **ZFC**, together with an associated canonical $\in$-isomorphism $* : \mathbf{V} \xrightarrow{\text{onto}} \mathbb{S} = \mathbb{S}^{(\mathbb{H})}$, where $\mathbf{V}$ is the set universe of **ZFC**. The classes $\mathbb{S}$ and $\mathbb{WF} = \mathbb{WF}^{(\mathbb{H})}$ are $\in$-isomorphic to each other in **HST**, hence, by superposition, there is an $\in$-isomorphism, say $\pi : \mathbf{V} \xrightarrow{\text{onto}} \mathbb{WF}$. Let $\kappa = \pi(\hat{\kappa})$, so that it is true in $\mathbb{H}$ that κ is a (well-founded) cardinal. However (see "second option" on p. 253), there is a class, say, $\mathbb{H}'_\kappa \subseteq \mathbb{H}$, satisfying $\mathbf{HST}'_\kappa$. It follows that $\mathbb{H}'_\kappa$ is a standard core interpretation of $\mathbf{DNST}_\varphi$ in **ZFC**, hence (see Proposition 4.1.9) $\mathbf{DNST}_\varphi$ is a conservative standard core extension of **ZFC**.

Exercise 8.4.1. Show that **DNST** is not a "realistic" nonstandard theory in the sense of Definition 4.1.8. (*Hint*. A minimal **ZFC** model M is not **DNST**-extendible since, as all sets in M are $\in$-definable in M, see Exercise 4.6.22, such an extension would be a model of the full standard size Saturation, contrary to the Hrbaček paradox.) □

8.4b Stratified nonstandard set theories

Under this title, we gathered three theories which have the common property of being focused on certain parts of the nonstandard universe rather than on the latter as a whole. We give here a rather sketchy review of the theories and refer the reader to original papers for details, in particular, regarding the proofs of their conservativity and equiconsistency with **ZFC**. (Reservation: Fletcher's presentation of **SNST** in [Fl 89] is very sketchy.)

Fletcher's stratified nonstandard set theory. The theory **SNST** defined in [Fl 89] sees the nonstandard universe as the union of a system of internal

[6] Actually Di Nasso's formalization uses $*$ as a primary notion, while $\mathtt{st}$ is a definable predicate, *i.e.*, $\mathtt{st}\, x$ iff $x = {}^*u$ for some well-founded u.

subuniverses I_α and external subuniverses E_α, where α is a cardinal in the standard universe S (which satisfies **ZFC**, as usual). This system of subuniverses looks rather similar to the system of classes $\mathbb{I}_\kappa$ and $\mathbb{L}[\mathbb{I}_\kappa]$ in **HST**, with some minor differences, for instance, κ-size Saturation rather than κ-deep Saturation is postulated.

Ballard's enlargement set theory. The nonstandard theory **EST** defined in [Bal 94] has a definite flavour of category – theoretical ideas: it essentially denies anything like a common "working" set universe, but instead postulates a conglomerate of universes connected via embeddings so that still each universe admits an elementary embedding into another, suitably saturated, universe. The whole picture can be compared to a system of transitive classes within a universe of Boffa's set theory **ZFBC**, which consists of those classes which satisfy a certain version of the von Neumann– Gödel – Bernays class theory **NBG** (Theorem 8.3.8 validates the existence of sufficiently saturated extensions.) The following citation from [Bal 94, p. 128] gives an impression of Ballard's philosophical position: *"In designing the vehicle EST, I have deliberately ignored the needs of practitioners and sought instead to decisively illustrate the full implications of this relativistic mathematical ontology."*

Theories of relative standardness. Péraire's system **RST** [Per 92, Per 95] utilizes st as a *binary* predicate, *i.e.*, in the form $x \,\mathtt{st}\, y$, which is understood as *x is standard relative to y*. This is a theory of *internal* kind, like **BST** or **IST**, and its universe has some semblance to a **BST** universe where $x \,\mathtt{st}\, y$ is defined by the st-$\in$-formula $x \in \mathbb{S}[y]$ (see Definition 3.1.13), but the whole structure of axioms is closer to **IST**. Note that the binary predicate $x \,\mathtt{st}\, y$ is atomic in **RST**, which allows to avoid the restrictions imposed by Theorem 6.1.15 and consistently add Inner Standardization for any class of the form $I_y = \{x : x \,\mathtt{st}\, y\}$.

Péraire demonstrated in a number of examples that the relative standardness gives an adequate treatment for phenomena connected with double and more complicated limits in topology and analysis.

Another approach to relative standardness, related rather to **BST**, has recently been proposed by Hrbaček [Hr 04, Hr **].

8.4c Nonstandard class theories

Hrbaček's idea, that any reasonable "standard" theory of set theoretic type admits a certain nonstandard version, was applied to the von Neumann–Gödel – Bernays class [7] theory **NBG** by Gordon in [Gor 97]. This resulted

[7] Recall that the common feature that distinguishes class theories from set theories (both standard and nonstandard) is that the former consider classes as primary objects, while sets are distinguished as those classes which are elements of other classes. Axiomatic systems of class theories look different from those of set theories even in the case when the theories are very close metamathematically as *e.g.* **ZFC** and **NBG**.

in the *nonstandard class theory* **NCT** (in more advanced form, see Andreev and Gordon [AnG 2001]), which is a standard core extension of **NBG** in approximately the same way as **BST** is a standard core extension of **ZFC**.

The universe of **NCT** consists of sets and classes, both types containing standard (satisfying st) and nonstandard objects, with appropriate Comprehension schemata which reflect the idea that the set universe is internal while classes are not necessarily internal. Internal classes are introduced as follows: if X is a standard class (*i.e.*, $\mathtt{st}\, X$) and p any set then the class $\{x : \langle p, x\rangle \in X\}$ is internal. A related Comprehension axiom postulates that any intersection of a set and an internal class is still a set, hence, any internal class X satisfying $X \subseteq x$ for a set x is itself a set (but there are non-internal classes with this property which are not sets).

As for metamathematical properties, being a standard core extension of **NBG**, the theory **NCT** is, at the same time, a *class extension* of **BST**, in the sense that the set universe of **NCT** satisfies **BST**, and conversely, any model of **BST** can be embedded, as the class of all sets, into a model of **NCT**. (The latter can be obtained by adjoining all st-$\in$-definable subclasses of the given **BST** universe; our Theorem 3.2.3 plays the key role to make such an extension procedure working.) Accordingly, any theorem of **NCT** which speaks only about sets is a theorem of **BST**, so that **NCT** is a *conservative* class extension of **BST**. It follows that **NCT** is a conservative (hence, equiconsistent) standard core extension of **ZFC**.

Nonstandard class theories can be expected to be useful in the treatment of those phenomena in the model theoretic version of nonstandard analysis which naturally lead to class-size objects in the frameworks of a nonstandard set theory, see, *e.g.*, Kanovei and Reeken [KanR 99b], where a version of SMA was considered in (a simplified version of) **NCT**. The same goals also can be achieved in Kawaï's theory **KST** (because $\mathbb{S} \subseteq \mathbb{I}$ are sets in **KST**, hence, $\mathbb{S}$-size and $\mathbb{I}$-size objects are sets in **KST**, so that there is no need for it), however, the use of **NCT** has a principal advantage here, because **NCT** provides what seems to be the minimal reasonable nonstandard universe containing $\mathbb{S}$-size objects.

It remains to briefly mention **THS**, *theory of hyperfinite sets* of Andreev and Gordon, see [AnG 2001] and especially a forthcoming paper [AnG **]. This nonstandard theory (actually, a class theory rather than set theory) shares some ideas with the alternative set theory **AST** of Vopenka, in particular, its set universe is intended to consist of sets with $*$-finite transitive closure (in the notation of **HST**). The main feature of **THS** is that it does not make use of standard sets. However to apply Saturation-like tools there should be a suitable notion of a "small" collection of sets — and this is achieved in **THS** by a careful combination of $\in$-definitions. Metamathematically, **THS** turns out to be as strong as the Zermelo theory **ZC**.

Historical and other notes to Chapter 8

Section 8.1. Kawaï's set theory was introduced in [Kaw 83] under the name: *nonstandard set theory*, **NST**. A weaker version was proposed earlier in [Kaw 81]. Theorem 8.1.5: Kawaï [Kaw 83].

Section 8.2. The theory **NST** was introduced, under the name $\mathbf{NS}_2(\mathbf{ZFC})$, in [Hr 78], where also its conservativity is established. A more comprehensive exposition was given in [Hr 79]. Our method to prove the conservativity of **NST** and its versions in § 8.2c by inner models in **KST** is, of course, rather anachronistic: Kawaï's paper [Kaw 83] was published later than Hrbaček's works. Theorem 8.2.10: Hrbaček [Hr 78].

Exercise 8.2.7: Hrbaček [Hr 78] and private communication.

Exercises 8.2.16, 8.2.17, 8.2.18: Kanovei and Reeken [KanR 00a].

Section 8.3. See [Kun 80, III.5] on the Mostowski collapse theorem in **ZFC**.

The Universality axiom is identified as BA_1 in [A 88]. See also [HrJ 98, p. 265]. Note that both Universality and SuperUniversality are different from (and in fact incompatible with) another rather popular axiom which implies the existence of ill-founded sets: *the antifoundation axiom*, or AFA, formulated as *every graph has a unique decoration*, see [A 88, Dev 98, HrJ 98]. AFA describes a set universe in a sense less ill-founded than those described by Universality and SuperUniversality, and apparently does not lead to applications to nonstandard set theories.

The content of § 8.3a – 8.3c is mainly due to Boffa [Bof 72] (regarding **ZFBC** in general and its relations to standard set theories) and Ballard and Hrbaček [BalH 92] (regarding applications to nonstandard analysis). The key axiom of SuperUniversality (or **BAFA** in [A 88]) was introduced in [Bof 72]. Aczel [A 88] gives a broad reference in the history of non-well-founded set theories which in fact goes back to the times of Zermelo and Fraenkel.

Di Nasso's theory $\mathbf{ZFC}[\alpha]$ first appeared in [DiN 99] with a slightly different (but equivalent) list of axioms. One of its motivations was to give rigorous treatment of a pre-Robinson attempt in nonstandard analysis, due to Schmieden and Laugwitz [SchmiedL 58].

9 "Hyperfinite" descriptive set theory

Descriptive set theory studies those subsets of topological spaces (called *pointsets*) which can be defined, by means of a list of specified operations including, *e. g.*, complement, countable union and intersection, projection, beginning with open sets of the space. Classical descriptive set theory (DST) considers mainly sets in Polish (that is, separable metric) spaces, this is why we shall identify it here as *Polish* descriptive set theory.

"Hyperfinite" descriptive set theory follows this scheme in a different setting: the construction of hierarchies begins with internal subsets of a fixed infinite internal set H as the basic sets. Note that internal subsets of an infinite internal H do not form a topology, moreover, the weakest topology where all internal sets are open, is discrete because all singletons are internal, hence, H is called rather *domain* than space. [1]

It turns out that many questions on the nature of pointsets, considered by Polish descriptive set theory, remain meaningful in the "hyperfinite" setting, sometimes directly sometimes in a more or less revised form. Accordingly, the results obtained are sometimes similar to those of Polish descriptive set theory, sometimes just the opposite. But in general "hyperfinite" descriptive set theory is much less developed than Polish DST. As for the methods, they can be very different. The following is a very rough classification of theorems of "hyperfinite" descriptive set theory from the point of view of the methods involved:

(A) results similar to "Polish" theorems and obtained by virtue of the substitution of Saturation for completeness (or compactness) in "Polish" proofs;

(B) corollaries of "Polish" theorems by means of shadow maps;

(C) results that appear stronger than their "Polish" counterparts because Saturation is in some cases stronger than completeness or compactness;

(D) results based on a kind of "hyperfinite" combinatorics, including plain pigeonhole-type arguments, sometimes w. r. t. non-internal objects.

---

[1] The domain H is sometimes taken to be a $*$-finite, that is hyperfinite set, which is essential for applications like Loeb measures, — this is why this direction is called *"hyperfinite"* DST. However most results will be true for all infinite internal domains H.

The content of this Chapter includes the following. We begin in Section 9.1 with the basic set-up including Borel, projective, Souslin subsets of internal sets. Operations of countable character, countably determined sets, and the related concept of shadows follow in Section 9.2. Closure properties of Borel and projective classes, based on the key shadow theorem (Theorem 9.3.3, which shows that shadow preimages keep a Borel or projective class in both directions), are considered in Section 9.3. The next Section 9.4 is central: we present main structural theorems of "hyperfinite" descriptive set theory, including Separation, Reduction, Uniformization, sets with special cross-sections, and some other theorems. Some questions related to Loeb measures (like the existence of liftings) are considered in Section 9.5. Section 9.6 presents studies on "Borel cardinals", that is, relations between Borel sets in terms of Borel injections and bijections, and "countably determined cardinals", that is, relations between countably determined sets in terms of countably determined injections and bijections. This research line continues in Section 9.7, where we study quotients over Borel and countably determined equivalence relations, a topic quite typical for modern works in Polish descriptive set theory.

We left aside such notable topics as some foundational issues in nonstandard real and functional analysis, topology, and Loeb measures (except for a brief Section 9.5 not at all covering the issue), which have some relevance to "hyperfinite" descriptive set theory. Unfortunately we have also to sacrifice our original plan to add a survey of the following topics:

1) nonstandard topologies generated by countably determined cuts in ${}^*\mathbb{N}$ (see [KL 91, Jin 01] and references in the second paper),
2) sets with special properties related to category and Loeb measure (as in [Mil 90]),
3) completeness properties of the $*$-reals (see [Jin 96] and references there),

— because of the limited space available for "hyperfinite" descriptive set theory in this book.

Our exposition will follow the standards of "hyperfinite" descriptive set theory in model theoretic nonstandard analysis, including its commitment to countable Saturation (see Blanket Agreement 9.1.2) and its stress on countably determined sets. In principle all results below (except those explicitly indicated) are true in the model theoretic setting.

We tried to make the exposition as self-contained as possible within a rather restricted space, yet some degree of aquaintance with Polish descriptive set theory and, to a lesser extent, with "hyperfinite" DST the in model-theoretic version will be assumed. Kechris [Kech 95] is given as a general reference in matters of Polish descriptive set theory.

9.1 Introduction to "hyperfinite" DST

Development of "hyperfinite" descriptive set theory in **HST** is quite similar to the model theoretic version, yet we have to pay attention to some essential details in the beginning of this introductory Section. Then we introduce Borel, projective, and Souslin sets in internal domains.

9.1a General set-up

The model-theoretic version of "hyperfinite" descriptive set theory deals with a basic "standard universe", that is a structure V in the **ZFC** world of sets, which models a fragment of **ZFC** (usually equal or weaker than **ZC**), and a "nonstandard universe", usually a nonstandard type-theoretic superstructure over an elementary extension ${}^*V$ of V, which contains, in particular, an element ${}^*x$ for any $x \in V$.

In **HST**, we change to the scheme "$\mathbb{WF} \xrightarrow{*} \mathbb{I}$ [in $\mathbb{H}$]" (see § 1.2a), which proposes the "standard" structure [2] $\mathbb{WF}$ and <u>the</u> "nonstandard extension" $\mathbb{I}$, both transitive classes in a wider external set universe $\mathbb{H}$ of **HST**. Thus, the multitude of "nonstandard universes", which model theorists are accustomed to in **ZFC**, apparently disappears — we have a uniquely defined pair of the "standard" (*i.e.*, well-founded) universe $\mathbb{WF}$ and the internal universe $\mathbb{I}$.

Problem 9.1.1. It is a challenging problem to utilize partially saturated subuniverses, of the type considered in Chapter 6, in a way emulating the ongoing study of the multitude of "nonstandard universes" in the **ZFC** set universe in model theoretic nonstandard analysis, especially w. r. t. questions related to "hyperfinite" descriptive set theory. □

Blanket agreement 9.1.2. Our development of "hyperfinite" descriptive set theory is compatible with the weakest reasonable version of **HST**, where

1°. Saturation is reduced to *countable* Saturation, that is, Saturation for <u>countable</u> f. i. p. families of internal sets.

2°. Countable Saturation is sufficient to prove *countable Extension*, that is,
for every sequence $\{x_n\}_{n\in\mathbb{N}}$ *of internal sets* x_n *there exists an internal function* f *with* $\mathbb{N} \subseteq \operatorname{dom} f$ *and* $f(n) = x_n$ *for all* n;
this is a particular case of Theorem 1.3.12.

3°. Standard Size Choice is reduced to $\mathfrak{c}$-size Choice (Choice in the case when the domain of a choice function is a set of cardinality $\le \mathfrak{c} = 2^{\aleph_0}$). [3] □

Any strengthening of these assumptions will be explicitly indicated.

[2] Isomorphic to the true class $\mathbb{S}$ of all standard sets in **HST**, but more convenient, in particular, $\mathbb{WF}$ is transitive and $\subseteq$-complete while $\mathbb{S}$ is not.

[3] Recall that κ-size Saturation, generally, corresponds to 2^κ-size Choice, see Footnote 22 on page 253, or Remark 3.3.6.

This is compatible with **HST** as well as with any of the partially saturated theories $\mathbf{HST}_\kappa$ and $\mathbf{HST}'_\kappa$ introduced in § 6.4a. In different terms, this is also compatible both with the "main" scheme "$\mathbb{WF} \xrightarrow{*} \mathbb{I}$ [in $\mathbb{H}$]" (*i.e.*, with $\mathbb{I}$ as the internal domain) and the partially saturated schemes of § 6.4h, *e.g.*, "$\mathbb{WF} \xrightarrow{*} \mathbb{I}_\kappa$ [in $\mathbb{L}[\mathbb{I}_\kappa]$]" (with $\mathbb{I}_\kappa$ as the internal domain, κ being any infinite cardinal, for instance, $\aleph_0$). This is also compatible with many nonstandard theories considered in Chapter 8.

And finally, 9.1.2 is compatible with the ordinary assumptions of model-theoretic nonstandard analysis. In fact all theorems below, except for those few explicitly marked as "full-**HST**" results, are valid in the set-up of model-theoretic nonstandard analysis, with countable Saturation.

Fortunately neither Power Set nor κ-size Choice for cardinals $\kappa > 2^{\aleph_0}$ is really of importance in "hyperfinite" descriptive set theory (as they are, generally, not important to Polish descriptive set theory, except for rather special issues). Similarly, it is customary not to assume more than countable Saturation in "hyperfinite" descriptive set theory.

9.1b Comments on notation

The remainder of this Chapter involves a special notation which deserves a few comments. Recall that if $X \subseteq \mathbb{I}$ and $\lambda \in \mathtt{Ord}$ then the collection X^λ of all functions $f : \lambda = \{\xi : \xi < \lambda\} \to X$ is a set (Theorem 1.3.14). In particular, the following collections are sets whenever $X \subseteq \mathbb{I}$:

$X^{\mathbb{N}}$ = all infinite sequences of elements of X, *i.e.*, maps $f : \mathbb{N} \to X$;

X^n = all finite sequences s of elements of X of length $\mathtt{lh}\, s = n \in \mathbb{N}$;

$X^{<\omega}$ = $\bigcup_{n\in\mathbb{N}} X^n$, all finite sequences of elements of X,
note that if $f \in X^{\mathbb{N}}$ and $n \in \mathbb{N}$ then $f \restriction n \in X^n \subseteq X^{<\omega}$;

in addition, $\mathscr{P}_{\mathtt{fin}}(X) = \{\text{all finite subsets of } X\}$ is a set.

If $X, Y \in \mathbb{WF}$ then X^Y, $X^{\mathbb{N}}$, X^n, $X^{<\omega}$, $\mathscr{P}(X)$ are sets and belong to $\mathbb{WF}$ since $\mathbb{N} \in \mathbb{WF}$ and $\mathbb{WF}$ is a transitive and $\subseteq$-complete class satisfying **ZFC** (Theorem 1.1.9). Typical cases below are $X = \mathbb{N}$ and $X = 2 = \{0, 1\}$. This leads to the following sets in $\mathbb{WF}$:

$\mathbb{N}^{\mathbb{N}}$ = all infinite sequences of natural numbers;

$\mathbb{N}^{<\omega}$ = $\bigcup_{k\in\mathbb{N}} \mathbb{N}^k$, all finite sequences of natural numbers;

$2^{\mathbb{N}}$ = all infinite dyadic sequences, or maps $\mathbb{N} \to 2 = \{0, 1\}$;

2^I = all maps $I \to 2$ (where I is a set in $\mathbb{WF}$);

$2^{<\omega}$ = $\bigcup_{k\in\mathbb{N}} 2^k$, all finite dyadic sequences.

A set X is *(at most) countable* if there is a bijection $f : \mathbb{N} \xrightarrow{\text{onto}} X$. Countable sets are sets of standard size (**exercise**: prove it !).

On the other hand, if H is an internal set then ${}^{\text{int}}2^H$ will be systematically used to denote the (internal) set of all internal maps $f : H \to 2$.

As usual, $\aleph_1 = \omega_1$ is the least uncountable ordinal. According to § 1.2c, ω_1 is the first uncountable ordinal and the least uncountable cardinal in $\mathbb{WF}$, hence, as $\mathbb{WF}$ is a model of **ZFC**, ω_1 behaves in **ZFC**-like manner in **HST**.

In parallel with notions and concepts of "hyperfinite" descriptive set theory, we shall systematically consider and refer to notions and results related to Polish spaces, especially compact spaces of the form 2^I where I is a countable set. A natural open base $\mathbf{Base}(2^I)$ of 2^I consists of all non-empty finite intersections of sets of the form $\{x \in 2^I : x(i) = \nu\}$, $i \in I$, $\nu = 0, 1$. By compactness any clopen $Y \subseteq 2^I$ is a finite union of sets in $\mathbf{Base}(2^I)$.

For any Polish space $\mathscr{X}$, for instance $\mathscr{X} = 2^I$, I being a countable set, $\mathbf{\Sigma}^0_\xi[\mathscr{X}]$, $\mathbf{\Pi}^0_\xi[\mathscr{X}]$, $\mathbf{\Delta}^0_\xi[\mathscr{X}]$ ($\xi < \omega_1$) and $\mathbf{\Sigma}^1_n[\mathscr{X}]$, $\mathbf{\Pi}^1_n[\mathscr{X}]$, $\mathbf{\Delta}^1_n[\mathscr{X}]$ ($n \in \mathbb{N}$) indicate the classes of resp. Borel and projective hierarchies of subsets of $\mathscr{X}$. We define separately $\mathbf{\Sigma}^0_0[\mathscr{X}] = \mathbf{\Pi}^0_0[\mathscr{X}] = \mathbf{\Delta}^0_0[\mathscr{X}] = \mathbf{Clop}(\mathscr{X})$ (clopen subsets of $\mathscr{X}$) for any space $\mathscr{X}$ of the form 2^I.

If U is a given set then we put $X^{\mathsf{C}} = U \setminus X$, the *complement* of X. As X^{C} obviously depends on U, too, this notation will be used only if it is clear from the context which basic domain U is considered.

9.1c Borel and projective sets in a nonstandard domain

Similarly to Polish DST, "hyperfinite" descriptive set theory classifies sets in accordance with the complexity of their definitions or constructions from certain initial sets, the latter being now just internal sets.

Borel sets over an algebra. Suppose that $\mathscr{A} \subseteq \mathscr{P}(H)$ is an algebra of subsets of a given set H, that is a collection of subsets of H closed under unions and intersections (of two sets) and complements to H. The *Borel hierarchy* over $\mathscr{A}$ consists of *Borel classes* [4] $\mathbf{\Sigma}^0_\xi(\mathscr{A})$, $\mathbf{\Pi}^0_\xi(\mathscr{A})$, $\mathbf{\Delta}^0_\xi(\mathscr{A})$ of subsets of H defined by induction on $\xi < \omega_1$ as follows:

$\mathbf{\Pi}^0_0(\mathscr{A}) = \mathbf{\Delta}^0_0(\mathscr{A}) = \mathbf{\Sigma}^0_0(\mathscr{A}) = \mathscr{A}$;

$\mathbf{\Pi}^0_\xi(\mathscr{A}) = \{X^{\mathsf{C}} : X \in \mathbf{\Sigma}^0_\xi(\mathscr{A})\}$ for any $\xi < \omega_1$, where $X^{\mathsf{C}} = H \setminus X$;

$\mathbf{\Delta}^0_\xi(\mathscr{A}) = \mathbf{\Sigma}^0_\xi(\mathscr{A}) \cap \mathbf{\Pi}^0_\xi(\mathscr{A})$ for each $\xi < \omega_1$, *i.e.* a set belongs to $\mathbf{\Delta}^0_\xi(\mathscr{A})$ iff it belongs to both $\mathbf{\Sigma}^0_\xi(\mathscr{A})$ and $\mathbf{\Pi}^0_\xi(\mathscr{A})$;

$\mathbf{\Sigma}^0_\xi(\mathscr{A})$ ($\xi \geq 1$) consists of all countable unions of sets in $\bigcup_{\eta<\xi} \mathbf{\Pi}^0_\xi(\mathscr{A})$; in other words, X belongs to $\mathbf{\Sigma}^0_\xi(\mathscr{A})$ iff there is a sequence $\{X_n\}_{n\in\mathbb{N}}$ of sets $X_n \in \bigcup_{\eta<\xi} \mathbf{\Pi}^0_\eta(\mathscr{A})$, such that $X = \bigcup_n X_n$.

Sets in $\mathbf{Borel}(\mathscr{A}) = \bigcup_{\xi<\omega_1} \mathbf{\Sigma}^0_\xi(\mathscr{A})$ are called *Borel sets over* $\mathscr{A}$. Obviously $\mathbf{Borel}(\mathscr{A})$ is the least σ-algebra of subsets of H containing all sets in $\mathscr{A}$.

[4] Actually, these classes are sets in **HST**, see Theorem 9.3.9(vii), but the word "class" is traditional and convenient here.

Borel sets in an internal domain. Let H be an internal set, usually infinite, called *the domain*. Hyperfinite, *i. e.* $*$-finite domains form a special category. A domain H will sometimes be fixed, but we shall often work with different domains, whose role will be rather similar to that of the underlying space in standard topology. However the basic structure is here not a topology but rather the (internal) algebra $\mathscr{P}_{\mathtt{int}}(H)$ of all internal subsets of H, whose role is similar, to some extent, to the role of clopen sets in spaces $\mathscr{X} = 2^I$.

Given an internal domain H, we define, for any $\xi < \omega_1$,

$$\mathbf{\Sigma}^0_\xi[H] = \mathbf{\Sigma}^0_\xi(\mathscr{A}), \quad \mathbf{\Pi}^0_\xi[H] = \mathbf{\Pi}^0_\xi(\mathscr{A}), \quad \mathbf{\Delta}^0_\xi[H] = \mathbf{\Delta}^0_\xi(\mathscr{A}),$$

— where $\mathscr{A} = \mathscr{P}_{\mathtt{int}}(H)$. Sets in $\mathbf{Borel}[H] = \bigcup_{\xi<\omega_1} \mathbf{\Sigma}^0_\xi[H] = \mathbf{Borel}(\mathscr{A})$ are called *Borel sets in H*. Thus $\mathbf{Borel}[H]$ is the least σ-algebra in $\mathscr{P}(H)$ containing all internal subsets of H. [5]

Projective sets. Define, by induction on $n \in \mathbb{N}$, *the projective hierarchy* of *projective classes* $\mathbf{\Sigma}^1_n[H]$, $\mathbf{\Pi}^1_n[H]$, $\mathbf{\Delta}^1_n[H]$ of subsets of a fixed internal domain H as follows:

$\mathbf{\Sigma}^1_{n+1}[H]$ consists of projections of sets in $\mathbf{\Pi}^1_n[H \times K]$ (if $n \geq 1$) or in $\mathbf{Borel}[H \times K]$ (if $n = 0$), where K is internal; thus a set $X \subseteq H$ belongs to $\mathbf{\Sigma}^1_{n+1}[H]$ iff there is an internal set K [6] and a set P in $\mathbf{\Pi}^1_n[H \times K]$ $(n \geq 1)$ or in $\mathbf{Borel}[H \times K]$ $(n = 0)$ such that

$$X = \operatorname{dom} P = \{x \in H : \exists y \in K\, (\langle x, y\rangle \in P)\}, \text{ the } \textit{projection} \text{ of } P;$$

$\mathbf{\Pi}^1_n[H] = \{X^{\complement} : X \in \mathbf{\Sigma}^1_n[H]\}$ for any n, where still $X^{\complement} = H \setminus X$;

$\mathbf{\Delta}^1_n[H] = \mathbf{\Sigma}^1_n[H] \cap \mathbf{\Pi}^1_n[H]$ for any n.

Sets in $\mathbf{Proj}[H] = \bigcup_{n\in\mathbb{N}} \mathbf{\Sigma}^1_n[H]$ are called *projective sets in H*.

Sets in $\mathbf{\Sigma}^1_1[H]$ are alternatively called *analytic*. This family of sets can be defined differently, see § 9.1e below.

That (in **HST**) all Borel and projective classes of sets in a fixed internal domain $H \in \mathbb{I}$ are sets rather than proper classes, can be demonstrated in a rather straightforward manner by induction, based on Theorem 1.3.14, since the construction can be carried out so that the axioms of Power Set and Choice (except for the $2^{\aleph_0}$-size Choice) are not involved. See Theorem 9.3.9 below for another proof based on operations over countable index sets.

Exercise 9.1.3 (absoluteness). Prove, by induction on ξ and n, that if H, K are two internal domains, $\mathbf{\Gamma}$ is a Borel or projective class, say, $\mathbf{\Sigma}^0_\xi$ or $\mathbf{\Sigma}^1_n$, and $X \subseteq H \cap K$, then $X \in \mathbf{\Gamma}[H]$ iff $X \in \mathbf{\Gamma}[K]$. Thus being of a certain Borel class is an intrinsic property of a set independent of the domain.

This allows us to omit specifications $[H]$ wherever possible. □

[5] Note that different parentheses are used to denote classes *over* an algebra and *in* a domain; in all cases $[H]$ means $(\mathscr{A})$, where $\mathscr{A} = \mathscr{P}_{\mathtt{int}}(H)$.

[6] The definition of $\mathbf{\Sigma}^1_{n+1}[H]$ does not depend on the choice of K in the category of infinite internal sets — see Theorem 9.3.9(iv).

Remark 9.1.4. There is another principal absoluteness issue. All uncountable Polish spaces are identical to each other with respect to basic properties of Borel and projective classes (except for questions related to low Borel classes) — simply because they are isomorphic to each other via Borel isomorphisms preserving all Borel classes beginning with the level ω.

On the contrary, it follows from Theorem 5.5.8(iv) that **HST** does not prove that all infinite internal (even all $*$-finite) sets are equinumerous, let alone Borel isomorphic. However we can extend the result of Corollary 7.2.5 to the elementary equivalence, in **HST**, of the Borel and projective hierarchies over any two different infinite internal sets. This is by virtue of the same argument as in the proof of 7.2.5, with the following additional note: countable sets (which participate in the definition of the hierarchies) do not vanish under the restriction to $\mathbb{L}[\mathbb{I}]$ by 6° of Theorem 5.5.4 and new countable sets do not appear in generic extensions by Theorem 7.1.20(ii). □

9.1d Some applications of countable Saturation

We present here several elementary results which demonstrate how countable Saturation works in rather simple cases; more applications are given below. Claim (i) of the next lemma shows that Saturation may work similarly to completeness or compactness, but according to (ii) and (iii) there is also a difference: neither of the two items follows from completeness or compactness.

Lemma 9.1.5. (i) *If* $X_0 \supseteq X_1 \supseteq X_2 \supseteq \dots$ *is a sequence of non-empty* $\mathbf{\Pi}^0_1$ *sets* (in particular, this applies for internal sets) *then* $X = \bigcap_{n \in \mathbb{N}} X_n \neq \varnothing$;

(ii) *if every* X_n, $n \in \mathbb{N}$, *contains at least two elements then so does* X;

(iii) *if every* X_n *is infinite then* X *contains an infinite internal subset.*

Proof. (i) Let $X_n = \bigcap_{k \in \mathbb{N}} X_{nk}$, X_{nk} being internal. To prove that $X \neq \varnothing$ apply Saturation to the countable family of sets X_{nk}.

(ii), (iii) Apply Saturation to the family of all sets $\mathscr{Y}_{nk}$, $n, k \in \mathbb{N}$, where $\mathscr{Y}_{nk}$ is the set of all internal $Y \subseteq X_{nk}$ with ≥ 2, resp., $\geq k$ elements. □

Lemma 9.1.6. *Let* I *be a countable set,* $\{P_i : i \in I\}$ *a family of* $\mathbf{\Pi}^0_1$ *subsets of* $H \times K$, *where* H, K *are internal sets, and* $P = \bigcap_{i \in I} P_i$. *Then* $\mathtt{dom}\, P = \bigcap_{u \in U} \mathtt{dom} \bigcap_{i \in u} P_i$, *where* $U = \mathscr{P}_{\mathtt{fin}}(I)$. *The same is true provided* $H, K \in \mathbb{WF}$ *are compact spaces and* P_i *closed sets in* $H \times K$.

Recall that $\mathtt{dom}\, P = \{x : \exists\, y\, P(x, y)\}$, the projection of $P \subseteq H \times K$ on H.

Proof. By definition, $P_i = \bigcap_{n \in \mathbb{N}} P_{ni}$, where $P_{ni} \subseteq X \times Y$ are internal sets. ($\aleph_0$-size Choice is applied.) Suppose that $x \in \bigcap_{u \in U} \mathtt{dom} \bigcap_{i \in u} P_i$. Then, for any u, $s \in U$ the internal set $S_{us} = \{y : \langle x, y \rangle \in \bigcap_{i \in u,\, n \in s} P_{ni}\}$ is non-empty, and the sets S_{us} form a f. i. p. family, because $S_{u \cup v,\, s \cup t} \subseteq S_{us} \cap S_{vt}$. As $U \times U$ is a countable set, the sets S_{us} contain a common element y by countable Saturation. However obviously $\langle x, y \rangle \in P$, and hence $x \in \mathtt{dom}\, P$.

Compactness substitutes Saturation in the second part of the lemma. □

Exercise 9.1.7. (1) Prove that any countable $\mathbf{\Pi}^0_1$ set $X = \{x_n : n \in \mathbb{N}\}$ is finite. (Apply Lemma 9.1.5 to the sequence of sets $X_n = \{x_k : k \geq n\}$.)

(2) Prove using Saturation that $\mathbf{\Delta}^0_1[H] = \mathscr{P}_{\mathtt{int}}(H)$ for any internal H. (See Theorem 1.4.2(ii).)

Lemma 9.1.8. *Suppose that H, K are internal sets and $P \subseteq H \times K$ is a $\mathbf{\Sigma}^0_1$ (resp., $\mathbf{\Pi}^0_1$) set. Then $X = \operatorname{dom} P$ is $\mathbf{\Sigma}^0_1$ (resp., $\mathbf{\Pi}^0_1$).*

Proof. If $P = \bigcup_n P_n \subseteq H \times K$, all P_n being internal, then $\operatorname{dom} P = \bigcup_n \operatorname{dom} P_n$, where each projection $\operatorname{dom} P_n$ is internal together with P_n — this proves the result for $\mathbf{\Sigma}^1_1$. As for $\mathbf{\Pi}^0_1$, apply Lemma 9.1.6 for the sequence of internal sets P_n such that $P = \bigcap_{n \in \mathbb{N}} P_n$. □

9.1e Operation A and Souslin sets

The *Souslin operation* over a family $\{X_s\}_{s \in \mathbb{N}^{<\omega}}$ of sets X_s of any kind produces a set $\mathsf{A}\{X_s\}_{s \in \mathbb{N}^{<\omega}} = \bigcup_{f \in \mathbb{N}^{\mathbb{N}}} \bigcap_{n \in \mathbb{N}} X_{f \restriction n}$. For any family $\mathscr{B}$ of sets, $\mathsf{A}{\bullet}\mathscr{B}$ will indicate the collection of all sets of the form $\mathsf{A}\{X_s\}_{s \in \mathbb{N}^{<\omega}}$ where all X_s belong to $\mathscr{B}$. Sets in $\mathsf{A}{\bullet}\mathscr{B}$ are called *Souslin sets over* $\mathscr{B}$. If H is an internal set then sets in $\mathsf{A}{\bullet}\mathscr{P}_{\mathtt{int}}(H)$ are called *Souslin sets in* H.

Exercise 9.1.9. Prove the following (well known) properties of Souslin sets: for any algebra of sets $\mathscr{B}$, $\mathbf{Borel}(\mathscr{B}) \subseteq \mathsf{A}{\bullet}\mathscr{B} = \mathsf{A}{\bullet}\mathsf{A}{\bullet}\mathscr{B}$. □

Yet the next result is somewhat more surprising: A substitutes the operation of projection in the definition of the first projective class $\mathbf{\Sigma}^1_1$.

Proposition 9.1.10. *Let H, K be two infinite internal sets. Then the following four classes of subsets of H coincide:*

(1) $\mathbf{\Sigma}^1_1[H]$; (2) $\mathsf{A}{\bullet}\mathscr{P}_{\mathtt{int}}(H)$; (3) $\mathsf{A}{\bullet}\mathbf{Borel}[H]$;

(4) *projections* $\operatorname{dom} P$ *of* $\mathbf{\Pi}^0_2$ *sets* $P \subseteq H \times K$.

Proof. To obtain (2) = (3) apply the result of 9.1.9.

(2) $\subseteq$ (4). Suppose that $X \in \mathsf{A}{\bullet}\mathscr{P}_{\mathtt{int}}(H)$, i.e. $X = \mathsf{A}\{X_s\}_{s \in \mathbb{N}^{<\omega}}$, where each $X_s \subseteq H$ is internal. To prove that X is of the form (4), we w. l. o. g. suppose that $K = k^k$ (the set of all internal functions $h : k \to k$), where $k \in {}^*\mathbb{N} \setminus \mathbb{N}$. The set P of all pairs $\langle x, h \rangle \in H \times K$ with $h(n) \in \mathbb{N}$ and $x \in X_{h \restriction n}$ for all $n \in \mathbb{N}$ is $\mathbf{\Pi}^0_2[H \times K]$ and $X = \operatorname{dom} P$.

(1) $\subseteq$ (2). Suppose that K is internal, $P \subseteq H \times K$ is Borel, and prove that $X = \operatorname{dom} P$ is of the form (2). First of all, as above, $P \in \mathsf{A}{\bullet}\mathscr{P}_{\mathtt{int}}(H)$, thus, let $P = \mathsf{A}\{P_s\}_{s \in \mathbb{N}^{<\omega}}$, where all sets $P_s \subseteq H$ are internal. We have

$$\begin{aligned} \operatorname{dom} P &= \textstyle\bigcup_{f \in \mathbb{N}^{\mathbb{N}}} \operatorname{dom} \bigcap_{n \in \mathbb{N}} P_{f \restriction n} = \\ &= \textstyle\bigcup_{f \in \mathbb{N}^{\mathbb{N}}} \bigcap_{k \in \mathbb{N}} \operatorname{dom} \bigcap_{n \leq k} P_{f \restriction n} = \mathsf{A}\{P_s\}_{s \in \mathbb{N}^{<\omega}}, \end{aligned}$$

by Lemma 9.1.6, where all sets $X_s = \operatorname{dom} \bigcap_{n \leq \mathtt{lh}\, s} P_{s \restriction n}$ are internal. □

9.2 Operations, countably determined sets, shadows

This family of sets does not have a reasonable counterpart in Polish descriptive set theory, yet it plays an extremely important role in the "hyperfinite" theory. To define countably determined sets in §9.2b, we introduce generalized operations and quantifiers, in particular those over countable index sets. Another approach to countably determined sets, connected with shadow (or standard part) maps, is considered in §9.2c.

9.2a Operations and quantifiers

Suppose that I is any (usually countable) set, called *the index set*. For any $u \subseteq I$, $\chi_u \in 2^I$ is *the characteristic function* of u, i.e. $\chi_u(i) = 1$ iff $i \in u$.

Let H be any set treated as a fixed domain; in most of applications H will be an infinite internal set or a space of the form 2^J, J a countable set.

Definition 9.2.1. Any set $U \subseteq 2^I$ is called a *base over* I.[7] Such a set U defines a *generalized quantifier* Q_U, acting as follows: if $R(i)$ is any relation then we let $\mathsf{Q}_U\, i \in I\, R(i)$, or briefly $\mathsf{Q}_U\, i\, R(i)$, mean that $\chi_{\{i \in I : R(i)\}} \in U$.

Also, Φ_U, *an operation over* I, acting on indexed families $\{A_i\}_{i\in I}$ of sets $A_i \subseteq H$, with the result being also a subset of H, is defined as follows:

$$\Phi_U\{A_i\}_{i\in I} = \{x \in H : \chi_{\{i\in I : x\in A_i\}} \in U\} = \{x \in H : \mathsf{Q}_U\, i\,(x \in A_i)\}$$
$$= \bigcup_{b\in U} \bigcap_{i\in I} A_i^{b(i)}, \text{ where } A_i^1 = A_i\,,\ A_i^0 = A_i{}^{\complement} = H \smallsetminus A_i\,. \quad \square$$

In these terms, the ordinary quantifiers $\exists$ and $\forall$ over I can be identified with resp. the bases of *I-union* $\cup(I) = 2^I \smallsetminus \{\mathbf{0}\}$ and *I-intersection* $\cap(I) = \{\mathbf{1}\}$, where $\mathbf{0}, \mathbf{1} \in 2^I$ are resp. the constant 0 and the constant 1. Obviously

$$\Phi_{\cup(I)}\{X_i\}_{i\in I} = \bigcup_{i\in I} X_i \qquad \text{and} \qquad \Phi_{\cap(I)}\{X_i\}_{i\in I} = \bigcap_{i\in I} X_i\,.$$

The Souslin operation A over the index set $\mathbb{N}^{<\omega}$ is the same as $\Phi_{\mathbf{sus}}$, where $\mathbf{sus}$ consists of all $b \in 2^{(\mathbb{N}^{<\omega})}$ such that the set $\{s : b(s) = 1\}$ contains a subset of the form $\{f \restriction k : k \in \mathbb{N}\}$ for some $f \in \mathbb{N}^{\mathbb{N}}$. This is a bit too complicated, but clearly $\mathsf{A}\{X_s\}_{s\in\mathbb{N}^{<\omega}} = \Phi_{\mathbf{sus}}\{X_s\}_{s\in\mathbb{N}^{<\omega}} = \bigcup_{f\in\mathbb{N}^{\mathbb{N}}} \bigcap_{m\in\mathbb{N}} X_{f\restriction m}$.

Generally speaking, quantifiers of the form $\mathsf{Q}_U\, i$ are not monotone, i.e. $\mathsf{Q}_U\, i\, A(i)$ and $\forall i\,(A(i) \Longrightarrow B(i))$ do not necessarily imply $\mathsf{Q}_U\, i\, B(i)$. U being a filter implies monotony, but this will not be assumed.

[7] It would be more convenient, in some respect, to consider bases formally as subsets of $\mathscr{P}(I)$ rather than subsets of 2^I — for instance this would eliminate a frequent passage from sets of indices to characteristic functions. Yet such a change would lead to its own inconveniences, in particular because it is common practice to consider spaces of the form 2^I rather than $\mathscr{P}(I)$.

Anyway, if the reader is going to identify subsets of the index set I with their characteristic functions in 2^I then any occurrence of χ below can be disregarded.

Exercise 9.2.2 (characterization of operations Φ_U). Suppose that Φ is an operation acting on indexed families $\{A_i\}_{i\in I}$ of sets $A_i \subseteq H$, with the result always being also a subset of H, and such that whether $x \in H$ belongs to $\Phi\{A_i\}_{i\in I}$ depends only on the set $I_x = \{i \in I : x \in A_i\}$, not on x itself. Prove that there exists a unique base $U \subseteq 2^I$ such that $\Phi = \Phi_U$. □

There are several general methods to define new bases and operations:

Example 1 (complementary operation). For any base $U \subseteq 2^I$ define the *complementary base* $U^{\mathsf{C}} = 2^I \setminus U$. [8] **Exercise**: Prove that $\mathsf{Q}_{(U^{\mathsf{C}})}\, i\, R(i)$ iff $\neg\, \mathsf{Q}_U\, i\, R(i)$, or, in terms of operations, $\Phi_{U^{\mathsf{C}}}\{X_i\}_{i\in I} = (\Phi_U\{X_i\})^{\mathsf{C}}$. □

Example 2 (superposition). Suppose that U is a base over an index set I and, for any $i \in I$, V_i is a base over a set J_i. Define a new index set $J = \{\langle i, j\rangle : i \in I \wedge j \in J_i\}$ and a new *superposition* base

$$V = U \circ \{V_i\}_{i\in I} = \{\beta \in 2^J : \mathsf{Q}_U\, i \in I\, \mathsf{Q}_{V_i}\, j \in J_i\, (\beta(i,j) = 1)\}\,.$$

Exercise. Prove that in this case $\mathsf{Q}_V\, \langle i, j\rangle\, R(i,j)$ iff $\mathsf{Q}_U\, i\, \mathsf{Q}_{V_i}\, j\, R(i,j)$, or, in terms of operations, $\Phi_V\{A_{ij}\}_{\langle i,j\rangle\in J} = \Phi_U\{\Phi_{V_i}\{A_{ij}\}_{j\in J_i}\}_{i\in I}$. Prove that $V = \Phi_U\{V_i'\}_{i\in I}$, where V_i' consists of all $\beta \in 2^J$ such that the "cross-section" map $\beta_i(j) = \beta(i,j)$, $j \in J_i$, belongs to V_i. □

Example 3 (projection). Assume that U is a base over I. Let Π be the set of all functions s such that $\mathtt{dom}\, s \subseteq I$ is finite and $\mathtt{ran}\, s \subseteq \{0,1\}$. Let V consist of all $\beta \in 2^\Pi$ such that the set $\{s \in \Pi : \beta(s) = 1\}$ includes a subset of the form $\Pi(b) = \{s \in \Pi : s \subset b\}$ for some $b \in U$. Such a base V over Π will be denoted by $\mathtt{prj}\, U$ (the *projection*). □

Lemma 9.2.3. *Under the conditions of Example 3, suppose that I is at most countable, H, K are internal sets, and $P = \Phi_U\{P_i\}_{i\in I}$, where $P_i \subseteq H \times K$ are also internal sets. Put $P_s = \bigcap_{i\in \mathtt{dom}\, s} P_i^{s(i)}$* [9] *for any $s \in \Pi$, where $P_i^1 = P_i$ and $P_i^0 = P_i^{\mathsf{C}}$. Then $\mathtt{dom}\, P = \Phi_V\{\mathtt{dom}\, P_s\}_{s\in\Pi}$. The same is true provided $H, K \in \mathbb{WF}$ are compact spaces and P_i are clopen sets in $H \times K$.*

Proof. Suppose that $\langle x, y\rangle \in P$, thus $b = \chi_{\{i : \langle x,y\rangle \in P_i\}} \in U$. Then obviously $x \in \mathtt{dom}\, P_s$ for any $s \in \Pi(b)$, hence $x \in \Phi_V\{\mathtt{dom}\, P_s\}_{s\in\Pi}$ by the definition of V. Conversely suppose that $x \in H$ and $\beta = \chi_{\{s : x \in \mathtt{dom}\, P_s\}} \in V = \mathtt{prj}\, U$, thus $\Pi(b) \subseteq \{s : \beta(s) = 1\}$ for some $b \in U$. It follows, by Lemma 9.1.6, that $x \in \mathtt{dom}\,(\bigcap_{i\in I} P_i^{b(i)})$, and hence immediately $x \in \mathtt{dom}\, P$. □

---

[8] A somewhat more complicated *dual base* $U^{\mathsf{D}} = \{b^{\mathsf{C}} : b \in U^{\mathsf{C}}\}$, is also considered, where $b^{\mathsf{C}} \in 2^I$ and $b^{\mathsf{C}}(i) = 1 - b(i)$, $\forall i$. Then $\mathsf{Q}_{(U^{\mathsf{D}})}\, i\, R(i)$ iff $\neg\, \mathsf{Q}_U\, i\, \neg\, R(i)$, and $\Phi_{U^{\mathsf{D}}}\{X_i\}_{i\in I} = (\Phi_U\{X_i{}^{\mathsf{C}}\}_{i\in I})^{\mathsf{C}}$. This version is more useful than U^{C} when rings (not necessarily algebras) are considered as initial families of sets — but not in our case. **Exercise**: show that $\cup(I)^{\mathsf{D}} = \cap(I)$ and $\cap(I)^{\mathsf{D}} = \cup(I)$. What is $\mathbf{sus}^{\mathsf{D}}$?

[9] Note that the sets $\mathtt{dom}\, P_\pi$ are internal together with the sets P_π; the latter are internal by Lemma 1.2.16.

9.2b Countably determined sets

Unlike the Borel and projective classes, the following definition has no reasonable analogy in Polish descriptive set theory:

Definition 9.2.4. A set $X \subseteq H$ is *countably determined*, or CD, *over* a family $\mathscr{A}$ of sets, in brief $X \in \mathbf{CD}(\mathscr{A})$, if there exist a countable set I, a base $U \subseteq 2^I$, and a family of sets $X_i \in \mathscr{A}$ such that $X = \Phi_U\{X_i\}_{i\in I}$.

Let H be an internal set. Sets in $\mathbf{CD}[H] = \mathbf{CD}(\mathscr{A})$, where $\mathscr{A} = \mathscr{P}_{\mathtt{int}}(H)$, are called countably determined (CD) *in* H. □

Exercise 9.2.5. (1) Prove that, similarly to 9.1.3, "to be countably determined" is an intrinsic property, *i.e.*, if H, K are two internal domains, and $X \subseteq H \cap K$, then $X \in \mathbf{CD}[H]$ iff $X \in \mathbf{CD}[K]$.

(2) Prove that if $\mathscr{A}$ is the family of all open sets in a Polish space $\mathscr{X}$ then $\mathbf{CD}(\mathscr{A})$ contains all subsets of $\mathscr{X}$. (*Hint*: if $X \subseteq 2^{\mathbb{N}}$ then $X = \Phi_U\{X_n\}_{n\in\mathbb{N}}$, where $U = X$ while $X_n = \{x \in 2^{\mathbb{N}} : x(n) = 1\}$.) □

Thus the notion of countably determined sets is vacuous in Polish spaces. But even countable Saturation makes countably determined sets an extremely interesting family, most likely the largest family that admits a fruitful study on the base of countable Saturation. The role of countably determined sets in "hyperfinite" descriptive set theory is, to some extent, similar to the role of projective sets in Polish descriptive set theory, while projective sets (except for $\mathbf{\Sigma}^1_1$ = analytic and $\mathbf{\Pi}^1_1$) do not seem to play a well-defined separate role here, compared with their role in Polish descriptive set theory.

Two modifications of the structure of operations, leading to the same collection of countably determined sets, deserve to be mentioned.

$$X = \bigcup_{u\in B}\bigcap_{n\in u} X_n \quad , \text{ where } B \subseteq \mathscr{P}(\mathbb{N}), \tag{1}$$

$$X = \bigcup_{f\in F}\bigcap_{n\in\mathbb{N}} X_{f\restriction n} \;, \text{ where } F \subseteq 2^{\mathbb{N}} \text{ and } X_v \subseteq X_u \text{ whenever } u, v \in 2^{<\omega} \text{ and } u \subset v \text{ (the } \textit{regularity}\text{)}, \tag{2}$$

where X_n and X_u are internal sets. Version (1) is called *δs-operation*. Version (2), very convenient technically, exploits the idea of Souslin operation A.

Lemma 9.2.6. *Each of the operations (1) and (2), applied to internal sets, produces all countably determined sets and only them.*

Proof. Let $X = \Phi_U\{X_n\}_{n\in\mathbb{N}}$, where $U \subseteq 2^{\mathbb{N}}$ and all sets $X_n \subseteq H$ are internal subsets of an internal set H. To get (2) define $X_s = \bigcap_{n<\mathrm{lh}\, s} X_n^{s(n)}$ for any $s \in 2^{<\omega}$, where $X_n^1 = X_n$, $X_n^0 = (X_n)^{\complement}$, and let $F = U$.

Suppose that X is defined by (2), where all sets $X_s \subseteq H$ ($s \in 2^{<\omega}$) are internal. Then $X = \bigcup_{u\in B}\bigcap_{s\in u} X_s$, where B consists of all sets $u \subseteq 2^{<\omega}$ of the form $u = \{f\restriction n : n \in \mathbb{N}\}$, where $f \in F$, as in (1).

Finally, suppose that X is defined as in (1). We can assume that $u \in B \implies u' \in B$ whenever $u \subseteq u' \subseteq \mathbb{N}$ (otherwise add to B all supersets of sets $b \in B$). Then $X = \Phi_U\{X_n\}_{n\in\mathbb{N}}$, where $U = \{\chi_u : u \in B\}$. □

Note that the operations involved in the definition of countably determined sets are, generally speaking, of uncountable character. Indeed, although the index set I is assumed to be countable, the union in the right-hand side of $\Phi_U\{A_i\}_{i\in I} = \bigcup_{b\in U}\bigcap_{i\in I} A_i^{b(i)}$ can be uncountable together with the base $U \subseteq 2^I$. However, since the operation applies to a countable family of sets, it still remains in the scope of the countable Saturation (to which we are committed). The following simple result demonstrates how countable Saturation works for countably determined sets; more applications see below.

Lemma 9.2.7 (Compare with Theorem 1.4.11!). *Any uncountable CD set contains an infinite internal subset.*

Proof. Let $X = \bigcup_{f\in F}\bigcap_n X_{f\restriction n}$, where $F \subseteq 2^{\mathbb{N}}$, all sets X_u are internal and $X_v \subseteq X_u$ whenever $u \subset v$. (We apply Lemma 9.2.6.) If there exists $f \in F$ such that $X_{f\restriction n}$ is infinite for each n then $\bigcap_n X_{f\restriction n}$ contains an infinite internal subset by Lemma 9.1.5. Otherwise for any $f \in F$ choose $m_f \in \mathbb{N}$ such that $X_{f\restriction m_f}$ is finite. Then the set $U = \{f\restriction m_f : f \in F\} \subseteq 2^{<\omega}$ is countable, and hence so is X because $X \subseteq \bigcup_{u\in U} X_u$, contradiction. □

Exercise 9.2.8 (a definable non-CD set). Prove that any countably determined set belongs to $\mathbf{\Delta}_2^{\mathtt{ss}}$ (recall § 1.4a). Prove, using Lemma 9.2.7, that the set ${}^*\mathbb{R} \cap \mathbb{S}$ of all standard $*$-reals is $\mathbf{\Delta}_2^{\mathtt{ss}}$ but not countably determined. □

We finally prove

Theorem 9.2.9. *If H is an internal set then* $\mathbf{CD}[H]$ *is a set, and is closed under complements, countable unions and intersections, projections, and operations over countable index sets.*

Proof. To show that $\mathbf{CD}[H]$ is a set, note that by Extension (see 2° of 9.1.2), $\mathbf{CD}[H]$ is the image of $D = \mathscr{P}(2^{\mathbb{N}}) \times \mathscr{P}_{\mathtt{int}}(H)^{\mathbb{N}}$ via the map $\langle U, f\rangle \longmapsto \Phi_U\{f(n)\}_{n\in\mathbb{N}}$, and hence the result follows by Replacement because D is a set in **HST** (see Remark 1.3.10 and Theorem 1.3.14).

That $\mathbf{CD}[H]$ is closed under operations over countable index sets follows from the fact that a superposition of operations Φ_U is still an operation of this type by the above. To see that $\mathbf{CD}[H]$ is closed under complements and projections apply transformations, resp., $U \mapsto U^{\mathsf{C}}$, $U \mapsto \mathtt{prj}\, U$ of § 9.3a. □

The closure properties in Theorem 9.2.9 can be reformulated. Let a *countably determined n-ary relation* ($n \in \mathbb{N}$) be any countably determined subset of a set of the form $H = H_1 \times \ldots \times H_n$, where $H_1, ..., H_n$ are internal sets.

Exercise 9.2.10. Prove, using Theorem 9.2.9, that the class of all countably determined relations is closed under $\wedge, \vee, \neg$, and also under:

(i) internal quantifiers, i.e. if X is internal, and $R(x, y, ...)$ is a CD relation, then the relations $\exists\, x \in X\, R(x, y, ...)$ and $\forall\, x \in X\, R(x, y, ...)$ are CD;

(ii) quantifiers over countable sets, i.e. if $R(i, x, ...)$ is a CD relation, $I \subseteq \mathbb{I}$ a countable set, and $\Phi \subseteq 2^I$, then $\Phi\, i\, R(i, x, ...)$ is a CD relation. □

9.2c Shadows or standard part maps

Let I be a countable index set and H an internal set.

Suppose that $\mathscr{A} = \{A_i\}_{i \in I}$ is a family of internal subsets of an internal set H. For any $x \in H$ define $\mathbf{sh}_{\mathscr{A}}(x) \in 2^I$ to be the characteristic function of the set $\{i \in I : x \in A_i\}$ — the *$\mathscr{A}$-shadow* of x. Thus $\mathbf{sh}_{\mathscr{A}} : H \to 2^I$.

We define $x \equiv_{\mathscr{A}} y$ iff $\mathbf{sh}_{\mathscr{A}}(x) = \mathbf{sh}_{\mathscr{A}}(y)$ iff $\forall i \in I\, (x \in A_i \iff y \in A_i)$ (here x, y are arbitrary elements of the domain H.)

These concepts allow us to approach the operations from another angle:

Exercise 9.2.11. Prove that $\mathbf{sh}_{\mathscr{A}}^{-1}(U) = \Phi_U\{A_i\}_{i \in \mathbb{N}}$ for any $U \subseteq 2^I$.

Prove that the following are equivalent for any $X \subseteq H$:

(1) X is countably determined over the family $\mathscr{A}$;

(2) $X = \Phi_U\{A_i\}_{i \in I}$, where $U = \mathbf{sh}_{\mathscr{A}}{}''X = \{\mathbf{sh}_{\mathscr{A}}(x) : x \in X\}$;

(3) $X = \mathbf{sh}_{\mathscr{A}}^{-1}(U)$ for a set $U \subseteq \mathbf{ran}\,\mathbf{sh}_{\mathscr{A}}$;

(4) $\mathbf{sh}_{\mathscr{A}}^{-1}(\mathbf{sh}_{\mathscr{A}}{}''X) = X$;

(5) the set X is $\equiv_{\mathscr{A}}$-invariant, that is, we have $x \in X \iff y \in X$ whenever $x, y \in H$ satisfy $x \equiv_{\mathscr{A}} y$. □

To compare $\mathbf{sh}_{\mathscr{A}}$ with the shadow defined in 2.1.10, suppose that $H = {}^{\mathtt{int}}2^h$, the set of all <u>internal</u> maps $x : h = \{0, 1, ..., h-1\} \to \{0, 1\}$, where $h \in {}^*\mathbb{N} \setminus \mathbb{N}$. Let $I = \mathbb{N}$ and $A_i = \{x \in H : x(i) = 1\}$ for all $i \in \mathbb{N}$. Then obviously $\mathbf{sh}_{\mathscr{A}}(x) = x \restriction \mathbb{N}$ for any $x \in H$. Such a map is called "the standard part map" and denoted by $\mathtt{st}$ in the model theoretic version of nonstandard analysis, which is impossible here since $\mathtt{st}$ is reserved for another purpose.

If, in this case, we identify elements of ${}^{\mathtt{int}}2^h$ with binary $*$-rationals and elements of $2^{\mathbb{N}}$ with reals in $[0, 1]$, by means of the binary expansion, then $\mathbf{sh}_{\mathscr{A}}(x)$ will coincide with ${}^{\circ}x$ in the sense of Definition 2.1.10.

There is a somewhat more general approach to shadows.

Definition 9.2.12. Suppose that $\mathscr{A}$ is a family of subsets of a set H.

A function $\varphi : H \to 2^I$ is *$\mathscr{A}$-measurable* if all φ-preimages $\varphi^{-1}(U)$ of open sets $U \subseteq 2^I$ belong to $\mathscr{A}$.

If H is internal then $\mathbf{\Sigma}^0_1[H]$-measurable maps are called *shadows*. [10] □

Lemma 9.2.13. *Let H be an internal set and I countable. If $\mathscr{A} = \{A_i\}_{i \in I}$ is a system of internal subsets of H then $\mathbf{sh}_{\mathscr{A}}$ is a shadow in the sense of 9.2.12. Conversely if $\varphi : H \to 2^I$ is a shadow then $\varphi = \mathbf{sh}_{\mathscr{A}}$, where $\mathscr{A} = \{A_i\}_{i \in I}$ and all sets $A_i = \{x \in H : \varphi(x)(i) = 1\}$ are internal.*

Proof. Obviously $\mathbf{sh}_{\mathscr{A}}$-preimages of the sets $U_{i\nu} = \{b \in 2^I : b(i) = \nu\}$, $i \in I$ and $\nu = 0, 1$, are the sets A_i and their complements ${A_i}^{\mathsf{C}}$. However any open set $U \subseteq 2^I$ is a countable union of finite intersections of sets $U_{i\nu}$.

[10] This is a narrower notion than the concept of shadow defined in § 2.3a, and hence the latter is abandoned for the remainder of this Chapter.

Thus $\mathbf{sh}_{\mathscr{A}}$-preimages of open subsets $U \subseteq 2^I$ are countable unions of finite intersections of internal sets — and hence sets in $\mathbf{\Sigma}^0_1[H]$.

The second part of the lemma is elementary. □

Corollary 9.2.14. *A subset X of an internal set H is countably determined iff there is a shadow $\varphi : H \to 2^{\mathbb{N}}$ and a set $U \subseteq 2^{\mathbb{N}}$ with $X = \varphi^{-1}(U)$.* □

Exercise 9.2.15. Prove that if $X \subseteq H$ is $\mathbf{\Pi}^0_1$ and $\varphi : H \to 2^I$ is a shadow then the set $\varphi"X = \{\varphi(x) : x \in X\}$ is closed in 2^I, in particular $\mathtt{ran}\,\varphi$ is closed. (*Hint.* If $X = \bigcap_k X_k$, all X_k internal, and $r \in 2^I$ is a limit point of $\varphi"X$ as a subset of 2^I, then apply Saturation to the family of internal sets $Y_{ku} = \{x \in X_k : \varphi(x) \restriction u = r \restriction u\}$, u being a finite subset of I.) □

The representation of countably determined sets as shadow preimages of subsets of 2^I, I being a countable set, will be quite useful in some arguments below. In particular, the next lemma, which proves that shadow preimages preserve classes, looks much more natural in this form.

Lemma 9.2.16. *Let $\mathbf{\Gamma}$ be a Borel or projective class $\mathbf{\Sigma}^0_\xi$, $\mathbf{\Pi}^0_\xi$, $\mathbf{\Sigma}^1_n$, or $\mathbf{\Pi}^1_n$, I a countable set, and $U \subseteq 2^I$ a set in $\mathbf{\Gamma}[2^I]$. Then*

(i) *$\varphi^{-1}(U) \in \mathbf{\Gamma}[\mathscr{A}]$, whenever $\mathbf{\Gamma}$ is a Borel class $\mathbf{\Sigma}^0_\xi$ or $\mathbf{\Pi}^0_\xi$, $\mathscr{A}$ is any algebra of subsets of a set H, and $\varphi : H \to 2^I$ is $\mathscr{A}$-measurable;*

(ii) *$\varphi^{-1}(U) \in \mathbf{\Gamma}[H]$ for any internal H and any shadow $\varphi : H \to 2^I$;*

(iii) *$\varphi^{-1}(U) \in \mathbf{\Gamma}[2^J]$ for any countable set J and continuous $\varphi : 2^J \to 2^I$;*

(iv) *for projective classes $\mathbf{\Gamma} = \mathbf{\Sigma}^1_n$, $\mathbf{\Pi}^1_n$, $n \geq 1$, $\varphi^{-1}(U) \in \mathbf{\Gamma}[H]$ for any internal H and any $\mathbf{Borel}[H]$-measurable $\varphi : H \to 2^I$, and $\varphi^{-1}(U) \in \mathbf{\Gamma}[2^J]$ for any countable set J and Borel-measurable $\varphi : 2^J \to 2^I$.*

Proof. (i) is trivial because 1) preimages commute with complements, unions, and intersections, 2) clopen sets of 2^I form the least algebra of subsets of 2^I containing all sets of the form $\{b \in 2^I : b(i) = \nu\}$, $i \in I$ and $\nu = 0, 1$, and 3) $\mathscr{A}$ is an algebra.

The case of Borel classes $\mathbf{\Gamma}$ in (ii) is covered by (i).

Prove (ii) for projective classes $\mathbf{\Gamma} = \mathbf{\Sigma}^1_n$, $\mathbf{\Pi}^1_n$, and also (iv) for H, by induction on n. The step $\mathbf{\Sigma}^1_n \to \mathbf{\Pi}^1_n$ is trivial as above. To carry out the principal step $\mathbf{\Pi}^1_n \to \mathbf{\Sigma}^1_{n+1}$, suppose that $U \subseteq 2^I$ is $\mathbf{\Sigma}^1_{n+1}$. We can w. l. o. g. assume that $I \cap \mathbb{N} = \varnothing$. Put $L = I \cup \mathbb{N}$. Then there is a set $W \subseteq 2^L$ of the class $\mathbf{\Pi}^1_n$ (if $n \geq 1$) or Borel (if $n = 0$) such that $U = \{w \restriction I : w \in W\}$.

Now let $\varphi : H \to 2^I$ be a $\mathbf{Borel}[H]$-measurable map. Take any $h \in {}^*\mathbb{N} \setminus \mathbb{N}$. Let $K = {}^{\mathrm{int}}2^h$ (the internal set of all internal maps $h \to 2$). Put $\psi(x, \alpha) = \varphi(x) \cup (\alpha \restriction \mathbb{N})$ for $x \in H$ and $\alpha \in K$. Clearly $\psi : H \times K \to 2^L$ is a $\mathbf{Borel}[H \times K]$-measurable map. If $n \geq 1$ then it follows by the inductive hypothesis that $P = \psi^{-1}(W)$ is a $\mathbf{\Pi}^1_n$ subset in $H \times K$, while if $n = 0$ then P is Borel by (i) for $\mathscr{A} = \mathbf{Borel}[H \times K]$. On the other hand, the set $X = \varphi^{-1}(U)$ is equal to the projection $\mathtt{dom}\,P$ of P.

(iii), and (iv) for 2^J, are considered the same way. □

9.3 Structure of the hierarchies

The definitions of Borel and projective hierarchies in "hyperfinite" descriptive set theory look rather similar to their Polish versions, thus, one would expect similar properties. In reality, both some similarities and some dissimilarities can be observed.

We present, in this section, proofs of some key theorems in "hyperfinite" descriptive set theory, related to Borel, projective, and countably determined sets — the last category, as we mentioned, has no proper analogy in Polish descriptive set theory. The content includes the definition of an operation $\Phi_{\boldsymbol{B}(\boldsymbol{\Gamma})}$, for any Borel or projective class $\boldsymbol{\Gamma}$ of $\boldsymbol{\Sigma}$- or $\boldsymbol{\Pi}$-type, which produces all sets in $\boldsymbol{\Gamma}$ (§ 9.3a), a class preservation theorem for shadow maps (§ 9.3b), and closure properties of Borel and projective classes (§ 9.3c).

9.3a Operations associated with Borel and projective classes

Our first task is to prove that, for any Borel or projective $\boldsymbol{\Sigma}$- or $\boldsymbol{\Pi}$-class $\boldsymbol{\Gamma}$ there is an operation which produces all sets in $\boldsymbol{\Gamma}$. If $\mathscr{A}$ is a family of sets and Φ a operation over a set I then let $\Phi{\bullet}\mathscr{A}$ be the collection of all sets of the form $\Phi\{X_i\}_{i\in I}$, where all X_i belong to $\mathscr{A}$. The nature of sets in $\Phi{\bullet}\mathscr{A}$ depends both on the choice of Φ and the nature of sets in $\mathscr{A}$, and Lemma 9.2.16 gives a useful upper estimate.

The next theorem of a complementary character is rather transparent for Borel classes while the result for projective classes is based on Lemma 9.2.3.

Recall that $\boldsymbol{\Delta}^0_1[\mathscr{X}]$ is the family of all clopen subsets of a space $\mathscr{X}$.

Theorem 9.3.1. *For any Borel or projective class* $\boldsymbol{\Gamma} = \boldsymbol{\Sigma}^0_\xi,\ \boldsymbol{\Pi}^0_\xi,\ \boldsymbol{\Sigma}^1_n,\ \boldsymbol{\Pi}^1_n$ *there exist a countable index set* $I(\boldsymbol{\Gamma})$ *and a base* $\boldsymbol{B}(\boldsymbol{\Gamma}) \subseteq 2^{I(\boldsymbol{\Gamma})}$ *such that*

(i) $\boldsymbol{B}(\boldsymbol{\Gamma})$, *as a subset of* $2^{I(\boldsymbol{\Gamma})}$, *is a set of the class* $\boldsymbol{\Gamma}$;

(ii) *for any internal* H : $\boldsymbol{\Gamma}[H] \subseteq \Phi_{\boldsymbol{B}(\boldsymbol{\Gamma})}{\bullet}\mathscr{P}_{\mathbf{int}}(H)$;

(iii) *for any countable* J : $\boldsymbol{\Gamma}[2^J] \subseteq \Phi_{\boldsymbol{B}(\boldsymbol{\Gamma})}{\bullet}\boldsymbol{\Delta}^0_1[2^J]$.

Proof. We define $\boldsymbol{B}(\boldsymbol{\Gamma})$ by induction on the index ξ (for Borel classes) and on the index $n \in \mathbb{N}$ (for projective classes):

1) we put $I(\boldsymbol{\Sigma}^0_0) = I(\boldsymbol{\Pi}^0_0) = \{0\}$ and let $\boldsymbol{B}(\boldsymbol{\Sigma}^0_0) = \boldsymbol{B}(\boldsymbol{\Pi}^0_0)$ consist of a single function $b(0) = 1$, thus $\Phi_{\boldsymbol{B}(\boldsymbol{\Sigma}^0_0)}\{X_0\} = \Phi_{\boldsymbol{B}(\boldsymbol{\Pi}^0_0)}\{X_0\} = X_0$;
2) we put $\boldsymbol{B}(\boldsymbol{\Sigma}^0_\xi) = \cup(\mathbb{N} \times \xi) \circ \{B_{k\eta}\}_{\langle k,\eta\rangle \in \mathbb{N}\times\xi}$, where $B_{k\eta} = \boldsymbol{B}(\boldsymbol{\Pi}^0_\eta)$ for all $\eta < \xi$, $k \in \mathbb{N}$ (see § 9.2a on $\cup(I)$);
3) we put $\boldsymbol{B}(\boldsymbol{\Pi}^0_\xi) = \boldsymbol{B}(\boldsymbol{\Sigma}^0_\xi)^{\mathsf{C}}$;
4) we put $\boldsymbol{B}(\boldsymbol{\Sigma}^1_1) = \mathbf{sus}$ (see § 9.2a on the Souslin operation $\mathsf{A} = \Phi_{\mathbf{sus}}$);
5) finally, we put $\boldsymbol{B}(\boldsymbol{\Pi}^1_n) = \boldsymbol{B}(\boldsymbol{\Sigma}^1_n)^{\mathsf{C}}$ and $\boldsymbol{B}(\boldsymbol{\Sigma}^1_{n+1}) = \mathtt{prj}\,\boldsymbol{B}(\boldsymbol{\Pi}^1_n)$.

See Examples 1 and 3 in § 9.2a on the definition of the complementary base U^{C} and the projection base $\mathtt{prj}\,U$ in 3) and 5).

Now (ii), (iii) for Borel classes easily follow by induction using the results of exercises in examples 1 and 2 in § 9.2a. The inductive step $\mathbf{\Pi}^1_n \to \mathbf{\Sigma}^1_{n+1}$ follows from Lemma 9.2.3. (Note that projections $\mathtt{dom}\, P_i$ of internal, resp., clopen sets P_i are internal, resp., clopen.) That (ii), (iii) hold for the class $\mathbf{\Sigma}^1_1$ follows from Proposition 9.1.10.

It remains to check the definability requirement (i).

In the case of Borel classes, $\boldsymbol{B}(\mathbf{\Sigma}^0_0) = \boldsymbol{B}(\mathbf{\Pi}^0_0)$ is by definition a subset of a finite discrete space, and hence is $\mathbf{\Sigma}^0_0$ (that is clopen). The step $\mathbf{\Sigma}^0_\xi \to \mathbf{\Pi}^0_\xi$ is easy: by definition $\boldsymbol{B}(\mathbf{\Pi}^0_\xi) = \boldsymbol{B}(\mathbf{\Sigma}^0_\xi)^{\complement} = 2^{I(\mathbf{\Sigma}^0_\xi)} \smallsetminus \boldsymbol{B}(\mathbf{\Sigma}^0_\xi)$. To carry out the step $\{\mathbf{\Pi}^0_\eta : \eta < \xi\} \to \mathbf{\Sigma}^0_\xi$, note that, in terms of Example 2 in § 9.2a, the base $U \circ \{V_i\}_{i \in I}$ belongs to $\mathbf{\Phi}_U \boldsymbol{\cdot} \mathbf{\Gamma}$ whenever a class $\mathbf{\Gamma}$ contains all sets V_i. Therefore $\boldsymbol{B}(\mathbf{\Sigma}^0_\xi)$ belongs to $\mathbf{\Phi}_{\mathsf{U}(\mathbb{N}\times\xi)} \boldsymbol{\cdot} \mathbf{\Delta}^0_\xi$ by the inductive hypothesis (indeed $\bigcup_{\eta<\xi} \mathbf{\Pi}^0_\eta \subseteq \mathbf{\Delta}^0_\xi$). However $\mathsf{U}(\mathbb{N} \times \xi)$ is obviously an open set, that is, $\mathbf{\Sigma}^0_1$ in the space $2^{\mathbb{N}\times\xi}$. It follows that $\boldsymbol{B}(\mathbf{\Sigma}^0_\xi)$ is $\mathbf{\Sigma}^0_1$ over the algebra $\mathbf{\Delta}^0_\xi$ by Lemma 9.2.16, and this is obviously $\mathbf{\Sigma}^0_\xi$.

To see that the base **sus** is a $\mathbf{\Sigma}^1_1$ set in $2^{(\mathbb{N}^{<\omega})}$, note that by definition **sus** is the projection $\mathtt{dom}\, P$ of the Borel set (in $2^{(\mathbb{N}^{<\omega})} \times \mathbb{N}^{\mathbb{N}}$)

$$P = \{\langle b, f\rangle : b \in 2^{(\mathbb{N}^{<\omega})} \wedge f \in \mathbb{N}^{\mathbb{N}} \wedge \forall m\, (b(f \restriction m) = 1)\},$$

and hence **sus** is a $\mathbf{\Sigma}^1_1$ set. As for the step $\mathbf{\Pi}^1_n \to \mathbf{\Sigma}^1_{n+1}$, note that, in the notation of Example 3 in § 9.2a, for any $\beta \in 2^{\Pi}$ the equivalence $\beta \in V \iff \exists\, b \in U\, (\Pi(b) \subseteq \{s : \beta(s) = 1\})$ holds, where the right-hand side indicates the projection of a certain $\mathbf{\Pi}^1_n$ set in $2^{\Pi} \times 2^{I}$, thus a $\mathbf{\Sigma}^1_{n+1}$ set. □

Definition 9.3.2. Fix bases $\boldsymbol{B}(\mathbf{\Gamma})$ satisfying Theorem 9.3.1. □

Theorem 9.3.1 fails for $\mathbf{\Delta}$-classes, the class of all Borel sets, the class of all projective sets, and the class of all countably determined sets. See [Das 96].

9.3b The "shadow" theorem

Here we introduce another technical tool, typically used to derive some claims in "hyperfinite" descriptive set theory from theorems of Polish descriptive set theory. Recall that if $\varphi : H \to 2^I$ is a shadow map then the class of a set $U \subseteq 2^I$ is an upper estimate of the class of the preimage $\varphi^{-1}(U)$ by Lemma 9.2.16. Let us show that it can be a lower estimate as well.

Theorem 9.3.3. *Let H be an internal set, $\varphi : H \to 2^{\mathbb{N}}$ a shadow (in the sense of 9.2.12), and $\mathbf{\Gamma}$ a Borel or projective class beginning with $\mathbf{\Pi}^0_1$. Then*

(i) *if $U \subseteq \mathtt{ran}\, \varphi$ then U is $\mathbf{\Gamma}$ in $2^{\mathbb{N}}$ iff $X = \varphi^{-1}(U)$ is $\mathbf{\Gamma}$ in H ;*

(ii) *claim (i) is also true for $\mathbf{\Gamma} = \mathbf{\Sigma}^0_1$ provided $\mathtt{ran}\, \varphi$ is clopen in $2^{\mathbb{N}}$;*

(iii) *if $\mathbf{\Gamma}$ is as above but not $\mathbf{\Pi}^0_1$ and $X \subseteq H$ belongs to $\mathbf{\Gamma}$ then there is a set $V \subseteq 2^{\mathbb{N}}$ in $\mathbf{\Gamma}$ with $V \subseteq \varphi"X$ and $(\mathtt{ran}\, \varphi) \smallsetminus V \subseteq \varphi"(H \smallsetminus X)$.*

We precede the proof by a warming-up lemma the proof of which resembles the proof of the theorem in the sense that an inverse function is the core of the argument, but is much more elementary.

Lemma 9.3.4. *Suppose that H, K are internal sets, $F : H \to K$ is an internal function, $\mathbf{\Gamma}$ is a Borel or projective class, $Y \subseteq K$, and $X = F^{-1}(Y)$. Then $Y \in \mathbf{\Gamma} \Longrightarrow X \in \mathbf{\Gamma}$. If moreover $\mathtt{ran}\, F = K$ then $Y \in \mathbf{\Gamma} \Longleftrightarrow X \in \mathbf{\Gamma}$.*

Proof. $\Longrightarrow$ follows by elementary induction based on the obvious fact that the operations involved commute with the preimage operation. To prove the "moreover" claim note that by Choice in the internal universe if $\mathtt{ran}\, F = K$ then there is an internal function $G : K \to H$ such that $F(G(y)) = y$ for all $y \in K$, and hence $Y = G^{-1}(X)$. This yields $\Longleftarrow$ in the lemma. □

Proof (Theorem 9.3.3). The implication $\Longrightarrow$ in the equivalence "iff" in (i), also for the class $\mathbf{\Sigma}^0_1$, holds by Lemma 9.2.16.

The implication $\Longleftarrow$ in (i) follows from (iii) for classes $\mathbf{\Gamma}$ other than $\mathbf{\Pi}^0_1$ because if $X = \varphi^{-1}(U)$ and $U \subseteq \mathtt{ran}\, \varphi$ then any set V as in (iii) must be equal to U. Separately for $\mathbf{\Pi}^0_1$ we have an even stronger result 9.2.15.

We observe that (ii) is a simple consequence of (i). Indeed if $X = \varphi^{-1}(U)$ is $\mathbf{\Sigma}^0_1$ then $Y = H \smallsetminus X$ is $\mathbf{\Pi}^0_1$ in H. Assuming (i) for $\mathbf{\Pi}^0_1$, the set $V = \varphi"Y$ is $\mathbf{\Pi}^0_1$ in $2^{\mathbb{N}}$. However V is equal to $\mathtt{ran}\, \varphi \smallsetminus U$, and hence U is $\mathbf{\Sigma}^0_1$ in $2^{\mathbb{N}}$ provided $\mathtt{ran}\, \varphi$ is clopen and $U \subseteq \mathtt{ran}\, \varphi$.

Thus we can concentrate on (iii) in the assumption that $\mathbf{\Gamma}$ is strictly higher than $\mathbf{\Pi}^0_1$, that is at least $\mathbf{\Sigma}^0_2$ or $\mathbf{\Pi}^0_2$. The proof is based on the next lemma, the idea of which is quite similar to the choice of G in the proof of Lemma 9.3.4, but the construction of G, trivial in 9.3.4, is here quite tricky.

Lemma 9.3.5. *Let $\{X_n\}_{n \in \mathbb{N}}$ be a sequence of $\mathbf{\Sigma}^0_2$ (resp., $\mathbf{\Pi}^0_2$) subsets of H. Then there is a function $G : 2^{\mathbb{N}} \to H$ such that $\varphi(G(r)) = r$ for all $r \in 2^{\mathbb{N}}$ and $G^{-1}(X_n)$ is $\mathbf{\Sigma}^0_2$ (resp., $\mathbf{\Pi}^0_2$) in $2^{\mathbb{N}}$ for any n.*

Proof. The result for $\mathbf{\Pi}^0_2$ follows by taking complements. As the class $\mathbf{\Sigma}^0_2$ is closed under countable unions, it suffices to prove the result for $\mathbf{\Sigma}^0_2$ under the assumption that all sets X_n belong to $\mathbf{\Pi}^0_1$. We define a system $\{Q_s\}_{s \in \mathbb{N}^{<\omega}}$ of nonempty $\mathbf{\Pi}^0_1$ subsets of H satisfying $Q_\Lambda = H$ and

(a) for each $s \in \mathbb{N}^{<\omega}$, the sets $Q_{s^\wedge k}$, $k \in \mathbb{N}$, are disjoint subsets of Q_s while the sets $\varphi"(Q_{s^\wedge k})$, $k \in \mathbb{N}$, form a disjoint partition of $\varphi"Q_s$;

(b) for each $s \in \mathbb{N}^{n+1}$, we have $p \restriction n = q \restriction n$ for all $p, q \in \varphi"Q_s$, and also either $Q_s \subseteq X_n$ or $Q_s \cap X_n = \varnothing$.

Suppose that a $\mathbf{\Pi}^0_1$ set $Q_s \subseteq H$ has been defined, $s \in \mathbb{N}^n$. Both $F = \varphi"Q_s$ and $U = \varphi"R \subseteq F$, where $R = X_n \cap Q_s$, are closed sets in $2^{\mathbb{N}}$ by 9.2.15, and hence F admits a partition $F = \bigcup_k F_k$, where each F_k is nonempty, and <u>either</u> $F_k \cap U = \varnothing$ and $F_k = \{r \in F : r \restriction m_k = u_k\}$ for some $m_k \in \mathbb{N}$, $m_k \geq n$, and $u_k \in 2^{m_k}$ — then we put

$$Q_{s^\wedge k} = \{x \in Q_s : \varphi(x) \restriction m_k = u_k\},$$

or $F_k = \{r \in U : r \restriction m_k = u_k\} \subseteq U$ for some $m_k \in \mathbb{N}$, $m_k \geq n$, and $u_k \in 2^{m_k}$ — then we put $Q_{s^\wedge k} = \{x \in R : \varphi(x) \restriction m_k = u_k\}$.

After the construction of $\{Q_s\}_{s \in \mathbb{N}^{<\omega}}$ is accomplished, note that by (a) for any $r \in 2^{\mathbb{N}}$ and n there is a unique $s = s_n(r) \in \mathbb{N}^{n+1}$ with $r \in \varphi"Q_s$, and $s_n(r) \subset s_{n+1}(r)$, so that $Q_r = \bigcap_{n \in \mathbb{N}} Q_{s_r(n)} \neq \varnothing$ (Lemma 9.1.5). Let $G(r)$ be an arbitrary element of Q_r. (The axiom of $2^{\aleph_0}$-size Choice is applied, recall 9.1.2.) We have $\varphi(G(r)) = r$ by (b). To show that $G^{-1}(X_n)$ is a $\mathbf{\Sigma}^0_2$ subset of $2^{\mathbb{N}}$, it suffices to prove that $G^{-1}(X_n) = \bigcup_{s \in S_n} \varphi"Q_s$, where $S_n = \{s \in \mathbb{N}^{n+1} : Q_s \subseteq X_n\}$.

If $r \in \varphi"Q_s$ for some $s = s_r(n) \in S_n$ then clearly $G(r) \in Q_s \subseteq X_n$. If $r \notin \bigcup_{s \in S} \varphi"Q_s$ then $s_r(n) \notin S$, therefore $Q_{s_r(n)} \cap X_n = \varnothing$, and we have $G(r) \notin X_n$ because $G(r) \in Q_{s_r(n)}$. □ *(Lemma)*

We come back to (iii) of Theorem 9.3.3.

Case 1: $\mathbf{\Gamma}$ is a Borel class $\mathbf{\Sigma}^0_m$ or $\mathbf{\Pi}^0_m$, $2 \leq m \in \mathbb{N}$. Suppose that a set $X \subseteq H$ belongs to $\mathbf{\Sigma}^0_m$. Then X can be obtained by $m-2$ consecutive alternating operations of countable union and intersection applied to sets either in $\mathbf{\Sigma}^0_2$ (m even) or in $\mathbf{\Pi}^0_2$ (m odd). For instance if X is $\mathbf{\Sigma}^0_4$ then $X = \bigcup_{i \in \mathbb{N}} \bigcap_{j \in \mathbb{N}} X_{ij}$, where all $X_{ij} \subseteq H$ are $\mathbf{\Sigma}^0_2$ sets. In this case Lemma 9.3.5 gives a function $G : 2^{\mathbb{N}} \to H$ such that $\varphi(G(r)) = r$ for all $r \in 2^{\mathbb{N}}$ and $G^{-1}(X_{ij})$ is $\mathbf{\Sigma}^0_2$ in $2^{\mathbb{N}}$ for all i, j. Then $V = \bigcup_{i \in \mathbb{N}} \bigcap_{j \in \mathbb{N}} G^{-1}(X_{ij})$ is a $\mathbf{\Sigma}^0_4$ set. Moreover $V = G^{-1}(\bigcup_{i \in \mathbb{N}} \bigcap_{j \in \mathbb{N}} X_{ij}) = G^{-1}(X)$ since the preimage commutes with the operations. Thus $Y \subseteq \varphi"X$ and $(\mathtt{ran}\,\varphi) \setminus Y \subseteq \varphi"(H \setminus X)$ because G satisfies $\varphi(G(r)) = r$, $\forall r$.

Case 2: $\mathbf{\Gamma}$ is a Borel class $\mathbf{\Sigma}^0_\xi$ or $\mathbf{\Pi}^0_\xi$, $\omega \leq \xi < \omega_1$. Suppose that a set $X \subseteq H$ belongs to $\mathbf{\Sigma}^0_\xi[H]$, where $\omega \leq \xi < \omega_1$. Then, by the choice of $\boldsymbol{B}(\mathbf{\Sigma}^0_\xi)$, $X = \mathbf{\Phi}_{\boldsymbol{B}(\mathbf{\Sigma}^0_\xi)}\{X_i\}_{i \in I}$, where $I = I(\mathbf{\Sigma}^0_\xi)$ is a (countable) index set of the operation $\mathbf{\Phi}_{\boldsymbol{B}(\mathbf{\Sigma}^0_\xi)}$ and all sets $X_i \subseteq H$ are internal. Let $G : 2^{\mathbb{N}} \to H$ be as in the lemma, so that $G^{-1}(X_i)$ is $\mathbf{\Sigma}^0_2$ in $2^{\mathbb{N}}$ for any $i \in I$. The set $V = \mathbf{\Phi}_{\boldsymbol{B}(\mathbf{\Sigma}^0_\xi)}\{G^{-1}(X_i)\}_{i \in I}$ satisfies $V \subseteq \varphi"X$ and $\mathtt{ran}\,\varphi \setminus V \subseteq \varphi"(H \setminus X)$ by the same reasons as above. In addition V belongs to the class $\mathbf{\Sigma}^0_\xi$ over the algebra $\mathbf{\Delta}^0_3[2^{\mathbb{N}}]$. (Indeed V is the ψ-preimage of $\boldsymbol{B}(\mathbf{\Sigma}^0_\xi)$, a $\mathbf{\Sigma}^0_\xi$ set, where $\psi = \mathtt{sh}_{\{G^{-1}(X_i)\}_{i \in I}}$, a $\mathbf{\Delta}^0_3[2^{\mathbb{N}}]$-measurable map $2^{\mathbb{N}} \to 2^I$.) Yet this is obviously the same as $\mathbf{\Sigma}^0_\xi$ over clopen sets because $\xi \geq \omega$.

Case 3: $\mathbf{\Gamma}$ is a projective class $\mathbf{\Sigma}^1_m$ or $\mathbf{\Pi}^1_m$, where $m \geq 1$. Argue as in Case 2, use the algebra **Borel** $[2^{\mathbb{N}}]$ instead of $\mathbf{\Delta}^0_3[2^{\mathbb{N}}]$, and apply Lemma 9.2.16(iv) in the final argument.

□ *(Theorem 9.3.3)*

The following simple corollary will be very important in some applications.

Corollary 9.3.6. *Suppose that $\xi < \omega_1$, H is an internal set, $\mathscr{B} \subseteq \mathscr{P}_{\mathtt{int}}(H)$ is an algebra of sets, and $X \subseteq H$ is a set both countably determined over $\mathscr{B}$ and $\mathbf{\Pi}^0_\xi[H]$. Then X is $\mathbf{\Pi}^0_\xi$ over $\mathscr{B}$. The same for $\mathbf{\Sigma}^0_\xi$.*

In particular ($\xi = 0$), any set both internal and $\mathbf{CD}(\mathscr{B})$ belongs to $\mathscr{B}$.

Proof. First suppose that $\xi \geq 1$. There is a base $U \subseteq 2^{\mathbb{N}}$ such that $X = \Phi_U\{A_n\}_{n\in\mathbb{N}} = \varphi^{-1}(U)$, where $\varphi = \mathbf{sh}_{\mathscr{A}}$ and $\mathscr{A} = \{A_n\}_{n\in\mathbb{N}} \subseteq \mathscr{B}$. Then $V = \varphi"X$ is $\mathbf{\Pi}^0_\xi$ in $2^{\mathbb{N}}$ by Theorem 9.3.3 and $X = \varphi^{-1}(V) = \Phi_V\{A_n\}_{n\in\mathbb{N}}$ by 9.2.11(2). It follows that X is $\mathbf{\Pi}^0_\xi$ over $\mathscr{A}$ by Lemma 9.2.16(i), and hence over $\mathscr{B}$ as well. The result for $\mathbf{\Sigma}^0_\xi$ can be obtained by taking complements. As for $\xi = 0$, we leave it as an easy **exercise** for the reader to demonstrate that any set both $\mathbf{\Sigma}^0_1$ over $\mathscr{B}$ and $\mathbf{\Pi}^0_1$ over $\mathscr{B}$ belongs to $\mathscr{B}$ since $\mathscr{B}$ is an algebra (use Saturation !). □

Exercise 9.3.7. Show that Theorem 9.3.3 fails for $\mathbf{\Gamma} = \mathbf{\Sigma}^0_1$. □

Exercise 9.3.8. In the conditions of Theorem 9.3.3 show that $\varphi"X$ is $\mathbf{\Sigma}^1_n$ in $2^{\mathbb{N}}$ whenever $n \geq 1$ and $X \subseteq H$ is $\mathbf{\Sigma}^1_n$ in H.

(*Hint.* $\varphi"X = \varphi"Y$, where $Y = \varphi^{-1}(\varphi"X)$, thus it suffices to show that Y is $\mathbf{\Sigma}^1_n$. Note that the relation $x \equiv_\varphi y$ iff $\varphi(x) = \varphi(y)$ is equivalent to $\forall n\, (x \in A_n \iff y \in A_n)$, where $A_n = \{x : \varphi(x)(n) = 1\}$. Now we have $Y = \{y : \exists x\, (x \in X \wedge x \equiv_\varphi y)\}$.) □

9.3c Closure properties of the classes

We have established all necessary instrumentarium for the following theorem. The theorem contains a list of assertions, mainly asserting a closure property. Generally speaking, the assertions resemble known results in Polish descriptive set theory, and some proofs are based on arguments also valid in the Polish case, but several claims need very different arguments.

Theorem 9.3.9. (i) *All Borel and projective classes, and the class of all Borel sets, are closed under finite unions and intersections* (of sets in the same internal domain) *and preimages by internal functions. In addition,*

(1) *the classes $\mathbf{\Pi}^0_\xi$ are closed under countable intersections, the classes $\mathbf{\Sigma}^0_\xi$ are closed under countable unions, and any set in $\mathbf{\Sigma}^0_\xi$ is the union of a* **pairwise disjoint** *countable family of sets in $\bigcup_{\eta<\xi} \mathbf{\Pi}^0_\eta$,*

(2) *the classes $\mathbf{\Sigma}^1_n$, $\mathbf{\Pi}^1_n$, $\mathbf{\Delta}^1_n$ ($n \geq 1$) are closed under countable unions, countable intersections, and Borel preimages,*

(3) *the classes $\mathbf{\Delta}^0_\xi$ and $\mathbf{\Delta}^1_n$ are closed under complements, that is, if H is internal and $X \in \mathbf{\Delta}^0_\xi[H]$ then the complementary set $X^{\mathsf{C}} = H \setminus X$ also belongs to $\mathbf{\Delta}^0_\xi[H]$, and the same for $\mathbf{\Sigma}^1_n[H]$,*

(4) *the class of all Borel sets is closed under both complements and countable unions and intersections,*

(5) *the classes* $\mathbf{\Sigma}^1_n$, $n \geq 1$, *are closed under projections, that is if* H, K *is internal and* $P \in \mathbf{\Sigma}^1_n[H \times K]$ *then the projection* $\operatorname{dom} P$ *is* $\mathbf{\Sigma}^1_n[H]$,

(6) *the classes* $\mathbf{\Sigma}^0_1$ *and* $\mathbf{\Pi}^0_1$ *are also closed under projections,*

In addition for any infinite internal H *the following holds:*

(ii) $\mathbf{\Sigma}^0_1[H] \cap \mathbf{\Pi}^0_1[H] = \mathbf{\Delta}^0_0[H] = \mathbf{\Delta}^0_1[H] = \mathscr{P}_{\mathtt{int}}(H)$.

(iii) (1) **(the Souslin theorem)** $\mathbf{\Sigma}^1_1[H] \cap \mathbf{\Pi}^1_1[H] = \mathbf{Borel}[H]$,

(2) **(Borel separation)** *any two disjoint* $\mathbf{\Sigma}^1_1$ *sets* $X, Y \subseteq H$ *can be separated by a Borel set* Z (meaning that $X \subseteq Z$ while $Y \cap Z = \varnothing$).

(iv) *If* K *is an infinite internal set and* $n \geq 1$ *then any* $\mathbf{\Sigma}^1_{n+1}$ *set* $X \subseteq H$ *is the projection of a set in* $\mathbf{\Pi}^1_n[H \times K]$.

(v) (1) *if* $1 \leq \xi < \omega_1$ *then* $\mathbf{\Sigma}^0_\xi[H] \not\subseteq \mathbf{\Pi}^0_\xi[H]$ *and* $\mathbf{\Pi}^0_\xi[H] \not\subseteq \mathbf{\Sigma}^0_\xi[H]$;

(2) *if* $n \geq 1$ *then* $\mathbf{\Sigma}^1_n[H] \not\subseteq \mathbf{\Pi}^1_n[H]$ *and* $\mathbf{\Pi}^1_n[H] \not\subseteq \mathbf{\Sigma}^1_n[H]$, *in particular, for* $n = 1$, $\mathbf{Borel}[H] \subsetneqq \mathbf{\Sigma}^1_1[H]$ *by* (iii).

(vi) (1) *if* $\xi < \eta < \omega_1$ *then* $\mathbf{\Sigma}^0_\xi[H] \cup \mathbf{\Pi}^0_\xi[H] \subsetneqq \mathbf{\Delta}^0_\eta[H]$;

(2) *if* $n < k$ *then* $\mathbf{\Sigma}^1_n[H] \cup \mathbf{\Pi}^1_n[H] \subsetneqq \mathbf{\Delta}^1_k[H]$.

(vii) *all Borel and projective subsets of* H *belong to* $\mathbf{CD}[H]$; *all Borel and projective classes in* H *and the class* $\mathbf{Borel}[H]$ *are sets.*

Proof. (i) Straightforward verification of (i)(1) – (i)(5), following patterns of Polish descriptive set theory, is left as an **exercise** for the reader. That $\mathbf{\Sigma}^0_1$ and $\mathbf{\Pi}^0_1$ are closed under projections follows from Lemma 9.1.8.

(ii) See 9.1.7(2) for the nontrivial part.

(iii) Prove (iii)(2), a stronger statement. We can present the sets X, Y in the Souslin form by 9.1.10, *i.e.* $X = \mathsf{A}\{X_s\}_{s \in \mathbb{N}^{<\omega}}$ and $Y = \mathsf{A}\{Y_s\}_{s \in \mathbb{N}^{<\omega}}$, all sets X_s, $Y_s \subseteq H$ being internal and $X_t \subseteq X_s$, $Y_t \subseteq Y_s$ whenever $s \subset t$.

As X, Y are disjoint, $\bigcap_m X_{f \restriction m} \cap \bigcap_m Y_{g \restriction m} = \bigcap_m (X_{f \restriction m} \cap Y_{g \restriction m}) = \varnothing$ for any $f, g \in \mathbb{N}^\mathbb{N}$, so that, by countable **Saturation** (in the Polish case completeness is used here), there is a number $m \in \mathbb{N}$ such that already $X_{f \restriction m} \cap Y_{g \restriction m} = \varnothing$. Let m_{fg} be the least of such numbers m. Then $T = \{\langle f \restriction k, g \restriction k \rangle : f, g \in \mathbb{N}^\mathbb{N} \wedge k \leq m_{fg}\}$ is a well-founded tree in $(\mathbb{N}^{<\omega})^2$ and $M = \{\langle f \restriction m_{fg}, g \restriction m_{fg} \rangle : f, g \in \mathbb{N}^\mathbb{N}\}$ is the set of all maximal elements of T. Note that if $\langle s, t \rangle \in T \smallsetminus M$ then $\langle s^\wedge k, t^\wedge l \rangle \in T$ for all k, $l \in \mathbb{N}$.

Define a set Z_{st} for any pair $\langle s, t \rangle \in T$ as follows:

1) $Z_{st} = X_s$ for all $\langle s, t \rangle \in M$;

2) if $\langle s, t \rangle \in T \smallsetminus M$ then $Z_{st} = \bigcup_{k \in \mathbb{N}} \bigcap_{l \in \mathbb{N}} Z_{s^\wedge k, t^\wedge l}$.

(This inductive definition is legitimate because T is a well-founded tree.) Clearly every set $Z_{st} \subseteq H$ is Borel. On the other hand, define

$$X'_s = \bigcup_{f \in \mathbb{N}^\mathbb{N},\, s \subset x} \bigcap_m X_{f \restriction m} \quad \text{and} \quad Y'_t = \bigcup_{g \in \mathbb{N}^\mathbb{N},\, t \subset g} \bigcap_m Y_{g \restriction m}$$

for all $s, t \in \mathbb{N}^{<\omega}$; then $X'_s \subseteq X_s \cap X$ and the same for Y'_t. Moreover, $X'_s = \bigcup_k X'_{s^\wedge k}$ and $Y'_t = \bigcup_l Y'_{t^\wedge l}$, so that, still by the well-founded induction within T, we prove that Z_{st} separates X'_s from Y'_t for any pair $\langle s, t\rangle \in T$. In particular, for $s = t = \Lambda$ (the empty sequence), we have $X'_\Lambda \subseteq Z_{\Lambda\Lambda}$ but $Y'_\Lambda \cap Z_{\Lambda\Lambda} = \varnothing$. It remains to note that $X'_\Lambda = X$ while $Y'_\Lambda = Y$.

(iv) We have $X = \Phi_U\{X_i\}_{i\in I}$, where $U = \boldsymbol{B}(\boldsymbol{\Sigma}^1_{n+1}) \subseteq 2^I$ is a $\boldsymbol{\Sigma}^1_{n+1}$ set, and all sets $X_n \subseteq H$ are internal (Theorem 9.3.1). There is $h \in {}^*\mathbb{N} \setminus \mathbb{N}$ such that K contains, in $\mathbb{I}$, at least 2^h elements; suppose that simply $K = {}^{\mathtt{int}}2^h$, the set of all internal maps $h \to 2$. Arguing as in the proof of Lemma 9.2.16(ii), we find a $\boldsymbol{\Pi}^1_n$ set $W \subseteq 2^L$, where $L = I \cup \mathbb{N}$, and a system $\{Y_\ell\}_{\ell\in L}$ of internal subsets of $H \times {}^{\mathtt{int}}2^h$, such that $X = \mathtt{dom}\, P$, where $P = \Phi_W\{Y_\ell\}_{\ell\in L}$. However P is $\boldsymbol{\Pi}^1_n$ in $H \times {}^{\mathtt{int}}2^h$ by Lemma 9.2.16.

(v) Polish descriptive set theory proves similar statements with the help of *universal sets* (or *parametrizations*) and Cantor's diagonal argument. In the "hyperfinite" setting this argument does not work [11], however, the shadow map leads to the result. There is $h \in {}^*\mathbb{N} \setminus \mathbb{N}$ such that H contains, in $\mathbb{I}$, at least 2^h elements. We can assume, by 9.1.3, that simply $H = {}^{\mathtt{int}}2^h$. Consider the family $\mathscr{A} = \{A_i\}_{i\in\mathbb{N}}$ of sets $A_i = \{x \in H : x(i) = 1\}$, $i \in \mathbb{N}$. Take any $\boldsymbol{\Sigma}^0_\xi$ set $X \subseteq 2^{\mathbb{N}}$ which does not belong to $\boldsymbol{\Pi}^0_\xi$ (the existence of such a set is a classical fact). Then $\mathbf{sh}^{-1}_{\mathscr{A}}(X)$ is, accordingly, $\boldsymbol{\Sigma}^0_\xi$ but not $\boldsymbol{\Pi}^0_\xi$ in H by Theorem 9.3.3 (applicable because $\mathtt{ran}\, \mathbf{sh}_{\mathscr{A}} = 2^{\mathbb{N}}$).

(vi) That $\boldsymbol{\Sigma}^0_\xi[H] \cup \boldsymbol{\Pi}^0_\xi[H] \subseteq \boldsymbol{\Delta}^0_\eta[H]$ $(\xi < \eta)$ can be proved by induction on η; we leave this as an **exercise** for the reader. To show $\subsetneqq$, let $H = H' \cup H''$ be a partition on two disjoint internal infinite sets. Take $X' \subseteq H'$ of class $\boldsymbol{\Sigma}^0_\xi[H'] \setminus \boldsymbol{\Pi}^0_\xi[H']$, and $X'' \subseteq H''$ of class $\boldsymbol{\Pi}^0_\xi[H''] \setminus \boldsymbol{\Sigma}^0_\xi[H'']$, then $X = X' \cup X''$ is $\boldsymbol{\Delta}^0_{\xi+1}$ but neither $\boldsymbol{\Sigma}^0_\xi$ nor $\boldsymbol{\Pi}^0_\xi$ in H.

(vii) That Borel and projective sets are countably determined follows from the closure properties of classes $\mathbf{CD}[H]$; alternatively, the operations introduced by 9.3.2 can be applied. To infer that all Borel and projective classes are sets, it suffices to apply the axiom of Separation because $\mathbf{CD}[H]$ is a set by Theorem 9.2.9.

□ (*Theorem 9.3.9*)

[11] To prove $\boldsymbol{\Sigma}^0_1[H] \subsetneqq \boldsymbol{\Pi}^0_1[H]$, we would need a $\boldsymbol{\Sigma}^0_1$ set $U \subseteq H \times H$, universal in the sense that for any $\boldsymbol{\Sigma}^0_1[H]$ set X there is $y \in H$ with $X = \{x : \langle y, x\rangle \in U\}$. That such a set does not exist follows from Theorem 1.4.9. A set with this property exists in $\boldsymbol{\Sigma}^0_1[{}^{\mathtt{int}}2^H \times H]$, which does not allow to carry out the diagonal argument, thus it is useless for (v). These difficulties underline a principal difference between the Polish and "hyperfinite" versions of DST: while all uncountable Polish spaces are Borel isomorphic, hyperfinite domains can be Borel isomorphic only if they have rather close numbers of elements, as in Theorem 1.4.9.

9.4 Some classical questions

“Hyperfinite” descriptive set theory inherits from the classical theory a list of questions on the nature of sets generally viewed as those worth to study. Some of those questions, commonly known as *the structure theory*, are considered in this section. This includes results related to “planar” sets with special cross-sections, Separation-like properties, Uniformization, and a remarkable decomposition theorem of the Louveau-like type.

9.4a Separation and reduction

For any class $\mathbf{\Gamma}$ of subsets of a given set H, the Separation and Reduction principles are defined as follows:

$\mathbf{\Gamma}$-Sep: for any two disjoint sets $X, Y \subseteq H$ in $\mathbf{\Gamma}$ there is a set Z which separates X from Y, that is $X \subseteq Z$ and $Y \cap Z = \varnothing$, and both X and its complement $X^{\mathsf{C}} = H \smallsetminus X$ belong to $\mathbf{\Gamma}$;

$\mathbf{\Gamma}$-Red: for any two sets $X, Y \subseteq H$ in $\mathbf{\Gamma}$ there exist disjoint sets $X' \subseteq X$ and $Y' \subseteq Y$, still in $\mathbf{\Gamma}$, such that $X \cup Y = X' \cup Y'$.

Exercise 9.4.1. Prove that $\mathbf{\Gamma}$-**Red** implies $\mathbf{\Gamma}^{\mathsf{C}}$-**Sep** for the class $\mathbf{\Gamma}^{\mathsf{C}} = \{X^{\mathsf{C}} : X \in \mathbf{\Gamma}\}$ of all complementary sets, in any domain H. □

Both Separation and Reduction trivially hold for any class $\mathbf{\Gamma}$ closed under the set difference $X \smallsetminus Y$, for instance for any class $\mathbf{\Delta}^0_\xi$ or $\mathbf{\Delta}^1_n$. As for $\mathbf{\Sigma}$- and $\mathbf{\Pi}$-classes, the following theorem reduces the question in the “hyperfinite” setting to the case of Polish spaces. Note also that if $\mathbf{\Gamma} = \mathbf{\Sigma}^1_1$ in $\mathbf{\Gamma}$-**Sep** then the separating set Z must be $\mathbf{\Delta}^1_1$, and hence Borel by Theorem 9.3.9(iii)(1), so that $\mathbf{\Sigma}^1_1$-**Sep** is just a reformulation of Theorem 9.3.9(iii)(2).

Theorem 9.4.2. *Suppose that $\mathbf{\Gamma}$ is a Borel or projective class of $\mathbf{\Sigma}$-type or $\mathbf{\Pi}$-type, beginning with $\mathbf{\Pi}^0_1$, and H an infinite internal set. Then:*

(i) *$\mathbf{\Gamma}[2^{\mathbb{N}}]$-**Sep** implies $\mathbf{\Gamma}[H]$-**Sep**, and the same for Reduction;*
(ii) *if $\mathbf{\Gamma}$ is as above but not $\mathbf{\Pi}^0_1$ then conversely $\mathbf{\Gamma}[H]$-**Sep** implies $\mathbf{\Gamma}[2^{\mathbb{N}}]$-**Sep**, and the same for Reduction.*

Proof. We consider $\mathbf{\Gamma}$-**Sep**; the arguments for $\mathbf{\Gamma}$-**Red** do not differ much.

(i) Let $X, Y \subseteq H$ be disjoint sets in $\mathbf{\Gamma}$. They are countably determined, and hence countably determined over one and the same countable algebra $\mathscr{A} = \{A_n\}_{n \in \mathbb{N}}$ of internal sets $A_n \subseteq H$. Then $\varphi = \mathbf{sh}_{\mathscr{A}} : H \to 2^{\mathbb{N}}$ is a shadow map by Lemma 9.2.13, and we have, by 9.2.11(4), $X = \varphi^{-1}(U)$, $V = \varphi^{-1}(V)$, where $U = \varphi"X$, $V = \varphi"Y$ (disjoint subsets of $2^{\mathbb{N}}$). Moreover U, V belong to $\mathbf{\Gamma}[2^{\mathbb{N}}]$ by Theorem 9.3.3(i). Therefore by $\mathbf{\Gamma}[2^{\mathbb{N}}]$-**Sep** there is a set $W \subseteq 2^{\mathbb{N}}$ such that $X \subseteq Z$, $Y \cap Z = \varnothing$, and both W and $2^{\mathbb{N}} \smallsetminus W$ belong to $\mathbf{\Gamma}$. Then $Z = \varphi^{-1}(Z)$ satisfies $X \subseteq Z$ and $Y \cap Z = \varnothing$. Finally both Z and $Z^{\mathsf{C}} = H \smallsetminus Z$ belong to $\mathbf{\Gamma}$ by Lemma 9.2.16.

(ii) Since H is infinite there is a sequence $\mathscr{A} = \{A_n\}_{n \in \mathbb{N}}$ of internal sets $A_n \subseteq H$ such that $\bigcap_{n \in u} A_n \cap \bigcap_{n \in v} (A_n{}^{\mathsf{C}})$ is an infinite (internal) set for any pair of disjoint finite sets $u, v \subseteq \mathbb{N}$. Put $\varphi = \mathbf{sh}_{\mathscr{A}}$. Note that $\mathtt{ran}\,\varphi = 2^{\mathbb{N}}$: indeed by **Saturation** the set $\bigcap_{n \in b} A_n \cap \bigcap_{n \in \mathbb{N} \smallsetminus b} (A_n{}^{\mathsf{C}})$ is non-empty for any $b \subseteq \mathbb{N}$. Consider any pair of disjoint sets $U, V \subseteq 2^{\mathbb{N}}$ in $\mathbf{\Gamma}$. The sets $X = \varphi^{-1}(U)$ and $Y = \varphi^{-1}(V)$ also belong to $\mathbf{\Gamma}$ (as subsets of H) by Lemma 9.2.16. Assuming $\mathbf{\Gamma}[H]$-**Sep**, there is a set $Z \subseteq H$ with $X \subseteq Z$ and $Y \cap Z = \varnothing$ such that both Z and Z^{C} belong to $\mathbf{\Gamma}$. It follows from Theorem 9.3.3(iii) that there is a set $W \subseteq 2^{\mathbb{N}}$ in $\mathbf{\Gamma}$ such that $W \subseteq \varphi"Z$ while $W^{\mathsf{C}} \subseteq \varphi"(Z^{\mathsf{C}})$.

However $\varphi"Z$ obviously does not intersect V, thus $W \cap V = \varnothing$, and by a similar reason $U \subseteq W$. □

Corollary 9.4.3. (i) $\mathbf{\Sigma}^0_\xi$-**Red** *and* $\mathbf{\Pi}^0_\xi$-**Sep** *for all* $1 \le \xi < \omega_1$, *and also* $\mathbf{\Sigma}^1_2$-**Red**, $\mathbf{\Pi}^1_1$-**Red**, $\mathbf{\Pi}^1_2$-**Sep**, $\mathbf{\Sigma}^1_1$-**Sep** *<u>hold</u> for any internal* H ;

(ii) $\mathbf{\Pi}^0_\xi$-**Red** *and* $\mathbf{\Sigma}^0_\xi$-**Sep** *for all* $2 \le \xi < \omega_1$, *and also* $\mathbf{\Pi}^1_2$-**Red**, $\mathbf{\Sigma}^1_1$-**Red**, $\mathbf{\Sigma}^1_2$-**Sep**, $\mathbf{\Pi}^1_1$-**Sep** *<u>fail</u> for any infinite internal* H.

Proof. All these results for Polish spaces are well-known theorems of classical descriptive set theory, in particular, they hold in the space $2^{\mathbb{N}}$. It remains to apply the theorem. □

The corollary says nothing about $\mathbf{\Pi}^0_1$-**Red** and $\mathbf{\Sigma}^0_1$-**Sep**. Differently from the Polish case we have:

Lemma 9.4.4. $\mathbf{\Pi}^0_1$-**Red** *and* $\mathbf{\Sigma}^0_1$-**Sep** *hold for any internal set* H.

Proof. We observe that $\mathbf{\Pi}^0_1$-**Red** implies $\mathbf{\Sigma}^0_1$-**Sep** by 9.4.1, but it occurs that to prove $\mathbf{\Pi}^0_1$-**Red** we first need $\mathbf{\Sigma}^0_1$-**Sep**.

To check $\mathbf{\Sigma}^0_1$-**Sep** consider a pair of disjoint $\mathbf{\Sigma}^0_1$ sets $X = \bigcup_{n \in \mathbb{N}} X_n$ and $Y = \bigcup_{n \in \mathbb{N}} Y_n$ in H; all X_n, Y_n being internal. By **Extension** (see 2° of 9.1.2) there exist internal functions $f, g : {}^*\mathbb{N} \to \mathscr{P}_{\mathbf{int}}(H)$ such that $X_n = f(n)$ and $Y_n = g(n)$ for all $n \in \mathbb{N}$. The set Ω of all $h \in {}^*\mathbb{N}$ such that $X_{<h} = \bigcup_{n<h} f(n)$ does not intersect $Y_{<h} = \bigcup_{n<h} g(n)$ is internal together with f, g, and $\mathbb{N} \subseteq \Omega$ because $X \cap Y = \varnothing$. Thus Ω contains a number $h \in {}^*\mathbb{N} \smallsetminus \mathbb{N}$. Then $X_{<h}$, $Y_{<h}$ are disjoint internal sets inluding resp. X, Y, as required.

To prove $\mathbf{\Pi}^0_1$-**Red** consider a pair of $\mathbf{\Pi}^0_1$ sets $X, Y \subseteq H$. The complementary sets $A = X^{\mathsf{C}}$ and $B = Y^{\mathsf{C}}$ (in H) are $\mathbf{\Sigma}^0_1$, and hence by $\mathbf{\Sigma}^0_1$-**Red** there exist disjoint $\mathbf{\Sigma}^0_1$ sets $A' \subseteq A$ and $B' \subseteq B$ with $A' \cup B' = A \cup B$. By $\mathbf{\Sigma}^0_1$-**Sep** there exists a Borel set C such that $A' \subseteq C$ but $B' \cap C = \varnothing$. The $\mathbf{\Pi}^0_1$ sets $X' = X \smallsetminus C$ and $Y' = Y \cap C$ witness $\mathbf{\Pi}^0_1$-**Red** for $X, Y \subseteq H$. □

Lemma 9.4.4 shows that the phenomenon of Polish descriptive set theory, saying that $\mathbf{\Gamma}$-**Red** and $\mathbf{\Gamma}$-**Sep** cannot hold for one and the same $\mathbf{\Sigma}$- or $\mathbf{\Pi}$-class $\mathbf{\Gamma}$, is not true in the "hyperfinite" setting any more. Recall that the proof of such an incompatibility in the Polish case involves a diagonal argument applied to universal sets, which fails in the "hyperfinite" case because of the non-existence of suitable universal sets, see Footnote 11.

There is a useful generalization of $\mathbf{\Sigma}^1_1$**-Sep** and $\mathbf{\Pi}^1_1$**-Red**:

Exercise 9.4.5. Prove the following for any internal H:

multiple $\mathbf{\Sigma}^1_1$**-Sep**: for any system $\{Y_n\}_{n\in\mathbb{N}}$ of pairwise disjoint $\mathbf{\Sigma}^1_1$ subsets of H there is a system of pairwise disjoint *Borel* sets $X_n \supseteq Y_n$;

multiple $\mathbf{\Pi}^1_1$**-Red**: for any system $\{Y_n\}_{n\in\mathbb{N}}$ of $\mathbf{\Pi}^1_1$ subsets of H there is a system of *pairwise disjoint* $\mathbf{\Pi}^1_1$ sets $X_n \subseteq Y_n$ with $\bigcup_n X_n = \bigcup_n Y_n$.

Hint. Either of the two claims can be derived from its Polish counterpart by the same methods as in the proof of Theorem 9.4.2(i). Alternatively multiple $\mathbf{\Sigma}^1_1$**-Sep** follows from $\mathbf{\Sigma}^1_1$**-Sep**: let, for any n, X_n be a Borel set separating Y_n from $\bigcup_{k\neq n} Y_k$. Such a trick, generally, does not work for multiple $\mathbf{\Pi}^1_1$**-Red**, but still works in the case when $\bigcup_n Y_n$ is a Borel set. □

The next corollary resembles a result in Polish descriptive set theory.

Exercise 9.4.6 (Souslin operation with disjoint summands). Let H be an infinite internal set. Prove that a set $X \subseteq H$ is Borel iff it can be presented in the form $X = \bigcup_{f\in\mathbb{N}^\mathbb{N}} \bigcap_m X_{f\restriction m}$, where all $X_s \subseteq H$, $s \in \mathbb{N}^{<\omega}$, are internal and all intersections $X_f = \bigcap_m X_{f\restriction m}$ (called *summands*) are pairwise disjoint.

Hint. $\Longrightarrow$ is a corollary of a very general theorem saying that sets Souslin over an algebra of sets, in the version of disjoint summands, form a σ-algebra. To prove $\Longleftarrow$ put $A_s = \bigcup_{f\in\mathbb{N}^\mathbb{N},\, s\subset f} \bigcap_m X_{f\restriction m}$ for any $s \in \mathbb{N}^{<\omega}$. The sets A_s are Souslin, therefore $\mathbf{\Sigma}^1_1$, and obviously $A_s \cap A_t = \varnothing$ whenever $s \neq t \in \mathbb{N}^{<\omega}$ satisfy $\mathtt{lh}\, s = \mathtt{lh}\, t$. Using multiple $\mathbf{\Sigma}^1_1$**-Sep**, find a system of Borel sets B_s, $s \in \mathbb{N}^{<\omega}$, such that $A_s \subseteq B_s \subseteq X_s$ for all s and still $B_s \cap B_t = \varnothing$ whenever $s \neq t \in \mathbb{N}^{<\omega}$ satisfy $\mathtt{lh}\, s = \mathtt{lh}\, t$. Show that $X = \bigcap_{n\in\mathbb{N}} \bigcup_{\mathtt{lh}\, s = n} B_s$. [12] □

9.4b Countably determined sets with countable cross-sections

Here we consider countably determined "planar" sets, that is subsets of a set of the form $H \times K$, where H, K are internal sets. Suppose that $P \subseteq H \times K$. For any $x \in H$, the *cross-section* P_x is the set $P_x = \{y : \langle x, y\rangle \in P\} \subseteq K$, and the *projection* of P is the set $\operatorname{dom} P = \{x \in H : P_x \neq \varnothing\}$.

There are several typical questions that can be addressed to sets P satisfying special requirements related to their cross-sections P_x, and their projections $\operatorname{dom} P$. The next theorem presents some results on *uniform* sets —

[12] In Polish descriptive set theory there is another proof: first, a "disjoint summand" Souslin set is the projection of a uniform Borel set, second, such a projection is necessarily Borel. But this argument fails in the "hyperfinite" setting in its first part: the intended Borel set $\{\langle x, f\rangle \in H \times {}^{\mathrm{int}}2^h : \forall n \in \mathbb{N}\, (x \in X_{f\restriction n})\}$ is not necessarily uniform. (h is an arbitrary number in ${}^*\mathbb{N} \setminus \mathbb{N}$.)

those P having at most one point in each cross-section P_x, and also sets with countable sections. In different words a uniform set $P \subseteq H \times K$ is (the graph of) a *partial* function $H \to K$, and if, in addition, $\operatorname{dom} P = H$ then P is a *total* function $H \to K$, so that $\langle x, y\rangle \in P$ iff $P(x) = y$.

Theorem 9.4.7. *Suppose that H, K are internal sets, and $P \subseteq H \times K$ is a countably determined set. Then there is a sequence of internal (total) functions $F_n : H \to K$ $(n \in \mathbb{N})$ satisfying the following:*

(i) *the set $P^{\aleph_0} = \{\langle x, y\rangle \in P : P_x$ is at most countable$\}$* (all points of P that belong to at most countable cross-sections P_x) *is covered by $\bigcup_n F_n$;*

(ii) *if all cross-sections P_x, $x \in H$, are at most countable then $P \subseteq \bigcup_n F_n$, and hence P is a countable pairwise disjoint union of uniform countably determined sets $P \cap (F_n \setminus \bigcup_{k<n} F_k)$;*

(iii) *if P is a Borel set in* (i) *then both the set $P^{\aleph_0}$ and its projection $\operatorname{dom} P^{\aleph_0}$ are $\boldsymbol{\Pi}^1_1$ sets, in particular, if all cross-sections P_x are at most countable then $\operatorname{dom} P^{\aleph_0} = \operatorname{dom} P$ is Borel;*

(iv) *if P is a Borel set and if all cross-sections P_x, $x \in H$, are at most countable then P is a countable pairwise disjoint union of uniform Borel sets, and $\operatorname{dom} P$ is a Borel set, too;*

(v) *if, in* (iv), *$\boldsymbol{\Gamma}$ is a class of the form $\boldsymbol{\Sigma}^0_\xi$, $\boldsymbol{\Sigma}^1_n$, $\boldsymbol{\Pi}^1_n$, $\xi < \omega_1$ and $1 \le n \in \mathbb{N}$, or the class $\boldsymbol{\Pi}^0_1$, but not $\boldsymbol{\Pi}^0_\xi$, $\xi \ge 2$, then $\operatorname{dom} P$ is $\boldsymbol{\Gamma}$ as well;*

(vi) *if P is uniform and $\operatorname{dom} P = H$ in* (v) *then P belongs also to the complementary class, resp., $\boldsymbol{\Pi}^0_\xi$, $\boldsymbol{\Pi}^1_n$, $\boldsymbol{\Sigma}^1_n$ — and hence to, resp., $\boldsymbol{\Delta}^0_\xi$, $\boldsymbol{\Delta}^1_n$, $\boldsymbol{\Delta}^1_n$, while if P is $\boldsymbol{\Pi}^0_1$ then it is internal;*

(vii) *if P is a Borel uniform set then $\operatorname{dom} P$ is Borel and there is a partition $\operatorname{dom} P = \bigcup_{n\in\mathbb{N}} B_n$ into Borel sets such that $F_n \restriction B_n = P \restriction B_n$ for all n.*

Note that assertions (v), (vi) of the theorem are in drastic contradiction to the results known from Polish descriptive set theory, where any Borel set is the projection of a suitable uniform $\boldsymbol{\Pi}^0_2$ (that is $\mathbf{G}_\delta$) set, any $\boldsymbol{\Sigma}^1_2$ set is the projection of a suitable uniform $\boldsymbol{\Pi}^1_1$ set, and, consistently with **ZFC**, any $\boldsymbol{\Sigma}^1_{n+1}$ set $(n \ge 2)$ is the projection of a suitable uniform $\boldsymbol{\Pi}^1_n$ set.

Proof. Let $P = \bigcup_{f\in F} \bigcap_k A_{f\restriction k}$, where $F \subseteq 2^{\mathbb{N}}$, all A_s are internal subsets of $H \times K$, and $A_t \subseteq A_s$ whenever $s \subset t$. (Lemma 9.2.6.) Let, for $s \in \mathbb{N}^{<\omega}$ and $m \in \mathbb{N}$, $A_s(m)$ be the set of all points $\langle x, y\rangle \in A_s$ such that the cross-section $(A_s)_x$ has $\le m$ elements; all sets $A_s(m)$ are internal together with A_s. Moreover we have $A_s(m) = \bigcup_{j<m} A^j_s(m)$, where all sets $A^j_s(m)$ are internal and uniform. (Indeed, in $\mathbb{I}$ any set with cross-sections containing $\le m$ elements is a union of m uniform sets.) Since the projection $\operatorname{dom} U$ of any internal $U \subseteq H \times K$ is internal as well, any internal uniform $U \subseteq H \times K$ can be covered by (the graph of) an internal function $F : H \to K$. It follows

that there is a countable sequence of internal maps $F_n : H \to K$ $(n \in \mathbb{N})$ such that every set of the form $A_s^j(m)$ is covered by at least one F_n.

(i) It suffices to show that $P^{\aleph_0}$ is a subset of the union of all sets $A_s^j(m)$. Suppose that $\langle x, y \rangle \in P$ and P_x is at most countable. There is $f \in 2^{\mathbb{N}}$ such that $\langle x, y \rangle$ belongs to the set $A_f = \bigcap_k A_{f \restriction k}$. Thus the cross-section $(A_f)_x = \bigcap_k (A_{f \restriction k})_x \subseteq P_x$ is an at most countable $\mathbf{\Pi}_1^0$ set, hence $(A_f)_x$ is finite by 9.1.7. It follows that there is a number $k = k_{fx}$ such that $(A_s)_x$ is equal to $(A_f)_x$ and hence finite, where $s = f \restriction k$. Thus $\langle x, y \rangle \in A_s(m)$, where m is the number of elements in $(A_s)_x$, as required.

(ii) follows from (i).

(iii) Suppose, by (i), that $P^{\aleph_0} \subseteq F = \bigcup_n F_n$. The sets $X_n = \mathtt{dom}\,(F_n \cap P)$ are Borel together with P : indeed, $X_n = G_n{}^{-1}(P)$, where $G_n(x) = \langle x, F_n(x) \rangle$ is still an internal function $H \to (H \times K)$, thus X_n is Borel by Theorem 9.3.9 (internal preimages). Then $x \in \mathtt{dom}\, P^{\aleph_0}$ is equivalent to

$$\exists n\, (x \in X_n) \wedge \forall y\, (\langle x, y \rangle \in P \Longrightarrow \exists n\, (\langle x, y \rangle \in F_n)),$$

which witnesses that $\mathtt{dom}\, P^{\aleph_0}$ is a $\mathbf{\Pi}_1^1$ set since X_n, P, F_n are Borel sets. ($\forall y$ is the key quantifier in this judgement.)

(iv) $\mathtt{dom}\, P$ is $\mathbf{\Sigma}_1^1$ by definition and $\mathtt{dom}\, P$ is $\mathbf{\Pi}_1^1$ by (iii) because $P = P^{\aleph_0}$ in this case. It remains to use Theorem 9.3.9(iii).

(v) Let, by (ii), $P \subseteq \bigcup_{n \in \mathbb{N}} F_n$. Then $\mathtt{dom}\, P = \bigcup_n X_n$, where each $X_n = \{x \in H : \langle x, F_n(x) \rangle \in P\}$ belongs to $\mathbf{\Gamma}$ by the same reasons as in the proof of (iii). However all classes $\mathbf{\Gamma}$ we deal with in (v), except for $\mathbf{\Gamma} = \mathbf{\Pi}_1^0$, are closed under countable unions by Theorem 9.3.9.

The result for $\mathbf{\Gamma} = \mathbf{\Pi}_1^0$ follows directly from Theorem 9.3.9(i)(6).

(vi) Define X_n as in the proof of (v). Then $\langle x, y \rangle \notin P$ is equivalent to $\exists n\, (x \in X_n \wedge y \neq F_n(x))$. It follows that the complement of P is still a set in $\mathbf{\Gamma}$.

Separately for $\mathbf{\Gamma} = \mathbf{\Pi}_1^0$, we have to prove that any map $P : H \to K$, $\mathbf{\Pi}_1^0$ as a set of pairs in $H \times K$, is internal. Suppose that $P = \bigcap_{n \in \mathbb{N}} P_n$, all $P_n \subseteq H \times K$ being internal. Since P is uniform, a routine application of Saturation shows that one of P_n must be uniform as well. But then $P = P_n$.

(vii) We have $P \subseteq \bigcup_n F_n$ by (ii). Note that the sets $X_n = \mathtt{dom}\,(F_n \cap P) \subseteq \mathtt{dom}\, P$ are Borel (see the proof of (iii)). Put $B_n = X_n \setminus \bigcup_{k<n} B_k$. □

The following result is complementary to (v), (vi) in the theorem.

Exercise 9.4.8. Let H, K be infinite internal sets, and $2 \le \xi < \omega_1$. Prove that there exist a uniform $\mathbf{\Pi}_\xi^0$ set $P \subseteq H \times K$ such that $\mathtt{dom}\, P$ is not $\mathbf{\Pi}_\xi^0$, and a uniform $\mathbf{\Pi}_\xi^0$ set $P \subseteq H \times K$ such that $\mathtt{dom}\, P = H$ and P is not $\mathbf{\Sigma}_\xi^0$. (*Hint*: see [KKML 89, Thm. 4.6].) □

The next exercise is an interesting variation of (iii) of the theorem.

Exercise 9.4.9. Prove that if $P \subseteq H \times K$ is Borel then the set P' of all points $\langle x, y \rangle \in P$, such that the cross-section P_x is a singleton, is $\mathbf{\Pi}^1_1$ and $\mathrm{dom}\, P'$ is $\mathbf{\Pi}^1_1$ as well. (See similar results in the next Subsection.)

Hint. P_x is a singleton iff, in terms of the proof of Theorem 9.4.7(iii),

$$\exists n\,(x \in X_n) \wedge \forall y\,(\langle x, y \rangle \in P \Longrightarrow \exists n\,(\langle x, y \rangle \in F_n)) \\ \wedge \forall m, n\,(x \in X_m \cap X_n \Longrightarrow F_m(x) = F_n(x))\,. \qquad \square$$

9.4c Countably determined sets with internal and $\mathbf{\Sigma}^0_1$ cross-sections

These are larger classes of "planar" sets: indeed, any uniform set is a set with internal cross-sections, and similarly, countable cross-sections are $\mathbf{\Sigma}^0_1$. Nevertheless the next theorem has obvious similarities with Theorem 9.4.7 in the sense of the parallel between internal and $\mathbf{\Sigma}^0_1$ cross-sections at the one hand and with singletons (or finite internal) and countable cross-sections at the other hand.

Theorem 9.4.10. *Suppose that H, K are internal sets, and $P \subseteq H \times K$ is a countably determined set. Then there is a sequence of internal sets $A^n \subseteq H \times K$ $(n \in \mathbb{N})$ satisfying the following, where $A^n_x = \{y \in K : \langle x, y \rangle \in A^n\}$:*

(i) *for any $x \in H$, if P_x is internal then $P_x = A^n_x$ for some n, while if P_x is $\mathbf{\Sigma}^0_1$ then $P_x = \bigcup_{n \in \{n : A^n_x \subseteq P_x\}} A^n_x$;*

(ii) *if all cross-sections P_x, $x \in H$, are internal then $P = \bigcup_n A^n \restriction D_n$, where $D_n = \{x : P_x = A^n_x\}$ and $A^n_x \restriction D_n = \{\langle x, y \rangle \in A^n_x : x \in D_n\}$, while if all cross-sections P_x, $x \in H$, are $\mathbf{\Sigma}^0_1$ then $P = \bigcup_n A^n \restriction E_n$, where $E_n = \{x : A^n_x \subseteq P_x\}$;*

(iii) *if P is a Borel set then the sets*

$$P^{(\mathtt{int})} = \{\langle x, y \rangle \in P : P_x \text{ is internal}\},\ P^{(\mathbf{\Sigma}^0_1)} = \{\langle x, y \rangle \in P : P_x \text{ is } \mathbf{\Sigma}^0_1\}$$

and their projections $\mathrm{dom}\, P^{(\mathtt{int})}$, $\mathrm{dom}\, P^{(\mathbf{\Sigma}^0_1)}$ are $\mathbf{\Pi}^1_1$ sets;

(iv) (a) *if P is a Borel set and all cross-sections P_x, $x \in H$, are internal then $P = \bigcup_{n \in \mathbb{N}} A^n \restriction X_n$, where $X_n \subseteq \mathrm{dom}\, P$ are Borel and pairwise disjoint sets,*

(b) *if P is a Borel set and all cross-sections P_x, $x \in H$, are $\mathbf{\Sigma}^0_1$ then $P = \bigcup_{n \in \mathbb{N}} P^n$, where each $P^n \subseteq P$ is a Borel set with internal cross-sections P^n_x,*

(c) *in both case* (iv)(a) *and case* (iv)(b) *the projection $\mathrm{dom}\, P$ is Borel.*

Proof. (i) Consider a countably determined set $P = \Phi_U\{A^n\}_{n \in \mathbb{N}} \subseteq H \times K$, where $U \subseteq 2^{\mathbb{N}}$ and $\mathscr{A} = \{A^n\}_{n \in \mathbb{N}}$ is an algebra of internal subsets in $H \times K$.

By 9.3.6, a cross-section $P_x = \Phi_U\{A^n_x\}_{n\in\mathbb{N}}$ is internal iff it is equal to some A^n_x, and is $\mathbf{\Sigma}^0_1$ iff it is equal to the union of all A^n_x satisfying $A^n_x \subseteq P_x$.

(ii) is a consequence of (i).

(iii) It immediately follows from (i) that P_x is internal and non-empty iff $\exists n\,(\varnothing \neq A^n_x = P_x)$, leading to a $\mathbf{\Pi}^1_1$ form provided P is Borel. Similarly P_x is $\mathbf{\Sigma}^0_1$ and non-empty iff $\forall y \in P_x\, \exists n\,(y \in A^n_x \wedge \varnothing \neq A^n_x \subseteq P_x)$, leading to a $\mathbf{\Pi}^1_1$ representation. (Note that $\varnothing \neq A^n_x$ is equivalent to $x \in \mathtt{dom}\, A^n$, where $\mathtt{dom}\, A^n$ is obviously an internal set because so is A^n itself.)

(iv)(c) follows from (iii) because any set both $\mathbf{\Sigma}^1_1$ (as all projections of Borel sets) and $\mathbf{\Pi}^1_1$ is Borel by Theorem 9.3.9(iii)(1).

(iv)(a) The sets D_n as in (ii) are $\mathbf{\Pi}^1_1$: indeed, $x \in D_n$ if and only if $\forall y\,(y \in P_x \iff y \in A^n_x)$, leading to a $\mathbf{\Pi}^1_1$ form. Therefore the sets $Y_n = D_n \cap \mathtt{dom}\, P$ are $\mathbf{\Pi}^1_1$ as well by (iv)(c). By multiple $\mathbf{\Pi}^1_1$-**Red** (Exercise 9.4.5) there exist *pairwise disjoint* $\mathbf{\Pi}^1_1$ sets $X_n \subseteq Y_n$ with $\bigcup_{n\in\mathbb{N}} X_n = \bigcup_{n\in\mathbb{N}} Y_n = \mathtt{dom}\, P$. However each X_n is then Borel by Theorem 9.3.9(iii)(1): indeed its complement $\bigcup_{k\neq n} X_k$ to $\mathtt{dom}\, P$ is $\mathbf{\Pi}^1_1$, too, but $\mathtt{dom}\, P$ is Borel by (iv)(c).

(iv)(b) Prove that the sets E_n as in (ii) are $\mathbf{\Pi}^1_1$ and apply multiple $\mathbf{\Pi}^1_1$-**Red** to the family of $\mathbf{\Pi}^1_1$ sets $Q^n = \{\langle x, y\rangle \in P : x \in E_n\}$. □

Thus by 9.4.10(iv)(b) any Borel set with $\mathbf{\Sigma}^0_1$ cross-sections is a countable union of Borel sets with internal cross-sections. It will be demonstrated in § 9.4e that this is a particular case of a much more general result. Assertion 9.4.10(iv)(c) also will be slightly improved: any Borel set with $\mathbf{\Sigma}^0_2$ cross-sections has a Borel projection, see § 9.4f.

9.4d Uniformization

Let $P \subseteq H \times K$. A set $Q \subseteq P$ is said to *uniformize* P if Q is uniform in the sense of § 9.4b and $\mathtt{dom}\, P = \mathtt{dom}\, Q$. Thus *uniformization* of a set $P \subseteq H \times K$ is equivalent to the choice of an element in every non-empty cross-section $P_x = \{y : \langle x, y\rangle \in P\}$ of P. The next lemma shows that even countably determined uniformizability of a "planar" set implies that its projection has roughly the same class as the set itself — which is very different from the state of affairs with Uniformization in Polish descriptive set theory.

Lemma 9.4.11. *Suppose that* $\mathbf{\Gamma}$ *is a class of the form* $\mathbf{\Sigma}^0_\xi$, $\mathbf{\Sigma}^1_n$, $\mathbf{\Pi}^1_n$, $\xi < \omega_1$ *and* $1 \le n \in \mathbb{N}$, *and a set* $P \subseteq H \times K$ *in* $\mathbf{\Gamma}$ *is uniformized by a countably determined set* $Q \subseteq P$. *Then*

(i) *the projection* $\mathtt{dom}\, P$ *is still a* $\mathbf{\Gamma}$ *set;*

(ii) P *is uniformizable by a set in* $\mathbf{\Delta}^1_{n+1}$ *provided* $\mathbf{\Gamma}$ *is* $\mathbf{\Sigma}^1_n$ *or* $\mathbf{\Pi}^1_n$, $n \ge 1$, *and by a set in* $\mathbf{\Sigma}^0_{\xi+1}$ *provided* $\xi < \omega_1$;.

(iii) P *is uniformizable by a set in* $\mathbf{\Pi}^1_1$ *provided* $\mathbf{\Gamma}$ *is* $\mathbf{\Pi}^1_1$.

Proof. (i) By Theorem 9.4.7(ii) we have $Q \subseteq \bigcup_n U_n$, where $U_n \subseteq H \times K$ are internal uniform sets. Then $Q_n = U_n \cap P$ are still sets in $\mathbf{\Gamma}$, and so are their projections $X_n = \mathtt{dom}\, Q_n$ still by Theorem 9.4.7(v). Moreover $\mathtt{dom}\, P = \bigcup_n X_n$, and hence $\mathtt{dom}\, P$ belongs to $\mathbf{\Gamma}$ because $\mathbf{\Gamma}$ is closed under countable unions under the conditions of the lemma.

(ii) Put $Y_n = X_n \smallsetminus \bigcup_{k<n} X_k$ and $R_n = \{\langle x, y\rangle \in Q_n : x \in Y_n\}$. Then the set $R = \bigcup_n R_n$ uniformizes P. Moreover if $\mathbf{\Gamma}$ is $\mathbf{\Sigma}^1_n$ or $\mathbf{\Pi}^1_n$, $n \geq 1$, then all sets Y_n belong to $\mathbf{\Delta}^1_{n+1}$, and so does R since $\mathbf{\Delta}^1_{n+1}$ is closed under countable unions. If $\mathbf{\Gamma}$ is $\mathbf{\Sigma}^0_\xi$ then all sets Y_n belong to $\mathbf{\Delta}^0_{\xi+1}$, and hence R, a countable union of them, is $\mathbf{\Sigma}^0_{\xi+1}$.

(iii) Multiple $\mathbf{\Pi}^1_1$-**Red** (see § 9.4a) enables us to choose pairwise disjoint $\mathbf{\Pi}^1_1$ sets $Y_n \subseteq X_n$ with $\bigcup_n Y_n = H$. Then argue as above. □

For any class $\mathbf{\Gamma}$ of subsets of a given set of the form $H \times K$, the Uniformization principle is defined as follows:

$\mathbf{\Gamma}$-**Unif**: any set $P \subseteq H \times K$ in $\mathbf{\Gamma}$ can be uniformized by a set $Q \subseteq P$ in $\mathbf{\Gamma}$.

In Polish descriptive set theory $\mathbf{\Gamma}$-**Unif** holds for $\mathbf{\Gamma} = \mathbf{\Pi}^1_1$ and $\mathbf{\Sigma}^1_2$ (the famous Novikov – Kondo uniformization theorem) but fails for $\mathbf{\Sigma}^1_1$, $\mathbf{\Pi}^1_2$, $\mathbf{\Delta}^1_1$, $\mathbf{\Delta}^1_2$, all Borel classes, and the class of all Borel sets. The negative part for the classes $\mathbf{\Sigma}^1_1$, $\mathbf{\Pi}^1_2$ can be derived from the negation of $\mathbf{\Pi}^1_1$-**Red** and $\mathbf{\Sigma}^1_2$-**Red** (see § 9.4a) because $\mathbf{\Gamma}$-**Unif** implies $\mathbf{\Gamma}$-**Red** for all reasonable classes $\mathbf{\Gamma}$. (Indeed to reduce sets $X, Y \subseteq H$ it suffices to uniformize the set

$$P = \{\langle x, z_1\rangle : x \in X\} \cup \{\langle y, z_2\rangle : y \in Y\} \subseteq H \times H,$$

where $z_1 \neq z_2$ – arbitrary elements of H.) The next theorem shows that the picture is rather different in the "hyperfinite" setting.

Theorem 9.4.12. *Suppose that H, K are infinite internal sets. Then*

(i) *any countably determined (resp., Borel, resp., $\mathbf{\Pi}^1_1$) set $P \subseteq H \times K$ with countable cross-sections P_x can be uniformized by a countably determined (resp., Borel, resp., $\mathbf{\Pi}^1_1$) set;*

(ii) *moreover any countably determined (resp., Borel, resp., $\mathbf{\Pi}^1_1$) set $P \subseteq H \times K$ with $\mathbf{\Sigma}^0_1$ cross-sections P_x can be uniformized by a countably determined (resp., Borel, resp., $\mathbf{\Pi}^1_1$) set;*

(iii) $\mathbf{\Sigma}^0_1$-**Unif**, $\mathbf{\Pi}^0_1$-**Unif**, $\mathbf{\Sigma}^0_2$-**Unif** *hold for sets in $H \times K$;*

(iv) *but there exists a $\mathbf{\Pi}^0_2$ set $P \subseteq H \times K$ with all cross-sections P_x being $\mathbf{\Pi}^0_1$ sets, which cannot be uniformized by a countably determined set.*

Proof. (i) Let, by Theorem 9.4.7(ii), $P = \bigcup_{n\in\mathbb{N}} P^n$, where each P^n is a countably determined uniform set. The projections $Y_n = \mathtt{dom}\, P^n \subseteq H$ are countably determined as well, and so are the sets $X_n = Y_n \smallsetminus \bigcup_{k<n} Y_k$ and finally the set $Q = \bigcup_n P^n \restriction X_n$. (We refer to Theorem 9.2.9, also in the form 9.2.10.) But it is clear that Q uniformizes P.

The Borel and $\mathbf{\Pi}^1_1$ versions now follow from Lemma 9.4.11.

(ii) Let, by Theorem 9.4.10(iii), $P = \bigcup_{n\in\mathbb{N}} P^n$, where each P^n is a countably determined set of the form $P^n = A^n \restriction E_n$, where the sets $A^n \subseteq H \times K$ are internal while the sets $E_n \subseteq H$ countably determined. Let $B^n \subseteq A^n$ be any internal set which uniformizes A^n. Then $P' = P \cap \bigcup_n B^n$ is a countably determined subset of P with at most countable cross-sections, satisfying $\mathtt{dom}\, P' = \mathtt{dom}\, P$. It remains to apply (i).

(iii) $\mathbf{\Sigma}^0_1$-**Unif** follows from (ii) and Lemma 9.4.11. To prove $\mathbf{\Pi}^0_1$-**Unif** for a $\mathbf{\Pi}^0_1$ set $P = \bigcap_n P_n$, where each $P_n \subseteq H \times K$ is internal, apply Saturation to the countable family of internal non-empty sets A_n, where each A_n consists of all internal uniform sets $U \subseteq H \times K$ such that $U \cap \bigcup_{k<n} P_k$ uniformizes $\bigcup_{k<n} P_k$. To get $\mathbf{\Sigma}^0_2$-**Unif** from $\mathbf{\Pi}^0_1$-**Unif** assemble a countably determined uniform subset $Q \subseteq P$ of a given $\mathbf{\Sigma}^0_2$ set $P = \bigcup_n P^n$ using uniform subsets of $\mathbf{\Pi}^0_1$ sets P^n as above. Then improve to $\mathbf{\Sigma}^0_2$ using Lemma 9.4.11.

(iv) We are going to define a required counterexample in the assumption that $H = K = {}^{\mathrm{int}}2^h$ (all *internal* maps $h \to 2$, as usual), where h is a number in ${}^*\mathbb{N} \smallsetminus \mathbb{N}$.[13] Put ${}^\circ x = \mathbf{sh}\, x = x \restriction \mathbb{N}$ for any $x \in H$ and $\psi(x, y) = \langle {}^\circ x, {}^\circ y\rangle$ for $x,\, y \in H$. Clearly $\psi : H \times K \to 2^{\mathbb{N}} \times 2^{\mathbb{N}}$ is a shadow map.

It is known from Polish descriptive set theory (and easily provable by induction on the Borel construction) that any Borel set in a Polish space is an image of a closed subset of the Baire space $\mathbb{N}^{\mathbb{N}}$ via a continuous bijection. Since $\mathbb{N}^{\mathbb{N}}$ is homeomorphic to a $\mathbf{\Pi}^0_2$ subset of $2^{\mathbb{N}}$, it follows that any Borel $X \subseteq 2^{\mathbb{N}}$ is the projection $\mathtt{dom}\, W$ of a uniform $\mathbf{\Pi}^0_2$ subset of $2^{\mathbb{N}} \times 2^{\mathbb{N}}$. Therefore there is a uniform $\mathbf{\Pi}^0_2$ set $U \subseteq 2^{\mathbb{N}} \times 2^{\mathbb{N}}$ such that the projection $E = \mathtt{dom}\, U$ is a (Borel) set **not** in the class $\mathbf{\Pi}^0_2$.

Then $P = \psi^{-1}(U)$ is a $\mathbf{\Pi}^0_2$ set in $H \times K$ by Lemma 9.2.16. Furthermore, by Theorem 9.3.3, the set $\mathtt{dom}\, P = \{x \in H : {}^\circ x \in E\}$ is a (Borel) set **not** of the class $\mathbf{\Pi}^0_2$ in H. (Indeed $x \mapsto {}^\circ x$ is a shadow map and $E \notin \mathbf{\Pi}^0_2$.) It follows that P is not uniformizable by a countably determined set by Lemma 9.4.11(i). Finally any cross-section P_x obviously coincides with $\{y \in K : {}^\circ y \in U_{{}^\circ x}\}$, and hence P_x is $\mathbf{\Pi}^0_1$ still by Lemma 9.2.16. □

Exercise 9.4.13. It is known from Polish descriptive set theory that there is a non-Borel uniform $\mathbf{\Pi}^1_1$ set $W \subseteq 2^{\mathbb{N}} \times 2^{\mathbb{N}}$ such that $\mathtt{dom}\, W = 2^{\mathbb{N}}$. (Apply $\mathbf{\Pi}^1_1$-**Unif** to a Borel-non-uniformizable Borel set $V \subseteq 2^{\mathbb{N}} \times 2^{\mathbb{N}}$ with $\mathtt{dom}\, W = 2^{\mathbb{N}}$.)

Prove that the set $P = \psi^{-1}(W) \subseteq H \times K$ (in terms of the proof of 9.4.12(iv)) is a $\mathbf{\Pi}^1_1$ set with $\mathbf{\Pi}^0_1$ cross-sections not equal to a countable intersection of $\mathbf{\Pi}^1_1$ sets with internal cross-sections. Accordingly the complement of W to $H \times K$ is a $\mathbf{\Sigma}^1_1$ set with $\mathbf{\Sigma}^0_1$ cross-sections not equal to a countable union of $\mathbf{\Sigma}^1_1$ sets with internal cross-sections. Thus Theorem 9.4.10(iv)(b) (in

[13] Why does this imply the general case? Indeed let X, Y be arbitrary infinite internal sets. There is a number $h \in {}^*\mathbb{N} \smallsetminus \mathbb{N}$ such that $2^h \le \mathtt{min}\{\mathtt{card}\, X, \mathtt{card}\, Y\}$. Let $H = K = {}^{\mathrm{int}}2^h$. We can assume that $H \subseteq X$ and $K \subseteq Y$. If $P \subseteq H \times K$ is a counterexample then it remains one in $X \times Y$ by 9.1.3.

the form that any Borel set with $\mathbf{\Sigma}^0_1$ cross-sections is a countable union of Borel sets with internal cross-sections) fails for $\mathbf{\Sigma}^1_1$ instead of Borel.

Hint: see Živaljević [Ziv 95, Example 11]. □

Exercise 9.4.14. Show that, in terms of the proof of 9.4.12(iv), there is an internal uniform set $P \subseteq H \times K$ such that $\psi"P = 2^{\mathbb{N}} \times 2^{\mathbb{N}}$. □

Exercise 9.4.15. Take a $\mathbf{\Pi}^0_2$ set $P \subseteq H \times K$ such that $\mathtt{dom}\, P$ is a non-Borel $\mathbf{\Sigma}^1_1$ set. (We refer to Theorem 9.3.9(v)(2), (iii) and Proposition 9.1.10.)

Show using Lemma 9.4.11 that P is not uniformizable by a countably determined set. (This is weaker than 9.4.12(iv) because we cannot claim that all cross-sections are $\mathbf{\Pi}^1_1$.) □

In spite of counterexamples like those in 9.4.12(iv), Uniformization theorems for Polish spaces still have some weak counterparts in "hyperfinite" setting in terms of sets with $\mathbf{\Pi}^0_1$ cross-sections rather than uniform sets, in the sense of the following theorem proved below as a corollary of the Novikov – Kondo $\mathbf{\Pi}^1_1$-**Unif** theorem for Polish spaces. Other "Polish" Uniformization theorems admit analogous "$\mathbf{\Pi}^0_1$-formization" versions in "hyperfinite" descriptive set theory.

Theorem 9.4.16 (see 9.4.22 for a proof). *If H, K are infinite internal sets then any $\mathbf{\Pi}^1_1$ set $P \subseteq H \times K$ has a $\mathbf{\Pi}^1_1$ subset $Q \subseteq P$ such that $\mathtt{dom}\, Q = \mathtt{dom}\, P$ and all cross-sections Q_x, $x \in H$, are $\mathbf{\Pi}^0_1$.*

Thus while Uniformization is the choice of an element in every non-empty cross-section, a set Q as in the theorem manifests the choice of a non-empty $\mathbf{\Pi}^0_1$ subset in every non-empty cross-section.

9.4e Variations on Louveau's theme

It is clear that, provided enough Choice is available, any set $P \subseteq H \times K$ such that all cross-sections P_x, $x \in H$, are $\mathbf{\Sigma}^0_{\xi+1}$ sets for a fixed ordinal $\xi < \omega_1$ can be decomposed into a union $P = \bigcup_n P^n$ such that all cross-sections $P^n_x = \{y : \langle x, y \rangle \in P^n\}$, $x \in H$, are $\mathbf{\Pi}^0_\xi$ sets for any n. The question becomes much more difficult if we require that the sets P^n belong to a certain class of sets provided so does P. Louveau proved the following theorem:

Theorem 9.4.17 (ZFC). *Suppose that $\mathscr{X}, \mathscr{Y}$ are Polish spaces and $P \subseteq \mathscr{X} \times \mathscr{Y}$ a Borel set, and $\xi < \omega_1$. Then*

(I) *the set $D = \{x \in \mathscr{X} : P_x \text{ is } \mathbf{\Pi}^0_\xi\}$ is of the class $\mathbf{\Pi}^1_1$, the same for $\mathbf{\Sigma}^0_\xi$;*

(II) *if $1 \le \xi < \omega_1$ and all cross-sections P_x, $x \in \mathscr{X}$, are $\mathbf{\Sigma}^0_\xi$ then there is a partition $P = \bigcup_{n \in \mathbb{N}} P^n$ into Borel sets P^n such that all cross-sections P^n_x belong to the class $\mathbf{\Pi}^0_{<\xi} = \bigcup_{\eta < \xi} \mathbf{\Pi}^0_\eta$.* □

The proof of this theorem for Borel sets P in Polish descriptive set theory in [Louv 80] involves both the instrumentarium of classical descriptive set theory and some recursion-theoretic ideas. We use the theorem to prove the following analogous result in "hyperfinite" descriptive set theory.

Theorem 9.4.18. *Suppose that H, K are internal sets, $P \subseteq H \times K$ is a countably determined set, and $\xi < \omega_1$. Then*

(i) *the set $D = \{x \in H : P_x$ is $\mathbf{\Pi}^0_\xi$ (or $\mathbf{\Sigma}^0_\xi$)$\}$ is countably determined;*

(ii) *if P is Borel then the set D as in (i) is $\mathbf{\Pi}^1_1$;*

(iii) *if $1 \le \xi < \omega_1$ and all cross-sections P_x, $x \in H$, are $\mathbf{\Sigma}^0_\xi$ sets then there is a partition $P = \bigcup_{n \in \mathbb{N}} P^n$ into countably determined sets P^n such that all cross-sections P^n_x belong to the class $\mathbf{\Pi}^0_{<\xi}$;*

(iv) *if P is Borel in (iii) then the sets P^n can be chosen to be Borel as well.*

Proof. There is a shadow $\varphi : H \times K \to 2^{\mathbb{N}}$ such that $P = \varphi^{-1}(U)$, where $U = \varphi"P \subseteq 2^{\mathbb{N}}$ (we refer to Corollary 9.2.14). Define a map $\varphi_x : K \to 2^{\mathbb{N}}$ for any $x \in H$ so that $\varphi_x(y) = \varphi(x, y)$ for all $y \in K$. Then φ_x is also a shadow, and $P_x = \varphi_x^{-1}(U) = \varphi_x^{-1}(U_x)$, where $U_x = \varphi_x"P_x = U \cap B_x$ and $B_x = \operatorname{ran} \varphi_x = \{\varphi(x, y) : y \in K\}$.

Define another map $\tau : H \to 2^{(2^{<\omega})}$ as follows: for all $x \in H$ and $s \in 2^{<\omega}$, $\tau(x)(s) = 1$ if and only if $\exists\, y \in K\, (s \subset \varphi_x(y))$. *We claim that τ is a shadow in the sense of 9.2.12.* It suffices to prove that the τ-preimage $\tau^{-1}(C_s)$ of any set of the form $C_s = \{c \in 2^{(2^{<\omega})} : c(s) = 1\}$, $s \in 2^{<\omega}$, is an internal set. Yet by definition $\tau^{-1}(C_s) = \operatorname{dom} Q_s$, where $Q_s = \{\langle x, y\rangle \in H \times K : s \subset \varphi(x, y)\}$ is internal because φ is a shadow while any set $\{b \in 2^{\mathbb{N}} : s \subset b\}$ is clopen.

Note that for any $x \in H$, if $s \in T_x = \{s \in 2^{<\omega} : \tau(x)(s) = 1\}$ then obviously at least one of the extensions $s^\wedge 0$, $s^\wedge 1$ belongs to T_x, and hence $T_x \subseteq 2^{<\omega}$ is a tree without endpoints, and accordingly $[T_x] = \{b \in 2^{\mathbb{N}} : \forall m\, (b \restriction m \in T_x)\}$ is a closed subset of $2^{\mathbb{N}}$. *We claim that $[T_x] = B_x = \{\varphi_x(y) : y \in K\}$.* Indeed, if $b = \varphi_x(y) \in B_x$ then for any $m \in \mathbb{N}$ y itself witnesses that $\tau(x)(b \restriction m) = 1$, that is $b \restriction m \in T_x$. To prove the converse suppose that $b \in [T_x]$, that is $b \in 2^{\mathbb{N}}$ and $b \restriction m \in T_x$ for any $m \in \mathbb{N}$. There is $y \in K$ such that $b \restriction m \subset \varphi_x(y)$, or, in different terms, $b \restriction m = \varphi_x(y) \restriction m$. As m is arbitrary, this implies that b belongs to the closure of the set $B_x = \{\varphi_x(y) : y \in K\}$ in $2^{\mathbb{N}}$. However B_x is closed by 9.2.15!

We conclude that $U_x = \varphi_x"P_x = U \cap B_x = U \cap [T_x]$ for any x. Put

$$\psi(x, y) = \langle \tau(x), \varphi_x(y)\rangle \quad \text{and} \quad W = \psi"P = \{\psi(x, y) : x \in H \wedge y \in P_x\}.$$

Lemma 9.4.19. (a) $P = \psi^{-1}(W) = \{\langle x, y\rangle \in H \times K : \psi(x, y) \in W\}$;
(b) $W_{\tau(x)} = \{\varphi_x(y) : y \in P_x\} = U_x = U \cap [T_x]$ for any $x \in H$.

Proof. (a) If $\psi(x, y) = \psi(x', y')$ then in particular $\varphi_x(y) = \varphi_{x'}(y')$. Thus we have $\langle x, y\rangle \in P$ iff $\langle x', y'\rangle \in P$ by 9.2.11.

(b) It suffices to show that $\{\varphi_x(y) : y \in P_x\} = \{\varphi_{x'}(y') : y' \in P_{x'}\}$ whenever $x, x' \in H$ satisfy $\boldsymbol{\tau}(x) = \boldsymbol{\tau}(x')$. To prove the equality, note that $\{\varphi_x(y) : y \in P_x\} = [T_x] \cap U$, but by definition the tree T_x depends rather on $\boldsymbol{\tau}(x)$ than on x itself. □

Now we are ready to prove the theorem.

(i) Note first of all that if we have the result for $\mathbf{\Pi}^0_\xi$ then it holds for $\mathbf{\Sigma}^0_\xi$ too: just take the complement of P instead of P. In particular the result for $\mathbf{\Pi}^0_1$ suffices to immediately derive it for $\mathbf{\Sigma}^0_1$ and subsequently for $\mathbf{\Sigma}^0_0 = \mathbf{\Pi}^0_0 = \mathbf{\Sigma}^0_1 \cap \mathbf{\Pi}^0_1$. Thus it suffices to prove (i), and also (ii) by the same reasons, only for classes $\mathbf{\Pi}^0_\xi$, $\xi \geq 1$. Under this assumption, let Ω be the set of all $\tau \in 2^{(2^{<\omega})}$ such that the cross-section $W_\tau = \{b : \langle \tau, b\rangle \in W\}$ is $\mathbf{\Pi}^0_\xi$. *We claim that*

$$D = \{x \in H : \boldsymbol{\tau}(x) \in \Omega\} = \boldsymbol{\tau}^{-1}(\Omega).$$

Indeed, $x \in D$ iff P_x is $\mathbf{\Pi}^0_\xi$ iff $U_x = W_{\boldsymbol{\tau}(x)}$ is $\mathbf{\Pi}^0_\xi$ (by Theorem 9.3.3 and Lemma 9.4.19(b)) iff $\boldsymbol{\tau}(x) \in \Omega$, as required.

We conclude that D is countably determined by Corollary 9.2.14.

(ii) If P is Borel then so is W by Theorem 9.3.3 and Lemma 9.4.19(b). Therefore the set Ω defined in the proof of (i) is $\mathbf{\Pi}^1_1$ in $2^{(2^{<\omega})}$ by (I) of Theorem 9.4.17. It follows that $D = \boldsymbol{\tau}^{-1}(\Omega)$ is $\mathbf{\Pi}^1_1$ in H by Lemma 9.2.16.

(iii) Suppose that all cross-sections P_x, $x \in H$, are $\mathbf{\Sigma}^0_\xi$ sets in K. Then the set $W_{\boldsymbol{\tau}(x)} = U_x = B_x \cap U$ is $\mathbf{\Sigma}^0_\xi$ together with P_x by Theorem 9.3.3 for any $x \in H$. Thus all cross-sections W_τ of W are $\mathbf{\Sigma}^0_\xi$ sets. We conclude that W can be decomposed in the form $W = \bigcup_{n \in \mathbb{N}} W^n$ so that all cross-sections $W^n_\tau = \{b : \langle \tau, b\rangle \in W^n\}$, $\tau \in 2^{(2^{<\omega})}$, are $\mathbf{\Pi}^0_{<\xi}$ sets in $2^{\mathbb{N}}$. ($\mathfrak{c}$-size Choice is applied to get such a decomposition $W = \bigcup_n W^n$ of W.)

Consider the sets $P^n = \psi^{-1}(W^n) = \{\langle x, y\rangle \in H \times K : \psi(x,y) \in W^n\}$. It follows from Lemma 9.4.19(a) that $P = \bigcup_n P^n$, and this is a pairwise disjoint union. Thus it remains to prove that all cross-sections P^n_x, $n \in \mathbb{N}$ and $x \in H$, are $\mathbf{\Pi}^0_{<\xi}$ in K. To prove this fact note that, by Lemma 9.4.19,

$$\begin{aligned} P^n_x = \{y : \langle x, y\rangle \in P^n\} \qquad &= \{y : \psi(x,y) \in W^n\} = \\ = \{y : \langle \boldsymbol{\tau}(x), \varphi_x(y)\rangle \in W^n\} &= \{y : \varphi_x(y) \in W^n_{\boldsymbol{\tau}(x)}\} = \varphi_x^{-1}(W^n_{\boldsymbol{\tau}_x}). \end{aligned}$$

Thus P^n_x belongs to $\mathbf{\Pi}^0_{<\xi}$ together with $W^n_{\boldsymbol{\tau}_x}$ by Lemma 9.2.16.

(iv) If P is Borel then the set W is also Borel (see the proof of (ii)), and with $\mathbf{\Sigma}^0_\xi$ cross-sections. It follows from Theorem 9.4.17 that there is a decomposition $W = \bigcup_{n \in \mathbb{N}} W^n$ where W^n are *Borel* sets, with all cross-sections W^n_τ, $\tau \in 2^{(2^{<\omega})}$, being $\mathbf{\Pi}^0_{<\xi}$ sets in $2^{\mathbb{N}}$. It remains to ckeck that the sets $P^n = \psi^{-1}(W^n)$, defined as above, are Borel — but this holds by Lemma 9.2.16 because ψ is a shadow. □

Exercise 9.4.20. There is a slightly stronger "Polish" result in [Louv 80]: if $\xi < \omega_1$, $P, Q \subseteq \mathscr{X} \times \mathscr{Y}$ are Borel sets in the product of two Polish spaces, and for any $x \in \mathscr{X}$ the cross-section P_x can be separated by a $\mathbf{\Sigma}^0_\xi$ set from Q_x then there is a Borel set $R \subseteq \mathscr{X} \times \mathscr{Y}$ such that all cross-sections R_x are $\mathbf{\Sigma}^0_\xi$ and R separates P from Q.

Exercise: formulate and prove the "hyperfinite" version of the result. □

9.4f On sets with $\mathbf{\Pi}^0_1$ cross-sections

The method of transformation of P to W in the proof of Theorem 9.4.18 can be used to derive some other results related to "planar" sets in "hyperfinite" descriptive set theory from similar theorems known in Polish descriptive set theory. The following theorem is an example.

Theorem 9.4.21. *Suppose that H, K be internal sets, and $P \subseteq H \times K$ is a countably determined set. Then*

(i) *the set $D = \{x \in H : P_x \neq \varnothing \wedge P_x$ is $\mathbf{\Pi}^0_1\}$ is countably determined;*

(ii) *if P is Borel then D as in* (i) *is $\mathbf{\Pi}^1_1$;*

(iii) *if P is Borel and all cross-sections P_x, $x \in H$, are $\mathbf{\Pi}^0_1$ then the projection $\mathtt{dom}\, P$ is $\mathbf{\Delta}^1_1$.*

The same is true for $\mathbf{\Sigma}^0_2$ instead of $\mathbf{\Pi}^0_1$ everywhere.

Note that (i), (ii) are not particular cases of (i), (ii) of Theorem 9.4.18 because the latter do not contain the requirement of $P_x \neq \varnothing$!

Proof. In terms of the proof of Theorem 9.4.18, let us consider the set $\Omega = \{x \in 2^{(2^{<\omega})} : W_x \neq \varnothing \wedge W_x$ is $\mathbf{\Pi}^0_1\}$. Then $D = \tau^{-1}(\Omega)$ is countably determined (as in the proof of 9.4.18(i)). If P is Borel then so is W, and hence the set Ω is $\mathbf{\Pi}^1_1$ by a classical theorem of Arsenin – Kunugui. It follows that $D = \tau^{-1}(\Omega)$ is $\mathbf{\Pi}^1_1$ as well. Finally (iii) follows from (ii).

The proof for $\mathbf{\Sigma}^0_2$ is the same because the abovementioned Arsenin – Kunugui theorem is known both for $\mathbf{\Sigma}^0_2$ and for $\mathbf{\Pi}^0_1$. □

Note that Theorem 9.4.21 fails for $\mathbf{\Pi}^0_2$. Indeed take a non-Borel $\mathbf{\Sigma}^1_1$ set $X \subseteq H$. It follows from Proposition 9.1.10 that there is a $\mathbf{\Pi}^0_2$ set $P \subseteq H \times K$ such that $\mathtt{dom}\, P = X$. Then all cross-sections P_x are $\mathbf{\Pi}^0_2$ in K but the projection $\mathtt{dom}\, P = X$ is non-Borel.

Exercise 9.4.22. Prove Theorem 9.4.16.

Hint. In terms of the proof of Theorem 9.4.18, if P is $\mathbf{\Pi}^1_1$ then so is $W = \psi\text{"}P \subseteq 2^{\mathbb{N}} \times 2^{\mathbb{N}}$, and hence there is a $\mathbf{\Pi}^1_1$ set $U \subseteq W$ uniformizing W. Then the set $Q = \psi^{-1}(U)$ satisfies $Q \subseteq P$, $\mathtt{dom}\, Q = \mathtt{dom}\, P$, and all cross-sections Q_x are $\mathbf{\Pi}^0_1$ sets. □

9.5 Loeb measures

This section is written to demonstrate how **HST** can be used to develop, in a fully adequate manner, such a remarkable topic in nonstandard analysis as Loeb measures. In fact our goal is not to give a comprehensive exposition of the theory and applications of Loeb measures, or even a comprehensive introduction into the topic — simply because this is not suitable in such a book of mainly foundational content. On the contrary we try rather to convince the reader that the existing development of Loeb measures can be successfully reproduced on the base of **HST** (or even weaker principles formulated in § 9.1a.

Two more "exotic" topics are considered at the end of the section: random elements and hyperfinite gambling.

9.5a Definitions and examples

Recall that $\mathbb{R}$ denotes the set of reals in $\mathbb{WF}$, the well-founded universe, while the $*$-extension ${}^*\mathbb{R} \in \mathbb{S}$ indicates the (standard) set of all $*$-reals. ${}^*[0,\infty)$ is understood similarly. (See Section 2.1, especially § 2.1a.)

Recall that a $*$-real $x \in {}^*\mathbb{R}$ is *infinitesimal* iff $|x| < {}^*\varepsilon$ for any real $\varepsilon > 0$, and $x \simeq y$ means that $|x - y|$ is infinitesimal. A $*$-real x is *limited* iff $|x| < {}^*c$ for a real $c \in \mathbb{R}$. By Lemma 2.1.9, every limited $*$-real x is near-standard, that is there is a unique real $r = {}^\circ x \in \mathbb{R}$ such that $x \simeq {}^*r$, the *shadow* of x.

Suppose that H is an internal set and μ is an internal measure on $\mathscr{P}_{\mathtt{int}}(H)$, that is, an internal function $\mu : \mathscr{P}_{\mathtt{int}}(H) \to {}^*[0,\infty)$ satisfying the requirement of finite additivity: $\mu(X \cup Y) = \mu(X) + \mu(Y) - \mu(X \cap Y)$ for all internal $X, Y \subseteq H$. By the above, if $X \in \mathscr{P}_{\mathtt{int}}(H)$ and $\mu(X)$ is a limited $*$-real then there exists a unique real number $r = {}^\circ\mu(X) \in \mathbb{R}$ such that ${}^*r \simeq \mu(X)$. If the value $\mu(X)$ is non-limited then we naturally put ${}^\circ\mu(X) = \infty$.

Exercise 9.5.1. Prove that ${}^\circ\mu : \mathscr{P}_{\mathtt{int}}(H) \to [0,\infty]$ is still a (non-internal) measure, but now with values in $\mathbb{R}$ rather than ${}^*\mathbb{R}$. □

Definition 9.5.2. The Loeb measure $\mathrm{L}\mu$ is a natural σ-additive extension of ${}^\circ\mu$ obtained as follows. If $D \subseteq H$ is internal, ${}^\circ\mu(D) < \infty$, $X \subseteq D$, and

$$\mathtt{sup}\{{}^\circ\mu(I) : I \subseteq X,\ I \text{ internal}\} = \mathtt{inf}\{{}^\circ\mu(I) : X \subseteq I \subseteq D,\ I \text{ internal}\}$$

then X is called *μ-approximable.* [14]

A set $X \subseteq H$ is *Loeb measurable*, or *$\mathrm{L}\mu$-measurable*, iff $X \cap D$ is μ-approximable for any $D \in \mathscr{P}_{\mathtt{int}}(H)$ with ${}^\circ\mu(D) < \infty$.

[14] Take notice that the sup and inf in the displayed formula do exist. Indeed it follows from Theorem 1.1.9 that every set $Y \subseteq \mathbb{R}$ belongs to $\mathbb{WF}$, thus one may execute the operations in $\mathbb{WF}$, a **ZFC** universe.

$\mathcal{L}(\mu)$ is the family of all $L\mu$-measurable sets.

$L\mu(X) = \sup\{{}^\circ\mu(I) : I \subseteq X,\ I \text{ internal}\} \in [0, +\infty]$ is the *Loeb measure* of any $X \in \mathcal{L}(\mu)$. The Loeb measure $L\mu$ is *finite* iff $L\mu(H) < \infty$ — and this the case is if and only if the original $\mu(H)$ is a limited $*$-real. □

Theorem 9.5.3. *Under the above assumptions $\mathcal{L}(\mu)$ is a σ-algebra containing all Borel subsets of H, $L\mu$ is a σ-additive measure on $\mathcal{L}(\mu)$ satisfying $L\mu(X) = {}^\circ\mu(X)$ for all $X \in \mathscr{P}_{\mathtt{int}}(H)$, and a set $X \subseteq H$ is $L\mu$-measurable iff $X \cap D$ is $L\mu$-measurable (or, equivalently, $L\mu$-approximable) for any $D \in \mathscr{P}_{\mathtt{int}}(H)$ with ${}^\circ\mu(D) < \infty$.*

Proof (sketch). There are some pitfalls in the argument, mostly related to the case of infinite measures, regarding those we refer the reader to manuals on Loeb measures like [StrB 86, Cut 94, Ross 97, Gol 98, Loeb 00].

Yet the core of the problem is to prove that a countable union of null sets is a null set. Thus suppose that sets $Z_n \subseteq H$ $(n \in \mathbb{N})$ satisfy $L\mu(Z_n) = 0$ for all $n \in \mathbb{N}$. We have to prove that $Z = \bigcup_{n\in\mathbb{N}} Z_n$ also satisfies $L\mu(Z) = 0$. In other words, for any real $\varepsilon > 0$, $\varepsilon \in \mathbb{R}$, we have to find an internal set X with $Z \subseteq X \subseteq H$ and $\mu(X) \le \varepsilon$.

By definition for any $n \in \mathbb{N}$ there exists an internal set $X_n \subseteq H$ such that $Z_n \subseteq X_n$ and $\mu X_n < {}^*\varepsilon\, 2^{-n-1}$. Note that $\mathbb{N}$ is a set of standard size (like any other set in $\mathbb{WF}$). By the Standard Size Choice axiom of **HST**, the family of sets X_n can be chosen as a whole. Now it follows from Extension (see 9.1.22°) that there exists an internal function ξ defined on ${}^*\mathbb{N}$ and satisfying $\xi(n) = X_n$ for all $n \in \mathbb{N}$. Since ξ is internal, there exists an infinitely large number $N \in {}^*\mathbb{N} \smallsetminus \mathbb{N}$ such that $\xi(n) \subseteq H$ and $\mu(\xi(n)) < {}^*\varepsilon\, 2^{-n-1}$ for all $n < N$. It remains to define $X = \bigcup_{n<N} \xi(n)$. □

The collection $\mathcal{L}(\mu)$ of all Loeb–measurable sets $Z \subseteq H$ is *not* a set in **HST**; it is too big to be a set. (Indeed take any infinite internal $X \subseteq Z$ with infinitesimal counting measure and hence Loeb measure 0. Then any $Y \subseteq X$ is Loeb measurable as well, but $\mathscr{P}(Y)$ is not a set in **HST** by Theorem 1.3.9.) For most applications this does not really matter because by 9.5.4 Loeb–measurable sets are sufficiently approximable by Borel sets, and on the other hand **Borel**$[H]$ is a set by Theorem 9.3.9. Another plausible solution is to argue in a universe of the form $\mathbb{L}[\mathbb{I}_\kappa]$ for a sufficiently large cardinal κ, in the frameworks of the scheme "$\mathbb{WF} \xrightarrow{*} \mathbb{I}_\kappa$ [in $\mathbb{L}[\mathbb{I}_\kappa]$]" of §6.4h: $\mathcal{L}(\mu) \cap \mathbb{L}[\mathbb{I}_\kappa]$ is a set because the Power Set axiom holds in $\mathbb{L}[\mathbb{I}_\kappa]$.

Exercise 9.5.4. Prove that in the assumptions above if $\mu(H)$ is limited then for any Loeb measurable set $X \subseteq H$ there exist a $\mathbf{\Sigma}^0_1$ set A and a $\mathbf{\Pi}^0_1$ set B such that $A \subseteq X \subseteq B$ and $L\mu(A) = L\mu(B)$ (and hence $= L\mu(X)$). □

Example 9.5.5. (1) Suppose that $H = \{1, 2, 3, ..., h\}$, where $h \in {}^*\mathbb{N} \smallsetminus \mathbb{N}$. Put $\mu(X) = \frac{\#X}{h}$ (where $\#X \in {}^*\mathbb{N}$ is the number of elements of X in $\mathbb{I}$). Then $L\mu$ is called *the uniform probability Loeb measure on* H.

(2) Let H be as in (1) and $\mu(X) = \#X$ for any internal $X \subseteq H$. Then all sets $X \subseteq H$ are $\mathrm{L}\mu$-measurable, with $\mathrm{L}\mu(X) = \#X$ for X finite, and $\mathrm{L}\mu(X) = \infty$ otherwise. In particular $\mathrm{L}\mu(\{x\}) = 1$ for any singleton.

(3) Let $K = \{1, 2, 3, ..., k\}$, where $k \in {}^*\mathbb{N} \setminus \mathbb{N}$ is a fixed number, and $\mu(X) = \frac{\#X}{k}$ for any $X \subseteq H = K \times K$. Then $\mathrm{L}\mu(H) = \infty$, of course. We observe that 9.5.4 fails in this case. Indeed let $Y \subseteq K$ be any non-Borel $\mathbf{\Sigma}^1_1$ set (we refer to Theorem 9.3.9(v)(2)). Then $X = K \times Y$ is still a $\mathbf{\Sigma}^1_1$ set in H, and hence Loeb measurable (see § 9.5b). Suppose on the contrary that $B \subseteq H$ is a Borel set with $\mathrm{L}\mu(B \triangle X) = 0$. There is internal $U \subseteq H$ with $B \triangle X \subseteq U$ and $\mathrm{L}\mu(U) < 0.5$, that is $\#U < \frac{k}{2}$. As $\#K = k$, there is a number $1 \le j \le k$ such that U does not contain a pair of the form $\langle x, j \rangle$, $x \in K$. Then the cross-sections X_j and B_j coincide. However $X_j = Y$ is non-Borel while B_j is Borel together with B.
Note that $\mathrm{L}\mu$ is point-vanishing in this case: $\mathrm{L}\mu(\{x\}) = 0$ for any $x \in H$. For a measure $\mathrm{L}\mu$ as in (2), the approximation of the type defined in 9.5.4 is possible only for internal X because for $\mathrm{L}\mu(A \triangle B) = 0$ it is necessary that $A = B$. □

9.5b Loeb measurability of projective sets

It follows from Theorem 9.5.3 that all Borel sets are Loeb measurable. What about more complicated, say projective, sets? The question of measurability of projective sets has been substantially studied by Polish descriptive set theory. In particular it is well known that

1°. All $\mathbf{\Sigma}^1_1$ and all $\mathbf{\Pi}^1_1$ subsets of Polish spaces are measurable in the sense of any finite Borel measure.

(A *finite Borel measure* on a Polish space $\mathscr{X}$ is a σ-additive measure λ on $\mathscr{X}$ such that any Borel set $B \subseteq \mathscr{X}$ is measurable with $\lambda(X) < \infty$, and any $U \subseteq \mathscr{X}$ is measurable if and only if there exist Borel sets $B, B' \subseteq \mathscr{X}$ such that $B \subseteq U \subseteq B'$ and $\lambda(B) = \lambda(B')$ — and then $\lambda(U) = \lambda(B) = \lambda(B')$.)

2°. The measurability of sets in higher projective classes is undecidable in the following sense:

(a) it is consistent with the axioms of **ZFC** that there is a $\mathbf{\Delta}^1_2$ set of reals X non-measurable in the sense of the Lebesgue measure of $\mathbb{R}$;

(b) it is also consistent with **ZFC** that all projective subsets of Polish spaces are measurable in the sense of any finite Borel measure.

Let us employ shadow maps to infer analogous results for Loeb measures. We keep here the notation of § 9.5a.

Exercise 9.5.6. Assume, to be on the safe side, that $\mathrm{L}\mu$ is a finite Loeb measure. Let $\varphi : H \to 2^{\mathbb{N}}$ be a shadow map in the sense of 9.2.12. Define a measure λ on $2^{\mathbb{N}}$ as follows: $\lambda(U) = \mathrm{L}\mu(\psi^{-1}(U))$ for any $U \subseteq 2^{\mathbb{N}}$ such that $\mathrm{L}\mu(\psi^{-1}(U))$ is defined. Prove that then λ is a σ-additive Borel measure defined on the σ-algebra $\mathcal{L}(\lambda) = \{U : \psi^{-1}(U) \in \mathcal{L}(\mu)\}$. □

Corollary 9.5.7. *All sets in* $\mathbf{\Sigma}^1_1$ *and* $\mathbf{\Pi}^1_1$ *are Loeb measurable.*

Proof (we consider only the case of finite Loeb measures $\mathrm{L}\mu$). Let $X \subseteq H$ be a $\mathbf{\Sigma}^1_1$ set. In particular X is countably determined, that is $X = \Phi_U\{B_n\}_{n\in\mathbb{N}} = \varphi^{-1}(U)$, where $U \subseteq 2^{\mathbb{N}}$, $\mathscr{B} = \{B_n\}_{n\in\mathbb{N}}$ is a system of internal subsets of H, and $\varphi = \mathbf{sh}_{\mathscr{B}}$. We can assume that $U = \varphi"X$, and then U is $\mathbf{\Sigma}^1_1$ together with X by Theorem 9.3.3.

Define a Borel σ-additive measure λ on $2^{\mathbb{N}}$ as in 9.5.6. Then U, as a $\mathbf{\Sigma}^1_1$ set, is λ-measurable by 1°, and hence X is $\mathrm{L}\mu$-measurable by the result of 9.5.6, as required. □

There is another proof of this result, not related to any facts in Polish descriptive set theory. Recall that the class $\mathbf{\Sigma}^1_1$ coincides with the class $\mathsf{A}\cdot\mathscr{P}_{\mathbf{int}}(H)$ of all sets obtained by action of the Souslin operation A on internal subsets of H (Proposition 9.1.10). However it is known that, by a very general theorem of measure theory due to Marczewski [15], under certain conditions (definitely satisfied in this case) the class of all measurable sets is closed w. r. t. the action of A, as required.

Exercise 9.5.8. By analogy with 9.5.7, prove a suitable consequence of 2° for Loeb measurability of projective sets in internal domains. □

9.5c Approximations almost everywhere

It is a typical question of measure theory whether a given rather complicated object can be "approximated" by a rather simple object so that the domain where the two differ from each other is "small", for instance, is a null set in the sense of a certain measure. The following theorem is an example of this kind in "hyperfinite" descriptive set theory.

Theorem 9.5.9. *Let* H, K *be infinite internal sets, and* $P \subseteq H \times K$ *is a set in* $\mathbf{\Sigma}^1_1$ *or in* $\mathbf{\Pi}^1_1$ *such that all cross-sections* P_x *are internal. Suppose also that* $\mathrm{L}\mu$ *is a Loeb measure on* H *such that* $\mathrm{L}\mu(H)$ *is finite. Then there is an internal set* $Q \subseteq H \times K$ *such that* $\mathrm{L}\mu(\{x \in H : P_x \neq Q_x\}) = 0$.

Proof. We can assume that P is $\mathbf{\Pi}^1_1$; the result for $\mathbf{\Sigma}^1_1$ follows by taking complements. It follows from Theorem 9.4.10(ii) that $P = \bigcup_n P^n$, where $P^n = A^n \restriction E_n$, $A^n \subseteq H \times K$ are internal sets, and

$$E_n = \{x : A^n_x \subseteq P_x\} = \{x : \forall y\, (\langle x, y\rangle \in A^n \Longrightarrow \langle x, y\rangle \in P)\}$$

are $\mathbf{\Pi}^1_1$ subsets of H since P is $\mathbf{\Pi}^1_1$, and hence $\mathrm{L}\mu$-measurable by Theorem 9.5.7. Fix a real $\varepsilon > 0$. Then for any $n \in \mathbb{N}$ there exists an internal

[15] See, for instance, [KurM 76, 7a in XII.8], [Cohn 80, 8.4.1], or [Rog 70, Thm. 26].

set $X_n(\varepsilon) \subseteq E_n$ with $\mathrm{L}\mu(R_n(\varepsilon)) < \varepsilon\, 2^{-n}$, where $R_n(\varepsilon) = E_n \smallsetminus X_n(\varepsilon)$. Thus we still have $P_x = \bigcup_n Q^n(\varepsilon)_x$ for all $x \in H(\varepsilon) = H \smallsetminus R(\varepsilon)$, where $R(\varepsilon) = \bigcup_{n\in\mathbb{N}} R_n(\varepsilon)$ and each $Q^n(\varepsilon) = A^n \restriction X_n(\varepsilon)$ is an internal set.

Thus by Saturation for any $x \in H(\varepsilon)$ there is a number $n(x) \in \mathbb{N}$ such that already $P_x = \bigcup_{n<n(x)} Q^n(\varepsilon)_x$. However, as all $Q^n(\varepsilon)$ are internal, the sets $C_m = \{x \in H(\varepsilon) : P_x = \bigcup_{n<m} Q^n(\varepsilon)_x\}$ are Borel, and $H(\varepsilon) = \bigcup_{m\in\mathbb{N}} C_m$ by the above. It follows that there is an internal set $Y(\varepsilon) \subseteq H(\varepsilon)$ and a number $m(\varepsilon) \in \mathbb{N}$ such that $\mathrm{L}\mu(H(\varepsilon) \smallsetminus Y(\varepsilon)) < \varepsilon$ and $Y(\varepsilon) \subseteq C_{m(\varepsilon)}$. Then $Q(\varepsilon) = \bigcup_{n<m(\varepsilon)} (A^n \restriction (X_n(\varepsilon) \cap Y(\varepsilon)))$ is an internal subset of P satisfying $P_x = Q(\varepsilon)_x$ for all $x \in Y(\varepsilon)$. We observe that $\mathrm{L}\mu(H \smallsetminus Y(\varepsilon)) \le 2\varepsilon$.

Taking $\varepsilon = \frac{1}{n}$, $n \in \mathbb{N}$, in this construction, we obtain an increasing sequence of internal sets $Y_n = Y(\frac{1}{n}) \subseteq H$ and an increasing sequence of internal sets $Q^n = Q(\frac{1}{n}) \subseteq P$ such that $\mathrm{L}\mu(H \smallsetminus Y_n) \le \frac{1}{n}$, $\mathtt{dom}\, Q^n \subseteq Y_n$, and $P_x = Q^n_x$ for all $x \in Y_n$, and hence $Q^{n+1}_x = Q^n_x$ for all $x \in Y_n$.

By Extension (see 9.1.2), the sequences $\{Q^n\}_{n\in\mathbb{N}}$ and $\{Y_n\}_{n\in\mathbb{N}}$ admit internal extensions $\{Q^n\}_{n\in{}^*\mathbb{N}}$ and $\{Y_n\}_{n\in{}^*\mathbb{N}}$. Then there is a number $h \in {}^*\mathbb{N} \smallsetminus \mathbb{N}$ such that the reduced sequences $\{Q^n\}_{n\le h}$ and $\{Y_n\}_{n\le h}$ are still increasing sequences of internal subsets of resp. $H \times K$ and H such that for any $n \le h$ the following holds: $\mathrm{L}\mu(H \smallsetminus Y_n) \le \frac{1}{n}$, $\mathtt{dom}\, Q^n \subseteq Y_n$, and $Q^{n+1}_x = Q^n_x$ for all $x \in Y_n$. It follows that $\mathrm{L}\mu(H \smallsetminus Y_h) \simeq 0$ and we have $Y_n \subseteq Y_h$, $Q^h_x = Q^n_x = P_x$ for all $n \in \mathbb{N}$ and $x \in Y_n$. Thus $Q^h_x = P_x$ for all $x \in Y = \bigcup_{n\in\mathbb{N}} Y_n$, therefore the set $Q = Q^h$ is as required. □

Corollary 9.5.10. *If $P : H \to K$ is a $\boldsymbol{\Pi}^1_1$ or $\boldsymbol{\Sigma}^1_1$ function* [16]*, partial or total, then there is a total internal function $F : H \to K$ such that the set $D = \{x \in \mathtt{dom}\, P : P(x) \ne F(x)\}$ satisfies $\mathrm{L}\mu(D) = 0$.* □

A set Q as in the theorem is called *an internal* $\mathrm{L}\mu$*-lifting* of P. The notion of $\boldsymbol{\Sigma}^0_1$*-lifting* (of a set with $\boldsymbol{\Sigma}^0_1$ cross-sections), as well as of $\boldsymbol{\Gamma}$-lifting for any other class $\boldsymbol{\Gamma}$, can be defined similarly.

Exercise 9.5.11. Prove, following the proof of the theorem, that any $\boldsymbol{\Pi}^1_1$ set with $\boldsymbol{\Sigma}^0_1$ cross-sections admits a $\boldsymbol{\Sigma}^0_1$ lifting. □

Problem 9.5.12 ("large sections"). In Polish descriptive set theory, any Borel set in a product of two Polish spaces, all of whose non-empty cross-sections have non-zero measure (different version: are non-meager) is Borel-uniformizable and hence its projection is Borel too (see [Kech 95, § 18B]).

Are there any reasonable analogies in "hyperfinite" descriptive set theory? For instance, if $P \subseteq H \times K$ is a Borel set and $\mathrm{L}\mu(P_x) > 0$ for every *non-empty* cross-section P_x, where $\mathrm{L}\mu$ is a fixed Loeb measure on K, does it follow that P is Borel-uniformizable (and then $\mathtt{dom}\, P$ is Borel, too)?

This question may be more interesting and difficult in the case of category since in general there exist difficulties with the Baire category notions in "hyperfinite" descriptive set theory, see [KL 91]. □

[16] We identify any function with its graph.

9.5d Randomness in a hyperfinite domain

It sometimes happens in mathematics that an intuitive notion cannot be easily formalized so that both the spirit and the letter is kept. The notion of a random object (for instance a random real) is among those notions. The approach determined by classical probability theory simply dismisses as nonsense the concept of a *single* random real.

Different attempts were made to introduce an adequate definition of a single random real, mostly in the frameworks of recursion theory. Their common denominator is as follows: a real x is defined to be random if no infinite amount of information about x is available. A similar notion is known in set theory: a real is called random, or Solovay–random, over a given model $\mathfrak{M}$ of **ZFC** if it avoids any Borel set of measure zero coded in $\mathfrak{M}$.

We attempt here to give a reasonable notion of randomness in **HST**. Our approach has some semblance of the Solovay randomness, but we employ the standard universe $\mathbb{S}$, and universes of the form $\mathbb{S}[w]$, $w \in \mathbb{I}$, in the role of a ground model in the Solovay randomness. Recall that $\mathbb{S}[w]$ consists of sets of the form $f(w)$, where f is a standard function such that $w \in \mathtt{dom}\, f$, and sets in $\mathbb{S}[w]$ are called *w-standard*.

Lemma 9.5.13. *Let Y, w be internal sets. Then the set $Y' = Y \cap \mathbb{S}[w]$ of all w-standard elements $y \in Y$ is a set of standard size.*

Proof. There exist standard sets W and S such that $Y \subseteq S$ and $w \in W$. Let F be the standard set of all internal functions $f : W \to S$. Each $y \in Y'$ has the form $f(w)$ for some $f \in F \cap \mathbb{S}$, which is a set of standard size. □

Definition 9.5.14. Let w be an internal set. Say that a $*$-real x is

w-infinitely large iff $x \geq c$ for some w-standard infinitely large $c > 0$;
w-infinitesimal iff $|x| \leq \varepsilon$ for some w-standard infinitesimal ε. [17]

Suppose that H is a hyperfinite internal set and $\mu : \mathscr{P}_{\mathtt{int}}(H) \to {}^*[0,1]$ an internal $*$-finitely additive probability measure on H. An element $x \in H$ is *w-random w. r. t. μ* iff x does not belong to any $\langle H, w\rangle$-standard set $X \subseteq H$ with H-infinitesimal value $\mu(X)$. □

(The postfix "w. r. t. μ" can be omitted if this does not lead to ambiguity.)

The following lemma shows that, in agreement with intuition, non-random elements form a scattered family.

Lemma 9.5.15. *In the assumptions of 9.5.14 the collection $\mathscr{R}^{\mathsf{C}}$ of all $x \in H$ non-w-random w. r. t. μ can be covered by an internal set $X \subseteq H$ with infinitesimal $\mu(X)$, and hence $\mathrm{L}\mu(\mathscr{R}^{\mathsf{C}}) = 0$.*

[17] The definition of w-infinitely large and w-infinitesimal reals makes sense iff there really exist w-standard infinitesimals and infinitely large numbers. In particular it does not make sense (and will not be used) in the case when w is standard.

It is not asserted here that $\mu(X)$ is e.g. H-infinitesimal. We cannot claim that $\mu(\mathscr{R}^{\mathsf{C}})$ itself is infinitesimal because μ is defined only for internal subsets of H while $\mathscr{R}$ is, generally speaking, external.

Proof. It follows from Lemma 9.5.13 that the collection $\mathscr{I}$ of all $\langle H, w\rangle$-standard sets $I \subseteq H$, such that $\mu(I)$ is H-infinitesimal, is a standard size subset of the internal power set $P = \mathscr{P}_{\mathtt{int}}(H)$ (which is a hyperfinite set). By Saturation, there is an infinitesimal ε bigger than all numbers $\mu(I)$, where $I \in \mathscr{I}$. Also by Saturation, there is an internal set $\mathscr{J} \subseteq P$ containing $\le 1/\sqrt{\varepsilon}$ elements and satisfying $\mathscr{I} \subseteq \mathscr{J}$. We can assume that $\mu(J) < \varepsilon$ for all $J \in \mathscr{J}$ (otherwise $\mathscr{J}$ can be accordingly restricted). Then the internal set $X = \bigcup \mathscr{J}$ satisfies $\mu(X) < \sqrt{\varepsilon}$. On the other hand $\mathscr{R}^{\mathsf{C}} \subseteq X$. □

In principle it can be required, in the lemma, that in addition $\mu(X) < \delta$, where δ is an arbitrary but fixed $*$-real (perhaps, infinitesimal) bigger than all h-infinitesimals.

Example 9.5.16. (1) Let $H = {}^{\mathtt{int}}2^h$ (the internal set of all internal maps $s : h \to 2$). Obviously $\#H = 2^h$ in $\mathbb{I}$, so we can define the counting measure η on H by $\eta(X) = \#(X)\, 2^{-h}$ for all internal $X \subseteq H$. Sequences $s \in H$ random w. r. t. η can be called *uniformly random*.

(2) Consider the set $\mathscr{B} = \{0, 1, ..., h\}$, with the *Bernoulli measure* β, defined on singletons by $\beta(\{k\}) = 2^{-h}\binom{h}{k}$. Numbers $k \in \{0, ..., h\}$ random w. r. t. β can be called *Bernoulli random*. □

Exercise 9.5.17. Let $s \in H$ be uniformly w-random. Prove that $k(s) = \#\{n : s(n) = 1\} \in \mathscr{B}$ is a Bernoulli w-random number. □

Theorem 9.5.18 (Fubini). *Suppose that w is an internal set, H, K are infinite $*$-finite sets, while μ, ν are internal finitely additive probability measures on resp. H, K, and ν is $\langle w, H, K\rangle$-standard. Then*

(i) *if $\langle x, y\rangle \in H \times K$ is w-random in $H \times K$ w. r. t. $\mu \times \nu$ then x is w-random in H w. r. t. μ while y is $\langle w, x\rangle$-random in K w. r. t. ν ;*

(ii) *if K is H-standard, H is K-standard, $x \in H$ is w-random in H w. r. t. μ, and $y \in K$ is $\langle w, x\rangle$-random in K w. r. t. ν, then $\langle x, y\rangle$ is w-random in $H \times K$ w. r. t. $\mu \times \nu$.*

Proof. (i) Let $X \subseteq H$ be a $\langle w, H\rangle$-standard set of measure $\mu(X) < \varepsilon$, where ε is H-infinitesimal. Assume on the contrary that $x \in X$. Then $\langle x, y\rangle \in P$, where $P = X \times K$ is $\langle w, H, K\rangle$-standard and satisfies $(\mu \times \nu)(P) < \varepsilon$, which is a contradiction. (Note, in passing by, that to be $\langle w, H, K\rangle$-standard and to be $\langle w, H \times K\rangle$-standard is one and the same.)

Let $Y \subseteq K$ be a $\langle w, x, K\rangle$-standard set of measure $\nu(Y) < \varepsilon$, where ε is K-infinitesimal. Suppose on the contrary that $y \in Y$. By definition we have $Y = f(w, x, K)$, where f is a standard function. Let P be the set of all pairs

$\langle x', y' \rangle \in H \times K$ such that $Y_{x'} = f(w, x', K)$ is a subset of K satisfying the inequality $\nu(Y_{x'}) < \varepsilon$, and $y' \in Y_{x'}$. Note that P is $\langle w, H, K \rangle$-standard by the assumptions above, and $(\mu \times \nu)(P) \leq \varepsilon$. On the other hand, $\langle x, y \rangle \in P$ by definition, which is a contradiction.

(ii) Consider a $\langle w, H, K \rangle$-standard set $P \subseteq H \times K$ with $(\mu \times \nu)(P) < \varepsilon$, where ε is a $\langle H, K \rangle$-infinitesimal; hence H-infinitesimal because K is H-standard. Put $P_{x'} = \{y \in K : \langle x', y \rangle \in P\}$ for any $x' \in H$. Under our assumptions the set $X = \{x' \in H : \nu(P_{x'}) \geq \sqrt{\varepsilon}\}$ is $\langle w, H \rangle$-standard, and $\mu(X) \leq \sqrt{\varepsilon}$ because $(\mu \times \nu)(P) < \varepsilon$. (A "discrete" version of Fubini theorem is applied.) Therefore $x \notin X$ by the randomness of x. Thus the $\langle w, H, x \rangle$-standard (therefore $\langle w, K, x \rangle$-standard) set $Y = P_x$ satisfies $\nu(Y) < \sqrt{\varepsilon}$. However $y \in Y$, which contradicts the randomness of y. □

Corollary 9.5.19 (Steinitz Exchange). *Under the assumptions of Theorem 9.5.18 suppose that K is H-standard, H is K-standard, and both μ and ν are $\langle w, H, K \rangle$-standard. Then, if $x \in H$ is w-random w. r. t. μ and $y \in K$ is $\langle w, x \rangle$-random w. r. t. ν then x is $\langle w, y \rangle$-random w. r. t. μ.* □

9.5e Law of Large Numbers

In classical probability theory, this is a common name for several important theorems saying that under some conditions the arithmetic mean $\frac{\xi_1 + \cdots + \xi_n}{n}$ of jointly independent random variables ξ_i is close to the arithmetic mean of their expectations $\frac{m_1 + \cdots + m_n}{n}$. (See [Sin 93], Section 12.)

Our goal here is to obtain a "hyperfinite" version, based on the notion of randomnes introduced in § 9.5d.

Suppose that μ an internal finitely additive probability measure on a $*$-finite set H, as above. Assume, in addition, that $H \subseteq {}^*\mathbb{R}$. We define

$$\begin{aligned} \mathbf{E}\,\mu &= \textstyle\sum_{x \in H} x\,\mu(\{x\}), && \text{the } \textit{expectation} \text{ of } \mu; \\ \mathbf{Var}\,\mu &= \textstyle\sum_{x \in H} (x - \mathbf{E}\,\mu)^2\,\mu(\{x\}), && \text{the } \textit{variance} \text{ of } \mu. \end{aligned}$$

Note that the expectation and variance are functions of the measure (= the probability distribution) rather than of random elements as we defined them.

Suppose that $h \in {}^*\mathbb{N} \setminus \mathbb{N}$, and for any $n = 1, 2, \ldots h$, $H_n \subseteq {}^*\mathbb{R}$ is a $*$-finite set and μ_n is an internal finitely additive probability measure on H_n, such that the maps $n \mapsto H_n$ and $n \mapsto \mu_n$ are internal. We put $m_n = \mathbf{E}\,\mu_n$ and $v_n = \mathbf{Var}\,\mu_n$ for all n. Define $H = \prod_{n=1}^{h} H_n$ (the product consists of all internal functions f defined on $\{1, 2, \ldots h\}$ so that $f(n) \in H_n$, $\forall n$) and let $\mu = \prod_{n=1}^{h} \mu_n$ be the internal product probability measure on H.

Theorem 9.5.20 (Hyperfinite Law of Large Numbers). *Under these assumptions, if $v = h^{-1} \sum_{n=1}^{h} v_n$ is a limited number and the measure μ is H-standard then, for any sequence $x = \{x_n\}_{n=1}^{h}$, random (i. e. 0-random) in H w. r. t. μ, the following difference is infinitesimal:*

$$\Delta(x) = \frac{x_1 + \cdots + x_h}{h} - \frac{m_1 + \cdots + m_h}{h}.$$

Proof. By Kolmogorov's inequality (see e.g. [Sin 93], Theorem 12.2), applied in the internal universe, we have

$$\mu(\{y \in H : \Delta(y) \geq s\}) \leq \frac{v}{hs^2}.$$

for any $s > 0$. By the assumption, vs^{-2} is a limited number whenever $s > 0$ is standard. Thus the set $X_s = \{y \in H : \Delta(y) \geq s\}$ has an h-infinitesimal measure $\mu(X_s)$ whenever $s > 0$ is standard. On the other hand, if s is standard then X_s is $\langle H, \mu \rangle$-standard, and hence H-standard because μ is assumed to be H-standard. We conclude, by definition, that $x \notin X_s$ for any standard $s > 0$, as required. □

9.5f Random sequences and hyperfinite gambling

There exists another idea of randomness. One may view a binary infinite sequence $a \in 2^\omega$ as random if a human cannot win an unlimited amount of money in gambling against a. In **HST**, this idea can be realized by a certain game of $*$-finite length.

Fix a number $h \in {}^*\mathbb{N} \setminus \mathbb{N}$. The set $\mathscr{S} = {}^{\text{int}}\{-1, 1\}^h$ of all internal sequences of the form $\alpha = \langle a_0, a_2, ..., a_{h-1} \rangle$, where each a_i is -1 or 1, is an internal $*$-finite set of $*$-cardinality $\#\mathscr{S} = 2^h$ in $\mathbb{I}$. Every set $A \subseteq \mathscr{S}$, not necessarily internal, defines a game $G(A)$ between two players, the `Gambler` and the `Casino`, which proceeds in $\mathbb{I}$, the internal universe, as follows.

`Gambler` has, at the beginning, an initial amount of money, $B_0 = \$1$.

A run in this game consists of h steps. At each step $n = 0, 1, 2, ..., h-1$:

1) `Gambler` bets an amount of money b_n, a $*$-real satisfying $|b_n| \leq B_n$, as to the result of `Casino`'s forthcoming move $a_n \in \{-1, 1\}$;
2) `Casino` observes b_n and moves $a_n = -1$ or 1;
3) `Gambler`'s next balance B_{n+1} is computed by $B_{n+1} = B_n + b_n a_n$. (In other words, say $b_n = -0.75$ means that `Gambler` bets $\$0.75$ on the move $a_n = -1$. If `Casino` in fact plays $a_n = -1$ then `Gambler` wins $\$0.75$ at this step, otherwise `Gambler` loses this amount.)

This results in an internal sequence $\alpha = \alpha_h = \langle a_0, ..., a_{h-1} \rangle \in \mathscr{S}$ of `Casino`'s moves, and the final `Gambler`'s balance B_h, a nonnegative $*$-real. The `Casino`'s goal in this game is to produce a sequence $\alpha \in \mathscr{S}$ which belongs to A; the `Gambler`'s aim is, by betting money, either to force `Casino` to produce $\alpha \notin A$ or to gain a large enough amount of money if `Casino` is willing to reach A by all means. Who wins the game in the case $\alpha \in A$ depends on a definition of what is the "large enough" final balance B_h to determine `Gambler`'s win. See Definition 9.5.22 below on possible version.

Thus the role of the set A in the game is the following: Casino *must play* so that $\alpha_h \in A$ in order not to lose independently of the balance score. Therefore Gambler can exploit the unability of Casino to play in an absolutely free way, and make reasonable predictions aiming at increasing the balance.

It is intuitively clear that the larger A is the easier Casino's task should be, and the other way around for Gambler. In quantitative terms, this is expressed by the following result of [KanR 96a] given here without a full proof.

Put $\eta(X) = \#X\, 2^{-h}$ for any internal set $X \subseteq \mathscr{S}$; this is a *counting measure* on $\mathscr{S}$. Let $\mathrm{L}\eta$ denote the corresponding Loeb measure.

Theorem 9.5.21. (i) *$A \subseteq \mathscr{S}$ is a set of Loeb measure $\mathrm{L}\eta(A) = 0$ if and only if* Gambler *has an internal strategy* [18] *in $G(A)$ which guarantees that B_h is infinitely large whichever way* Casino *plays.*

(ii) *Let r be a positive $*$-real. Then A has an internal superset of counting measure r^{-1} if and only if* Gambler *has an internal strategy in $G(A)$ which guarantees $B_h \geq r$ in $G(A)$.*

(iii) *Let r be a positive $*$-real. Then A has an internal subset of counting measure r^{-1} if and only if* Casino *has an internal strategy in $G(A)$ which guarantees $B_h \leq r$ in $G(A)$.* □

Proof (sketch). (i) Assume, for the sake of simplicity, that A is internal. Let $\varepsilon = \eta(A)$. For any $t \in {}^{\mathrm{int}}\{-1,1\}^n$ $(n \leq h)$ put $\mathscr{S}_t = \{\alpha \in \mathscr{S} : t \subset \alpha\}$ and $d_t = \eta(\mathscr{S}_t \cap A)\, 2^{-n}$, the *density* of A on $\mathscr{S}_t$. Thus for instance $d_\Lambda = \varepsilon$. (Λ is the empty sequence.) Suppose that $n < h$ and $t = \langle a_0, \ldots, a_{n-1}\rangle$ is the sequence of Casino's n initial moves. Obviously $d_{t^\wedge -1} = d_t - r$ and $d_{t^\wedge 1} = d_t + r$ for some (positive or negative) $*$-real $r = r_t$, $|r_t| \leq d_t$. Gambler's optimal strategy is to bet $b_n = B_n \frac{r_t}{d_t}$, so that $\frac{B_{n+1}}{d_{t^\wedge a_n}} = \frac{B_n}{d_t}$ independently of the ensuing Casino's move a_n. Playing this way, Gambler has $\frac{B_h}{d_\alpha} = \frac{B_0}{d_0} = \varepsilon^{-1}$ in the end of the run. That is, if Casino has played $\alpha \in A$ (and otherwise Casino has lost) then the final density d_α is equal to 1, and hence B_h, the final balance, is equal to ε^{-1}.

In other words Gambler has a strategy that guarantees $B_h \geq \varepsilon^{-1}$. This proves $\Longrightarrow$ in (i). To prove $\Longleftarrow$ note that by similar reasons Casino has a strategy providing $B_n \leq \varepsilon^{-1}$. □

If e.g. $A = \{\alpha\}$ contains only one sequence $\alpha \in \mathscr{S}$ then all Casino's moves are forced, and Gambler can upgrade the initial balance $B_0 = 1$ to $B_h = 2^h$. Yet to realize this plan, Gambler needs to know α.

[18] A *strategy* for Gambler is a "rule" σ which tells them how to play, for each n, their nth move $b_n = \sigma(a_0, \ldots, a_{n-1})$, given Casino's previous moves $a_0, \ldots, a_{n-1}$. Technically this can be defined as an **internal** function σ mapping binary sequences into nonnegative $*$-reals. The notion of a strategy for Casino is defined similarly, see [KanR 96a]. Henson [Hen 92] proposed a wider category of strategies, not necessarily internal themselves but rather those which preserve the internality of the result provided the opponent plays internally.

Definition 9.5.22. Say that `Gambler` *wins the game* $G(A)$ iff either $\alpha \notin A$ or the final balance B_h is h-infinitely large (that is, B_h is bigger than some infinitely large h-standard $*$-real).

Let w be an internal set. A sequence $\alpha \in \mathscr{S}$ is *w-invincible* iff `Gambler` does not have a $\langle h, w\rangle$-standard winning strategy in $G(\{\alpha\})$. □

Theorem 9.5.23. *Let w be an internal set. A sequence $\alpha \in \mathscr{S}$ is w-invincible if and only if it is w-random w. r. t. η.*

Proof. Assume that α is not w-random, so that $\alpha \in A$, where $A \subseteq \mathscr{S}$ is a $\langle h, w\rangle$-standard set such that the measure $\varepsilon = \eta(A)$ is an h-infinitesimal $*$-real. Now `Gambler` simply forgets α and starts gambling in $G(A)$. The set A is bigger than $\{\alpha\}$, of course, but it is still rather small, and it is for sure $\langle h, w\rangle$-standard, unlike α. `Gambler`'s strategy described in the proof of Theorem 9.5.21(i) is A-standard and guarantees the final balance $B_h \geq \varepsilon^{-1}$ in $G(A)$. But ε^{-1} is h-infinitely large by the choice of ε. In other words `Gambler` has a $\langle h, w\rangle$-standard winning strategy in $G(\{\alpha\})$.

Suppose that α is not w-invincible, so that `Gambler` has a $\langle h, w\rangle$-standard winning strategy σ in $G(\{\alpha\})$. Let, for each $\alpha' \in \mathscr{S}$, $B(\alpha')$ denote the final balance in the run where `Gambler` plays following σ while `Casino` plays the sequence α'. Then $B(\alpha)$ is h-infinitely large, and hence $B(\alpha) > b$, where b is an infinitely large h-standard $*$-integer. Thus α belongs to the $\langle h, w\rangle$-standard set $A = \{\alpha' \in \mathscr{S} : B(\alpha') > b\} \subseteq \mathscr{S}$. Finally, $\eta(A) < b^{-1}$ because $\sum_{\alpha' \in \mathscr{S}} B(\alpha') = 2^h$. It follows that $\eta(A)$ is h-infinitesimal, as required. □

Recall that a game is *determined* if one of the players has a winning strategy. For instance for any $*$-real $r \geq 0$, if $A \subseteq \mathscr{S}$ is a Lη-measurable set then the game $G(A)_r$, that is $G(A)$ specified so that `Gambler` wins whenever $\alpha_H \notin A$ or $B_h \geq r$, is determined by Theorem 9.5.21. In Polish descriptive set theory any Borel game of length $\mathbb{N}$ is determined (see e.g. [Kech 95, 20.C]). 'Borel' here means that the set $A \subseteq \mathbb{N}^{\mathbb{N}}$, which defines the result in the sense that player I wins iff the final sequence $\alpha = \{a_n\}_{n \in \mathbb{N}}$ of moves belongs to A. But in "hyperfinite" descriptive set theory such a Borel determinacy badly fails.

Exercise 9.5.24. Fix $h \in {}^*\mathbb{N} \setminus \mathbb{N}$. Any set $A \subseteq H = {}^{\text{int}}2^{2h}$ (internal functions $\{0, 1, ..., 2h-1\} \to 2$) defines a game G_A in which player I makes moves $a_0, a_2, ..., a_{2h-2} \in \{0, 1\}$, player II makes moves $a_1, a_3, ..., a_{2h-1} \in \{0, 1\}$, and I wins iff the sequence $\{a_k\}_{k<2h}$ belongs to A.

(1) Prove that G_A is determined provided $A \subseteq H$ is a set in $\mathbf{\Sigma}^0_1 \cup \mathbf{\Pi}^0_1$.

(2) Fix a number $n \in {}^*\mathbb{N} \setminus \mathbb{N}$, $n < h$. Let A be the set of all $\alpha = \{a_k\}_{k<2h} \in H$ such that there is $j \in \mathbb{N}$ such either $a_{n-j} = 0$ for all $j \in \mathbb{N}$ or there is an **odd** number $j \in \mathbb{N}$ such that $a_{2n} = a_{2n-1} = ... = a_{2n-j} \neq a_{2n-j-1}$. Prove that neither of the players has an internal winning strategy in G_A.

Hint. See [Cut 84] on (1) and [Hen 92] on (2). □

9.6 Borel and countably determined cardinalities

The cantorian notion of cardinal has been the *sine qua non* of modern mathematical foundations since more than a century ago. However a variety of phenomena, first of all in modern descriptive set theory and related areas, clearly demonstrate that it does not fully reflect all sides and angles of a quantitative measure of infinite sets in mathematics. Studies in 1990s revealed the importance of the notion of *Borel cardinality*, arising when only Borel maps are allowed to participate in the comparison of two pointsets. In this sense, it was found that there are strictly more Vitali equivalence classes than reals themselves, and even more countable sets of the reals, and a multitude of other inequalities of this kind — in the case when the cantorian cardinality of all sets considered is the continuum $\mathfrak{c} = 2^{\aleph_0}$ by very transparent reasons.

The goal of this section is to study Borel and countably determined cardinalities of Borel and countably determined subsets of ${}^*\mathbb{N}$. Even the Borel case here is highly different from the "Polish" case: while all uncountable Borel sets in Polish spaces are Borel bijective, even bounded Borel subsets of ${}^*\mathbb{N}$ present a quite nontrivial structure of Borel cardinalities.

9.6a Preliminaries

We begin with the following definition.

Definition 9.6.1. A *Borel* (resp., *countably determined*) *map* is a function whose graph is a Borel (resp., countably determined) set.

Suppose that H is an infinite internal set and $X, Y \subseteq H$.

(1) Let $X \leq_{\mathtt{B}} Y$ mean that there is a Borel injective map ϑ with $X \subseteq \operatorname{dom} \vartheta$ and $\vartheta''X \subseteq Y$. (If X is Borel then just $\vartheta : X \to Y$.)

(2) Accordingly, let $X \equiv_{\mathtt{B}} Y$ mean that both $X \leq_{\mathtt{B}} Y$ and $Y \leq_{\mathtt{B}} X$, and $X <_{\mathtt{B}} Y$ will mean that $X \leq_{\mathtt{B}} Y$ but not $Y \leq_{\mathtt{B}} X$.

(3) Changing "Borel" (map) to "countably determined" in this definition, we obtain the relations $\leq_{\mathtt{CD}}$, $\equiv_{\mathtt{CD}}$, $<_{\mathtt{CD}}$. □

Lemma 9.6.2. *Let X, Y be Borel sets. Then*

(i) *$X \equiv_{\mathtt{B}} Y$ iff there is a Borel bijection of X onto Y.*

(ii) *$X \equiv_{\mathtt{CD}} Y$ iff there is a countably determined bijection of X onto Y.*

Proof. Apply the Schröder – Bernstein argument. To see that it yields a Borel bijection in the Borel case note that the image $\operatorname{ran} \vartheta$ of a Borel injection ϑ is equal to $\operatorname{dom}(\vartheta^{-1})$, hence, is still a Borel set by Theorem 9.4.7(iii). As for the countably determined case, this class is generally closed under projections, images, preimages and the like by Theorem 9.3.9. □

Thus, the relation $X \equiv_{\mathsf{B}} Y$ can be interpreted as saying that the sets X, Y have the same *Borel cardinality*; the latter then can be defined as the $\equiv_{\mathsf{B}}$-class of X. Accordingly, the relation $X \equiv_{\mathsf{CD}} Y$ is interpreted as having the same *CD* [19] *cardinality*.

Exercise 9.6.3. Say that a set $X \subseteq {}^*\mathbb{N}$ is *bounded* if there is $c \in {}^*\mathbb{N}$ such that $X \subseteq [0, c)$. Prove that any unbounded countably determined set $X \subseteq {}^*\mathbb{N}$ contains an internal unbounded subset. (Apply the methods used in the proofs of 9.1.5 and 9.2.7.) Infer that any unbounded countably determined set $X \subseteq {}^*\mathbb{N}$ satisfies $X \equiv_{\mathsf{CD}} {}^*\mathbb{N}$, and even $X \equiv_{\mathsf{B}} {}^*\mathbb{N}$ whenever it is Borel. □

This result allows us to restrict attention to <u>bounded</u> countably determined and Borel subsets of ${}^*\mathbb{N}$.

The next theorem describes Borel and CD cardinalities of the most elementary subsets of ${}^*\mathbb{N}$ — those of the form $[0, x)$, $x \in {}^*\mathbb{N}$. Note that any $x \in {}^*\mathbb{N}$ is equal to the set $[0, x) = \{z : z < x\}$ (an interval, and hence bounded subset of ${}^*\mathbb{N}$), thus $x \equiv_{\mathsf{B}} y$ is the same as $[0, x) \equiv_{\mathsf{B}} [0, y)$, and $x \equiv_{\mathsf{CD}} y$ is the same as $[0, x) \equiv_{\mathsf{CD}} [0, y)$.

Recall that ${}^\circ x \in \mathbb{R}$ is the shadow of $x \in {}^*\mathbb{R}$.

Theorem 9.6.4. *Let* $x, y \in {}^*\mathbb{N} \setminus \mathbb{N}$. *Then*

(i) $x \equiv_{\mathsf{B}} y$ *iff* ${}^\circ(\frac{x}{y}) = 1$; (ii) $x \equiv_{\mathsf{CD}} y$ *iff* $0 < {}^\circ(\frac{x}{y}) < \infty$.

Equivalence (i) obviously holds also in the case when one or both of x, y belong to $\mathbb{N}$, but $\Longleftarrow$ in (ii) fails whenever $x \neq y \in \mathbb{N}$.

Proof. (i) If $x < y$ and $x \simeq y$ then the map

$$f(z) = \begin{cases} z\,, & \text{provided } \frac{y-x}{x} \text{ is infinitesimal} \\ z + y - x\,, & \text{otherwise} \end{cases}$$

is a bijection of $[0, y)$ onto $[0, x)$, thus we have $\Longleftarrow$. To prove the implication $\Longrightarrow$ assume towards the contrary that $x < y$, $\frac{y}{x} \not\simeq 1$, but $G : [0, y) \to [0, x)$ is a Borel injection. It follows from Theorem 9.4.7(vii) that there is a partition $[0, y) = \bigcup_{n \in \mathbb{N}} Y_n$ into Borel sets Y_n, and also a sequence of <u>internal</u> maps $F_n : [0, y) \to [0, x)$ such that $F_n \restriction Y_n = G \restriction Y_n$ for all n. Let $\eta(X) = \frac{\# X}{y}$ be the counting probability measure on $[0, y)$ and $\mathrm{L}\eta$ the associated Loeb measure. Thus $\mathrm{L}\eta([0, y)) = \eta([0, y)) = 1$ while $\mathrm{L}\eta([0, x)) = d < 1$. Then there is at least one n such that $a = \mathrm{L}\eta(X_n) < b = \mathrm{L}\eta(Y_n)$ where $X_n = F_n{}''Y_n$ (also a Borel set). Put $\varepsilon = b - a$. There exist internal sets $A \subseteq Y_n$ and B, $X_n \subseteq B \subseteq [0, x)$ such that $\mathrm{L}\eta(A) > b - \frac{1}{3\varepsilon}$ and $\mathrm{L}\eta(B) < a + \frac{1}{3\varepsilon} < \mathrm{L}\eta(A)$, and hence F injectively maps an internal set A into an internal set B of a strictly smaller number of elements, which is impossible.

[19] Recall that CD is a routine shortcut for "countably determined".

(ii) The implication $\Longleftarrow$ will be proved below (Lemma 9.6.12(ii)). Let us concentrate on $\Longrightarrow$ [20]. Thus suppose, towards the contrary, that $x < y$ and $G : [0, x) \overset{\text{onto}}{\longrightarrow} [0, y)$ is a countably determined map, and $y > nx$ for all $n \in {}^*\mathbb{N}$. Then $y > hx$ also for some $h \in {}^*\mathbb{N} \setminus \mathbb{N}$. Theorem 9.4.7(i) implies that there is a countable sequence of internal functions $F_n : [0, x) \to [0, y)$ such that $G(r) \in \{F_n(r) : n \in \mathbb{N}\}$ for all $r < x$. Let, by Extension, $\{F_n\}_{n \in {}^*\mathbb{N}}$ be any internal extension of this sequence. There is $\nu \in {}^*\mathbb{N} \setminus \mathbb{N}$, $\nu < h$, such that F_n is a map $[0, x) \to [0, y)$ for all $n < \nu$. Then $Y = \{F_n(r) : n < \nu \wedge r < x\}$ contains $\le \nu x < hx < y$ elements, but $[0, y) \subseteq Y$, contradiction. □

Exercise 9.6.5. Prove, using the theorem, that $\equiv_{\mathsf{B}}$ and $\equiv_{\mathsf{CD}}$ are Borel equivalence relations on ${}^*\mathbb{N}$. □

9.6b Borel cardinals and cuts

The following theorem describes the structure of Borel cardinalities of Borel sets $X \subseteq {}^*\mathbb{N}$, in terms of *cuts*, that is initial segments in ${}^*\mathbb{N}$.

Theorem 9.6.6. *For any Borel set $X \subseteq {}^*\mathbb{N}$ there is a Borel cut $U \subseteq {}^*\mathbb{N}$ with $X \equiv_{\mathsf{B}} U$, actually, there is a minimal Borel cut U satisfying $X \equiv_{\mathsf{B}} U$.*

The proof follows below in this section. We begin here with some preparatory materials. Thus, *cuts* are initial segments of ${}^*\mathbb{N}$ (including $\varnothing, \mathbb{N}, {}^*\mathbb{N}$). A cut U is *additive* if $x + y \in U$ whenever $x, y \in U$. Given a cut U, the sets

$$U\mathbb{N} = \bigcup\nolimits_{n \in \mathbb{N},\, x \in U} [0, xn] \quad \text{and} \quad U/\mathbb{N} = \bigcap\nolimits_{n \in \mathbb{N},\, x \in U} [0, \mathcal{E}(\tfrac{x}{n})] ,$$

where $\mathcal{E}(r) \in {}^*\mathbb{N}$ is the entire part of r, are additive cuts, obviously the least additive cut including U and the largest additive cut included in U. In particular, the cuts $c/\mathbb{N} = [0, c)/\mathbb{N}$ and $c\mathbb{N} = [0, c)\,\mathbb{N}$ are defined for any $c \in {}^*\mathbb{N}$.

Internal cuts are $\varnothing$, ${}^*\mathbb{N}$, and those of the form $c = [0, c)$, $c \in {}^*\mathbb{N}$. Some non-internal cuts can be obtained with the following general procedure. If $\{a_n\}_{n \in \mathbb{N}}$ is a strictly increasing, resp., decreasing sequence in ${}^*\mathbb{N}$ then define a *countably cofinal* cut $\bigcup_n [0, a_n)$, resp., *countably coinitial* cut $\bigcap_n [0, a_n)$. Both types consist of Borel sets of classes resp. $\mathbf{\Sigma}^0_1$ and $\mathbf{\Pi}^0_1$. It occurs that these types exhaust all countably determined (including Borel) cuts in ${}^*\mathbb{N}$:

Lemma 9.6.7. *Any countably determined cut $\varnothing \ne U \subsetneqq {}^*\mathbb{N}$ it either countably cofinal or countably coinitial or contains a maximal element (and then is internal).*

Proof. [21] Let $U = \bigcup_{f \in F} \bigcap_{m \in \mathbb{N}} X_{f \restriction m}$, where F and the sets X_s are as in (2) of § 9.2b. Put $\operatorname{cut} X = \bigcup_{x \in X} [0, x]$ for any set $X \subseteq {}^*\mathbb{N}$, the least cut

[20] Compare the implication $\Longrightarrow$ here with Theorem 1.4.9.

[21] The result is a corollary of Theorem 1.4.6(i), of course, but we are interested to give a proof using only countable Saturation.

which includes X. By (countable) Saturation, $U = \mathsf{cut}\, U = \bigcup_{f \in F} \bigcap_m U_{f \restriction m}$, where $U_s = \mathsf{cut}\, X_s$, hence, $U_s = [0, \mu_s]$, where $\mu_s = \max X_s \in {}^*\mathbb{N}$ for all $s \in 2^{<\omega}$. If there is $f \in F$ with $U = \bigcap_m U_{f \restriction m}$ then the sequence $\{h_{f \restriction m}\}_{m \in \mathbb{N}}$ witnesses that U is countably coinitial, or contains a maximal element if the sequence is eventually constant. Otherwise, by Saturation, for any $f \in F$ there is $m_f \in \mathbb{N}$ such that $h_{f \restriction m_f} \in U$. Let $S = \{f \restriction m_f : f \in F\}$; this is a countable set and easily $U = \bigcap_{s \in S} [0, \mu_s]$, so that U is countably cofinal. □

Lemma 9.6.8. *Suppose that A_n, B_n are $*$-finite internal sets, and $b_n = \#B_n \le a_n = \#A_n$ for each n. Then*

(i) *if $A_{n+1} \subseteq A_n$ and $B_{n+1} \subseteq B_n$ for each n then $\bigcap_n B_n \le_{\mathsf{B}} \bigcap_n A_n$;*

(ii) *if $A_n \subseteq A_{n+1}$ and $B_n \subseteq B_{n+1}$ for each n then $\bigcup_n B_n \le_{\mathsf{B}} \bigcup_n A_n$.*

Thus $\le_{\mathsf{B}}$ is sometimes preserved under unions and intersections!

Proof. (i) For any n there is an internal bijection $f : A_0 \xrightarrow{\text{onto}} [0, a_0)$ such that $f"A_k = [0, a_k)$ for all $k \le n$. By Saturation, there is an internal bijection $f : A_0 \xrightarrow{\text{onto}} [0, a_0)$ with $f"A_n = [0, a_n)$ for all $n \in \mathbb{N}$. We conclude that $\bigcap_n A_n \equiv_{\mathsf{B}} A = \bigcap_n a_n$. Also, $\bigcap_n B_n \equiv_{\mathsf{B}} B = \bigcap_n b_n$. However $B \subseteq A$.

(ii) Arguing the same way, we prove that $\bigcup_n A_n \equiv_{\mathsf{B}} A$ and $\bigcup_n B_n \equiv_{\mathsf{B}} B$, where $B = \bigcup_n b_n \subseteq A = \bigcup_n a_n$. □

If $U \subseteq V \subseteq {}^*\mathbb{N}$ are cuts then we write $U \approx V$ iff $\frac{x}{y} \simeq 1$ for all $x, y \in V \setminus U$. (For instance if $U = [0, a)$ and $V = [0, b)$ then $U \approx V$ iff $\frac{a}{b} \simeq 1$.) This turns out to be a necessary and sufficient condition for $U \equiv_{\mathsf{B}} V$.

Lemma 9.6.9. (i) *If U, V are Borel cuts then $U \equiv_{\mathsf{B}} V$ iff $U \approx V$.*

(ii) *Any $\approx$-class of Borel cuts contains a $\subseteq$-minimal cut.*

(iii) *Any <u>additive</u> Borel cut is $\approx$-isolated, i.e., $U \not\approx V$ for any cut $V \ne U$.*

Proof. (i) Let, say, $U \subseteq V$. Suppose that $U \equiv_{\mathsf{B}} V$. Take any $x < y$ in $V \setminus U$. Then $x \equiv_{\mathsf{B}} y$, hence, $\frac{x}{y} \simeq 1$ by Theorem 9.6.4. To prove the converse suppose that $U \approx V$. Take any $c \in V \setminus U$. Let $A = \{a \in {}^*\mathbb{N} : \frac{a}{c} \simeq 0\}$. We observe that $A \subsetneqq U$ (indeed, the entire part of $\frac{c}{2}$ belongs to $U \setminus A$). Put $X^+ = \{c + a : a \in A\}$ and $X^- = \{c - a : a \in A\}$. Obviously $U' = [0, c) \setminus X^- \subseteq U$ and $V \subseteq V' = [0, c) \cup X^+$, and hence it suffices to define a Borel bijection $U' \xrightarrow{\text{onto}} V'$. Let $Z = [0, c) \setminus A$. Then $U' = Z \cup A$ and $V' = Z \cup A \cup X^+ \cup X^-$, where the unions are pairwise disjoint. **Exercise**: prove that the map

$$F(z) = \begin{cases} z\,, & \text{whenever } z \in Z \\ x\,, & \text{whenever } z = 3x \in A \\ c - x\,, & \text{whenever } z = 3x + 1 \in A \\ c + x\,, & \text{whenever } z = 3x + 2 \in A \end{cases}$$

is a Borel bijection $U' \xrightarrow{\text{onto}} V'$.

(ii) Let $\widetilde{U}$ be the set of all $x \in U$ such that there is $y \in U$, $y > x$ with $\frac{x}{y} \not\simeq 1$. This is a cut, moreover, a projective set, hence, countably determined, which implies that $\widetilde{U}$ is actually Borel by Lemma 9.6.7. Moreover $\widetilde{U} \approx U$. Finally, note that for any $x \in \widetilde{U}$ there exists $x' \in \widetilde{U}$, $x' > x$, with $\frac{x'}{x} \not\simeq 1$: indeed, let $x' = \frac{x+y}{2}$, where $y \in U$, $y > x$, $\frac{y}{x} \not\simeq 1$. This suffices to infer that $V \not\approx \widetilde{U}$ for any cut $V \subsetneqq \widetilde{U}$. In other words, $\widetilde{U}$ is the $\subseteq$-least cut $\equiv_{\mathsf{B}}$-equivalent to U, as required.

(iii) That $\widetilde{U} = U$ for any <u>additive</u> cut U is a simple **exercise**. □

9.6c Proof of the theorem on Borel cardinalities

Here we prove Theorem 9.6.6. Lemma 9.6.9 allows us to concentrate on the first assertion of the theorem.

Since all Borel sets are countably determined, we can present a given Borel set $X \subseteq {}^*\mathbb{N}$ in the form $X = \bigcup_{f \in F} \bigcap_n X_{f \restriction n}$, where F and the internal sets $X_s \subseteq {}^*\mathbb{N}$ are as in (2) of § 9.2b. In accordance with 9.6.3, we can assume that X is bounded in ${}^*\mathbb{N}$ — then it can be assumed that all sets X_s are also bounded, and hence $*$-finite. Let $\nu_s = \#X_s$.

Let C be the set of all $c \in {}^*\mathbb{N}$ such that there is $f \in F$ and an internal injection $\varphi : [0, c) \to X_f = \bigcap_n X_{f \restriction n}$. Then C is a cut and a countably determined set. (By Saturation, for any internal Y to be internally embeddable in X_f it suffices that $\#Y \le \nu_{f \restriction m}$ for any m.)

We claim that $C \le_{\mathsf{B}} X$. Indeed if there is $f \in F$ such that $C \subseteq [0, \nu_{f \restriction n})$ for all n then immediately $C \le_{\mathsf{B}} X_f$ by Lemma 9.6.8(i). Otherwise for any $f \in F$ there is $n_f \in \mathbb{N}$ such that $\nu_{f \restriction n_f} \in C$. As $X_{f \restriction n_f}$ is an internal set with $\#X_{f \restriction n_f} = \nu_{f \restriction n_f}$, no internal set Y with $\#Y > \nu_{f \restriction n_f}$ admits an internal injection in X_f. Thus the countable set $\{\nu_{f \restriction n_f} : f \in F\}$ is cofinal in C, and hence $C = \bigcup_k z_k$, where all z_k belong to C. However for any k there is an internal $R_k \subseteq X$ with $\#R_k = z_k$. Lemma 9.6.8(ii) implies $C \le_{\mathsf{B}} \bigcup_k R_k$.

In continuation of the proof of the theorem, we have the following cases.

Case 1: C is not additive. Then there is $c \in C$ such that $c\mathbb{N} = U$ and $2c \notin C$. Prove that $X \le_{\mathsf{B}} c\mathbb{N}$. By Lemma 9.6.8(ii), it suffices to cover X by a countable union $\bigcup_j Y_j$ of internal sets Y_j with $\#Y_j \le 2c$ for all j. For this it suffices to prove that for any $f \in F$ there is m such that $\nu_{f \restriction m} = \#X_{f \restriction m} \le 2c$. To prove this, assume, on the contrary, that $f \in F$ and $\nu_{f \restriction m} \ge 2c$ for all m; we obtain, by Saturation, an internal subset $Y \subseteq X_f$ with $\#Y = 2c \notin C$, contradiction. We return to this case below.

In the remainder, we assume that C is additive.

Case 2: C is countably cofinal. Arguing as in Case 1, we find that for any $f \in F$ there is m such that $\nu_{f \restriction m} = \#X_{f \restriction m} \in C$. (Otherwise, using Saturation and the assumption of countable cofinality, we obtain an internal subset $Y \subseteq X_f$ with $\#Y \notin C$, contradiction.) Thus, X can be covered by a countable union $\bigcup_j Y_j$ of internal sets Y_j with $\#Y_j \in C$ for all j. It follows, by Lemma 9.6.8(ii), that $X \le_{\mathrm{B}} C$. Since $C \le_{\mathrm{B}} X$ has been established, we have $X \equiv_{\mathrm{B}} C$, so that $U = C$ proves the theorem.

Case 3: C is countably coinitial, and there exists a decreasing sequence $\{h_k\}_{k\in\mathbb{N}}$, coinitial in ${}^*\mathbb{N}\setminus U$, such that $\frac{h_k}{h_{k-1}}$ is infinitesimal for all $k \in \mathbb{N}$. For any $k \in \mathbb{N}$, if $f \in F$ then there is m with $\nu_{f\restriction m} \le h_{k+1}$ (otherwise, by Saturation, X_f contains an internal subset Y with $\#Y > h_{k+1}$, contradiction), so that X is covered by a countable union of internal sets Y_j with $\#Y_j \le h_{k+1}$ for all j. It follows, by Saturation and because $\frac{h_k}{h_{k-1}}$ is infinitesimal, that, for any k, X can be covered by an internal set R_k with $\#R_k \le h_k$. Now $X \le_{\mathrm{B}} C$ by Lemma 9.6.8(i), hence, $U = C$ proves the theorem.

Case 4: finally, $C = c/\mathbb{N}$ for some $c \notin U$. We have $c/\mathbb{N} \le_{\mathrm{B}} X \le_{\mathrm{B}} c\mathbb{N}$ (similarly to Case 2).

We finish the proof. Cases 2 and 3 led us directly to the result required, while cases 1 and 4 can be summarized as follows: there is a number $c \in {}^*\mathbb{N}\setminus\mathbb{N}$ such that $c/\mathbb{N} \le_{\mathrm{B}} X \le_{\mathrm{B}} c\mathbb{N}$. We can assume that $X \subseteq c\mathbb{N}$.

Let $\eta_n(Y) = \frac{\#Y}{c}$ be the counting measure on the interval $[nc, nc+c)$ for any $n \in \mathbb{N}$, and $\mathrm{L}\eta_n$ the corresponding Loeb measure. For any $Z \subseteq c\mathbb{N}$ such that $Z_n = Z \cap [nc, nc+c)$ is $\mathrm{L}\eta_n$-measurable for all $n \in \mathbb{N}$, put $\mathrm{L}\eta(Z) = \sum_{n\in\mathbb{N}} \mathrm{L}\eta_n(Z_n)$. The set X is Borel, thus $\mathrm{L}\eta(X)$ is defined. If $\mathrm{L}\eta(X) = \infty$ then there is a sequence $\{X_n\}$ of internal subsets of X with $\#X_n = nc$, $\forall\, n$. It follows that $c\mathbb{N} \le_{\mathrm{B}} X$ by Lemma 9.6.8, hence, $X \equiv_{\mathrm{B}} U = c\mathbb{N}$, as required.

Suppose that $\mathrm{L}\eta(X) = r < \infty$. There is an increasing sequence $\{A_n\}_{n\in\mathbb{N}}$ of internal subsets of X and a decreasing sequence $\{B_n\}_{n\in\mathbb{N}}$ of supersets of X such that $\eta(B_n) - \eta(A_n) \to 0$ as $n \to \infty$ (*i.e.*, the difference is eventually less than any fixed real $\varepsilon > 0$). If $r = 0$ then $\frac{\#B_n}{c} \to 0$, thus $\bigcap_n B_n \equiv_{\mathrm{B}} c/\mathbb{N}$ by Lemma 9.6.8, which implies $X \equiv_{\mathrm{B}} c/\mathbb{N}$ since $c/\mathbb{N} \le_{\mathrm{B}} X$, therefore, $U = c/\mathbb{N}$ proves the theorem.

Finally, assume that $r > 0$. Prove that then $X \equiv_{\mathrm{B}} \mathcal{E}(cr)$ (the entire part of cr). We have $\frac{\#A_n}{c} \to r$ from below and $\frac{\#B_n}{c} \to r$ from above. Let $U = \bigcup_{n\in\mathbb{N}} \#A_n$ and $V = \bigcap_{n\in\mathbb{N}} \#B_n$; then $\bigcup_n A_n \equiv_{\mathrm{B}} U$ and $\bigcap_n B_n \equiv_{\mathrm{B}} V$ by Lemma 9.6.8, while $\mathcal{E}(cr) \in V \setminus U$, hence, in remains to prove that $U \equiv_{\mathrm{B}} V$. By Lemma 9.6.9, it suffices to show that $U \approx V$. Let $x < y$ belong to $V \setminus U$. If $\frac{y}{x} \not\simeq 1$ then $\frac{y}{c} - \frac{x}{c}$ is not infinitesimal, which contradicts the fact that $\frac{\#B_n}{c} - \frac{\#A_n}{c} \to 0$ because $\frac{\#A_n}{c} \le \frac{x}{c}$ and $\frac{y}{c} \le \frac{\#B_n}{c}$ for all n.

□ (*Theorem 9.6.6*)

Corollary 9.6.10. (i) *Any two Borel sets* $X, Y \subseteq {}^*\mathbb{N}$ *are* $\le_{\mathrm{B}}$*-comparable.*

(ii) *If* $c \in {}^*\mathbb{N}\setminus\mathbb{N}$ *and* $X, Y \subseteq c\mathbb{N}$ *are Borel sets of non-*0 *measure* $\mathrm{L}\eta$ (see the proof of the theorem) *then* $X \equiv_{\mathrm{B}} Y$ *iff* $\mathrm{L}\eta(X) = \mathrm{L}\eta(Y)$.

Proof. (ii) See the last paragraph of the proof of the theorem. □

9.6d Complete classification of Borel cardinalities

Call a Borel cut $U \subseteq {}^*\mathbb{N}$ *minimal* if $V \not\equiv_{\mathrm{B}} U$ for any cut $V \subsetneqq U$. It follows from Theorem 9.6.6 that any $\equiv_{\mathrm{B}}$-class of Borel subsets of ${}^*\mathbb{N}$ contains a unique minimal Borel cut, so that minimal Borel cuts can be viewed as **Borel cardinals** (of Borel subsets of ${}^*\mathbb{N}$).

For instance, any additive Borel cut is minimal by Lemma 9.6.9, hence such a cut is a Borel cardinal. But if U is a non-additive minimal Borel cut, then there is a number $c \in U$ with $2c \notin U$, so that $c/\mathbb{N} \subsetneqq U \subsetneqq c\mathbb{N}$, and, accordingly, $c/\mathbb{N} <_{\mathrm{B}} U <_{\mathrm{B}} c\mathbb{N}$, because $c/\mathbb{N}$ and $c\mathbb{N}$ are minimal cuts themselves. (Note that $c\mathbb{N}$ is the least attitive cut bigger than $c/\mathbb{N}$.)

To study the structure of minimal Borel cuts between $c/\mathbb{N}$ and $c\mathbb{N}$ for a fixed nonstandard $c \in {}^*\mathbb{N}$, put $y_{cr} = \mathcal{E}(cr)$ for any real $r \in \mathbb{R}$, $r > 0$, where, we recall, $\mathcal{E}(\cdot)$ is the entire part in the internal universe. Then $\widetilde{y_{cr}}$ is the least Borel cut $\equiv_{\mathrm{B}}$-equivalent to $y_{cr} = [0, y_{cr})$, see the proof of Lemma 9.6.9(ii).

Exercise 9.6.11. Prove that any minimal Borel cut U satisfying $c/\mathbb{N} <_{\mathrm{B}} U <_{\mathrm{B}} c\mathbb{N}$ is equal to $\widetilde{y_{cr}}$ for a positive real r. Prove that $\widetilde{y_{cr}} \neq \widetilde{y_{cr'}}$ for different r, r' (and one and the same c). □

Thus, Borel cardinals of Borel subsets of ${}^*\mathbb{N}$ are either additive Borel cuts or those of the form $\widetilde{y_{cr}}$, or, finally, (finite) natural numbers.

9.6e Countably determined cardinalities

It can be expected that different Borel cardinalities are identified by countably determined maps. The following lemma shows the exact measure of this phenomenon.

Lemma 9.6.12. (i) *If $U \subseteq {}^*\mathbb{N}$ is an infinite Borel cut then $U \equiv_{\mathrm{CD}} \mathbb{N} \times U$ (the Cartesian product) and $U \equiv_{\mathrm{CD}} U\mathbb{N}$ (a cut).*

(ii) *If $x, y \in {}^*\mathbb{N} \setminus \mathbb{N}$, $n \in \mathbb{N}$, and $y \le nx$, then $y \le_{\mathrm{CD}} x$.* [22]

(iii) *On the other hand, if $U \subsetneqq V$ are Borel cuts, and U is additive, then there is no CD map $\varphi : U$ onto V.*

Note that Claim (ii) is sufficient for the implication $\Longleftarrow$ in (ii) of Theorem 9.6.4. Indeed $0 < {}^\circ(\frac{x}{y}) < \infty$ means that there is $n \in \mathbb{N}$ such that both $x \le ny$ and $y \le nx$.

Proof. (i) Theorem 9.7.10 below implies that there exists a CD set $W \subseteq U$ such that for any $x \in U$ there is a unique $w_x \in W$ with $|x - w_x| \in \mathbb{N}$. Let $a \mapsto \langle z_a, n_a \rangle$ be a recursive bijection of $\mathbb{Z}$ (the integers) onto $\mathbb{Z} \times \mathbb{N}$. Now, if $x \in U \setminus \mathbb{N}$ then put $a = x - w_x$ and $\vartheta(x) = \langle w_x + z_a, n_a \rangle$. If $x = m \in \mathbb{N}$ then let $\Phi(x) = \langle i_m, j_m \rangle$, where $m \mapsto \langle i_m, j_m \rangle$ is a fixed bijection of $\mathbb{N}$ onto

[22] Recall that any $z \in {}^*\mathbb{N}$ is identified with the set $[0, z) = \{h \in {}^*\mathbb{N} : h < z\}$.

$\mathbb{N} \times \mathbb{N}$. Also, if U has a maximal element μ and $x = \mu - m$, $m \in \mathbb{N}$, then let $\vartheta(x) = \langle \mu - j_m, i_m \rangle$. Then ϑ is a countably determined bijection of U onto $U \times \mathbb{N}$.

In the second equivalence, if U is an additive cut then $U = U\mathbb{N}$ and there is nothing to prove. Otherwise there is $c \in U$ such that $U\mathbb{N} = c\mathbb{N}$. Note that $U\mathbb{N} = \bigcup_{n \in \mathbb{N}} U_n$, where $U_n = cn + U$, hence, there is a Borel bijection of $U \times \mathbb{N}$ onto $U\mathbb{N}$.

(ii) It suffices to consider $U = [0, x)$ in (i).

(iii) Let, on the contrary, $P = \bigcup_{f \in F} \bigcap_m P_{f \restriction m}$ be such a map ($P_s \subseteq {}^*\mathbb{N} \times {}^*\mathbb{N}$ are internal sets and $P_s \subseteq P_t$ whenever $t \subset s$.) Then any $P_f = \bigcap_m P_{f \restriction m}$ is still a function, hence, by **Saturation**, there is a number m_f such that $P_{f \restriction m_f}$ is a function. Thus, there is a countable family of *internal* functions Φ_i, $i \in \mathbb{N}$, with $U \subseteq \operatorname{dom} \Phi_i$, such that $V \subseteq \bigcup_i \Phi_i{}''U$. We can assume that $V = [0, c)$, where $c \in {}^*\mathbb{N} \smallsetminus U$. Put $c_0 = c$ and, by induction, let c_{n+1} be the entire part of $c_n/2$. Then still $c_n \notin U$ for any n as U is an additive cut, therefore, $V \subseteq \bigcup_i \Phi_i{}''[0, c_{i+2}]$. Yet every $V_i = \Phi_i{}''[0, c_{i+2})$ is an internal set with $\#V_i \le c/2^{i+2}$, hence, by **Saturation**, $\bigcup_i V_i$ can be covered by an internal set with $c/2$ elements, and cannot cover V. □

Thus, for any $c \in {}^*\mathbb{N} \smallsetminus \mathbb{N}$, all Borel cardinals (as defined in § 9.6d) between $c/\mathbb{N}$ and $c\mathbb{N}$ are $\equiv_{\mathrm{CD}}$-equivalent to each other and to $c\mathbb{N}$. It follows that for any Borel set $X \subseteq {}^*\mathbb{N}$ there is a unique *additive* Borel cut U with $X \equiv_{\mathrm{CD}} U$, so that we can define *CD-cardinals* of Borel sets to be just additive Borel cuts in ${}^*\mathbb{N}$. What about CD-cardinalities of countably determined sets? The next theorem gives a partial answer.

Theorem 9.6.13. *If $X \subseteq {}^*\mathbb{N}$ is an infinite countably determined set then either there is a unique additive Borel cut $U \equiv_{\mathrm{CD}} X$ or there is $c \in {}^*\mathbb{N} \smallsetminus \mathbb{N}$ such that $c/\mathbb{N} <_{\mathrm{CD}} X <_{\mathrm{CD}} c\mathbb{N}$.*

Proof. We leave it as an **exercise** for the reader to verify that the arguments in § 9.6c are partially applicable to any countably determined, not necessarily Borel, set $X \subseteq {}^*\mathbb{N}$. More exactly. If X is unbounded in ${}^*\mathbb{N}$ then $X \equiv_{\mathrm{CD}} {}^*\mathbb{N}$. If X is bounded in ${}^*\mathbb{N}$ then either X is $\equiv_{\mathrm{CD}}$-equivalent to an additive Borel cut (cases 2 and 3) or there is a number $c \in {}^*\mathbb{N} \smallsetminus \mathbb{N}$ with $c/\mathbb{N} <_{\mathrm{CD}} X <_{\mathrm{CD}} c\mathbb{N}$ (cases 1 and 4). The Loeb measurability of Borel sets enables us to further study the "or" case in § 9.6c provided X is a Borel set, but the method does not seem to apply for CD sets in general. □

While the "either" case is realized in simple examples (for instance, additive Borel cuts themselves), the "or" case needs further study.

Suppose that $c \in {}^*\mathbb{N} \smallsetminus \mathbb{N}$. Then ${}^{\mathrm{int}}2^c$ is the (internal) set of all internal functions $x : c \to 2$, and $\#({}^{\mathrm{int}}2^c) = 2^c$. Put $\varphi_c(x) = x \restriction \mathbb{N}$ for any $x \in {}^{\mathrm{int}}2^c$. Then $\varphi_c : {}^{\mathrm{int}}2^c \to 2^{\mathbb{N}}$ is a shadow map in the sense of 9.2.12. We put $M_U^c = \varphi_c^{-1}(U)$ for any $U \subseteq 2^{\mathbb{N}}$.

Theorem 9.6.14. *Suppose that* $h = 2^c \in {}^*\mathbb{N} \setminus \mathbb{N}$. *If* $X \subseteq h\mathbb{N}$ *is a countably determined set then either* $X <_{\mathtt{CD}} h/\mathbb{N}$ *or* $X \equiv_{\mathtt{CD}} M^c_U$ *for some* $U \subseteq 2^{\mathbb{N}}$. □

This theorem (see [KanR 03] for a proof) brings some order into CD cardinalities of countably determined subsets of ${}^*\mathbb{N}$. In particular it implies that there exist at most $\mathfrak{c}$-many different CD cardinalities between $h/\mathbb{N}$ and $h\mathbb{N}$. The following is the main open problem:

Problem 9.6.15. Find a reasonable necessary and sufficient condition for $M^c_U \leq_{\mathtt{CD}} M^c_V$ to hold for any two sets $U, V \subseteq 2^{\mathbb{N}}$. Does the relation $M^c_U \leq_{\mathtt{CD}} M^c_V$ for a given pair of sets U, V depend on the choice of c? □

It follows from a theorem of Čuda and Vopenka [CV 79, p. 651] (see also Theorems 1.5 and 4.14 in [KalZ 89a]) that, under the assumption of **CH**, there exists a pair of sets $U, V \subseteq 2^{\mathbb{N}}$ such that the sets M^c_U and M^c_V are $\leq_{\mathtt{CD}}$-incomparable. Beside of this result little is known, even regarding the order $\leq_{\mathtt{B}}$ on sets of the form M^c_U, or even a much simpler order $M^c_U \leq_{\mathtt{int}} M^c_V$ iff there is an internal map $F : {}^{\mathtt{int}}2^c \to {}^{\mathtt{int}}2^c$ such that $F \upharpoonright M^c_U$ is an injection of M^c_U into M^c_V.

Note that the relation $U \leq^c V$ iff $M^c_U \leq_{\mathtt{CD}} M^c_V$ is a binary relation on the "standard" power set $\mathscr{P}(2^{\mathbb{N}})$ (let us identify here $\mathbb{WF}$ with $\mathbb{S}$), that is a "standard" object itself. Does it depend on the choice of $c \in {}^*\mathbb{N} \setminus \mathbb{N}$? The same question can be addressed to the Borel and internal variations.

Exercise 9.6.16. Fix $c \in {}^*\mathbb{N} \setminus \mathbb{N}$. Prove that if $U \subseteq 2^{\mathbb{N}}$ is Borel (as a subset of $2^{\mathbb{N}}$) then M^c_U is Borel in ${}^{\mathtt{int}}2^c$, and if U is measurable in the sense of the natural σ-additive probability measure on $2^{\mathbb{N}}$ then M^c_U is Loeb measurable in the sense of $\mathrm{L}\eta$, where η is the internal counting measure on ${}^{\mathtt{int}}2^c$.

Show that any $\mathrm{L}\eta$-measurable countably determined set $X \subseteq {}^{\mathtt{int}}2^c$ is $\equiv_{\mathtt{CD}}$-equivalent to a Borel cut in ${}^*\mathbb{N}$. □

9.7 Equivalence relations and quotients

While the cardinality problems for Borel and Souslin sets in Polish spaces were solved in the early era of Polish descriptive set theory, the cardinality problems for Borel quotient structures, *i.e.*, quotients over Borel equivalence relations became the focal point of studies in descriptive set theory only in the early 1990s — especially in the form of *Borel reducibility* of the quotients. It is far too early to mention anything comparable in "hyperfinite" descriptive set theory, yet the obvious importance of the topic forces us to take some space to present some of the few results already obtained. Those with an experience in Polish descriptive set theory may be interested to recognize similarities and differences with the set-up they are accustomed to.

9.7a Silver's theorem for countably determined relations

Aleksandrov and Hausdorff independently proved in 1916 that any uncountable Borel set (in a Polish space) contains a perfect subset — and then obviously has the cardinality of the continuum $\mathfrak{c} = 2^{\aleph_0}$. It had taken few decades until Silver proved in mid-70s a similar result for Borel quotients: if E is a Borel, or even $\mathbf{\Pi}^1_1$ equivalence relation (ER, for brevity) on a Polish space $\mathscr{X}$ then either the quotient $\mathscr{X}/\mathsf{E}$ is at most countable or there is a perfect set of pairwise E-inequivalent elements.

The following theorem can be considered as a counterpart of the Silver theorem for countably determined quotients, in the same sense as Lemma 9.2.7 is a counterpart of the Aleksandrov – Hausdorff theorem.

Theorem 9.7.1. *If E is a countably determined equivalence relation on a countably determined set X then exactly one of the following holds:*

(I) *there is a shadow* (in the sense of Definition 9.2.12) $\varphi : H \to 2^{\mathbb{N}}$ *such that $\varphi(x) = \varphi(y)$ implies $x \,\mathsf{E}\, y$ for all $x, y \in H$ — and then the quotient X/E has cardinality $\le \mathfrak{c}$;*

(II) *there is an infinite internal subset of X of pairwise E-inequivalent elements.*

Equivalence relations of type (I) can be viewed as those with a small number of equivalence classes, while ERs of type (II) as those with a large number of equivalence classes. The incompatibility of (I) and (II) means that this is a sound dichotomy.

Say that an equivalence relation E is *thin* if there does not exist a pairwise E-inequivalent infinite internal set. Then ERs of type (II) are non-thin. Moreover it follows from the incompatibility of (I) and (II) that equivalence relations of type (I) are thin.

Example 9.7.2. Note that $\mathfrak{c}$ cannot be improved in (I) to any smaller cardinality. Indeed take any $z \in {}^*\mathbb{N} \setminus \mathbb{N}$ and put $H = {}^{\mathtt{int}}2^z$. Define, for $x, y \in H$, $x \,\mathsf{E}\, y$ iff $\varphi(x) = \varphi(y)$, where $\varphi(x) = x \restriction \mathbb{N}$. Then E is a countably determined, even $\mathbf{\Pi}^0_1$ equivalence relation, φ witnesses (I), and $\mathtt{card}\, H/\mathsf{E} = \mathfrak{c}$. □

Proof (theorem). [23] Let H be an internal set. Consider a countably determined equivalence relation E on a countably determined set $X \subseteq H$.

[23] The proof is, generally speaking, based on the arguments in the proof of Theorem 1.4.11, but is not really a copy because the setting is somewhat different: any operation of the form of § 9.2a consists of two operations, one of which can be uncountable – but we have committed ourselves to use not more than countable Saturation. However it is a special property of operations (over countable index sets) that they are applied to *countable* families of sets (which can be combined in uncountably many ways) — this is why countable Saturation is sufficient to prove many deep results on countably determined sets despite the in principle uncountable character of operations.

We can assume that $X = H$, otherwise extend E so that $x \mathrel{\mathsf{E}} y$ for all $x, y \in H \smallsetminus X$. It follows from Lemma 9.2.6 that E can be presented in the form $\mathsf{E} = \bigcup_{f \in F} \bigcap_{m \in \mathbb{N}} P_{f \restriction m}$, where $F \subseteq 2^{\mathbb{N}}$, P_u $(u \in 2^{<\omega})$ are internal subsets of H and $P_v \subseteq P_u$ whenever $u \subset v$. We can w. l. o. g. assume that the sets P_u are symmetric, that is $P_u = {P_u}^{-1}$, where ${P_u}^{-1} = \{\langle y, x\rangle : x \mathrel{P_u} y\}$. Indeed otherwise we can "symmetrize" the sets by changing each P_u to ${P_u}^{-1} \cup P_u$. (We systematically write $x \mathrel{P} y$ instead of $\langle x, y\rangle \in P$ whenever P is a subset of $H \times H$, that is, a binary relation on H.)

It follows from transitivity of E that, for any $x, y \in H$,

$$\exists f \in F \,\exists z \in H \,\forall m \,(x \mathrel{P_{f \restriction m}} z \wedge y \mathrel{P_{f \restriction m}} z) \Longrightarrow x \mathrel{\mathsf{E}} y\,.$$

(All quantifiers over m and k in the course of the proof are restricted to the domain $\mathbb{N}$.) Therefore, by countable Saturation and the assumption that $P_v \subseteq P_u$ whenever $u \subset v$, we have

$$\exists f \in F \,\forall m \,\exists z \in H \,(x \mathrel{P_{f \restriction m}} z \wedge y \mathrel{P_{f \restriction m}} z) \Longrightarrow x \mathrel{\mathsf{E}} y\,.$$

Since the two leftmost quantifiers are over sets in $\mathbb{WF}$ (F and $\mathbb{N}$), the displayed formula is equivalent to

$$\forall T \in A(F) \,\exists u \in T \,\exists z \in H \,(x \mathrel{P_u} z \wedge y \mathrel{P_u} z) \Longrightarrow x \mathrel{\mathsf{E}} y\,, \qquad (*)$$

where $A(F)$ is the family of all sets $T \subseteq 2^{<\omega}$ intersecting any set of the form $\{f \restriction m : m \in \mathbb{N}\}$, $f \in F$. (And this holds for arbitrary $x, y \in H$.)

Suppose now that there is no infinite internal pairwise E-inequivalent set $Y \subseteq H$. The assumption can formally be expressed as

$$\forall Y \in \mathscr{P}_{\mathbf{int}}(H) \,\big(\forall k \,(\#Y > k) \Longrightarrow \exists x \neq y \in Y \,\exists f \in F \,\forall m \,(x \mathrel{P_{f \restriction m}} y)\big).$$

The subformula to the right of $\Longrightarrow$ can be consecutively transformed (using countable Saturation and the assumption that $P_v \subseteq P_u$ provided $u \subset v$) to

$$\exists f \in F \,\forall m \,\exists x \neq y \in Y \,(x \mathrel{P_{f \restriction m}} y)\,,$$

and then to

$$\forall T \in A(F) \,\exists u \in T \,\exists x \neq y \in Y \,(x \mathrel{P_u} y)\,,$$

so that we obtain

$$\forall Y \in \mathscr{P}_{\mathbf{int}}(H) \,\big(\forall k \,(\#Y > k) \Longrightarrow \exists u \in T \,\exists x \neq y \in Y \,(x \mathrel{P_u} y)\big)$$

for every $T \in A(F)$. Applying countable Saturation once again, we get, for any $T \in A(F)$, a number $k_T \in \mathbb{N}$ and a finite set $T' \subseteq T$ such that

$$\forall Y \,\big(\#Y > k_w \Longrightarrow \exists u \in T' \,\exists x \neq y \in Y \,(x \mathrel{P_u} y)\big)\,.$$

Since the sets P_u are assumed to be symmetric, this leads us to a finite set $Z_T \subseteq H$, with at most k_T elements, satisfying

$$\forall x \in H\, \exists z \in Z_T\, \exists u \in T'\, (x\ P_u\ z)\,. \tag{†}$$

Note that the displayed formula, as a property of Z_T, depends only on T', a finite subset of $2^{<\omega}$, not on T itself, and hence the sets Z_T can be chosen so that there are only countably many really different sets among them. Then $Z = \bigcup_{T \in A(F)} Z_T$ is a countable subset of H.

We put $\varphi(x) = \{\langle z, u\rangle : z \in Z \wedge u \in 2^{<\omega} \wedge x\ P_u\ z\}$ for any $x \in H$. Thus $\varphi : H \to Z \times 2^{<\omega}$, where both Z and $2^{<\omega}$ are countable sets. For any $\langle z, u\rangle \in Z \times 2^{<\omega}$ the set $\{x \in H : \langle z, u\rangle \in \varphi(z)\}$ is equal to $\{x : x\ P_u\ z\}$, and hence is internal. Thus, modulo the natural identification of $\mathscr{P}(Z \times 2^{<\omega})$ with $2^{Z \times 2^{<\omega}}$, φ is a shadow map. It remains to show that, for $x, y \in H$, $\varphi(x) = \varphi(y)$ implies that $x \mathsf{E} y$. To prove this we are going to employ $(*)$. Let $T \in A(F)$. Choose, by (†), $z \in Z_T \subseteq Z$ and $u \in T' \subseteq T$ such that $x\ P_u\ z$ — then $\langle z, u\rangle \in \varphi(x) = \varphi(y)$, thus we also have $y\ P_u\ z$. As T is arbitrary, this ends the proof of $x \mathsf{E} y$.

Finally, let us prove that clauses (I) and (II) are incompatible. Otherwise there exist an infinite internal set X and an **injective** shadow $\varphi : X \to 2^{\mathbb{N}}$. As X is infinite there is $f \in 2^{\mathbb{N}}$ such that the set $X_m = \{x \in X : \varphi(x) \restriction m = f \restriction m\}$ is infinite for any $m \in \mathbb{N}$. However the sets X_m are internal (because φ is a shadow). Thus the intersection $X_f = \bigcap_m X_m$ must be infinite by Lemma 9.1.5. But X_f must be a singleton because φ is injective and clearly $\varphi(x) = f$ for all $x \in X_f$. □

Exercise 9.7.3. Prove the following, somewhat stronger, results for any $\mathbf{\Sigma}^0_1$ equivalence relation E on a set X, and infer Henson's result 9.2.7 from (2):

(1) if X is $\mathbf{\Pi}^0_1$ then either the quotient X/E is finite or there is an infinite internal pairwise E-inequivalent set $C \subseteq X$;

(2) if X is countably determined then either X/E is at most countable or there is an infinite pairwise E-inequivalent internal set $C \subseteq X$.

Hint: see [KanR 03, 7.2]. □

Exercise 9.7.4. Prove, following the last paragraph in the proof of the theorem, that if E is a thin countably determined ER on an internal set H, then any countably determined set $X \subseteq H$ which intersects each E-class in an at most countable set is finite. □

9.7b Application: nonstandard partition calculus

For the sake of brevity, let $R(k, \ell, m, n)$ denote the following statement where k, ℓ, m, n are natural numbers (in a standard setting) or numbers in ${}^*\mathbb{N}$, and as usual any number n is identified with the set $\{0, 1, ..., n-1\}$ of all smaller numbers:

$R(k,\ell,m,n)$: for any partition of the set $[n]^k = \{u \subseteq n : \#u = k\}$ into ℓ-many parts there is a set $A \subseteq n$, $\#A = m$, *homogeneous* for the partition in the sense that all sets $u \in [A]^k$ belong to one and the same class of the partition.

The next result belongs to the classics of combinatorics:

Theorem 9.7.5 (the Ramsey theorem, ZFC). *For any numbers $k,\ell,m \in \mathbb{N}$ there is $n \in \mathbb{N}$ satisfying $R(k,\ell,m,n)$.* □

Note the logical structure of the theorem:

$$\forall\, k,\ell,m\, \exists\, n\, \forall \text{ partition } \exists\, A\, \forall\, u,v \in [A]^k \tag{1}$$

As usual nonstandard analysis yields less complicated formulations.

Lemma 9.7.6 (HST). *For any $k,\ell \in \mathbb{N}$ and $\nu \in {}^*\mathbb{N} \setminus \mathbb{N}$ and any internal partition of $[\nu]^k$ into ℓ parts there is an infinite internal homogeneous set.*

Note the logical structure of this result, two quantifiers less than (1):

$$\forall\, k,\ell,\nu\, \forall \text{ partition } \exists\, A\, \forall\, u,\, v \in [A]^k, \tag{2}$$

Proof. Arguing in $\mathbb{H}$, put $\rho(m) = \mathtt{min}\{n : R(m,m,m,n)\}$. By Theorem 9.7.5 ρ is a total function $\mathbb{N} \to \mathbb{N}$, and it is known that ρ is a very fast growing computable function. (Note that $R(m,m,m,n)$ implies $R(k,\ell,m,n)$ whenever $k,\ell \le m$.) By Transfer, ${}^*\!\rho$ is a standard total function ${}^*\mathbb{N} \to {}^*\mathbb{N}$. Let $m \in {}^*\mathbb{N}$ be the largest number with ${}^*\!\rho(m) \le \nu$. Note that $m \in {}^*\mathbb{N} \setminus \mathbb{N}$, because otherwise by Transfer ${}^*\!\rho(m) = {}^*(\rho(m)) \in \mathbb{N}$, but $\nu \notin \mathbb{N}$! Thus, still by Transfer, we have $R(m,m,m,\nu)$ in the internal universe $\mathbb{I}$, and hence $R(k,\ell,m,\nu)$ because $k,\ell < m$. This implies the result required. □

However nonstandard analysis leads to much more general results:

Theorem 9.7.7. *Suppose that $k \in \mathbb{N}$, $\nu \in {}^*\mathbb{N} \setminus \mathbb{N}$, and E is a countably determined equivalence relation on $[\nu]^k$ such that there is no infinite internal pairwise E-inequivalent set $I \subseteq [\nu]^k$. Then the E-partition admits an internal infinite homogeneous set $A \subseteq [\nu]^k$.*

Proof. First of all, arguing as in the proof of Lemma 9.7.6, choose $c \in {}^*\mathbb{N} \setminus \mathbb{N}$ such that $\nu \ge {}^*\!\rho(2^c)$.

It follows from Theorem 9.7.1 that there is a shadow $\varphi : [\nu]^k \to 2^{\mathbb{N}}$ such that $\varphi(u) = \varphi(v)$ implies $u \,\mathsf{E}\, v$ for all $u,\, v \in [\nu]^k$. For any $j \in \mathbb{N}$ the set $U_j = \{u \in [\nu]^k : \varphi(u)(j) = 1\}$ is an internal subset of $[\nu]^k$. By Extension there is an internal sequence $\{U_j\}_{j \in {}^*\mathbb{N}}$ of (internal) subsets of $[\nu]^k$ extending $\{U_j\}_{j \in \mathbb{N}}$. Define an internal map $\vartheta : [\nu]^k \to {}^{\mathtt{int}}2^c$ so that $\vartheta(u)(j) = 1$ iff $u \in U_j$ for all $j < c$. Then obviously $\varphi(u) = \vartheta(u) \restriction \mathbb{N}$ for any $u \in [\nu]^k$, and hence $\vartheta(u) = \vartheta(v)$ implies $u \,\mathsf{E}\, v$ for all $u,\, v \in [\nu]^k$.

However by the choice of c the internal partition of $[\nu]^k$ defined by ψ (that is the equivalence relation $u \mathsf{E}_\vartheta v$ iff $\vartheta(u) = \vartheta(v)$) has $\le 2^c$ classes, therefore, admits an infinite internal homogeneous set $A \subseteq [\nu]^k$. But E_ϑ is finer than E, thus A is E-homogeneous as well. □

9.7c Generalization

Quantitatively, the difference between (I) and (II) of Theorem 9.7.1 can be presented as the difference between sets which do contain an infinite internal subset (II) and those which do not (I). But this in not the only possible quantitative dichotomy: the division line can be drawn differently! The following general result of [KanR 03] is given here without a proof.

Theorem 9.7.8. *Let* E *be a countably determined equivalence relation on an infinite internal set* H, *and* U *a countably cofinal additive cut in* $^*\mathbb{N}$. *Then* **at least one** *of the following two statements holds*:

(I) *there is a number* $c \in {}^*\mathbb{N} \setminus U$ *and an internal map* $\vartheta : H \to {}^{\text{int}}2^c$ *such that* $\vartheta(x) \restriction U = \vartheta(y) \restriction U \Longrightarrow x \mathsf{E} y$;

(II) *there is an internal pairwise* E*-inequivalent set* $Y \subseteq H$ *with* $\#Y \notin U$.

Moreover, if (II) *holds and* U *satisfies* $x \in U \Longrightarrow 2^x \in U$ *then* (I) *fails even for countably determined maps* ϑ.

The theorem yields a true dichotomy only for "exponential" cuts U, *i.e.*, those satisfying $x \in U \Longrightarrow 2^x \in U$. If this condition fails then (I) and (II) are compatible, for take E to be the equality on $[0, 2^x)$ but $y \mathsf{E} z$ for all y, $z \ge 2^x$. It is an open problem to obtain a true dichotomy in the general case.

Note that the case $U = \mathbb{N}$ in this theorem is equivalent to Theorem 9.7.1. Indeed, for the less trivial direction, a shadow map φ as in 9.7.1(I) can be transformed to an internal map $\vartheta : H \to {}^{\text{int}}2^c$ such that $\varphi(x) = \vartheta(x) \restriction \mathbb{N}$ for any $x \in H$, see the proof of Theorem 9.7.7. Then $\vartheta(x) = \vartheta(y)$ implies $x \mathsf{E} y$ for all x, $y \in H$, as required.

Exercise 9.7.9 (difficult!). Prove Theorem 9.7.8 following the proof of Theorem 9.7.1 with appropriate corrections. □

There is a somewhat different approach to maps ϑ as in (I), which may lead to new insights. Put $\varphi(x) = \vartheta(x) \restriction U$. Then φ is a map defined on H, with values in 2^U, and $\varphi(x) = \varphi(y)$ implies $x \mathsf{E} y$. The values of φ are not just arbitrary external maps $U \to 2$. Say that a function $\xi : U \to 2$ is *internally extendable*, in symbols $\xi \in (2^U)_{\mathtt{iex}}$, if there exist an internal set Z with $U \subseteq Z$ and a map $f \in {}^{\text{int}}2^Z$ (that is $f : Z \to 2$ is an internal function) such that $\xi = f \restriction U$. (If U itself is internal then this is the same as an internal function.) This definition obviously does not depend on the choice of an internal set $Z \supseteq U$, that is, we can take $Z = [0, c)$, $c \in {}^*\mathbb{R} \setminus U$.

In these terms, we have $x \mathsf{E} y \iff \varphi(x) \mathsf{R} \varphi(y)$, where $\xi \mathsf{R} \eta$ iff $\xi = \eta$, or there exist $x, y \in H$ with $x\mathsf{E}y$, is an equivalence relation on $(2^U)_{\mathtt{iex}}$. However it is difficult to study the relation R and its connection with E by means of "hyperfinite" descriptive set theory simply because the domain $(2^U)_{\mathtt{iex}}$ of R consists of non-internal objects. Yet we can define the *lifting* $\mathsf{F} = \mathsf{R}^{\uparrow c}$ of R to $c = [0, c)$, that is an equivalence relation $f \mathsf{F} g$ iff $(f \restriction U) \mathsf{R} (g \restriction U)$ on the internal set ${}^{\mathtt{int}}2^c$. Clearly F is countably determined. Then a similar equivalence $x \mathsf{E} y \iff \vartheta(x) \mathsf{F} \vartheta(y)$ holds. This means that ϑ is a reduction of E to R (see § 9.7f on a general definition). Moreover F is *concentrated on* U in the sense that whether $f \mathsf{F} g$ $(f, g \in {}^{\mathtt{int}}2^c)$ depends only on $f \restriction U$, $g \restriction U$. We conclude that (I) of Theorem 9.7.8 can be reformulated as follows:

(I′) *there exist* $c \in {}^*\mathbb{N} \setminus U$, *a countably determined equivalence relation* F *on* ${}^{\mathtt{int}}2^c$ *concentrated on* U, *and an internal reduction* ϑ *of* E *to* F.

Note that (I′) does not depend on the choice of c, that is if it holds for some $c \notin U$ then it also holds for any other $c' \notin U$.

9.7d Transversals of "countable" equivalence relations

An equivalence relation E is called *"countable"* if all of its equivalence classes $[x]_{\mathsf{E}} = \{y : x \mathsf{E} y\}$, $x \in \mathtt{dom}\,\mathsf{E}$, are at most countable. In Polish descriptive set theory, "countable" Borel ERs form a rather rich category whose full structure in terms of Borel reducibility is a topic of deep investigations (see [JackKL 02]). In nonstandard setting, the structure of "countable" ERs is much more elementary due to the next theorem. This is another side of the same phenomenon making "planar" sets with countable cross-sections look simpler in "hyperfinite" descriptive set theory than in Polish spaces, see § 9.4b.

Recall that a transversal for an equivalence relation is any set having exactly one element in common with each equivalence class.

Theorem 9.7.10. *Any "countable" countably determined equivalence relation* E *on an internal set* H *admits a countably determined transversal.*

Proof. Note that E, as a subset of $H \times H$, is a countably determined set with all cross-sections $\mathsf{E}_x = [x]_{\mathsf{E}} = \{y : x \mathsf{E} y\}$ being at most countable sets. It follows from Theorem 9.4.7(ii) that there exists a countable sequence $\{F_k\}_{k \in \mathbb{N}}$ of internal functions $F_k : H \to H$ such that $\mathsf{E} \subseteq \bigcup_k F_k$, or in different terms $[x]_{\mathsf{E}} \subseteq \{F_k(x) : n \in \mathbb{N}\}$ for all $x \in H$. The sets $D_k = \mathtt{dom}\,(\mathsf{E} \cap F_k) = \{x \in H : x \mathsf{E} F_k(x)\}$ are countably determined (as internal preimages of E, a countably determined set).

Let us fix any internal well-ordering $\prec$ of H. In other words, it is true in $\mathbb{I}$ that $\prec$ is a well-ordering of H.

Let $n \in \mathbb{N}$. For any $x \in H$ we carry out the following construction, called *the* n*-construction for* x. Define an internal $\prec$-decreasing sequence

$\{x_{(a)}\}_{a \le a(x)}$ with $a(x)+1 \in {}^*\mathbb{N}$ terms, by internal induction on a; the length $a(x) \in {}^*\mathbb{N}$ will be determined in the course of the construction. Put $x_{(0)} = x$. Suppose that $x_{(a)}$ is defined. If $z = F_n(x_{(a)}) \prec x_{(a)}$ then put $x_{(a+1)} = z$, otherwise put $a(x) = a$ and end the construction. Note that eventually the construction stops simply because $x_{(a+1)} \prec x_{(a)}$ for all a. Put $\nu_n(x) = 0$ if $a(x)$ is even and $\nu_n(x) = 1$ otherwise.

Put $\psi(x)(n) = \nu_n(x)$ for $x \in H$, $n \in \mathbb{N}$; thus $\psi : H \to 2^{\mathbb{N}}$. We observe that the n-construction is internal. Therefore the sets $X_n = \{x \in H : \psi(x)(n) = 0\}$, $n \in \mathbb{N}$, are internal, and hence ψ is a shadow in the sense of 9.2.12. Note that the sets $X_{nk} = \{x \in H : \psi(F_k(x))(n) = 0\}$ are internal as well, and by the same reasons.

Lemma 9.7.11. *If $x \ne y \in H$ and $x \mathsf{E} y$ then $\psi(x) \ne \psi(y)$.*

Thus, while it is, generally speaking, possible that different elements $x \in H$ have equal "profiles" $\psi(x)$, this cannot happen if they are E-equivalent.

Proof. We can assume that $y \prec x$. There is $n \in \mathbb{N}$ such that $y = F_n(x)$. Then $y = x_{(1)}$, in the sense of the n-construction for x. It follows that the n-construction for y has exactly one step less than the n-construction for x. We conclude that $\nu_n(x) \ne \nu_n(y)$. □ (*Lemma*)

Coming back to the theorem, choose, in $\mathbb{WF}$, an element $r_A \in A$ in any non-empty set $A \subseteq 2^{\mathbb{N}}$. (The Choice axiom of **ZFC** is applied in the universe $\mathbb{WF}$.) For any $x \in {}^*\mathbb{N}$, the set $A(x) = \{\psi(y) : y \in [x]_{\mathsf{E}}\}$ is a non-empty countable subset of $2^{\mathbb{N}}$. Then $X = \{x \in H : \psi(x) = r_{A(x)}\}$ is a transversal for E by Lemma 9.7.11.

To prove that X is countably determined consider the least algebra $\mathscr{X}$ of subsets of H containing all sets D_k, X_n, X_{nk} defined above. Note that $\mathscr{X}$ is countable and consists of countably determined sets. (Indeed D_k are CD while X_n, X_{nk} are just internal.) Let C be the set of all at most countable non-empty sets $A \subseteq 2^{\mathbb{N}}$. We observe that $X = \bigcup_{A \in C} X(A)$, where

$$X(A) = \{x \in X : A(x) = A\} = \{x \in H : A(x) = A \wedge \psi(x) = r_A\}.$$

Lemma 9.7.12. *Any set $X(A)$, $A \in C$, is countably determined over $\mathscr{X}$ in the sense of Definition 9.2.4.*

Proof. Note that $X(A) = X' \cap X''$, where $X' = \{x \in H : \psi(x) = r_A\}$ is obviously Borel over $\mathscr{X}$, and $X'' = \{x \in H : A(x) = A\}$. It remains to prove that X'' is countably determined over $\mathscr{X}$. Let $A = \{g_m : m \in \mathbb{N}\}$. Note that $A(x) = \{\psi(F_n(x)) : n \in \mathbb{N} \wedge x \in D_n\}$, thus $A(x) = A$ is equivalent to

$$\forall n \, \exists m \, (x \in D_n \Longrightarrow x \in S_{mn}) \wedge \forall m \, \exists n \, (x \in D_n \wedge x \in S_{mn}),$$

where $S_{mn} = \{x : \psi(F_n(x)) = g_m\}$. It is clear that every set S_{mn} is Borel over $\mathscr{X}$ (even over the sets X_{mn}). It follows that $X(A)$ is Borel, and hence countably determined, over $\mathscr{X}$, as required. □

On the other hand, the class of all sets countably determined over a fixed countable algebra $\mathscr{X}$ of sets is closed under any unions (as well as under complements and intersections): to show this take the set theoretic union of the associated bases under the assumption that the assignment of sets in $\mathscr{X}$ to indices $i \in \mathbb{N}$ is fixed once and for all. (Note that the class of all countably determined sets is closed only under countable unions and intersections!) It follows that $A(X)$ is countably determined over $\mathscr{X}$. But $\mathscr{X}$ itself consists of countably determined sets, therefore it remains to cite Theorem 9.2.9.

□ (*Theorem 9.7.10*)

Example 9.7.13. The equivalence relation $x \mathsf{M}_{\mathbb{N}} y$ iff $|x - y| \in \mathbb{N}$ on ${}^*\mathbb{N}$ is "countable", and hence it has a countably determined transversal by Theorem 9.7.10. Note that $\mathsf{M}_{\mathbb{N}}$ is a $\mathbf{\Pi}^0_1$ relation. Thus it is natural to ask whether $\mathsf{M}_{\mathbb{N}}$ has a transversal of a type simpler than CD. It clearly does not admit a Borel transversal — by the same "shift" argument as in the proof of the fact that the Vitali equivalence on $\mathbb{R}$ does not admit a Borel (generally, Lebesgue measurable) transversal. Theorem 9.7.18 below contains an even stronger result.

Whether $\mathsf{M}_{\mathbb{N}}$ has a projective transversal depends on the Loeb measurability of projective sets in segments $[0, a)$ of ${}^*\mathbb{N}$ (and hence on the Lebesgue measurability in $2^{\mathbb{N}}$, see § 9.5b) for the negative direction, or projective choice of an element in an arbitrary countable subset of $2^{\mathbb{N}}$ for the positive direction — and hence is independent of **HST**. □

9.7e Equivalence relations of monad partitions

This class of equivalence relations was defined in § 1.4c: any additive cut $U \subseteq {}^*\mathbb{N}$ induces an equivalence relation $x \mathsf{M}_U y$ iff $|x - y| \in U$ on ${}^*\mathbb{N}$, which divides ${}^*\mathbb{N}$ into M_U-equivalence classes $[x]_U = \{y : x \mathsf{M}_U y\} = \{y : |x - y| \in U\}$, called *U-monads*. It follows from Lemma 9.6.7 that the case of additive countably determined cuts $U \subseteq {}^*\mathbb{N}$ splits into two subcases: *countably cofinal* and *countably coinitial* cuts. (Let alone $\varnothing$ and ${}^*\mathbb{N}$, the only internal additive cuts.) Accordingly, this leads to the following classes of ERs:

countably cofinal equivalence relations: those of the form M_U, where $U \subseteq {}^*\mathbb{N}$ is a countably cofinal additive cut — all of them belong to $\mathbf{\Sigma}^0_1$;

countably coinitial equivalence relations: those of the form M_U, where $U \subseteq {}^*\mathbb{N}$ is a countably coinitial additive cut — all of them belong to $\mathbf{\Pi}^0_1$.

We proved (Theorem 1.4.7) that among these ERs, only those of the form $\mathsf{M}_{h\mathbb{N}}$, $h \in {}^*\mathbb{N}$, in particular, $\mathsf{M}_{\mathbb{N}}$, and those of the form $\mathsf{M}_{h/\mathbb{N}}$, $h \in {}^*\mathbb{N} \setminus \mathbb{N}$ admit transversals in the class $\mathbf{\Delta}^{\mathbf{ss}}_2$. As the latter is strictly bigger than the class of all countably determined sets (see 9.2.8), it is a natural question whether the ERs of the form $h\mathbb{N}$ and $h/\mathbb{N}$ have countably determined transversals. The answer turns out to be different for the two subfamilies.

Theorem 9.7.14. (i) *If* $1 \le h \in {}^*\mathbb{N}$ *then* $\mathsf{M}_{h\mathbb{N}}$ *admits a CD transversal*; (ii) *if* $h \in {}^*\mathbb{N} \smallsetminus \mathbb{N}$ *then* $\mathsf{M}_{h/\mathbb{N}}$ *does not have a CD transversal.*

Proof. (i) Note that $\mathsf{M}_{\mathbb{N}}$ admits a countably determined transversal T, see 9.7.13, or alternatively, the transversal defined in the proof of Theorem 1.4.7(i) is CD. It follows that any $\mathsf{M}_{h\mathbb{N}}$ has a countably determined transversal: just take $\{hx : x \in T\}$.

(ii) Let us prove that even the restricted relation $\mathsf{E}_h = \mathsf{M}_{h/\mathbb{N}} \restriction [0, h)$ does not admit a CD transversal. In the notation of the proof of (ii) of Theorem 1.4.7 (the "if" part), it suffices to show that the equivalence $\xi \,\mathsf{R}\, \eta$ iff $\xi \restriction \mathbb{N} = \eta \restriction \mathbb{N}$ on ${}^{\mathrm{int}}2^c$ (where $c \in {}^*\mathbb{N} \smallsetminus \mathbb{N}$) does not have a countably determined transversal.

We claim that moreover any countably determined set $X \subseteq {}^{\mathrm{int}}2^c$ intersecting every R-class in a countable set is countable. (Note that $2^{\mathbb{N}}$ is uncountable, and hence so is the set of all R-equivalence classes.) As usual, we have $X = \bigcup_{f \in F} \bigcap_{m \in \mathbb{N}} X_{f \restriction m} \subseteq {}^{\mathrm{int}}2^c$, where $F \subseteq 2^{\mathbb{N}}$ and all sets X_s, $s \in 2^{<\omega}$, are internal subsets of ${}^{\mathrm{int}}2^c$. Thus, for any $f \in F$ and $g \in 2^{\mathbb{N}}$, the intersection of $X_f = \bigcap_{m \in \mathbb{N}} X_{f \restriction m}$ (a subset of X) and $Y_g = \{\xi \in {}^{\mathrm{int}}2^c : \xi \restriction \mathbb{N} = g\}$ (an equivalence class of R) is at most countable. It follows that $X_f \cap Y_g$, a $\mathbf{\Pi}^0_1$ set, is in fact finite by 9.1.7. Therefore, by Lemma 9.1.5(iii), for any pair of $f \in F$ and $g \in 2^{\mathbb{N}}$ there is a number $m \in \mathbb{N}$ such that $D_{f \restriction m,\, g \restriction m} = X_{f \restriction m} \cap Y_{g \restriction m}$ is a finite set. On the other hand, we have $X \subseteq \bigcup_{s,t \in 2^{<\omega},\, \mathtt{lh}\, s = \mathtt{lh}\, t} D_{st}$, so that X is countable, as required. □

Here follow several corollaries and related results.

Exercise 9.7.15. Let $h \in {}^*\mathbb{N} \smallsetminus \mathbb{N}$ and $\mathsf{E}_h = \mathsf{M}_{h/\mathbb{N}} \restriction [0, h)$. Prove the following:

(1) For all $x, y < h$, the equivalence holds: $x \,\mathsf{M}_{h/\mathbb{N}}\, y$ iff $\frac{x-y}{h} \simeq 0$, and hence E_h has exactly $\mathfrak{c}$-many equivalence classes

(2) Any internal set $X \subseteq [0, h)$ which intersects every E_h-class in a finite set is finite. (*Hint.* If X is infinite then there is a decreasing chain $[0, h) = I_0 \supseteq I_1 \supseteq I_2 \dots$ of subintervals whose length tends to 0, but $X \cap I_k$ still infinite for any finite k. Now apply Lemma 9.1.5.)

(3) The relation $x \equiv_{\mathsf{B}} y$ on ${}^*\mathbb{N}$ does **not** admit a countably determined transversal. (*Hint.* $x \equiv_{\mathsf{B}} y$ is equivalent to $x/y \simeq 1$ by Theorem 9.6.4, and hence $\equiv_{\mathsf{B}}$ coincides with E_h on $[\frac{h}{2}, h)$.)

(4) On the contrary, the relation $x \equiv_{\mathsf{CD}} y$ on ${}^*\mathbb{N}$ admits a countably determined transversal. For instance, take, using Theorem 9.6.4(ii), a countably determined transversal for $\mathsf{M}_{\mathbb{N}}$ — then $\{2^x : x \in T\}$ is a countably determined transversal for $\equiv_{\mathsf{CD}}$. □

The next exercise shows that equivalence relations of the form M_U can be presented differently. Such an alternative form will be useful in some applications.

Exercise 9.7.16. Let $\mathbf{S} = \{\xi \in 2^{\mathbb{N}} : \{k : \xi(k) = 1\}$ is finite$\}$, accordingly, ${}^*\mathbf{S}$ is the set of all internal functions $\xi \in 2^{{}^*\mathbb{N}}$ such that $\{\nu \in {}^*\mathbb{N} : \xi(\nu) = 1\}$ is $*$-finite; ${}^*\mathbf{S}$ is internal. Suppose that $C \subsetneqq {}^*\mathbb{N}$ is a non-internal cut. Put, for $\xi, \eta \in {}^{\mathtt{int}}2^{{}^*\mathbb{N}}$, $\xi \,\mathsf{R}_C\, \eta$ iff $\xi \restriction ({}^*\mathbb{N} \smallsetminus C) = \eta \restriction ({}^*\mathbb{N} \smallsetminus C)$.

(1) Prove that $2^C = \bigcup_{c \in C} [0, 2^c)$ is an additive cut and the map $\xi \mapsto s(\xi) = \sum_{\nu \in {}^*\mathbb{N}} 2^\nu \xi(\nu)$ is an isomorphism of $\langle {}^*\mathbf{S}; \mathsf{R}_C \rangle$ onto $\langle {}^*\mathbb{N}; \mathsf{M}_{2^C} \rangle$.

To restrict this phenomenon to a $*$-finite domain, let us choose any $m \in {}^*\mathbb{N}$ satisfying $C \subseteq [0, m)$.

(2) Prove that the map $\xi \mapsto s(\xi) = \sum_{\nu < m} 2^\nu \xi(\nu)$ is a bijection of ${}^{\mathtt{int}}2^m$ onto $[0, 2^m)$ satisfying $\xi \restriction ([0, m) \smallsetminus C) = \eta \restriction ([0, m) \smallsetminus C)$ iff $s(\xi) \,\mathsf{M}_{2^C}\, s(\eta)$.

Finally, we put $W = \{m - x : x \in [0, m) \smallsetminus C\}$, the dual cut. Recall that ${\mathsf{D}_W}^{\uparrow m}$ is the lifting of the equality D_W on the set $(2^W)_{\mathtt{iex}}$ of all internally extendable maps $\sigma : W \to 2$ to m, that is an equivalence relation on ${}^{\mathtt{int}}2^m$ such that $\xi \,{\mathsf{D}_W}^{\uparrow m}\, \eta$ iff $\xi \restriction W = \eta \restriction W$ (see §9.7c).

(3) Prove that $\xi \mapsto s'(\xi) = \sum_{\nu < m} 2^\nu \xi(m - \nu)$ sends ${\mathsf{D}_W}^{\uparrow m}$ onto M_{2^C}.

Thus we may consider the quotient $[0, m)/\mathsf{M}_{2^C}$ as an adequate "model" of $(2^W)_{\mathtt{iex}}$ in "hyperfinite" descriptive set theory. □

9.7f Borel and countably determined reducibility

The following definition is copied from studies in Polish DST.

Suppose that E, F are equivalence relations (ERs, for brevity) on Borel sets X, Y. We write $\mathsf{E} \le_{\mathtt{B}} \mathsf{F}$, in words: E *is Borel-reducible to* F, iff there is a Borel map (called: *reduction*) $\vartheta : X \to Y$ such that $x \,\mathsf{E}\, x' \iff \vartheta(x) \,\mathsf{F}\, \vartheta(x')$ for all x, $x' \in X$. We can also write $X/\mathsf{E} \le_{\mathtt{CD}} Y/\mathsf{F}$ in this case. Define $\mathsf{E} \equiv_{\mathtt{B}} \mathsf{F}$ if both $\mathsf{E} \le_{\mathtt{B}} \mathsf{F}$ and $\mathsf{F} \le_{\mathtt{B}} \mathsf{E}$, and $\mathsf{E} <_{\mathtt{B}} \mathsf{F}$ iff $\mathsf{E} \le_{\mathtt{B}} \mathsf{F}$ but not $\mathsf{F} \le_{\mathtt{B}} \mathsf{E}$.

Changing "Borel" to "countably determined" in these definitions, we obtain the relations $\le_{\mathtt{CD}}$, $\equiv_{\mathtt{CD}}$, $<_{\mathtt{CD}}$ of *countably determined* reducibility. The reducibility of sets (that is, subsets of internal sets), as considered above in §9.6, corresponds to the case when $\mathsf{E} = \mathsf{D}_X$ is the equality on X considered as an equivalence relation. (D from "diagonal".)

Exercise 9.7.17. Suppose that $h \in {}^*\mathbb{N} \smallsetminus \mathbb{N}$. Let E_h be defined as in 9.7.15. Prove that, in terms of Theorem 9.7.1, E is of type (I) iff $\mathsf{E} \le_{\mathtt{B}} \mathsf{E}_h$. □

The informal meaning of $\mathsf{E} \le_{\mathtt{B}} \mathsf{F}$ and $\mathsf{E} \le_{\mathtt{CD}} \mathsf{F}$ is that F *has at least as many equivalence classes as* E, *and this is witnessed by a Borel, resp., CD map.* Quantitative study in terms of Borel and countably determined reductions is also possible in some cases when the sets considered consist of non-internal elements, see Exercise 9.7.16(3).

The following definition also has an obvious prototype in Polish descriptive set theory. An equivalence relation E is *B-smooth* if $\mathsf{E} \le_{\mathtt{B}} \mathsf{D}_R$ for an internal set R, and *CD-smooth* if $\mathsf{E} \le_{\mathrm{CD}} \mathsf{D}_R$ for an internal set R. Thus, E is, say, B-smooth iff there is a Borel map ϑ of $\mathtt{dom}\,\mathsf{E}$ into an internal set R such that $x \,\mathsf{E}\, y \Longleftrightarrow \vartheta(x) = \vartheta(y)$ for all x, y.

Theorem 9.7.18. *The equivalence relation* $\mathsf{M}_{\mathbb{N}}$ *on* $^*\mathbb{N}$ *is CD-smooth but not B-smooth.*

Proof. To show that $\mathsf{M}_{\mathbb{N}}$ is CD-smooth take a countably determined transversal $T \subseteq {}^*\mathbb{N}$ (by Theorem 9.7.10) and let $\vartheta(x)$ be equal to the only element $y \in T$ such that $x \,\mathsf{M}_{\mathbb{N}}\, y$. The map ϑ withnesses the CD-smoothness of $\mathsf{M}_{\mathbb{N}}$.

To show that $\mathsf{M}_{\mathbb{N}}$ is not B-smooth fix $h \in {}^*\mathbb{N} \smallsetminus \mathbb{N}$ and prove that even the restriction $\mathsf{M}_{\mathbb{N}} \restriction [0,h)$ is not B-smooth. Suppose towards the contrary that R is an internal set and $\vartheta : [0,h) \to R$ is a Borel map satisfying $|x-y| \in \mathbb{N} \Longleftrightarrow \vartheta(x) = \vartheta(y)$. We consider the counting measure $\mu(Z) = \frac{\#Z}{h}$ on $[0,h)$ and the corresponding Loeb measure $\mathrm{L}\mu$.

As any Borel set, (the graph of) ϑ is $\mathbf{\Sigma}^1_1$, and hence it admits a presentation in the form $\vartheta = \bigcup_{f\in\mathbb{N}^{\mathbb{N}}} \bigcap_m C_{f\restriction m}$, where all $C_s \subseteq [0,h) \times R$ are internal, and $C_t \subseteq C_s$ whenever $s \subset t$. (In the sequel m, n, k indicate only numbers in $\mathbb{N}$.) By a simple measure-theoretic argument using the σ-additivity of $\mathrm{L}\mu$, there is a sequence of numbers $\{j_m\}_{m\in\mathbb{N}}$ in $\mathbb{N}$ such that $X = \mathtt{dom}\,\vartheta'$ has $\mathrm{L}\mu$-measure $\ge \frac{1}{2}$, where $\vartheta' = \bigcup_{f\in F} \bigcap_m C_{f\restriction m}$ and $F = \{f \in \mathbb{N}^{\mathbb{N}} : \forall m\,(f(m) \le j_m)\}$. By Koenig's lemma, $\vartheta' = \bigcap_m C_m$, where $C_m = \bigcup_u C_u$ and the union is taken over all sequences $u \in \mathbb{N}^{<\omega}$ of length m such that $u(k) \le j_k$, $\forall\, k < m$. Thus each C_m is internal, and hence ϑ' is a $\mathbf{\Pi}^0_1$ set. (Note that $C_{m'} \subseteq C_m$ whenever $m \le m'$.) Also, $\vartheta' = \vartheta \restriction X$, where $X = \mathtt{dom}\,\vartheta' \subseteq [0,h)$ is a $\mathbf{\Pi}^0_1$ set (by Lemma 9.1.8) satisfying $\mathrm{L}\mu(X) \ge \frac{1}{2}$.

Since ϑ is a reduction, we have, for all x, $x' \in X$ and y, $y' \in R$:

$$\forall m\,(x\,C_m\,y \wedge x'\,C_m\,y') \Longrightarrow \big(\exists\,k\,(|x-x'|<k) \Longleftrightarrow y = y'\big). \qquad (1)$$

Changing $\Longleftrightarrow$ to $\Longleftarrow$ in the subformula in the right-hand side of (1), we re-write the formula obtained as

$$y = y' \Longrightarrow \big(\forall m\,(x\,C_m\,y \wedge x'\,C_m\,y') \Longrightarrow \exists\,k\,(|x-x'|<k)\big).$$

In other words the $\mathbf{\Pi}^0_1$ set $W = \{\langle x,x',y,y'\rangle : x,x' \in X \wedge y = y' \in R\}$ is covered by the union of all (internal) sets of the form

$$U_{mk} = \{\langle x,x',y,y'\rangle : x\,C_m\,y \wedge x'\,C_m\,y' \Longrightarrow |x-x'|<k\}\,, \quad m,k \in \mathbb{N}\,.$$

Note that $U_{m'k'} \subseteq U_{mk}$ whenever $m \le m'$, $k \le k'$. Thus by Saturation in the form of Lemma 9.1.5(i), we find numbers m_0, $k_0 \in \mathbb{N}$ such that, for all x, $x' \in X$ and y, $y' \in R$:

$$x\,C_{m_0}\,y \wedge x'\,C_{m_0}\,y' \wedge y = y' \Longrightarrow |x-x'| < k_0\,. \qquad (2)$$

Further, changing $\Longleftrightarrow$ to $\Longrightarrow$ in the subformula in (1), we have

$$\forall m\,(x\ C_m\ y \wedge x'\ C_m\ y') \implies y = y'$$

whenever $x,\ x' \in X$ and $y,\ y' \in R$ satisfy $|x - x'| < 4k_0$. Applying Saturation as above, we find a number $m_1 \geq m_0$ such that

$$x\ C_{m_1}\ y \wedge x'\ C_{m_1}\ y' \wedge |x - x'| < 4k_0 \implies y = y' \tag{3}$$

holds for all $x,\ x' \in X$ and $y,\ y' \in R$. Now we claim that

$$|x - x'| < k \vee |x - x'| \geq 4k \qquad \text{for all} \quad x,\ x' \in X\,, \tag{4}$$

which obviously contradicts the assumption $\mathrm{L}\mu(X) \geq \frac{1}{2}$. To prove (4) put $y = \vartheta(x) = \vartheta'(x)$ and $y' = \vartheta(x') - \vartheta'(x')$, thus $x\ C_m\ y \wedge x'\ C_{m1}\ y'$ for all m. If $y = y'$ it then follows from (2) that $|x - x'| < k_0$. If $y = y'$ then $|x - x'| > 4k_0$ by (3). □

Exercise 9.7.19. (1) Show that Theorem 9.7.18 implies that there exists no Borel transversal for $\mathsf{M}_{\mathbb{N}}$.

(2) Prove $\mathsf{D}_{{}^*\mathbb{N}} \leq_{\mathsf{B}} \mathsf{M}_{\mathbb{N}}$, and hence $\mathsf{D}_{{}^*\mathbb{N}} <_{\mathsf{B}} \mathsf{M}_{\mathbb{N}}$ by the Theorem 9.7.18. In other words, $\mathsf{M}_{\mathbb{N}}$ has strictly more than ${}^*\mathbb{N}$ equivalence classes in Borel sence, while exactly ${}^*\mathbb{N}$-many classes in CD-sense.

(3) Let E be a $\mathbf{\Pi}^0_1$ equivalence relation on an internal set. Prove that $\mathsf{M}_{\mathbb{N}} \not\leq_{\mathsf{B}} \mathsf{E}$. In fact this implies that $\mathsf{M}_U \not\leq_{\mathsf{B}} \mathsf{E}$ for any countably cofinal additive cut U by Theorem 9.7.20 below.

(4) Prove that $\mathsf{M}_U \not\leq_{\mathsf{B}} \mathsf{E}$ whenever U is a countably coinitial additive cut and E is a $\mathbf{\Sigma}^0_1$ equivalence relation.

Hints for (3), (4): argue approximately as in the proof of Theorem 9.7.18. See details in [KanR 03], Lemma 14.1 and part 3 in Section 13.

9.7g Reducibility structure of monad partitions

According to the next theorem, mutual reducibility of countably cofinal and countably coinitial ERs is largely determined by a quantitative factor, *the rate of* U, which we define as follows:

$$\mathtt{rate}\,U \;=\; \textstyle\bigcap_{a \in \log U,\ a' \notin \log U} [0\,,\ a' - a)\ ^{24}, \quad \text{where} \ \ \log U = \{a : 2^a \in U\}.$$

Note that if U is a countably cofinal (coinitial) *additive* cut then $\log U$ is still a countably cofinal, (resp., coinitial) cut, but not necessarily additive, and $U = \bigcup_{a \in \log U} [0, 2^a)$. Additive cuts of lowest possible rate are obviously those of the form $U = c\mathbb{N}$, $c \in {}^*\mathbb{N}$ and $U = c/\mathbb{N}$, $c \in {}^*\mathbb{N} \smallsetminus \mathbb{N}$, which we call *slow*; they satisfy $\mathtt{rate}\,U = \mathbb{N}$. Other additive cuts will be called *fast*.

---

[24] The right-hand side of the displayed formula, as a function of $\log U$, is known, for instance, from papers on **AST**; it can be called *the thickness* of $\log U$.

Theorem 9.7.20. *Suppose that U, V are additive countably determined cuts in ${}^*\mathbb{N}$ other than $\varnothing$ and ${}^*\mathbb{N}$. Then $\mathsf{D}_{{}^*\mathbb{N}} \le_{\mathsf{B}} \mathsf{M}_U$. In addition,*

(i) *if both U, V are countably cofinal or both countably coinitial then M_U and M_V are $\le_{\mathsf{B}}$-comparable, and $\mathsf{M}_U \le_{\mathsf{B}} \mathsf{M}_V$ iff $\mathsf{M}_U \le_{\mathsf{CD}} \mathsf{M}_V$ iff $\mathtt{rate}\, U \subseteq \mathtt{rate}\, V$, in particular, if U is slow then $\mathsf{M}_U \le_{\mathsf{CD}} \mathsf{M}_V$;*

(ii) *if U is countably cofinal and V countably coinitial then $\mathsf{M}_V \not\le_{\mathsf{CD}} \mathsf{M}_U$ and $\mathsf{M}_U \not\le_{\mathsf{B}} \mathsf{M}_V$, while $\mathsf{M}_U \le_{\mathsf{CD}} \mathsf{M}_V$ holds if and only if U is slow;*

(iii) *M_U is not B-smooth, and M_U is CD-smooth if and only if U is countably cofinal and slow (which means that $U = c\mathbb{N}$ for some $c \in {}^*\mathbb{N}$).*

Thus either of the two subclasses is linearly $\le_{\mathsf{CD}}$-(pre)ordered and has slow equivalence relations as the $\le_{\mathsf{CD}}$-least element, and there is no $\le_{\mathsf{CD}}$-connection between them except that any slow countably cofinal equivalence relation (it is necessarily CD-smooth) is $\le_{\mathsf{CD}}$-reducible to any countably coinitial ER. And, with the same exception, $\le_{\mathsf{B}}$ and $\le_{\mathsf{CD}}$ coincide on either subclass.

Exercise 9.7.21. Prove that for any countable sequence of countably cofinal fast cuts U_n there are countably cofinal fast cuts U, V with $\mathsf{M}_U <_{\mathsf{B}} \mathsf{M}_{U_n} <_{\mathsf{B}} \mathsf{M}_V$ for all $n \in \mathbb{N}$, and the same for countably coinitial cuts.

Prove that, in addition, either of the two classes is dense and countably saturated, *i.e.* contains no gaps of countable character. □

Proof (selected items of Theorem 9.7.20 [25]**).** To check that $\mathsf{D}_{{}^*\mathbb{N}} \le_{\mathsf{B}} \mathsf{M}_U$ for any additive countably determined cut U choose a number $c \notin U$; then $x \mapsto xc$ is an internal, hence, Borel reduction of $\mathsf{D}_{{}^*\mathbb{N}}$ to M_U, in other words, $x = x'$ iff $xc \,\mathsf{M}_U\, x'c$. This argument works for both countably cofinal and countably coinitial cuts U.

Incomparability in (ii) follows from (3) and (4) of Exercise 9.7.19.

Now we are going to prove the part of (i) of the theorem related to countably cofinal cuts and associated monadic equivalence relations.

In the course of the proof, k, ℓ, m, n will denote only numbers in $\mathbb{N}$.

Thus let U, V be countably cofinal cuts. Choose increasing sequences $\{a_n\}$, $\{b_k\}$ cofinal in resp. $\log U = \{c : 2^c \in U\}$ and $\log V = \{c : 2^c \in V\}$. Then the sequences $\{2^{a_n}\}$, $\{2^{b_k}\}$ are cofinal in the cuts U, V themselves.

Part 1. Suppose that $\mathsf{M}_U \le_{\mathsf{CD}} \mathsf{M}_V$. Then $\mathsf{R}_{\log U} \le_{\mathsf{CD}} \mathsf{R}_{\log V}$ by 9.7.16(1). Let $\vartheta : {}^*\mathbf{S} \to {}^*\mathbf{S}$ be a countably determined reduction of $\mathsf{R}_{\log U}$ to $\mathsf{R}_{\log V}$, thus $\xi \,\mathsf{R}_{\log U}\, \xi'$ iff $\vartheta(\xi) \,\mathsf{R}_{\log V}\, \vartheta(\xi')$ for all $\xi, \xi' \in {}^*\mathbf{S}$. By Lemma 9.2.6 we have $\vartheta = \bigcup_{f \in F} C_f$, where $F \subseteq 2^{\mathbb{N}}$ and $C_f = \bigcap_m C_{f \restriction m}$ for any $f \in F$, the sets C_s, $s \in 2^{<\omega}$, are internal, and $C_t \subseteq C_s \subseteq {}^*\mathbf{S} \times {}^*\mathbf{S}$ for $s \subset t$.

Consider any $f \in F$. Then C_f is a subset of the graph of ϑ, hence, by the choice of ϑ, for any $k \in \mathbb{N}$ we have, for all $\xi, \xi', \eta, \eta' \in {}^*\mathbf{S}$,

$$\forall m\, (\xi \, C_{f\restriction m}\, \eta \wedge \xi' \, C_{f\restriction m}\, \eta') \wedge \eta\restriction_{\ge b_k} = \eta'\restriction_{\ge b_k} \implies \exists n\, (\xi\restriction_{\ge a_n} = \xi'\restriction_{\ge a_n}),$$

[25] See a rather lengthy full proof of the theorem in [KanR 03].

where $\sigma\restriction_{\geq c} = \sigma \restriction ({}^*\mathbb{N} \smallsetminus [0,c))$ for $\sigma \in {}^*\mathbf{S}$ and $c \in {}^*\mathbb{N}$. Then, by Saturation,

$$\begin{aligned}&\forall k \;\; \exists n \, \exists m \quad \forall \xi, \xi', \eta, \eta' \in {}^*\mathbf{S} : \\ &\quad \xi \; C_{f\restriction m} \; \eta \wedge \xi' \; C_{f\restriction m} \; \eta' \wedge \eta\restriction_{\geq b_k} = \eta'\restriction_{\geq b_k} \implies \xi\restriction_{\geq a_n} = \xi'\restriction_{\geq a_n} . \end{aligned} \tag{1}$$

A similar (symmetric) argument also yields the following:

$$\begin{aligned}&\forall n \;\; \exists k \, \exists m \quad \forall \xi, \xi', \eta, \eta' \in {}^*\mathbf{S} : \\ &\quad \xi \; C_{f\restriction m} \; \eta \wedge \xi' \; C_{f\restriction m} \; \eta' \wedge \xi\restriction_{\geq a_n} = \xi'\restriction_{\geq a_n} \implies \eta\restriction_{\geq b_k} = \eta'\restriction_{\geq b_k} . \end{aligned} \tag{2}$$

Suppose, towards the contrary, that $\mathtt{rate}\, U \not\subseteq \mathtt{rate}\, V$, thus $\mathtt{rate}\, V \subsetneqq \mathtt{rate}\, U$. Then $\mathbb{N} \subsetneqq U$, and hence U is a fast cut. We can suppose that $a_{n+1} - a_n$ is infinitely large for all n. As $\mathtt{rate}\, V \subsetneqq \mathtt{rate}\, U$, there is an index k such that the sequence $\{b_{k'} - b_k\}_{k'>k}$ is not cofinal in $\mathtt{rate}\, U$. Let n, m be a pair of numbers satisfying (1) for this k. By the choice of k, there exists a number $n' > n$ such that $a_{n'} - a_n > b_{k'} - b_k$ for any $k' > k$, hence, in fact, $a_{n'} - a_n > \ell + b_{k'} - b_k$ for any $m' > m$ and any $\ell \in \mathbb{N}$. Finally, choose $k' > k$ and $m' > m$ according to (2) but w. r. t. n'. Put $C(f) = C_{f\restriction m'}$. Then we have, for all $\langle \xi, \eta \rangle$, $\langle \xi', \eta' \rangle$ in $C(f)$:

$$\left.\begin{aligned} \eta\restriction_{\geq b_k} = \eta'\restriction_{\geq b_k} &\implies \xi\restriction_{\geq a_n} = \xi'\restriction_{\geq a_n} \\ \eta\restriction_{\geq b_{k'}} \neq \eta'\restriction_{\geq b_{k'}} &\implies \xi\restriction_{\geq a_{n'}} \neq \xi'\restriction_{\geq a_{n'}} \end{aligned}\right\} . \tag{3}$$

Note that ${}^*\mathbf{S} = \mathtt{dom}\, \vartheta = \bigcup_{f \in F} X(f)$, where $X(f) = \mathtt{dom}\, C(f)$. Thus by Saturation there is a finite set $F' \subseteq F$ such that ${}^*\mathbf{S} = \bigcup_{f \in F'} X(f)$. Let us show that all sets $X(f)$ are too small for a finite union of them to cover ${}^*\mathbf{S}$.

Call an internal set $X \subseteq {}^*\mathbf{S}$ *small* iff

(∗) there is a number $z \in {}^*\mathbb{N} \smallsetminus \mathbb{N}$ such that, for any internal map $\sigma \in {}^{\mathrm{int}}2^{{}^*\mathbb{N} \smallsetminus [0,z)}$, the set $X_\sigma = \{\xi \in X : \xi\restriction_{\geq z} = \sigma\}$ satisfies $2^{-z} \# X_\sigma \simeq 0$.

Clearly ${}^*\mathbf{S}$ is not a union of finitely many small internal sets. To get a contradiction, it remains to show that any set $X(f)$ is small, with $z = a_{n'}$ in the notation above. (Note that $a_{n'}$ depends on f, of course.)

Take any $\langle \xi, \eta \rangle \in C(f)$ and let $\sigma = \xi\restriction_{\geq a_{n'}}$, $\tau = \eta \restriction_{\geq b_{k'}}$. By (3), each $\langle \xi', \eta' \rangle \in C(f)$ with $\xi'\restriction_{\geq a_{n'}} = \sigma$ satisfies $\eta'\restriction_{\geq b_{k'}} = \tau$. Divide the domain $\Psi = \{\eta' \in {}^*\mathbf{S} : \eta'\restriction_{\geq b_{k'}} = \tau\}$ into subsets $\Psi_w = \{\eta' \in \Psi : \eta' \restriction [b_k, b_{k'}) = w\}$, where $w \in {}^{\mathrm{int}}2^{[b_k, b_{k'})}$, totally $2^{b_{k'} - b_k}$ of the sets Ψ_w. For any such Ψ_w, the set $\Phi_w = \{\xi' : \exists \eta' \in \Psi_w \, \langle \xi', \eta' \rangle \in C(f)\}$ contains at most 2^{a_n} elements by the first implication in (3). Thus $X(f)_\sigma = \{\xi' \in X(f) : \xi'\restriction_{\geq a_{n'}} = \sigma\}$ contains at most $2^{a_n + b_{k'} - b_k}$ elements of the set $X(f)$, which is less than $2^{a_{n'} - \ell}$ for any $\ell \in \mathbb{N}$. We conclude that $X(f)$ is small, as required.

Part 2. Suppose that $\mathtt{rate}\, U \subseteq \mathtt{rate}\, V$. In this case it does not take much effort to redefine the sequences $\{a_n\}$, $\{b_k\}$, cofinal in resp. $\log U$, $\log V$, so that $a_{n+1} - a_n \leq b_{n+1} - b_n$ for all $n \in \mathbb{N}$. By Robinson's lemma (Theorem 2.2.12), there exist a number $h \in {}^*\mathbb{N} \smallsetminus \mathbb{N}$ and internal extensions

$\{a_\nu\}_{\nu\le h}$ and $\{b_\nu\}_{\nu\le h}$ of sequences $\{a_n\}_{n\in\mathbb{N}}$ and $\{b_n\}_{n\in\mathbb{N}}$, both being increasing hyperfinite sequences satisfying $a_{\nu+1}-a_\nu \le b_{\nu+1}-b_\nu$ for all $\nu < h$.

Now to prove that $\mathsf{R}_{\log U} \le_{\mathsf{B}} \mathsf{R}_{\log V}$ we define a Borel reduction ϑ of $\mathsf{R}_{\log U}$ to $\mathsf{R}_{\log V}$. If $\xi \in {}^*\mathbf{S}$ then let $\vartheta(\xi) = \eta \in {}^*\mathbf{S}$ be defined as follows:

1) $\eta \restriction [0, b_0)$ is constant 0 (not important);
2) $\eta(b_\nu + j) = \xi(a_\nu + j)$ whenever $\nu < h$ and $j < a_{\nu+1} - a_\nu$;
3) $\eta \restriction [b_\nu + a_{\nu+1} - a_\nu, b_{\nu+1})$ is constant 0 for any $\nu < h$;
4) $\eta(b_h + z) = \xi(a_h + z)$ for all $z \in {}^*\mathbb{N}$.

Thus, to define η, we move each piece $\xi \restriction [a_\nu, a_{\nu+1})$ of ξ so that it begins with b_ν-th position in η, and fill in the rest of $[b_\nu, b_{\nu+1})$ by 0s; in addition, $\eta\restriction_{\ge b_h}$ is a shift of $\xi\restriction_{\ge a_h}$. That such a map ϑ is a Borel reduction of $\mathsf{R}_{\log U}$ to $\mathsf{R}_{\log V}$ is a matter of routine verification. □

Beyond Theorem 9.7.20 little is known on the structure of the reducibility relations $\le_{\mathsf{B}}$ and $\le_{\mathsf{CD}}$. The following problem can be proposed with a moderate hope for a positive answer.

Problem 9.7.22. Is it true that the equivalence relations of the form M_U, where U is a countably determined cut in ${}^*\mathbb{N}$, and their restrictions to $*$-finite subsets of ${}^*\mathbb{N}$ form an initial segment within Borel (variant: countably determined) equivalence relations on ${}^*\mathbb{N}$ in the sense of $\le_{\mathsf{B}}$ (variant: $\le_{\mathsf{CD}}$)?

Is it true that for any non-B-smooth Borel equivalence relation E there is an equivalence relation of the form M_U and a number $h \in {}^*\mathbb{N} \smallsetminus \mathbb{N}$ such that the restriction $\mathsf{E}' = \mathsf{M}_U \restriction [0, h)$ is still non-B-smooth and $\mathsf{E}' \le_{\mathsf{B}} \mathsf{E}$? □

There is, still, an additional result.

Exercise 9.7.23. Let $P = \mathscr{P}_{\mathtt{fin}}(\mathbb{N})$, accordingly ${}^*P$ is the (internal) set of all $*$-finite subsets of ${}^*\mathbb{N}$. For $u, v \in$ define $u \,\mathsf{FD}\, v$ iff the symmetric difference $u \Delta v$ is finite. (FD from "finite difference".) Prove the following:

(1) if $U \subseteq {}^*\mathbb{N}$ is an additive countably cofinal cut then $\mathsf{M}_U \le_{\mathsf{B}} \mathsf{FD}$;
(2) if $V \subseteq {}^*\mathbb{N}$ is an additive countably coinitial cut then $\mathsf{M}_V \not\le_{\mathsf{CD}} \mathsf{FD}$.

Hints. (1) Suppose that $U = \bigcup_n [0, 2^{a_n})$, where $\{a_n\}$ is an increasing sequence in ${}^*\mathbb{N}$; accordingly, $\log U = \bigcup_n [0, a_n]$. It suffices to prove that $\mathsf{R}_{\log U} \le_{\mathsf{B}} \mathsf{FD}$. The sequence $\{a_n\}$ admits an internal $*$-extension $\{a_\nu\}_{\nu\le h}$, where $h \in {}^*\mathbb{N} \smallsetminus \mathbb{N}$, still an increasing internal sequence of elements of ${}^*\mathbb{N}$. Let, for any $\xi \in {}^*\mathbf{S}$, $\vartheta(\xi)$ be the (internal, hyperfinite) set of all restricted maps $\xi\restriction_{\ge a_\nu}$, $\nu \le h$. By definition, $\xi \,\mathsf{R}_{\log U}\, \eta$ iff the symmetric difference $\vartheta(\xi)\Delta\vartheta(\eta)$ is finite. Yet ϑ takes values in the set of all hyperfinite subsets of a certain internal $*$-countable set (because ${}^*\mathbf{S}$ itself is $*$-countable) which can be identified with ${}^*\mathbb{N}$.

(2) That $\mathsf{M}_V \not\le_{\mathsf{CD}} \mathsf{FD}$ follows from (4) of Exercise 9.7.19. □

In fact we have $\mathsf{M}_U <_{\mathsf{B}} \mathsf{FD}$ in (1) of the lemma because there is no $\le_{\mathsf{B}}$-largest one among all countably cofinal ERs by 9.7.21.

Historical and other notes to Chapter 9

Section 9.1. Borel and projective sets in a nonstandard domain: [KKML 89] (where earlier references can be found). That all infinite internal sets give rise to the same descriptive set theoretic properties (Remark 9.1.4) was observed in [HenR 93].

The Souslin operation A : [Sous 17]. δs-operations, as in (1) of § 9.2b, were introduced independently by Hausdorff [Haus 27] and Kolmogorov [Kol 28, Kol 93]. Operations Φ_U as in Definition 9.2.1 were introduced and studied in [KantL 32]. They are known as *set theoretic operations*, and their characteristic property is that whether an element x belongs to the result $\Phi_U\{X_i\}_{i\in I}$ is fully determined by the set $\{i \in I : x \in X_i\}$.

Section 9.2. Countably determined sets: Henson [Hen 79b]. Lemma 9.2.7: [Hen 79b] and [CV 79], see also 2.5 in [KKML 89].

In parallel "hyperfinite" descriptive set theory was developed by **AST** followers, with many results obtained independently and in a totally different notational system. For instance countably determined sets were independently introduced by Čuda and Vopenka [CV 79] under the name: *real classes*. See [MlZ 02, Prop. 2] on the equivalence of the definitions in [Hen 79b] and [CV 79]. See [Vop 79, Soc 76] on the **AST** system in general.

The history of shadows (or standard part maps) goes back to the earliest days of nonstandard analysis (standard parts of hyperreals, Definition 2.1.10). In a more general setting (see *e.g.* Henson [Hen 79a]) a shadow is a partial map $x \mapsto {}^\circ x$, defined in the set ${}^*\mathscr{X}_{\mathbf{ns}}$ of all near-standard elements of the $*$-extension ${}^*\mathscr{X}$ of a space $\mathscr{X} \in \mathbb{WF}$ so that ${}^\circ x = r \in \mathscr{X}$ iff $x \simeq {}^*r$ in ${}^*\mathscr{X}$. This is not exactly how we define shadows in § 9.2c, but the key property is the same: preimages of open subsets of $\mathscr{X}$ are $\mathbf{\Sigma}^0_1$ sets in ${}^*\mathscr{X}_{\mathbf{ns}}$.

Section 9.3. The operations $\Phi_{B(\mathbf{\Gamma})}$ for Borel and projective classes $\mathbf{\Gamma}$ were defined in [KantL 32]. Proposition 9.1.10 for Polish descriptive set theory is essentially due to Souslin [Sous 17], in the nonstandard setting due to Henson [Hen 79a, Hen 79b], see also [KKML 89, 1.2].

Theorem 9.3.3: see [KKML 89, § 1] where some earlier references are given. Lemma 9.3.5 is a slight generalization of Lemma 1.10 in [KKML 89].

Exercise 9.3.8: Henson [Hen 79a] for $n = 1$.

Corollary 9.3.6: see Schilling and Živaljević [SchilZ 97], where also references to early partial results are given.

Theorem 9.3.9: mainly from [KKML 89], where earlier references can be found. The results gathered in this theorem can be roughly divided into two groups. The first group contains claims that retain proofs achieved in Polish descriptive set theory possibly corrected so that completeness (or compactness) as the method to guarantee the non-emptiness of countable intersections is substituted by Saturation, for instance (iii). (The corresponding classical results are due to Souslin and Luzin [Sous 17, Luz 27]. Hausdorff gave a systematic exposition of those early proofs in [Haus 27].) The second group contains assertions proved by reduction to the Polish case by means of the shadow map, like (iv), (v), (vi), but classical proofs designed for the Polish case just do not go through directly.

Section 9.4. Separation and reduction results in § 9.4a are taken from [KKML 89, § 3]. Exercise 9.4.6 is due to Živaljević [Ziv 90]. See [KurM 76, XIII.1, XIII.4] or [Kech 95, 28.5] on Separation and Reduction in Polish spaces (including multiple principles) in classical descriptive set theory.

The content of § 9.4b — mainly [KKML 89, § 4].

Exercise 9.4.9 and § 9.4c: mainly Živaljević [Ziv 90, Ziv 91, Ziv 93].

Theorem 9.4.12 (uniformization) and § 9.4d in general: mainly § 4 in [KKML 89]. Uniformization of Borel and $\mathbf{\Pi}^1_1$ sets with $\mathbf{\Sigma}^0_1$ cross-sections: [Ziv 93, Ziv 95]. See [Kech 95, 36.D] on the Novikov – Kondo uniformization theorem in Polish spaces (originally due to Kondo [Kon 38]).

See Louveau [Louv 80] for earlier partial results related to Theorem 9.4.17. Claim (iv) of Theorem 9.4.18 was proved by Schilling and Živaljević [SchilZ 97].

Theorem 9.4.21(ii), (iii) is from Živaljević [Ziv 93] (Prop. 2.10), however a short proof given there (based on the claim that the sets D, E defined there belong to $\mathbf{\Pi}^1_1$) is not convincing. See also [Ziv 95], where Živaljević proves some other interesting results, mainly on $\mathbf{\Pi}^1_1$ sets.

See Larman [Lar 72] or Kechris [Kech 95, 35.H] on Arsenin – Kunugui results on Borel sets with $\mathbf{\Pi}^0_1$ and $\mathbf{\Sigma}^0_2$ cross-sections. (Original results were published in [Ar 40, Ar 40a, Kunug 40].)

Section 9.5. Loeb measures were introduced by Loeb [Loeb 75]. See [StrB 86, Lin 88, Cut 94, Ross 97, Gol 98, Cut 00] on the theory and applications of Loeb measures in differend fields of nonstandard analysis.

Example 9.5.5(3) is presented in [Ziv 96] together with a more advanced construction of a non-approximable $\mathbf{\Sigma}^1_1$ set.

Liftings (Theorem 9.5.9, Exercise 9.5.11): see Živaljević [Ziv 93, Ziv 95], Corollary 9.5.10 for $\mathbf{\Sigma}^1_1$ is due to Henson and Ross see [HenR 93].

See [LiV 93, US 93, vLam 96, DKUV 03] on different notions of randomness (of a single real), and [Sol 70] on the Solovay randomness. Definition 9.5.14: [KanR 97]. Gordon considered a slightly different notion of randomness: a $*$-real x is *random* in [Gor 91] iff it avoids any standard set of null Lebesgue measure. However this approach does not seem to prove Steinitz Exchange; the natural argument (as in Theorem 9.5.19) works only under the assumption that all sets are Lebesgue measurable, incompatible with **ZFC**. On Steinitz Exchange see [vLam 92, vLam 96] where van Lambalgen summarized reasonable requirements for the notion of randomness.

Theorem 9.5.20: [KanR 00b].

The content of § 9.5f: [KanR 96a]. See [Cut 84, Hen 92] on some other games in a "hyperfinite" domain.

Section 9.6. See [Kech 95, 15.7] on a Polish version of Lemma 9.6.2(i).

See [KKML 89, § 2] for references to different parts of Theorem 9.6.4.

Theorem 9.6.6 and Corollary 9.6.10(i) were established, in the frameworks of **AST**, by Kalina and Zlatoš [KalZ 89a] (Theorems 4.10 and 4.12). Kalina and Zlatoš obtained other related results in [KalZ 88, KalZ 89, KalZ 90], in particular, under the assumptions of the theorem, there also exists a *maximal* Borel cut U satisfying $X \equiv_{\mathsf{B}} U$ [KalZ 89, Theorem 1.10]. Those arguments in **AST** do not involve the two-cardinal hypothesis (essentially, the continuum hypothesis **CH**), thus can be considered as valid proofs in hyperfinite descriptive set theory. On the other hand, Schilling [Schil 98] (based in part on earlier results of Živaljević [Ziv 90]) gave a proof of Theorem 9.6.6 in modern terms of “hyperfinite” descriptive set theory.

Lemma 9.6.8 and Corollary 9.6.10(ii): Živaljević [Ziv 90].

Theorem 9.6.13: [KanR 03]; some relevant results were obtained in the frameworks of **AST**, for instance, in [KalZ 89] (Thms 2.5, 2.6).

Section 9.7. See [Hj 00, Kech 99] on the ongoing study of “Borel cardinals” in Polish descriptive set theory.

Silver’s theorem (for Polish spaces) was published in [Sil 80]. Theorem 9.7.1 first appeared in explicit form in [KanR 03], but P. Zlatos informed us that a similar unpublished result was earlier obtained by Vencovská in the frameworks of **AST**. Exercise 9.7.4: [KKML 89, 2.6].

Theorem 9.7.7: Mlček and Zlatoš [MlZ 02] (with several other interesting results). A weaker result, for countably determined partitions into countably many classes, was obtained in [KKML 89, 2.8].

Theorem 9.7.8 and § 9.7c in general: [KanR 03].

Theorem 9.7.10 on “countable” ERs: [KanR 03], but in the particular case $\mathsf{E} = \mathsf{M}_{\mathbb{N}}$ (Exercise 9.7.13) by [Jin 01]. Jin’s original proof is presented in the “if” part of the proof of Theorem 1.4.7 in Chapter 1.

Monads and monad partitions (§ 9.7e) were employed in [KL 91] to define and study a family of topologies on ${}^*\mathbb{N}$ with various properties. See also [Jin 01] and references there.

Theorem 9.7.14: [Jin 01]. The key claim in the proof of (ii) Theorem 9.7.14 in contained in [KKML 89, 2.6].

Borel reducibility of equivalence relations in Polish spaces has been extensively studied in classical descriptive set theory since the late 1980s. The version for “hyperfinite” descriptive set theory (§ 9.7f, including the concept of smoothness) was introduced in [KanR 03]. Theorem 9.7.18: [KanR 03].

The content of § 9.7g: [KanR 03], where full proofs are given and some open problems are discussed in the final section.

References

[A 88] P. Aczel, *Non-well-founded sets*, CSLI Lecture Notes, vol. 14, Stanford University, 1988.

[An 97] P. Andreev, On the standardization principle in the theory of bounded sets, *Moscow Univ. Math. Bull*, 1997, 52, 1, pp. 46 – 48.

[An 99] P. Andreev, On definable predicates of standardness in internal set theory, *Math. Notes*, 1999, 66, 5 – 6, pp. 665 – 669.

[AnG 2001] P. Andreev and E. Gordon, An axiomatics for nonstandard set theory, based on von Neumann-Bernays-Gödel theory. *J. Symbolic Logic*, 2001, 66, 3, pp. 1321 – 1341.

[AnG **] P. Andreev and E. Gordon, *A theory of hyperfinite sets*. Preprint.

[AnH 04] P. Andreev and K. Hrbaček, Standard sets in nonstandard set theory. *J. Symbolic Logic*, 2004, 69, 1, pp. 165 – 182.

[Ar 40] V. Arsenin, Sur les projections de certains ensembles mesurables B, *C. R. (Dokl.) Acad. Sci. URSS (N. S.)* 1940, 27, pp. 107 – 109.

[Ar 40a] V. Arsenin, Sur la nature des projections de certains ensembles mesurables B, *Bull. Acad. Sci. URSS Ser. Math. [Izvestia]* 1940, 4, pp. 403 – 410.

[Bal 94] D. Ballard, Foundational aspects of "non"standard mathematics, *Contemporary Math.* 1994, 176.

[BalH 92] D. Ballard and K. Hrbaček, Standard foundations for nonstandard analysis, *J. Symbolic Logic* 1992, 57, pp. 741 – 748.

[Bell 77] J. L. Bell, *Boolean valued models and independence proofs in set theory*, Oxford, Clarendon, 1977.

[BenDN 03] V. Benci and M. Di Nasso, Alpha–theory: an elementary axiomatics for nonstandard analysis, *Expositiones Math.* 2003, 21, pp. 355 – 386.

[vdBerg 87] I. van den Berg, *Nonstandard asymptotic analysis*, Lecture Notes in Mathematics 1249, Springer 1987.

[vdBerg 92] I. van den Berg, Extended use of IST, *Ann. Pure Appl. Log.* 1992, 58, pp. 73 – 92.

[vdBergK 95] I. van den Berg and F. Koudjeti, Neutrices, external numbers, and external calculus. *Nonstandard analysis in practice*, Springer, Berlin, 1995, pp. 145 – 170.

[BertC 94] N. Bertoglio, R. Chuaqui, An elementary geometric nonstandard proof of the Jordan curve theorem, *Geometriae Dedicata*, 1994, 51 , pp. 14 – 27.

[Bof 72] M. Boffa, *Forcing et Négation de l'Axiome de Fondement*, *Mém. Acad. Sci. Belg.* 1972, t. XL, fasc. 7.

[Bur 79] J. Burgess, Effective enumeration of classes in a $\mathbf{\Sigma}^1_1$ equivalence relation, *Indiana Univ. Math. J.* 1979, 28, pp. 353 – 364.

[CK 92] C. C. Chang and H. J. Keisler, *Model Theory, 3rd ed.*, North Holland, Amsterdam, 1992, xiv + 650 pp.

[CherH 70] G. Cherlin and J. Hirschfeld, Ultrafilters and ultraproducts in non-standard analysis. *Contributions to non-standard analysis (Sympos., Oberwolfach, 1970)*, pp. 261 – 279. Studies in Logic and Found. Math., Vol. 69, North-Holland, Amsterdam, 1972.

[Coh 66] P. J. Cohen, *Set Theory and the Continuum Hypothesis*, Benjamin, 1966.

[Cohn 80] D. L. Cohn, *Measure Theory*, Birkhäuser, 1980.

[CV 79] K. Čuda and P. Vopenka, Real and imaginary classes in the AST, *Comment. Math. Univ. Carol.* 1979, 20, pp. 639 – 653.

[Cut 84] N. Cutland, A question of Borel hyperdeterminacy, *Z. Math. L. Grundl. Math.* 1984, 30, pp. 313 – 316.

[Cut 94] N. Cutland, Loeb measure theory, *Developments in nonstandard mathematics (Aveiro, 1994)*, pp. 151 – 177, Pitman Res. Notes Math. Ser., 336, Longman, Harlow, 1995.

[Cut 97] N. Cutland, Nonstandard real analysis. *Nonstandard Analysis: Theory and Applications (Edinburg 1996)*, pp. 51 – 76. NATO Adv. Sci. Inst. Ser. C: Math. Phys. Sci. 493 Eds. L. O. Arkeryd, N. J. Cutland and C. W. Henson. Kluwer, Dordrecht, 1997.

[Cut 00] N. Cutland, *Loeb measures in practice: recent advances*, Lecture Notes in Mathematics, 1751. Springer-Verlag, Berlin, 2000. xii+111 pp.

[Das 96] A. Dasgupta, Boolean operations, Borel sets, and Hausdorff's question. *J. Symbolic Logic* 1996, 61, no. 4, pp. 1287 – 1304.

[Dev 98] K. J. Devlin, *The joy of sets: fundamentals of contemporary set theory*, 2nd Edition. Springer, New York *et. al.*, 1998, X + 192 p. (Undergraduate texts in mathematics).

[DienD 95] F. Diener and M. Diener (eds.), *Nonstandard analysis in practice*, Springer, Berlin, 1995, xiv + 250 pp.

[DienR 89] F. Diener and G. Reeb, *Analyse non standard*, Herrmann Editeurs, Paris, 1989, 196 pp.

[DienS 88] F. Diener and K. D. Stroyan, Syntactical methods in infinitesimal analysis, in: N. Cutland (ed.) *Nonstandard analysis and its applications* (London Math. Soc. Student Texts 10, Cambridge Univ. Press, 1988), pp. 258 – 281.

[DiN 97] M. Di Nasso, ***ZFC** : an axiomatic *approach to nonstandard methods, *C. R. Acad. Sci. Paris* 1997, 324, Série 1, pp. 963 – 967.

[DiN 98] M. Di Nasso, Pseudo-superstructures as nonstandard universes, *J. Symbolic Logic* 1998, 63, pp. 222 – 236.

[DiN 99] M. Di Nasso, On the foundations of nonstandard mathematics, *Math. Japonica* 1999, 50, pp. 131 – 160.

[DiNH 03] M. Di Nasso and K. Hrbaček, Combinatorial principles in nonstandard analysis, *Ann. Pure Appl. Logic* 2003, 119, no. 1-3, pp. 265 – 293.

[DKUV 03] B. Durand, V. Kanovei, V. A. Uspensky, N. Vereshchagin, Do stronger definitions of randomness exist? *Theoret. Comput. Sci.* 2003, 290, no. 3, pp. 1987 – 1996.

[E:Inf] L. Euler, *Introduction to Analysis of the Infinite*, Springer, 1988, xiv + 327 pp.

[E:Dif] L. Euler, *Foundations of differential calculus*. Springer-Verlag, New York, 2000. xvi+194 pp.

[Fel 71] U. Felgner, Comparision of the axioms of local and universal choice, *Fund. Math.* 1971, 71, pp. 43 – 62.

[Fl 89] P. Fletcher, Nonstandard set theory. *J. Symbolic Logic* 1989, 54, pp. 1000 – 1008.

[FS 87] H. Friedman and M. Sheard, An axiomatic approach to self-referential truth, *Ann. Pure Appl. Logic* 1987, 33, no. 1, pp. 1 – 21.

[Gol 98] R. Goldblatt, *Lectures on the hyperreals. An introduction to nonstandard analysis*. Graduate Texts in Mathematics, 188. Springer-Verlag, New York, 1998. xiv+289 pp.

[Gor 89] E. Gordon, Relatively standard elements in the theory of internal sets of E. Nelson. *Siberian Math. J.* 1989, 30, pp. 89 – 95.

[Gor 91] E. Gordon, On Loeb measures, *Russian Mathematics (Izv. VUZ)* 1991, 2, pp. 25 – 33.

[Gor 97] E. Gordon, *Nonstandard methods in commutative harmonic analysis*, AMS, 1997.

[HKL 88] L. A. Harrington, A. S. Kechris, A. Louveau, A Glimm – Effros dichotomy for Borel equivalence relations, *J. Amer. Math. Soc.* 1988, 310, pp. 293 – 302.

[Haus 27] F. Hausdorff, *Mengenlehre*, zweite neubearbeitete Aufl., Berlin: W. de Gruyter & Co, 1927.

[Haus 37] Ф. Хаусдорф, *Теория множеств,* М. ОНТИ, 1937. (Russian translation of [Haus 27] edited by P. Aleksandrov and A. Kolmogorov. Contains some cuts and insertions written by the editors.)

[Hen 74] C. W. Henson, The isomorphism property in nonstandard analysis and its use in the theory of Banach spaces. *J. Symbolic Logic* 1974, 39, pp. 717 – 731.

[Hen 75] C. W. Henson, When do two Banach spaces have isometrically isomorphic nonstandard hulls. *Israel J. Math.* 1975, 22, pp. 57 – 67.

[Hen 79a] C. W. Henson, Analytic sets, Baire sets, and the standard part map, *Canadian J. of Math.* 1979, 31, pp. 663 – 672.

[Hen 79b] C. W. Henson, Unbounded Loeb measures, *Proc. Amer. Math. Soc.* 1979, 64, pp. 143 – 160.

[Hen 92] C. W. Henson, Strong counterexamples to Borel hyperdeterminacy, *Arch. Math. Log.*, 1992, 31, pp. 215 – 220.

[Hen 97] C. W. Henson, Foundations of nonstandard analysis: a gentle introduction to nonstandard extension. *Nonstandard Analysis: Theory and Applications (Edinburg 1996)*, pp. 1 – 49. NATO Adv. Sci. Inst. Ser. C: Math. Phys. Sci. 493, Eds. L. O. Arkeryd *et. al.*, Kluwer, Dordrecht, 1997.

[HenK 86] C. W. Henson and H. J. Keisler, On the strength of nonstandard analysis, *J. Symbolic Logic* 1986, 51, pp. 377 – 386.

[HenR 93] C. W. Henson and D. Ross, Analytic mappings on hyperfinite sets, *Proc. Amer. Math. Soc.* 1993, 118, pp. 587 – 596.

[Hj 00] G. Hjorth, *Classification and Orbit Equivalence Relations* (Mathematical surveys and monographs, 75), AMS, 2000.

[Hr 78] K. Hrbaček, Axiomatic foundations for nonstandard analysis, *Fundamenta Mathematicae* 1978, 98, pp. 1 – 19.

[Hr 79] K. Hrbaček, Nonstandard set theory, *Amer. Math. Monthly* 1979, 86, pp. 659 – 677.

[Hr 01] K. Hrbaček, Realism, nonstandard set theory, and large cardinals, *Annals of Pure and Applied Logic*, 2001, 109, pp. 15 – 48.

[Hr 04] K. Hrbaček, Internally iterated ultrapowers. *Proceedings of the Special Session on Nonstandard Methods, Baltimore 2003.* Eds. A. Enayat and R. Kossak. To appear in *Contemporary Math.*, AMS, Providence, R.I., 2004.

[Hr **] K. Hrbaček, *Nonstandard objects in set theory*. Preprint, 2003. (Submitted to Proceedings of "Nonstandard Methods and Applications in Math.", Pisa 2002.)

[HrJ 98] K. Hrbaček, T. Jech, *Introduction to set theory*, 3rd edition (Marcel Dekker, 1998).

[HuL 85] A. E. Hurd and P. A. Loeb, *An introduction to nonstandard real analysis* (Academic Press, 1985).

[JackKL 02] S. Jackson, A. S. Kechris, and A. Louveau, Countable Borel equivalence relations, *J. Math. Logic* 2002, 2, no. 1, pp. 1 – 80.

[Jech 78] T. Jech, *Set theory* (Academic Press, 1978).

[Jin 92] R. Jin, The isomorphism property versus special model axiom, *J. Symbolic Logic* 1992, 57, pp. 975 – 987.

[Jin 92a] R. Jin, A theorem on the isomorphism property. *J. Symbolic Logic* 1992, 57, pp. 1011 – 1017.

[Jin 96] R. Jin, Better nonstandard universes with applications. *Nonstandard Analysis: Theory and Applications (Edinburg 1996)*, pp. 183 – 208, NATO Adv. Sci. Inst. Ser. C: Math. Phys. Sci. 493, Eds. L. O. Arkeryd *et. al.*, Kluwer, Dordrecht, 1997.

[Jin 99] R. Jin, Distinguishing three strong saturation properties in nonstandard analysis. *Ann. Pure Appl. Logic* 1999, 98, no. 1-3, pp. 157 – 171.

[Jin 01] R. Jin, Existence of some sparse sets of nonstandard natural numbers, *J. Symbolic Logic* 2001, 66, 2, pp. 959 – 973.

[Jin **] R. Jin, On some questions of Hrbacek and Di Nasso, to appear.

[JinK 93] R. Jin and H. J. Keisler, Game sentences and ultrapowers, *Ann. Pure Appl. Logic* 1993, 60, pp. 261 – 274.

[JinS 94] R. Jin and S. Shelah, The strength of the isomorphism property, *J. Symbolic Logic* 1994, 59, pp. 292 – 301.

[Jor 1893] C. Jordan, *Cours d'analyse*, 2nd ed. 1893, Gauthier-Villars.

[KalZ 88] M. Kalina and P. Zlatoš, Arithmetic of cuts and cuts of classes, *Comment. Math. Univ. Carol.* 1988, 29, pp. 435 – 456.

[KalZ 89] M. Kalina and P. Zlatoš, Cuts of real classes, *Comment. Math. Univ. Carol.* 1989, 30, pp. 129 – 136.

[KalZ 89a] M. Kalina and P. Zlatoš, Borel classes in AST, measurability, cuts, and equivalence, *Comment. Math. Univ. Carol.* 1989, 30, pp. 357 – 372.

[KalZ 90] M. Kalina and P. Zlatoš, Some connections between measure, indiscernibility and representation of cuts, *Comment. Math. Univ. Carol.* 1990, 31, pp. 751 – 763.

[Kan 88] V. Kanovei, The correctness of Euler's method for the factorization of the sine function into an infinite product. *Russian Math. Surveys* 1988, 43:4, pp. 65 – 94.

[Kan 90] V. Kanovei, Ограниченные множества в теории внутренних множеств Эдварда Нельсона. (Bounded sets in Edvard Nelson's internal set theory), *Nonstandard analysis. 3rd USSR seminar*. Saratov, 1990, pp. 15 – 23. [Russian]

[Kan 91] V. Kanovei, Undecidable hypotheses in Edward Nelson's Internal Set Theory. *Russian Math. Surveys* 1991, 46:6, pp. 1 – 54.

[Kan 94a] V. Kanovei, **IST** *is more than an algorithm to prove* **ZFC** *theorems*, University of Amsterdam, Research report ML-94-05, June 1994.

[Kan 94b] V. Kanovei, *A course on foundations of nonstandard analysis*, (With a preface by M. Reeken.) IPM, Tehran, Iran, 1994.

[Kan 95] V. Kanovei, Uniqueness, Collection, and external collapse of cardinals in **IST** and models of Peano arithmetic. *J. Symbolic Logic* 1995, 57, 1, pp. 1 – 7.

[KanR 95] V. Kanovei and M. Reeken, Internal approach to external sets and universes. Appeared in *Studia Logica*, in three parts:
Part 1, Bounded set theory, *SL* 1995, 55, no. 2, pp. 229 – 257.
Part 2, External universes over the universe of bounded set theory, *SL* 1995, 55, no. 3, pp. 347 – 376.
Part 3, Partially saturated universes, *SL* 1996, 56, no. 3, pp. 293 – 322.

[KanR 96a] V. Kanovei and M. Reeken, Loeb measure from the point of view of coin flipping game, *Math. Logic Quarterly* 1996, 42, no 1, pp. 19 – 26.

[KanR 96b] V. Kanovei and M. Reeken, Summation of divergent series from the nonstandard point of view. *Real Analysis Exchange*, 1996, 21, 2, pp. 453 – 477.

[KanR 97] V. Kanovei and M. Reeken, Mathematics in a nonstandard world. (In two parts.) *Math. Japonica*, 1997, vol. 45, part 1: no. 2, pp. 369 – 408, part 2: no. 3, pp. 555 – 571.

[KanR 97a] V. Kanovei and M. Reeken, Isomorphism property in nonstandard extensions of a **ZFC** universe, *Ann. Pure Appl. Logic*, 1997, 88, pp. 1 – 25.

[KanR 98] V. Kanovei and M. Reeken, Elementary extensions of external classes in a nonstandard universe. *Studia Logica*, 1998, 60, 2, pp. 253 – 273.

[KanR 99a] V. Kanovei and M. Reeken, A nonstandard proof of the Jordan curve theorem, *Real Analysis Exchange*, 1999, 24, no 1, pp. 161–170.

[KanR 99b] V. Kanovei and M. Reeken, Special model axiom in nonstandard set theory. *Math. Logic Quarterly*, 1999, 45, 3, pp. 371 – 384.

[KanR 99c] V. Kanovei and M. Reeken, Extension of standard models of ZFC to models of nonstandard Nelson's set theory IST. *Math. Notes*, 1999, 66, 2, pp. 160 – 166.

[KanR 00a] V. Kanovei and M. Reeken, Extending standard models of ZFC to models of nonstandard set theories. *Studia Logica*, 2000, 64, pp. 37 - -59.

[KanR 00b] V. Kanovei and M. Reeken, A nonstandard set theory in the $\in$-language. *Archive for Math. Logic*, 2000, 39, 4, pp. 403 – 416.

[KanR 03] V. Kanovei and M. Reeken, Borel and countably determined reducibility in nonstandard domain. *Monatshefte für Mathematik*, 2003, 140, 3, pp. 197–231.

[KanS 04] V. Kanovei and S. Shelah, A definable nonstandard model of the reals. *J. Symbolic Logic*, to appear.

[KantL 32] L. Kantorovich, E. Livenson, Memoir on the analytical operations and projective sets, *Fund. Math.*, Part 1: 1932, 18, pp. 214 – 279, Part 2: 1933, 20, pp. 54 – 97.

[Kaw 81] T. Kawaï, Axiom systems for nonstandard set theory, in: *Logic Symposia Hakone* 1979, 1980 (Lecture Notes Math. 891, Springer, 1981), pp. 57 – 65.

[Kaw 83] T. Kawaï, Nonstandard analysis by axiomatic methods, in: *Southeast Asia Conference on Logic, Singapore* 1981 (Studies in Logic and Foundations of Mathematics, 111, North Holland, 1983), pp. 55 – 76.

[Kech 95] A. S. Kechris. *Classical Descriptive Set Theory*. (Graduate Texts in Mathematics, vol. 156.) Springer, 1995.

[Kech 99] A. S. Kechris, New directions in descriptive set theory, *Bull. Symbolic Logic*, 1999, 5(2), pp. 161 – 174.

[Keis 94] H. J. Keisler, The hyperreal line, in P. Erlich (ed.) *Real numbers, generalizations of reals, and theories of continua*, Kluwer Academic Publishers, 1994, pp. 207 – 237.

[KKML 89] H. J. Keisler, K. Kunen, A. Miller, and S. Leth, Descriptive set theory over hyperfinite sets, *J. Symbolic Logic* 1989, 54, pp. 1167 – 1180.

[KL 91] H. J. Keisler and S. Leth, Meager sets on the hyperfinite time line, *J. Symbolic Logic* 1991, 56, pp. 71 – 102.

[Kol 28] A. Kolmogoroff, Operations sur des ensembles, *Matem. Sbornik* 1928, 35, pp. 414 – 422. (Dated January 1922.)

[Kol 93] A. Kolmogorov, On operations on sets, II, *In*: A. Shiryaev ed. *Selected Works of A. N. Kolmogorov, Vol. III: Information Theory and the Theory of Algorithms*, Kluwer, 1993, pp. 266 – 274. (Dated February 1922.)

[Kon 38] M. Kondô, Sur l'uniformization des complémentaires analytiques et les ensembles projectifs de la seconde classe, *Japan J. Math.* 1938, 15, pp. 197 – 230.

[Kr 69] G. Kreisel, Axiomatizations of nonstandard analysis that are conservative extensions of formal systems for classical standard analysis, in: *Applications of model theory to algebra, analysis and probability* (Holt, Rinehart and Winston 1969), pp. 93 – 106.

[Kun 80] K. Kunen, *Set theory, an introduction to independence proofs*, Studies in Logic and the Foundations of Mathematics, vol. 102 (Elsevier, 1980).

[Kunug 40] K. Kunugui, Sur un problème de M. E. Szpilrajn, *Proc. Imp. Acad. Tokyo* 1940, 16, pp. 73 – 78.

[KurM 76] K. Kuratowski and A. Mostowski, *Set Theory, with an introduction to descriptive set theory*, North-Holland, 1976.

[KusK 94] A. G. Kusraev and S. S. Kutateladze, *Nonstandard Methods of Analysis*, Kluwer, 1994.

[vLam 92] M. van Lambalgen, Independence, randomness, and the axiom of choice, *J. Symbolic Logic* 1992, 57, pp. 1274 – 1304.

[vLam 96] M. van Lambalgen, Randomness and foundations of probability: von Mises' axiomatisation of random sequences, *Statistics, probability and game theory*, pp. 347 – 367. (IMS Lecture Notes Monogr. Ser., 30.) Inst. Math. Statist., Hayward, CA, 1996.

[Lar 72] D. G. Larman, Projecting and uniformising Borel sets with $\mathcal{K}_\sigma$ sections. *Mathematika* 1972, 19, pp. 231 – 244 and 1973, 20, pp. 233 – 246.

[LiV 93] Ming Li, P. Vitanyi, *An introduction to Kolmogorov complexity and its applications*, Springer, 1993.

[Lin 88] T. Lindstrøm, An invitation to nonstandard analysis, in: N. Cutland (ed.) *Nonstandard analysis and its applications* (London Math. Soc. Student Texts 10, Cambridge Univ. Press, 1988), pp. 1 – 105.

[Loeb 75] P. Loeb, Conversion from nonstandard to standard measure spaces and applications in probability theory, *Trans. Amer. Math. Soc.* 1975, 211, pp. 113 – 122.

[Loeb 97] P. Loeb, Nonstandard analysis and topology. *Nonstandard Analysis: Theory and Applications (Edinburg 1996)*, pp. 77 – 89. NATO Adv. Sci. Inst. Ser. C: Math. Phys. Sci. 493, Eds. L. O. Arkeryd *et. al.*, Kluwer, Dordrecht, 1997.

[Loeb 00] P. Loeb, An introduction to nonstandard analysis, *Nonstandard analysis for the working mathematician*, pp. 1 – 95. Mathematics and its Applications, 510. Kluwer, Dordrecht, 2000.

[LoebW 00] P. Loeb and M. Wolff (eds.), *Nonstandard analysis for the working mathematician*, Mathematics and its Applications, 510. Kluwer, Dordrecht, 2000.

[Louv 80] A. Louveau, A separation theorem for Σ^1_1 sets, *Trans. Amer. Math. Soc.* 1980, 260, no. 2, pp. 363–378.

[LutG 81] R. Lutz and M. Goze, *Nonstandard analysis. A practical guide with applications* (Lecture Notes Math. 881, Springer, 1981).

[Lux 62] W. A. J. Luxemburg, *Non–standard analysis, lecture notes* (Dept. Math. Cal. Inst. Tech. 1962).

[Lux 73] W. A. J. Luxemburg, What is nonstandard analysis? *Papers in the foundations of mathematics, Amer. Math. Monthly* 1973, 80, no. 6, part II, pp. 38 – 67.

[Lux 77] W. A. J. Luxemburg, Non-standard analysis. *Logic, foundations of mathematics and computability theory (Proc. Fifth Internat. Congr. Logic, Methodology and Philos. of Sci., 1975)*, Part I, pp. 107 – 119. Reidel, Dordrecht, 1977.

[Luz 27] N. N. Luzin, Sur les ensembles analytiques, *Fundamenta Math.* 1927, 10, pp. 1 – 95.

[LyanK 97] W. Lyantse, T. Kudryk, *Introduction to nonstandard analysis*. Mathematical Studies Monograph Series, 3. VNTL Publishers, L'viv, 1997. 255 pp.

[Mat 01] A. R. D. Mathias, The strength of Mac Lane set theory, *Ann. Pure Appl. Logic* 2001, 110, no. 1-3, pp. 107 – 234.

[Mil 90] A. W. Miller, Set-theoretic properties of Loeb measure, *J. Symbolic Logic* 1990, 55, no. 3, pp. 1022 – 1036.

[MlZ 00] J. Mlček and P. Zlatoš, The Ramsey structure of A-determined sets in a κ-saturated universe, *Logic Colloquium '98 (Prague)*, ASL, 2000, pp. 316–333.

[MlZ 02] J. Mlček and P. Zlatoš, Some Ramsey-type theorems for countably determined sets, *Arch. Math. Logic* 2002, 41, no. 7, pp. 619–630.

[Nar 71] L. Narens, A nonstandard proof of the Jordan curve theorem, *Pacific J. Math.* 1971, 36, 1 pp. 219 – 229.

[Nel 77] E. Nelson, Internal set theory; a new approach to nonstandard analysis, *Bull. Amer. Math. Soc.* 1977, 83, 6, pp. 1165 – 1198.

[Nel 88] E. Nelson, The syntax of nonstandard analysis, *Ann. Pure Appl. Log.* 1988, 38, pp. 123 – 134.

[Nel **] E. Nelson, *Unfinished book on nonstandard analysis, Chapter 1, Internal Set Theory*, unpublished.

[Os 12] W. F. Osgood, *Funktionentheorie*, vol. 1, p. 161, 1912, Teubner.

[Per 92] Y. Péraire, Théorie relative des ensembles internes, *Osaka J. Math.* 1992, 29, pp. 267 – 297.

[Per 95] Y. Péraire, Some extensions of the principles of idealization, transfer, and choice in the relative internal set theory, *Arch. Math. Logic* 1995, 34, pp. 269 – 277.

[Pr 98] M. Prokhorova, On relative near-standardness in IST, *Siberian Math. J.* 1998, 39, 3, pp. 518 – 521.

[R 92] M. Reeken, On external constructions in Internal Set Theory. *Expositiones Mathematicae* 1992, 10, pp. 193 – 247.

[Rob 88] A. Robert, *Nonstandard analysis* (Wiley, 1988).

[RobinZ 69] A. Robinson and E. Zakon, A set-theoretic characterization of enlargements, in: W. A. J. Luxemburg (ed.) *Applications of model theory to algebra, analysis, and probability* (Holt, Rinehart, and Winston, 1969), pp. 109 – 122.

[Rog 70] C. A. Rogers, *Hausdorff Measures*, Cambridge Univ. Press, 1970.

[Ross 90] D. Ross, The special model axiom in nonstandard analysis, *J. Symbolic Logic* 1990, 55, pp. 1233 – 1242.

[Ross 97] D. Ross, Loeb measure and probability, *Nonstandard analysis: Theory and Applications (Edinburgh, 1996)*, pp. 91 – 120. NATO Adv. Sci. Inst. Ser. C: Math. Phys. Sci. 493, Eds. L. O. Arkeryd *et. al.*, Kluwer, Dordrecht, 1997.

[Schil 98] K. Schilling, Vanishing Borel sets. *J. Symbolic Logic* 1998, 63, no. 1, pp. 262 – 268.

[SchilZ 97] K. Schilling and B. Živaljević, Louveau's theorem for the descriptive set theory of internal sets, *J. Symbolic Logic*, 1997, 62, 2, pp. 595 – 607.

[Schm 95] J. H. Schmerl, The isomorphism property for nonstandard universes, *J. Symbolic Logic* 1995, 60, pp. 512 – 516.

[SchmiedL 58] C. Schmieden and D. Laugwitz, Eine erweiterung der Infinitesimalrechnung, *Math. Zeitschr.* 1958, 69, pp. 1 – 39.

[Sh 94] M. Sheard, A guide to truth predicates in the modern era, *J. Symbolic Logic* 1994, 59, no. 3, pp. 1032 – 1054.

[Shoen 67] J. R. Shoenfield, *Mathematical logic*, Addison–Wesley, 1967.

[Shoen 71] J. R. Shoenfield, Unramified forcing, *Proc. Symp. Pure Math.* 13, part 1 (1971), pp. 357 – 380.

[Sil 80] J. Silver, Counting the number of equivalence classes of Borel and coanalytic equivalence relations, *Ann. Math. Log.* 1980, 18, pp. 1 – 28.

[Sin 93] Y. G. Sinai, *Probability theory*, Springer, 1992.

[Soc 76] A. Sochor, The alternative set theory, *Lecture Notes in Math.* 537, pp. 259 – 273, Springer 1976.

[Sol 70] R. M. Solovay, A model of set theory in which every set of reals is Lebesgue measurable, *Ann. of Math.* 1970, 92, pp. 1 – 56.

[Sous 17] M. Souslin, Sur une définition des ensembles mesurables B sans nombres transfinis, *C. r. Acad. sci. Paris* 1917, 164, pp. 88 – 91.

[StrB 86] K. D. Stroyan and J. M. Bayod, *Foundations of infinitesimal stochastic analysis* (North Holland, 1986).

[Suz 99] A. Suzuki, No elementary embedding from V into V is definable from parameters, *J. Symbolic Logic* 1999, 64, 4, pp. 1591 – 1594.

[Tz **] A. Tzouvaras, *Forcing and antifoundation*, under review.

[US 93] V. Uspensky and A. Semenov, *Algorithms: Main Ideas and Applications*. Kluwer Acad. Publ., 1993.

[Veb 05] O. Veblen, Theory of plane curves in non-metrical Analysis Situs, *Trans. Amer. Math. Soc* 1905, 6, pp. 83 – 98.

[Vop 79] P. Vopěnka, *Mathematics in the alternative set theory*. Teubner Verlagsgesellschaft, Leipzig, 1979. 120 pp.

[VopH 72] P. Vopěnka and P. Hájek, *The theory of Semisets*, North-Holland, Amsterdam, 1972.

[Zak 74] E. Zakon, A new variant of non-standard analysis, in A. Hurd (ed.) *Victoria Symposium on Nonstandard Analysis*, (Lecture Notes in Math. 369) Springer-Verlag, 1974, pp. 313 – 339.

[Ziv 90] B. Živaljević, Some results about Borel sets in descriptive set theory of hyperfinite sets, *J. Symbolic Logic* 1990, 55, no. 2, pp. 604 – 614.

[Ziv 91] B. Živaljević, The structure of graphs all of whose Y-sections are internal sets, *J. Symbolic Logic* 1991, no. 1, pp. 50 – 66.

[Ziv 93] B. Živaljević, Graphs with $\Pi^0_1(\kappa)$ Y-sections, *Arch. Math. Logic* 1993, 32, no. 4, pp. 259 – 273.

[Ziv 95] B. Živaljević, Π^1_1 functions are almost internal, *Trans. Amer. Math. Soc.* 1995, 347, no. 7, pp. 2621 – 2632.

[Ziv 96] B. Živaljević, Π^1_1 sets of unbounded Loeb measure, *Proc. Amer. Math. Soc.* 1996, 124, no. 7, pp. 2205 – 2210.

Index

absolute formula, 23
absolute operation, 23
absolute set, 23
A-code, 192
alephs, 25
algebra, 321
– $\mathscr{D}$, 170
asterisks, 16
atom, 304
axiom
– Basic Enlargement, 87
– Bounded Inner Transfer, 125
– Basic Idealization, 86
– $\mathbf{BST}^{\mathbf{int}}$, 182
– Choice
– – 2^{κ}-size Choice, 241
– Choice, 21
– – $\mathfrak{c}$-size, 319
– Collection, 21
– Comprehension, 20
– Dependent Choice, 19
– Enlargement, 86
– Extensionality, 20
– Global Choice, 148, 303
– "$\mathbb{H} = \mathbb{L}[\mathbb{I}]$", 214
– Idealization, 85
– Ill-founded Mostowski Collapse, 303
– Infinity, 21
– Inner Boundedness, 86
– Inner $\boldsymbol{\kappa}$-Boundedness, 105
– Inner Standardization, 85
– Inner Strong $\boldsymbol{\kappa}$-Boundedness, 105
– Inner Transfer, 84
– Internal Size Collection, 298
– $\boldsymbol{\kappa}$-Boundedness, 241
– $\boldsymbol{\kappa}$-deep BI, 105
– $\boldsymbol{\kappa}$-size BE, 138
– $\boldsymbol{\kappa}$-size BI, 105
– Minimality, 159
– of costructibility, 159
– Pair, 20
– Parametrization, 182
– Power Set, 21
– Regularity, 21
– Regularity over $\mathbb{I}$, 15
– Replacement, 21
– Saturation, 19
– – κ-deep, 241
– – κ-size, 241
– Separation, 20
– SMA, 294
– Special Model Axiom, 294
– Standardization, 15
– Standard Condensation, 297
– Standard Size Choice, 19
– Standard Size Collection, 297
– SuperUniversality, 303
– Transfer, 15
– Transitivity of $\mathbb{I}$, 15
– Union, 21
– Universality, 303
– $\mathbf{ZFC}^{\mathbf{st}}$, 14, 84

base, 325
– $\boldsymbol{B}(\boldsymbol{\Pi}^0_\xi)$, 331
– $\boldsymbol{B}(\boldsymbol{\Pi}^1_n)$, 331
– $\boldsymbol{B}(\boldsymbol{\Sigma}^0_\xi)$, 331
– $\boldsymbol{B}(\boldsymbol{\Sigma}^1_n)$, 331
base $\mathbf{Base}(2^I)$, 321
bijection
– internal-preserving, 272
– locally internal, 272
bisimulation, 198
Borel
– map, 362
– set, 321

Borel cardinality, 363
Borel class, 321
Borel hierarchy, 321
B-smooth, 381

cardinal, 10, 24
– $*$-cardinal, 25
– $*$-cardinality, 25
– $\mathbb{I}$-cardinal, 25
– regular, 106
– $\mathbb{S}$-cardinal, 25
– singular, 106
cardinality, 10, 29
cardinality of continuum, 371
CD (countably determined), 327
CD-smooth, 381
chain
– elementary continuous, 144
characteristic function, 325
class, 10, 45
– $\equiv$-class, 47
– $\mathbf{A}$, 197
– $\mathbb{A}$ of all atoms, 304
– $\mathbf{A}(\mathscr{I})$, 245
– almost universal, 306
– $\underline{\mathbf{A}}$, 197
– $\underline{\mathbf{A}}(\mathscr{I})$, 245
– $\mathbb{B}$ of all bounded sets in **IST**, 111
– Borel, 321
– $\mathbf{Borel}(\mathscr{A})$, 321
– $\mathbf{Borel}[H]$, 322
– $\mathtt{Card}$ of all cardinals, 24
– $^*\mathtt{Card}$ of all $*$-cardinals, 25
– $\mathbf{CD}(\mathscr{A})$, 327
– $\mathbf{CD}[H]$, 327
– complete over, 237
– $\mathbf{\Delta}_1^{\mathtt{ss}}$, 34
– $\mathbf{\Delta}_2^{\mathtt{ss}}$, 34
– $\mathbf{\Delta}_n^1[H]$, 322
– $\mathbf{\Delta}_n^1[\mathscr{X}]$, 321
– $\mathbf{\Delta}_0^0[\mathscr{X}]$, 321
– $\mathbf{\Delta}_\xi^0(H)$, 321
– $\mathbf{\Delta}_\xi^0[H]$, 322
– $\mathbf{\Delta}_\xi^0[\mathscr{X}]$, 321
– $\mathbb{E}$, "external sets"
–– in **BST**, 102
– $\mathbb{E}$, elementary external sets
–– in external theories, 186
– $\mathbf{E}$, 184
– $\mathbb{E}[\mathscr{I}]$, 245
– extensional, 220, 237
– external subuniverse, 237
– formally $\mathcal{L}_{\in,\mathbf{G}}$-definable, 168
– $\mathbb{H}$ of all sets in a nonstandard universe, 12
– $\mathbb{H}$-class, 260
– $\mathbb{H}[\mathbf{G}]$, generic extension, 261
– $\mathbb{I}$ of all internal sets, 12, 258
–– in internal theories, 84
–– in **IST**, 111
– $\mathbb{I}_\kappa$, 230
–– in **IST** and **BST**, 111
– $\mathbb{I}'_\kappa$, 255
– internal subuniverse, 220
– κ-deep saturated, 230
– κ-size saturated, 230
– $\mathbb{L}[\mathbb{I}]$ of all sets constructible from internal sets, 202, 211
– $\mathbb{L}[\mathbb{I}_\kappa]$, 248
– $\mathbb{L}[\mathscr{I}]$, 245
– $\mathtt{Nms}(\mathbf{P})$ of names, 261
– $\mathtt{Ord}$ of all ordinals, 10, 24
– $^*\mathtt{Ord}$ of all $*$-ordinals, 25
– $\mathtt{Ord}^{(\mathbb{H})}$ of all $\mathbb{H}$-ordinals, 258
– $\mathbb{P}$ of all sub-internal sets, 203
– $\mathbf{\Pi}_1^{\mathtt{ss}}$, 34
– $\mathbf{\Pi}_2^{\mathtt{ss}}$, 34
– $\mathbf{\Pi}_n^1[H]$, 322
– $\mathbf{\Pi}_n^1[\mathscr{X}]$, 321
– $\mathbf{\Pi}_{<\xi}^0$, 348
– $\mathbf{\Pi}_0^0[\mathscr{X}]$, 321
– $\mathbf{\Pi}_\xi^0(H)$, 321
– $\mathbf{\Pi}_\xi^0[H]$, 322
– $\mathbf{\Pi}_\xi^0[\mathscr{X}]$, 321
– $\mathbf{Proj}[H]$, 322
– projective, 322
– proper, 45
– $\mathbb{S}$ of all standard sets, 12, 258
– self-definable, 246
– $\mathbf{\Sigma}_1^{\mathtt{ss}}$, 34
– $\mathbf{\Sigma}_2^{\mathtt{ss}}$, 34
– $\mathbf{\Sigma}_n^1[H]$, 322
– $\mathbf{\Sigma}_n^1[\mathscr{X}]$, 321
– $\mathbf{\Sigma}_0^0[\mathscr{X}]$, 321
– $\mathtt{SOrd}$ of all $\mathbb{S}$-ordinals, 190
– $\mathbf{\Sigma}_\xi^0[\mathscr{X}]$, 321
– $\mathbf{\Sigma}_\xi^0(H)$, 321

- $\mathbf{\Sigma}^0_\xi[H]$, 322
- $\subseteq$-complete, 16
- thin, 222
- transitive, 16
- transitive over, 237
- $\mathbb{V}$, the universe of **ZFBC**$^-$, 304
- universal class $\mathbb{V}$, 304
- $\mathbb{WF}$ of all well-founded sets, 16
- $\mathbb{WF}$ in Boffa's theory, 304
- $\mathbb{WF}[\mathscr{I}]$, 239
-
- $\mathbb{WF}$ of all well-founded sets, 258
clopen sets, 321
code
- A-code, 192
- E-code, 102
- $\mathscr{I}$-regular, 245
- regular, 195
coded formula, 118, 166
collection
- set-size collection, 104
complementary base, 326
complementary operation, 326
complement, $X^{\mathbf{C}}$, 321
complete
- over, 237
- $\subseteq$-complete, 16
concatenation, 191
core
- internal, 180
- standard, 132
- sub-internal, 203
- well-founded, 134
countably determined
- map, 362
- set, 327
cross-section, 340
cut, 35, 364
- additive, 37, 364
- countably cofinal, 35, 364
- countably coinitial, 35, 364
- minimal, 368
- standard size cofinal, 35
- standard size coinitial, 35
cut (initial segment), 37

descriptive set theory
- "hyperfinite", 317
- Polish, 317

direct limit, 145
domain, 317, 322
dual operation, 326

E-code, 102
$\in$-isomorphism, 132
elementary continuous chain, 144
embedding
- standard core embedding, 132
- $\in$-embedding, 132
- elementary, 133
- internal core embedding, 180
- natural, $*$, 141
- standard core embedding, 132
- st-$\in$-embedding, 180
entire part, 364
equinumerous, 10, 24
equivalence class
- $[\mathbf{x}]^{\mathbf{a}}$, 211
- $[p]^{\bullet}$, 185
equivalence relation
- B-smooth, 381
- CD-smooth, 381
- "countable", 376
- countably cofinal, 378
- countably coinitial, 378
- M_U, 378
- thin, 371
ER (equivalence relation, 371
$\in$-structure, 46, 132
extension, 132
- $*$-extension, 16, 59
- wf-core
-- conservative, 134
- conservative, 49
- elementary, 133
- internal core, 180, 181, 237
-- conservative, 181
- of an $\in$-structure, 132
- regular (of a model), 259
- standard core, 132, 133
-- conservative, 133
- transitive, 237
extensional class, 237
extensional relation, 303
"external power set", 104
external subuniverse, 237

filter, 140

– C-adequate, 143
finite intersection property, f. i. p., 30
finite support, 149
forcing
– condition, 260
–– stronger, 260
– notion, 260
– set-size, 260
– forc, 262
forcing relation, 263
Form, 118
$\mathtt{Form}_{\mathbf{G}}$, 166
formula
– absolute, 23
– bounded, 42, 113
– bounded st-∈-formula, 126
– coded, 118, 166
– directed, 57
– ∈-formula, 12
– external, 12
– formally false (f. false), 119, 169
– formally true (f. true), 119, 169
– int y, 12
– internal, 12
– Π_n formula, 42
– $\Sigma_2^{\mathtt{st}}$ formula, 94
– Σ_n formula, 42
– st-∈-formula, 12
– subint x, 203
foundations
– model-theoretic, V
Fubini product, 148, 150
function
– $\mathscr{A}$-measurable, 329
– Borel, 362
– continuous, uniformly continuous, 61
– countably determined, 362
– $\mathscr{I}$-extendible, 32
– of finite support, 150
– validation function, 48

gap, 36
graph
– locally of set size, 305
ground model, 258

halo, 56
$\mathbb{H}$-class, 260
height, 258
hierarchy
– Borel, 321
– projective, 322
– von Neumann, 42
–– relative, 44
hyperrational, 54
hyperreal (∗-real), 54
– appreciable, 55
– bounded, 55
– infinitely large, 55
– infinitesimal, 55
– limited, 55
– near-standard, 55
– standard, 55
– unbounded, 55
– unlimited, 55

index set, 140
induction, 26
– ∈-induction, 28
– on the ≺-rank, 265
– transfinite, 26
Inf. Large Exchange, 58
internal core, 180
internal core embedding, 180
internal core extension, 180, 237
internal core interpretability, 181
internal core interpretation, 181
internal power set, 23, 32
internal subuniverse, 220
internally presented, 273
interpretability
– internal core, 181
– standard core, 133
interpretation, 47
– internal core, 181
– standard core, 133
invariant, 45
invariant $\mathcal{L}$-structure, 45
isomorphism
– ∈-isomorphism, 132

language
– ∈-language, 12
– $\mathcal{L}_{\in,\mathtt{st},\mathtt{int}}$, 290
– extended, $\mathcal{L}^{\infty}$, 275
– $\mathcal{L}_{\in,\mathbf{G}}$, 148
– $\mathcal{L}_{\in,\mathbf{G},\mathbf{T}}$, 166
– of standard size, 273

– st-∈-language, 12
lifting, 355, 376
Loeb measure, 352
– finite, 352

map
– $\mathscr{A}$-measurable, 329
– Borel, 362
– countably determined, 362
measure
– finite Borel, 353
membership predicate ∈, 12
membership relation
– $\in_{\mathbf{G}}$ on $\mathbb{H}[\mathbf{G}]$, 261
model, 48
– extendible, 134
– ground model, 258
– $\mathbb{H}$, 258
– of a theory, 49
– T-extendible, 134
model-theoretic foundations, V
monad, 56, 95
– U-monad, 37, 378

name, 261
natural embedding, *, 141
natural number, 10, 26
– *-natural number, 26
– in **EEST**, 190
– in "internal" theories, 90
Nelson's algorithm, 94, 127
$\mathbb{N}$-sequence, 307
number
– infinitely large, 27

operation, 325
– absolute, 23
– complementary operation, 326
– dual, 326
– Souslin, 324, 325
– superposition, 326
order
– $<_{\mathbf{L}}$, 161
ordinal, 10, 24
– *-ordinal, 25
– good, 232
– in **EEST**, 190
– $\mathbb{I}$-ordinal, 25
– $\mathbb{S}$-ordinal, 25, 190
ordinals $\boldsymbol{\alpha}_k$, 160

P-forcing relation, 263
"planar" (set), 340
Polish space, 317
power class, $\mathscr{P}(X)$ 10, 21
– in **HST**, 21
power set, $\mathscr{P}(X)$ 10, 21
– "external", 104
– in **HST**, 21
– internal, 23, 32
– in **ZFC**, 21
predicate
– well-foundedness wf, 16
– membership ∈, 12
– standardness st, 12
principle
– *-Transfer
– – in **HST**, 17
– Boundedness, 18
– Choice
– – $\mathfrak{c}$-size, 319
– Σ_n-Collection, 43
– Compactness, 30
– Formal Truth Completeness, 122
– Extension, 32
– – countable, 319
– Inner Collection, 98
– Inner Dependent Choice, 99
– Internal Definitions, 56
– Inner Extension, 99
– Inner S. S. Choice, 99
– – restricted, 127
– Internal Induction, 56
– Inner Saturation, 93
– Isomorphism Property, 279
– Local Idealization, 93
– Map-Standardization, 99
– $\mathbf{\Pi}^1_1$-**Red** multiple, 340
– $\mathbf{\Sigma}^1_1$-**Sep** multiple, 340
– Overflow, 56
– Permanence, 57
– $\mathbf{\Gamma}$-**Red**, 338
– Reflection, 43
– Saturation, 30
– – countable, 319
– – κ-deep, 241
– – κ-size, 241
– $\mathbf{\Gamma}$-**Sep**, 338

– SMA, 294
– Special Model Axiom, 294
– $\mathbb{S}$-Size Choice, 122
– $\mathbb{S}$-Separation, 122
– Underflow, 57
– **Γ-Unif**, 345
– Uniqueness, 99
– Well-Ordering, 21, 32
prj U, 326
problem of external sets, 5
– in **BST**, 102
– in **IST**, 117
projection, 322, 340

quantifier
– $\forall^{\mathtt{bd}}$, 111
– $\forall^{\mathtt{int}}$, 14
– $\forall^{\infty \mathtt{lg}}$, 58
– $\forall^{\mathtt{st}}$, 14
– $\forall^{\mathtt{stfin}}$, 85
– $\forall^{\mathtt{wf}}$, 16
– bounded, 42
– $\exists^{\mathtt{bd}}$, 111
– $\exists^{\mathtt{int}}$, 14
– $\exists^{\infty \mathtt{lg}}$, 58
– $\exists^{\mathtt{st}}$, 14
– $\exists^{\mathtt{wf}}$, 16
– $\mathsf{Q}_U\, i\, R(i)$, 325
– U-many, 141
quotient power, 141
– set-indexed, 141
quotient structure, 47
– $\overline{\mathfrak{e}} = \langle \mathbf{E}/{}^{\bullet}{=}\,; {}^{\bullet}{\in}, {}^{\bullet}\mathtt{st} \rangle$, 185
$U, \mathscr{F}$-quotient power
– "definable", 163
$U, \mathscr{F}$-quotient power, 141

rank
– in a wf tree, 192, 197
– $\mathtt{irk}\, x$, 44
– $\mathtt{nrk}\, x$, 261
– $\mathtt{rank}\, x$, 42
rationals, $\mathbb{Q}$, 54
real, 54
– hyperreal, 54
–– appreciable, 55
–– bounded, 55
–– infinitely large, 55
–– infinitesimal, 55
–– limited, 55
–– near-standard, 55
–– standard, 55
–– unbounded, 55
–– unlimited, 55
"realistic" theory, 135
reals, $\mathbb{R}$, 54
reduction to true equality, 46
reflects, 43
regular extension, 259
relation
– well-founded, 16
–– externally, 104
– extensional, 303
– invariant, 45
– transitive, 215, 303
relativization, 10, 46, 141
– ${}^{\mathtt{a}}\Phi$, 202
– $\Phi^{\mathtt{bd}}$, 111
– ${}^{\bullet}\Phi$, 185
– $\Phi^{\mathtt{int}}$ of an $\in$-formula, 14
– φ^V, 43
– $\Phi^{\mathtt{st}}$, 14
– $\Phi^{\mathtt{wf}}$, 16
– to a $\in$-structure, 132
– to a $\mathtt{st}$-$\in$-structure, 132
Robinson's lemma, 62

saturation
– D-Saturation, 57
scheme
– "$\mathbb{S} \subseteq \mathbb{I}$", 83
– "$\mathbb{WF}^{\mathtt{feas}} \xrightarrow{*} \mathbb{I}$ [in $\mathbb{H}$]", 291
– "$\mathbb{WF} \xrightarrow{*} \mathbb{I}$ [in $\mathbb{H}$]", 22
– "$\mathbb{WF} \xrightarrow{*} \mathbb{I}_\kappa$ [in $\mathbb{L}[\mathbb{I}_\kappa]$]", 253
– "$\mathbb{WF} \xrightarrow{*} \mathscr{I}$ [in $\mathbb{WF}[\mathscr{I}]$]", 254
sequence
– $\mathbb{N}$-sequence, 307
set
– absolute, 23
– analytic, 322
– Borel, 321, 322
–– in H, 322
–– over $\mathscr{A}$, 321
– bounded, 111, 363
– CD, 327
– closed, 58
– cofinal, 35
– coinitial, 35

– compact, 58
– $\subseteq$-complete, 16
– condensable, 291
– condensed, 17
– constructible, 159
– countable, 320
– countably determined, 327
– Def(V), the collection of all sets definable in V, 161
– dense, 266
– elementary external, 4, 186
– extendible, 134
– external, 12
– "external", in **BST**, 101
– feasible well-founded, 291
– finite set, 10, 26
– – in **EEST**, 190
– – in internal theories, 90
– $*$-finite set, 26
– generic, 266
– hereditarily finite, 28
– – in **EEST**, 190
– homogeneous, 374
– hyperfinite set, 26
– index set, 325
– inductive, 192
– internal, 12, 84
– internal set in **IST**, 111
– κ-closed, 215
– $(<\kappa)$-closed, 215
– κ-distributive, 215
– κ-specially distributive, 215
– $\mathscr{I}$-wrong, 237
– large, 40
– λ-complete, 233
– Loeb measurable, 351
– $\mathbb{N}$ of all natural numbers, 10, 26
– – in **EEST**, 190
– – in "internal" theories, 90
– $^*\mathbb{N}$ of all $*$-natural numbers, 26
– of $\mathbb{S}$-size, 290
– open dense, 215
– $\cap$-closed, 19
– "planar", 340
– projective, 322
– – in H, 322
– relative standard, 89, 221, 223
– set of standard size, 19
– small, 40
– Souslin, 324
– Souslin over $\mathscr{B}$, 324
– Souslin in H, 324
– standard, 12
– standard size closed, 267
– standard size distributive, 268
– sub-internal, 203
– T-extendible, 134
– transitive, 16
– truth set
– – good, 122
– U-measurable, 141
– uniform, 340
– well-founded over $\mathscr{I}$, 239
– well-founded, 16
– – externally, 104
– well-founded over U, 44
– w-standard, 89, 223
– – in the modified sense, 224
set-size collection, 104
set-like collection, 189
sets
– equinumerous, 10, 24, 270
– external
– – problem of, 102
sets $\mathbf{y}_k$, 160
shadow, 56, 329
– $\mathscr{A}$-shadow, 329
shadow map, 64, 329
Souslin operation, 325
standard core, 132
standard core embedding, 132
standard core interpretability, 133
standard core interpretation, 133
standard part, 56
st-$\in$-structure, 46, 132
structure, 45
– $\mathfrak{a} = \langle \mathbf{A}\,; {}^{\mathfrak{a}}\!\in, {}^{\mathfrak{a}}\mathtt{st}\,; {}^{\mathfrak{a}}\!= \rangle$, 202
– domain of, 45
– $\mathbf{e} = \langle \mathbf{E}\,; {}^{\mathbf{e}}\!\in, {}^{\mathbf{e}}\mathtt{st}\,; {}^{\mathbf{e}}\!= \rangle$, 184
– $\in$-structure, 46, 132
– internally presented, 273
– invariant, 45
– κ-saturated, 138
– – strongly, 140
– $\mathcal{L}$-structure, 45
– quotient, 47

– set size, 48
– st-∈-structure, 46, 132
– strongly κ-saturated, 140
– underlying, 47
– universe of, 45
– with true equality, 46
sub-internal core, 203
submodel
– elementary, 43
superposition of operations, 326
support, 149, 150

theorem
– Collection
–– for set-like classes, 189
– Collection
–– in **BST**, 98
–– in **EEST**, 189
–– in **IST**, 114
– Dependent Choice
–– in **EEST**, 190
– Induction, 90
– Map-Standardization in **BST**, 100
– parametrization in **BST**, 103
– Reduction to Σ_2^{st} in **BST**, 94
– Reduction to Σ_2^{st} in **IST**, 113
– Saturation
–– in **BIST**, 92
–– in **EEST**, 190
– Separation
–– for set-like classes, 189
– Inner S. S. Choice
–– in **BST**, 100
– Standardization
–– in **EEST**, 189
– Standard Size Choice
–– in **EEST**, 189
– Uniqueness
–– in **BIST**, 89
–– in **BST**, 100
–– in **IST**, 117
theory
– ***ZC**, 125
– ***ZCN**, 125
– **BIST**, 88
– **BST**, 86
– $\mathbf{BST}'_\kappa$, 105
– $\mathbf{BST}_\kappa$, 105
– **BST**[T], 87
– $\mathbf{BST}\vartheta$, 157
– Δ_0-**ZC**, 126
– **DNST**, 312
– **EEST**, 182
– **EST**, 313
– **HST**, 13, 20
– ^{0}**HST**, 237
– $\mathbf{HST}_\kappa$, 241
– $\mathbf{HST}_\kappa^-$, 241
– $\mathbf{HST}'_\kappa$, 241
– internal, 84
– internal core interpretable, 181
– **IST**, 84
– **IST**′, 127
– $\mathbf{IST}^+$, 169
– **IST**[T], 87
– **IST**[**ZC**], 87
– **KST**, 290
– **KSTI**, 300
– **NBG**, 313
– **NCT**, 314
– nonstandard, 133
– **NST**, 295
– $\mathbf{NST}^+$, 298
– "realistic", 135
– reducible, 133
– **RST**, 313
– **SNST**, 312
– standard, 133
– standard core interpretable, 133
– **THS**, 314
– T^{st}, 87
– **ZC**, 21
– **ZCN**, 124
– $\mathbf{ZC}\vartheta$, 124
– **ZFC**, 21
– **ZFCI**, 299
– **ZFC**[**J**], 307, 309
– $\mathbf{ZFC}_\kappa[*]$, 311
– Σ_n-**ZFC**, 43
– $\mathbf{ZFC}\vartheta^+$, 160
– $\mathbf{ZFC}\vartheta$, 49
– **ZFGC**, 148
– **ZFGT**, 166
– $\mathbf{ZFGC}^{weak}$, 148
thin (equivalence relation), 371
transitive, 16
– over, 237

transitive closure, 44
transitive extension, 237
transversal, 37
tree, 191
– well-founded, wf, 192
truth
– truth set, 119, 169
`Truth` T, truth predicate, 119
$\mathtt{Truth}_G\, T$, modified truth predicate, 169
type, 274

ultrafilter, 140
– C-adequate, 143
– countably incomplete, 143
– Fubini product, 150
– good, 143
– regular, 143
– ultrafilter U, 163
uniformize, 344
union, $\bigcup X$, 21
universe
– **V**, universe of **ZFC**, 42

validation function, 48

well-founded core, 134

$\mathtt{card}\, X$, 10, 29
$\mathfrak{M} \models \Phi$, 10, 48
ω, $\aleph_0$, $\aleph_1$, 10, 25
`Ord`, 10, 24
$\mathscr{P}(X)$, 10, 21, 23
X^Y, 10
${}^Y X$, 10
$\# X$, 10, 26

`st`, 12
$\mathtt{int}\, y$, 12
$\forall^{\mathtt{int}}$, 14
$\forall^{\mathtt{st}}$, 14
$\exists^{\mathtt{int}}$, 14
$\exists^{\mathtt{st}}$, 14
$\Phi^{\mathtt{st}}$, 14
$\Phi^{\mathtt{int}}$, 14
${}^\sigma X$, 14, 85
${}^{\mathtt{s}}X$, 15
${}^*w$, 16
$\exists^{\mathtt{wf}}$, 16
$\forall^{\mathtt{wf}}$, 16
$\Phi^{\mathtt{wf}}$, 16
$\mathbb{WF}$, 16
$\bigcup X$, 21
$\mathscr{P}_{\mathtt{int}}(X)$, 23, 32
`Card`, 24
$\aleph_\xi$, 25
ω_ξ, 25
${}^*\mathtt{card}\, X$, 25
$\mathbb{N}$, 26
${}^*\mathbb{N}$, 26
$\mathscr{P}_{\mathtt{fin}}(X)$, 27
$X^{<\omega}$, 27, 220
$\mathbb{HF}$, 28
X^λ, 33
$X^{<\lambda}$, 33
$[X]^\lambda$, 33
$[X]^{<\lambda}$, 33
$\mathbf{\Pi}_1^{\mathtt{ss}}$, 34
$\mathbf{\Pi}_2^{\mathtt{ss}}$, 34
$\mathbf{\Delta}_1^{\mathtt{ss}}$, 34
$\mathbf{\Delta}_2^{\mathtt{ss}}$, 34
$\mathbf{\Sigma}_1^{\mathtt{ss}}$, 34
$\mathbf{\Sigma}_2^{\mathtt{ss}}$, 34
M_U, 37, 378
$[x]_U$, 37, 378
$\mathtt{rank}\, x$, 42
$\mathtt{sup}\, X$, 42
$\mathbf{V}_\xi$, 42
$\mathtt{irk}\, x$, 44
$\mathtt{TC}(x)$, 44
$\mathbb{WF}[U]$, 44
$\mathbf{V}_\xi[U]$, 44
$\Phi^{\mathtt{q}}$, 46
$\mathtt{Form}(\mathcal{L}, \mathfrak{M}, \Phi)$, 48
$\mathrm{TRUE}(\mathfrak{M}, \Phi)$, 48

$*$-rational, 54
$*$-real, 54
$\mathbb{Q}$, 54
$\mathbb{R}$, 54
${}^*\mathbb{Q}$, 54
${}^*\mathbb{R}$, 54
$x \simeq y$, 54
${}^\circ x$, 56
$\exists^{\infty\mathtt{lg}}$, 58

$\forall^{\infty \mathtt{lg}}$, 58

$\forall^{\mathtt{stfin}}$, 85
$^{\mathsf{s}}\{x \in X : \Phi(x)\}$, 85
$\mathbb{S}[c_1, ..., c_n]$, 89, 221
$\mathbb{S}[w]$, 89, 221, 222
$^{\sigma}\mathbb{N}$, 90
$\mathbf{mon}\, U$, 95
E_p, 102
$^{\bullet}x$, 103
$\mathscr{P}_{\mathtt{ext}}(X)$, 104
$p \doteq q$, 106
$\mathbb{B}$, 111
$\Phi^{\mathtt{bd}}$, 111
$\mathbb{I}_\kappa$ in **IST** and **BST**, 111
$\forall^{\mathtt{bd}}$, 111
$\exists^{\mathtt{bd}}$, 111
$\boldsymbol{\theta}(Z, x, Y)$, 116
$\ulcorner ... \urcorner$, 118

$\mathbb{S}^{(\mathtt{v})}$, 132
$\mathbb{WF}^{(^*V)}$, 134
$\mathbf{f}_x$, 140
$^*x$, 141
$^*=$, 141
$[f]$, 141
$\mathscr{F}/U$, 141
$^*\in$, 141
$\varphi^{\mathtt{v}}$, 141
$^*\mathtt{st}$, 141
$U i\, \Phi(i)$, 141
$U i \in I\, \Phi(i)$, 141
$^*V = \langle ^*V ; {}^*\varepsilon, {}^*\mathtt{st} \rangle$, 141
$\Phi[i]$, 142
$[\Phi]$, 142
$C^{\mathtt{fin}}$, 143
$I(C, i)$, 143
∞, 144
$\|X\|$, 149
$\mathscr{F}^P$, 150
$\mathscr{F}^s$, 150
$\|f\|$, 150
$\bigtimes_{p \in P} U_p$, 150
$<_\xi$, 151
$\mathbf{exp}\, \kappa$, 151
S_κ, 151
"$\mathbf{V} = \mathbf{L}$", 159
$\boldsymbol{\alpha}_k$, 160
$\mathbf{y}_k$, 160
$<_{\mathbf{L}}$, 161
$\mathtt{Fun}\, f$, 164
$\Xi(n, x)$, 164
x_α, 167
$<_{\mathbf{G}}$, 168
$\mathscr{D}$, 170

$\mathbb{I}^{(\mathtt{v})}$, 180
$^{\bullet}=$, 184
$\mathbf{e}$, 184
$\mathbf{E}$, 184
$^{\bullet}\in$, 184
$^{\bullet}\mathtt{int}$, 184
$^{\bullet}\mathtt{int}\, p$, 184
$^{\bullet}\mathtt{st}$, 184
$^{\bullet}\mathtt{st}\, p$, 184
$p \;{}^{\bullet}\!= q$, 184
$p \;{}^{\bullet}\!\in q$, 184
ε, 184
$x\; \varepsilon\; p$, 184
$^{\bullet}\Phi$, 185
$[p]^{\bullet}$, 185
$\mathbb{HF}$ in **EEST**, 190
$\mathtt{SOrd}$, 190
$\mathbf{sup}^{\mathsf{S}}\, X$, 190
$\mathtt{Seq}$, 191
$s\,{}^\wedge t$, 191
$\mathtt{Max}\, T$, 191
$\mathtt{Min}\, T$, 191
$\mathtt{Succ}_T(t)$, 191
$\mathsf{A}_{\mathbf{x}}$, 192
$\mathsf{F}_{\mathbf{x}}(t)$, 192
$|t|_T$, 192
$|T|$, 192
$T_{\mathbf{x}}$, 192
$T|_a$, 193
$T|_t$, 193
$^{\mathtt{a}}x$, 193
$\mathbf{x}|_t$, 193
$T[y]$, 194
$\mathbf{c}[y]$, 194
$\mathtt{DI}_{\mathbf{x}}$, 195
$\mathbf{x}^{\mathtt{R}}$, 195
$|t|^*_T$, 197
$|T|^*$, 197
$\underline{\mathbf{A}}$, 197
$\mathbf{A}$, 197
$^{\mathtt{a}}\mathtt{st}$, 199
$^{\mathtt{a}}\mathtt{st}\, \mathbf{x}$, 199

$^{\mathtt{a}}{=}$, 199
$^{\mathtt{a}}{\in}$, 199
$^{\mathtt{a}}\mathtt{int}$, 199
$^{\mathtt{a}}\mathtt{int}\,\mathbf{x}$, 199
$\mathbf{j}_{\mathbf{xy}}$, 199
$^{\mathtt{a}}\Phi$, 202
$\mathfrak{a}$, 202
$\mathbb{L}[\mathbb{I}]$, 202, 211
$\mathbb{P}^{(\mathfrak{a})}$, 203
$\mathtt{subint}\,x$, 203
$[\mathbf{x}]^{\mathtt{a}}$, 211

$\mathrm{Def}^{\mathbb{I}}_{\in,\mathtt{st}}(P)$, 220
$\mathbb{S}(X)$, 221
$\mathbb{N}[w]$, 223, 224
$w\text{-}\mathtt{st}\,x$, 223
$w\text{-}\mathtt{st}_{\mathrm{M}}\,x$, 224
$\mathbb{N}_{\mathrm{M}}[w]$, 224
$\mathbb{S}_{\mathrm{M}}[w]$, 224
$\mathtt{st}_{\mathscr{I}}\,x$, 225
$\mathbb{I}_{\kappa}$, 230
$\mathbb{WF}[\mathscr{I}]$, 239
$\mathbb{E}[\mathscr{I}]$, 245
$\mathbf{A}(\mathscr{I})$, 245
$\underline{\mathbf{A}}(\mathscr{I})$, 245
$\mathbb{L}[\mathscr{I}]$, 245
$\mathbb{L}[\mathbb{I}_{\kappa}]$, 248
$\mathbb{I}'_{\kappa}$, 255

$\mathbb{H}$, a model of **HST**, 258
$\mathtt{ht}_{\mathbb{H}}$, 258
$x^{(\mathbb{H})}$, 258
$\mathbb{I}_0$, 258
$\breve{x}$, 258
$\mathtt{Nms}_0$, 261
$\mathtt{Nms}(\mathbf{P})$, 261
$\mathtt{Nms}_{\xi}(\mathbf{P})$, 261
$\mathtt{nrk}\,x$, 261
$a[\mathbf{G}]$, 261
$a^{\mathtt{put}}[\mathbf{G}]$, 261
$\mathbb{H}[\mathbf{G}]$, 261
Δ_a, 263
$\Phi[\mathbf{G}]$, 266
$\diamond(t_1,...,t_k)$, 274
$\mathfrak{C}[D]$, 275
D^t, 275
$\mathcal{L}^{\infty}$, 275
$p\,\Phi$, 275
$p^t(x^t)$, 275
$\mathbb{P}_{\mathcal{L}}(\mathfrak{A},\mathfrak{B})$, 276
$\mathfrak{A}_i$, 280
$\mathfrak{B}_i$, 280
$\mathcal{L}_i$, 280
$\mathcal{L}(\kappa)$, 280
$\mathbf{L}(\kappa)$, 280
$|\pi|$, 280
$\mathtt{Ind}$, 280
$\mathbb{P}$, product forcing, 280
$\mathbb{P}_i$, 280
$\mathbf{P}_i$, 281
$\|a\|$, 282
$\Vdash_C$, 282
$\|\Phi\|$, 283

$\mathcal{L}_{\in,\mathtt{st},\mathtt{int}}$, 290
$\mathfrak{i}$, 291
$\mathfrak{s}$, 291
$\mathsf{A}^{\mathsf{S}}_{\mathbf{x}}$, 296
$\mathbf{S}$, 296
$|t|^{\mathsf{S}}_{\mathbf{x}}$, 296

$\aleph_1$, 321
ω_1, 321
$\mathbf{Base}(2^I)$, 321
$\mathbf{Clop}(\mathscr{X})$, 321
$X^{\mathtt{C}}$, 321
${}^{\mathtt{int}}2^H$, 321
$\mathbf{\Sigma}^0_0[\mathscr{X}]$, 321
$\mathbf{\Pi}^0_0[\mathscr{X}]$, 321
$\mathbf{\Delta}^0_0[\mathscr{X}]$, 321
$\mathbf{\Sigma}^0_{\xi}[\mathscr{X}]$, 321
$\mathbf{\Pi}^0_{\xi}[\mathscr{X}]$, 321
$\mathbf{\Delta}^0_{\xi}[\mathscr{X}]$, 321
$\mathbf{\Sigma}^1_n[\mathscr{X}]$, 321
$\mathbf{\Pi}^1_n[\mathscr{X}]$, 321
$\mathbf{\Delta}^1_n[\mathscr{X}]$, 321
$\mathbf{\Sigma}^0_{\xi}(H)$, 321
$\mathbf{\Pi}^0_{\xi}(H)$, 321
$\mathbf{\Delta}^0_{\xi}(H)$, 321
$\mathbf{\Sigma}^0_{\xi}[H]$, 322
$\mathbf{\Pi}^0_{\xi}[H]$, 322
$\mathbf{\Delta}^0_{\xi}[H]$, 322
$\mathbf{\Sigma}^1_n[H]$, 322
$\mathbf{\Pi}^1_n[H]$, 322
$\mathbf{\Delta}^1_n[H]$, 322
$\mathtt{dom}\,P$, 322
A, 324, 325
$\mathsf{A}{\bullet}\mathscr{B}$, 324

χ_u, 325
$\Phi\{A_i\}$, 325
$\cap(I)$, 325
$\cup(I)$, 325
$U^{\complement}$, 326
$Q_U\, i\, R(i)$, 325
$\circ$, 326
$U \circ \{V_i\}_{i \in I}$, 326
$\mathbf{CD}(\mathscr{A})$, 327
$\mathbf{CD}[H]$, 327
${}^{\mathtt{int}}2^h$, 329
$\mathbf{sh}_{\mathscr{A}}{}''X$, 329
$\mathbf{sh}_{\mathscr{A}}(x)$, 329
$\boldsymbol{B}(\mathbf{\Pi}^0_\xi)$, 331
$\boldsymbol{B}(\mathbf{\Pi}^1_n)$, 331
$\boldsymbol{B}(\mathbf{\Sigma}^0_\xi)$, 331
$\boldsymbol{B}(\mathbf{\Sigma}^1_n)$, 331
$\Phi \cdot \mathbf{K}$, 331
$\mathbf{\Pi}^0_{<\xi}$, 348
$\mathcal{L}(\mu)$, 352
$\mathrm{L}\mu$, 352
$\equiv_{\mathtt{B}}$, 362, 380
$<_{\mathtt{B}}$, 362, 380
$\leq_{\mathtt{B}}$, 362, 380
$\equiv_{\mathtt{CD}}$, 362, 380
$<_{\mathtt{CD}}$, 362, 380
$\leq_{\mathtt{CD}}$, 362, 380
$c\mathbb{N}$, 364
$c/\mathbb{N}$, 364
$\mathcal{E}(r)$, 364
$U\mathbb{N}$, 364
$U/\mathbb{N}$, 364
$\mathfrak{c}$, 371
M_U, 378

Springer Monographs in Mathematics

This series publishes advanced monographs giving well-written presentations of the "state-of-the-art" in fields of mathematical research that have acquired the maturity needed for such a treatment. They are sufficiently self-contained to be accessible to more than just the intimate specialists of the subject, and sufficiently comprehensive to remain valuable references for many years. Besides the current state of knowledge in its field, an SMM volume should also describe its relevance to and interaction with neighbouring fields of mathematics, and give pointers to future directions of research.

Abhyankar, S.S. **Resolution of Singularities of Embedded Algebraic Surfaces** 2nd enlarged ed. 1998
Andrievskii, V.V.; Blatt, H.-P. **Discrepancy of Signed Measures and Polynomial Approximation** 2002
Ara, P.; Mathieu, M. **Local Multipliers of C*-Algebras** 2003
Armitage, D.H.; Gardiner, S.J. **Classical Potential Theory** 2001
Arnold, L. **Random Dynamical Systems** corr. 2nd printing 2003 (1st ed. 1998)
Aubin, T. **Some Nonlinear Problems in Riemannian Geometry** 1998
Auslender, A.; Teboulle M. **Asymptotic Cones and Functions in Optimization and Variational Inequalities** 2003
Bang-Jensen, J.; Gutin, G. **Digraphs** 2001
Baues, H.-J. **Combinatorial Foundation of Homology and Homotopy** 1999
Brown, K.S. **Buildings** 3rd printing 2000 (1st ed. 1998)
Cherry, W.; Ye, Z. **Nevanlinna's Theory of Value Distribution** 2001
Ching, W.K. **Iterative Methods for Queuing and Manufacturing Systems** 2001
Crabb, M.C.; James, I.M. **Fibrewise Homotopy Theory** 1998
Dineen, S. **Complex Analysis on Infinite Dimensional Spaces** 1999
Elstrodt, J.; Grunewald, F. Mennicke, J. **Groups Acting on Hyperbolic Space** 1998
Fadell, E.R.; Husseini, S.Y. **Geometry and Topology of Configuration Spaces** 2001
Fedorov, Y.N.; Kozlov, V.V. **A Memoir on Integrable Systems** 2001
Flenner, H.; O'Carroll, L. Vogel, W. **Joins and Intersections** 1999
Gelfand, S.I.; Manin, Y.I. **Methods of Homological Algebra** 2nd ed. 2003
Griess, R.L.Jr. **Twelve Sporadic Groups** 1998
Gras, G. **Class Field Theory** 2003
Ivrii, V. **Microlocal Analysis and Precise Spectral Asymptotics** 1998
Jech, T. **Set Theory** (3rd revised edition 2002)
Jorgenson, J.; Lang, S. **Spherical Inversion on SLn (R)** 2001
Kanamori, A.; **The Higher Infinite** (2nd edition 2003)
Khoshnevisan, D. **Multiparameter Processes** 2002
Koch, H. **Galois Theory of *p*-Extensions** 2002
Kozlov, V.; Maz'ya, V. **Differential Equations with Operator Coefficients** 1999
Landsman, N.P. **Mathematical Topics between Classical & Quantum Mechanics** 1998
Lebedev, L.P.; Vorovich, I.I. **Functional Analysis in Mechanics** 2002
Lemmermeyer, F. **Reciprocity Laws: From Euler to Eisenstein** 2000
Malle, G.; Matzat, B.H. **Inverse Galois Theory** 1999
Mardesic, S. **Strong Shape and Homology** 2000
Margulis, G.A. **On Some Aspects of the Theory of Anosov Systems** 2004
Murdock, J. **Normal Forms and Unfoldings for Local Dynamical Systems** 2002
Narkiewicz, W. **The Development of Prime Number Theory** 2000
Parker, C.; Rowley, P. **Symplectic Amalgams** 2002
Peller, V. (Ed.) **Hankel Operators and Their Applications** 2003
Prestel, A.; Delzell, C.N. **Positive Polynomials** 2001
Puig, L. **Blocks of Finite Groups** 2002
Ranicki, A. **High-dimensional Knot Theory** 1998
Ribenboim, P. **The Theory of Classical Valuations** 1999
Rowe, E.G.P. **Geometrical Physics in Minkowski Spacetime** 2001
Rudyak, Y.B. **On Thom Spectra, Orientability and Cobordism** 1998

Ryan, R.A. **Introduction to Tensor Products of Banach Spaces** 2002
Saranen, J.; Vainikko, G. **Periodic Integral and Pseudodifferential Equations with Numerical Approximation** 2002
Schneider, P. **Nonarchimedean Functional Analysis** 2002
Serre, J-P. **Complex Semisimple Lie Algebras** 2001 (reprint of first ed. 1987)
Serre, J-P. **Galois Cohomology** corr. 2nd printing 2002 (1st ed. 1997)
Serre, J-P. **Local Algebra** 2000
Serre, J-P. **Trees** corr. 2nd printing 2003 (1st ed. 1980)
Smirnov, E. **Hausdorff Spectra in Functional Analysis** 2002
Springer, T.A. Veldkamp, F.D. **Octonions, Jordan Algebras, and Exceptional Groups** 2000
Sznitman, A.-S. **Brownian Motion, Obstacles and Random Media** 1998
Taira, K. **Semigroups, Boundary Value Problems and Markov Processes** 2003
Tits, J.; Weiss, R.M. **Moufang Polygons** 2002
Uchiyama, A. **Hardy Spaces on the Euclidean Space** 2001
Üstünel, A.-S.; Zakai, M. **Transformation of Measure on Wiener Space** 2000
Yang, Y. **Solitons in Field Theory and Nonlinear Analysis** 2001
Kanovei, V.; Reeken, M. **Nonstandard Analysis, Axiomatically** 2004

Printed by Books on Demand, Germany